Principles of Engineering Thermodynamics

Second Edition, SI Edition

John R. Reisel
University of Wisconsin – Milwaukee

CENGAGE

Australia • Brazil • Canada • Mexico • Singapore • United Kingdom • United States

Principles of Engineering Thermodynamics,
Second Edition, SI Edition
John R. Reisel

Senior Vice President, Higher Education & Skills Product: Erin Joyner

Product Director: Mark Santee

Senior Product Manager: Timothy L. Anderson

Learning Designer: MariCarmen Constable

Associate Content Manager: Alexander Sham

Senior Product Assistant: Anna Goulart

Senior Marketing Director: Jennifer Fink

Executive Marketing Manager: Tom Ziolkowski

Senior Digital Delivery Lead: Nikkita Kendrick

IP Analyst: Deanna Ettinger

IP Project Manager: Nick Barrows

Text and Image Permissions Researcher: Kristiina Paul and Lumina Datamatics

Production Service: RPK Editorial Services, Inc.

Compositor: SPi Global

Designer: Nadine Ballard

Manufacturing Planner: Ron Montgomery

Cover Designer: Nadine Ballard

Cover Image: iStockPhoto.com/Striker77s

For product information and technology assistance, contact us at **Cengage Customer & Sales Support, 1-800-354-9706** or **support.cengage.com.**

For permission to use material from this text or product, submit all requests online at **www.cengage.com/permissions**.

Library of Congress Control Number: 2020948721

ISBN: 978-0-357-11179-6

Cengage
200 Pier 4 Boulevard
Boston, MA 02210
USA

Cengage is a leading provider of customized learning solutions with employees residing in nearly 40 different countries and sales in more than 125 countries around the world. Find your local representative at **www.cengage.com**.

To learn more about Cengage platforms and services, register or access your online learning solution, or purchase materials for your course, visit **www.cengage.com**.

Printed at CLDPC, USA, 02-21

Preface

Mission

Why should an engineering student want to study thermodynamics? The answers are all around you. Look at all of the devices that use energy—electric lights, automobiles, computers, smartphones, and so many more. Things that don't directly use energy likely were made by machines that do use energy. In today's world, energy is being used everywhere, and Thermodynamics is the study of energy. Engineers need to know how to use energy effectively. As such, the goal of this book is to prepare students to be practicing engineers with an intuitive understanding of how energy-related systems work and how the performance of such systems is affected by variations in its operational parameters.

While the basic principles and concepts of thermodynamics were developed well over 100 years ago, they are still being used today to analyze and explain how things work in the world. Thermodynamics makes all of today's technology possible, and will continue to power the development of new technology in the future. Therefore, engineers need to have a strong fundamental understanding of thermodynamics in order to continue to develop technology to improve the world.

This textbook is written with the philosophy that it is most important to prepare engineers to understand how to use thermodynamics in professional practice. Engineers must gain an intuitive understanding of how changes in a parameter of a system will impact the energy-related performance of a process. The approach taken in this book is to help students develop this understanding. Throughout the book, students will be asked to use modern computational tools of their choosing to quickly vary the parameters of a system so that they can recognize interactions between features in systems. A historical problem with learning thermodynamics is that it is often taught as a subject that involves only solving individual problems with tabulated property data. Students who learn thermodynamics in such a manner often end up as practicing engineers who do not recognize how various parameters may impact the energy consumption of a piece of equipment or a process. For example, an engineering student may learn in thermodynamics how to solve for the power required to operate an air compressor. However, as a working engineer they may not realize that the energy consumption can be reduced by compressing cold air rather than hot air. As a result, their company may continue to pull hot air from the interior of a factory into the compressor intake, rather than using cold exterior air in the winter, which would waste both energy and money. This book aims to correct this deficiency that has plagued thermodynamics education in the past by encouraging and directing the students to explore the relationships between system parameters.

Special Features of the Book

Emphasis on Computer-based Properties and Equation Solving Platforms

To aid in their understanding of energy relationships, students are strongly encouraged to develop computer-based models of devices, processes, and cycles, and to take advantage of the plethora of Internet-based programs and computer apps for rapidly finding thermodynamic data; these are things that practicing engineers do regularly. Students who are comfortable with a particular equation-solving platform are encouraged to use that platform for their equation development. Some platforms may directly connect into thermodynamic property data, making such platforms potentially easier to use for students already familiar with the platform. Alternatively, an external property-data program can be used to find values for the properties, and these values can be directly input into an equation-solving program. This approach allows students to spend more time focusing on thermodynamics and less time learning a new piece of software.

Parametric Analysis-Based Problems

In keeping with the goal of developing an intuitive understanding of how thermodynamic systems work, many examples and problems throughout the book guide students to perform parametric analyses. These problems are designed to isolate a particular quantity and allow the students to learn how variation of that quantity affects the rest of the system.

Streamlining of Thermodynamic Topics

Another philosophical difference behind this textbook is a streamlining of the material presented in this book in comparison to other thermodynamics books. The content of this book focuses on what is most important for most students to learn about thermodynamics as they strive to become practicing engineers. This is not to say that there are not many other important topics in the subject of thermodynamics. However, the author believes that these topics are more suited for a higher-level engineering thermodynamics course—primarily a course to be taken by students engaged in graduate studies in energy-related areas.

Course Organization

The content of this book is suitable for either a one-semester course or a two-semester course sequence for thermodynamics. For a one-semester course, it is suggested that the material covered be Chapters 1-6, and if time permits some coverage of basic cycles (such as the basic Rankine cycle or the Otto cycle in Chapter 7, or the vapor-compression refrigeration cycle in Chapter 8) may be included. A two-semester sequence would include the remainder of the material in Chapters 7-11 in the second semester of thermodynamics. This second course focuses on applying the basic principles covered in the first course in practical systems. Students who complete only the first course will have a strong understanding of the basic engineering principles of thermodynamics and have some knowledge of the interrelationship between parameters impacting thermodynamic systems. A student who completes two courses will have a much deeper understanding of the relationship between thermodynamics parameters and will be capable of applying thermodynamics in a wide variety of mechanical systems.

This book aims to make thermodynamics enjoyable for students and help them understand the importance of thermodynamics in today's world. Many of the problems facing the world today revolve around the use of energy. Through the use of this book, it is expected that many more engineers will be prepared and eager to help solve these energy-related problems by properly applying the classic concepts of thermodynamics.

New in the Second Edition

A key concept behind this book is to keep the amount of content in the textbook manageable for today's students. As such, many of the changes in the second edition involve editorial changes to the content, with the intent of these changes being to improve the pedagogy of the text.

A new feature found throughout the textbook is "Question for Thought/Discussion." The purpose of these questions is to act as a stimulus for students to think about topics related to Thermodynamics and engineering. The questions are often centered on non-technical concepts. By thinking about and discussing these questions, students will gain insights into how energy use impacts individuals and the world. This will help them understand how engineers may use this information as they design and build. Instructors can either ask students to think about these questions on their own, or they can have class discussions on the topics. An added benefit is that attention paid to these questions should help programs meet the ABET student outcomes.

To address their growing use in many internal combustion engines in hybrid vehicles, a section on the Atkinson and Miller cycles has been added to Chapter 7. While the analysis of these cycles is not tremendously different from more traditional engine cycles, this section will draw attention to the evolving nature of engine design.

Finally, over 100 new end-of-chapter problems have been added to the textbook, offering up a new array of exercises through which students can learn thermodynamics.

Supplements for the Instructor

Supplements to the text include a Solution and Answer Guide that provides complete solutions to the end-of-chapter problems, Lecture Note PowerPoint™ slides, and an image library of all figures in the book. These can be found on the password-protected Instructor's Resources website for the book at login.cengage.com.

Acknowledgments

I would like to acknowledge the helpful comments and suggestions of the many reviewers of this book as it underwent development, including

- Edward E. Anderson, *Texas Tech University*
- Sarah Codd, *Montana State University*
- Gregory W. Davis, *Kettering University*
- Elizabeth M. Fisher, *Cornell University*
- Sathya N. Gangadharan, *Embry-Riddle Aeronautical University*
- Dominic Groulx, *Dalhousie University*
- Fouad M. Khoury, *University of Houston*
- Kevin H. Macfarlan, *John Brown University*
- Kunal Mitra, *Florida Institute of Technology*
- Patrick Tebbe, *Minnesota State University, Mankato*
- Kenneth W. Van Treuren, *Baylor University*

I also wish to thank the reviewers of the first edition for their thoughtful feedback:

- Paul Akangah, *North Carolina A&T State University*
- Emmanuel Glakpe, *Howard University*
- James Kamm, *University of Toledo*
- Chaya Rapp, *Yeshiva University*
- Francisco Ruiz, *Illinois Institute of Technology*
- David Sawyers, *Ohio Northern University*
- Keith Strevett, *University of Oklahoma*
- Victor Taveras, *West Kentucky Community and Technical College*

Their work has helped immensely in improving this book.

I also wish to thank those who taught me thermodynamics, including Charles Marston at Villanova University and Normand Laurendeau at Purdue University. Without their outstanding abilities to teach thermodynamics, I may never have developed the passion I have for thermodynamics. I also would like to acknowledge my colleagues, both at the University of Wisconsin—Milwaukee and elsewhere, for their thoughtful discussions on thermodynamics education over the years, which helped to form my perspective on how thermodynamics should be taught. This perspective has reached fruition in this book. The students whom I have taught over the years also deserve recognition for their patience as I've experimented with different pedagogical approaches.

Finally, I would like to thank all the individuals at Cengage who have been instrumental in the creation of this book. In particular, I would like to thank Timothy Anderson, Senior Product Manager; MariCarmen Constable, Learning Designer; Alexander Sham, Associate Content Manager; and Anna Goulart, Senior Product Assistant. Special thanks are also due to Rose Kernan of RPK Editorial Services, Inc. Their assistance has been invaluable in bringing this project to completion.

John R. Reisel

About the Author

John R. Reisel is a Professor in the Mechanical Engineering Department at the University of Wisconsin—Milwaukee (UWM). He received his B.M.E., with a minor in Mathematics, from Villanova University, and his M.S. and Ph.D. in Mechanical Engineering from Purdue University. His areas of research interest include combustion, energy usage modeling, energy efficiency, fuel production, sustainable engineering, and engineering education. He has received numerous awards in engineering education, including the UWM Distinguished Undergraduate Teaching Award and the UWM College of Engineering and Applied Science Outstanding Teaching Award. He was also a recipient of the SAE Ralph R. Teetor Educational Award.

Dr. Reisel is a member of the American Society for Engineering Education (ASEE), the American Society of Mechanical Engineers (ASME), the Combustion Institute, the European Society for Engineering Education (SEFI), and the Society of Automotive Engineers (SAE). He has served as division chair of the Engineering and Public Policy Division of ASEE, and program chair of the Technological and Engineering Literacy/Philosophy of Engineering of ASEE. Dr. Reisel is a registered Professional Engineer in the state of Wisconsin.

Digital Resources

New Digital Solution for Your Engineering Classroom

WebAssign is a powerful digital solution designed by educators to enrich the engineering teaching and learning experience. With a robust computational engine at its core, WebAssign provides extensive content, instant assessment, and superior support.

WebAssign's powerful question editor allows engineering instructors to create their own questions or modify existing questions. Each question can use any combination of text, mathematical equations and formulas, sound, pictures, video, and interactive HTML elements. Numbers, words, phrases, graphics, and sound or video files can be randomized so that each student receives a different version of the same question.

In addition to common question types such as multiple choice, fill-in-the-blank, essay, and numerical, you can also incorporate robust answer entry palettes (mathPad, chemPad, calcPad, physPad, Graphing Tool) to input and grade symbolic expressions, equations, matrices, and chemical structures using powerful computer algebra systems.

WebAssign Offers Engineering Instructors the Following

- The ability to create and edit algorithmic and numerical exercises.
- The opportunity to generate randomized iterations of algorithmic and numerical exercises. When instructors assign numerical WebAssign homework exercises (engineering math exercises), the WebAssign program offers them the ability to generate and assign their students differing versions of the same engineering math exercise. The computational engine extends beyond and provides the luxury of solving for correct solutions/answers.
- The ability to create and customize numerical questions, allowing students to enter units, use a specific number of significant digits, use a specific number of decimal places, respond with a computed answer, or answer within a different tolerance value than the default.

Visit www.webassign.com/instructors/features/ to learn more. To create an account, instructors can go directly to the signup page at www.webassign.net/signup.html.

WebAssign Features for Students

Review Concepts at Point of Use

Within WebAssign, a "Read It" button at the bottom of each question links students to corresponding sections of the textbook, enabling access to the MindTap Reader at the precise moment of learning. A "Watch It" button allows a short video to play. These videos help students understand and review the problem they need to complete, enabling support at the precise moment of learning.

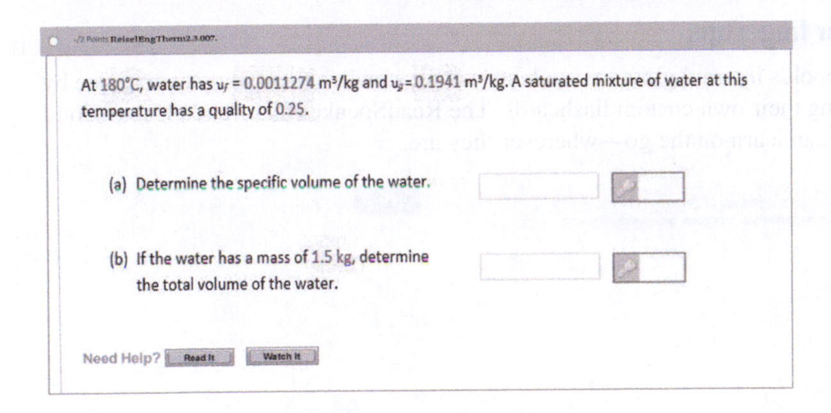

At 180°C, water has v_f = 0.0011274 m³/kg and v_g = 0.1941 m³/kg. A saturated mixture of water at this temperature has a quality of 0.25.
(a) Determine the specific volume of the water.
(b) If the water has a mass of 1.5 kg, determine the total volume of the water.
Need Help?
Read It
Watch It

My Class Insights

WebAssign's built-in study feature shows performance across course topics so that students can quickly identify which concepts they have mastered and which areas they may need to spend more time on.

Ask Your Teacher

This powerful feature enables students to contact their instructor with questions about a specific assignment or problem they are working on.

MindTap Reader

Available via WebAssign, **MindTap Reader** is Cengage's next-generation eBook for engineering students.

The MindTap Reader provides more than just text learning for the student. It offers a variety of tools to help our future engineers learn chapter concepts in a way that resonates with their workflow and learning styles.

Personalize their experience

Within the MindTap Reader, students can highlight key concepts, add notes, and bookmark pages. These are collected in My Notes, ensuring they will have their own study guide when it comes time to study for exams.

2.3 Transport of Energy

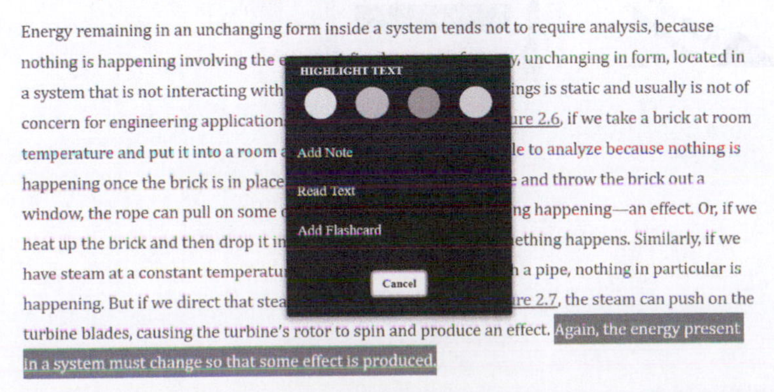

Energy remaining in an unchanging form inside a system tends not to require analysis, because
nothing is happening involving the
y, unchanging in form, located in
a system that is not interacting with
ings is static and usually is not of
concern for engineering application
ure 2.6, if we take a brick at room
temperature and put it into a room
le to analyze because nothing is
happening once the brick is in place
and throw the brick out a
window, the rope can pull on some
ng happening—an effect. Or, if we
heat up the brick and then drop it in
ething happens. Similarly, if we
have steam at a constant temperatu
a pipe, nothing in particular is
happening. But if we direct that stea
re 2.7, the steam can push on the
turbine blades, causing the turbine's rotor to spin and produce an effect. Again, the energy present in a system must change so that some effect is produced.
HIGHLIGHT TEXT
Add Note
Read Text
Add Flashcard
Cancel

Flexibility at their fingertips

With access to the book's internal glossary, students can personalize their study experience by creating and collating their own custom flashcards. The ReadSpeaker feature reads text aloud to students, so they can learn on the go—wherever they are.

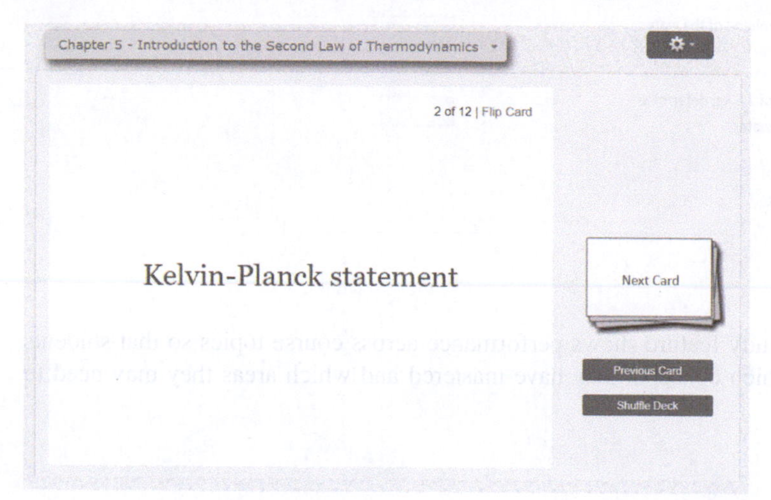

The Cengage Mobile App

Available on iOS and Android smartphones, the Cengage Mobile App provides convenience. Students can access their entire textbook anyplace and anytime. They can take notes, highlight important passages, and have their text read aloud whether they are online or off.

To learn more and download the mobile app, visit www.cengage.com/mobile-app/.

To my wife Jennifer, and my children Theresa and Thomas—may Thermodynamics continue to provide them with modern wonders.

Contents

Preface to the SI Edition viii
Preface ix
About the Author xiii
Digital Resources xiv

Chapter 1 INTRODUCTION TO THERMODYNAMICS AND ENERGY 1
1.1 Basic Concepts: Systems, Processes, and Properties 6
1.2 An Introduction to Some Common Properties 16
1.3 Zeroth Law of Thermodynamics 24
1.4 Phases of Matter 25
Summary 27
Problems 28

Chapter 2 THE NATURE OF ENERGY 35
2.1 What Is Energy? 35
2.2 Types of Energy 36
2.3 Transport of Energy 40
2.4 Heat Transfer 41
2.5 Work Transfer 48
2.6 Energy Transfer via Mass Transfer 57
2.7 Analyzing Thermodynamics Systems and Processes 59
2.8 Platform for Performing Thermodynamics Analysis 60
Summary 61
Problems 62

Chapter 3 THERMODYNAMIC PROPERTIES AND EQUATIONS OF STATE 69
3.1 Introduction 69
3.2 Phase Diagrams 69
3.3 The State Postulate 78
3.4 Internal Energy, Enthalpy, and Specific Heats 78
3.5 Equations of State for Ideal Gases 80
3.6 Incompressible Substances 91
3.7 Property Determination for Water and Refrigerants 92
Summary 97
Problems 98

Chapter 4 THE FIRST LAW OF THERMODYNAMICS 107
4.1 Introduction 107
4.2 Conservation of Mass 108
4.3 First Law of Thermodynamics in Open Systems 112
4.4 First Law of Thermodynamics in Closed Systems 144

Chapter 9 IDEAL GAS MIXTURES 351

9.1 Introduction 351
9.2 Defining the Composition of a Gas Mixture 352
9.3 Ideal Gas Mixtures 357
9.4 Solutions of Thermodynamic Problems Incorporating Ideal Gas Mixtures 364
9.5 Introduction to Real Gas Mixture Behavior 370
Summary 372
Problems 372

Chapter 10 PSYCHROMETRICS: THE STUDY OF "ATMOSPHERIC AIR" 383

10.1 Introduction 383
10.2 Basic Concepts and Terminology of Psychrometrics 385
10.3 Methods of Determining Humidity 389
10.4 Comfort Conditions 397
10.5 Cooling and Dehumidifying of Moist Air 399
10.6 Combining the Cooling and Dehumidifying Process with Refrigeration Cycles 404
10.7 Heating and Humidifying Air 406
10.8 Mixing of Moist Air Streams 410
10.9 Cooling Tower Applications 413
Summary 416
Problems 417

Chapter 11 COMBUSTION ANALYSIS 427

11.1 Introduction 427
11.2 The Components of the Combustion Process 429
11.3 A Brief Description of the Combustion Process 431
11.4 Balancing Combustion Reactions 432
11.5 Methods of Characterizing the Reactant Mixture 437
11.6 Determining Reactants from Known Products 440
11.7 Enthalpy of a Compound and the Enthalpy of Formation 443
11.8 Further Description of the Combustion Process 445
11.9 Heat of Reaction 446
11.10 Adiabatic Flame Temperature 458
11.11 Entropy Balance for Combustion Processes 462
11.12 The Gibbs Function 465
11.13 Fuel Cells 465
11.14 Introduction to Chemical Equilibrium 468
11.15 The Water–Gas Shift Reaction and Rich Combustion 472
Summary 474
Problems 476

Appendices

A.1 Properties of Some Ideal Gases 489
A.2 Values of the Specific Heats at Different Temperatures for Common Ideal Gases (kJ/kg · K) 490

4.5	Thermal Efficiency of Heat Engines, Refrigerators, and Heat Pumps	150
	Summary	155
	Problems	156
Chapter 5	**INTRODUCTION TO THE SECOND LAW OF THERMODYNAMICS**	**173**
5.1	The Nature of the Second Law of Thermodynamics	173
5.2	Summary of Some Uses of the Second Law	175
5.3	Classical Statements of the Second Law	176
5.4	Reversible and Irreversible Processes	179
5.5	A Thermodynamic Temperature Scale	181
5.6	Carnot Efficiencies	182
5.7	Perpetual Motion Machines	185
	Summary	186
	Problems	187
Chapter 6	**ENTROPY**	**195**
6.1	Entropy and the Clausius Inequality	195
6.2	Entropy Generation	198
6.3	Evaluating Changes in the Entropy of a System	201
6.4	The Entropy Balance	205
6.5	Isentropic Efficiencies	218
6.6	Consistency of Entropy Analyses	228
6.7	Entropy Generation and Irreversibility	230
	Summary	234
	Problems	236
Chapter 7	**POWER CYCLES**	**251**
7.1	Introduction	251
7.2	The Ideal Carnot Power Cycle	253
7.3	The Rankine Cycle	255
7.4	Gas (Air) Power Cycles and Air Standard Cycle Analysis	287
7.5	Brayton Cycle	288
7.6	Otto Cycle	297
7.7	Diesel Cycle	303
7.8	Dual Cycle	307
7.9	Atkinson/Miller Cycle	310
	Summary	310
	Problems	310
Chapter 8	**REFRIGERATION CYCLES**	**327**
8.1	Introduction	327
8.2	The Vapor-Compression Refrigeration Cycle	330
8.3	Absorption Refrigeration	337
8.4	Reversed Brayton Refrigeration Cycle	338
	Summary	342
	Problems	342

A.3 Ideal Gas Properties of Air 491
A.4 Ideal Gas Properties of Nitrogen, Oxygen, Carbon Dioxide, Carbon Monoxide, Hydrogen, and Water Vapor 492
A.5 Thermodynamic Properties of Select Solids and Liquids 498
A.6 Properties of Saturated Water (Liquid-Vapor)—Temperature 499
A.7 Properties of Saturated Water (Liquid-Vapor)—Pressure 501
A.8 Properties of Superheated Water Vapor 503
A.9 Properties of Compressed Liquid Water 508
A.10 Enthalpy of Formation, Gibbs Function of Formation, Entropy, Molecular Mass, and Specific Heat of Common Substances at 25°C and 1 atm 509
A.11 Values of the Natural Logarithm of the Equilibrium Constant, ln K_p, for Various Chemical Equilibrium Reactions 510

Index **511**

Preface to the SI Edition

This edition of *Principles of Engineering Thermodynamics* has been adapted to incorporate the International System of Units (*Le Système International d'Unités or* SI) throughout the book.

Le Système International d'Unités

The United States Customary System (USCS) of units uses FPS (foot–pound–second) units (also called English or Imperial units). This system is also referred to as the English Engineering (EE) system of units. SI units are primarily the units of the MKS (meter–kilogram–second) system. However, CGS (centimeter–gram–second) units are often accepted as SI units, especially in textbooks.

Using SI Units in this Book

In this book, we have used both MKS and CGS units. USCS (U.S. Customary Units) or FPS (foot-pound-second) units used in the U.S. Edition of the book have been converted to SI units throughout the text and problems. However, in case of data sourced from handbooks, government standards, and product manuals, it is not only extremely difficult to convert all values to SI, but it also encroaches upon the intellectual property of the source. Some data in figures, tables, and references, therefore, remain in FPS units. Chapters 1 and 2 in particular contain USCS and FPS units; these chapters introduce you to these systems of units and allow you to practice converting between the most common systems. For readers unfamiliar with the relationship between the USCS and the SI systems, a conversion table has been provided inside the front cover.

To solve problems that require the use of sourced data, the sourced values can be converted from FPS units to SI units just before they are to be used in a calculation. To obtain standardized quantities and manufacturers' data in SI units, readers may contact the appropriate government agencies or authorities in their regions.

CHAPTER 1

Introduction to Thermodynamics and Energy

Learning Objectives

Upon completion of Chapter 1, you will be able to:

1.1 Describe what the subject of thermodynamics studies, and identify the types of engineering applications that involve thermodynamics;

1.2 Discuss concepts such as thermodynamic systems, processes, thermodynamic equilibrium, and properties;

1.3 Manipulate different temperature scales;

1.4 Recognize the difference between mass and force;

1.5 Use the basic properties of volume and pressure;

1.6 Explain and apply the Zeroth Law of Thermodynamics; and

1.7 Identify the different phases of matter.

Look around you. What do you see? You probably see people and things in motion, electrical devices in operation, and comfortable buildings. What you are seeing is energy being used. The use of energy is so commonplace that most people don't even notice it until it is unavailable, such as during an electrical power outage in a storm, or when an automobile runs out of fuel, or when someone is weak due to a lack of food. Anything that moves needs some energy to do so. Anything that is powered by electricity needs energy. Even the earth as a whole needs energy from the sun to stay warm and for life to flourish. The world as we know it exists because of energy in action.

If you have been even casually following the news in recent years, you have probably seen stories involving energy in the world. Stories on the rising costs and stressed supplies of petroleum or electricity are common. There are serious concerns over the effects on the environment caused by the rising levels of carbon dioxide (CO_2) in the atmosphere produced through the burning of fossil fuels. As people seek new, cleaner sources of energy, we see wind turbines rising across the world and solar panels appearing in locations that have rarely seen the use of such technology. The price and availability of energy, as well as how its use impacts the environment, is increasingly important to society.

Yet the demand that humans have for energy is at its greatest level ever. Not only are there more people than ever before, but people worldwide want the goods and standard of living associated with the wealthiest nations. People want ready access to transportation, and transportation systems require energy. People want buildings heated and cooled to the desired comfort level, and this requires energy. Factories use prodigious amounts of energy to produce the products that people want. Growing and transporting food requires energy. The demand for energy has never been higher, and it is likely to keep growing.

It can even be said that the harnessing of energy sources has been a key element in the development of civilization. **Figure 1.1** shows a number of examples of how the use of energy by people has developed over time. People learned how to control fire for heating and cooking. People harnessed the power of wind to pump water. Engines were developed to allow the chemical energy in a fuel to do useful work for us. Humans even learned how to unleash the power locked inside atoms to generate electricity. In the future, we don't know how humans will use the energy present in nature all around us, but it is likely that new means of harnessing energy will be needed to keep the development of civilization moving forward.

Engineers play a key role in creating systems that convert energy from one form to another, usually taking energy in a form that is otherwise rather useless and transforming

Dutourdumonde Photography/Shutterstock.com

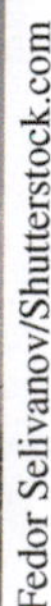

Fedor Selivanov/Shutterstock.com

Oliver Sved/Shutterstock.com

SpaceKris/Shutterstock.com

FIGURE 1.1 Various images showing the application of energy by humans throughout history: a fire, a windmill, an automobile engine, and a nuclear power plant.

it so that people can use it to do something productive. For example, the energy bound up in the molecules that make up the fluid called "gasoline" is of little use as is. But if the gasoline is ignited with air in a combustion process, large amounts of heat can be released, and this heat can be used to create a high-temperature, high-pressure gas that can push on a piston in an engine, as shown in **Figure 1.2**; the work produced can be used to propel a vehicle forward. Engineers also play a key role in designing systems that use energy efficiently. Engineers can create devices that use less energy to accomplish the same task, and by using less energy these devices save consumers money. Furthermore, more efficient devices reduce the overall demand for more energy. For example, even a technology as widespread as lighting, illustrated in **Figure 1.3**, has seen dramatic improvements in energy efficiency. Light-emitting diode (LED) lights can be six times more efficient than incandescent bulbs and 40% more efficient than compact fluorescent lights, and last much longer than either technology. If engineers are to develop means of using energy efficiently while benefiting humankind significantly, they must understand the basic science behind energy.

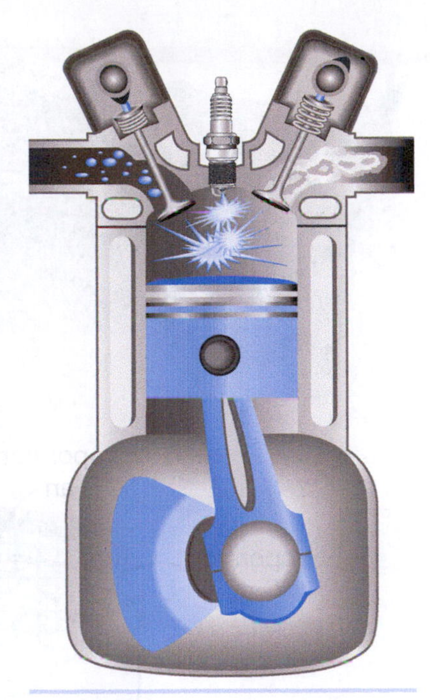

FIGURE 1.2 A cutaway image of an internal-combustion engine cylinder.

Thermodynamics is the science of energy. If you look at the original Greek roots of the word, *thermos* means "heat" and *dynamikos* means "power"—and power applied to an object produces movement. Thermodynamics is the power of heat, or the movement of heat. As our understanding of energy has evolved, we recognize that energy involves more than just heat, and, as such, thermodynamics is considered the science of all energy. In engineering, we use thermodynamics to understand how energy is transformed from one form to another to accomplish a given purpose. Thus, we will be exploring not only the basic laws that describe thermodynamics but also the technology that is employed to accomplish tasks using energy.

FIGURE 1.3 Examples of lighting technology: an incandescent light bulb, a compact fluorescent fixture, and an LED bulb.

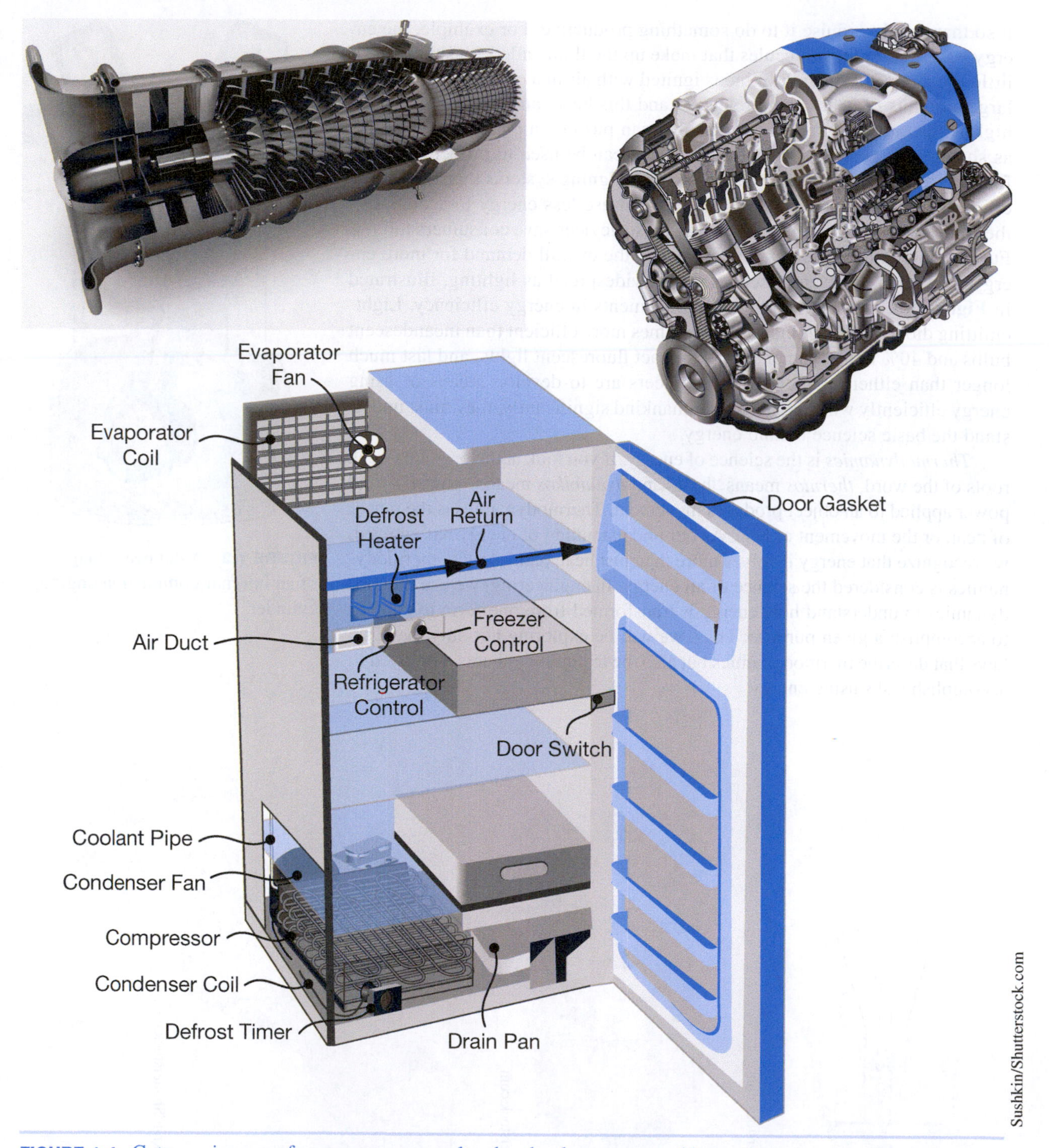

FIGURE 1.4 Cutaway images of common energy-related technology: a gas turbine engine, a reciprocating engine, and a refrigerator.

Figure 1.4 shows many applications in the world today, developed by engineers, that use energy and for which thermodynamics is an integral design component. Turbines are used to transform the energy in a working fluid into a rotational motion of a shaft that in turn produces electricity in a generator. The turbines take in a high-energy gas or vapor (which generally has a high temperature and pressure) and extract energy from that fluid to produce the work needed to turn the shaft (also known as a rotor). A low-energy fluid is exhausted from the turbine. An automobile engine takes high-temperature, high-pressure gases (formed from the combustion of fuel in air, which releases the chemical energy bound up in the fuel) and has these gases push on a piston. The piston then drives the crankshaft, which transmits power to the wheels, thus moving the vehicle forward. Cooler, lower pressure gases exit from the cylinder after their energy had been extracted.

A refrigerator is used to keep food cool. It accomplishes this by taking electrical power and using this power to operate a compressor that increases the pressure of a vapor. Before the vapor is compressed, it is cooler than the interior of the refrigerator, and so it can remove heat from inside the refrigerator to cool the interior. After it is compressed, the vapor is hotter than the room temperature, and it is able to release the excess heat to the air outside the refrigerator. So, in this case, electrical power is used to change the state of the refrigerant so that it can accomplish the task of moving energy from a cooler space inside the refrigerator to a warmer space outside the refrigerator.

There are a vast number of devices and systems that are encountered every day that use thermodynamics to some extent, some of which are shown in **Figure 1.5**. A furnace to heat a building, an air conditioner to cool a building, the radiator on an engine, the sun heating the earth, a light bulb illuminating (and heating) a room, a bicycle being ridden, and a computer generating heat while performing its tasks are all such examples. All around you, energy is changing forms. Energy is moving throughout the world. Thermodynamics describes these energy motions and transformations. Although conventional thermodynamic analysis is not needed to analyze many things in everyday life, keep in mind that thermodynamics is fundamental to how the world functions.

Oleksiy Mark/Shutterstock.com

liseykina/Shutterstock.com

EcoPrint/Shutterstock.com

FIGURE 1.5 Examples of thermodynamics in the world today: an industrial furnace, the sun heating the earth, and human-powered bicycles.

Thermodynamics, and thermodynamic analysis, is defined by four scientific laws. These laws will be introduced when appropriate throughout the book. Scientific laws are not absolutely proven principles, but rather are concepts that are well established through observation and have never been shown to be incorrect. Occasionally, a law may need to be refined as our knowledge of the world deepens, but the basic principle usually remains unchanged. (For example, the law of the conservation of energy had to be modified to include mass for nuclear processes when Einstein showed that mass and energy were equivalent.) If one of the four laws upon which thermodynamics is based is ever shown to be incorrect, future scientists and engineers will need to reformulate the basics upon which thermodynamics rests. However, this is extremely unlikely.

Thermodynamics is one of the basic sciences that engineers use daily as they apply scientific principles to solve problems to aid humanity. This doesn't mean that every engineer will be performing thermodynamic analyses every day of his or her career, but some engineers do frequently design and analyze devices and processes by relying on the principles of thermodynamics. Others will only occasionally need to invoke thermodynamic principles in their careers. Still others rarely use thermodynamics directly, but thermodynamics still informs their work and may influence their work in ways that are not immediately apparent. As such, it is important for all engineers to be fluent in the basics of thermodynamics.

Before we can explore the principles of thermodynamics, we must first define and describe a number of basic concepts upon which our subsequent presentations will be based. This is the focus of the next section.

1.1 BASIC CONCEPTS: SYSTEMS, PROCESSES, AND PROPERTIES

1.1.1 The Thermodynamic System

At the basis of all thermodynamic analysis is a construct known as the ***thermodynamic system***. A thermodynamic system is the volume of space that contains the object(s) that are the focus of the thermodynamic analysis. The system is defined by the person performing the analysis and should be made as simple as possible. Unnecessary complexity should be avoided because it will either result in an incorrect analysis or will lead to significant amounts of additional work on the part of the person performing the analysis.

As shown in **Figure 1.6**, a thermodynamic system is delineated by a system boundary; everything inside the system boundary (which we will represent with a dashed line) is the system, and everything outside the boundary is considered the *surroundings*. **Figure 1.7** shows several possible systems that could all be considered the system for analyzing a particular problem. The quantity to be determined is the amount of heat needed to heat liquid water in a kettle on a stove. In Figure 1.7a, the system is proposed to be only the water in the kettle. In Figure 1.7b, the system is proposed to be the water and the kettle. In Figure 1.7c, the system is proposed to be the water, kettle, and the air above the water inside the pot. All three of these systems could be used to analyze the problem. However, the systems in (b) and (c) add complexity to the fundamental problem of determining the amount of heat that is added to the

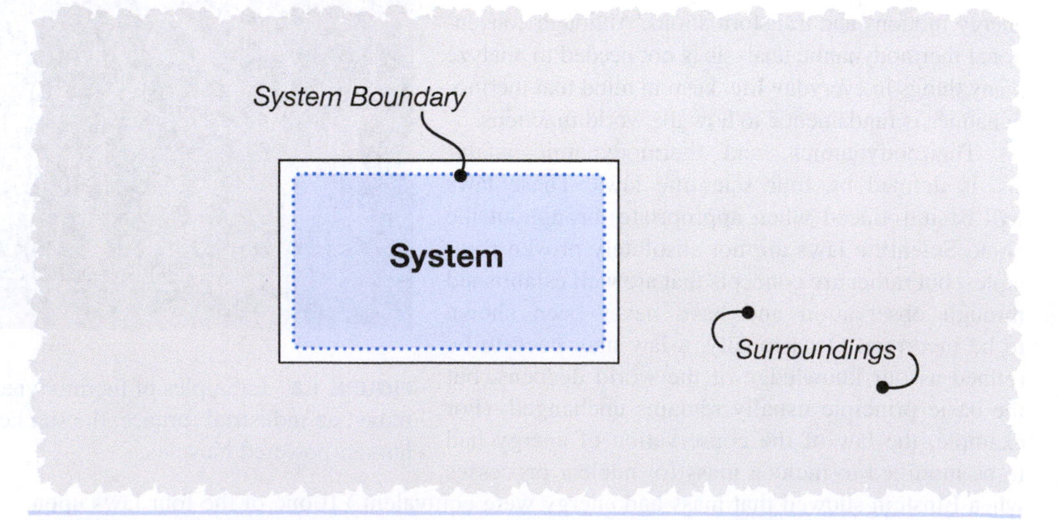

FIGURE 1.6 Example of a thermodynamic system, the system boundary (represented by a dashed line), and the surroundings.

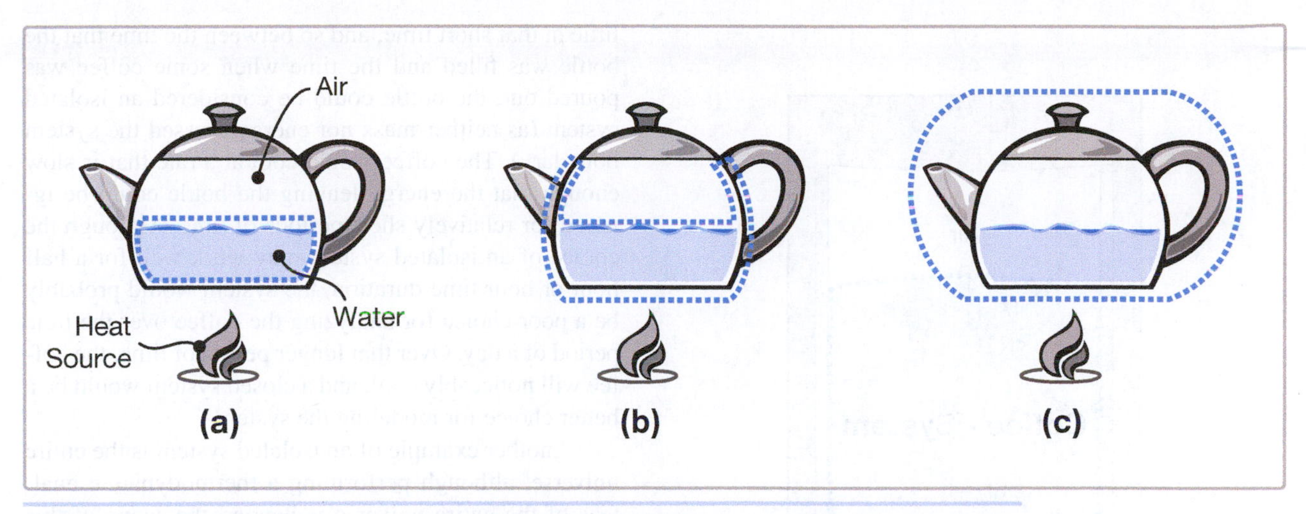

FIGURE 1.7 Examples of how the choice of the thermodynamic system changes the problem to be analyzed: (a) the system as only the water, (b) the system as the water and kettle, and (c) the system as the kettle, its contents, and its immediate surroundings.

water alone. In system (b), we would need to determine how much heat was added to both the water and the kettle, and then further analysis would be needed to determine how much heat was added to the water itself. In system (c), the problem would be further complicated by needing to determine how much heat was also added to the air, and then again separating out only the heat added to the water. So, although all three systems could be used for the problem, it is best to take care when defining the extent of the system so that the smallest volume possible is used in the analysis. This will reduce the amount of extra work that is necessary for finding the solution.

There are three types of systems; the types of systems are differentiated by whether or not mass and/or energy can cross the system boundary:

Isolated System: Neither mass nor energy can cross the system boundary.
Closed System: Energy can cross the system boundary, but mass cannot.
Open System: Both mass and energy can cross the system boundary.

Sometimes, a closed system is referred to as a "control mass," whereas sometimes an open system is referred to as a "control volume." However, in this book we will refer to each by the name of "closed" or "open" system, respectively.

In determining what type of system is to be employed for an analysis, we need to consider the important characteristics of the problem on the time scale likely to be employed. It is likely that given enough time, some small amount of mass may diffuse across the solid boundary of a closed system; however, if the amount of mass flowing across the system boundary is negligible over the time frame of the problem, it is likely best to view the system as a closed system. For example, if we have a bicycle tire filled with air, and the tire has no obvious leaks, it is safe to treat the system as a closed system if the time period under consideration is no more than a day or two. But if we are considering the tire over the course of a year, we may need to consider the impact of air very slowly leaving the tire—which would make it an open system.

Isolated systems will play a minor role in this book but are important in more advanced thermodynamics studies. One type of system that could be considered an isolated system is an insulated thermos bottle, after it has been closed, as shown in Figure 1.8. Suppose the bottle is filled with hot coffee at the start of the day. The bottle is closed, and then a short time later the bottle is opened and some coffee is poured out. The conditions of the coffee would change

FIGURE 1.8 A cutaway diagram of an insulated bottle containing coffee and air, with the system being only the coffee.

little in that short time, and so between the time that the bottle was filled and the time when some coffee was poured out, the bottle could be considered an isolated system (as neither mass nor energy crossed the system boundary). The coffee would cool at a rate that is slow enough that the energy leaving the bottle could be ignored for relatively short periods of time. Although the choice of an isolated system may work well for a half hour or hour time duration, the system would probably be a poor choice for analyzing the coffee over the time period of a day. Over that longer period of time, the coffee will noticeably cool, and a closed system would be a better choice for modeling the system.

Another example of an isolated system is the entire universe, although performing a thermodynamic analysis of the entire universe is beyond the scope of this book. As far as we understand the universe, it contains all of the mass and energy that exists, and so mass and energy cannot cross the system boundary.

There are many more practical examples of closed systems, and some of these are shown in **Figure 1.9**. Solid objects generally are considered closed systems, unless they are specifically losing mass. A liquid or gas inside a closed container also is typically viewed as a closed system. The contents inside objects such as sealed

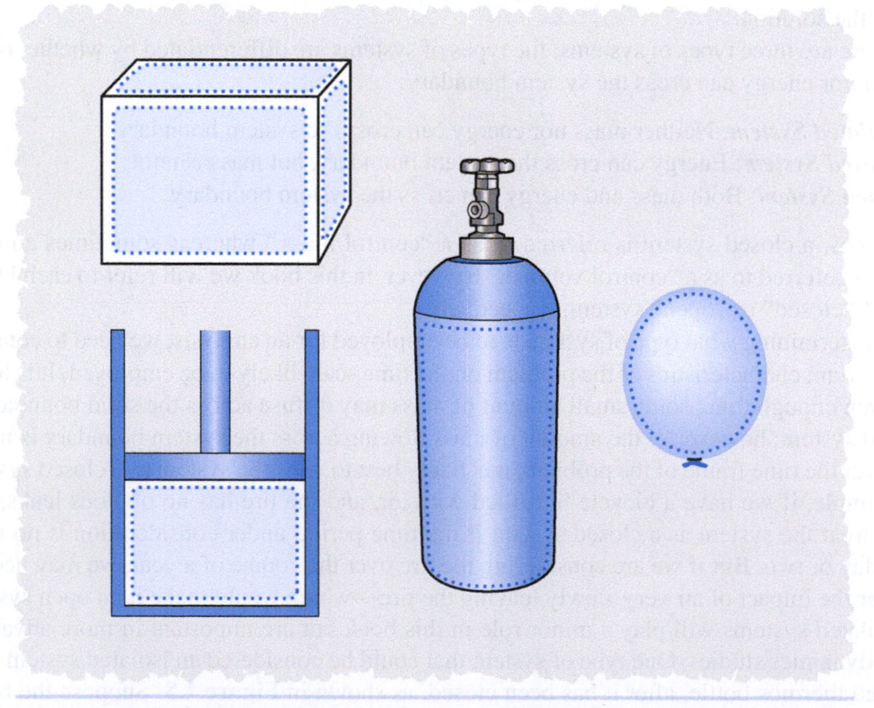

FIGURE 1.9 Examples of closed systems: a solid block of metal, the gas inside a tank, the air inside a balloon, and the gas in a piston–cylinder device.

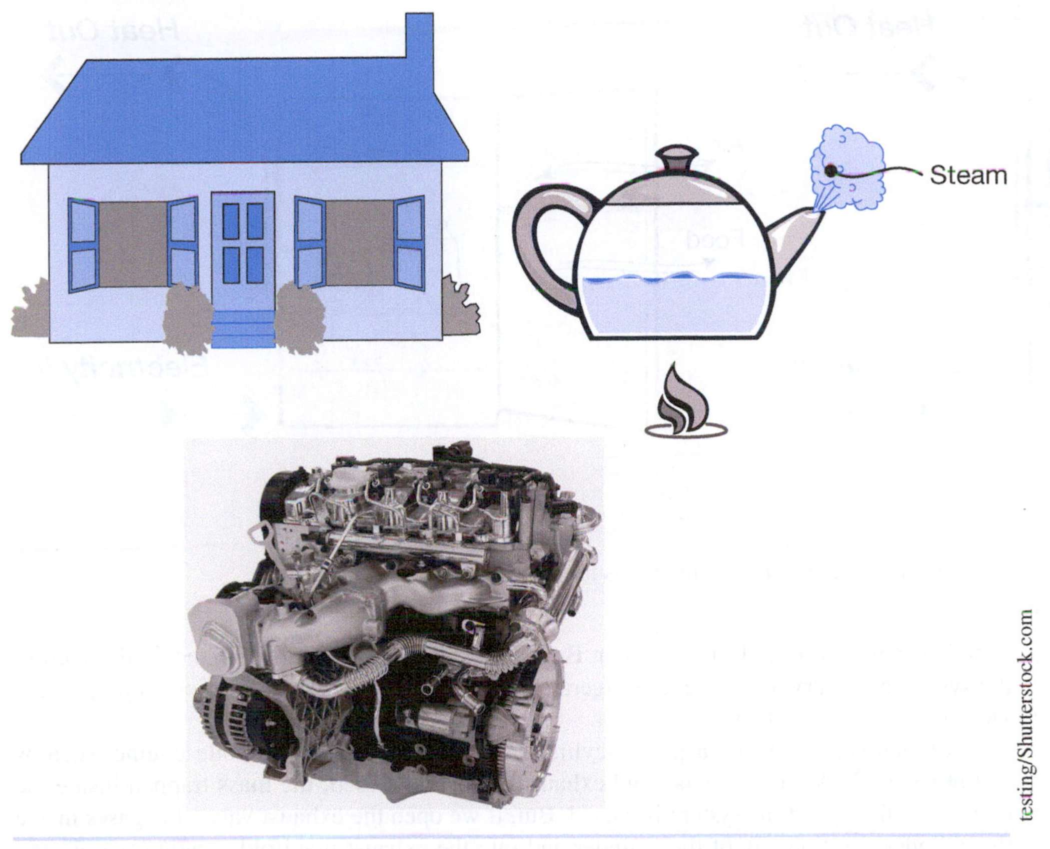

FIGURE 1.10 Examples of open systems: a house with open windows, steam escaping from a kettle, and an automobile engine.

balloons are considered closed systems because, even though their volume changes, the mass inside the balloon doesn't change. Similarly, a piston–cylinder assembly containing a fixed mass of a gas or liquid will be considered a closed system.

Any system that clearly has mass being added to or removed from it is viewed as an open system. Figure 1.10 provides a few of the many possible examples of open systems. A garden hose, a house with open windows, an air compressor, an automobile engine, and a kettle containing escaping boiling water are all examples of open systems. It should be noted that closed systems can be viewed as special applications of open systems. A thermodynamic analysis of a generic open system contains all the elements that would be seen in a closed system analysis; however, the closed system analysis will allow terms involving mass flow into and out of the system to be eliminated.

When determining the type of system to be used, it is important to consider the application of the object under consideration. For example, if we are in the act of filling or emptying a thermos bottle, we are dealing with what is clearly an open system rather than a potentially isolated system. Or, consider a kitchen refrigerator as shown in Figure 1.11. Assuming that the door is well sealed, the mass inside the refrigerator is fixed when the refrigerator door is closed. Energy in the form of electricity is still flowing into the refrigerator, and energy in the form of heat is being rejected to the environment as the refrigerant cools. (Heat is also slowly flowing through the walls to the interior of the refrigerator, but this heat can often be neglected for a well-built refrigerator.) Because energy can cross the system boundary but mass cannot, this would be viewed as a closed system. Now, consider if the door is open so that food can be

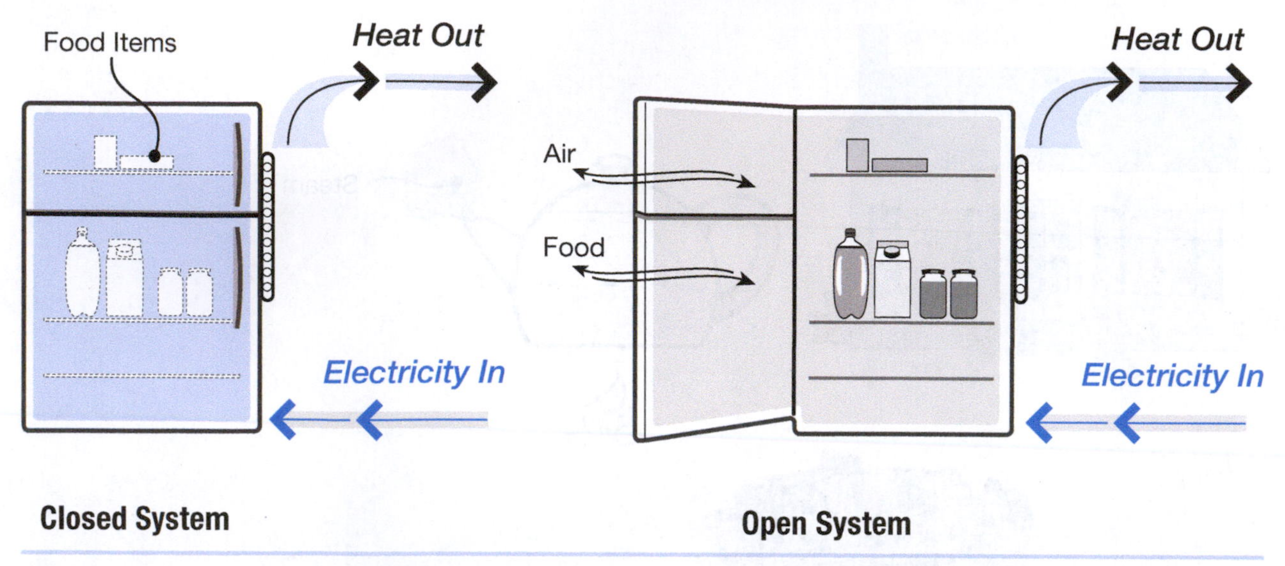

FIGURE 1.11 A refrigerator viewed as a closed system and as an open system.

added or removed from the refrigerator. Both air and the mass contained in the food can cross the system boundary; therefore, a refrigerator with an open door would be more appropriately viewed as an open system.

Alternatively, consider a piston–cylinder device inside an automobile engine, such as in **Figure 1.12**. When the intake and exhaust valves are closed, the mass trapped inside the cylinder is fixed, and the system is closed. But, if we open the exhaust valve, the gases inside the cylinder can flow out of the cylinder and into the exhaust manifold, making the piston–cylinder device an open system. Therefore, the choice of the type of system to use depends on the portion of the engine cycle under consideration.

Previously, we said that a garden hose is an open system. However, suppose that the valve that allows water to flow into the hose is shut. At this point, as long as the hose does not leak, there is a fixed mass of water that resides inside the hose. Some water may slowly evaporate, but because little mass if actively flowing into or out of the hose, the hose would be best considered a closed system. So, keep in mind that we cannot always determine whether a system is open or closed just by identifying the object under consideration, but rather we should also learn the nature of the process impacting the system.

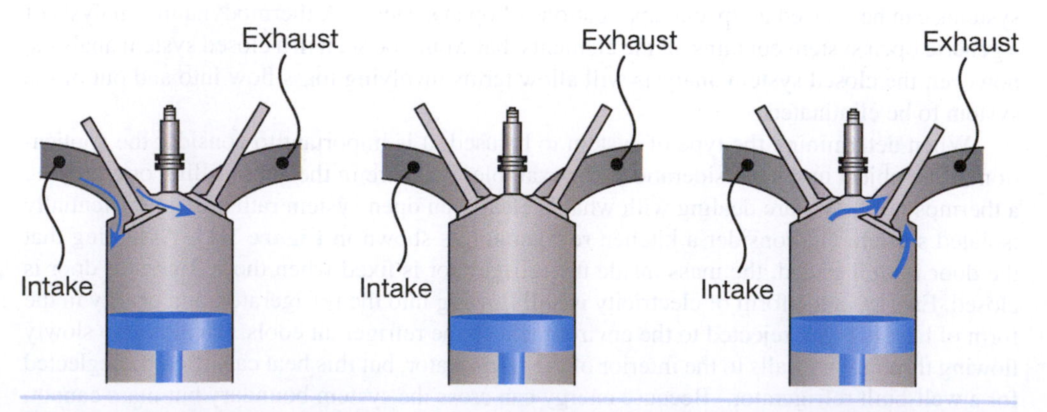

FIGURE 1.12 An engine cylinder as an open system (with intake valve open), as a closed system (with both valves closed), and as an open system (exhaust valve open).

1.1.2 The Thermodynamic Process

A thermodynamic system exists in a particular state, as described by the properties of the system, at a particular time. If the system is undergoing a change in state, the system is undergoing a thermodynamic process. A ***thermodynamic process*** is the action of changing a thermodynamic system from one state to another and is described by the series of thermodynamic states that a system experiences as it transforms from the "initial state" to the "final state." This is a rather formal definition of what is intuitively a rather simple concept. Essentially, a process is what happens to a system when mass and/or energy is added to or removed from the system, or as the mass or energy inside the system undergoes an internal transformation. Examples of processes include the heating or cooling a system, the compressing of a system or its expansion, and the changes in a system as electricity flows into it. Having an object accelerate as it falls from a height is an example of a process involving an internal transformation of energy, as will be described in Chapter 2. Normally, we show the progression of a thermodynamic process by drawing a diagram that illustrates how two of the system's properties vary during the process. Figure 1.13, for example, shows the thermodynamic process of heating air in a rigid (fixed volume) container by visually illustrating the relationship between the pressure and temperature for this process.

Often systems undergo a series of processes. As an example, consider the gases inside a piston–cylinder assembly in an automobile engine with the intake and exhaust valves closed. These processes are shown in Figure 1.14. First, the air and fuel mixture is compressed by the piston. Then a spark is used to ignite

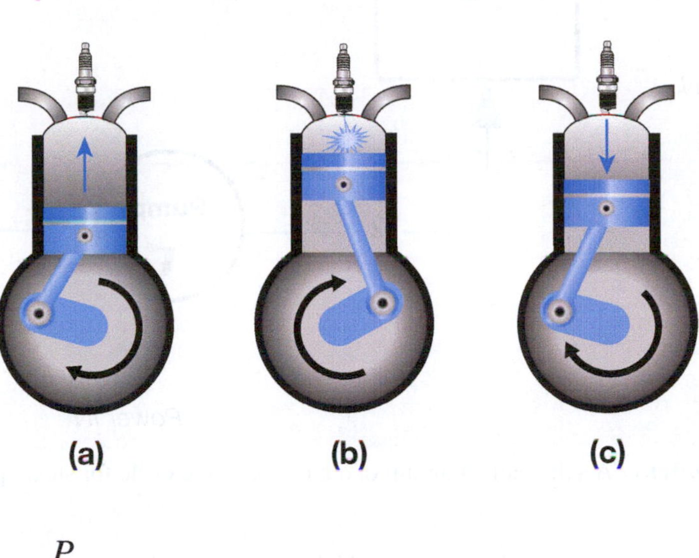

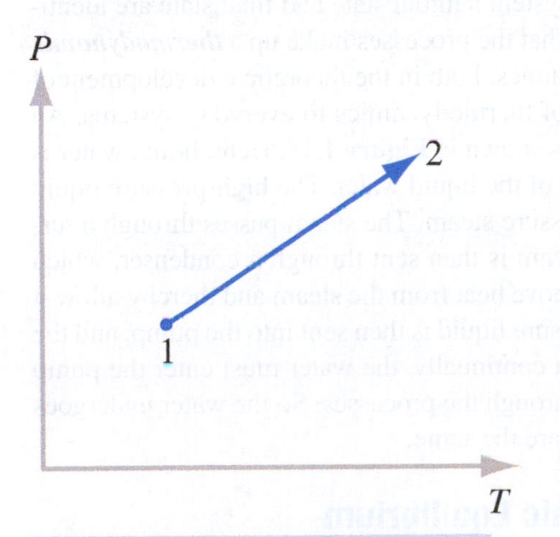

FIGURE 1.13 A pressure vs. temperature (P-T) diagram of a thermodynamic process.

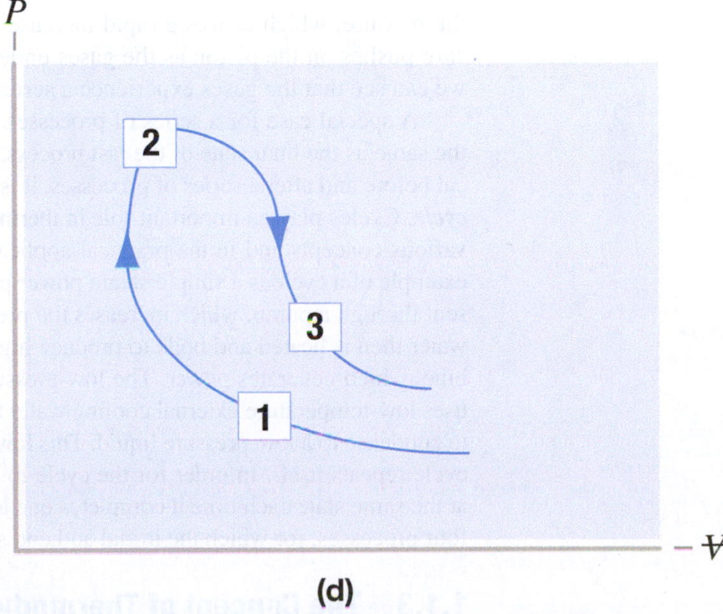

FIGURE 1.14 Three processes that occur during automobile engine operation: (a) the compression stroke, (b) the combustion process, and (c) the expansion stroke. (d) The processes are then shown on a pressure vs. volume (P-V) diagram.

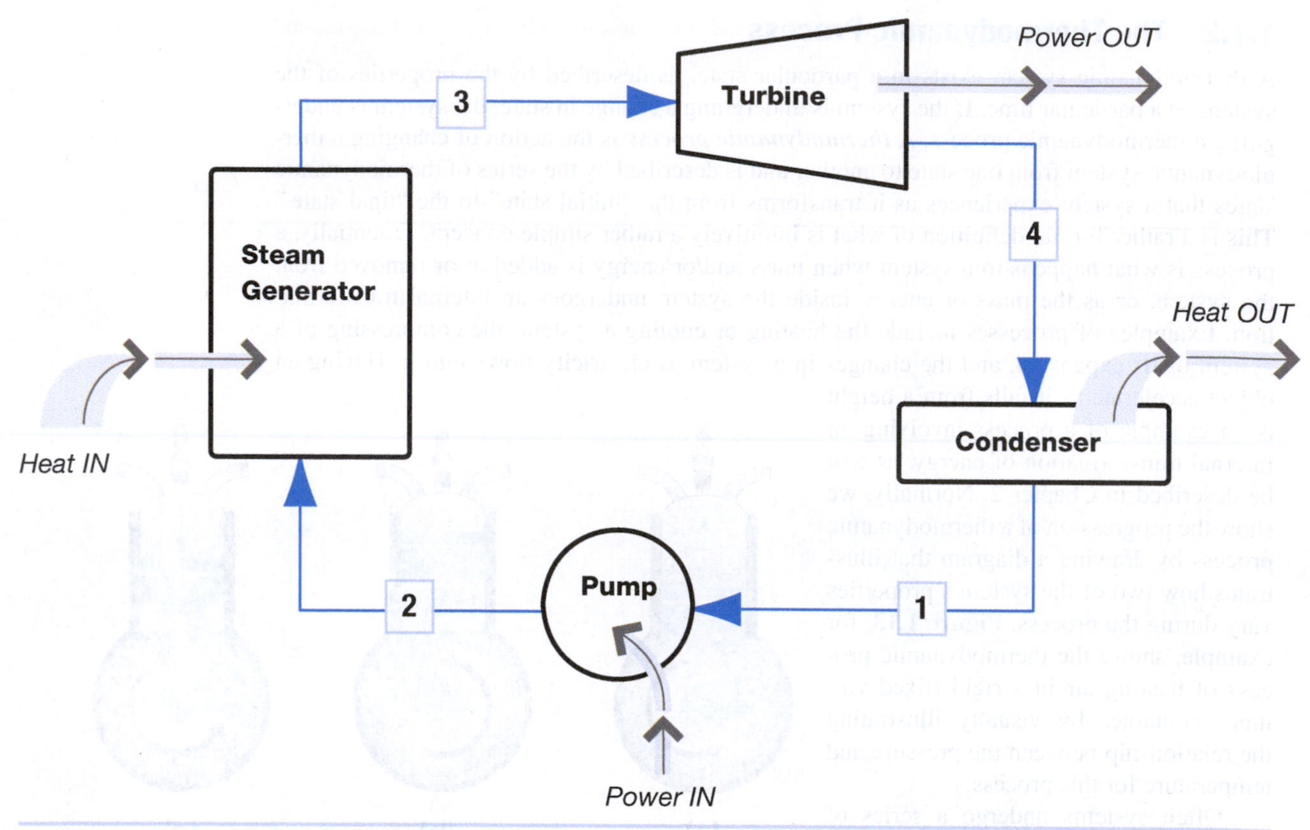

FIGURE 1.15 A schematic diagram of a basic Rankine cycle for steam power plants.

the mixture, which causes a rapid increase in temperature and pressure. The increased pressure pushes on the piston as the gases undergo an expansion process. From this description, we can see that the gases experience a series of three processes with the valves closed.

A special case for a series of processes occurs when the initial state of the first process is the same as the final state of the last process. If a system's initial state and final state are identical before and after a series of processes, it is said that the processes make up a ***thermodynamic cycle***. Cycles play an important role in thermodynamics, both in the theoretical development of various concepts and in the practical application of thermodynamics to everyday systems. An example of a cycle is a simple steam power plant as shown in Figure 1.15. Here, liquid water is sent through a pump, which increases the pressure of the liquid water. The high-pressure liquid water then is heated and boils to produce high-pressure steam. The steam passes through a turbine, which generates power. The low-pressure steam is then sent through a condenser, which uses low-temperature external cooling water to remove heat from the steam and thereby allow it to condense to a low-pressure liquid. This low-pressure liquid is then sent into the pump, and the cycle repeats itself. In order for the cycle to repeat continually, the water must enter the pump at the same state each time it completes one loop through the processes. So the water undergoes four processes, for which the initial and end states are the same.

1.1.3 The Concept of Thermodynamic Equilibrium

Equilibrium, in general, is defined as a state of balance due to the canceling of the actions by opposing forces. There are many different types of equilibrium, some of which you may already be familiar with. Mechanical equilibrium is the state that exists when a system undergoes no acceleration because the mechanical forces acting on the system are balanced. In thermodynamics, we often will view mechanical equilibrium as the state at which the pressure in a system is uniform. Thermal equilibrium is the state when two or more systems are at the same temperature, or when a single system has a uniform temperature.

Thermodynamic equilibrium is the state that exists when a system is in a combination of thermal, mechanical, chemical, and phase equilibriums. A system that is in thermodynamic equilibrium does not have the capacity to spontaneously change its state; for a system in thermodynamic equilibrium to change its state, it must experience a driving force from outside the system.

In this book, we will study equilibrium thermodynamics. There is another branch of thermodynamics known as nonequilibrium thermodynamics that is beyond the scope of this book and that does not have as many practical applications for most engineers. In equilibrium thermodynamics, we assume that our systems are in a state of thermodynamic equilibrium. You may note, though, that as a system proceeds from one state to another during a process, it may not be at equilibrium every instant of time. For example, if a container of water is heated on a stove, the water at the bottom of the container near the heat source may be warmer than the water at the top of the container. Given sufficient time, the system will all be at the same temperature, but at a given instant in time, the system may not technically be at thermal equilibrium. In such situations, we normally consider the system to be passing through a series of quasi-equilibrium states. Although such states are not strictly in equilibrium, the deviations from equilibrium experienced in these states are insignificant in terms of the overall analysis of the problem, and the deviations exist for relatively short periods of time. Although these deviations from equilibrium may cause errors in a very detailed analysis of a system, they generally do not cause significant errors in an engineering analysis or design of a system.

1.1.4 Thermodynamic Properties

We need a way to describe the state of a thermodynamic system. The thermodynamic properties of a system are the tools that are used to describe the state of a system. A ***thermodynamic property*** is a quantity whose numerical value is independent of how the state of a system was achieved and only depends on the system's local thermodynamic equilibrium state. This is an important distinction, because it means that anything whose value depends on a specific process that a system undergoes is not a property of the system but is rather a description of the process.

You are undoubtedly already familiar with many properties, such as temperature, pressure, mass, volume, and density. As an example of how temperature fits our definition of a property, consider air in a room. If air in a room is at a particular temperature, a description of the system does not need to explain how that temperature was achieved (via heating or cooling); the system is simply at that temperature. Other properties of note are viscosity, thermal conductivity, emissivity, and many more that will be introduced when appropriate throughout the book. Color also fits the definition of a property, because a system's color can be described numerically through a spectrum describing the wavelengths of light composing the color.

Some properties are considered ***extensive*** properties and some are considered ***intensive*** properties. An extensive property is a property whose value depends on the mass of the system, whereas an intensive property is one whose value is independent of a system's mass. A quick method to use to determine whether a property is extensive or intensive is to mentally divide the system in half and determine if the value of the property in half the system would change. Doing this should yield the result that temperature, pressure, and density are intensive properties, because a system of half the size will have the same value of those properties as the original system (i.e., dividing a system in half should have no impact on its temperature). However, mass and volume are extensive properties because each would have different values in a system half the size of the initial system (i.e., the volume of a system half the size of the initial system is clearly half of the initial volume).

Extensive properties can be transformed into intensive properties by dividing the extensive property by the mass of the system. Such transformed properties are given the name "specific." For example, the volume of a system, $\forall$, can be divided by the mass, m, yielding the specific volume, v:

$$v = \forall/m \tag{1.1}$$

You may note that the specific volume is the inverse of the density, ρ, which is defined as the mass divided by the volume ($\rho = m/\forall$).

As will be discussed below, in general it is easier to work with intensive properties of a system in a thermodynamic analysis, and then multiply the value of the property by the system's mass to get the total value of the extensive quantity for that particular system. Working with intensive properties allows us to avoid needing to perform calculations for every possible mass of a system undergoing a process. We can solve the process on an intensive (per unit mass) basis in general and then apply that solution to any specific mass undergoing the process.

Properties of a substance are related to each other through ***equations of state***. Some equations of state, as we will see, are very simple relationships, whereas others are so complicated that they are better calculated through computer programs. An example of a simple equation of state that you are probably familiar with is the ideal gas law:

$$P\forall = mRT \tag{1.2}$$

where P is the pressure, T is the temperature, and R is the gas-specific ideal gas constant, which is equal to the universal ideal gas constant, $\overline{R}$, divided by the molecular mass of the gas, M. The ideal gas law can also be written in terms of the specific volume:

$$Pv = RT \tag{1.3}$$

We will explore the ideal gas law in more detail later.

Considering Eq. (1.3), we can see that for a particular gas, three properties are related through the given equation of state. Two of these properties can be independently set, and the third property will be calculated from knowledge of those other two properties. The properties that can be arbitrarily chosen for a particular system are called ***independent properties***, whereas the properties whose values are subsequently determined through an equation of state are known as ***dependent properties***. In Eq. (1.3), if the pressure, P, and specific volume, v, are chosen as the independent properties, the temperature, T, is a dependent property whose value is determined through the ideal gas law. Similarly, if P and T are known as the independent properties, then v is the dependent property.

Equations (1.2) and (1.3) also provide a concrete example of the benefit of using specific properties. Suppose you were asked to compile a list of information that could be used to find the total volume occupied by a particular gas for a set of pressures and temperatures. If you were to use Eq. (1.2) to calculate the total volume directly, you would need to compile a list for every possible mass of the gas. However, if you used Eq. (1.3), you could prepare one list of specific volumes, and then ask the user of the data to just multiply the given number by the system's particular mass. Clearly, the second approach is simpler.

Below, we will formally introduce the basic, easily measured properties that are commonly used to describe a system. But, first, we need to comment on the nature of the unit systems involved in thermodynamics.

1.1.5 A Note on Units

In engineering practice, there are generally two broad systems of units that are employed. One system, known as the International System (SI), is common throughout much of the world, whereas the other system, known as the English Engineering (EE) system, is mostly limited to use in the United States, although the SI system is becoming more commonly used there as well. The SI system of units is a system that is based upon scientific principles and strongly employs a decimal numbering system. The EE system developed haphazardly over time using measurements of convenience that often lacked consistent universal standards. Although standards have since been developed, unit conversions in the EE system are often nonintuitive and today may appear to have been assigned arbitrary values.

Although this book will primarily make use of the SI system of units, we will spend some time in this chapter and Chapter 2 introducing you to the units of the EE system. This will also allow you the opportunity to practice converting between these commonly used systems of units. Conversion tables between the SI and EE systems can be found within the inside covers of this book.

To illustrate the intuitive simplicity of the SI system, let us compare the units used in the two systems for a measurement of length. The SI system uses the meter, m, as the base unit of length. If a system is substantially larger than or smaller than a meter, we can apply one of the prefixes shown in **Table 1.1** for the SI system. So, 1 millimeter (mm) equals one thousandth of a meter: 1 mm = 0.001 m. One kilometer (km) equals one thousand meters: 1 km = 1000 m. In common practice, the centimeter (cm) is also used and is equal to one hundredth of a meter: 1 cm = 0.01 m.

For the EE system, the base unit of length is the foot (ft). To convert to larger or smaller sizes, we would employ conversion factors such as 1 ft = 12 inches, and 1 mile = 5280 ft. Although such conversions are possible, they can introduce unnecessary complexity to calculations and can more easily introduce mistakes into the calculations than the SI unit conversions.

The unit of time for both the SI and EE system of units is the second (s). Both systems of units will employ concepts such as milliseconds (ms) for 0.001 s. In addition, both systems consider 1 minute equal to 60 seconds, and 1 hour to be 3600 seconds. These conversions are the only situations where the SI system of units fails to deal exclusively with factors of 10 in unit conversions.

Remember that once the thermodynamic principles are learned, they are applied in the same way no matter which unit system is employed. It is important in both unit systems to be certain that the units employed in equations appropriately cancel so as to give the correct final units.

QUESTION FOR THOUGHT/DISCUSSION

Most of the world uses the SI system of units, whereas the United States generally employs the EE system of units. Would it be a good idea for the United States to adopt the SI system, and what might be needed for this to occur?

TABLE 1.1 Common Prefixes for the SI System of Units

Prefix	Factor	Symbol
Pico	10^{-12}	p
Nano	10^{-9}	n
Micro	10^{-6}	μ
Milli	10^{-3}	m
Centi	10^{-2}	c
Kilo	10^{3}	k
Mega	10^{6}	M
Giga	10^{9}	G
Tera	10^{12}	T

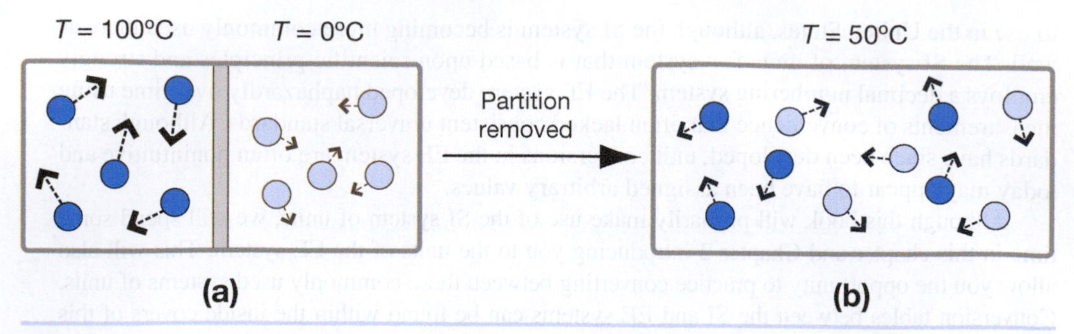

FIGURE 1.16 (a) A box partitioned in half with some molecules at 100°C and some molecules at 0°C. (b) After the hot molecules collide with the cold molecules, the molecules are all at 50°C.

1.2 AN INTRODUCTION TO SOME COMMON PROPERTIES

As mentioned previously, some thermodynamic properties are familiar, commonly encountered, and (rather) easily measured. These are the properties that we often use to describe a system. In this section, we will discuss some of these common properties and the units associated with them.

1.2.1 Temperature

When we think of concepts such as "heat," the first property we often think of is temperature. And although we discuss temperatures freely, we may find it difficult to define the word *temperature*. Temperature is a measure of the molecular motion and the energy associated with the motion inside a system. Systems with low temperatures have relatively slow motions associated with their atoms and molecules, whereas systems with high temperatures have relatively fast motion of their atoms and molecules. In one sense, this is why a hot system will tend to cool when it comes in contact with a cool system (and why the cool system heats up). The fast-moving molecules in the hot system will collide with the slow-moving molecules in the cold system, transferring some energy to the slower-moving molecules. This causes the fast-moving molecules to slow (leading to a decrease in temperature for the hot system) and the slow-moving molecules to speed up (leading to an increase in temperature for the cold system); this process is illustrated in Figure 1.16.

There are many methods for measuring the temperature of a system, but the one most commonly recognized is the thermometer (see Figure 1.17). If we look at a thermometer, we will see a scale showing a series of marks indicating the "degrees" of the substance. In the SI system of units, the unit of temperature used is the degree Celsius (°C). This system is given logical scientific reference points to define the scale:

0°C corresponds to the freezing point of water at atmospheric pressure.
100°C corresponds to the boiling point of water at atmospheric pressure.

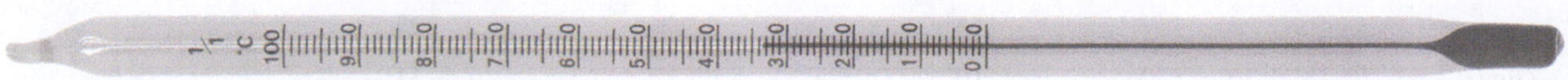

TAGSTOCK1/Shutterstock.com

FIGURE 1.17 A common thermometer for measuring temperature.

In the EE system of units, the unit of temperature is the degree Fahrenheit (°F). The set points used to define this scale are as follows, and are compared to the Celsius scale in **Figure 1.18**:

32°F corresponds to the freezing point of water at a pressure of 1 atmosphere.
212°F corresponds to the boiling point of water at a pressure of 1 atmosphere.

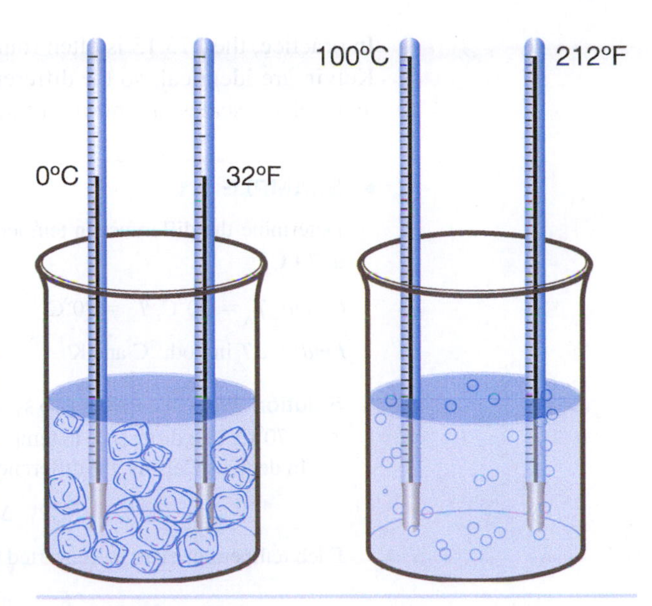

FIGURE 1.18 The image on the left shows the temperature in Celsius and Fahrenheit for ice water, whereas the image on the right shows the same for boiling water at atmospheric pressure.

Clearly, these values do not make much sense scientifically. However, the temperature scale is rather convenient to use for everyday temperature descriptions. If we consider most of the world, a temperature range of 0°F to 100°F is good for describing the range of temperatures that people are likely to experience. Although some people encounter temperatures outside this range, most people live in temperatures that fall within this range for most of the year.

Both scales have fixed temperatures at the same physical points, so a relationship can be derived between the two scales:

$$T(°C) = \frac{5}{9}[T(°F) - 32] \tag{1.4}$$

or, equivalently,

$$T(°F) = \frac{9}{5}T(°C) + 32 \tag{1.5}$$

Both of these temperature scales are based on an arbitrarily set 0 point (zero point), which makes them *relative* scales. Although relative scales are perfectly adequate for comparing the temperatures of different systems, and although they are adequate for determining a change in temperature, relative scales can cause significant problems when performing calculations involving a specific temperature. Consider the ideal gas law, Eq. (1.2), written in a form to solve for the mass of a system:

$$m = P\forall/RT$$

Now, suppose we wish to determine the mass of a system whose temperature is that of the freezing point of water at atmospheric pressure. If we used the SI system of units and degrees Celsius, we would obtain an infinite mass for the system, whereas the EE system of units and degrees Fahrenheit would give a finite mass. Clearly, there is a problem if an infinite mass and a finite mass can be calculated for the same system just because of the units used for temperature. Similarly, if the temperature scale gave a temperature below zero, the mass of the system would be negative. To avoid the problems that can result from using a relative temperature scale, an absolute temperature scale can be used.

If we return to the idea of what temperature is—a measure of molecular motion in a system—then we can develop a concept of a temperature scale that would not be based on an arbitrary zero point. If we cool a system, the molecules in the system will slow. The cooler the system becomes, the slower the molecules will move. At some point, all molecular-level motion will cease. At the point where all motion ceases, we can define a temperature scale to have a temperature of zero. Such a scale is called an absolute temperature scale, because the scale is based on a true definition of zero motion. It is not possible to have a lower temperature than this "absolute zero" because motion cannot become slower once the molecules have stopped.

The absolute temperature scale in the SI system of units is the Kelvin (K) scale. One Kelvin (1 K) is equal in magnitude to 1°C. Absolute zero occurs at −273.15°C, and this corresponds to 0 K. Therefore, the relationship between the Celsius and Kelvin temperature scales is

$$T(K) = T(°C) + 273.15 \tag{1.6}$$

In practice, the 273.15 is often rounded to 273. The size of a degree Celsius and the size of a Kelvin are identical, so the difference in temperature is identical numerically for the Kelvin and Celsius scales, as shown in Example 1.1.

▶ EXAMPLE 1.1

Determine the difference in temperature in both °C and K between a system at 30°C and a system at 70°C.

Given: $T_A = 30°C$, $T_B = 70°C$

Find: ΔT in both °C and K

Solution: We will consider two systems, A and B, with the given indicated temperatures: $T_A = 30°C$, $T_B = 70°C$. The difference in temperature is $\Delta T = T_B - T_A$.

In degrees Celsius, the difference in temperature is equal to

$$\Delta T = 70°C - 30°C = \mathbf{40°C}$$

Each temperature can be converted to the Kelvin scale.

$$T_A = 30 + 273.15 = 303.15\ \text{K} = 303\ \text{K}$$

$$T_B = 70 + 273.15 = 343.15\ \text{K} = 343\ \text{K}$$

The difference in temperature in Kelvin is

$$\Delta T = 343\ \text{K} - 303\ \text{K} = \mathbf{40\ K}$$

Analysis: Numerically the difference in temperature is the same because the size of each unit is the same. In practice, this means that either scale can be used when considering differences in temperature. In addition, if the value of some quantity has units of "unit/K," the value is the same in terms of "unit/°C," and vice versa.

In the EE system of units, the absolute temperature scale is the Rankine (R) scale. As in the SI system, the magnitude of the unit 1 R equals the magnitude of 1°F. Absolute zero occurs at −459.67°F, which corresponds to 0 R. Therefore, the relationship between the Fahrenheit and Rankine temperature scales is

$$T(\text{R}) = T(°\text{F}) + 459.67 \tag{1.7}$$

The 459.67 is often rounded to 460 in practice. The same concepts discussed for the SI temperature scales regarding the implications of the size of the units hold true for the Rankine and Fahrenheit temperature scales.

It should be noted that the temperature in Kelvin and the temperature in Rankine are related through

$$T(\text{K}) = \frac{5}{9}T(\text{R}) \tag{1.8}$$

▶ EXAMPLE 1.2

A system has a temperature of 25°C. Determine the value of this temperature in °F, K, and R.

Given: $T(°C) = 25°C$

Find: $T(°F)$, $T(K)$, $T(R)$

Solution: With $T(°C) = 25°C$, Eq. (1.5) can be used to find the temperature of the system in °F:

$$T(°\text{F}) = \frac{9}{5}T(°\text{C}) + 32 = \frac{9}{5}(25) + 32 = \mathbf{77°F}$$

Equation (1.6) can be used to find the temperature of the system in Kelvin:

$$T(\text{K}) = T(°\text{C}) + 273.15 = 25 + 273.15 = 298.15\ \text{K} = \mathbf{298\ K}$$

Equation (1.8), rewritten to solve for T(R), can be used to find the temperature of the system in Rankine:

$$T(\text{R}) = \frac{9}{5}T(\text{K}) = \frac{9}{5}(298.15) = 536.67\ \text{R} = \mathbf{537\ R}$$

These different values are illustrated in **Figure 1.19.**

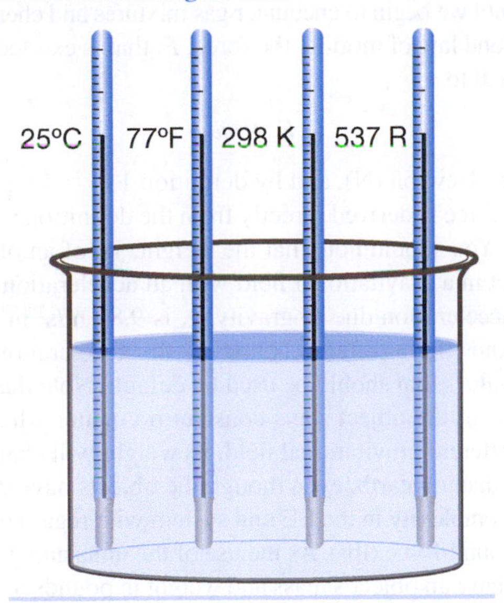

FIGURE 1.19 Four thermometers showing the same temperature in four temperature scales.

Today, the absolute temperature scales are defined through the absolute 0 point and the triple point of water. (The triple point of water is the temperature at which the solid, liquid, and vapor phase can all coexist.) This temperature is 0.01°C, and so the size of a unit of K in the Kelvin temperature scale is specified through the absolute 0 point and the triple point of water at a temperature of 273.16 K—there are 273.16 evenly sized units of Kelvin between absolute 0 and the triple point of water.

QUESTION FOR THOUGHT/DISCUSSION

Think of some possible scenarios that would give a nonsensical result if you divided by the temperature using a relative temperature scale. How can you use this to remind others to use only absolute temperatures when multiplying or dividing by a temperature?

1.2.2 Mass, Moles, and Force

The property known as ***mass*** specifies the amount of a substance. Mass is represented by the letter m. Somewhat like temperature, mass represents a concept that is easy to understand but difficult to define. The SI unit used for mass is the kilogram (kg), and the EE unit used for mass is the pound-mass (lbm).

A mole represents the amount of a substance that contains an Avogadro's number worth of atoms or molecules of that substance, with Avogadro's number being equal to 6.022×10^{23}. More formally, Avogadro's number is defined as the number of carbon atoms present in 0.012 kg of carbon-12. The number of moles of a substance is represented by the letter n. The ***molecular mass***, M, of a substance is the mass of one mole of the substance:

$$M = m/n$$

However, because we often will be using mass in kilograms, we are normally more concerned with the number of kilomoles of a substance; in such a case, the molecular mass represents the mass in kg of 1 kmole (one thousand moles) of the substance. In this book, we will be concerned primarily with mass rather than moles until we begin to encounter gas mixtures and chemically reacting systems.

From Newton's second law of motion, the force, F, that is exerted by a mass experiencing an acceleration, a, is equal to

$$F = ma \tag{1.9}$$

The SI unit of force is the Newton (N), and by definition 1 N = 1 kg · m/s². So, in the SI system of units, the unit for force is derived directly from the definition of force, because the units of acceleration are m/s². You should note that the weight, W, of an object is the force exerted by the mass of the object in a gravitational field with an acceleration equal to g: $W = mg$. At sea level on earth, the acceleration due to gravity, g, is 9.81 m/s² in SI units, and 32.17 ft/s² in EE units. In lieu of knowing a different value for the acceleration due to gravity at other locations, these are the values that should be used by default. Note that the weight of an object can change, but the mass of the object stays constant no matter where it is located. So if an object is moved into a different gravitational field, its weight will change—this is why objects weigh less on the moon than on earth, even though the objects have the same mass.

There is additional complexity in the EE unit system with regard to force. The unit of force in the EE system is the pound-force (lbf). As the use of the units and the concept of mass developed, it was desirable to have an object's mass and weight in pounds be numerically equal at sea level on earth, even though mass and weight are represented by different units. So, if an object has a mass of 50 lbm, it should have a weight of 50 lbf at sea level. However, considering that the acceleration due to gravity is 32.17 ft/s², this will clearly not occur if the equation $W = mg$ is used. To account for this, in the EE system of units, a unit conversion factor, g_c, is introduced:

$$g_c = 32.174 \text{ lbm} \cdot \text{ft/(lbf} \cdot \text{s}^2)$$

In turn, this changes Eq. (1.9) to

$$F = ma/g_c$$

for EE units. In fact, in any equation where there needs to be a conversion between lbm and lbf, the conversion factor g_c must be introduced; this is one of the complications of trying to learn thermodynamics using EE units. In SI units, $g_c = 1$. In this book, because we emphasize SI units, we will be ignoring the use of g_c in the equations, but be aware that the unit conversion will be needed for EE unit calculations.

▶ EXAMPLE 1.3

An object on a distant planet has a weight of 58.5 N. The acceleration due to gravity on the planet is 31.5 m/s². Determine the object's mass.

Given: $W = 58.5$ N

$g = 31.5$ m/s²

Find: m

Solution: Because $W = mg$, we can find

$$m = 58.5 \text{ N}/31.5 \text{ m/s}^2 = 58.5 \text{ kg} \cdot \text{m/s}^2/31.5 \text{ m/s}^2 = \mathbf{1.86\ kg}$$

▶ **EXAMPLE 1.4**

On earth at sea level, a block of metal has a weight of 95 kg. The block is placed on a rocket and delivered to the moon, where the acceleration due to gravity is 1.62 m/s². Determine the weight of the block of metal on the moon.

Given: Earth: $W_e = 95$ kg

Moon: $g_m = 1.62$ m/s²

Find: W_m (Weight on the moon)

Solution: The mass of the object can be found from the given information on the block's weight on the Earth.

$$m = W_e/g_e = (95 \text{ kg})/(9.81 \text{ m/s}^2) = 9.68 \text{ kg}$$

The mass is constant, so using the acceleration due to gravity on the moon, the weight of the block on the moon can be found:

$$W_m = mg_m = (9.68 \text{ kg})(1.62 \text{ m/s}^2) = \mathbf{15.7 \text{ kg}}$$

These different weights are illustrated in Figure 1.20.

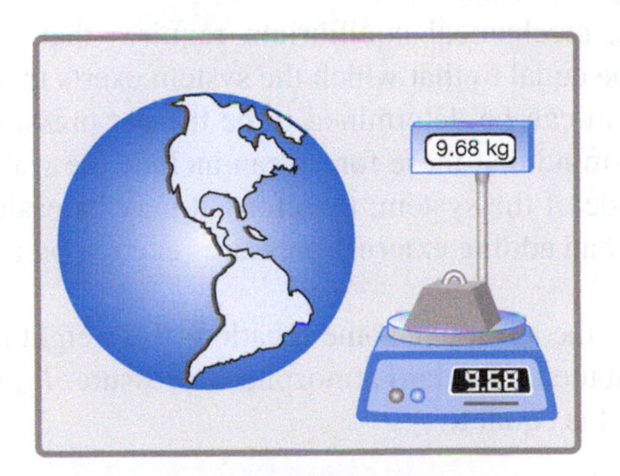

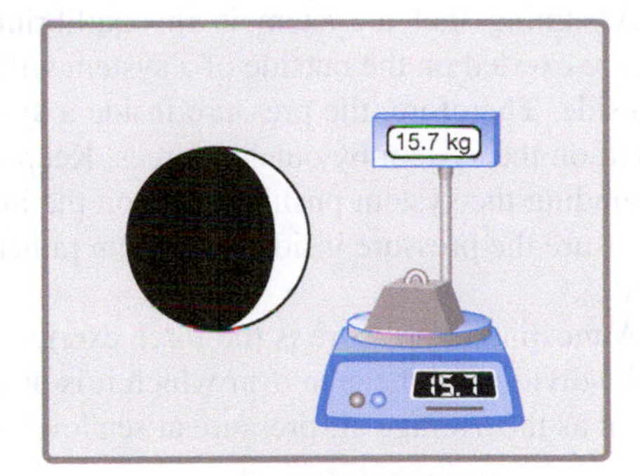

FIGURE 1.20 Scales showing the same object's weight on the earth and on the moon.

1.2.3 Volume and Specific Volume

The ***volume***, $\forall$, of a system is the physical space occupied by the system. Volume represents a three-dimensional spatial region, so the SI unit for volume is the cubic meter: m³. The EE unit for volume is the cubic foot: ft³.

As mentioned previously, thermodynamic calculations are typically performed using intensive properties, and the result can then be scaled by multiplying by the system's mass. Volume is an extensive property, whereas the ***specific volume***, v, is the corresponding intensive property. The specific volume is the total volume of the system divided by the system's mass:

$$v = \forall/m$$

The units for specific volume in the SI system are m³/kg, and the units are ft³/lbm in the EE system. As noted, the specific volume is the inverse of the density, ρ: $v = 1/\rho$.

1.2.4 Pressure

The last of the easily measured common properties of a system that we often use to describe a system is the pressure, P. The pressure is the force exerted divided by the area, A, over which the force acts. At a particular point in space, the pressure is found as the limit of the force divided by the area as the area approaches that of a point:

$$P = \lim_{A \to A'} \frac{F}{A} \tag{1.10}$$

where A' is the area of the point. In practice, the pressure exerted on a system or by a system will be uniform on a surface, and so in general the pressure will be calculated simply as

$$P = F/A \tag{1.11}$$

In SI units, the unit of pressure is the Pascal (Pa), where 1 Pa = 1 N/m^2. However, if you consider the size of this unit, you would see that it would be equal to the weight on earth of a roughly 0.1 kg object spread over a square meter of space: this is a very small unit of pressure. As such, in thermodynamics we typically will be concerned with units of pressure in kilopascals (kPa) (1000 Pa) and megapascals (MPa) (10^6 Pa). Occasionally, you may see units of pressure in "bar," where 1 bar = 100 kPa = 10^5 Pa. The reason for this unit will become apparent shortly. The standard unit of pressure in EE units is the pound-force per square foot: lbf/ft^2. Often, the pound-force per square inch (lbf/in^2 or psi) will be also used, where 1 lbf/ft^2 = 144 lbf/in^2 due to there being 12 inches in a foot.

Assuming that a system is in equilibrium, mechanical equilibrium requires that the pressure exerted on the outside of a system will be equal to that which the system exerts from the inside. Therefore, the pressure inside a system can be determined to be the net pressure exerted on the system by outside forces. Keep in mind that these forces can include the walls surrounding the system pushing back on the inside of the system; therefore, it may be easier to measure the pressure inside the system rather than adding external pressures exerted on the system.

Atmospheric pressure is the force exerted by the air above some location (the weight of the air) divided by the area over which it is distributed. Standard atmospheric pressure, P_0, is defined as the average air pressure at sea level and is equal to

$$P_0 = 101.325 \text{ kPa} = 14.696 \text{ lbf/in}^2 = 2116.2 \text{ lbf/ft}^2$$

The standard atmospheric pressure is sometimes also referred to as 1 atm, and pressures can be reported in terms of a number of atmospheres. As can be seen, standard atmospheric pressure is approximately 100 kPa, indicating that a pressure given in bar is approximately equal to the number of atmospheres. The local atmospheric pressure, P_{atm}, is often different from standard atmospheric pressure, particularly at an elevation significantly higher than sea level. The air pressure in mountainous areas is considerably less than that at sea level; the difference is significant enough that its effect on the boiling properties of water will change cooking times for some foods.

Measuring a difference in pressure is a relatively easy task. Manometers or other simple pressure gages can be used for this purpose. As shown in **Figure 1.21**, in a manometer, a tube containing a liquid is placed between the two systems whose pressures are to be compared; in this case, a cylinder containing a gas and the atmosphere. The cylinder contains pressurized gas, and this exerts a greater force on one side of the fluid in comparison to the force exerted by the atmosphere on the other side. This causes a difference in height, L, of the liquid between the two legs of the tube. Multiplying L by the density of the liquid and the local acceleration due to gravity

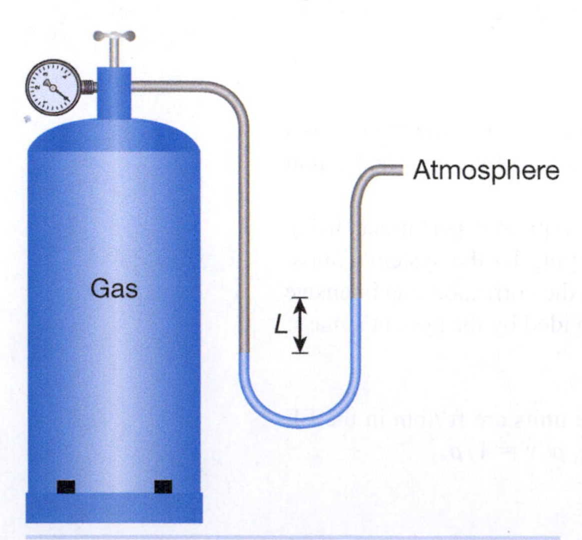
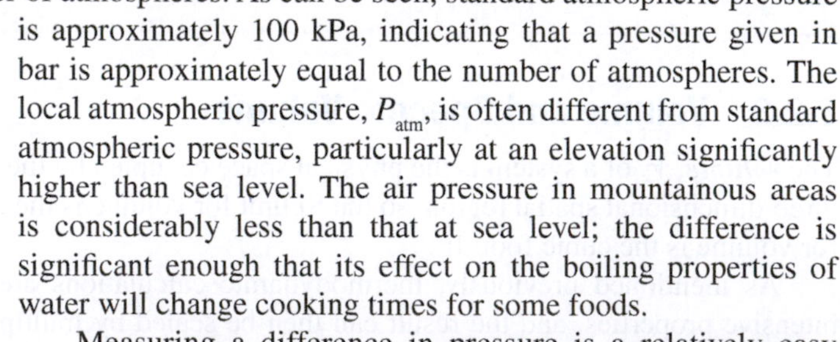

FIGURE 1.21 A manometer as set up to measure the pressure inside a tank of gas.

yields the pressure difference between the two systems. This pressure difference is called the gage pressure, P_g:

$$P_g = \rho g L \tag{1.12}$$

The gage pressure can be measured in other ways as well. Consider the state of the manometer if the gas inside the cylinder has a pressure equal to that of the local atmosphere. In this case, both ends of the tube will experience the same force, and there will be no height difference between the fluid in both legs of the manometer. This indicates that the gage pressure is zero. But is the pressure inside the cylinder equal to zero? No, the pressure is equal to the atmospheric pressure. Therefore, to get the actual pressure of the system, we must combine the gage pressure and the local atmospheric pressure to obtain the absolute pressure, P:

$$P = P_g + P_{atm} \tag{1.13}$$

The absolute pressure is the pressure that is needed for calculations involving equations of state to find other properties of a system.

Another way to envision the need for the absolute pressure is to consider the problem of determining the mass of air inside a flat bicycle tire. A flat bicycle tire is one whose internal air pressure is identical to the local atmospheric pressure: a tire gage would read zero for a flat tire. But is there still air inside the tire? Yes, there is still air inside the tire, because a vacuum condition has not been formed by a tire losing its air. Recall the ideal gas law (rewritten from Eq. (1.2)):

$$m = P\forall/RT$$

Clearly, the gage pressure of zero cannot be used as the pressure of the air inside the tire, because that would lead to a mass of zero. Instead, the absolute pressure is required in this calculation, where in this case the absolute pressure is equal to the local atmospheric pressure.

As mentioned, the local atmospheric pressure is not necessarily equal to standard atmospheric pressure, although standard atmospheric pressure can be used as a good approximation if other information is not available. It tends to be more difficult, and expensive, to measure the local atmospheric pressure. A typical device used for such a measurement is a barometer, and so the local atmospheric pressure is often called the barometric pressure.

QUESTION FOR THOUGHT/DISCUSSION

What would happen if you placed a well-sealed, fully-inflated bicycle tire into a chamber with a pressure of 1.5 MPa?

▶ EXAMPLE 1.5

A pressure gage on a tank filled with compressed helium reads that the pressure inside the tank is 352 kPa. A barometer in the room containing the tank indicates that the local barometric pressure is 100.2 kPa. Determine the absolute pressure of the helium inside the tank.

Given: $P_g = 352$ kPa (The pressure gage is providing a gage pressure in the tank.)

$P_{atm} = 100.2$ kPa (The barometer provides the local atmospheric pressure.)

Find: P

Solution: The absolute pressure is the sum of the gage pressure and atmospheric pressure:

$$P = P_g + P_{atm} = 352 \text{ kPa} + 100.2 \text{ kPa} = \mathbf{452.2\ kPa = 452\ kPa}$$

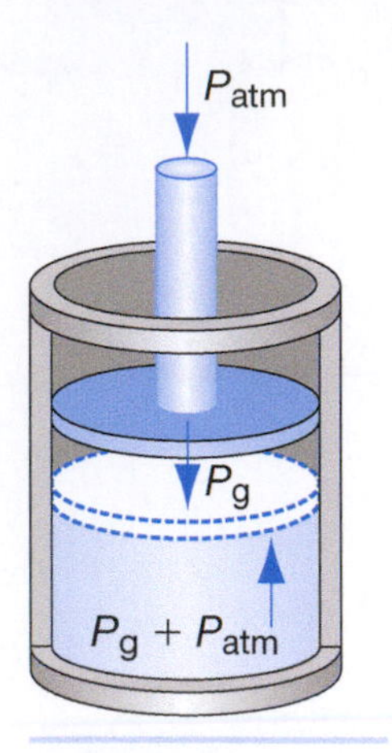

FIGURE 1.22 A diagram showing the pressures exerted on a gas in a piston–cylinder device.

▶ EXAMPLE 1.6

A circular piston in a cylinder has a diameter of 5 cm. The cylinder contains air. A pressure gage inside the cylinder reads a pressure of 300 kPa. The piston is at sea level, where the acceleration due to gravity is 9.81 m/s^2. Determine the mass of the piston.

Given: $D = 5$ cm $= 0.05$ m, $P_g = 300$ kPa, $g = 9.81$ m/s^2

Find: m

Solution: As shown in **Figure 1.22**, the absolute pressure inside the cylinder is equal to the pressure exerted by the weight of the piston plus the local atmospheric pressure. Therefore, the difference in pressure between the air inside the cylinder and outside the cylinder (i.e., the gage pressure) is that caused by the weight of the piston. As such, we do not need to find the absolute pressure in the tank to find the mass of the piston; we just need to equate the gage pressure to what is produced by the piston:

$$P_g = F/A = mg/A$$

The area of the piston is that of a circle:

$$A = \pi D^2/4 = \pi(0.05 \text{ m})^2/4 = 0.00196 \text{ m}^2$$

Keep in mind that the pressure unit that is naturally derived from a force in Newtons is the Pascal. Therefore, $P_g = 300$ kPa $= 300{,}000$ Pa.

$$m = P_g A/g = (300{,}000 \text{ Pa})(0.00196 \text{ m}^2)/(9.81 \text{ m/s}^2)$$

$$= (300{,}000 \text{ N/m}^2)(0.00196 \text{ m}^2)/(9.81 \text{ m/s}^2) = 60.0 \text{ N} \cdot \text{s}^2/\text{m}$$

$$= 60.0 (\text{kg} \cdot \text{m/s}^2)(\text{s}^2/\text{m}) = \mathbf{60.0 \text{ kg}}$$

Analysis: This problem is a good illustration of the benefit of tracking units in a calculation. Keeping track of the units and canceling the units appropriately will allow for the avoidance of careless mistakes. For example, if we had kept the gage pressure in kPa, the final answer would clearly have been incorrect because the units would not have properly canceled to give a mass in kilograms.

The two previous examples illustrate the use of "engineering accuracy" in the answers for thermodynamics calculations. In general, engineering accuracy is considered a precision of three significant figures in the value. A significant figure is a non-placeholding 0 number in an answer. Numbers such as 10,300, 431, 2.04, and 0.00352 all have three significant figures. In general, it is considered that most quantities can be measured to three significant figures of precision and that most objects can be built to such specifications without an excessive amount of effort. Clearly, some applications need more precision, and some need less. But in general in this course, we will be seeking to provide answers to three significant figures of precision and will assume that given quantities were known to that precision, even if they are not given to that level (such as the diameter of "5 cm" in Example 1.6).

There are some special thermodynamic processes that have one property held constant. A constant-temperature process is also known as an ***isothermal process***. A constant-pressure process is also known as an ***isobaric process***. Less commonly, a constant-volume process is sometimes referred to as an ***isochoric process***.

1.3 ZEROTH LAW OF THERMODYNAMICS

Now that some of the basic concepts and properties in thermodynamics have been introduced, we are ready to consider one of the four laws that govern thermodynamics. The law is known as the ***Zeroth Law of Thermodynamics***, as it was formally stated after the First Law of Thermodynamics, but was subsequently deemed as more fundamental.

Consider three systems: A, B, and C. The Zeroth Law of Thermodynamics states that if system A is in thermal equilibrium (i.e., has the same temperature) with system B, and system B is in thermal equilibrium with system C, then system A is in thermal equilibrium with system C. This law should seem logical, and because it is so logical it was not formally stated as early as other scientific laws. However, as we shall discuss, the Zeroth Law is at the heart of temperature measurement, and we have already considered that temperature is a very important property in thermodynamics. Therefore, the Zeroth Law was formally stated, and the other laws of thermodynamics rest upon it and rely upon it to provide a basis for correct temperature measurements.

As an example of a temperature measurement device, consider the mercury thermometer shown in Figure 1.23. The thermometer is placed in a glass containing water and is made of glass containing mercury. Once the thermometer reaches a steady temperature, we assume we are measuring the temperature of the water. But, in actuality, we are reading the temperature of the mercury inside the glass. To assume that we are reading the temperature of the water, we must assume that the temperature of the mercury (T_{Hg}) is equal to the temperature of the glass (T_g), which in turn is equal to the temperature of the water (T_w). For this to be the case, the mercury must be in thermal equilibrium with the glass, and the glass must be in thermal equilibrium with the water. From the Zeroth Law, we now can state that the mercury is in thermal equilibrium with the water, and indeed measuring the temperature of the mercury is the same as measuring the temperature of the water:

$$T_{Hg} = T_g$$

and

$$T_g = T_w$$

so

$$T_{Hg} = T_w$$

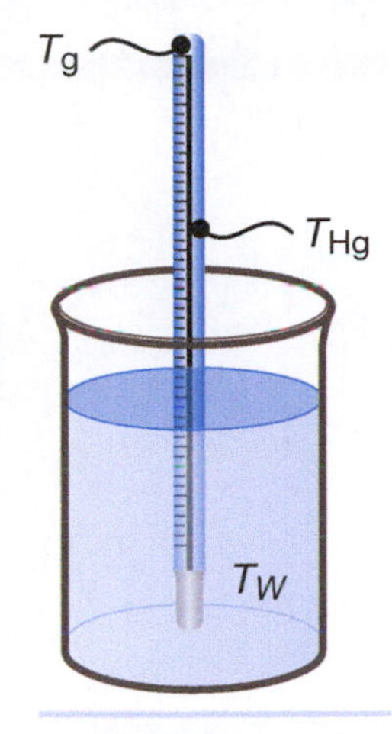

FIGURE 1.23 A mercury thermometer measuring the temperature of water. The three systems (the mercury, the glass, and the water) are all in thermal equilibrium, so the temperature of the mercury (what is being read) is the same as the temperature of the water.

Notice that reaching thermal equilibrium is a requirement for this temperature measurement concept to work. If we stored the mercury thermometer in a refrigerator and then took it out and placed it into a glass of boiling water, looking at the temperature of the mercury as shown by the thermometer immediately after placing it into the water would not yield the water temperature. Thermal equilibrium between the three systems would not have been reached at that point, and the Zeroth Law indicates that the mercury temperature is not the water temperature until the thermal equilibrium conditions are met.

1.4 PHASES OF MATTER

A ***pure substance*** is a substance that is chemically homogeneous—a substance that has a uniform chemical composition throughout. A pure substance can be either a single-molecule substance (such as water, nitrogen, and oxygen) or it can be a mixture of substances that has a constant composition throughout (such as air, which is a mixture of gases that has the same composition throughout a reasonably sized system). A ***phase*** of matter is a quantity of matter of a pure substance that is physically homogeneous. So a phase is chemically and physically uniform throughout. There are a relatively small number of phases of matter, three of which are of primary concern to engineers: solid, liquid, and gas. The plasma phase is important in some applications, and phases such as a Bose-Einstein condensate are primarily of interest at a scientific, but not necessarily practical, level.

The solid phase is characterized by a quantity of matter whose atoms or molecules are in a fixed lattice structure. The atoms or molecules are spaced closely together and held in place by intermolecular attractive forces. A solid maintains its shape without the aid of a container. As a solid is heated, the molecules will oscillate more and more inside the lattice structure and eventually will have enough energy to break free of the lattice. This is a melting process if the molecules move into the liquid phase, and a sublimation process if the molecules move into the gas phase.

The liquid phase still has molecules spaced closely together, but the molecules are not in a fixed lattice structure. The molecules are free to move inside the phase and are not forced into a position next to another particular molecule: the molecules can translate and rotate. The intermolecular forces are not as strong as with the solid phase, but are strong enough to keep the molecules in a somewhat orderly and structured environment. As more energy is added to the liquid, the molecules can gain enough energy to overcome the intermolecular forces binding the liquid together, and the boiling process to a gas occurs.

The gas phase is characterized by free atoms or molecules moving in random directions unattached to other molecules. The spacing between the molecules is great, and intermolecular forces are very small. Interactions between molecules in a gas occur primarily through random collisions as the molecules move in different directions. Sometimes, the gas phase will be referred to as the vapor phase. These phases are identical, with the term ***vapor*** being more typically applied to substances that are relatively close to their boiling/condensation point with the liquid phase. Gas is a term that is more commonly reserved for substances that are commonly experienced as a gas, even though we know that they could exist in other phases. So, we often will refer to hydrogen gas, or nitrogen gas, or carbon dioxide gas, because these substances are typically, in everyday life, experienced only in the gas phase. However, the gaseous phase of water is often called "water vapor" because we experience water in liquid and gaseous (and solid) forms often.

The phases can also change as a result of energy being removed from the material. A gas will condense into a liquid (or directly to a solid under the appropriate conditions), and a liquid will freeze into a solid. If energy is added to the material, the phase changes can occur in the opposite direction. For a substance, the temperatures at which these processes occur are dependent on the pressure of the system and do not depend on the direction of the phase change. So, at a pressure of 101.3 kPa, pure water will change from the solid phase (ice) to the liquid phase through melting at 0°C, and liquid water will change to ice through freezing at 0°C as well. Similarly, liquid water at 101.3 kPa will boil to the gas phase (water vapor) at 100°C, and water vapor will condense to liquid water at 100°C. These temperatures are different at different pressures, as illustrated in Figure 1.24. For example, the boiling/condensation temperature of water at 200 kPa is 120.2°C, and at 1000 kPa is 179.9°C.

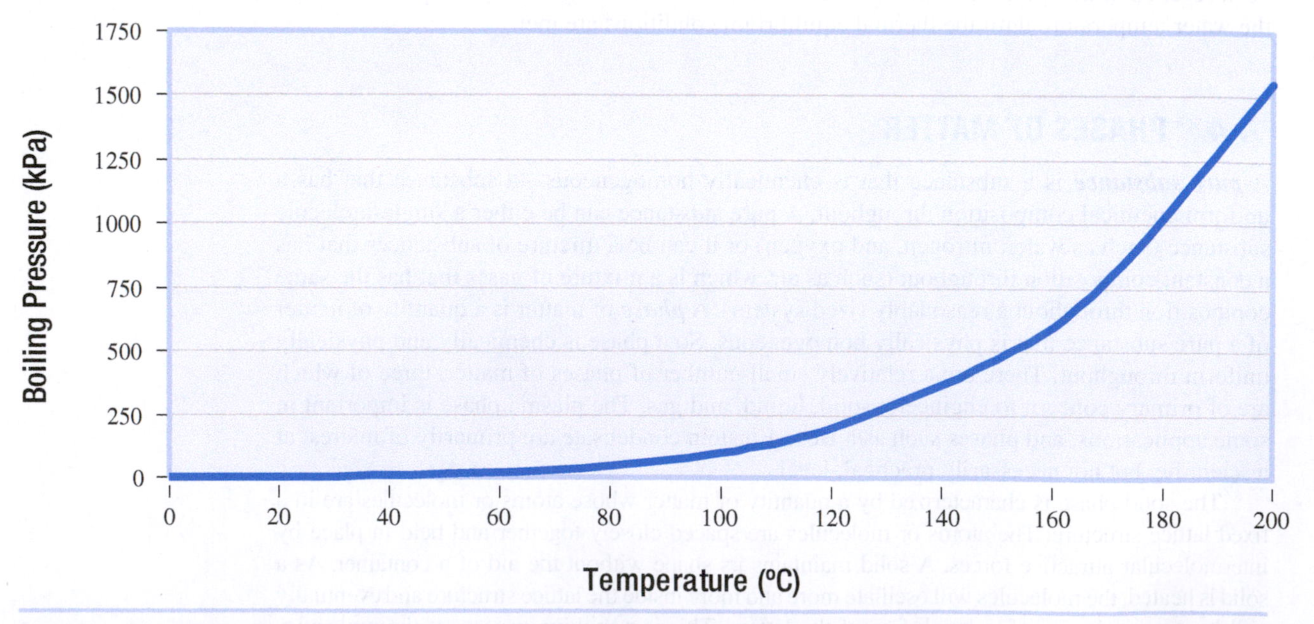

FIGURE 1.24 The boiling pressure of water as a function of temperature.

As will be discussed in greater detail in Chapter 3, a pure substance can exist in a system in more than one phase. A common example of this is a glass of ice water, containing both liquid water and ice, at the same temperature if thermal equilibrium has been reached between the two. When multiple phases of a substance exist inside a system and the masses of each phase are unchanging, the system is considered to be in ***phase equilibrium***. Phase equilibrium is one of the requirements necessary for a system to be considered to be in thermodynamic equilibrium.

QUESTION FOR THOUGHT/DISCUSSION

What are two ways you could stop liquid water at 101.3 kPa and 100°C from boiling?

Summary

In this chapter, we have established the importance of energy in the world and the need for engineers to understand the science that describes energy: thermodynamics. We have described some of the fundamental concepts involving thermodynamic analysis, such as the system, the process, and the concept of thermodynamic properties. The SI and EE unit systems have been introduced, and it should be noted that most of the focus of this book is on using SI units. The basic properties that are used to typically describe systems, such as temperature and pressure, have been discussed. The Zeroth Law of Thermodynamics was presented, and its application to temperature measurement systems was explored. Finally, the idea of a phase of matter was presented.

These are all fundamental concepts, many of which may already be familiar to you through other courses. However, the rest of this book relies on these concepts, and you should be certain you thoroughly understand these principles before moving forward. Some of the principles will be explicitly referred to at future points in the book, but other concepts (such as thermodynamic equilibrium or the Zeroth Law) will be implicitly assumed as being a fundamental component of the development of the principles of thermodynamics.

KEY EQUATIONS

Temperature Scale Conversions:

$$T(°C) = \frac{5}{9}[T(°F) - 32] \quad (1.4)$$

$$T(°F) = \frac{9}{5}T(°C) + 32 \quad (1.5)$$

$$T(K) = T(°C) + 273.15 \quad (1.6)$$

$$T(R) = T(°F) + 459.67 \quad (1.7)$$

Force:

$$F = ma/g_c \quad (1.9)$$

where $g_c = 1$ (SI units) or $g_c = 32.174$ lbm · ft/(lbf · s²) (EE units)

Absolute Pressure:

$$P = P_g + P_{atm} \quad (1.13)$$

PROBLEMS

1.1 For the following systems, determine whether the system described is best modeled as an isolated, closed, or open system:

(a) steam flowing through a turbine
(b) an incandescent light bulb
(c) a fuel pump in a moving automobile
(d) an anchor of a sunken ship resting 3000 m below the surface of the ocean
(e) the roof of a house

1.2 For the following systems, determine whether the system described is best modeled as an isolated, closed, or open system:

(a) a tree growing in a forest
(b) a television
(c) a laptop computer
(d) the *Voyager 2* spacecraft in its current state
(e) the *Messenger* spacecraft as it moved into orbit around Mercury

1.3 For the following systems, determine whether the system described is best modeled as an isolated, closed, or open system:

(a) an inflated tire
(b) a lawn sprinkler actively in use
(c) a cup filled with liquid water
(d) an engine's radiator
(e) a rock formation 200 m below the surface of the earth

1.4 For the following systems, determine whether the system described is best modeled as an isolated, closed, or open system:

(a) a pump supplying water to a building
(b) a tea kettle containing boiling water
(c) an active volcano
(d) a solid gold bar placed inside a very well-insulated box
(e) a chair

1.5 For the following systems, determine whether the system described is best modeled as an isolated, closed, or open system:

(a) a pulley on an elevator
(b) a bathtub
(c) a human being
(d) a piece of metal being shaped on a lathe
(e) a comet orbiting the sun in the Oort cloud (the cloud of inactive comets located well beyond the orbits of the planets)

1.6 Consider a closed bottle half-filled with water and placed in a refrigerator. Draw diagrams showing the most appropriate system for a thermodynamic analysis that

(a) only considers the water
(b) considers only the water and the air inside the bottle
(c) considers the water and air inside the bottle, and the bottle itself
(d) considers only the bottle and not the contents
(e) considers all the contents of the refrigerator, but not the physical refrigerator

1.7 Consider a fire hose with water flowing through the hose and then through a nozzle at the end of the hose. Draw diagrams showing the most appropriate system for a thermodynamic analysis that

(a) considers only the water in the nozzle of the system
(b) considers the water flowing through the hose and the nozzle
(c) considers both the water flowing through the nozzle and the nozzle itself

1.8 A basketball is about to leave a player's hand for a shot. Draw diagrams showing the most appropriate system for a thermodynamic analysis that

(a) considers only the air inside the basketball
(b) considers only the material making up the basketball, and not the air inside the ball
(c) considers the basketball and the air inside
(d) considers the basketball, the air inside, and the player's hand
(e) considers the entire arena in which the basketball is located

1.9 To condense a flow of steam, liquid cooling water is sent through a pipe, and the steam is passed over the exterior of the pipe. Draw diagrams showing the most appropriate system for a thermodynamic analysis that

(a) considers only the water flowing through the pipe
(b) considers only the steam condensing on the exterior of the pipe
(c) considers only the pipe
(d) considers the pipe, the internal cooling water, and the external condensing steam

1.10 Draw a schematic diagram of the place where you live. Identify any places where mass or energy may flow into or out of the room or building.

1.11 Draw a schematic diagram of an automobile engine. Identify any locations where mass or energy may flow into or out of the engine.

1.12 Draw a schematic diagram of a desktop computer. Identify any locations where mass or energy may flow into or out of the computer.

1.13 Draw a schematic diagram of a highway bridge over a river. Identify any mechanisms that may cause mass or energy to flow into or out of the system of the bridge.

1.14 Draw a schematic diagram of an airplane in flight. Identify any locations where mass or energy may flow into or out of the airplane.

1.15 A closed system undergoes a constant volume (isochoric) process at 0.25 m^3/kg, as the pressure changes from 100 kPa to 300 kPa. Draw this process on a P-v (pressure vs. specific volume) diagram.

1.16 A system undergoes an isothermal process at 30°C as the specific volume changes from 0.10 m^3/kg to 0.15 m^3/kg. Draw this process on a T-v (temperature vs. specific volume) diagram.

1.17 A system undergoes an isobaric process from 50°C to 30°C, at a pressure of 200 kPa. Draw this process on a P-T (pressure vs. temperature) diagram.

1.18 A system undergoes a process described by Pv = constant, from an initial state of 100 kPa and 0.25 m^3/kg, to a final specific volume of 0.20 m^3/kg. Determine the final pressure, and draw this process on a P-v (pressure vs. specific volume) diagram.

1.19 Draw on a P-v diagram the following three sequential processes that a system undergoes:

(a) a constant-pressure expansion from an initial state of 500 kPa and 0.10 m^3/kg to a specific volume of 0.15 m^3/kg
(b) a constant-specific-volume depressurization to a pressure of 300 kPa
(c) a process following Pv = constant to a final pressure of 400 kPa

1.20 Draw a T-v diagram of the following three sequential processes that a system undergoes:

(a) a constant-specific-volume heating from 300 K and 0.80 m^3/kg to a temperature of 450 K
(b) an isothermal compression to a specific volume of 0.60 m^3/kg
(c) an isochoric cooling to 350 K

1.21 Draw a P-T diagram of a system undergoing the following two sequential processes:

(a) an isothermal compression from 500 K and 250 kPa to a pressure of 500 kPa
(b) an isobaric cooling to a temperature of 350 K

1.22 Draw a P-v diagram of a closed system undergoing the following four sequential processes:

(a) an isobaric compression from 200 kPa and 0.50 m^3/kg to a specific volume of 0.20 m^3/kg
(b) a constant-volume expansion to a specific volume of 0.30 m^3/kg
(c) a constant-volume depressurization to a pressure of 125 kPa
(d) a constant-pressure expansion to a specific volume of 0.30 m^3/kg

1.23 A thermodynamic cycle consists of the following three processes. Draw the cycle on a T-v diagram.

(a) a constant-volume heating from 0.10 m^3/kg and 300 K to 500 K
(b) an isothermal expansion to a specific volume of 0.15 m^3/kg
(c) a linear process returning the process to its initial state

1.24 A thermodynamic cycle consists of the following three processes. Draw the cycle on a P-v diagram.

(a) an isobaric compression from 300 kPa and 1.20 m^3/kg to a specific volume of 0.80 m^3/kg
(b) a process for which Pv = constant to a specific volume of 1.20 m^3/kg
(c) a constant-volume process resulting in a pressure of 300 kPa

1.25 A thermodynamic cycle involves the following four processes. Draw the cycle on a P-T diagram.

(a) an isobaric heating from 500 K and 400 kPa to a temperature of 700 K
(b) an isothermal compression to a pressure of 800 kPa
(c) an isobaric cooling to a temperature of 500 K
(d) an appropriate isothermal expansion

1.26 The melting point of lead at atmospheric pressure is 601 K. Determine this temperature in °C, °F, and R.

1.27 The melting point of gold at atmospheric pressure is 1336 K. Determine this temperature in °C.

1.28 At a pressure of 517 kPa, carbon dioxide will condense into a liquid at –57°C. Determine this temperature in K.

1.29 The "normal" temperature for a human being is 37°C. Determine this temperature in K.

1.30 The boiling point of ammonia at atmospheric pressure is 239.7 K. Determine this temperature in °C.

1.31 The melting point of aluminum at atmospheric pressure is 660°C. Determine this temperature in K.

1.32 At atmospheric pressure, the boiling point of methanol is 337.7 K and the boiling point of ethanol is 351.5 K. Convert both of these temperatures to degrees Celsius, and determine the difference in these temperatures in both K and °C.

1.33 At atmospheric pressure, the melting point of pure platinum is 2045 K, and the melting point of silver is 1235 K. Convert both of these temperatures to degrees Celsius, and determine the difference in these temperatures in both K and °C.

1.34 You wish to drop an ice cube into a cup of hot water to cool the water. The temperature of the ice cube is –10°C, and the water temperature is 92°C. Convert both of these temperatures to Kelvin, and determine the difference between the temperatures in both K and °C.

1.35 Oxygen, O_2, has a molecular mass of 32 kg/kmole. How many moles does 17 kg of O_2 represent?

1.36 You determine that 1.2 kmole of a substance has a mass of 14.4 kg. Determine the molecular mass of the substance.

1.37 You are asked if you would like to have a box which contains 3.5 kmole of gold. The only condition of the deal is that you must carry the box away using only your own strength. What is the mass of the gold in the box if the molecular mass of the gold is 197 kg/kmole? Do you think you will be able to accept this deal?

1.38 Suppose that one kilomole of any gaseous substance at a given temperature and pressure occupies a volume of 24 m^3. The density of a particular gas at these conditions is 1.28 kg/m^3. How much mass of the gas is present if you have a 2.0-m^3 container full of the gas at the given temperature and pressure, and what is the molecular mass of the gas?

1.39 Burning a hydrocarbon fuel will convert the carbon in the fuel to carbon dioxide. For every kmole of carbon to be burned, you need 1 kmole of oxygen (O_2). This produces 1 kmole of CO_2. If you originally have 2 kg of carbon to be burned, what is the mass of the CO_2 that will be produced? The molecular mass of carbon is 12 kg/kmole, of oxygen is 32 kg/kmole, and of CO_2 is 44 kg/kmole.

1.40 A rock at sea level on earth (where g = 9.81 m/s^2) has a mass of 25 kg. What is the weight of the rock in Newtons?

1.41 On a distant planet, the acceleration due to gravity is 6.84 m/s^2. The weight of an object on that planet is 542 N. What is the mass of the object? If that object is moved to earth, where g = 9.81 m/s^2, what is the weight of the object?

1.42 How much force is needed to accelerate a ball with a mass of 0.5 kg at a rate of 25 m/s^2?

1.43 How much force is needed to accelerate a block with a mass of 0.72 kg at a rate of 11 m/s^2?

1.44 An object has a mass of 66 kg. This object is sent into space and is placed onto the surface of a planet where the acceleration due to gravity is 7.6 m/s^2. What is the weight of the object in N on the other planet?

1.45 The acceleration due to gravity on Mars is 3.71 m/s^2. At sea level on earth, an astronaut can lift an object that weighs 555 N. What is the mass of an object that the astronaut could lift on Mars?

1.46 A club applies a force of 50 N to a rubber ball that has a mass of 700 g. What is the acceleration experienced by the ball as it encounters the force?

1.47 What force is required to accelerate a 2.4 kg rock at a rate of 11 m/s^2?

1.48 The specific volume of steam at 500°C and 500 kPa is 0.7109 m^3/kg. You have a container whose volume is 0.57 m^3, which is full of the steam at 500°C and 500 kPa. Determine the mass of the steam in the container.

1.49 A solid block of unknown composition has dimensions of 0.5 m in length, 0.25 m in width, and 0.1 m in height. The weight of the block at sea level (g = 9.81 m/s^2) is 45 N. Determine the specific volume of the block.

1.50 A mixture of liquid water and water vapor occupies a cylindrical tube whose diameter is 0.05 m and whose length is 0.75 m. If the specific volume of the water is 0.00535 m^3/kg, determine the mass of the water present.

1.51 The density of several metals is as follows: lead: 11,340 kg/m^3; tin: 7310 kg/m^3; aluminum: 2702 kg/m^3. You are given a small box (0.1 m × 0.1 m × 0.075 m) and are told that it is filled with one of these metals. Unable to open the box and unable to read the label on the box, you decide to weigh the box to determine the metal inside. You find that the weight of the box is 53.8 N. Determine the density and specific volume of the box, and choose the likely metal inside.

1.52 A person with a mass of 81 kg stands on a small platform whose base is 0.25 m × 0.25 m. Determine the pressure exerted on the ground below the platform by the person.

1.53 A wall of area 2.5 m^2 is hit by a gust of wind. The force exerted by the wind on the wall is 590 kN. Determine the pressure exerted by the wind on the wall.

1.54 A press applies a pressure of 800 kPa uniformly over an area of 0.025 m^2. What is the total force applied by the press?

1.55 A manometer is used to determine the pressure difference between the atmosphere and a tank of liquid. The fluid used in the manometer is water, with a density of 1000 kg/m^3. The manometer is located at sea level, where g = 9.81 m/s^2. The difference in height between the liquid in the two legs of the manometer is 0.25 m. Determine the pressure difference.

1.56 A mercury (ρ = 13,500 kg/m^3) manometer is used to measure the pressure difference between two tanks containing fluids. The difference in height of the mercury in the two legs is 10 cm. Determine the difference in pressure between the tanks.

1.57 You choose to use a mercury (ρ = 13,500 kg/m^3) manometer to check the accuracy of a pressure gage on a compressed nitrogen gas tank. The manometer is set up between the tank and the atmosphere, and the height difference for the mercury in the two legs is 1.52 m. The pressure gage to be checked reads a pressure of 275 kPa for the gage pressure of the tank. Is the pressure gage accurate?

1.58 Compressed gas tanks often have gage pressures of at least 1 MPa. Suppose you wished to use a manometer to measure the gage pressure of a compressed air tank whose pressure was at least 1 MPa. The manometer would be set up between the tank and the atmosphere. What is the minimum length of tube needed for such a measurement if the liquid in the manometer is (a) mercury (ρ = 13,500 kg/m^3), (b) water (ρ = 1000 kg/m^3), and (c) engine oil (ρ = 880 kg/m^3)? Do these seem to be practical devices for such a measurement?

1.59 A manometer using a liquid with a density of 1750 kg/m^3 is set up to measure the pressure difference between two locations in a flow system. The height of the manometer liquid is 0.12 m. What is the pressure difference between the two locations?

1.60 The pressure gage on a tank of compressed nitrogen reads 785 kPa. A barometer is used to measure the local atmospheric pressure as 99 kPa. What is the absolute pressure in the tank?

1.61 The pressure gage on a tank of compressed air reads 872 kPa. The local atmospheric pressure is measured as 100.0 kPa. What is the absolute pressure in the tank?

1.62 A pressure gage is used to measure the pressure of air inside a piston–cylinder device. The diameter of the cylinder is 8 cm. While the piston is at rest, the gage measures the pressure to be 40 kPa. A barometer measures the atmospheric pressure to be 100 kPa. A weight with a mass of 20 kg is placed on the top of the piston, and the piston moves until it reaches a new equilibrium point. What is the new gage pressure and the new absolute pressure of the air in the cylinder when this new equilibrium is reached?

1.63 The absolute pressure of air in a piston–cylinder device is 220 kPa. The local atmospheric pressure is 99 kPa. If the acceleration due to gravity is 9.79 m/s^2, and if the diameter of the cylinder is 0.10 m, what is the mass of the piston?

1.64 Air is located in a piston–cylinder device. The diameter of the cylinder is 12 cm, the mass of the piston is 5 kg, and the acceleration due to gravity is 9.80 m/s^2. The local atmospheric pressure is 100.5 kPa. Determine the mass of a set of weights that needs to be added to the top of the piston so that the absolute pressure of the air in the cylinder is 250 kPa.

1.65 A tank of liquid exerts a pressure of 300 kPa on a plug on the bottom of the tank. The local atmospheric pressure is 99 kPa. The diameter of the circular plug is 2.5 cm. What is the additional force that needs to be applied to the plug to keep the plug in place?

1.66 What is the absolute pressure of air located in a piston–cylinder device for a cylinder of diameter 15 cm, with a piston mass of 70 kg, and with local atmospheric pressure of 101.01 kPa? The device is located at sea level on earth.

1.67 Consider a piston–cylinder device initially at equilibrium with the air pressure inside the cylinder being 150 kPa. It is desired to raise the pressure of the air to 300 kPa by adding air to the cylinder, without changing the location of the piston. If the piston has a diameter of 8 cm, how much mass needs to be added to the piston to keep the piston in the same location with the higher pressure? Assume standard acceleration due to gravity at sea level on earth.

CHAPTER 2

The Nature of Energy

Learning Objectives

Upon completion of Chapter 2, you will be able to

2.1 Explain the nature of energy and the different forms that energy can take;

2.2 Identify the methods of transporting energy into or out of a system;

2.3 Recognize the three modes of heat transfer;

2.4 Describe and compute the many modes of work; and

2.5 Express an approach and framework for solving thermodynamic problems.

2.1 WHAT IS ENERGY?

Energy is something that most people inherently understand, but it is very hard to define. The *Oxford English Dictionary* defines *energy* as the "ability or capacity to produce an effect." Considering this definition, a substance or object with energy has the ability to change itself or its surroundings. For example, a substance with energy can move, or it can do work, or it can heat some other substance. Energy is a property of a substance. The quantity of energy that a substance has is not related to how the state of the substance was reached, but rather is only a function of the local thermodynamic equilibrium state. In this chapter, we will describe some of these concepts in more detail.

Energy is also of great interest in the world today. Modern society has developed because of the ability of humans to harness the energy in the world around us. The use of energy surrounds us every day. Energy is used to heat or cool buildings. Energy is used by vehicles to transport people and goods. Energy is used in computers, lights, and every electrical device in our homes and offices. Energy is used to manufacture everything we see around us. As a result, the supply and cost of energy is often paramount in our minds. Many engineers deal with energy on an everyday basis. Some engineers work on producing electricity or transportation fuels. Some engineers seek new ways to harness nature's energy. Other engineers try to make devices more efficient users of energy so that the energy that is available can last longer. Although not every engineer works in such areas, the efficient use of energy touches the jobs of most engineers at least intermittently, and not necessarily in obvious ways. Consider that it might be natural for a bridge designer to seek to use less material in their bridges in order to keep costs down. How does using less material reduce costs? As one consideration, it requires energy to turn raw materials into the steel and concrete used in the bridge, and that energy costs money—the less material used, the lower the energy costs. It also costs money to transport materials to the building site, and transportation by trucks or ships requires energy. Therefore, the less material used, the lower the energy used in transporting the materials.

The standard SI unit of energy is the Joule, which is equal to a Newton-meter:

$$1\ \text{J} = 1\ \text{N} \cdot 1\ \text{m}$$

Often, we will deal with kilojoules or megajoules: 1 kJ = 1000 J; 1 MJ = 1000 kJ. When we deal with rates of energy use or change (generally when discussing energy flows or transports), the standard SI unit is a Watt, which equals a Joule per second:

$$1\ \text{W} = 1\ \text{J/s}$$

Similarly, we will often encounter kilowatts and megawatts: 1 kW = 1000 W; 1 MW = 1000 kW.

Users of the EE system of units will primarily encounter two units for energy. These units are the foot-pound (force) (ft · lbf) and the British thermal unit (Btu). The conversion between these two units is

$$1\ \text{Btu} = 778.1693\ \text{ft} \cdot \text{lbf}$$

We can also convert between SI and EE units, as 1 Btu = 1.055 kJ. The two units in the EE system also lead to two units of power, the ft · lbf/s and the Btu/s, although different time durations can be used, leading to a unit such as the Btu per hour: Btu/h. Furthermore, a traditional unit of power in the EE system is the horsepower (hp), where

$$1\ \text{hp} = 550\ \text{ft} \cdot \text{lbf/s} = 2544.43\ \text{Btu/h}$$

Because horsepower is a unit with considerable history, it may be useful to know its conversion to SI units:

$$1\ \text{hp} = 0.7457\ \text{kW}$$

Finally, when refrigeration processes are involved, we may encounter the unit of a "ton of refrigeration," where

$$1\ \text{ton of refrigeration} = 12{,}000\ \text{Btu/h}$$

Let's now consider the different forms that energy can take, and how energy can be transported from one system to another.

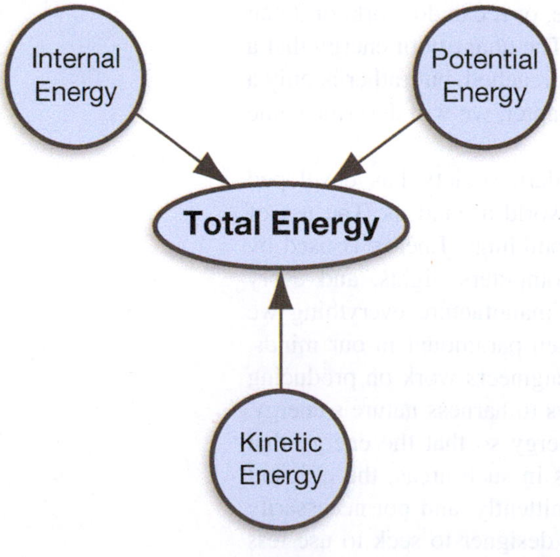

Internal Energy
Potential Energy
Total Energy
Kinetic Energy

FIGURE 2.1 The total energy of a system is the combination of the system's internal energy, kinetic energy, and potential energy.

2.2 TYPES OF ENERGY

Just as Julius Caesar said that all of Gaul is divided into three parts, for our purposes we can say that all of the energy of a substance is divided into three parts: internal energy, kinetic energy, and potential energy. Although each of these forms plays a prominent role in certain applications, they are usually not equally important. As we will see later, and as illustrated in Figure 2.1, we can neglect various forms of energy in a system depending on the application; this will simplify our thermodynamic problems.

2.2.1 Potential Energy

Potential energy is the energy in a system resulting from the system being in a gravitational field, and it is due to the mass of the system being higher than some reference point. There is the potential for the substance to produce an effect, and this potential will be unleashed if the substance is allowed to fall under the effect of gravity toward the reference point—typically the ground or floor. A substance that does not change its height during a process retains its initial potential energy. For example, as shown in Figure 2.2, a ball that is being held out a window has potential energy, but no changes are produced until the ball is released. Until the ball is released, the potential energy contained in the ball is essentially useless.

The potential energy, PE, of a system can be found from

$$PE = mgz \tag{2.1}$$

where m is the mass of the system, g is the acceleration due to gravity, and z is the height of the system above a reference point.

Potential energy is an important driving force for devices such as hydroelectric power plants. Any situation involving moving a solid object up or down a significant distance will generally require a consideration of changes in potential energy. But we will see that, for many typical thermodynamic applications, the presence of potential energy will have little impact on the engineering calculations.

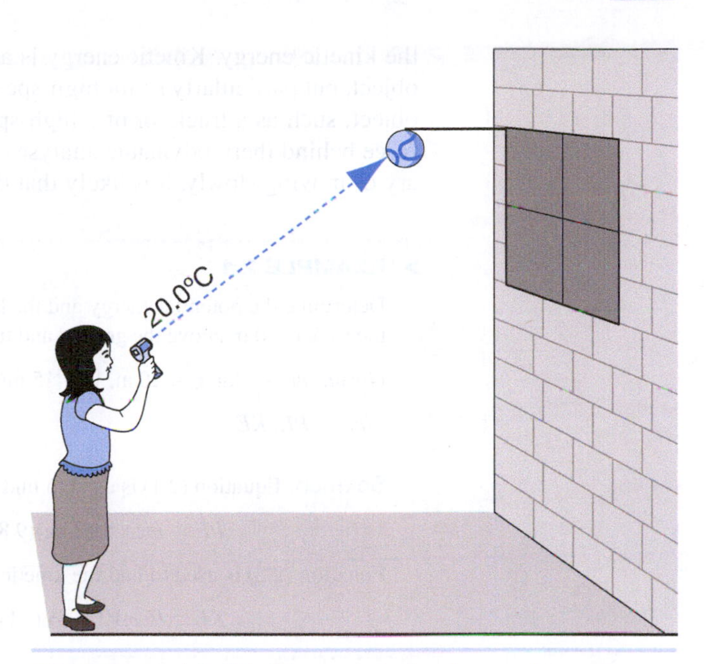

FIGURE 2.2 A ball held out a window has an internal energy corresponding to a temperature of 20°C, no kinetic energy, and some value of potential energy.

2.2.2 Kinetic Energy

Kinetic energy is the energy present in an object as a result of its motion. A moving object has the capacity to cause an effect. For example, a truck driving down the highway will cause a stationary car to move if the truck hits the car. The kinetic energy of the truck causes the car to move. Kinetic energy is usually the form of energy that is initially produced when an object with potential energy begins to fall. If we release a ball being held out a window, such as in Figure 2.3, the ball will begin to accelerate toward the ground. The stationary ball's potential energy is being converted into kinetic energy as the moving ball gains speed.

The kinetic energy, KE, of a system can be found from

$$KE = \tfrac{1}{2} mV^2 \tag{2.2}$$

where V is the velocity of the system. We can readily see that the faster an object is moving, the greater

FIGURE 2.3 As the ball falls to the ground, the potential energy is converted to kinetic energy. Ignoring frictional heating from the surrounding air, the internal energy of the ball is unchanged.

the kinetic energy. Kinetic energy is an important component of the energy of any moving object, but particularly so for high-speed applications. The kinetic energy of a moving heavy object, such as a truck, or of a high-speed gas or liquid flow will sometimes be the driving force behind thermodynamic analyses of such applications. However, if an object is stationary or moving slowly, it is likely that the impact of its kinetic energy is negligible.

▶ EXAMPLE 2.1

Determine the potential energy and the kinetic energy of a 2-kg rock that is falling from a cliff when the rock is 20 m above the ground and traveling at 15 m/s. Assume the cliff is located at sea level.

Given: $m = 2$ kg, $z = 20$ m, $V = 15$ m/s, $g = 9.81$ m/s^2 (acceleration due to gravity at sea level)

Find: PE, KE

Solution: Equation (2.1) is used to find the potential energy:

$$PE = mgz = (2 \text{ kg})(9.81 \text{ m/s}^2)(20 \text{ m}) = \mathbf{392\ J}$$

Equation (2.2) is used to find the kinetic energy:

$$KE = \tfrac{1}{2} mV^2 = \tfrac{1}{2}(2 \text{ kg})(15 \text{ m/s})^2 = \mathbf{225\ J}$$

▶ EXAMPLE 2.2

A parachutist, with a mass of 65 kg, is falling at a speed of 31 m/s at a height of 1800 m. Determine the kinetic and potential energy of the parachutist. Consider the acceleration due to gravity to be 9.79 m/s^2.

Given: $m = 65$ kg, $V = 31$ m/s, $z = 1800$ m/s, $g = 9.79$ m/s^2

Find: PE, KE

Solution: Equation (2.1) is used to find the potential energy:

$$PE = mgz = (65 \text{ kg})(9.79 \text{ m/s}^2)(1800 \text{ m})$$

$$= 1{,}145{,}000 \text{ N} \cdot \text{m} = 1{,}145{,}000 \text{ J} = 1145 \text{ kJ}$$

Equation (2.2) is used to find the kinetic energy:

$$KE = \tfrac{1}{2} mV^2 = \tfrac{1}{2}(65 \text{ kg})(31 \text{ m/s})^2$$

$$= 31{,}200 \text{ J} = 31.2 \text{ kJ}$$

2.2.3 Internal Energy

Although most people have a good intuitive feel for potential and kinetic energy because such energies are on a macroscopic scale and can be visualized easily, the internal energy of a substance is harder for people to grasp. Yet, for most systems, more energy is present in the form of internal energy than in the forms of kinetic or potential energy. The internal energy of a substance is all of the energy present at the molecular level. Molecules move, vibrate, and rotate. Electrons have energy as they move around a nucleus. All of this motion, which can be thought of as kinetic and potential energy at the molecular level, contributes to a substance's internal energy. Importantly, this energy will increase as the temperature of a substance increases, and so temperature can be closely related to the internal energy of a substance and can even be interpreted as a measure of the molecular motion of a substance. As such, in this text, we won't be developing an equation to determine a system's internal energy but we will use relationships (equations of state) that allow us to find changes in a system's internal energy as the system's temperature changes. Internal energy is also a much weaker function of a system's pressure—so much weaker that in some cases the effects of pressure on the internal energy are ignored. We will represent the internal energy of a substance with the symbol U.

As an example of internal energy, think about the ball that was dropped from a window. The moving ball eventually hits the ground and stops, as shown in **Figure 2.4**. At that point, its kinetic energy is back to 0, but its potential energy has decreased because it is at a lower height. Where did the energy go? The energy was converted from kinetic energy to internal energy by the ball increasing in temperature, and by the ground and air heating up as well due to friction. Although the amount of a temperature increase may be small, it is present, and the higher temperature indicates that there is an increase in the molecular motion inside the ball, ground, and air (i.e., there is more internal energy).

FIGURE 2.4 When the ball hits the ground and stops, both the potential and kinetic energy of the ball are 0, with the ball's potential and kinetic energy just before contact being converted to internal energy—which corresponds to an increased temperature of the ball.

As mentioned, all of the energy of a system can be divided into these three parts. We can write the total energy of the system, E, as

$$E = U + KE + PE = U + \tfrac{1}{2} mV^2 + mgz \tag{2.3}$$

In Chapter 1, we discussed how many problems will be solved using specific properties. The specific energy, e, of a system can be found by dividing Eq. (2.3) by the mass:

$$e = E/m = u + \tfrac{1}{2} V^2 + gz \tag{2.4}$$

where u is the specific internal energy ($u = U/m$).

2.2.4 Magnitudes of the Types of Energy

It is helpful to understand the relative sizes of each type of energy. As an example, let us consider 1 kg of liquid water at 20°C moving at 50 m/s through a pipe 20 m above the ground. Notice that this is a large velocity for liquid water (the water would move the length of an American football field in about 2 seconds), and the height of the pipe is also large. In Chapter 3, you will learn how to find the specific internal energy of liquid water, but for now, let us consider the specific internal energy of liquid water at 20°C to be u = 83.9 kJ/kg. Therefore, the various contributions of each type of energy can be found to be

$$U = mu = (1\text{ kg})(83.9\text{ kJ/kg}) = 83.9\text{ kJ}$$

$$KE = \tfrac{1}{2} mV^2 = (0.5)(1\text{ kg})(50\text{ m/s})^2 = 1250\text{ J} = 1.25\text{ kJ}$$

$$PE = mgz = (1\text{ kg})(9.81\text{ m/s}^2)(20\text{ m}) = 196\text{ J} = 0.196\text{ kJ}$$

These values are shown graphically in **Figure 2.5**.

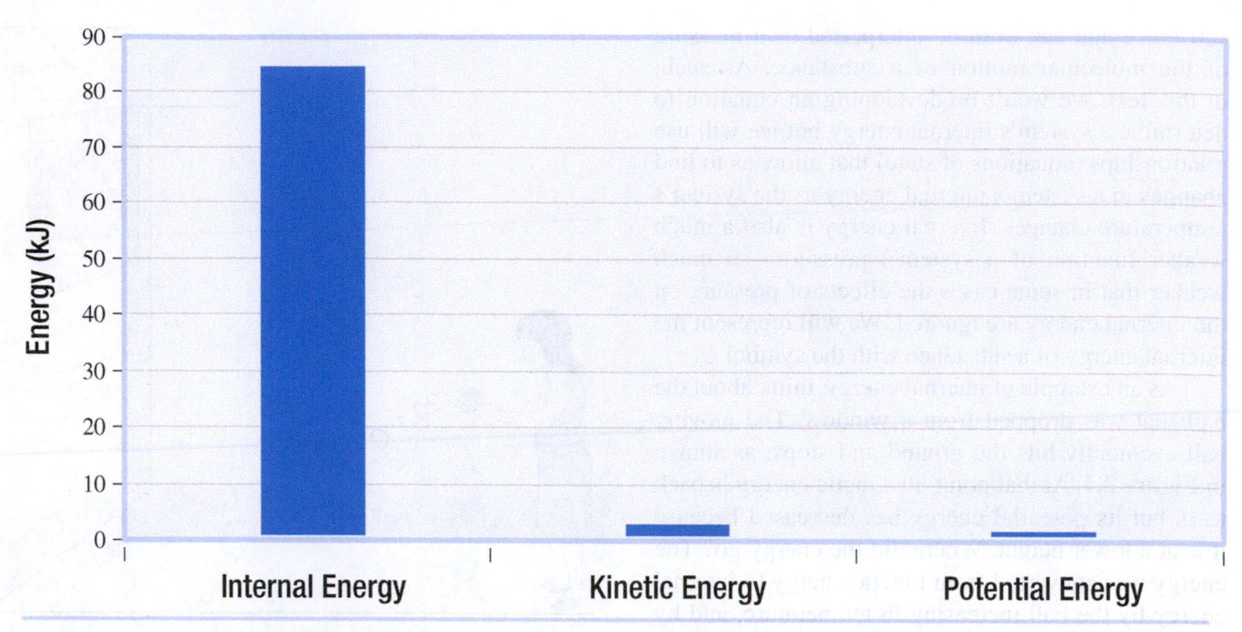

FIGURE 2.5 The internal energy of an object is usually much larger than its kinetic energy or potential energy.

The internal energy of the water is much larger than the kinetic energy, which in turn is much larger than the potential energy. As substances get hotter, and when they become gases or vapors, their specific internal energy increases rapidly, making the kinetic and potential energy contributions even less significant in many applications than shown for liquid water above.

> **QUESTION FOR THOUGHT/DISCUSSION**
> How has the use of energy by humans changed over the last 100 years? What are some uses of energy today that would have been unthought of 200 years ago?

2.3 TRANSPORT OF ENERGY

Energy remaining in an unchanging form inside a system tends not to require analysis, because nothing is happening involving the energy. A fixed amount of energy, unchanging in form, located in a system that is not interacting with other systems or the surroundings is static and usually is not of concern for engineering applications. For example, as shown in **Figure 2.6**, if we take a brick at room temperature and put it into a room at that temperature, there is little

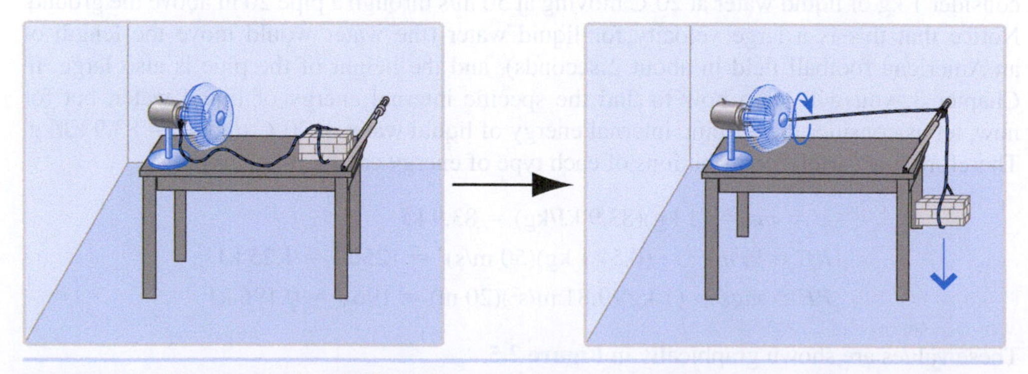

FIGURE 2.6 The energy in a falling object can be used to cause another effect—in this case, a falling brick can cause a fan blade to spin.

to analyze because nothing is happening once the brick is in place. But if we tie the brick to a rope and throw the brick out a window, the rope can pull on some other object, leading to something happening—an effect. Or, if we heat up the brick and then drop it into a bucket of water, again, something happens. Similarly, if we have steam at a constant temperature and pressure flowing through a pipe, nothing in particular is happening. But if we direct that steam into a turbine such as in Figure 2.7, the steam can push on the turbine blades, causing the turbine's rotor to spin and produce an effect. Again, the energy present in a system must change so that some effect is produced.

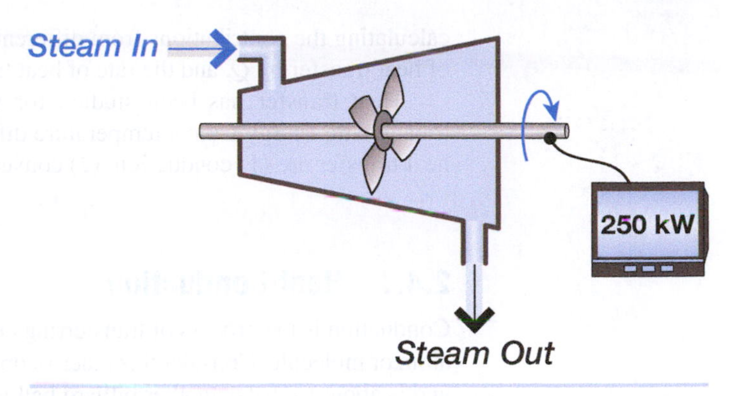

FIGURE 2.7 Flowing steam can cause a turbine shaft to spin as it impinges on the turbine's blades.

How does a substance with energy produce an effect? Effects are accomplished either by converting the energy in a substance to a different form (such as in the preceding extended description of the ball) or by transferring energy from one substance or system to another. By transferring energy into or out of a system, we can rapidly produce changes in the system. For example, by transferring energy out of a cup of boiling water, the water may cool to room temperature. Or, by transferring energy into a pulley on an elevator, the elevator may rise to a higher floor of a building. Having already considered the possibility of converting energy to different forms inside a system, let us now focus on transferring energy into or out of a system. This can also be called the transport of energy.

There are three general ways that energy can be transported: (1) heat transfer, (2) work, and (3) mass transfer. Notice that these three quantities are not thermodynamic properties. Remember that a thermodynamic property is a quantity whose numerical value only depends on the local thermodynamic equilibrium state, and not on how that state was achieved. Although energy is a property, these three transport mechanisms are processes, and their numerical value depends entirely on the process used. Therefore, it makes no sense to say that a system has a certain amount of heat or a certain amount of work. Rather, the system has a certain amount of energy, and the system underwent a process that delivered or removed a certain amount of heat or work. Below, we will consider each of these mechanisms in greater detail.

2.4 HEAT TRANSFER

The transport of energy via heat transfer is the movement of energy caused by a temperature difference between two systems (or one system and the surroundings). Intuitively, we know that if we have a block of metal at 100°C (system 1) and we bring it into contact with water at 20°C (system 2), the block of metal will cool whereas the water will become warmer, as shown in Figure 2.8. The process by which the internal energy of the metal (as represented by the lowering of the metal's temperature) was transferred to the internal energy of the water (as represented by the increase in its temperature) is heat transfer. The detailed analysis of how heat transfer can occur is the topic of another course, typically called "Heat Transfer," but here we will describe the basic mechanisms and provide simple equations for

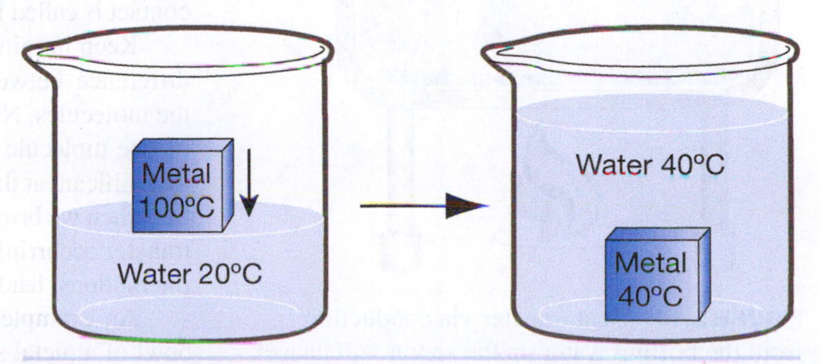

FIGURE 2.8 The placement of a hot object into cool water causes energy to be transferred in the form of heat from the hot object to the water, until thermal equilibrium is reached.

calculating the contributions from different heat transfer modes. We will represent the amount of heat transfer by Q, and the rate of heat transfer by $\dot{Q}$.

Heat transfer has been studied for well over 100 years. Yet only three methods for transferring energy from a temperature difference have been identified. These three modes of heat transfer are (1) conduction, (2) convection, and (3) radiation.

2.4.1 Heat Conduction

Conduction is the process of transferring energy from one atom or molecule directly to another atom or molecule. Consider this macroscopic example: a billiard ball is rolling rapidly on a table and is about to strike another billiard ball moving slowly. What will happen as contact is made between the two balls? The first ball will slow down, and the second ball will speed up as a result of the transfer of kinetic energy from the fast-moving ball to the slow-moving ball. Would there have been any change in the speeds of the balls if they did not hit? Neglecting friction, no, there would have been no speed changes in the balls. The transfer of energy required contact between the two balls.

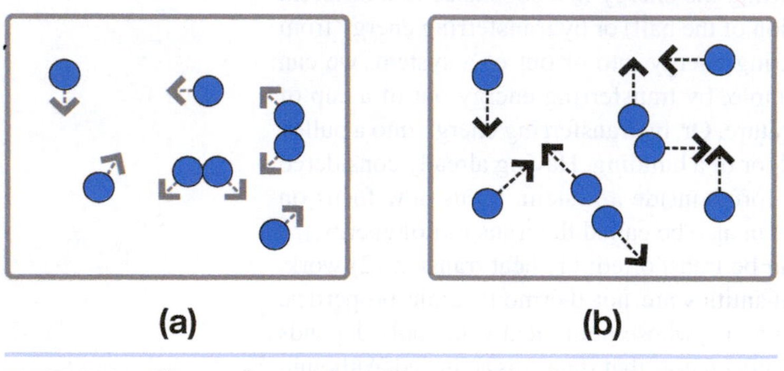

FIGURE 2.9 Energy is transferred between moving molecules. (a) Slow-moving (low temperature) molecules have lower energy, and (b) fast-moving (high temperature) molecules have higher energy.

This same concept can be seen at the molecular level, as shown in **Figure 2.9**. As mentioned previously, all molecules are moving, and the higher the temperature, the more rapid the motion of the molecules. If a molecule that is moving fast hits another molecule that is moving more slowly, energy will be transferred from the first molecule to the second molecule. The first molecule had a higher temperature due to its faster velocity than the second molecule, and then after the collision, the velocity (and temperature) of the first molecule was reduced whereas the velocity (and temperature) of the second molecule was increased. If the two molecules initially had the same temperature (and therefore velocity), the velocities would not have been altered in the collision and both molecules would have retained the same energy. If the two molecules had not collided, there would have been no transfer of energy. The process of transferring energy between atoms and molecules by direct contact is called heat conduction.

FIGURE 2.10 Heat transfer via conduction from the boiling water up the spoon will cause the spoon to get hot, and may injure the person about to grab the spoon.

Keep in mind what is needed for conduction: a temperature difference between the molecules and direct contact between the molecules. Now, if the only transfer of energy occurs because of one molecule contacting another molecule, the result will be insignificant at the macroscopic level of thermodynamic systems. So, when we bring two systems into direct contact, we see energy transfer occurring through very large numbers of direct molecular interactions, leading to the heat transfer between systems.

An example of heat conduction can be seen by placing the bowl of a metal spoon at room temperature into a pot of boiling water, as shown in **Figure 2.10**. If we wait a few minutes and then grab the handle of the spoon, we risk being burned as a result of the bowl of the spoon having increased in temperature by being in direct contact with the water, and then the handle of the spoon heating up as energy was conducted up the length of the spoon.

But does every spoon behave the same way? Would every spoon reach the same temperature at the same time? No. A spoon with a thinner handle may heat up differently than a spoon with a wider handle. And different materials affect the rate at which heat is conducted up the handle of the spoon. These features are represented in the basic equation for one-dimensional heat conduction, also known as Fourier's law:

$$\dot{Q}_{\text{cond}} = -\kappa A \frac{dT}{dx} \tag{2.5}$$

where κ is the thermal conductivity of the material (given in W/m · K), A is the area perpendicular to the direction of heat transfer, and dT/dx is the rate of change of temperature with respect to position (x). Often, dT/dx can be approximated as $\Delta T/\Delta x$, where ΔT is the temperature difference over a length Δx.

▸ EXAMPLE 2.3

On a cold winter day, the outside temperature of a wall is –5°C, while the inside temperature of the wall is 20°C, as shown in **Figure 2.11**. The wall thickness is 10 cm, and the wall is made of a type of brick that has a thermal conductivity of 0.70 W/m · °C. The surface area of the wall is 8 m².
Determine the rate of conductive heat transfer through the wall.

Given: $T_1 = 20°\text{C}$, $T_2 = -5°\text{C}$, $x_2 - x_1 = 10\text{ cm} = 0.10\text{ m}$, $A = 8\text{ m}^2$, $\kappa = 0.70\text{ W/m} \cdot °\text{C}$

Find: $\dot{Q}_{\text{cond}}$

Solution: The rate of heat transfer via conduction can be found with Eq. (2.5):

$$\dot{Q}_{\text{cond}} = -\kappa A \frac{dT}{dx}$$

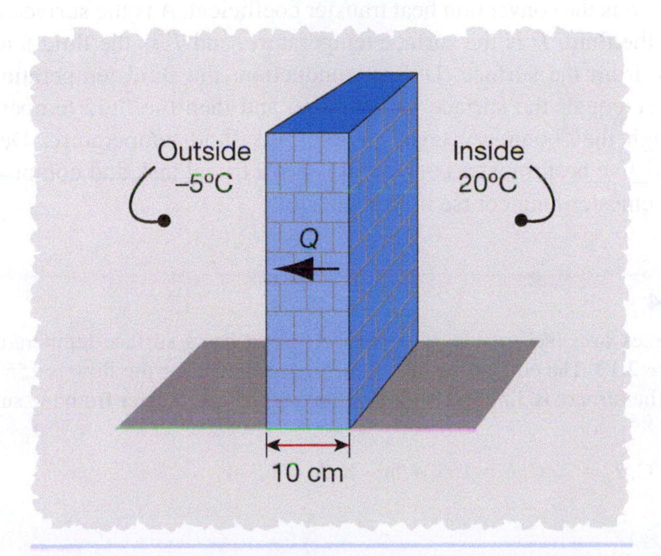

FIGURE 2.11 Heat will transfer via conduction through a wall from an area with a higher temperature to an area with a cooler temperature.

For the wall, we can approximate the temperature gradient by

$$\frac{dT}{dx} = \frac{T_2 - T_1}{x_2 - x_1} = 250°\text{C/m}$$

Therefore, $\dot{Q}_{cond} = (-0.70 \text{ W/m} \cdot °\text{C})(8 \text{ m}^2)(250°\text{C/m}) = -1400 \text{ W} = \mathbf{-1.40 \text{ kW}}$

Analysis: The negative sign indicates that the flow of energy is from the warm area to the cold area.

2.4.2 Heat Convection

A second way that energy can be transferred via a temperature difference is known as heat convection. In convection, energy is transferred between a solid surface and a moving fluid that is in contact with the surface. An example of heat convection occurs when we stand outside on a cold, windy day, as shown in Figure 2.12. Heat is transferred rapidly away from your skin to the moving air, cooling you down more quickly than would happen if you were in stationary air. Another example can be seen with the pot of boiling water used as an example of conduction. If steam is coming off the boiling water and you put your face into the rising steam, you will quickly feel your skin warm as the steam passes over your skin—this too is an example of convection.

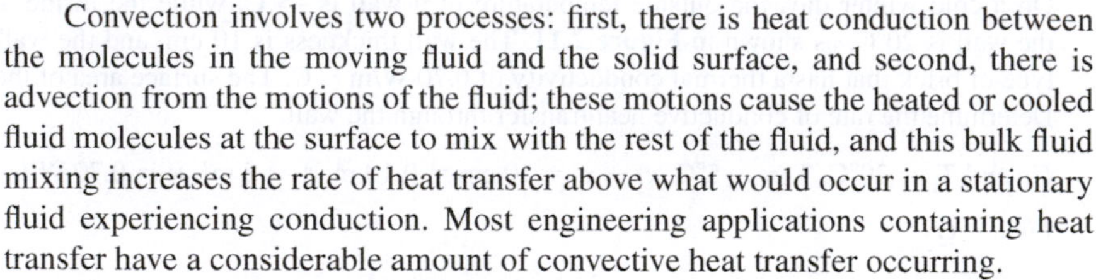

Convection involves two processes: first, there is heat conduction between the molecules in the moving fluid and the solid surface, and second, there is advection from the motions of the fluid; these motions cause the heated or cooled fluid molecules at the surface to mix with the rest of the fluid, and this bulk fluid mixing increases the rate of heat transfer above what would occur in a stationary fluid experiencing conduction. Most engineering applications containing heat transfer have a considerable amount of convective heat transfer occurring.

FIGURE 2.12 Wind blowing cold air over a person is an example of heat convection.

The basic equation for heat convection is Newton's law of cooling (or heating, depending on the fluid and surface temperatures):

$$\dot{Q}_{conv} = hA(T_f - T_s) \tag{2.6}$$

where h is the convection heat transfer coefficient, A is the surface area in contact with the fluid, T_s is the surface temperature, and T_f is the fluid temperature well away from the surface. Due to conduction, the fluid temperature right at the surface equals the surface temperature, and then the fluid temperature changes through the "boundary layer" to the bulk fluid temperature. Determining the convective heat transfer coefficient is not a trivial task and comprises much time in a semester-long course in heat course.

▶ EXAMPLE 2.4

Air at 20°C passes over the outside of a furnace, which has a surface temperature of 70°C, as shown in Figure 2.13. The convective heat transfer coefficient for the flow is 155 W/m² · K. The surface area of the furnace is 3.5 m². Determine the rate of heat transfer from the surface from heat convection.

Given: $T_f = 20°\text{C}$, $T_s = 70°\text{C}$, $h = 155 \text{ W/m}^2 \cdot \text{K}$, $A = 3.5 \text{ m}^2$

Find: $\dot{Q}_{conv}$

Solution: The convective heat transfer rate can be found from Eq. (2.6):

$$\dot{Q}_{conv} = ha(T_f - T_s)$$

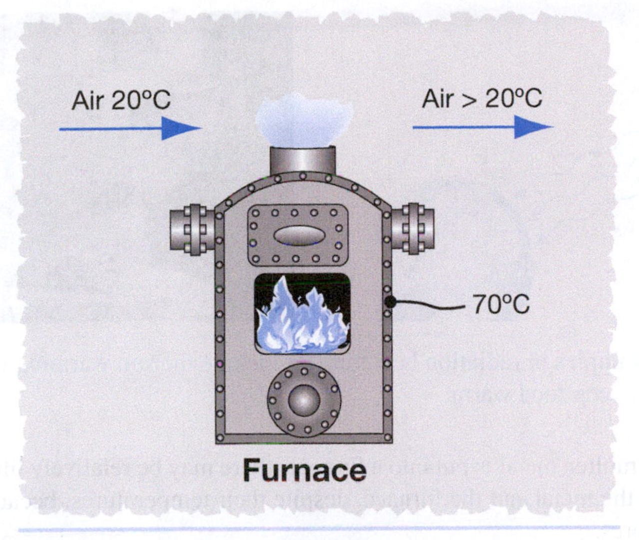

FIGURE 2.13 Cool air will heat up as it flows over a warm surface, via heat convection.

Substituting: $\dot{Q}_{conv} = (155\ \text{W/m}^2\cdot\text{K})(3.5\ \text{m}^2)(20 - 70)°\text{C} = -27{,}100\ \text{W} = \mathbf{-27.1\ kW}$

Analysis: The negative sign indicates that heat is leaving the surface. Furthermore, writing the heat transfer coefficient as "per K" is equivalent to saying "per °C." Alternatively, the temperature difference is 50°C = 50 K. Knowing this, you can see how the units appropriately cancel.

2.4.3 Radiation Heat Transfer

The third mechanism for transferring energy due to a temperature difference is radiation heat transfer. In this mechanism, energy is transferred via photons (or electromagnetic waves). The heat transfer is highly dependent on the temperature of the object, as can be seen in the Stefan-Boltzmann law, which describes the maximum rate of heat transfer from a surface at temperature T_s via radiation:

$$\dot{Q}_{emit} = \sigma A T_s^4 \tag{2.7}$$

where A is the surface area and σ is the Stefan-Boltzmann constant ($\sigma = 5.67 \times 10^{-8}\ \text{W/m}^2\cdot\text{K}^4 = 0.1714 \times 10^{-8}\ \text{Btu/h}\cdot\text{ft}^2\cdot\text{R}^4$). The net amount of radiation heat transfer to or from a surface depends on the surface temperature, the temperature of the surroundings, T_{surr} (because the surroundings are radiating heat to the surface as well), and the emissivity of the surface, ε. The emissivity of the surface can be thought of as the efficiency of a surface at emitting radiation, with an emissivity of 1 corresponding to a perfect emitter, also known as a "blackbody." The net rate of radiation heat transfer between a surface and the surroundings is given by

$$\dot{Q}_{rad} = -\varepsilon\sigma A\left(T_s^4 - T_{surr}^4\right) \tag{2.8}$$

The negative sign is used to maintain consistency with the heat transfer sign convention described later. Examples of situations where radiation heat transfer can be the dominant heat transfer mechanisms are the sun heating the earth and a heat lamp shining on a cold object, as shown in **Figure 2.14**. To have large amounts of radiation heat transfer, you need both a very hot surface (due to the dependence of the radiation on the surface temperature to the fourth power) and a large temperature difference between the surface and the surroundings.

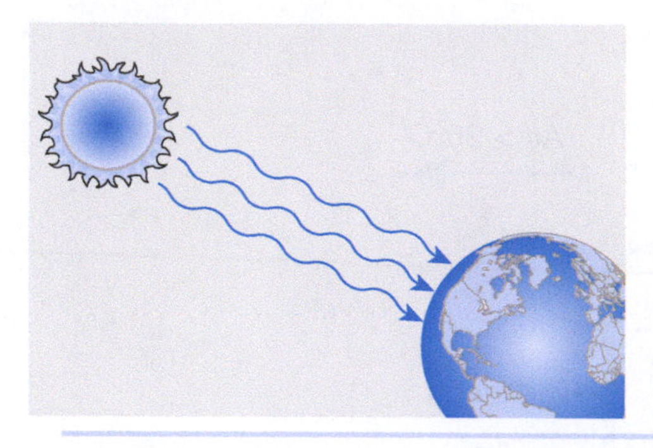

FIGURE 2.14 Examples of radiation heat transfer include the sun warming the earth and a heat lamp used to keep food warm.

For instance, if a molten metal is put into a furnace, there may be relatively little radiation heat transfer between the metal and the furnace, despite their temperatures, because they are both at high temperatures.

▶ EXAMPLE 2.5

The heat lamp shown in **Figure 2.15**, with a surface area of 0.25 m², is used for heating a train platform on winter days. The lamp has a surface temperature of 250°C, and the surroundings are at −2°C. The emissivity of the lamp is 0.92. Determine the net rate of radiation heat transfer leaving the heat lamp.

Given: $A = 0.25\ \text{m}^2$, $T_s = 250°\text{C} = 523\ \text{K}$, $T_{surr} = -2°\text{C} = 271\ \text{K}$, $\varepsilon = 0.92$

Find: $\dot{Q}_{rad}$

Solution: To calculate the heat transfer rate via radiative heat transfer for the heat lamp, Eq. (2.8) can be used.

$$\dot{Q}_{rad} = -\varepsilon\sigma A\left(T_s^4 - T_{surr}^4\right)$$

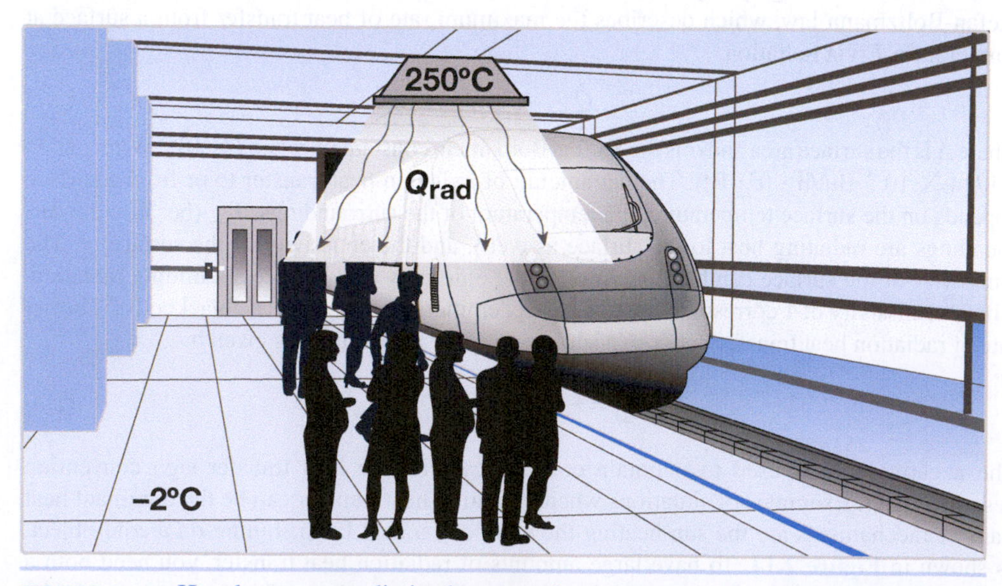

FIGURE 2.15 Heat lamps use radiative heat transfer to keep people warm in outdoor settings in winter.

Substituting: $\dot{Q}_{rad} = -(0.92)(5.67 \times 10^{-8}\ \text{W/m}^2 \cdot \text{K}^4)(0.25\ \text{m}^2)((523\ \text{K})^4 - (271\ \text{K})^4) = \mathbf{-905\ W}$

Analysis: The negative sign indicates that heat is leaving the lamp.

2.4.4 Total Heat Transfer and the Sign Convention

The total amount of heat transfer experienced by a system is the sum of the contributions from the three heat transfer mechanisms, as illustrated in Figure 2.16:

$$\dot{Q} = \dot{Q}_{cond} + \dot{Q}_{conv} + \dot{Q}_{rad} \tag{2.9}$$

As described previously, all three mechanisms are not necessarily important for all engineering applications. Three examples are as follows. Measuring the heat loss from a person standing outside on a windy summer day will likely involve only consideration of convection. Determining the rate of heat transfer through a wall likely involves only conduction. Determining the rate of heat loss from a spacecraft in orbit will likely involve only radiation. Sometimes two modes are important, such as determining the rate of heat loss from an engine cylinder to engine coolant (conduction and convection). And sometimes all three may be important, such as for determining the rate of heat loss from an incandescent light bulb. For most of this text, we will be concerned with the overall heat transfer, and we will not be particularly concerned with the heat transfer mechanisms. Nonetheless, it is helpful to understand how heat can be transferred to or from a system, so that we can recognize how to reduce or increase the heat transfer as desired.

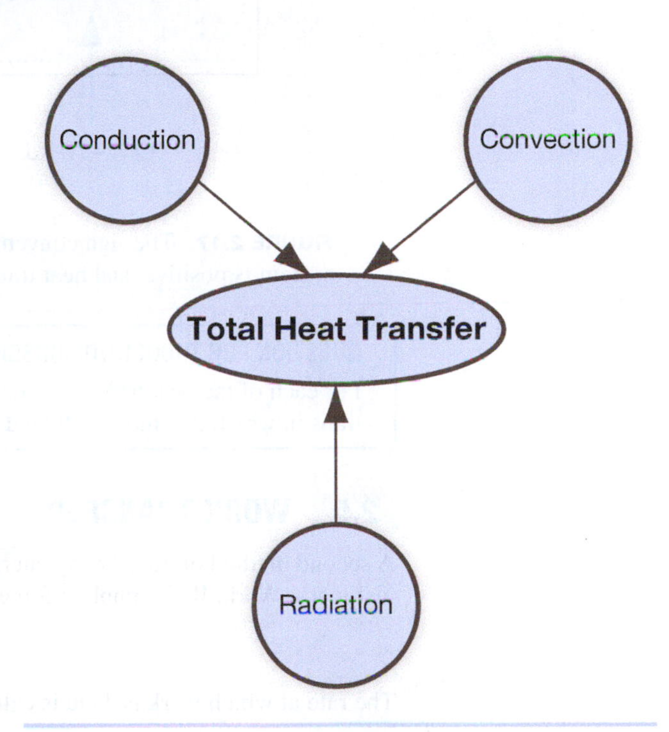

FIGURE 2.16 The total heat transfer experienced by a system is the sum of the conduction, convection, and radiation.

The overall heat transfer for a process can be determined, if necessary, by integrating the heat transfer rate over time, t:

$$Q = \int_{t_1}^{t_2} \dot{Q} \cdot dt \tag{2.10}$$

At times you will be concerned with the rates of processes, and at other times you will be concerned with the total quantities involved with a process.

An important element to solving thermodynamic problems is the establishment of a sign convention to describe the transfers of energy. For heat transfer, we will use the following sign convention:

Heat Transfer to a system $\Rightarrow$ Q is positive

Heat Transfer from a system $\Rightarrow$ Q is negative

This sign convention is shown graphically in Figure 2.17. Finally, there will be applications with no heat transfer. A process that does not involve any heat transfer (i.e., $Q = 0$) is called an ***adiabatic process***.

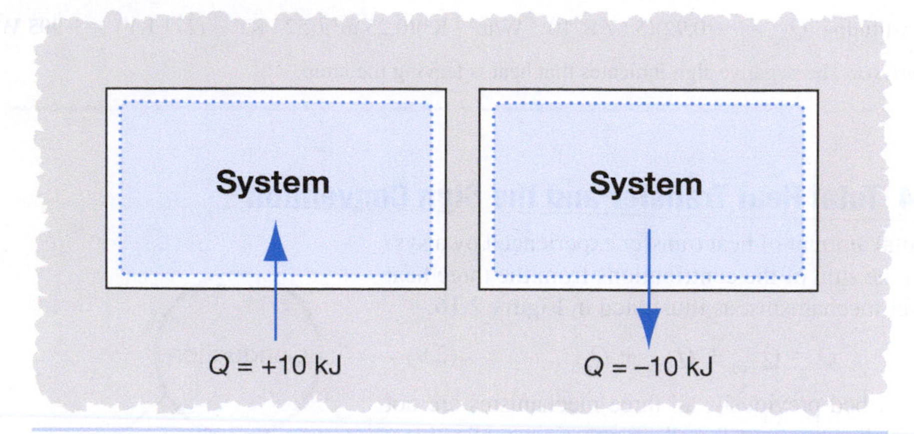

FIGURE 2.17 The sign convention for heat transfer is that heat transfer into a system is positive, and heat transfer out of a system is negative.

QUESTION FOR THOUGHT/DISCUSSION

For each of the heat transfer modes, what are two or three commonly encountered situations in which that mode will be dominant?

2.5 WORK TRANSFER

A second method of transferring energy into or out of a system is called work transfer, or often just work. Work, W, is simply a force, F, acting through a displacement, dx:

$$W = \int F \cdot dx \tag{2.11}$$

The rate at which work is done is called the **power**, $\dot{W}$, and is equal to

$$\dot{W} = \frac{\delta W}{\delta t} \tag{2.12}$$

As stated before, work is not a property of a system, but rather is a description of the thermodynamic process—therefore, the symbol used for the differential in Eq. (2.12) is a δ rather than a d; this indicates that the work is a *path function* (dependent on the thermodynamic process taken) rather than a *point function* (dependent on only the thermodynamic state, such as a property).

Any combination of a type of force acting through a type of displacement yields a mode of work. Unlike heat transfer, therefore, it is possible that the number of possible work transfer modes is limitless. However, there are only a few work transfer modes with which we will be concerned with in basic thermodynamics. In the following subsections, we describe several of the more commonly used work modes in basic thermodynamics.

2.5.1 Moving Boundary Work

When the physical size of a system changes during a process, the expansion or contraction of the system occurs as a result of moving boundary work. An example of moving boundary work is the expansion of a balloon as air inside the balloon is heated, such as in Figure 2.18. The increase in the temperature of the air will correspond to an increase in volume, as we will see in Chapter 3, and the expansion will cause the system to do work on the surroundings. Another example is the air located inside a piston–cylinder assembly shown in Figure 2.19, such as in an

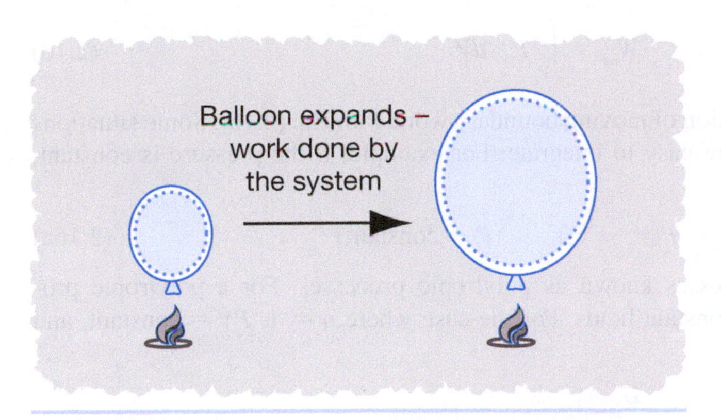

FIGURE 2.18 The change of the volume of a system is a form of work—moving boundary work.

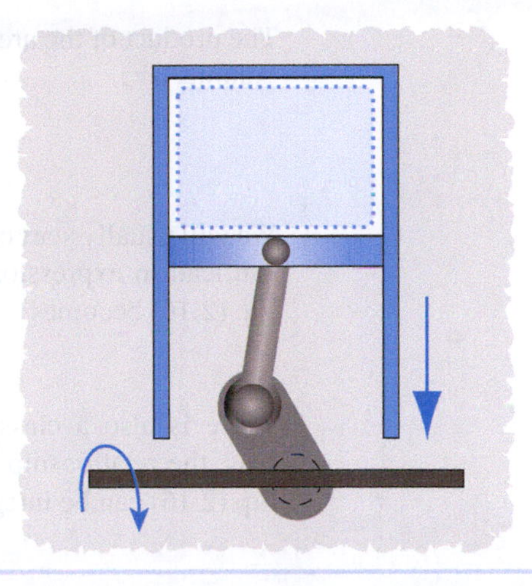

FIGURE 2.19 Through mechanical linkages, the moving boundary work in a piston–cylinder device can be converted to rotating shaft work.

automobile's engine. During the compression stroke, the piston moves, compressing the gas, and this adds energy to the system through moving boundary work. Conversely, during the expansion (power) stroke, the gas is pushing against the piston, causing the volume occupied by the gas to expand—in this case, energy leaves the system and moves into the surroundings through the moving boundary work done by the gas on the piston.

For a one-dimensional displacement, such as in a piston–cylinder device, the moving boundary work, W_{mb}, can be found by applying Eq. (2.11) between the initial and final position:

$$W_{mb} = \int_{x_1}^{x_2} F \cdot dx \tag{2.13}$$

where F is the force acting on the system boundary, x_1 is the initial location of the boundary, and x_2 is the final location of the boundary after the process is complete, as shown in Figure 2.20. Although this expression works for one-dimensional movement and for when the force is known, in thermodynamic systems we often deal with multidimensional expansion and contraction (such as the balloon expanding or contracting) and more readily measure the pressure of a gas or liquid rather than the force. Equation (2.11) can be applied to such a situation. Considering that the pressure, P, is equal to the force divided by the area, A, across which the force is applied,

$$P = F/A \tag{2.14}$$

we can substitute into Eq. (2.13) an expression for the force:

$$W_{mb} = \int_{x_1}^{x_2} P \cdot A \cdot dx \tag{2.15}$$

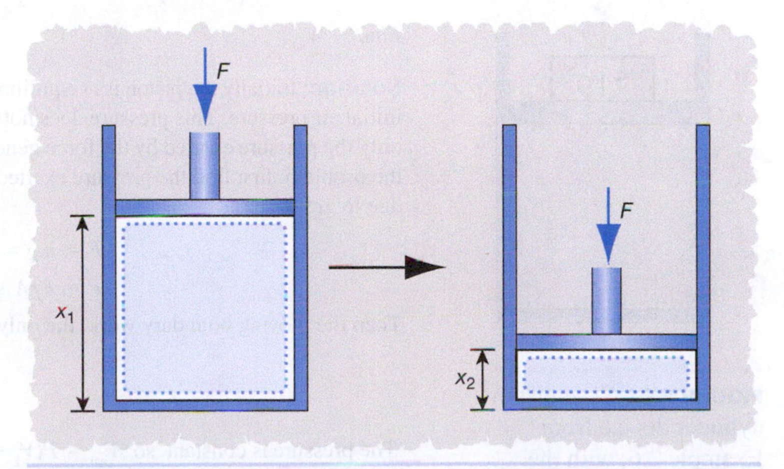

FIGURE 2.20 The force applied on the system by the piston causes the volume to decrease, requiring an input of energy through moving boundary work.

The product of the area and a differential linear displacement change is simply the differential volume, $d\forall$:

$$W_{mb} = \int_{\forall_1}^{\forall_2} P \cdot d\forall \tag{2.16}$$

We will usually start consideration of moving boundary work with Eq. (2.16). Some situations can lead to expressions that are easy to integrate. For example, if the pressure is constant, Eq. (2.16) becomes

$$W_{mb} = P(\forall_2 - \forall_1) \qquad (P = \text{constant}) \tag{2.16a}$$

There is also a class of processes known as polytropic processes. For a polytropic process, the relationship $P\forall^n$ = constant holds. For the case where $n = 1$, $P\forall$ = constant, and Eq. (2.16) can be integrated to

$$W_{mb} = P_1\forall_1 \ln \frac{\forall_2}{\forall_1} \qquad (P\forall^n = \text{constant}, n = 1) \tag{2.16b}$$

For polytropic processes when $n \neq 1$, Eq. (2.16) becomes

$$W_{mb} = \frac{P_2\forall_2 - P_1\forall_1}{1 - n} \qquad (P\forall^n = \text{constant}, n \neq 1) \tag{2.16c}$$

Finally, where the volume is constant, $d\forall = 0$ and $W_{mb} = 0$.

▶ EXAMPLE 2.6

Air is located inside a cylinder, under a piston, as shown in Figure 2.21. The cylinder has a cross-sectional area of 0.008 m^2. Initially, the air occupies a volume of 0.01 m^3, and the pressure of the air is 150 kPa. A weight with a mass of 5 kg is placed on top of the piston, and the piston moves down, compressing the gas to a volume of 0.007 m^3. Determine the amount of moving boundary work done in the process.

Given: $A = 0.008 \text{ m}^2$, $\forall_1 = 0.01 \text{ m}^3$, $\forall_2 = 0.007 \text{ m}^3$, $m_w = 5 \text{ kg}$

Find: W_{mb}

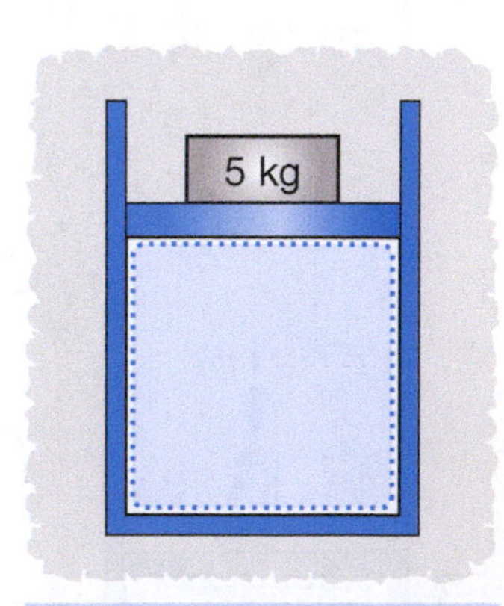

FIGURE 2.21 The piston–cylinder device from Example 2.6, with the 5-kg weight placed on top of the piston.

Solution: Initially the piston is at equilibrium, so the pressure exerted by the piston equals the initial air pressure. This pressure does not contribute to the work done by the process. Rather, only the pressure exerted by the force generated by the added weight produces work. To solve the problem, first find the pressure exerted by the 5-kg weight. Assume standard acceleration due to gravity.

$$F = mg = (5 \text{ kg})(9.81 \text{ m/s}^2) = 49.05 \text{ N}$$
$$P = F/A = 49.05 \text{ N}/0.008 \text{ m}^2 = 6131 \text{ Pa}$$

Then the moving boundary work, the only work mode present, is found from Eq. (2.16):

$$W_{mb} = \int_{\forall_1}^{\forall_2} P \cdot d\forall$$

The pressure is constant, so $W_{mb} = P(\forall_2 - \forall_1) = (6131 \text{ Pa})(0.007 - 0.01) \text{ m}^3 = \mathbf{-18.4 \text{ J}}$.

Analysis: As we will see, the negative sign on the work indicates that energy is being added to the system through this work mode.

▸ **EXAMPLE 2.7**

An unknown gas is heated inside a balloon. Initially, the gas occupies a volume of 0.21 m³, and its pressure is 320 kPa. The gas is heated until the volume expands to 0.34 m³. During the expansion process, the pressure and volume are related through $P\forall^{1.2}$ = constant. Find the moving boundary work done in the process.

Given: $\forall_1 = 0.21\text{ m}^3$, $\forall_2 = 0.34\text{ m}^3$, $P_1 = 320\text{ kPa}$, $P\forall^{1.2} = \text{constant}$

Find: W_{mb}

Solution: As this moving boundary work is a result of a polytropic process with $n = 1.2$, Eq. (2.16c) can be used to find the moving boundary work:

$$W_{mb} = \frac{P_2\forall_2 - P_1\forall_1}{1 - n}$$

Everything is known in this equation except the final pressure, P_2. P_2 is found from the relationship between the pressure and volume:

$$P_1\forall_1^{1.2} = P_2\forall_2^{1.2}$$

$$P_2 = (320\text{ kPa})(0.21\text{ m}^3/0.34\text{ m}^3)^{1.2} = 179.5\text{ kPa}$$

Substituting:

$$W_{mb} = \frac{(179.5\text{ kPa})(0.34\text{ m}^3) - (320\text{ kPa})(0.21\text{ m}^3)}{1 - 1.2} = \mathbf{30.9\text{ kJ}}$$

Analysis: The positive sign on the work indicates that the system is transferring energy to the surroundings through this expansion process.

▸ **EXAMPLE 2.8**

0.50 kg of air undergoes a constant-temperature expansion from a pressure of 1000 kPa and a volume of 0.06 m³ to a volume of 0.11 m³. The process follows a polytropic relationship of $P\forall$ = constant. Determine the final pressure and the moving boundary work done by the air in the expansion process.

Given: $m = 0.50\text{ kg}$, $P_1 = 1000\text{ kPa}$, $\forall_1 = 0.06\text{ m}^3$, $\forall_2 = 0.11\text{ m}^3$

Polytropic process with $n = 1$

Find: P_2, W_{mb}

Solution: First, the final pressure can be found from the relationship $P\forall$ = constant:

$$P_1\forall_1 = P_2\forall_2$$

Solving for P_2:

$$P_2 = P_1\forall_1/\forall_2$$

$$P_2 = (1000\text{ kPa})(0.06\text{ m}^3)/(0.11\text{ m}^3) = 545.5\text{ kPa}$$

To find the moving boundary work for a polytropic process with $n = 1$, Eq. (2.16b) can be used.

$$W_{mb} = P_1\forall_1 \ln\frac{\forall_2}{\forall_1} = (1000\text{ kPa})(0.06\text{ m}^3)\left(\ln\frac{0.11\text{ m}^3}{0.06\text{ m}^3}\right) = \mathbf{36.4\text{ kJ}}$$

2.5.2 Rotating Shaft Work

The work mode that is present when a machine has energy being transferred to or from a system through a revolving shaft is called rotating shaft work, W_{rs}. Examples of such work are the rotor on a turbine being used as the mechanical work input to an electrical generator and the crankshaft on an internal combustion engine moving power down the drive train, as shown in **Figure 2.22**. In this case, the force is represented by the torque, T, and the displacement is the angular displacement of the shaft, $d\theta$:

$$W_{rs} = \int_{\theta_1}^{\theta_2} T \cdot d\theta \tag{2.17}$$

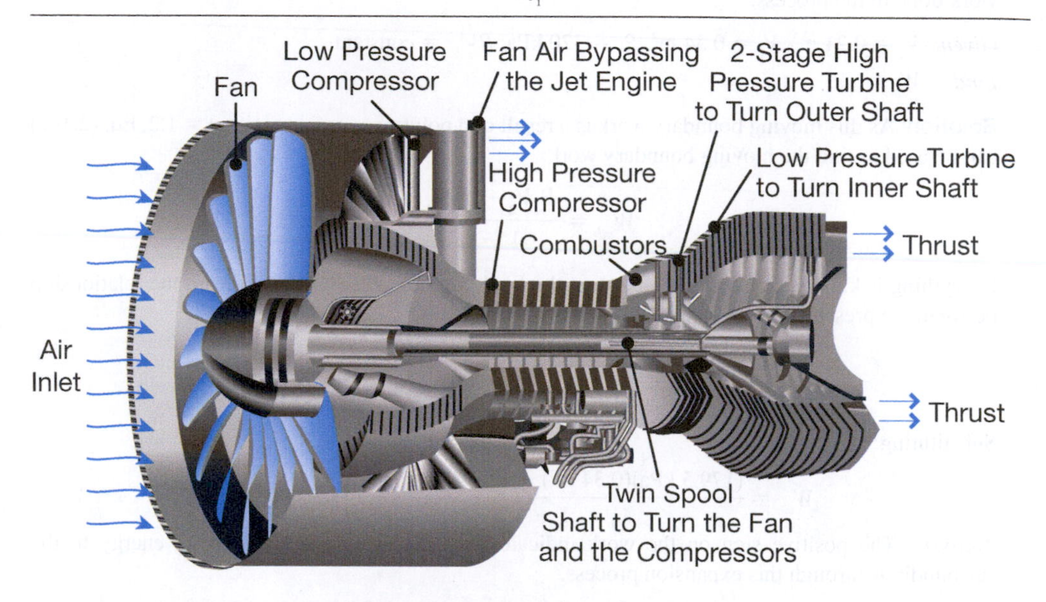

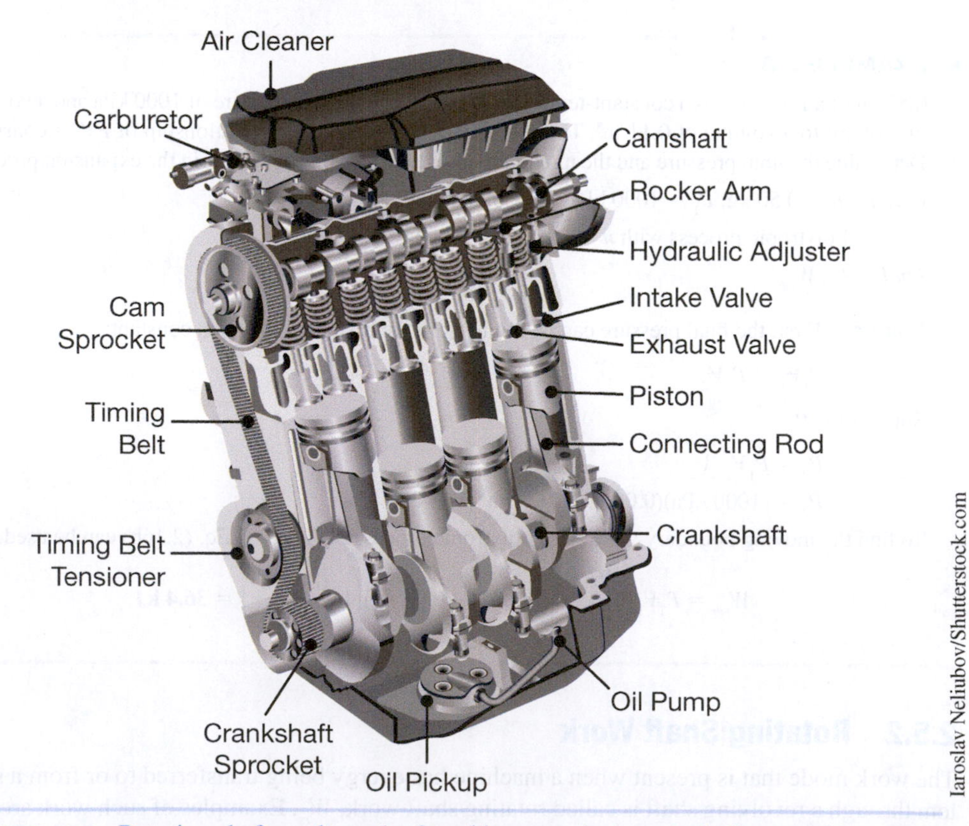

FIGURE 2.22 Rotating shaft work can be found in many devices, including a turbine and the crankshaft of a reciprocating engine.

Although rotating shaft work can be calculated from this equation, we will generally use the First Law of Thermodynamics to find the rotating shaft work present in applications such as turbines, compressors, and pumps, as shown in Chapter 4.

▶ EXAMPLE 2.9

An engine applies a torque of 200 N · m to an outlet shaft. The shaft rotates at 500 rpm. What is the rotating shaft power delivered by the engine?

Given: $T = 200$ N · m, $\dot{n} = 500$ rpm (rotational speed)

Find: $\dot{W}_{rs}$

Solution: The power is the rate of work, and we can find the rotating shaft power by taking the derivative of the rotating shaft work in Eq. (2.17) with respect to time:

$$\dot{W}_{rs} = \frac{\delta W_{rs}}{\delta t} = \int_{\theta_1}^{\theta_2} T \cdot \frac{d\theta}{dt}$$

As the rotational speed is constant, the number of radians that will be traversed in a given time is equal to the number of radians in a circle (2π) times the number of circles traversed. Extending this to a rate:

$$\int_{\theta_1}^{\theta_2} \frac{d\theta}{dt} = 2\pi\dot{n}$$

where $\dot{n}$ is the rotational speed. With a constant torque, the rotating shaft power is

$$\dot{W}_{rs} = 2\pi\dot{n}T$$

Substituting:

$$\dot{W}_{rs} = 2\pi\left(\frac{\text{rad}}{\text{rev}}\right)\left(500\,\frac{\text{rev}}{\text{min}}\right)\left(\frac{1\text{ min}}{60\text{ s}}\right)(200\text{ N}\cdot\text{m}) = 10{,}470\text{ W} = \mathbf{10.5\ kW}$$

Analysis: This is power out of the engine to the rotating shaft. You may note that this is a rather small amount of power in comparison to many engines. This is due primarily to the slow rotational speed rather than the torque.

2.5.3 Electrical Work

Electrons can move in a wire under the influence of an electromotive force. Considering the electrons moving in a certain period of time as a displacement, we can see that this represents a work mode, known as electrical work, as illustrated in Figure 2.23. For the electromotive force being represented as a potential difference (i.e, voltage, E), and the electrical charges in a unit of time being the current, I, the electrical work, W_e, can be written as

$$W_e = \int_{t_1}^{t_2} -EI \cdot dt \tag{2.18}$$

where t is the time. On a rate basis, the electrical power, $\dot{W}_e$, is

$$\dot{W}_e = -EI \tag{2.19}$$

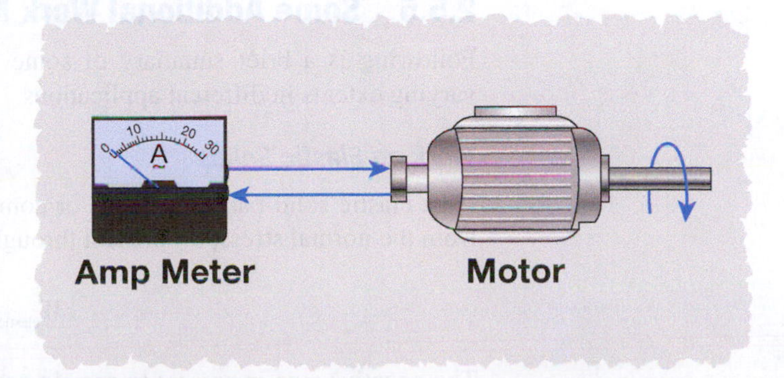

FIGURE 2.23 Electrical work can be converted by motors into rotating shaft work.

The negative signs are present to remain consistent with the sign convention for work that will be established later. An electrical input to a system is correctly modeled as electrical work. Alternatively, if the electrical input is flowing into an electrical resistance heater, it may be modeled as a heat input.

▶ EXAMPLE 2.10

A sump pump in a home uses 350 W of power. The pump is plugged into a standard 120-volt outlet. What current is drawn while the pump is operating?

Given: $\dot{W}_e = -350 \text{ W}$

$E = 120 \text{ V}$

Find: I

Solution: The current can be found by rearranging Eq. (2.19):

$$\dot{W}_e = -EI$$

So,

$$I = -\frac{\dot{W}_e}{E}$$

Substituting:

$$I = -\frac{(-350 \text{ W})}{120 \text{ V}} = \mathbf{2.92\ A}$$

2.5.4 Spring Work

As a force is applied to a spring, the spring will decrease in length, and when a spring increases in length, a force is released. Clearly, this is again a case of a force acting through a displacement, and the result is the work mode known as spring work, W_{spring}. For a linear elastic spring, the displacement is linearly proportional to the displacement, $F = kx$, where k is the spring constant. Substituting into Eq. (2.11), we find

$$W_{\text{spring}} = \int_{x_1}^{x_2} F \cdot dx = \int_{x_1}^{x_2} kx \cdot dx = \frac{1}{2}k\left(x_2^2 - x_1^2\right) \tag{2.20}$$

More complex expressions can be derived out in a similar fashion for nonlinear springs.

2.5.5 Some Additional Work Modes

Following is a brief summary of some additional work modes which are encountered to varying extents in different applications.

Work on Elastic Solids

If an elastic solid bar is stretched or compressed, work transfer occurs. It can be calculated from the normal stress, σ_n, present through

$$W_{\text{elastic}} = \int_{x_1}^{x_2} -\sigma_n A\, dx \tag{2.21}$$

The negative sign is present to remain consistent with the sign convention for work that will be established below.

Work from Surface Tension on a Film

A liquid film experiences surface tension, and this surface tension results in work transfer when the film is stretched or compressed. The surface tension, σ_s, leads to surface tension work through

$$W_{\text{surface}} = \int_{A_1}^{A_2} -\sigma_s \, dA \tag{2.22}$$

Magnetic Work

Work is done to change the magnetic field of a system. This can be found with the magnetic field strength, H, the magnetic susceptibility of the material, χ_m, and the constant known as the magnetic permeability ($\mu_0 = 4\pi \times 10^{-7}\,\text{V} \cdot \text{s/(A} \cdot \text{m)}$):

$$W_{\text{magnetic}} = -\mu_0 \forall (1 + \chi_m)\left(\frac{H_2^2 - H_1^2}{2}\right) \tag{2.23}$$

Chemical Work

When a chemical species is added to or removed from a system, chemical work is present. This work can be found using the chemical potential for species i, μ_i, and the changes in mass of species, i, summed for all the species added to or removed from, k, the system:

$$W_{\text{chemical}} = -\sum_{i=1}^{k} \mu_i (m_2 - m_1)_i \tag{2.24}$$

In Eq. (2.24), the chemical potential of each species has been assumed to be constant during the transfer of mass.

There are still other forms of work. These are some of the more commonly experienced forms, but remember that any force acting through a displacement will lead to a work mode.

2.5.6 Total Work and Sign Convention

Just as for the total heat transfer, the total work in a process is the sum of all the contributions from all of the work modes that are present:

$$W = \sum_i W_i \tag{2.25}$$

where W_i represents each type of work mode. This is illustrated in Figure 2.24.

We will use a sign convention for the work that is common in engineering thermodynamics; this sign is the opposite of the sign convention for heat. The sign convention we will use for work is

Work Transfer to a system $\Rightarrow$ W is negative

Work Transfer from a system $\Rightarrow$ W is positive

This sign convention is graphically shown in Figure 2.25. If a system undergoes a process that produces work, the sign convention indicates that the work term should be positive. This has already been established in our equations for the various work modes. The sign convention is easy to see in Eq. (2.16) for moving boundary work. In moving boundary work, if the volume is expanding, the system is doing work on the surroundings and thus there is a transfer of energy out of the system in the form of work. If the volume expands, Eq. (2.16) yields a positive quantity for the work.

It may seem odd that the sign convention for the work is opposite that of the heat transfer, and it should be noted that some people will use a sign convention that has the same direction of energy flow for both heat and work. The sign convention that we use comes from practical engineering applications. For example, consider an automobile engine. The purpose of the engine

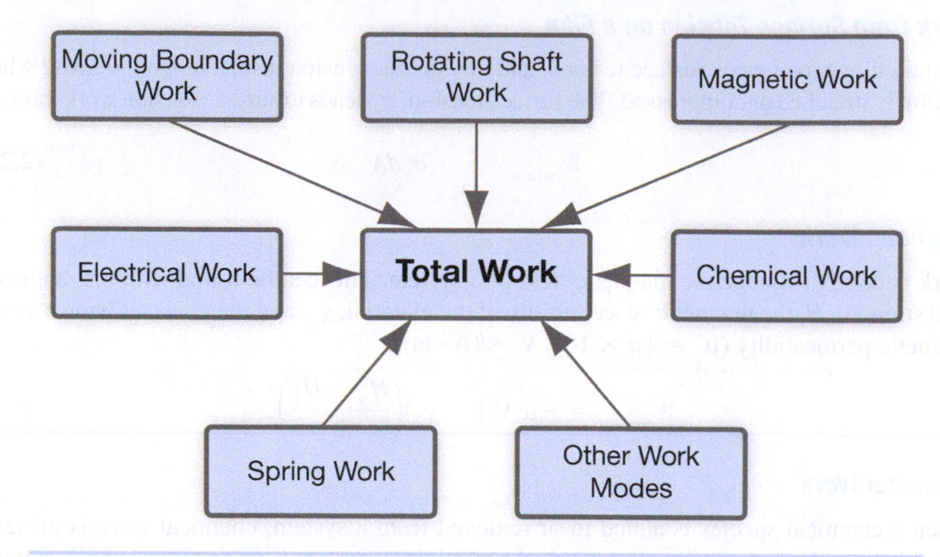

FIGURE 2.24 The total work experienced by a system is the sum of the contributions from all possible work modes.

is to produce work, and in order to get this work, heat is added to the gases in the cylinder through the combustion process. So for an engine, the processes can be interpreted as energy being added to the gases through heat (heat in is positive) and the gases in the engine cylinders then expanding and doing work (work out is positive). From a simple analysis, heat into the system produced work out of the system—and with the sign convention used here, both terms are positive quantities. There is no right or wrong sign convention to be used as long as the same convention is used consistently to develop equations that use the derived quantities appropriately. Keep this idea in mind in Chapter 4 when we develop equations for the First Law of Thermodynamics.

QUESTION FOR THOUGHT/DISCUSSION

While the use of electricity is properly considered work, the energy transfer involved with electrical work may be thought of as a heat input. When might a heat transfer interpretation of the energy transfer associated with electrical work be reasonable?

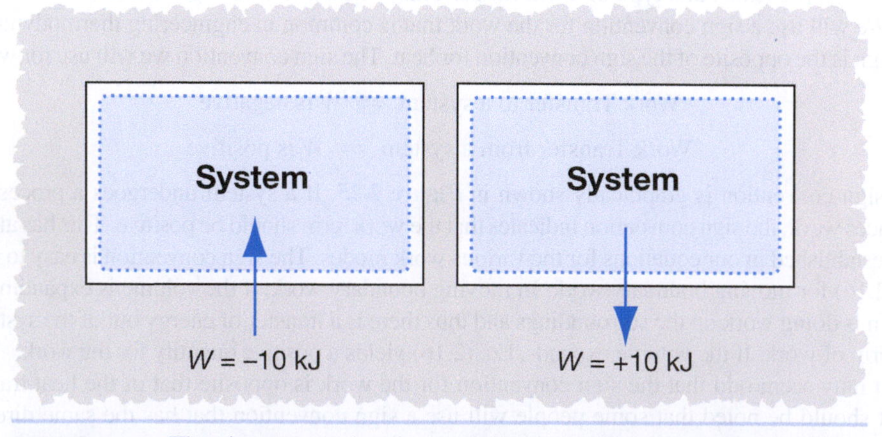

FIGURE 2.25 The sign convention for work is that work into a system is negative and work out of a system is positive.

2.6 ENERGY TRANSFER VIA MASS TRANSFER

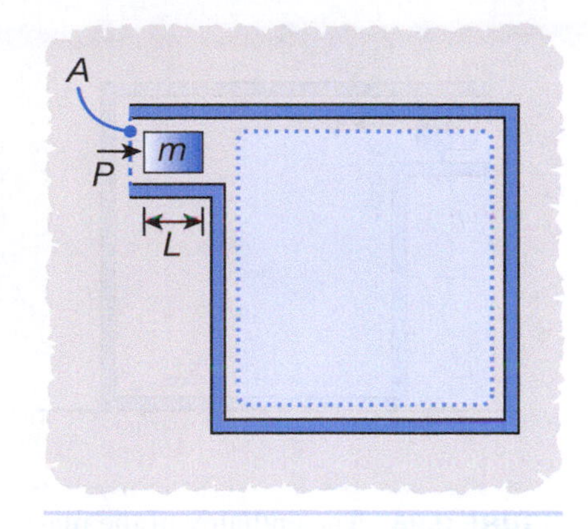

FIGURE 2.26 Mass flowing into or out of a system results in a method of transferring energy into or out of a system.

The third method of transferring energy into or out of a system is energy transfer via mass transfer. As discussed earlier in this chapter, a substance has internal energy, kinetic energy, and potential energy. If a substance flows into or out of a system, it carries the substance's energy along with it. For example, if you have a bucket half-filled with water and you pour a glass of water into the bucket, the water present in the bucket now has gained the energy that was present in the water in the glass initially. Or, as another example, consider a classroom full of students. If one student leaves the classroom, the energy that is present in the classroom is reduced because the student took their internal, kinetic, and potential energy with them as they exited the room.

There is a complication to this simple idea, however. As a mass crosses the boundary of a system, it does require energy to pass into or out of the system. This transfer of energy is a work mode, because it is a force acting through a displacement, and it is called the ***flow work***. The force pushing the mass can be found from

$$F = PA \tag{2.26}$$

where P is the pressure of the fluid and A is the cross-sectional area of the inlet or outlet through which the fluid passes. As shown in Figure 2.26, the mass has some length of space, L, that it occupies in the inlet or outlet, and so the flow work is the application of this force through this length:

$$W_{\text{flow}} = FL = (PA)(L) = P\forall = P(mv) = m(Pv) \tag{2.27}$$

This flow work is present anytime a mass flows into or out of a system and is only meaningful for such an occurrence.

Combining the concept of the energy of the mass with the flow work, we can develop an expression for the energy transfer into or out of a system that occurs during mass transfer:

$$E_{\text{mass flow}} = m\left(u + \frac{1}{2}V^2 + gz\right) + m(Pv) \tag{2.28}$$

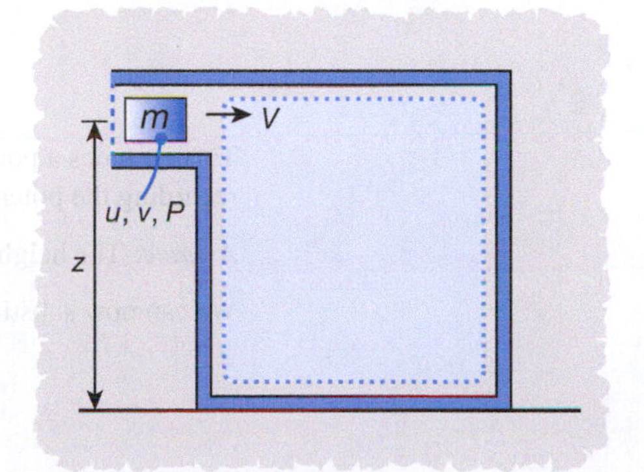

FIGURE 2.27 As mass enters a system, it adds its internal energy, kinetic energy, and potential energy to the system. In addition, energy is added to the system through work to push the mass across the system boundary.

where the first term on the right-hand side represents the energy contained in the mass, and the second term is the flow work. Equation (2.28) can be rewritten as

$$E_{\text{mass flow}} = m\left(u + Pv + \frac{1}{2}V^2 + gz\right) \tag{2.29}$$

The components of this energy transport from mass flow is shown in Figure 2.27.

For any mass flow into or out of a system, the expression $u + Pv$ will always appear. Because this expression occurs so frequently, and because it is a combination of three properties, it has been given its own name: specific enthalpy. The enthalpy, H, of a substance is a thermodynamic property,

$$H = U + P\forall \tag{2.30}$$

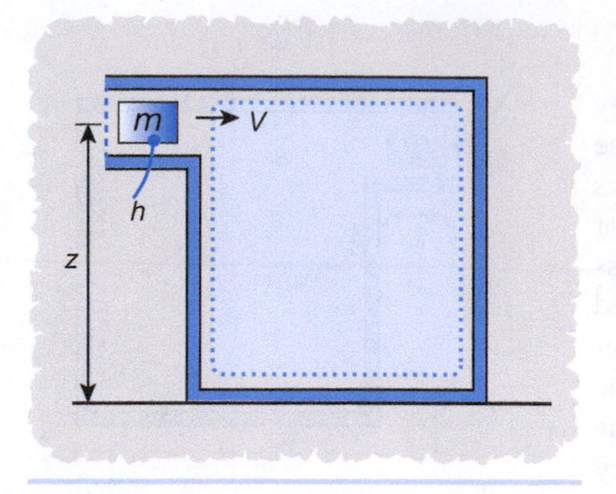

FIGURE 2.28 The enthalpy of the mass flowing into the system is the sum of its internal energy and the flow work.

and the specific enthalpy, h, is the resulting intensive property when the enthalpy is divided by the mass:

$$h = H/m = u + Pv \tag{2.31}$$

It should be noted that the enthalpy value of most substances is often dominated by the contribution of the internal energy in the system, but the difference between the enthalpy and the internal energy can be significant for substances with high specific volumes (such as gases) or at very high pressures.

Equation (2.29) can be rewritten in terms of specific enthalpy:

$$E_{\text{mass flow}} = m\left(h + \frac{1}{2}V^2 + gz\right) \tag{2.32}$$

On a rate basis, the rate of energy transfer via mass transfer is

$$\dot{E}_{\text{mass flow}} = \dot{m}\left(h + \frac{1}{2}V^2 + gz\right) \tag{2.33}$$

which assumes that the properties of the fluid are constant during the duration of the rate analysis. The combination of properties in a system is shown in Figure 2.28.

▶ EXAMPLE 2.11

Steam at 300°C and 2 MPa enters a turbine with a velocity of 105 m/s. The specific enthalpy of the steam at this state is 3024.2 kJ/kg. The mass flow rate of the steam is 3.5 kg/s. Determine the rate of energy transfer via mass transfer for the turbine from the entering steam.

Given: $h = 3024.2$ kJ/kg

$\dot{m} = 3.5$ kg/s

$V = 105$ m/s

Find: $\dot{E}_{\text{mass flow}}$

Solution: To determine the energy transfer via mass transfer, Eq. (2.33) can be used.

$$\dot{E}_{\text{mass flow}} = \dot{m}\left(h + \frac{V^2}{2} + gz\right)$$

This includes a potential energy term, for which we have no information. As a result, an assumption regarding the potential energy is needed.

Assume: The height of the steam is negligible ($PE = 0$).

We can now substitute in the known information:

$$\dot{E}_{\text{mass flow}} = \left(3.5\,\frac{\text{kg}}{\text{s}}\right)\left(3024.2\,\frac{\text{kJ}}{\text{kg}} + \frac{(105\,\frac{\text{m}}{\text{s}})^2}{2}\left(\frac{1}{1000\,\frac{\text{J}}{\text{kJ}}}\right) + 0\right)$$

$$= 10{,}604 \text{ kW} = \mathbf{10.6\ MW}$$

Analysis: The contribution from the kinetic energy is very small in comparison to the contribution from the enthalpy—which is primarily internal energy and is closely related to the temperature of the steam. In all likelihood, the potential energy contribution would be even smaller than the kinetic energy contribution, making it reasonable to assume the contribution to be 0 if no information is known about the height of the system.

We will not assign a specific sign convention to the energy transfer via mass transfer, and instead we will develop future equations with the knowledge that mass entering a system will increase the energy of the system, whereas mass leaving a system will decrease the system's energy. Finally, keep in mind that energy transfer via mass transfer only occurs in open systems, because by definition there is no mass flow into or out of a system in closed systems.

2.7 ANALYZING THERMODYNAMICS SYSTEMS AND PROCESSES

At this point, we have introduced many of the fundamental concepts that will be needed for performing thermodynamic analyses of devices and systems. Solving equations that develop out of the analyses is usually not difficult from a mathematical perspective, but many students struggle to assimilate the information given in a problem and recognize how to connect that information to the variables in the equations. There is no single analysis procedure that everyone finds to be satisfactory; however, the following is a solution framework that should help you organize your thoughts and guide you to a solution procedure that works best for you.

Suggested Problem Analysis and Solution Procedure:

1. Carefully read the problem statement.
2. List all the information that is known about the device or process (i.e., what is known).
3. List the quantities that are sought in the analysis (i.e., what is being solved for).
4. Draw a schematic diagram of the device, and indicate the system chosen for analysis.
5. Determine whether the thermodynamic system is open, closed, or isolated. Determine the type of substance to be analyzed (i.e., solid, liquid, gas, two-phase) and what equations of state describe the behavior of the substance (these will be identified primarily in Chapter 3).
6. Write the appropriate general equations that describe the type of system to be analyzed (these will be derived primarily in Chapters 4 and 6).
7. Determine what simplifying assumptions are appropriate for this particular analysis: state the assumptions clearly and apply the assumptions to the general equations.
8. Apply the equations of state to connect the given thermodynamic property data to the properties required in the modified equations.
9. Solve the equations for the desired quantities.
10. Thoughtfully consider the results to determine if the answers make sense from an engineering viewpoint.

Throughout this solution process, it is beneficial to keep track of the units of each quantity, because doing so reduces the chances of a numerical mistake and serves as a check to be certain that your equations are correct. For example, if you are expecting a power calculation to provide an answer in kW, but your units are ending as kJ/kg, you neglected to multiply by a mass flow rate of kg/s. Or, if you are forgetting to convert a given volumetric flow rate into a mass flow rate, your omission will be obvious if you are carrying the units through the calculation rather than just assuming that the units will work with the numbers given.

2.8 PLATFORM FOR PERFORMING THERMODYNAMICS ANALYSIS

Traditionally, thermodynamics has been taught by relying on hand calculation of individual problems, with property data being extracted from tables or charts. There are some benefits to hand calculations: when studying a device for the first time, it is helpful to think through the system and analysis carefully to understand the details of the thermodynamic analysis. However, after analyzing a device a few times, what you learn through hand calculations is limited. You may begin to view thermodynamics as a series of unrelated problems, and you may learn how to solve individual problems rather than getting a deep understanding of how different factors impact the performance of a device. For example, in solving individual problems, you may never understand why a compressor requires more power to compress high-temperature air than low-temperature air; as a result, you may not realize that simply changing the location of an air compressor intake can save a company significant amounts of money in reduced electricity bills.

As will be seen throughout this book, the individual equations to be solved in basic thermodynamics are not particularly difficult; what becomes time-consuming is the use of tables or charts to obtain necessary thermodynamic property data. So, although you could easily solve an equation for the power required to compress air at several different inlet temperatures, finding the necessary property data can be time-consuming and a deterrent to performing such an analysis. Furthermore, in the 21st century, it has become commonplace for engineers to perform much analysis on computers. Therefore, in this book, we will encourage you to develop computer-based models for various common devices and thermodynamic systems. Doing so will allow you to explore the parameters that affect the performance of a system, and it will allow you to learn how great those effects are—such understanding can be vital to a practicing engineer. We encourage you to work some problems initially by hand, and then to develop computer-based models once you understand how to set up appropriate equations for a process.

There are many software packages that are used for equation solving; examples include MATLAB, Mathcad, *Mathematica*, and Engineering Equation Solver (EES). Even spreadsheets such as Microsoft Excel can be set up to solve thermodynamics problems. Engineering students are frequently introduced to one or more of these packages before taking a course in thermodynamics. In addition, software is available that is designed specifically for solving thermodynamics problems. In today's world, what is available in terms of computer software or tablet apps changes frequently, as does how a specific piece of software or an app works. Therefore, this book will not promote the use of one particular software platform for solving thermodynamic problems. Instead, we will leave it to you (or your instructor) to choose a platform that you are most comfortable using and to use your knowledge of that platform to develop models for different devices and processes.

As will be discussed in Chapter 3, there are many options for finding thermodynamic properties via software. Again, the choice of what to use is left to you, because the speed at which those programs and apps change may result in a vastly superior option becoming available even during the duration of a course. Some of the common equation solvers are either directly connected to thermodynamic property solvers or can be connected through available plug-ins. You may find such integrated approaches convenient.

QUESTION FOR THOUGHT/DISCUSSION

What are some advantages to using a computer-based approach to solving thermodynamics problems in comparison to traditional hand-calculations?

Summary

In this chapter, we have learned about the different forms of energy that are present in substances. We have also learned about the three methods of transferring energy into or out of a system: heat transfer, work, and energy transfer via mass transfer. In Chapter 4, we will use this knowledge to develop expressions for the First Law of Thermodynamics. Before we do that, we will need to learn how to evaluate and relate various thermodynamic properties. That is the subject of Chapter 3.

KEY EQUATIONS

Potential Energy:

$$PE = mgz \tag{2.1}$$

Kinetic Energy:

$$KE = \tfrac{1}{2} mV^2 \tag{2.2}$$

Total Energy:

$$E = U + KE + PE = U + \tfrac{1}{2} mV^2 + mgz \tag{2.3}$$

Heat Conduction Rate:

$$\dot{Q}_{cond} = -\kappa A \frac{dT}{dx} \tag{2.5}$$

Convective Heat Transfer Rate:

$$\dot{Q}_{conv} = hA(T_f - T_s) \tag{2.6}$$

Net Radiative Heat Transfer Rate:

$$\dot{Q}_{rad} = -\varepsilon\sigma A\,(T_s^4 - T_{surr}^4) \tag{2.8}$$

Moving Boundary Work:

$$\text{General:}\quad W_{mb} = \int_{\mathcal{V}_1}^{\mathcal{V}_2} P \cdot d\mathcal{V} \tag{2.16}$$

$$P = \text{constant:}\quad W_{mb} = P\,(\mathcal{V}_2 - \mathcal{V}_1) \tag{2.16a}$$

$$\text{Polytropic process, } n = 1:\quad W_{mb} = P_1 \mathcal{V}_1 \ln \frac{\mathcal{V}_2}{\mathcal{V}_1} \tag{2.16b}$$

$$\text{Polytropic process, } n \neq 1:\quad W_{mb} = \frac{P_2\mathcal{V}_2 - P_1\mathcal{V}_1}{1 - n} \tag{2.16c}$$

Electrical Power:

$$\dot{W}_e = EI \tag{2.19}$$

Rate of Energy Transfer via Mass Flow

$$\dot{E}_{\text{mass flow}} = \dot{m}\left(h + \frac{1}{2}V^2 + gz\right) \tag{2.33}$$

PROBLEMS

2.1 A 5.00-kg light fixture is suspended above a theater's stage 15.0 m above the ground. The local acceleration due to gravity is 9.75 m/s^2. Determine the potential energy of the light fixture.

2.2 In 1908, Charles Street of the Washington, D.C., baseball team caught a baseball thrown from the top of the Washington Monument. If the height of the ball before it was thrown was 165 m and the mass of the baseball was 0.145 kg, what was the potential energy of the baseball before it was thrown? Assume that the acceleration due to gravity was 9.81 m/s^2.

2.3 10.0 kg of water is about to fall over a cliff in a waterfall. The height of the cliff is 115 m. Determine the potential energy of the mass of water considering standard gravity on earth.

2.4 A jet airplane is flying at a height of 10,300 m at a velocity of 240 m/s. If the mass of the airplane is 85,000 kg and if the acceleration due to gravity is 9.70 m/s^2, determine the potential energy and the kinetic energy of the airplane.

2.5 A piece of space debris with a mass of 7 kg is falling through the atmosphere toward earth at a height of 4.5 km with a velocity of 230 m/s. Determine the potential and kinetic energy of the object.

2.6 A person is taking a picture with a camera over the edge of a cliff which towers 125 m above an ocean beach. The person is bumped by someone else, causing them to drop the camera. The camera leaves the person's hands at a velocity of 1.5 m/s. The camera has a mass of 450 g. Determine the kinetic and potential energy of the camera with respect to the beach as the camera leaves the person's hands.

2.7 A golf ball with a mass of 45.8 g reaches a maximum height of 15 m on a particular shot. At that height, the velocity of the golf ball is 31 m/s. Determine the kinetic and potential energy of the golf ball at its maximum height on this shot.

2.8 A baseball with a mass of 0.145 kg is thrown by a pitcher at a velocity of 42.0 m/s. Determine the kinetic energy of the baseball.

2.9 A brick with a mass of 2.50 kg, which was dropped from the roof of a building, is about to hit the ground at a velocity of 27.0 m/s. Determine the kinetic energy of the brick.

2.10 An 80.0-kg rock that was hurled by a catapult is about to hit a wall while traveling at 7.50 m/s. What is the kinetic energy of the rock just before contacting the wall?

2.11 Steam flows through a pipe located 4 m above the ground. The velocity of the steam in the pipe is 80.0 m/s. If the specific internal energy of the steam is 2765 kJ/kg, determine the total internal energy, the kinetic energy, and the potential energy of 1.50 kg of steam in the pipe.

2.12 For Problem 2.11, replace the steam with liquid water. If the specific internal energy of the liquid water is 120 kJ/kg, determine the total internal energy, the kinetic energy, and the potential energy of 1.50 kg of water in the pipe.

2.13 Steam, with a specific internal energy of 2803 kJ/kg, is flowing through a pipe located 5 m above the ground. The velocity of the steam in the pipe is 75 m/s. Determine the total internal energy, the kinetic energy, and the potential energy of 1 kg of steam in the pipe.

2.14 Replace the steam in Problem 2.13 with liquid water. If the specific internal energy of the liquid water is 105 kJ/kg, determine the total internal energy, the kinetic energy, and the potential energy of 1 kg of water in the pipe.

2.15 Determine the specific energy of air at 25°C, traveling at 35.0 m/s at a height of 10.0 m. Consider the specific internal energy of the air to be 212.6 kJ/kg.

2.16 Determine the total energy of water vapor with a mass of 2.50 kg and a specific internal energy of 2780 kJ/kg traveling at a velocity of 56.0 m/s at a height of 3.50 m.

2.17 Determine the total energy of liquid water with a mass of 2.5 kg if it is traveling at 40 km/h at a height of 6 m. Consider the specific internal energy of the water to be 85 kJ/kg.

2.18 A house has outside walls made of brick with a thermal conductivity of 1.20 W/m · K. The wall thickness is 0.20 m.

- **(a)** If the inside temperature of the wall is 20.0°C and the outside temperature of the wall is −10.0°C, determine the heat transfer rate per unit area from conduction (the conductive heat flux) for the wall.
- **(b)** For an inside wall temperature of 20.0°C, plot the conductive heat flux through the wall for outside wall temperatures ranging from 30.0°C to −15.0°C.
- **(c)** For an outside wall temperature of −10.0°C, plot the conductive heat flux for inside wall temperatures ranging from 15.0°C to 25.0°C. Discuss how this relates to gaining cost savings by heating a home to a lower temperature in the winter.

2.19 To reduce heat transfer losses through a glass window, you consider replacing the window with one of three alternatives: a sheet of tin 0.50 cm thick, a layer of brick 8.0 cm thick, and a combination of wood and insulation 4.0 cm thick. The original thickness of the glass is 1.0 cm. Consider the thermal conductivities of each to be the following: glass—1.40 W/m · K; tin—66.6 W/m · K; brick—1.20 W/m · K; wood/insulation combination—0.09 W/m · K. Considering the inside temperature of the surface to be 20.0°C and the outside temperature to be −5.0°C, determine the heat conduction rate per unit area for the glass and the three alternatives, and discuss the relative merits of the three alternatives (ignoring material and installation costs).

2.20 A factory has a sheet metal wall with a thermal conductivity of 180 W/m · K. The wall thickness is 2.5 cm.

- **(a)** If the inside temperature of the wall is 25.0°C and the outside temperature of the wall is −12°C, determine the heat transfer rate per unit area from conduction (the conductive heat flux) for the wall.
- **(b)** For an inside wall temperature of 25°C, plot the conductive heat flux through the wall for outside wall temperatures ranging from 30°C to −30°C.
- **(c)** For an outside wall temperature of −12°C, plot the conductive heat flux for inside wall temperatures ranging from 10°C to 30°C. Discuss how this relates to gaining cost savings by heating a building to a lower temperature in the winter.

2.21 You are stoking a fire in a furnace, and you leave the stoker inside the furnace for several minutes. The end of the stoker inside the furnace reaches a temperature of 300°C, while the other end of the stoker in the air is cooled sufficiently to maintain a temperature of 60°C. The length of the stoker between these two ends is 2.0 m. If the cross-sectional area of the stoker is 0.0010 m^2, determine the rate of heat transfer that occurs if (a) the stoker is made of aluminum ($\kappa = 237$ W/m · K), (b) iron ($\kappa = 80.2$ W/m · K), and (c) granite ($\kappa = 2.79$ W/m · K).

2.22 A conductive heat transfer rate of 15 W is applied to a metal bar whose length is 0.50 m. The hot end of the bar is at 80°C. Determine the temperature at the other end of the bar for (a) a copper bar ($\kappa = 401$ W/m · K) with a cross-sectional area of 0.0005 m^2,

(b) a copper bar ($\kappa = 401$ W/m · K) with a cross-sectional area of 0.005 m^2, and (c) a zinc bar ($\kappa = 116$ W/m · K) with a cross-sectional area of 0.005 m^2.

2.23 A block of material with a cross-sectional area of 0.10 m^2 and a thickness of 0.15 m has one side maintained at 90°C and the other end maintained at 20°C. Determine the rate of heat transfer via conduction if the material is (a) nickel ($\kappa = 90.7$ W/m · K), (b) common brick ($\kappa = 0.72$ W/m · K), and (c) fiberglass ($\kappa = 0.04$ W/m · K).

2.24 A heat exchanger is a device that allows energy to be transferred via heat transfer from a hot fluid to a cold fluid. Suppose a heat exchanger is designed such that a hot fluid passes over one side of the flat plate whose surface area is 0.75 m^2, while the cold fluid passes over the other side. The fluid on the hot side of the flat plate maintains the surface temperature on the hot side at 100°C, while the cold fluid maintains the cold side of the plate at 30°C. The plate is made of a carbon steel ($\kappa = 60$ W/m · K). What is the thickness of the plate if the necessary conductive heat transfer rate across the plate is (a) 25 kW, (b) 100 kW, and (c) 300 kW?

2.25 Hot water is sprayed onto a countertop while washing it. The temperature of the water is 75°C, while the countertop is 20°C. The surface area of the countertop in contact with the water is 0.05 m^2. If the convective heat transfer coefficient is 30 W/m^2 · K, determine the heat transfer rate from the water to the countertop.

2.26 A wind blows over the face of a person on a cold winter day. The temperature of the air is -5.0°C, while the person's skin temperature is 35.0°C. If the convective heat transfer coefficient is 10.0 W/m^2 · K and the exposed surface area of the face is 0.008 m^2, determine the rate of heat loss from the skin via convection.

2.27 Cooling water flows over a hot metal plate with a surface area of 0.5 m^2 in a manufacturing process. The temperature of the cooling water is 15°C, and the metal plate's surface temperature is maintained at 200°C. If the convective heat transfer coefficient is 68.0 W/m^2 · K, determine the rate of convective heat transfer from the plate.

2.28 Air passes over a hot metal bar which has a surface area of 0.25 m^2. The temperature of the air is 20°C, while the bar has a temperature maintained at 140°C. If the convective heat transfer coefficient is 23 W/m^2 · K, determine the rate of convective heat transfer from the plate.

2.29 A cool, early spring breeze passes over the roof of a poorly insulated home. The surface area of the roof is 250 m^2, and the temperature of the roof is maintained at 20°C from heat escaping the house. The air temperature is 5.0°C, and the convective heat transfer coefficient is 12.0 W/m^2 · K. Determine the rate of convective heat transfer from the roof.

2.30 Air is to be blown over the top of a cup of coffee to cool the coffee. The temperature of the coffee is 90°C, and the air temperature is 25°C. The surface area of the coffee in contact with the air is 0.005 m^2. It is necessary to determine the velocity of the air needed to achieve the desired cooling rate; this could be determined from the heat transfer coefficient with the appropriate correlations available in more advanced heat transfer courses. If the desired convective heat transfer rate for the coffee is -10 W, what is the corresponding desired heat transfer coefficient?

2.31 A hot water bottle is placed inside an empty box. The walls of the box are initially at 17°C. The surface area of the bottle is 0.0024 m^2. If the surface temperature of the water bottle is 75°C and the emissivity of the bottle is 0.70, what is the initial rate of radiation heat transfer from the bottle to the walls?

2.32 An electric space heater has a metal coil at a temperature of 250°C and is used to heat a space with an air temperature of 15°C. If the surface area of the space heater is 0.02 m^2 and the emissivity of the heating element is 0.95, what is the rate of radiation heat transfer from the heater to the air?

2.33 An electrical resistance heater is placed inside a hollow cylinder. The heater has a surface temperature of 260°C, while the inside of the cylinder is maintained at 25°C. If the surface area of the heater is 0.12 m^2 and the emissivity of the heater is 0.90, what is the rate of radiation heat transfer from the heater?

2.34 A baker has developed a marvelous cookie recipe that works best if the cookie is cooled in a vacuum chamber through radiation heat transfer alone. The cookie is placed inside the chamber with a surface temperature of 125°C. Assume that the cookie is thin enough so that the temperature is uniform throughout during the cooling process. The walls of the chamber are maintained at 10.0°C. If the surface area of the cookie is 0.005 m^2 and if the emissivity of the cookie is 0.80, determine the initial radiation heat transfer rate from the cookie to the walls of the chamber.

2.35 A building is losing heat to the surrounding air. The inside walls of the building are maintained at 22.0°C, and the walls have a thermal conductivity of 0.50 W/m · K. The wall thickness is 0.10 m, and the outside wall temperature is held at 2.0°C. The air temperature is −10.0°C. Consider the emissivity of the outside of the walls to be 0.85. Calculate the conductive heat flux and the radiative heat flux. Considering that the heat flux entering the outside of the wall from the inside must balance the heat flux leaving the outside of the wall via convection and radiation, determine the necessary convective heat transfer coefficient of the air passing over the outside wall.

2.36 A long cylindrical rod with a diameter of 3.0 cm is placed in the air. The rod is heated by an electrical current so that the surface temperature is maintained at 200°C. The air temperature is 21°C. Air flows over the rod with a convective heat transfer coefficient of 5.0 $W/m^2 \cdot K$. Consider the emissivity of the rod to be 0.92. Ignoring the end effects of the rod, determine the heat transfer rate per unit length of the rod (a) via convection and (b) via radiation.

2.37 A heat lamp is used to treat material samples in a vacuum oven in the absence of air. The heat lamp is maintained at 400°C, while the walls are initially maintained at 20°C (before any material samples have been inserted into the oven, but after all the air has been removed from the oven). The emissivity of the lamp is 0.95. The lamp is attached to the walls by a solid titanium rod ($\kappa = 21.9$ W/m · K) whose diameter is 0.025 m and whose length is 0.20 m. The surface area of the lamp is 0.02 m^2. Determine the heat transfer rate from the lamp to the walls (a) via conduction and (b) via radiation.

2.38 A weight is attached to a horizontal load through a frictionless pulley. The weight has a mass of 25.0 kg, and the acceleration due to gravity is 9.80 m/s^2. The weight is allowed to fall 2.35 m. What is the work done by the falling weight as it pulls the load on a horizontal plane?

2.39 How much work is needed to lift a rock with a mass of 58.0 kg a distance of 15.0 m, where the acceleration due to gravity is 9.81 m/s^2?

2.40 How much work is needed to lift a 140 kg rock a distance of 7.5 m, where the acceleration due to gravity is 9.81 m/s^2?

2.41 A piston–cylinder device is filled with air. Initially, the piston is stabilized such that the length of the cylinder beneath the piston is 0.15 m. The piston has a diameter of 0.10 m.

A weight with mass of 17.5 kg is placed on top of the piston, and the piston moves 0.030 m, such that the new length of the air-filled cylinder beneath the piston is 0.12 m. Determine the work done by the newly added mass.

2.42 250 kPa of pressure is applied to a piston in a piston–cylinder device. This pressure causes the piston to move 0.025 m. The diameter of the piston is 0.20 m. Determine the work done on the gas inside the piston.

2.43 A piston–cylinder device is filled with 5 kg of liquid water at 150°C. The specific volume of the liquid water is 0.0010905 m^3/kg. Heat is added to the water until some of the water boils, giving a liquid–vapor mixture with a specific volume of 0.120 m^3/kg . The pressure of the water is 475.8 kPa. How much work was done by the expanding water vapor?

2.44 Air at 700 kPa and 25°C fills a piston–cylinder assembly to a volume of 0.015 m^3. The air expands, in a constant-temperature process, until the pressure is 205 kPa. (The constant-temperature process with a gas can be modeled as a polytropic process with $n = 1$.) Determine the work done by the air as it expands.

2.45 Air at 100 kPa and 35°C fills a piston–cylinder assembly to a volume of 0.001 m^3. The air is then compressed following the relationship $P\forall^{1.4}$ = constant, until the volume is 0.0001 m^3. Determine the work done in the process.

2.46 Air at 450 kPa and 20°C fills a piston–cylinder assembly to a volume of 0.075 m^3. The air expands, in a constant-temperature process, until the pressure is 150 kPa. (The constant-temperature process with a gas can be modeled as a polytropic process with $n = 1$.) Determine the work done by the air as it expands.

2.47 1.5 kg of water vapor at 500 kPa fills a balloon. The specific volume of the water vapor is 0.3749 m^3/kg. The water vapor condenses at constant pressure until a liquid–vapor mixture with a specific volume of 0.0938 m^3/kg is present. Determine the work done in the process.

2.48 Air at 1200 kPa and 250°C fills a balloon with a volume of 2.85 m^3. The balloon cools and expands until the pressure is 400 kPa. The pressure and volume follow a relationship given by $P\forall^{1.3}$ = constant. Determine the work done by the air as it expands, and the final temperature of the air.

2.49 Oxygen gas expands in a flexible container following a relationship of $P\forall^{1.15}$ = constant. The mass of the oxygen is 750 g, and initially the pressure and temperature of the oxygen are 1 MPa and 65°C, respectively. The expansion continues until the volume is double the original volume. Determine the final pressure and temperature of the oxygen, and the work done by the oxygen during the expansion.

2.50 Nitrogen gas is compressed in a flexible container following a relationship of $P\forall^{1.2}$ = constant. The mass of the nitrogen is 1.5 kg, and initially the pressure and temperature of the nitrogen are 120 kPa and 15°C, respectively. The compression continues until the volume reaches 0.10 m^3. Determine the final pressure and temperature of the nitrogen, and the work done on the nitrogen in the process.

2.51 Air expands in an isothermal process from a volume of 0.5 m^3 and a pressure of 850 kPa, to a volume of 1.2 m^3. The temperature of the air is 25°C. Determine the work done by the air in this expansion process.

2.52 3.0 kg of air is initially at 200 kPa and 10°C. The air is compressed in a polytropic process, following the relationship $P\forall^{n}$ = constant. The air is compressed until the volume is

0.40 m^3. Determine the work done and the final temperature of the air for values of n of 1.0, 1.1, 1.2, 1.3, and 1.4, and plot the work as a function of n.

2.53 You are told that a particular device compresses air following the relationship $P\mathcal{V}^{1.25}$ = constant. The device initially has a volume of 0.20 m^3, holding 0.24 kg of air at 101 kPa and 20°C. You need to determine the pressure of the air and the work required to compress the air to various volumes.

(a) Determine the pressure of the air and the work required to compress the air for a final volume of 0.10 m^3.

(b) Plot the pressure and the work for a range of final volumes from 0.18 m^3 to 0.025 m^3.

2.54 The following data are available for the compression/expansion of 0.025 kg of O_2 in a device:

P (kPa)	$\mathcal{V}$ (m^3)
125	0.0157
250	0.00859
500	0.00470
750	0.00331
1000	0.00257
1500	0.00181

Determine the exponent n in the expression $P\mathcal{V}^n$ = constant if this is a polytropic process, and determine the work done by the O_2 as it expands from 1500 kPa to 125 kPa.

2.55 A torque of 250 N · m is applied to a rotating shaft. Determine the work delivered for one revolution of the shaft.

2.56 A torque of 510 N · m is applied to a rotating shaft. Determine the power used if the shaft rotates at 1500 revolutions per minute (rpm).

2.57 An engine delivers 55 kW of power to a rotating shaft. If the shaft rotates at 2500 rpm, determine the torque exerted on the shaft by the engine.

2.58 A steam turbine operates at 1800 rpm. The turbine delivers 65.0 MW of power to the shaft of an electrical generator. Determine the torque on the steam turbine's shaft.

2.59 An electrical generator receives a torque of 850 N · m from a rotating shaft connected to a turbine. The shaft operates at 3600 rpm. Determine the power produced by the electrical generator.

2.60 An electrical oven operates on a 208-V line. The power drawn by the oven is 1.25 kW. What is the current drawn by the oven?

2.61 A room fan operates on 120 V and draws a current of 1.5 amps. What is the power used by the fan?

2.62 A 23-W Compact Fluorescent Lamp (CFL) bulb is plugged into a 120-V source. What is the current drawn by the bulb?

2.63 An air compressor is plugged into a 208-V outlet and requires 10 kW of power to complete its compression process. What is the current drawn by the air compressor?

2.64 You are placed in an unfamiliar environment, with many odd pieces of electrical equipment that use nonstandard plugs. You need to determine the voltage of a particular outlet. The information on the machine plugged into the outlet reveals that the machine uses 2.50 kW of power and draws a current of 20.8 amps. What is the voltage of the outlet?

2.65 A 100-W incandescent light bulb, a 25-W CFL bulb, and an 18-W LED bulb all produce approximately the same number of lumens of light. If each operate on 120 V, what is the current drawn by each lighting device?

2.66 A linear elastic spring requires 95 J of work to be compressed from a length of 0.15 m to a length of 0.11 m. Determine the spring constant of the spring.

2.67 A linear elastic spring with a spring constant of 250 N/m is compressed from a length of 0.25 m to a length of 0.17 m. Determine the work done in this compression process.

2.68 A linear elastic spring with a spring constant of 300 N/m is compressed from a length of 45 cm to a length of 40 cm. Determine the work done in this compression process.

2.69 A linear elastic spring with a spring constant of 1.25 kN/m is initially at a length of 0.25 m. The spring expands, doing 40 J of work in the process. What is the final length of the spring?

2.70 Air, with a mass flow rate of 8.0 kg/s, enters a gas turbine. The enthalpy of the air entering the turbine is 825 kJ/kg, the velocity of the air is 325 m/s, and the height of the air above the ground is 2.5 m. The acceleration due to gravity is 9.81 m/s^2. Determine the rate at which energy is being transferred into the gas turbine via mass flow by this air.

2.71 A jet of liquid water with a velocity of 42 m/s and an enthalpy of 62 kJ/kg enters a system at a mass flow rate of 210 kg/s. The potential energy of the water is negligible. What is the rate of energy transfer to the system via mass flow for the water jet?

2.72 Steam enters a steam turbine at a volumetric flow rate of 3 m^3/s. The enthalpy of the steam is 3070 kJ/kg, and the specific volume of the steam is 0.162 m^3/kg. The velocity of the steam as it enters the steam turbine is 75 m/s, and the height of the entrance above the ground is 3 m. The acceleration due to gravity is 9.81 m/s^2. Determine the rate at which energy is being transferred via mass flow for the steam.

2.73 You can use available steam to add energy to a system via mass flow. The enthalpy of the steam is 2750 kJ/kg, and the velocity of the steam is 120 m/s. The potential energy of the steam is negligible. If you must add 33.1 MW of energy to the system with this steam, what is the required mass flow rate of steam entering the system?

2.74 A hose sprays liquid water at 20°C into a container at a velocity of 15 m/s. The mass flow rate of the water is 0.15 kg/s, and the specific enthalpy of the water is 83.92 kJ/kg. Assuming the change in height of the water is negligible, what is the rate at which energy is being transferred via mass flow into the container for the water?

2.75 A hot-water line and a cold-water line both are used to add water to a bowl. The total rate at which water can be added to the bowl is fixed at 2.5 kg/s. The diameter of the hot-water line is 1.5 cm, and the diameter of the cold-water line is 2.0 cm. The velocity of the water can be found from $V = \frac{\dot{m}}{\rho A}$, where $\rho = 1000$ kg/m^3 for liquid water. The specific enthalpy of the cold water is 21.0 kJ/kg and the specific enthalpy of the hot water is 335 kJ/kg. Assume no contribution from potential energy changes for the system. Plot the rate at which energy is being transferred via mass flow into the bowl as the percentage of hot water added varies from 0% to 100% of the total flow (2.5 kg/s).

CHAPTER 3

Thermodynamic Properties and Equations of State

Learning Objectives

Upon completion of Chapter 3, you will be able to

3.1 Describe the structure of phase diagrams and the associated terminology;

3.2 Express the state postulate;

3.3 Summarize the relationship between the two specific heats and the properties of specific enthalpy and specific internal energy;

3.4 Apply the equations of state for ideal gases and incompressible substances, and employ more complex equations of state for real gases;

3.5 Compute thermodynamic properties for substances near phase-change regions, such as water and the refrigerants.

3.1 INTRODUCTION

Designing and analyzing thermodynamic systems requires being able to describe the state of the substance in the system. As discussed in Chapter 1, we describe the state of a substance by identifying various properties of the substance. Not all properties must be measured for a substance to specify its state. As we will see in this chapter, the state of a pure substance can often be adequately described by two independent intensive properties; all other properties can then be derived through equations of state. In this chapter, we will first explore a method to visually illustrate the state of a substance and the processes it undergoes, and then learn how to use equations of state to find all the needed properties of a well-defined system.

3.2 PHASE DIAGRAMS

As described in Chapter 1, matter exists in different phases, and the phases that we are most concerned with are solid, liquid, and gas (or vapor). At a particular set of conditions, a substance at equilibrium will always exist in the same phase. So, at a pressure of 101 kPa and a temperature of 20°C, water at equilibrium is always in liquid form. Similarly, at the same conditions, gold will always be a solid. There are more advanced thermodynamic concepts that can be used to determine the state of a particular substance at a given set of independent properties, but we will not explore those methods at this time. Suffice it to say, there are fundamental thermodynamic principles as to why a substance is in a specific phase at a particular set of conditions—it is not a random process.

We are able to generate three-dimensional (3-D) diagrams of a pure substance's pressure, temperature, and specific volume. Pressure and specific volume, and temperature and specific volume are two sets of properties that are always independent. If the pressure and specific volume of a substance are known, the temperature can be calculated (or if we know T and v, then P can be calculated). This leads to a 3-D surface on which the substance must rest. Examples of these P-v-T diagrams are shown in **Figures 3.1** and **3.2**. Figure 3.1 is for water, and Figure 3.2 is descriptive of most pure substances: water is an unusual substance in that it expands upon freezing, whereas most substances contract—this difference in behavior leads to a different shape for the P-v-T diagram.

In Figures 3.1 and 3.2, the phase of the substance is given in the appropriate regions of P, v, and T. In addition, regions where two or three phases coexist are shown. You may notice that in those regions, the surface becomes flat, and the pressure and temperature are not independent properties. So, although pressure and temperature are often useful properties for specifying the state of a system, it is insufficient to thoroughly define the state of a multiphase region with pressure and temperature alone—another property that is independent, such as specific volume or specific internal energy, is needed. In multiphase regions, knowing either the pressure or temperature is sufficient for knowing the other value, but not other properties of a system. For example, if you know you have boiling water at 101 kPa (i.e., a mixture of a liquid and a vapor), you know that the temperature is 100°C but have no idea what the specific

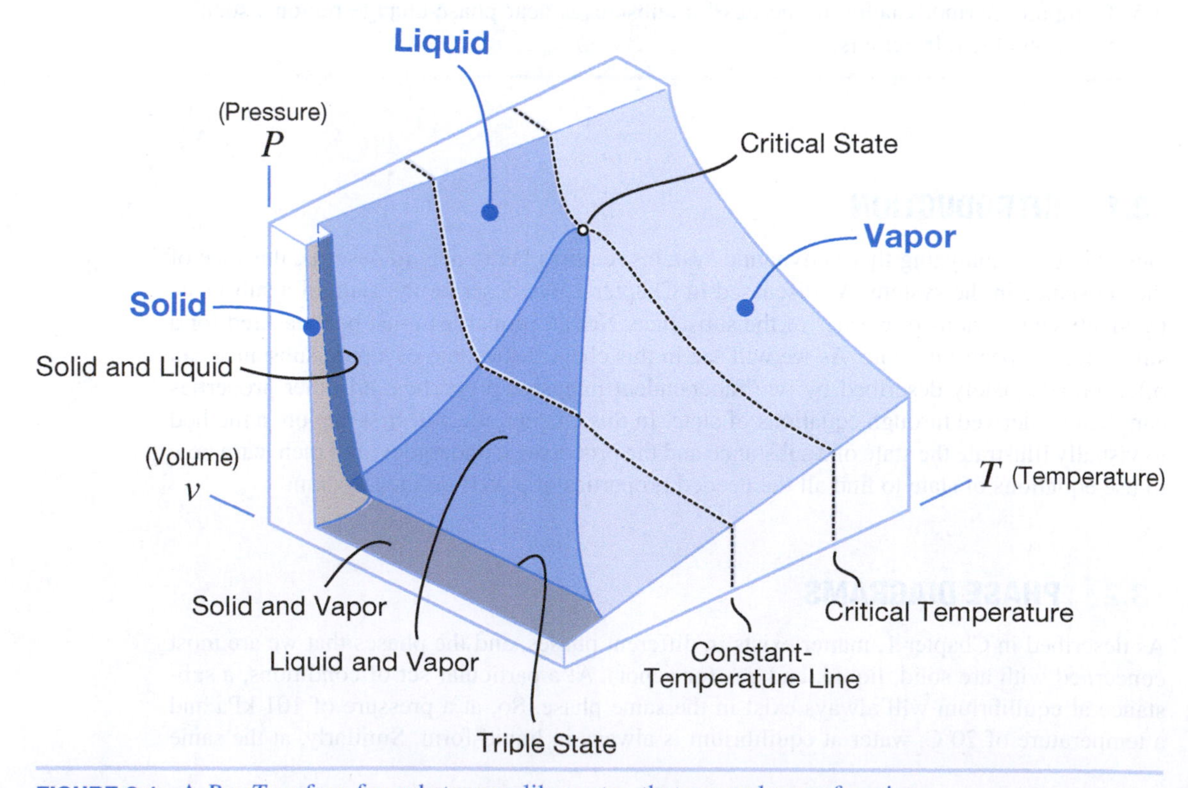

FIGURE 3.1 A P-v-T surface for substances, like water, that expand upon freezing.

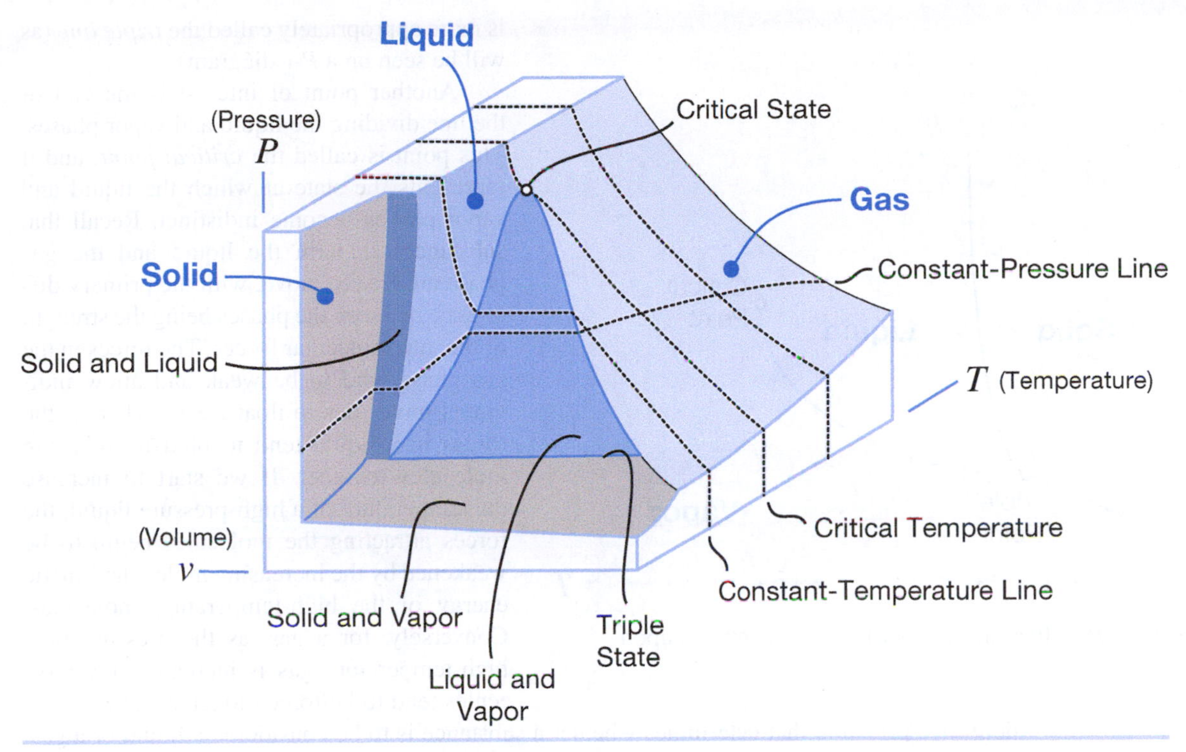

FIGURE 3.2 A *P-v-T* surface for substances that contract upon freezing (most substances).

volume or specific internal energy is. Those quantities depend on the relative amount of the water that is in liquid and in vapor form.

Although *P-v-T* diagrams are useful in helping understand phases of a substance, they are often more cumbersome than necessary for illustrating a process that the substance is undergoing. As such, we usually will rely on two-dimensional projections of this surface to illustrate thermodynamic processes. A *P-T* projection of water (such that the volume is into the page) is shown in **Figure 3.3**, and a *P-T* projection that is indicative of the behavior of most other substances is shown in **Figure 3.4**. As can be seen in Figures 3.3 and 3.4, the diagrams are similar, with the exception being the slope of the line dividing the solid and liquid phases: this is a consequence of the feature of water expanding upon freezing, whereas most substances contract upon freezing. The lines between two phases represent states where both states exist. At the intersection of all three phases, there is a pressure and temperature where all three phases coexist. On a *P-T* diagram, this is called the *triple point*; however, it should be noted that this state exists at a range of specific volumes and

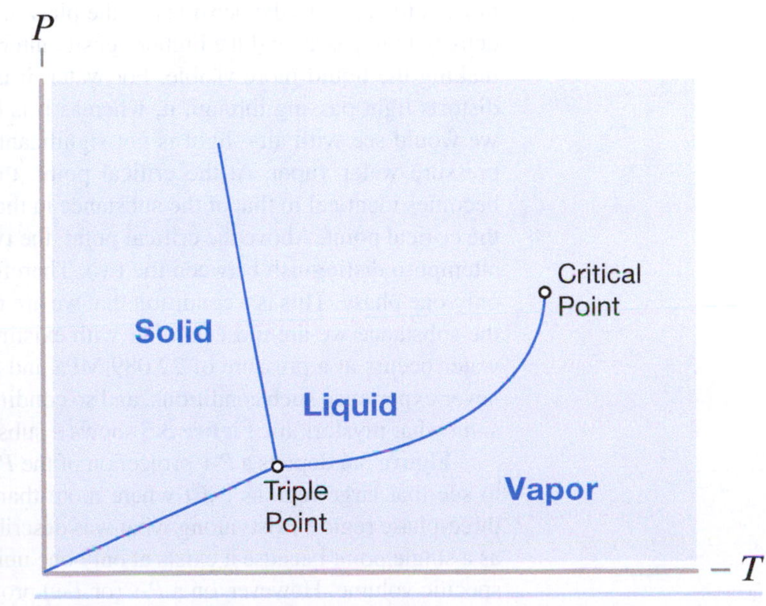

FIGURE 3.3 A *P-T* diagram of water.

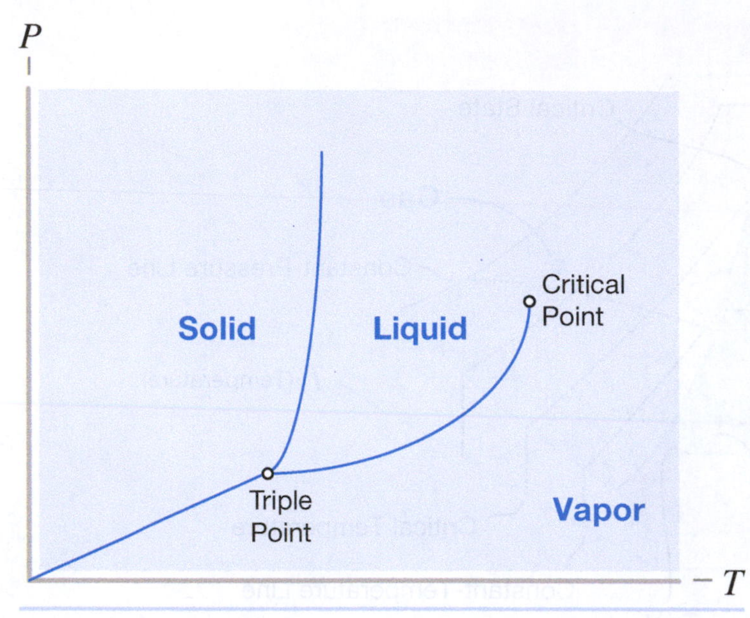

FIGURE 3.4 A *P-T* diagram for a surface that contracts upon freezing.

is more appropriately called the *triple line* (as will be seen on a *P-v* diagram).

Another point of interest is the end of the line dividing the liquid and vapor phases. This point is called the *critical point*, and it represents the state at which the liquid and vapor phases become indistinct. Recall that substances in both the liquid and the gas phase are free to move, with the primary difference between the phases being the strength of the intermolecular forces. The forces in the gas phase tend to be weak and allow individual molecules to float freely, whereas the forces in a liquid tend to bind neighboring molecules together. If we start to increase the temperature of a high-pressure liquid, the forces attracting the molecules begin to be weakened by the increasing molecular kinetic energy of the high-temperature molecules. Conversely, for a gas, as the pressure of a high-temperature gas is increased, the molecules tend to be forced together. There is no definitive principle that determines whether a substance is to be considered a liquid or a gas. Although you might regard a liquid as a substance that gravitates toward pooling at the bottom of a container, you can also have liquid droplets suspended in a gas such that they are floating.

The primary method of determining whether a substance is a liquid or a gas at such conditions is to compare the densities of the phases. We are accustomed to liquids having a higher density than gases, and the higher density alters the passage of light more through the liquid, making the liquid more visible. For water, it is easy to see the liquid phase at 101 kPa as it distorts light passing through it, whereas it is difficult to distinguish water vapor from what we would see with air—light is not significantly impacted by passing through atmospheric-pressure water vapor. At the critical point, the density of a substance in the liquid phase becomes identical to that of the substance in the gas phase; the phases are indistinguishable at the critical point. Above the critical point, the two phases become the same, and we no longer attempt to distinguish between the two. Therefore, the line on the *P-T* diagram ends: there is only one phase. This is a condition that we are not very familiar with in practical life. Water is the substance we are most familiar with existing in different phases, and the critical point of water occurs at a pressure of 22.089 MPa and a temperature of 374.14°C—most people will never experience such conditions, and so conditions of substances at their critical point remain somewhat mysterious. **Figure 3.5** shows a substance at its critical point.

Figure 3.6 depicts a *P-v* projection of the *P-v-T* diagram for water. In Figure 3.6, it is easy to see that large regions exist where more than one phase exists; it is also apparent that the three-phase region exists along what was described as the triple line. The critical point remains as a single point because it exists at only one unique combination of pressure, temperature, and specific volume. However, on a *P-v* (or *T-v*) projection, it is much easier to see that the density (which, recall, is the inverse of the specific volume) of the liquid and gas phases becomes identical at that state. It should also be noted that some people will reserve the term *gas* for substances that exist at pressures and temperatures above the critical point, whereas they use the term *vapor* for gas-phase substances below the critical point; however, this is not a universal distinction.

In this book, most of our concern will be with liquid and gas phase substances. Solids will be considered as well, but mostly in terms of substances that begin and end processes as solids;

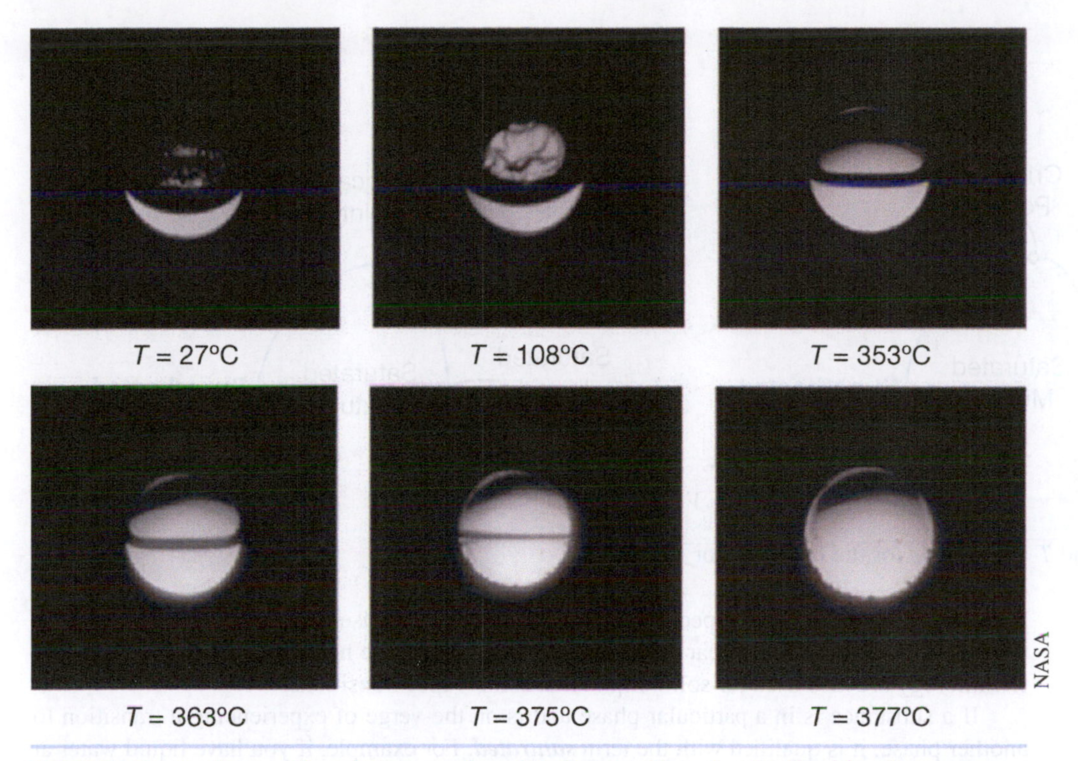

FIGURE 3.5 A picture of a substance at its critical point. Note that the liquid and gaseous phases become indistinguishable at the critical point.

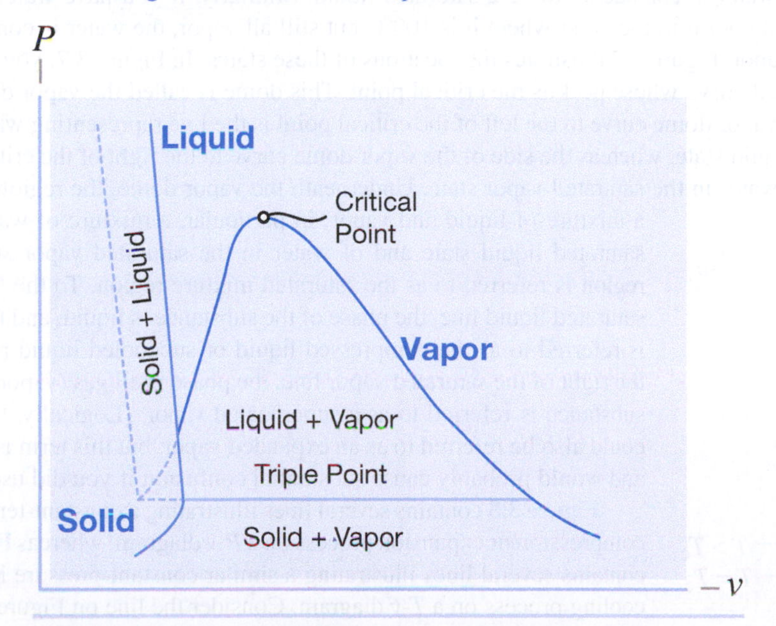

FIGURE 3.6 A *P-v* diagram for water, showing the solid, liquid, and vapor phases along with the phase transition regions.

we will not often be concerned with solid–liquid and solid–vapor transitions. Therefore, we will normally be concerned with only the parts of a *P-v* or *T-v* diagram around the liquid–vapor transition, such as shown in **Figure 3.7**. (Note that we will also use diagrams consisting of only the gas phase, because such diagrams are useful for visualizing what is happening to different

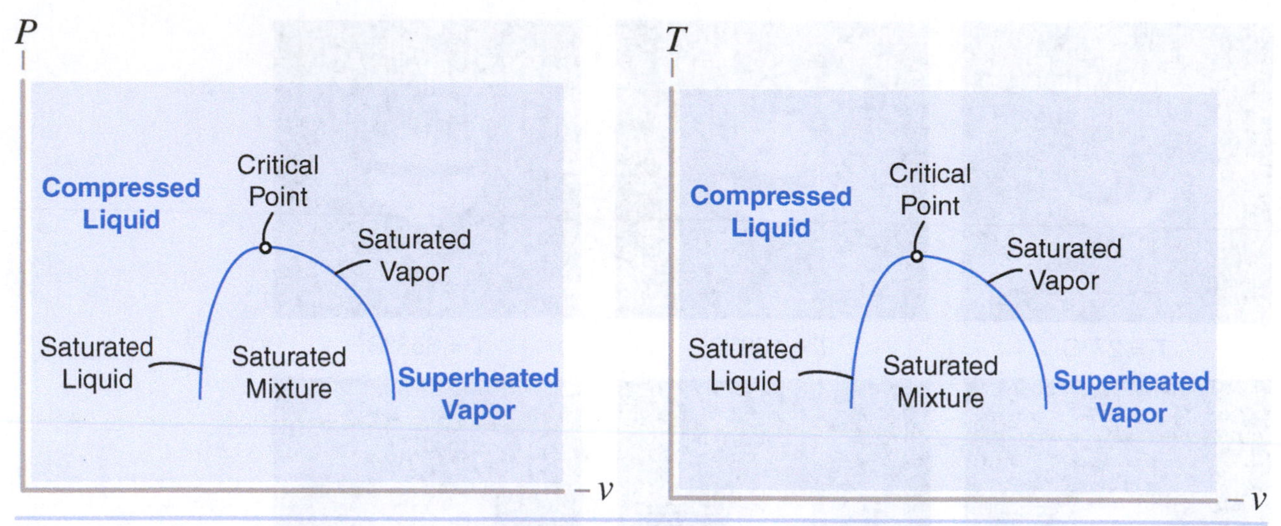

FIGURE 3.7 P-v and T-v diagrams for the liquid–vapor region.

properties as gases undergo specific processes.) We can also use Figure 3.7 to introduce new terminology for substances near a phase-change region. Please note that similar concepts and terminology are also used for solid–liquid and solid–vapor transitions.

If a substance is in a particular phase and is on the verge of experiencing a transition to another phase, it is qualified with the term ***saturated***. For example, if you have liquid water at 101 kPa pressure and add heat to it so that it reaches a temperature of 100°C but it is still all liquid, the water is considered to be a saturated liquid. Similarly, if you have water vapor at 101 kPa and cool it to the point where it is 100°C but still all vapor, the water is considered a saturated vapor. Figure 3.7 illustrates the locations of these states. In Figure 3.7, you can see a dome-shaped curve whose peak is the critical point. This dome is called the vapor dome. The side of the vapor dome curve to the left of the critical point is the line representing water in the saturated liquid state, whereas the side of the vapor dome curve to the right of the critical point represents water in the saturated vapor state. Underneath the vapor dome, the region contains

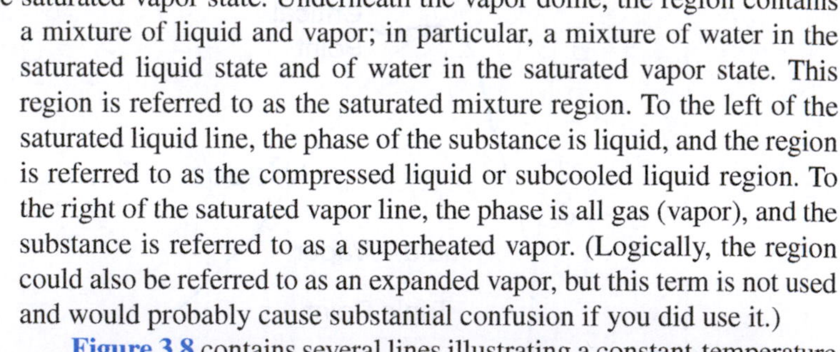

a mixture of liquid and vapor; in particular, a mixture of water in the saturated liquid state and of water in the saturated vapor state. This region is referred to as the saturated mixture region. To the left of the saturated liquid line, the phase of the substance is liquid, and the region is referred to as the compressed liquid or subcooled liquid region. To the right of the saturated vapor line, the phase is all gas (vapor), and the substance is referred to as a superheated vapor. (Logically, the region could also be referred to as an expanded vapor, but this term is not used and would probably cause substantial confusion if you did use it.)

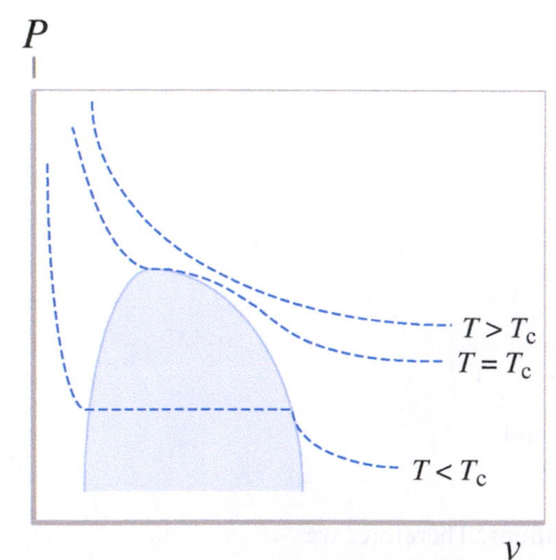

FIGURE 3.8 P-v diagram showing various lines of constant temperature.

Figure 3.8 contains several lines illustrating a constant-temperature compression or expansion process on a P-v diagram, whereas **Figure 3.9** contains several lines illustrating a similar constant-pressure heating or cooling process on a T-v diagram. Consider the line on Figure 3.9 for a subcritical pressure, and relate this to your own experiences with boiling water. If heat is added to liquid water at constant pressure, its temperature rises, with little change in its specific volume. The temperature continues to rise until the temperature reaches the boiling temperature. (The phase-change temperature for a given pressure is called the ***saturation temperature***, T_{sat}, whereas the phase-change pressure for a given temperature is called the ***saturation pressure***, P_{sat}.) Then the temperature remains constant (as illustrated in Figure 3.9) as the liquid becomes a gas (vapor). As this is happening, the volume occupied by the liquid slowly decreases, but the volume occupied by the gas rapidly expands. This results in an increase in the specific volume, with no change in temperature. Finally, once the water is all in vapor form, the temperature again begins to increase as more heat is added. These processes are illustrated in **Figure 3.10**.

Differences with this process can be seen at the critical pressure. Again, the temperature increases as heat is added until the water is at the critical point. Near the critical point, the rate of temperature increase with more heat addition slows, reaching 0 at the critical point, but then immediately begins to increase after the critical point is passed. At supercritical pressures, there is no plateau of the temperature; rather, the system experiences a fairly steady temperature increase as it changes from a very high-pressure liquid to a very dense gas.

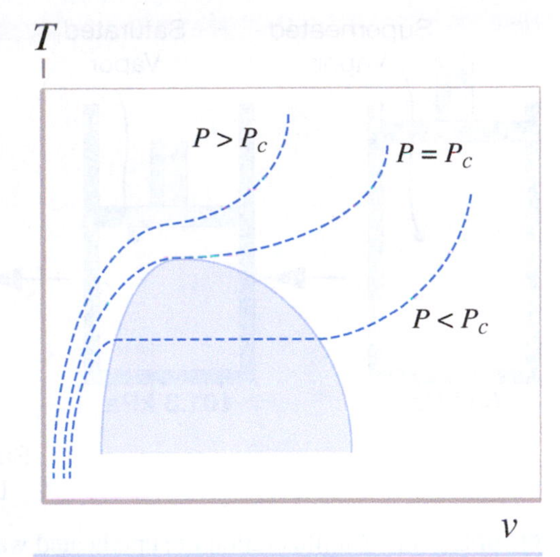

FIGURE 3.9 *T*-*v* diagram showing lines of constant pressure.

A similar description can be made for a constant-temperature process for the compression of water vapor, as shown in **Figure 3.11**. Initially, as we increase the pressure of a gas at constant, subcritical, temperature, its pressure increases slowly whereas its specific volume decreases. Once the saturation pressure is reached for that temperature, the water begins to condense. This continues at constant pressure until the water is all in the saturated liquid state. As the water is compressed further, the pressure of the compressed liquid begins to rise rapidly with relatively small changes in specific volume.

As you can see based on this discussion and the corresponding diagrams, substances that undergo a phase change exist in a range of specific volumes for a particular saturation pressure

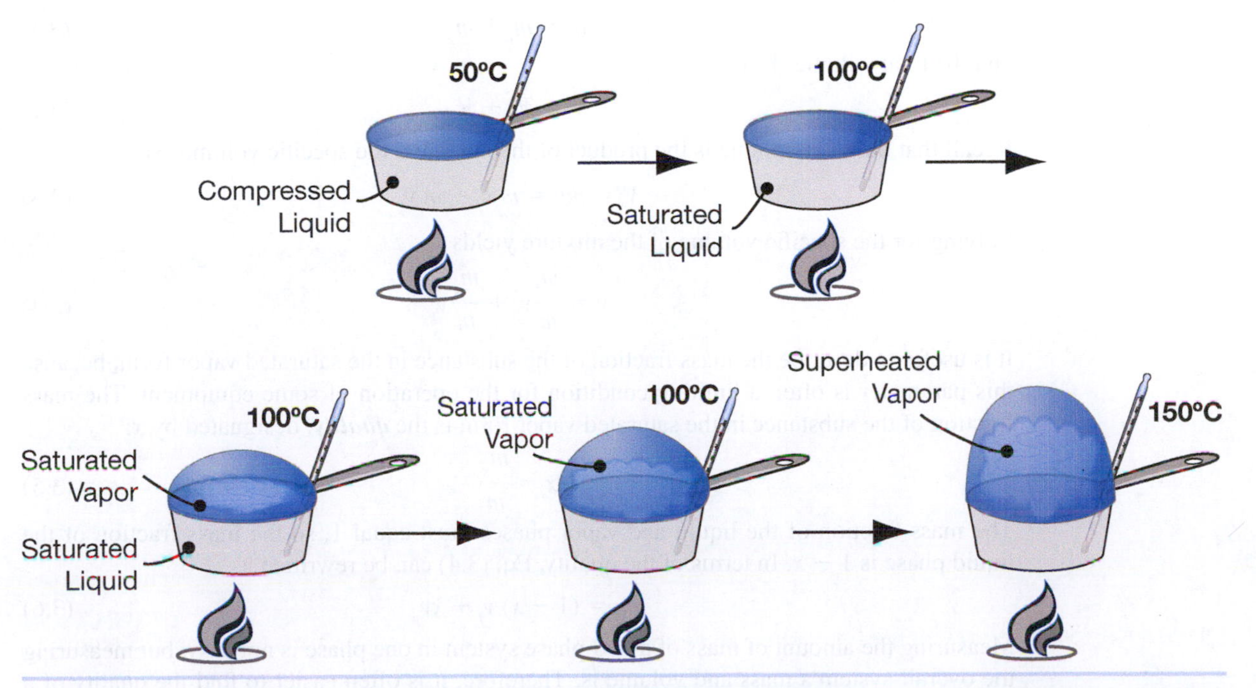

FIGURE 3.10 An illustration of the change in temperature and volume as water is heated at atmospheric pressure from a compressed liquid to a superheated vapor. Initially, the water is heated with little volume change, and then the volume changes significantly as the water boils at constant temperature. Once the water is completely vaporized, both the temperature and volume continue to increase with additional heating.

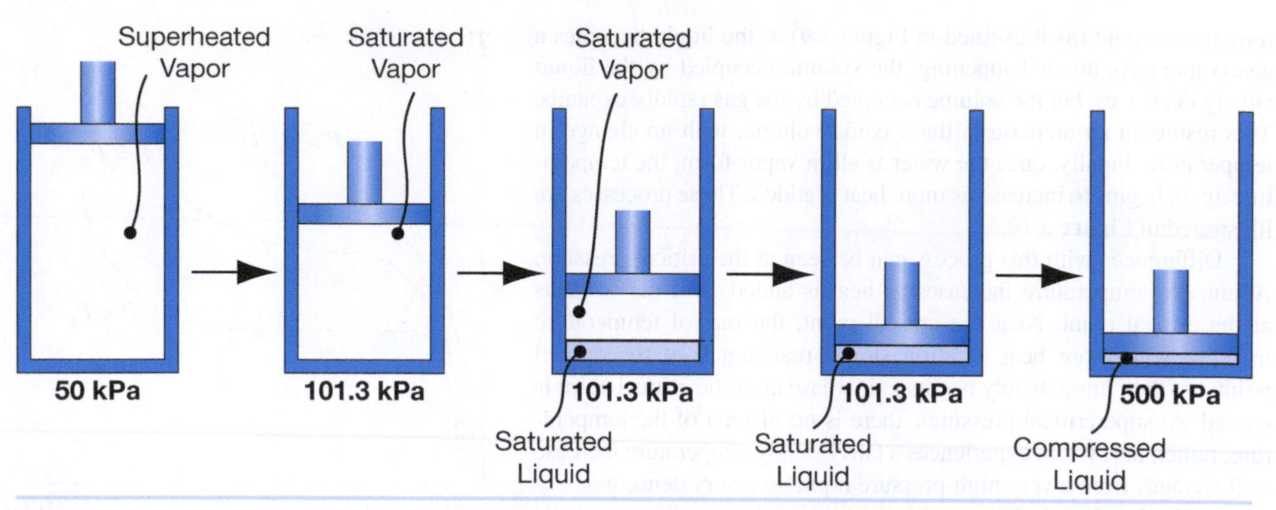

FIGURE 3.11 An illustration of superheated water vapor being compressed into a compressed liquid at a constant temperature of 100°C. At 101.3 kPa, the water transitions from being completely vapor to completely liquid.

or saturation temperature. It is important to recognize that for such a system, the entire system consists of a fraction of the substance in the saturated liquid phase and a fraction of the substance in the saturated vapor phase. The specific volume of the system is the mass fraction-weighted average of the specific volumes of the saturated liquid and saturated vapor states.

For reference purposes, saturated liquid properties are designated with a subscript "f," whereas saturated vapor properties are designated with a subscript "g." (We will learn more about finding values for these saturated liquid and vapor properties in Section 3.6.) Therefore, the total mass in the system, m, is

$$m = m_f + m_g \tag{3.1}$$

and the total volume, $\forall$, is

$$\forall = \forall_f + \forall_g \tag{3.2}$$

Recall that the total volume is the product of the mass and the specific volume. So

$$\forall = mv = m_f v_f + m_g v_g \tag{3.3}$$

Solving for the specific volume of the mixture yields

$$v = \frac{m_f}{m} v_f + \frac{m_g}{m} v_g \tag{3.4}$$

It is useful to describe the mass fraction of the substance in the saturated vapor form, because this parameter is often a limiting condition for the operation of some equipment. The mass fraction of the substance in the saturated vapor form is the ***quality***, designated by x:

$$x = \frac{m_g}{m} \tag{3.5}$$

The mass fraction of the liquid and vapor phases must equal 1, so the mass fraction of the liquid phase is $1 - x$. In terms of the quality, Eq. (3.4) can be rewritten as

$$v = (1 - x)\, v_f + x v_g \tag{3.6}$$

Measuring the amount of mass of a two-phase system in one phase is not easy, but measuring the overall system's mass and volume is. Therefore, it is often easier to find the quality of a mixture by solving Eq. (3.6) for x rather than using Eq. (3.5) to calculate x:

$$x = \frac{v - v_f}{v_g - v_f} \tag{3.7}$$

where the saturated liquid and vapor properties have been determined for most pure substances as a function of saturation temperature or saturation pressure.

It should be noted that the quality of a saturated liquid is 0.0, and the quality of a saturated vapor is 1.0. Any other saturated mixture has $0 < x < 1$:

$x = 0.0$	Saturated liquid
$0.0 < x < 1.0$	Saturated mixture
$x = 1.0$	Saturated vapor

The quality is not defined outside of the saturated (two-phase) region. (Similar descriptions do exist for solid–liquid and solid–vapor transitions). If a quality of something greater than 1 is calculated for a system, then the system is a superheated vapor and the quality is undefined. So, if you calculate a quality to be 1.25 using an equation such as (3.7), it does not mean that 125% of the substance is a vapor; rather, it is all a superheated vapor and the quality is undefined. Similarly, a calculated quality less than 0 indicates that the substance is a compressed liquid and the quality is again undefined. So, a calculated quality of −0.25 does not mean that −25% of the substance is a vapor, but rather that it is all a compressed liquid and the quality is undefined.

Keep in mind that the quality is defined as a mass fraction of saturated vapor in a system. For most subcritical pressures or temperatures, the density of a saturated liquid is much higher than the density of a saturated vapor (or, the specific volume of a saturated liquid is much lower than that of a saturated vapor). Therefore, a system that visually appears to have half of its volume being liquid and half being vapor will still have a very low quality. Even systems with just small amounts of pooled liquid will tend to have low qualities. Systems with high values of the quality will tend to have small droplets of liquid suspended in vapor. This is illustrated in Figure 3.12.

QUESTION FOR THOUGHT/DISCUSSION

What type of an experiment would you design to determine the mass fractions of water vapor and of liquid water in a saturated water mixture?

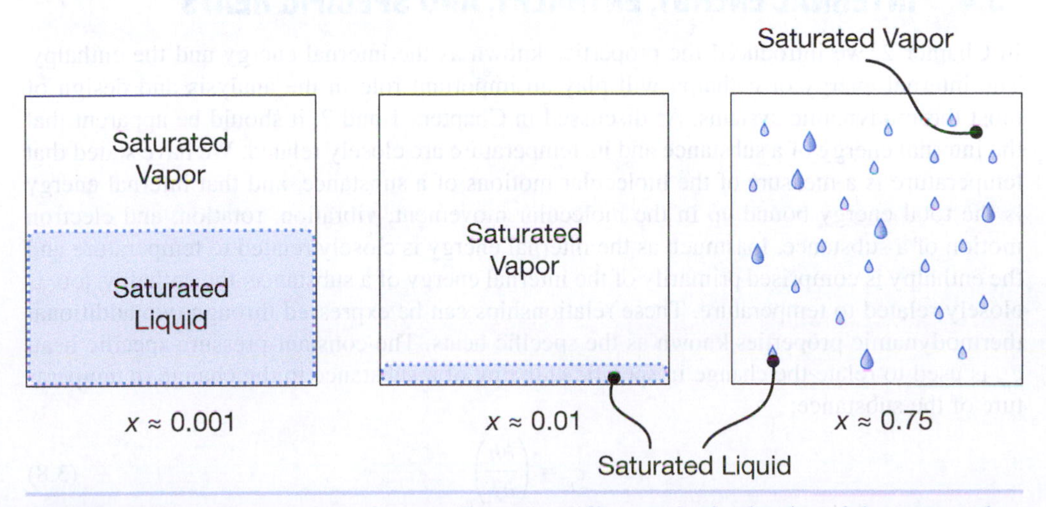

FIGURE 3.12 A visualization of the quality. The difference in the density of the saturated liquid and saturated vapor states results in little volume being needed to occupy a significant portion of the mass of the system that is in the liquid phase.

3.3 THE STATE POSTULATE

In the preceding section we stated that normally two independent intensive properties are required to fix the state of a system; the concept that leads to this conclusion is the state postulate, which reads as follows:

The number of independent, intensive properties required to completely specify the state of a pure substance is equal to the number of nonchemical work modes present plus 1.

For most applications that are encountered in engineering practice, the system can be considered a *simple compressible system.* A simple compressible system is a system that does not have substantial magnetic, motion, gravitational, electrical, or surface tension effects. This does not exclude the likelihood that a system is experiencing the pull of gravity, or receives an electrical work input that is the equivalent of being heated. Rather, a simple compressible system is not experiencing additional work interactions as a result of being acted on by an external force field. Most engineering applications have negligible impact from these external forces, and so most engineering applications are considered to be simple compressible systems.

Simple compressible systems experience only one nonchemical work mode. Therefore, for pure substances in simple compressible systems, the state postulate can be restated as follows:

The number of independent, intensive properties required to completely specify the state of a pure substance in a simple compressible system is 2.

If additional work modes are present, there must be additional properties known to specify the state. For example, if there are significant gravitational effects in the system, then we would need to specify the height of the substance in addition to two thermodynamic properties that describe the substance.

Furthermore, additional properties are necessary if the system consists of more than one pure substance. In such a situation, we typically need to specify the composition of the system so that we can determine each component's contributions to the total properties of the system.

3.4 INTERNAL ENERGY, ENTHALPY, AND SPECIFIC HEATS

In Chapter 2, we introduced the properties known as the internal energy and the enthalpy. The internal energy or enthalpy will play an important role in the analysis and design of most thermodynamic systems. As discussed in Chapters 1 and 2, it should be apparent that the internal energy of a substance and its temperature are closely related. We have stated that temperature is a measure of the molecular motions of a substance, and that internal energy is the total energy bound up in the molecular movement, vibration, rotation, and electron motion of a substance. Inasmuch as the internal energy is closely related to temperature and the enthalpy is comprised primarily of the internal energy of a substance, the enthalpy, too, is closely related to temperature. These relationships can be expressed through two additional thermodynamic properties known as the specific heats. The constant-pressure specific heat, c_p, is used to relate the change in specific enthalpy of a substance to the change in temperature of the substance:

$$c_p = \left(\frac{\partial h}{\partial T}\right)_p \tag{3.8}$$

The constant-pressure specific heat is equal to the partial derivative of the enthalpy with respect to temperature in a constant-pressure process.

The constant-volume specific heat, c_v, is equal to the partial derivative of the internal energy with respect to temperature in a constant-volume process:

$$c_v = \left(\frac{\partial u}{\partial T}\right)_v \tag{3.9}$$

As we will see, in some cases the values of the specific heats can be assumed to be constant, and in fact they are constant for monatomic gases. However, for most substances, the relationship between the specific heat and the temperature is complicated; in such cases it is unlikely that you will directly integrate the relationship by hand. In these cases, you will typically either make assumptions to simplify the calculation or use a computer program to solve for the appropriate property values. The complex relationships are found either using advanced statistical thermodynamic tools to theoretically predict the specific heat, or through experimental techniques followed by curve fitting of the data.

The specific heats can also be given on a molar basis:

$$\overline{c}_p = \left(\frac{\partial \overline{h}}{\partial T}\right)_p \tag{3.10}$$

and

$$\overline{c}_v = \left(\frac{\partial \overline{u}}{\partial T}\right)_v \tag{3.11}$$

where $\overline{h}$ is the molar-specific enthalpy ($\overline{h} = H/n$) and $\overline{u}$ is the molar-specific internal energy ($\overline{u} = U/n$).

It should also be noted that because the specific heats are properties, their use is not restricted to only constant-pressure or constant-volume processes. Remember, properties are independent of how the state was achieved, and so the changes in the enthalpy or internal energy are the same whether or not the process encountered was a constant-pressure or constant-volume process. The changes in the enthalpy and internal energy between states A and B are the same whether the process was constant pressure, constant volume, or neither. The qualifying terms describing the specific heats refer to how the properties were originally determined, either experimentally or theoretically. Therefore, if we need to find the change in the specific enthalpy for a given change in temperature, we can use Eq. (3.8) regardless of whether or not the change in temperature is occurring at constant pressure. The same can be said with the change in specific internal energy being unbound by the process being at constant volume: we can use Eq. (3.9) to find the change in specific internal energy for any type of process.

A value that will be used frequently as we proceed through our study of thermodynamics is the specific heat ratio, k. The specific heat ratio is defined as

$$k = \frac{c_p}{c_v}$$

The complexity of the equations of state that relate the specific heats to other properties can also appear in other equations of state relating different properties. As such, it is important to know when simplifying assumptions about the material can be made, or when more elaborate relationships are truly necessary to achieve good engineering design and analysis. For substances that are either undergoing a phase change or have properties relatively near a phase-change region, it is usually necessary to consider the complex property relationships, even in relatively simple processes. Water and various refrigerants are substances that we normally group into this category. But there are two categories of idealized substances for which the approximations made in developing equations of state do not substantially impact the calculations involved in engineering thermodynamic analysis. These categories are ideal gases and incompressible substances. Because many substances can be adequately modeled as one of these two types, there are many practical engineering systems where the assumption of idealized substance behavior provides sufficient accuracy to allow for an acceptable design or analysis. Next, we consider some of the equations of state for these substances.

3.5 EQUATIONS OF STATE FOR IDEAL GASES

An ideal gas is one for which it is assumed that there are no interactions between molecules or atoms in the system for a substance in the gas phase, such as shown in **Figure 3.13**. Outside of systems consisting of only a single molecule, such a substance does not exist because molecules in any real system will collide and intermolecular forces will exist. However, in engineering practice, substances that are in the gas phase and not near a phase transition are usually considered to be ideal gases. (This assumption should not be made when an exceptional degree of accuracy is required for a calculation, or when the density of the gas is very high. For such situations, other equations of state exist, some of which will be briefly introduced later.) A general rule of thumb for the use of the ideal gas assumption is that it is usually acceptably used for substances that are normally thought of as a gas (such as air, nitrogen, oxygen, hydrogen, etc.), provided that the gas is at a temperature well above its saturation temperature at its given pressure, and provided that the density of the gas is not very high.

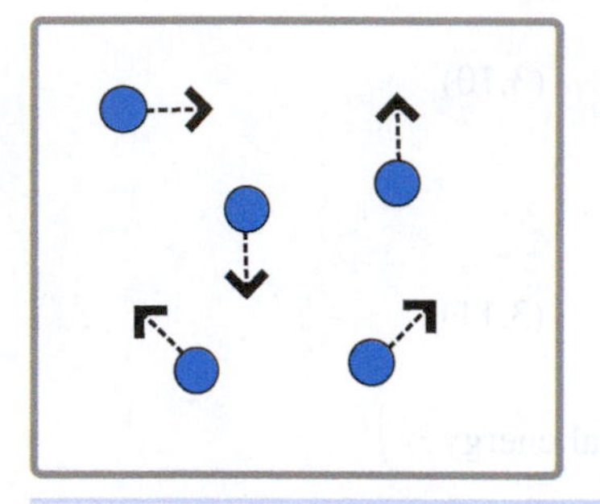

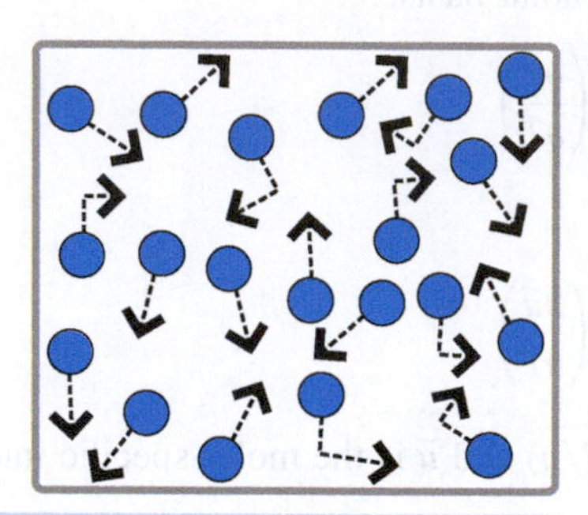

FIGURE 3.13 On the left, a gas at low molecular density will behave like an ideal gas, with no interactions between molecules. On the right, a gas with a high density will have many interactions between molecules, resulting in nonideal gas behavior.

3.5.1 The Ideal Gas Law

The ideal gas law is used to relate the pressure, volume, and temperature of an ideal gas. Several forms of the ideal gas law can be used, based on whether we want to deal with mass or molar units and whether or not we wish to deal with specific or total volumes:

$$P\mathcal{V} = mRT \tag{3.12a}$$

$$Pv = RT \tag{3.12b}$$

$$P\mathcal{V} = n\overline{R}T \tag{3.12c}$$

$$P\overline{v} = \overline{R}T \tag{3.12d}$$

where $\overline{R}$ is the universal ideal gas constant ($\overline{R}$ = 8.314 kJ/kmole · K), R is the gas-specific ideal gas constant ($R = \overline{R}/M$, where M is the molecular mass of the gas), and $\overline{v}$ is the molar-specific volume ($\overline{v} = \mathcal{V}/n$). We will be using Eqs. (3.12a) and (3.12b) the most until we begin to study reacting gas mixtures in Chapter 11.

Keeping in mind that ideal gases are to have no molecular interactions, the ideal gas law is most accurate when the number of molecular interactions is minimized. Such minimization occurs when the gas density is low. Considering that the density is the inverse of the specific volume, it can be seen from Eq. (3.12b) that

$$\rho = P/RT$$

Therefore, typically the ideal gas law will be most accurate for gases that are at a low pressure and/or a high temperature. Conversely, the ideal gas law becomes most inaccurate for gases at high pressure and/or low temperature, which is often the case near a phase-change region.

▶ EXAMPLE 3.1

A tank of compressed oxygen gas contains O_2 at 20°C and 1000 kPa. The volume of the tank is 0.120 m³. Using the ideal gas law, determine the mass of oxygen in the tank.

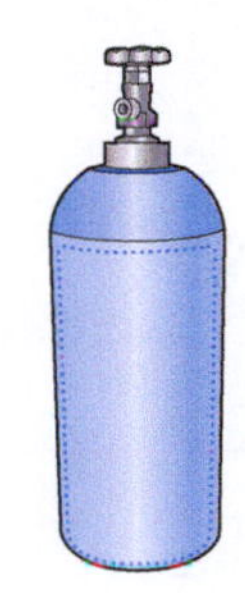

Given: $\forall = 0.120\ \text{m}^3$, $P = 1000$ kPa, $T = 20°\text{C} = 293$ K

Find: m

Solution: We will use the ideal gas law to find the mass of the O_2. From Eq. (3.12a),

$$m = \frac{P\forall}{RT}$$

From Table A.1, $R = 0.2598$ kJ/kg · K for O_2.

So

$$m = \frac{(1000\ \text{kPa})(0.120\ \text{m}^3)}{(0.2598\ \text{kJ/kg}\cdot\text{K})(293\ \text{K})} = \mathbf{1.58\ kg}$$

Analysis: Remember to use absolute temperatures in the ideal gas law, because we are multiplying or dividing by a temperature—use of a relative temperature scale in such mathematical operations will provide an erroneous result. Also, if you do not see how the units cancel, expand the kPa into kN/m², and the kJ into kN · m.

The ideal gas law can also be written on a rate basis, so that the volumetric flow rate, $\dot{V}$, can be related to the mass flow rate, $\dot{m}$, or the molar flow rate, $\dot{n}$:

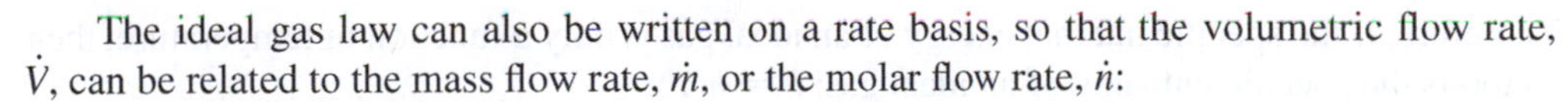

$$P\dot{\forall} = \dot{m}RT \tag{3.12e}$$

$$P\dot{\forall} = \dot{n}\overline{R}T \tag{3.12f}$$

▶ EXAMPLE 3.2

3.5 kg/s of air at 100 kPa and 25°C is to be inducted into an air compressor. The air will exit the compressor at 800 kPa and 300°C, with the same mass flow rate. Using the ideal gas law, determine the volumetric flow rates at the entrance and exit of the compressor.

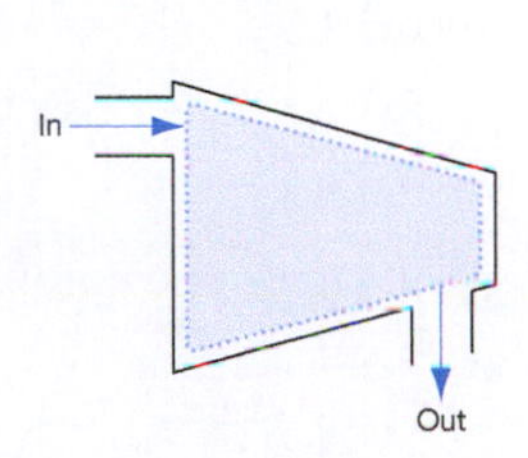

Given: $\dot{m} = 3.5$ kg/s, $T_{in} = 25°\text{C} = 298$ K, $P_{in} = 100$ kPa, $P_{out} = 800$ kPa,

$T_{out} = 300°\text{C} = 573$ K

Find: $\dot{\forall}_{in}, \dot{\forall}_{out}$

Solution: We will treat the air as an ideal gas, and use the ideal gas law to find the requested volumetric flow rates. From Eq. (3.12e),

$$\dot{\forall} = \frac{\dot{m}RT}{P}$$

The mass flow rate is identical at the entrance and exit of the compressor, as is typically seen in such devices, but the volumetric flow rate will change as the density of the gas changes.

From Table A.1, $R = 0.287$ kJ/kg · K for air. Therefore,

$$\dot{V}_{in} = \frac{\dot{m}RT_{in}}{P_{in}} = \frac{\left(3.5\ \frac{kg}{s}\right)\left(0.287\ \frac{kJ}{kg \cdot K}\right)(298\ K)}{(100\ kPa)} = \mathbf{2.99\ m^3/s}$$

$$\dot{V}_{out} = \frac{\dot{m}RT_{out}}{P_{out}} = \frac{\left(3.5\ \frac{kg}{s}\right)\left(0.287\ \frac{kJ}{kg \cdot K}\right)(573\ K)}{(800\ kPa)} = \mathbf{0.719\ m^3/s}$$

Analysis: Keep in mind that the volumetric flow rate will often change during an open system process, but the mass flow rate often stays constant. Although the outlet volumetric flow rate is considerably smaller than the inlet volumetric flow rate, in practice the air is often cooled before leaving the compressor, resulting in an even smaller volumetric flow rate.

3.5.2 Changes in Internal Energy and Enthalpy for Ideal Gases

In general, the specific enthalpy and specific internal energy can both be viewed as functions of temperature and specific volume: $h = h(T, v)$, $u = u(T, v)$. However, an ideal gas assumes that there are no molecular interactions present. Therefore, each molecule acts as though there are no other molecules present, and if there are no other molecules present, the specific volume (or density) of the system should be irrelevant. Logically, the properties of specific enthalpy and specific internal energy should be functions of temperature alone. It has been shown experimentally and theoretically that the specific internal energy of an ideal gas is a function of temperature alone: $u = u(T)$. Considering the definition of the specific enthalpy and substituting Eq. (3.12b), it can be shown that

$$h = u + Pv = u + RT$$

Therefore, if the specific internal energy of an ideal gas is only a function of temperature, then so too is the specific enthalpy of an ideal gas: $h = h(T)$.

Because u and h only depend on one property, temperature, the partial derivatives in the definitions of the specific heats (Eqs. (3.8) and (3.9)) can be reduced to ordinary derivatives. So, for ideal gases,

$$c_p = \frac{dh}{dT} \tag{3.13}$$

and

$$c_v = \frac{du}{dT} \tag{3.14}$$

Such relationships are, in general, easier to evaluate than partial derivatives. First, however, the nature of the functional relationships between c_p (and c_v) and T needs to be established. For an ideal gas, these relationships are typically given in terms of the molar constant-specific heat and take a form

$$\bar{c}_p = a + bT + cT^2 + dT^3 + \cdots \tag{3.15}$$

where the coefficients $a, b, c, d, \ldots$ terms represent constants for a particular gas.

For an ideal gas, the values of c_p and c_v can be easily related:

$$dh = du + d(Pv) = du + d(RT) = du + RdT$$

$$c_p = \frac{dh}{dT} = \frac{du}{dT} + R\frac{dT}{dT} = c_v + R \tag{3.16}$$

Therefore, a functional relationship between c_p and T can quickly yield a relationship between c_v and T. As can be seen, the specific heats differ for ideal gases by the gas-specific ideal gas constant. Additionally, it can be shown that for ideal gases

$$c_p = \frac{kR}{k-1} \quad \text{and} \quad c_v = \frac{R}{k-1} \tag{3.17a, b}$$

Ideal Gases Assuming Constant Specific Heats

Although the specific heats do vary with temperature, the variation can be rather slow, as shown graphically in **Figure 3.14**. Therefore, there are conditions for which the values of the specific heat can be safely assumed to be constant. Such conditions include the following:

(a) relatively small temperature changes ($\Delta T < \sim 200$ K) for a process;
(b) a need for only an approximate solution over a larger temperature change (for instance, when performing a quick estimate);
(c) somewhat larger temperature changes when using an average value for the specific heat.

It can also be noted that the specific heats of the noble gases (He, Ne, Ar, Kr, Xe, Rn) are constant, with $c_p = (5/2)$ R and $c_v = (3/2)$ R for each.

For ideal gases for which the specific heat is assumed to be constant, the changes in specific enthalpy and specific internal energy can be found through integration to be

$$h_2 - h_1 = c_p(T_2 - T_1) \tag{3.18a}$$

and

$$u_2 - u_1 = c_v(T_2 - T_1) \tag{3.18b}$$

Values of the specific heats at a temperature of 300 K for many gases can be found in Table A.1 in the appendix. Table A.2 contains values for the specific heats of six common gases at different temperatures.

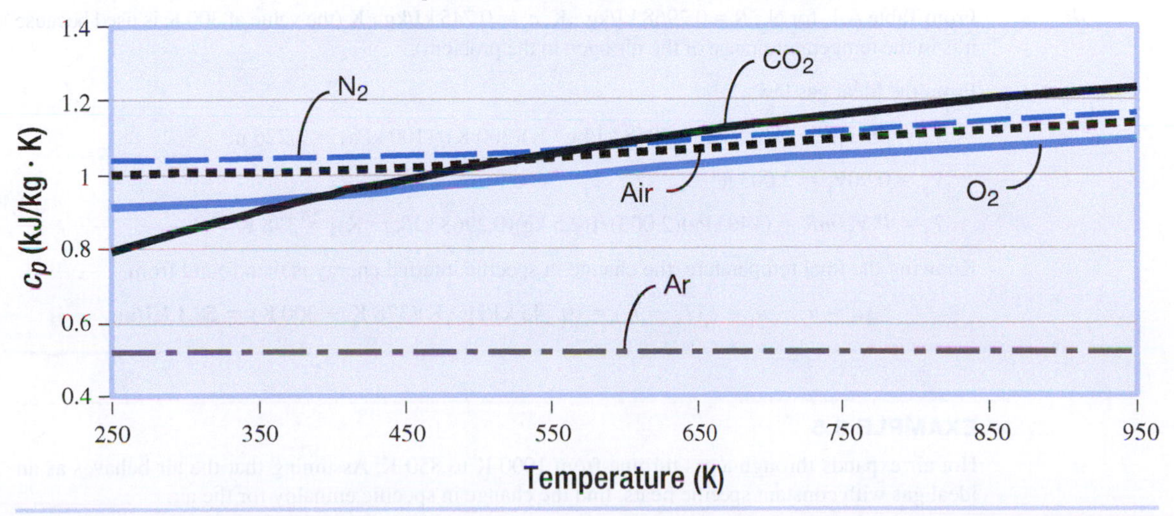

FIGURE 3.14 A plot of the constant-pressure specific heat, c_p, as a function of temperature for many gases. The specific heats increase with temperature, except for monatomic gases, which have constant specific heats.

▶ EXAMPLE 3.3

Air undergoes a heating process from 20°C to 100°C. Assuming that air behaves as an ideal gas with constant specific heats, find the change in the specific enthalpy and specific internal energy for the process.

Given: $T_1 = 20°C$, $T_2 = 100°C$

Find: Δh, Δu

Solution: We will be treating the air as an ideal gas with constant specific heats. The values of the specific heats for air at 300 K can be found from Table A.1. This temperature falls in the temperature range of the air and is appropriate for use. From this table, we find $c_p = 1.005$ kJ/kg · K and $c_v = 0.718$ kJ/kg · K.

The change in specific enthalpy is

$$\Delta h = h_2 - h_1 = c_p(T_2 - T_1) = (1.005 \text{ kJ/kg} \cdot \text{K})(100°\text{C} - 20°\text{C})$$

$$= (1.005 \text{ kJ/kg} \cdot °\text{C})(100°\text{C} - 20°\text{C}) = \mathbf{80.4\ kJ/kg}$$

$$\Delta u = u_2 - u_1 = c_v(T_2 - T_1) = (0.718 \text{ kJ/kg} \cdot \text{K})(100°\text{C} - 20°\text{C}) = \mathbf{57.4\ kJ/kg}$$

Analysis: Remember that the size of a Kelvin and a °C are identical, so a unit given "per Kelvin" is the same as "per °C" so that kJ/kg · K is the same as kJ/kg · °C. This is what allows the unit substitution in the solution.

▶ EXAMPLE 3.4

2.5 kg of nitrogen gas initially at a temperature of 300 K and a pressure of 100 kPa is compressed to a volume equal to 90% of the initial volume and at a pressure of 140 kPa. Determine the change in specific internal energy of the nitrogen during the process, assuming that the nitrogen acts as an ideal gas with constant specific heats.

Given: $m = 2.5$ kg, $T_1 = 300$ K, $P_1 = 100$ kPa, $V_2 = 0.90V_1$, $P_2 = 140$ kPa

Find: Δu

Solution: The N_2 is to be treated as an ideal gas with constant specific heats for this problem. First, we will need to use the ideal gas law to find the initial volume. The final volume is 90% of the initial volume. Once the final volume is found, we again use the ideal gas law to find the final temperature of the nitrogen.

From Table A.1, for N_2: $R = 0.2968$ kJ/kg · K, $c_v = 0.745$ kJ/kg · K (the value at 300 K is used because it is in the temperature range of the nitrogen in the problem).

From the ideal gas law:

$$V_1 = mRT_1/P_1 = (2.5 \text{ kg})(0.2968 \text{ kJ/kg} \cdot \text{K})(300 \text{ K})/(100 \text{ kPa}) = 2.226 \text{ m}^3$$

$$V_2 = 0.90V_1 = 2.003 \text{ m}^3$$

$$T_2 = P_2V_2/mR = (140 \text{ kPa})(2.003)/[(2.5 \text{ kg})(0.2968 \text{ kJ/kg} \cdot \text{K})] = 378 \text{ K}$$

Knowing the final temperature, the change in specific internal energy is then found from

$$\Delta u = u_2 - u_1 = c_v(T_2 - T_1) = (0.745 \text{ kJ/kg} \cdot \text{K})(378 \text{ K} - 300 \text{ K}) = \mathbf{58.1\ kJ/kg}$$

▶ EXAMPLE 3.5

Hot air expands through a gas turbine from 1000 K to 850 K. Assuming that the air behaves as an ideal gas with constant specific heats, find the change in specific enthalpy for the air.

Given: $T_1 = 1000$ K, $T_2 = 850$ K

Find: Δh

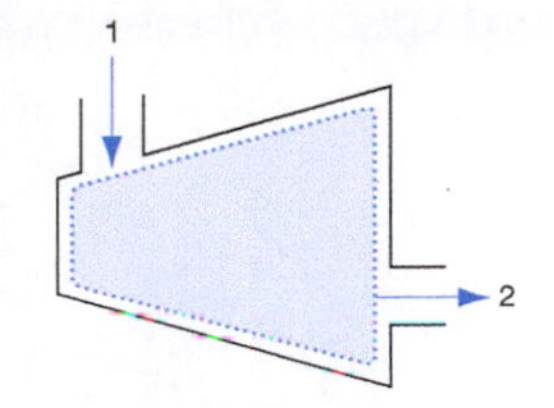

Solution: We will treat the air as an ideal gas with constant specific heats. Although we could use a value of the specific heat for air at the standard temperature, a more accurate value for the change in specific enthalpy will be found using a specific heat value in the temperature range of the process. Therefore, choose the value of the specific heat of air to be at 900 K (which is a quantity tabulated in Table A.2).

At 900 K, $c_p = 1.121$ kJ/kg · K

Therefore,

$$\Delta h = h_2 - h_1 = c_p(T_2 - T_1) = (1.121 \text{ kJ/kg} \cdot \text{K})(850 \text{ K} - 1000 \text{ K}) = \mathbf{-168 \text{ kJ/kg}}$$

Analysis: If c_p had been taken at 300 K, then $c_p = 1.005$ kJ/kg · K, and $h_2 - h_1 = -151$ kJ/kg. This is a substantial difference, and it illustrates a potential problem with assuming constant specific heats and then subsequently using values that correspond to a temperature from well outside the temperature range of interest.

Ideal Gases Considering Variable Specific Heats

When large temperature ranges are considered, or when more precise calculations of changes in specific internal energy or specific enthalpy are needed, it is often necessary to consider the change of the values of the specific heats as the temperature changes. As mentioned, this can be achieved through integration of a functional relationship such as Eq. (3.15). Table 3.1 provides examples of the functional relationship between $\bar{c}_p$ and temperature for different ideal gases. However, for more than one or two calculations, such integration can become tedious. As a result, for more commonly used gases the integration is normally performed through a computer-based calculation of the properties.

Vast amounts of information have become readily available electronically, and engineers are now able to use computer-based programs and apps to find property values for many ideal gases. Programs and apps are frequently updated and modified, so when looking for apps that will calculate thermodynamic properties for ideal gases, it's a good idea to perform either Internet searches for Internet-based programs or tablet app store searches for tablet-based apps. You should choose a program that allows you to input a temperature and that will give values for the specific enthalpy and specific internal energy. Finding a program that works for air is most important, but finding programs for other gases can also be beneficial. Some programs will allow you to enter both a temperature and a pressure, with the resulting properties being more representative of real gases. Typically, the pressure input will have little effect on

TABLE 3.1 Relationships between $\bar{c}_p$ and Temperature for Various Ideal Gases

$$\frac{\bar{c}_p}{R} = a + bT + cT^2 + dT^3 + eT^4$$

T is in Kelvin (K), and the coefficients are valid for 300 K $\leq T \leq$ 1000 K.

Gas	a	$b \times 10^3$	$c \times 10^6$	$d \times 10^9$	$e \times 10^{12}$
Air	3.653	−1.337	3.294	−1.913	0.2763
CH_4	3.826	−3.979	24.558	−22.733	6.963
CO_2	2.401	8.735	−6.607	2.002	0
H_2	3.057	2.677	−5.810	5.521	−1.812
N_2	3.675	−1.208	2.324	−0.632	−0.226
O_2	3.626	−1.878	7.055	−6.764	2.156
Monatomic Gases (i.e., He, Ne, Ar)	2.5	0	0	0	0

the properties. As of this writing, two sources worth considering for air properties are Wolfram Alpha™ (program and app) and Peacesoftware™ (peacesoftware.de). The National Institute of Standards and Technology (NIST) also provides property data for many gases. You may find that you will need to convert between molar units and mass-based units using the molecular mass (M) of the gas.

One limitation of many of the programs available is that they do not allow for the input of a value of the specific enthalpy (for example) and give the temperature as output. In such a case, you may need to enter several temperatures until you find the one that gives the desired specific enthalpy. Furthermore, note that you should use the same source for property values for a particular problem, because some sources may use different reference states, which could result in different values for these relative quantities; however, the change in a property value between two states should remain the same no matter what particular reference state is used. Finally, if you find an accurate plug-in for ideal gas property data for the solution platform you are using, you should strongly consider integrating the plug-in directly into the models you develop with that platform.

An alternative method for finding properties of ideal gases considering the variability of specific heats is to use gas tables, such as the air table (Table A.3) in the appendix or other gas tables (Table A.4). In these tables, values for the specific enthalpy, specific internal energy, and the temperature-dependent portion of the specific entropy (to be introduced in Chapter 6) are given as a function of temperature. Not all temperatures are included, so you may need to perform linear interpolation on the data to find the desired data point.

To perform linear interpolation, consider two data points (x_1, y_1) and (x_2, y_2). We will assume that a linear relationship exists between these two points. Such an assumption is most accurate when as small of an increment as possible is used between the data points (such as the nearest sequential values on the tables surrounding the unknown quantity). If we transform the x-values to shift the x-axis origin to the value x_1, the y_1 value becomes the y-intercept of the line. This x-value transformation is accomplished by $x \rightarrow x - x_1$. The equation of the line, in the $y = mx + b$ format, is

$$y = \left[\frac{(y_2 - y_1)}{(x_2 - x_1)}\right](x - x_1) + y_1 \tag{3.19}$$

For a value of x_3 between x_1 and x_2 (i.e., $x_1 < x_3 < x_2$), the corresponding value of y_3 can be found through substitution into the interpolation equation.

▶ EXAMPLE 3.6

Using linear interpolation with Table A.3, find the specific enthalpy of air at a temperature of 327 K.

Given: The data incorporating variable specific heats for air in Table A.3

Find: h at 327 K

Solution: This example uses the air properties given in Table A.3, which include the variability of the specific heats of air. Using Eq. (3.19), consider the specific enthalpy quantities to be y and the temperatures to be x. Therefore, Eq. (3.19) can be rewritten as

$$h_3 = \left[\frac{(h_2 - h_1)}{T_2 - T_1}\right](T_3 - T_1) + h_1$$

where state 3 is the condition with $T_3 = 327$ K.

For choosing states 1 and 2, use the table entries immediately surrounding $T_3 = 327$ K in Table A.3. State 1 will be the state at $T_1 = 325$ K, and state 2 will be the state at $T_2 = 330$ K. From Table A.3:

$T_1 = 320$ K $\quad h_1 = 320.29$ kJ/kg

$T_2 = 340$ K $\quad h_2 = 340.42$ kJ/kg

Substituting:

$$h_3 = \left[\frac{(340.42 - 320.29)\ \frac{\text{kJ}}{\text{kg}}}{(340 - 320)\ \text{K}} \right] (327 - 320)\ \text{K} + 320.29\ \frac{\text{kJ}}{\text{kg}}$$

$$\mathbf{h_3 = 327.32\ kJ/kg}$$

Analysis: Many hand-held calculators can perform linear interpolation upon entry of the two data points.

3.5.3 Other Equations of State for Gases

Engineers frequently make assumptions in order to make their design and analysis more manageable. It is important for engineers to be cognizant of whether or not the assumptions they have made are valid. Although small errors introduced by valid assumptions are easily accounted for in a robust design, poor assumptions can lead to design failure or a completely inaccurate analysis. The assumption of a gas behaving as an ideal gas is usually acceptable in engineering, but it is important for an engineer to know just how close to ideal behavior a gas is displaying.

A quick method to judge the accuracy of an ideal gas assumption is to determine the compressibility factor, Z, of the gas. Gases, when normalized by their critical state properties, display uniform behavior, and this behavior can be represented on a generalized compressibility chart, as shown in **Figure 3.15**. The uniform behavior of the gases is called *the principle of corresponding states*. On such a chart, the reduced temperature, T_R, and the reduced pressure, P_R, are defined as

$$T_R = T/T_c \quad \text{and} \quad P_R = P/P_c$$

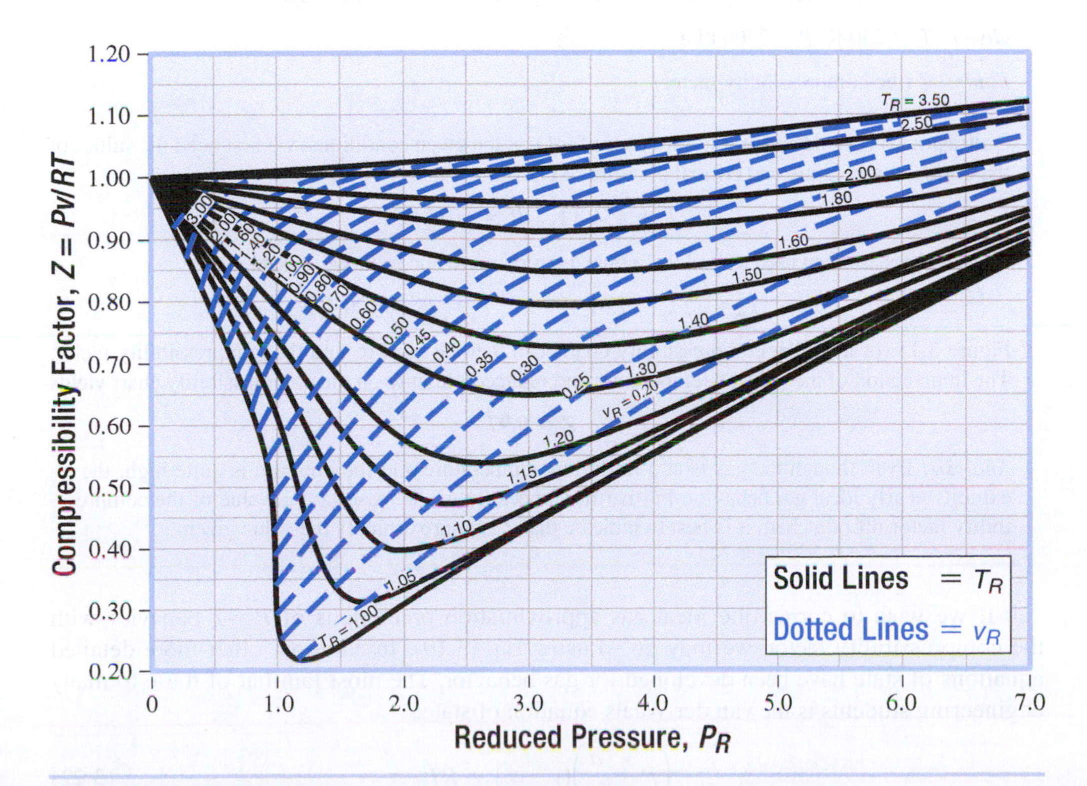

FIGURE 3.15 An example of a compressibility chart.

where T_c and P_c are the critical temperature and the critical pressure of the gas, respectively. Values for these quantities are found for many gases in Table A.1. These values can be used with the chart to find the compressibility factor, Z, which is defined as

$$Z = Pv/RT \tag{3.20}$$

The compressibility factor equals 1 ($Z = 1$) for a gas displaying ideal gas behavior. Use of a compressibility chart, such as in Figure 3.15, allows you to quickly approximate how close to ideal the behavior of a nonideal gas is. For temperatures well above the critical temperature of the gas, the behavior is often within 10% of ideal, and pressures much below or much above the critical pressure also drive gases to more ideal behavior. Engineers must recognize what levels of nonideal gas behavior are acceptable for their work if the ideal gas assumption is used. For many applications, if the behavior is within 20%, the ideal gas assumption is probably acceptable.

If the specific volume of a gas is known, the pseudo-reduced specific volume of the gas can be used to help determine the compressibility factor. The pseudo-reduced specific volume is defined as

$$v_R = \frac{v}{\dfrac{RT_c}{P_c}} \tag{3.21}$$

The pseudo-reduced specific volume is used rather than simply the reduced specific volume ($v_R = v/v_c$, where v_c is the critical state specific volume) because the correlation with corresponding states is found to be much better using the relationship given in Eq. (3.21).

▶ EXAMPLE 3.7

Determine the compressibility factor for oxygen gas at 250 K and 2500 kPa.

Given: $T = 250$ K, $P = 2500$ kPa

Find: Z (the compressibility factor)

Solution: To find the compressibility factor for O_2 at the given conditions, we first need the values of the critical temperature and critical pressure of O_2. From Table A.1, for O_2,

$$T_c = 154.8 \text{ K} \quad P_c = 5080 \text{ kPa}$$

These values are then used to find the reduced temperature and reduced pressure:

$$T_R = T/T_c = 1.61 \qquad P_R = P/P_c = 0.492$$

Figure 3.15 (or a similar compressibility chart) can then be used to find the compressibility factor. The intersection of the reduced temperature and reduced pressure on the compressibility chart yields

$$\mathbf{Z \approx 0.97}$$

Analysis: Even though the gas is at a fairly low temperature and the pressure is quite high, the O_2 exhibits nearly ideal gas behavior. Furthermore, because we are reading the value of the compressibility factor off of a chart, it is best to indicate that Z is approximately the value given.

If we wish to correct the ideal gas approximation predictions of P-v-T behavior with the compressibility factor, we may do so using Eq. (3.19). In addition, other more detailed equations of state have been developed for gas behavior. The most familiar of these to many engineering students is the van der Waals equation of state:

$$\left(P + \frac{a}{v^2}\right)(v - b) = RT \tag{3.22}$$

where a and b are constants particular to a specific gas. This model improves the ideal gas model by (a) accounting for intermolecular forces through the a/v^2 term and (b) accounting for the gas molecules occupying a finite volume through the b term. The need for this latter term is particularly straightforward to understand; clearly a molecule cannot occupy the space filled by another molecule, and so that volume should not be available for it to occupy. The use of the intermolecular force correction was a result of the recognition that gases do condense to liquids, which requires a force to draw molecules together.

The constants a and b can be determined through values at the critical point. At the critical point, the line of the critical temperature on a P-v diagram reaches a horizontal inflection point, and so the first and second derivatives of the pressure with respect to volume must be 0. As such, it can be found that

$$a = \frac{27R^2T_c^2}{64P_c} \quad \text{and} \quad b = \frac{RT_c}{8P_c} \tag{3.23}$$

The difference that can be experienced in a prediction of v between the van der Waals equation and the ideal gas equation is shown in **Figure 3.16**, where the value of v for oxygen is plotted as a function of temperature for three pressures, using both Eq. (3.22) and Eq. (3.12b).

▶ EXAMPLE 3.8

Determine the specific volume of O_2 at 250 K and 2500 kPa using (a) the ideal gas law, (b) the compressibility factor, and (c) the van der Waals equation.

Given: $T = 250$ K and $P = 2500$ kPa

Find: v using three different methods

Solution: We will use three methods for finding the specific volume. For all three methods, we need the gas-specific ideal gas constant for O_2. From Table A.1, $R = 0.2598$ kJ/kg · K.

(a) From the ideal gas law:

$$v = \frac{RT}{P} = \mathbf{0.0260\ m^3/kg}$$

(b) From Example 3.7, $Z \approx 0.97$ for this situation. So using Eq. (3.20):

$$v = \frac{ZRT}{P} = \mathbf{0.0252\ m^3/kg}$$

(c) Solution of the van der Waals equation is a bit more complex. First,

$$a = \frac{27R^2T_c^2}{64P_c} = 0.1343\ \text{kPa} \cdot \text{m}^6$$

and

$$b = \frac{RT_c}{8P_c} = 0.0009896\ \text{m}^3$$

Solving the van der Waals equation for v may be best done numerically. A numerical solution of

$$\left(P + \frac{a}{v^2}\right)(v - b) = RT$$

yields

$$v = \mathbf{0.0249\ m^3/kg}$$

If you are unfamiliar with numerical methods, v can be found via trial-and-error, with the ideal gas law value being a good first guess.

Analysis: All three methods give similar values for the specific volume, and none of them are likely to be exact. Fortunately, for most engineering purposes, any of the values are probably going to be acceptable, except when extreme accuracy is needed.

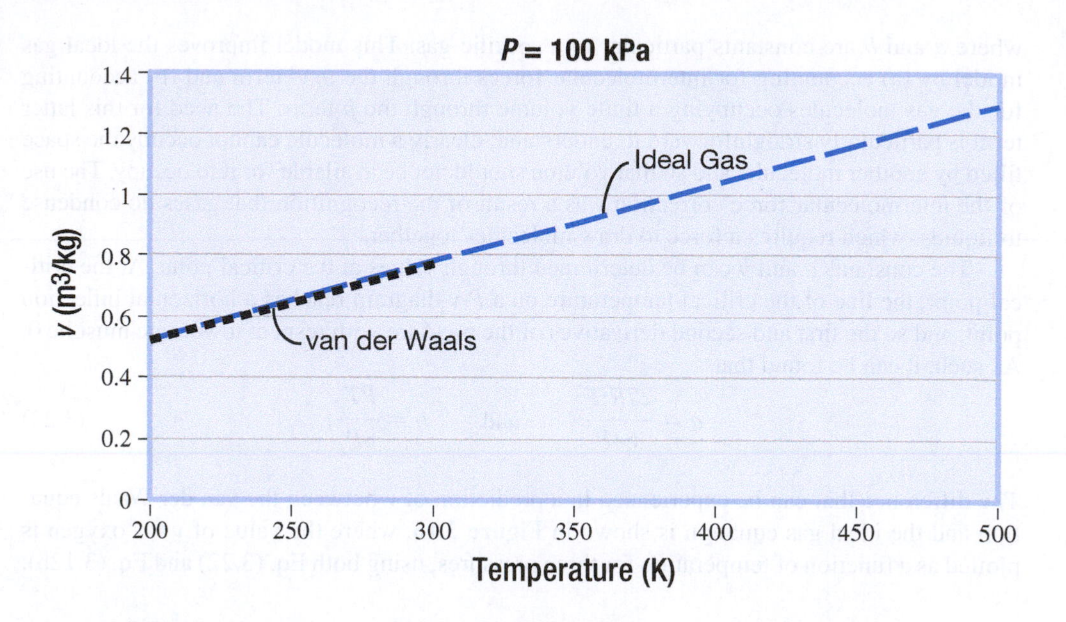

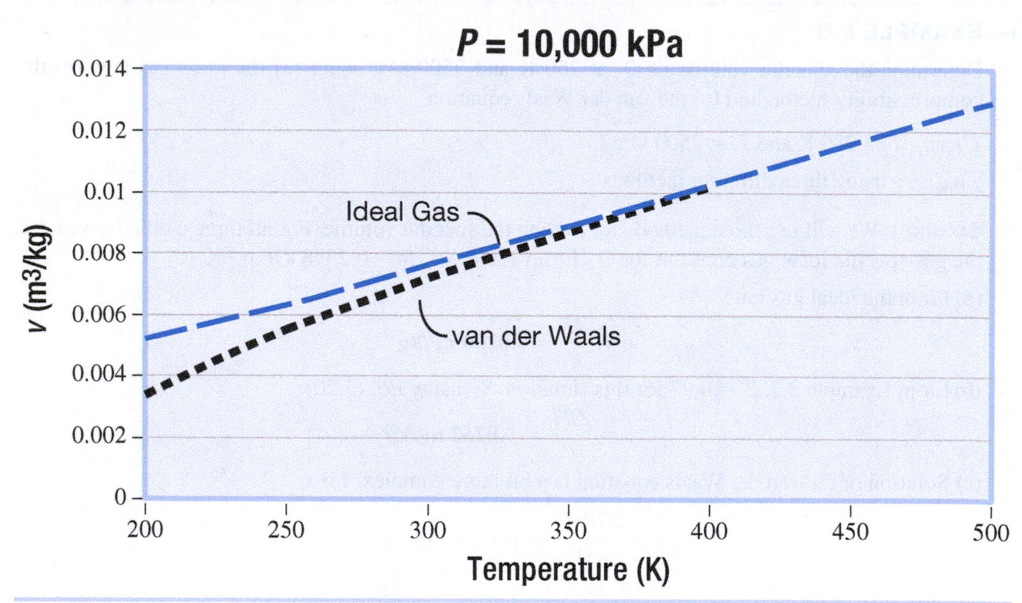

FIGURE 3.16 A comparison of the calculated specific volume of O_2 found using the ideal gas law and the van der Waals equation. The differences between the two calculations are greater at higher pressures and lower temperatures, where the gas density is larger.

▶ EXAMPLE 3.9

Experimentally, the specific volume of water vapor at 6.0 MPa and 280°C is found to be 0.03317 m³/kg. What is the specific volume of water at this state as predicted by the ideal gas law, and as predicted using the compressibility factor?

Given: $T = 280°C = 553$ K, $P = 6.0$ MPa $= 6000$ kPa

Find: v, using the ideal gas law and the compressibility factor

Solution: ***This is for illustrative purposes only. In practice, we do not want to use the ideal gas law for water vapor.***

For water, the gas-specific ideal gas constant is $R = 0.4615$ kJ/kg · K

From the ideal gas law, $v = RT/P = (0.4615 \text{ kJ/kg} \cdot \text{K})(553\text{K})/(6000 \text{ kPa}) = \mathbf{0.04253\ m^3/kg}$

For water, $T_c = 647.3$ K, $P_c = 22{,}090$ kPa

Therefore, $T_R = T/T_c = 0.854$ $P_R = P/P_c = 0.272$

Using the compressibility charts, $Z \approx 0.82$.

Then $v = ZRT/P = \mathbf{0.0349\ m^3/kg}$.

Analysis: For this condition of a substance near a phase change, it can be seen that the ideal gas law gives a specific volume that is 28.2% in error—an unacceptably large amount for most engineering applications. This is an excellent illustration of why we don't want to treat water vapor as an ideal gas. Using the compressibility factor does bring the error to approximately 5%, which is an error that can be typically expected for a compressibility factor read off of a chart. Although this level of error is acceptable in most cases, it is generally better to use other methods to find properties of substances near a phase change, such as water. These methods are discussed later.

More complex equations of state have been proposed and used in practice. Such equations include the Beattie-Bridgeman equation, the Benedict-Webb-Rubin equation, and virial equations of state. The equations are marked by more constants designed to best match observed behavior in real gases, and some equations are designed to work closer to the phase-change region between liquid and gases—even to the point of being used inside the vapor dome itself. In general, most engineers do not need to use these more complex equations of state.

QUESTION FOR THOUGHT/DISCUSSION

Any real gas does not perfectly follow the ideal gas law. How much deviation from the ideal gas law do you think is acceptable in typical engineering applications? What applications might demand using more complex equations of state for gases?

3.6 INCOMPRESSIBLE SUBSTANCES

A second type of idealized substance is called an incompressible substance. An incompressible substance is defined as a substance with an unchanging specific volume:

$$v = \text{constant}$$

This also indicates that an incompressible substance has a constant density. The incompressible substance approximation is best for solids and liquids, particularly those that do not experience large changes in pressure and temperature. Solids and liquids will expand and contract with temperature changes, but these changes in volume lead to a relatively small change in the specific volume, thereby making the approximation reasonably accurate. In addition, it often requires a very large change in pressure to produce a significant volume change in solids or liquids. As a result, we will usually assume solids and liquids to be incompressible substances, except for some liquids undergoing large pressure changes.

More advanced property relationships can be used to show that for incompressible substances, $c_p = c_v$. Although real solids and liquids have small amounts of compressibility, the difference between c_p and c_v is typically small enough to be ignored. Because incompressible substances have $c_p = c_v$, we will refer to a single specific heat of an incompressible substance: c. Using this single specific heat, for incompressible substances we can state

$$\Delta u = \int c\, dT \tag{3.24}$$

While the specific heat for an incompressible substance is a function of temperature, it can be treated as constant for small and moderate temperature ranges. With an assumption of a constant specific heat:

$$u_2 - u_1 = c(T_2 - T_1) \tag{3.25}$$

For changes in specific enthalpy, we first realize that $dh = du + vdP$, as $v =$ constant. If constant specific heat is assumed, integration yields

$$h_2 - h_1 = c(T_2 - T_1) + v(P_2 - P_1) \tag{3.26}$$

However, for solids and most liquids, the $v(P_2 - P_1)$ is very small except for extremely large pressure differences. As a result, for incompressible substances with constant specific heats,

$$h_2 - h_1 \approx u_2 - u_1 = c(T_2 - T_1) \tag{3.27}$$

If the specific heat is not assumed to be constant, the appropriate functional relationship between the specific heat and temperature must be integrated. The specific heats of some common solids and liquids are found in Table A.5.

▶ **EXAMPLE 3.10**

Iron, at atmospheric pressure, is heated from 10°C to 50°C. Determine the change in specific internal energy of the iron during the process.

Given: $T_1 = 10°C$, $T_2 = 50°C$

Find: Δu

Solution: We will assume that the iron is an incompressible substance and that the specific heats of the iron are constant. For an incompressible substance with constant specific heats, we can use Eq. (3.25) to find the change in specific internal energy:

$$\Delta u = u_2 - u_1 = c(T_2 - T_1)$$

For iron at 300 K (a reasonable temperature for this problem), $c = 0.447$ kJ/kg · K (from Table A.5).

$$\Delta u = u_2 - u_1 = (0.447 \text{ kJ/kg} \cdot \text{K})(50°\text{C} - 10°\text{C}) = \mathbf{17.9\ kJ/kg}$$

3.7 PROPERTY DETERMINATION FOR WATER AND REFRIGERANTS

As discussed previously, for engineering purposes it is often satisfactory to model substances that are gases as ideal gases, and substances that are liquids and solids as incompressible substances. The equations of state for such substances are rather simple, although we can also model gas P-v-T behavior more accurately through more complex equations. But, for substances that are either undergoing a phase change or whose thermodynamic state is close to the vapor dome, the equations of state describing the substance become much more complicated. Accurate determination of the thermodynamic properties of these substances requires different approaches.

First, however, there is a question as to what constitutes "close" with respect to the vapor dome. There is no absolute answer to this question. A good rule of thumb is to consider if the substance is readily experienced in the system as both a vapor and a liquid (we will ignore the solid phase in this discussion, although the same reasoning applies between solid–liquid and solid–vapor transitions). For example, in regular activities in a factory we can experience water in liquid form as well as in vapor form (either as water vapor in the air or as steam in a process). Therefore, in everyday usage, we would consider water as potentially a substance that is close to the vapor dome. The same can be said of refrigerants used in refrigeration or air conditioning applications. However, although nitrogen can become a liquid at atmospheric pressure, it must be cooled to $-196°C$ to do so. Therefore, we would not expect N_2 to be commonly experienced as a liquid in applications, and it is therefore modeled as a gas (and often as an ideal gas); but if we were working with cryogenic operations where liquid N_2 is used, we would treat it similarly to how water and refrigerants are treated, and we would not treat it as an ideal gas.

As noted, when close to or under the vapor dome, the equations of state become much more complicated and are usually not presented in equation form. For the purposes of this text, we will consider that water and refrigerants fall into this category of substances; however, keep in mind that the methods for finding properties of water and the refrigerants can be applied to other substances that are close to or experiencing a phase change in any particular application.

Engineers commonly use computers to assist in their design analysis of systems, and we encourage you to use computer programs or tablet/smartphone apps that will provide property data for water and the refrigerants at any state. As discussed previously with regard to ideal gases with variable specific heats, Internet sites and programs and computer apps change frequently, so you should conduct a search for "steam properties" or "refrigerant properties" to find currently available applications. Again, at the time of this writing, Wolfram Alpha™, Peacesoftware™, and NIST all have good tools available. In addition, some plug-ins are available for MATLAB users; these plug-ins may be beneficial for use in modeling systems in future chapters. Some programs may give values in molar units that will need conversion to mass-based units through the molecular mass. And because the values of such quantities as specific enthalpy are generally given relative to an arbitrary 0 point for a substance, the actual number provided by different sources may be different, but a change in a property should be the same.

Traditionally, finding properties for water and refrigerants was done using property tables; when used for water, these are commonly called the "steam tables." A set of steam tables for water is given in the appendix as Tables A.6–A.9. Upon inspection, we can see that there are two tables covering water in the saturation region (Tables A.6 and A.7), a set of tables covering water as superheated steam (Table A.8), and a set of tables covering water as a compressed liquid (Table A.9). Note that, for the saturation region, the tables contain similar data. The difference between the tables is that Table A.6 is designed around an orderly progression of saturation temperatures, whereas Table A.7 is built on an orderly progression of saturation pressures.

To use these tables, the first task is to determine which table to use based on the given data. It is typically best to use the saturated state tables (Tables A.6 and A.7) to assist in this determination. When given the temperature and pressure (T and P) of a substance, we can do one of the following two analyses to determine the state of the substance:

> ***Option 1***: Let $T = T_{sat}$. Using Table A.6: If $P > P_{sat}(T)$, then the substance is a compressed liquid. If $P < P_{sat}(T)$, then the substance is a superheated vapor. If $P = P_{sat}(T)$, the substance is saturated, and an additional property is needed to fix its state.

> ***Option 2***: Let $P = P_{sat}$. Using Table A.7: If $T < T_{sat}(P)$, then the substance is a compressed liquid. If $T > T_{sat}(P)$, then the substance is a superheated vapor. If $T = T_{sat}(P)$, then the substance is saturated and an additional property is needed to fix its state.

▶ EXAMPLE 3.11

Use the steam tables to determine the phase of water with a pressure of 1.0 MPa and a temperature of (a) 30°C, (b) 179.9°C, and (c) 300°C.

Given: Water, $P = 1.0$ MPa, (a) $T = 30$°C, (b) $T = 179.9$°C, (c) $T = 300$°C

Find: The phase of water at each temperature

Solution: We will use the steam tables for this problem, although a computer program or app could also be used. The pressure is given as one tabulated in Table A.7, so we should use Option 2 above to determine the phase of the water in each case. From Table A.7, for a saturation pressure of 1.0 MPa,

$$T_{sat} = 179.9°C$$

Therefore,

(a) for $T = 30$°C, $T < T_{sat}$ and the water is a **compressed liquid**;

(b) for $T = 179.9$°C, $T = T_{sat}$ and the water is **saturated**; either a saturated liquid, a saturated vapor, or a saturated mixture—an additional property is needed to fix the state;

(c) for $T = 300$°C, $T > T_{sat}$ and the water is a **superheated vapor**.

Often, a property other than the pressure or temperature is known. For example, we may know the temperature and the specific volume. In these types of situations (combinations of known properties of T, v; T, u; T, h; T, s [s is the specific entropy, which is introduced in Chapter 6]; P, v; P, u; P, h; and P, s), the following approach can be used to determine the phase:

(1) Using the saturation table appropriate for the known properties (Table A.6 if T is known, Table A.7 if P is known), set either T or P to the saturation state: $T = T_{sat}$ or $P = P_{sat}$.

(2) Compare the other known property to the value of the property as a saturated liquid and as a saturated vapor at the given saturation state. For example, if v is known, compare v to v_f and v_g:

if $v < v_f$, the substance is a compressed liquid
if $v = v_f$, the substance is a saturated liquid
if $v_f < v < v_g$, the substance is a saturated mixture
if $v = v_g$, the substance is a saturated vapor
if $v > v_g$, the substance is a superheated vapor

This comparison also works for u, h, and s.

▶ EXAMPLE 3.12

Determine the phase of water if (a) $T = 200$°C and $u = 1500$ kJ/kg, and (b) $P = 500$ kPa and $h = 3000$ kJ/kg.

Given: Water, (a) $T = 200$°C, $u = 1500$ kJ/kg; (b) $P = 500$ kPa, $h = 3000$ kJ/kg

Find: The phase of the water at each state

Solution: For water, we will use the steam tables to solve this problem. A computer program that provides water properties could also be used.

(a) For $T = 200$°C, use Table A.6. From Table A.7 at $T = T_{sat} = 200$°C,

$$u_f = 850.65 \text{ kJ/kg and } u_g = 2595.3 \text{ kJ/kg}$$

For the value $u = 1500$ kJ/kg, from inspection it can be seen that $u_f < u < u_g$; therefore, the water is a **saturated mixture**. The quality of the water can be found to be

$$x = \frac{u - u_f}{u_g - u_f} = \mathbf{0.372}$$

indicating that 37.2% of the water is a saturated vapor, and 62.8% of the water is a saturated liquid.

(b) For $P = 500$ kPa, use Table A.7, and set $P = P_{sat} = 500$ kPa. At this saturation pressure,

$$h_f = 640.23 \text{ kJ/kg and } h_g = 2748.7 \text{ kJ/kg}$$

Upon inspection, $h = 3000$ kJ/kg is $h > h_g$, so the water is a **superheated vapor**.

Once the phase of the fluid is known, the appropriate tables can be used.

3.7.1 Saturated Property Tables

The layout of Tables A.6 and A.7 is straightforward. The pressure (or temperature) corresponding to the saturation temperature (or saturation pressure) is given. The saturated liquid and saturated vapor values of the specific volume, specific internal energy, specific enthalpy, and specific entropy are then listed. Note that some tables list values of a property such as h_{fg}. These values represent the change in that property during evaporation (i.e., $h_{fg} = h_g - h_f$) and are provided to enable quicker calculation of quality values. If a value is not tabulated, you may need to perform linear interpolation between the two most closely surrounding values in the table. For example, if you needed the specific internal energy of saturated liquid water at a pressure of 130 kPa, you would interpolate between the values of u_f at 100 kPa and 150 kPa in Table A.7.

3.7.2 Superheated Vapor Tables

Table A.8 contains the superheated vapor tables for water. Because both temperature and pressure are independent properties in the superheated region, the tables take on a significantly different form. As can be seen, blocks of data are presented for specific values of the pressure. For each of these blocks, the specific volume, specific internal energy, specific enthalpy, and specific entropy are given for a series of temperatures. If we seek property information that falls directly on the table, we can quickly read off the value. For example, if the specific enthalpy of water vapor at 1.0 MPa and 400°C is needed, we can find the block of data for $P = 1.0$ MPa and then look at the row containing the temperature of 400°C to find that $h = 3263.9$ kJ/kg. However, if property data do not fall exactly on the table, linear interpolation is required. Such a process is relatively simple if one known property appears in the tables, but it will require multiple interpolations if neither known property appears. When three variables are involved in the interpolation, it is important to hold one of the values constant when selecting values from the tables. These ideas are illustrated in Example 3.13.

▶ EXAMPLE 3.13

Using the steam tables, determine the specific internal energy of water vapor for (a) $P = 1.0$ MPa and $T = 310$°C, and (b) $P = 2.20$ MPa and $T = 365$°C.

Given: Water (steam), (a) $P = 1.0$ MPa, $T = 310$°C; (b) $P = 2.20$ MPa, $T = 365$°C

Find: u for each case.

Solution: We will use the steam tables for this problem. You may want to check your answers with computer programs or apps.

(a) Data for $P = 1.0$ MPa are tabulated in Table A.8, so a single linear interpolation using T and u values surrounding $T = 310°C$ is possible. From Table A.8,

$T_1 = 300°C$, $u_1 = 2793.2$ kJ/kg

$T_2 = 350°C$, $u_2 = 2875.2$ kJ/kg

Using Eq. (3.19) as a model for linear interpolation,

$$u_3 = \left[\frac{u_2 - u_1}{T_2 - T_1}\right](T_3 - T_1) + u_1$$

With $T_3 = 310°C$, we can substitute and solve for

$$\mathbf{u_3 = 2809.6\ kJ/kg}$$

(b) Neither the pressure nor temperature is given in Table A.8, so multiple interpolations are necessary. To do this, we can either find the values of the specific internal energy at both 350°C and 400°C at $P = 2.2$ MPa and then interpolate between those two values for the value at 365°C, or we can find the specific internal energy values at 2.0 MPa and 2.5 MPa at $T = 365°C$ and then interpolate between those values for the value at 2.2 MPa. Notice that each interpolation involves holding one variable constant. In this example, we will use the first approach.

Holding $T = 350°C$, the following values for P and u can be found from Table A.8:

$P_1 = 2.0$ MPa, $u_1 = 2859.8$ kJ/kg

$P_2 = 2.5$ MPa, $u_2 = 2851.9$ kJ/kg

$$u_3 = \left[\frac{u_2 - u_1}{P_2 - P_1}\right](P_3 - P_1) + u_1$$

For $P_3 = 2.2$ MPa, this yields $u_3 = 2856.6$ kJ/kg.

Next, at $T = 400°C$,

$P_4 = 2.0$ MPa, $u_4 = 2945.2$ kJ/kg

$P_5 = 2.5$ MPa, $u_5 = 2939.1$ kJ/kg

$$u_6 = \left[\frac{u_5 - u_4}{P_5 - P_4}\right](P_6 - P_4) + u_4$$

For $P_6 = 2.2$ MPa, $u_6 = 2942.8$ kJ/kg.

This now gives us two data points, both at $P = 2.2$ MPa:

$T_3 = 350°C$ $\quad u_3 = 2856.6$ kJ/kg

$T_5 = 400°C$ $\quad u_6 = 2942.8$ kJ/kg

Interpolating between these values,

$$u_7 = \left[\frac{u_6 - u_3}{T_6 - T_3}\right](T_7 - T_3) + u_3$$

With $T_7 = 365°C$, this equation yields

$$\mathbf{u_7 = 2882.5\ kJ/kg}$$

Analysis: An online calculator such as Wolfram Alpha™ quickly yields 2883 kJ/kg. Clearly, use of computer resources can greatly reduce the need to manipulate data on the steam tables through linear interpolation.

3.7.3 Compressed Liquids Property Data

Table A.9 provides a format similar to Table A.8, but for compressed liquid water. Table A.9 can be used in the same manner as the superheated vapor tables, although you should note that the lowest pressure on the table is still quite high. Upon inspection of the values, you can also see that the values of v, u, h, and s do not change rapidly with pressure—these values are primarily a function of temperature. Finally, note that detailed tables for many substances as compressed liquids are not readily available.

As a result, approximations are often used when dealing with "slightly" compressed liquids—those liquids at a pressure of less than 5 MPa. If more detailed data are not available, these assumptions can be applied at higher pressures as well, without the loss of much accuracy from an engineering perspective. These approximations involve setting the value of the property to that of a saturated liquid at the substance's temperature:

$$v(T, P) \approx v_f(T)$$

$$u(T, P) \approx u_f(T)$$

$$s(T, P) \approx s_f(T)$$

However, some prefer to add a pressure correction term for the specific enthalpy:

$$h(T, P) \approx h_f(T) + v_f(P - P_{sat}(T))$$

You will find, however, that the pressure correction term will be small for most applications involving a slightly compressed liquid and can often be safely ignored.

QUESTION FOR THOUGHT/DISCUSSION

Why is it wrong to calculate (and report) a value for the quality of a compressed liquid or superheated vapor?

Summary

In this chapter, we have explored the *P-v-T* behavior of substances, and we have described phase changes and the relationships between properties. We presented the state principle, from which we know that for pure substances in simple compressible systems, two properties are needed to specify the state. From these properties, all other properties can be found using appropriate equations of state. In addition, we introduced the thermodynamic properties of the specific heats.

We described the concept of an ideal gas, and we presented the ideal gas law. We also explored how to find values for the changes in specific enthalpy and specific internal energy for ideal gases, both assuming constant specific heats and incorporating the variability of the specific heats into the calculations. We discussed incompressible substances, as well as the appropriate relationships between properties for such substances. Finally, we discussed methods for finding property data for substances near a phase-change region, such as water and refrigerants. We emphasized the importance of learning to find property data for water and refrigerants from computer-based sources.

With your knowledge of heat, work, and thermodynamic properties, you are now ready to learn the First Law of Thermodynamics. The tools that you have learned so far will enable you to apply the first law to the design and analysis of many thermodynamic systems.

KEY EQUATIONS

Ideal Gas Equations of State:

Ideal Gas Law:

$$P\forall = mRT \tag{3.12a}$$

$$Pv = RT \tag{3.12b}$$

Change in Enthalpy and Internal Energy If Assuming Constant Specific Heats:

$$h_2 - h_1 = c_p (T_2 - T_1) \tag{3.18a}$$

$$u_2 - u_1 = c_v (T_2 - T_1) \tag{3.18b}$$

Compressibility Factor:

$$Z = Pv/RT \tag{3.20}$$

van der Waals Equation of State:

$$\left(P + \frac{a}{v^2}\right)(v - b) = RT \tag{3.22}$$

Changes in Enthalpy and Internal Energy for Incompressible Substances with Constant Specific Heat:

$$u_2 - u_1 = c(T_2 - T_1) \tag{3.25}$$

$$h_2 - h_1 = c(T_2 - T_1) + v(P_2 - P_1) \tag{3.26}$$

$$h_2 - h_1 \approx u_2 - u_1 = c(T_2 - T_1) \tag{3.27}$$

PROBLEMS

3.1 Describe the process of cooling water at 101 kPa pressure from 125°C to 50°C, at constant pressure. What happens to the water as heat is removed? Show the process on a *T*-*v* diagram, and relate the description to the diagram.

3.2 Describe the process of heating water at 101 kPa pressure from −10°C to 150°C, at constant pressure. What happens to the water as heat is added? Show the process on a *T*-*v* diagram, and relate the description to the diagram.

3.3 Water is initially at 150°C and a pressure of 1000 kPa. The volume in the container is slowly increased, causing the pressure to decrease, until the pressure is 80 kPa. The saturation pressure of water at 150°C is 475.8 kPa. The temperature of the water is maintained at 150°C throughout the process. Show the process on a *P*-*v* diagram, and describe what happens to the water during the process.

3.4 A rigid container is completely full of compressed liquid water at 500 kPa. Liquid water is then slowly drained from the container, while the pressure is maintained at 500 kPa in the container. The liquid water is removed until all of the liquid water has been removed. Show this process on a *P*-*v* diagram, and describe what happens to the water during the process.

3.5 Water at 200°C occupies a volume of 0.005 m^3. The mass of the water is 0.0525 kg. The specific volume of saturated liquid water, v_f, at 200°C is 0.001156 m^3/kg, while the specific volume of saturated water vapor, v_g, is 0.1274 m^3/kg. Determine the quality of the saturated mixture.

3.6 0.5 kg of water at 120°C occupies a volume of 0.12 m^3. The specific volume of saturated liquid water, v_f, at 120°C is 0.0010603 m^3/kg, and the specific volume of saturated water vapor, v_g, at 120°C is 0.8919 m^3/kg. Find the quality of the saturated mixture.

3.7 At 80°C, water has v_f = 0.0010291 m^3/kg and v_g = 3.407 m^3/kg. The water occupies a volume of 0.05 m^3. For each of the following mass values, determine whether the water is a compressed liquid, saturated liquid, saturated mixture, saturated vapor, or superheated vapor, and determine the quality of the water: (a) 60 kg, (b) 5 kg, (c) 0.025 kg, (d) 0.0147 kg, and (e) 0.0095 kg.

3.8 At 400 kPa, water has v_f = 0.0010836 m^3/kg and v_g = 0.4625 m^3/kg. The mass of the water is 0.5 kg. For each of the following volumes, determine whether the water is a compressed liquid, saturated liquid, saturated mixture, saturated vapor, or superheated vapor, and determine the quality of the water: (a) 0.00045 m^3, (b) 0.000542 m^3, (c) 0.025 m^3, (d) 0.20 m^3, (e) 0.40 m^3.

3.9 At 180°C, water has v_f = 0.0011274 m^3/kg and v_g = 0.1941 m^3/kg. A saturated mixture of water at this temperature has a quality of 0.25. (a) Determine the specific volume of the water. (b) If the water has a mass of 1.5 kg, determine the total volume of the water.

3.10 At 1.7 MPa, water has v_f = 0.001164 m^3/kg and v_g = 0.1152 m^3/kg. A saturated mixture of water at 1.7 MPa has a quality of 0.71. (a) Determine the specific volume of the water. (b) If the water occupies a volume of 0.30 m^3, determine the mass of the water.

3.11 At –12°C, refrigerant R-134a has v_f = 0.0007498 m^3/kg and v_g = 0.1068 m^3/kg. 3 kg of a saturated mixture of R-134a at –12°C has a quality of 0.93. Determine the specific volume and the total volume of the R-134a.

3.12 Refrigerant R-134a at 140 kPa has v_f = 0.0007381 m^3/kg and v_g = 0.1395 m^3/kg. The R-134a at 140 kPa has a mass of 5.0 kg. Plot the volume occupied by the R-134a as a function of quality for values of the quality ranging between 0.0 and 1.0.

3.13 At 200°C, water has v_f = 0.0011565 m^3/kg and v_g = 0.1274 m^3/kg. The water at 200°C occupies a volume of 0.05 m^3. Plot the mass of the water as a function of quality for values of the quality between 0.0 and 1.0.

3.14 At 0°C, ammonia has v_f = 0.001556 m^3/kg and v_g = 0.2892 m^3/kg. A container of ammonia at 0°C has a volume of 0.012 m^3. Plot the quality of the ammonia as a function of mass for values of the mass between 0.042 kg and 7 kg.

3.15 Saturated water vapor at 14 MPa has a specific volume of 0.010 m^3/kg. The water is depressurized at a constant specific volume from this state. Considering the data below, determine the quality of the mixture as its pressure falls to 10 MPa, 5 MPa, 1 MPa, and 500 kPa.

P (MPa)	**v_f (m^3/kg)**	**v_g (m^3/kg)**
14	0.001611	0.01149
10	0.001452	0.01803
5	0.001286	0.04111
1	0.001127	0.1944
0.5	0.001093	0.3891

3.16 The specific volume of water at the critical point is 0.003155 m^3/kg. The water is cooled at a constant specific volume from the critical point. Considering the data given here, determine the quality of the mixture as its temperature falls to 340°C, 300°C, 200°C, and 100°C.

T (°C)	v_f (m^3/kg)	v_g (m^3/kg)
340	0.0016379	0.01080
300	0.0014036	0.02167
200	0.0011565	0.1274
100	0.0010435	1.673

3.17 Air has a pressure of 200 kPa and occupies a volume of 0.25 m^3. Determine the mass of the air present, assuming ideal gas behavior, for air temperatures of (a) 25°C, (b) 100°C, (c) 250°C, and (d) 500°C.

3.18 Air has a temperature of 350 K. Assume the air behaves as an ideal gas. Plot the pressure of the air as a function of specific volume for specific volumes ranging between 0.100 m^3/kg and 1.00 m^3/kg.

3.19 Nitrogen gas with a mass of 1.5 kg has a pressure of 690 kPa and a temperature of 390 K. Determine the volume occupied by the nitrogen gas, assuming ideal gas behavior.

3.20 An unknown gas occupies a volume of 0.075 m^3, has a pressure of 0.500 MPa, and has a temperature of 47°C. You determine that the gas is either CO_2, O_2, N_2, or CH_4. Using the volume, pressure, and temperature given, determine the mass that would be expected for these four gases. If you measure the mass to be 0.441 kg, which of the four gases is the unknown gas likely to be?

3.21 You are given a flask containing an unknown gas. The volume of the flask is 0.025 m^3, the mass of the gas is 0.0773 kg, and the temperature of the gas is 70°C. You know that the gas is either CO_2, Ar, O_2, or CH_4. If you measure the pressure of the gas to be 275 kPa, which of the four gases is the unknown gas likely to be?

3.22 Plot the pressure of CH_4 gas with a mass of 0.035 kg in a volume of 0.005 m^3 for temperatures varying between 0°C and 200°C. Assume the methane behaves as an ideal gas.

3.23 Plot the total volume as a function of temperature for 0.5 kg of hydrogen gas at 500 kPa, for temperatures varying between −50°C and 300°C. Assume the hydrogen behaves as an ideal gas.

3.24 Assuming ideal gas behavior, determine the mass of each of the following gases if the gas occupies 0.015 m^3 at 690 kPa and 300 K: (a) hydrogen, (b) methane, (c) nitrogen, (d) argon, and (e) carbon dioxide.

3.25 Air is heated from 20°C to 100°C. Determine the change in specific internal energy and the change in specific enthalpy for the air, assuming ideal gas with constant specific heat behavior.

3.26 The internal energy of 2 kg of air is reduced by 50 kJ during a cooling process. Initially, the air has a temperature of 40°C. Assuming ideal gas behavior and constant specific heats, determine the final temperature of the air.

3.27 The enthalpy of 1.5 kg of nitrogen gas is increased by 28 kJ during a heating process. Initially, the nitrogen temperature is 25°C. Assuming the nitrogen behaves as an ideal gas with constant specific heats, determine the final temperature of the nitrogen.

3.28 0.33 kg of oxygen is initially at 200 kPa and 280 K. The oxygen expands in an isobaric process until its volume is 0.175 m^3. Assuming the oxygen behaves as an ideal gas with constant specific heats, determine the change in total internal energy and total enthalpy of the oxygen during the process.

3.29 0.5 kg of oxygen is in a piston–cylinder device, initially occupying a volume of 0.15 m^3 at 70°C. The oxygen is cooled at constant pressure to a volume of 0.12 m^3. Assuming the oxygen behaves as an ideal gas with constant specific heats, determine the change in total internal energy and total enthalpy of the oxygen during the process.

3.30 Plot the change in specific internal energy as a function of final temperature for the following gases, each of which is initially at 10°C: (a) helium, (b) methane, (c) air, and (d) argon. Assume each gas behaves as an ideal gas with constant specific heats, and consider a final temperature range of 0°C to 100°C.

3.31 Plot the change in specific enthalpy as a function of final temperature for the following gases, each of which is initially at 25°C: (a) hydrogen, (b) nitrogen, (c) air, and (d) carbon dioxide. Assume each gas behaves as an ideal gas with constant specific heats, and consider a final temperature range of 0°C to 100°C.

3.32 0.5 kg of methane is initially at 500 kPa and 25°C. The methane expands in an isothermal process until its final volume is twice its initial volume. Assuming the methane behaves as an ideal gas with constant specific heats, determine (a) the initial volume, (b) the final pressure, (c) the change in specific internal energy, and (d) the change in specific enthalpy of the methane.

3.33 Argon, which has a constant specific heat as a noble gas, has a specific heat ratio of 1.667 and an ideal gas constant of 0.2081 kJ/kg · K. Determine the values of c_p and c_v for argon.

3.34 Neon has constant specific heats whose ratio is 1.667. The molecular mass of neon is 20.183 kg/kmole. Determine the values of c_p and c_v for neon.

3.35 Xenon is a noble gas and as such has constant specific heats, whose ratio is a constant 1.667. For xenon, c_p is 0.1583 kJ/kg · K. Determine the values of c_v, the molecular mass, and the gas-specific ideal gas constant for xenon.

3.36 Air is heated from an initial temperature of 300 K to a final temperature of 1200 K. Considering the variability of the specific heats, determine the change in specific internal energy and the change in specific enthalpy for the process.

3.37 Air is cooled from 610 K to 280 K. Considering the variability of the specific heats, determine the change in specific internal energy and the change in specific enthalpy for the process.

3.38 1.5 kg of air is initially at 25°C. The enthalpy of the air is increased by 850 kJ. Considering the variability of the specific heats, determine the final temperature of the air.

3.39 2.75 kg of air is initially at 800°C. The internal energy of the air is reduced by 525 kJ in a cooling process. Considering the variability of the specific heats, determine the final temperature of the air.

3.40 0.7 kg of air is initially at 17°C. Heat is added to the air, increasing its internal energy by 85 kJ. Considering the variability of the specific heats, determine the final temperature of the air.

3.41 Considering the variability of the specific heats of air, plot the changes in specific internal energy and specific enthalpy of air as it is cooled from an initial temperature of 1300 K to a final temperature ranging from 300 K to 1200 K.

3.42 Considering the variability of the specific heats of air, plot the change in specific enthalpy of air as it is heated from an initial temperature of 0°C to final temperatures ranging from 100°C to 1000°C.

3.43 Considering the variability of the specific heats of air, plot the change in specific internal energy of the air as it is cooled from an initial temperature of 1250 K to final temperatures ranging from 300 K to 1200 K.

3.44 Air is initially at 25°C. Plot the change in specific internal energy of the air as it is heated to final temperatures ranging between 30°C and 1200°C, considering (a) the specific heats to be constant with $c_v = 0.718$ kJ/kg · K, and (b) the specific heats to be variable.

3.45 Determine the change in specific enthalpy for air that is heated from 25°C to 1000°C, considering (a) the specific heats to be constant, with a value taken at 25°C, (b) the specific heats to be constant, with a value taken at 1000°C ($c_p = 1.185$ kJ/kg · K), (c) the specific heats to be constant, with a value taken at 500°C, and (d) the specific heats to be variable.

3.46 Air, with a mass of 0.7 kg, is initially at 1100°C. In a cooling process, the internal energy of the air is reduced by 370 kJ. Determine the final temperature of the air if (a) the specific heat is considered constant, with its value taken at 1100°C, (b) the specific heat is considered constant, with its value taken at 40°C, and (c) the specific heat is considered to be variable.

3.47 Determine the compressibility factor for nitrogen gas at the following conditions: (a) $T = 298$ K, $P = 101$ kPa; (b) $T = 200$ K, $P = 10{,}000$ kPa; (c) $T = 600$ K, $P = 10{,}000$ kPa; and (d) $T = 600$ K, $P = 50$ kPa.

3.48 Determine the specific volume of oxygen at 250 K and 3500 kPa (a) for ideal gas behavior, and (b) using the compressibility factor.

3.49 Using the ideal gas law and the compressibility factor, determine the pressure of methane gas at a temperature of 300 K and specific volumes of (a) 0.005 m^3/kg, (b) 0.05 m^3/kg, and (c) 0.5 m^3/kg.

3.50 Using the ideal gas law and the compressibility factor, determine the temperature of oxygen gas at a specific volume of 0.010 m^3/kg and pressures of (a) 5000 kPa, (b) 10,000 kPa, and (c) 20,000 kPa.

3.51 Using the ideal gas law and the compressibility factor, determine the pressure of nitrogen gas at a specific volume of 0.010 m^3/kg and a temperature of (a) 600 K, (b) 300 K, and (c) 150 K.

3.52 If you consider the ideal gas law to be an acceptable approximation for substances that have compressibility factors within 10% of 1 (i.e., between 0.90 and 1.10), determine the range of acceptable pressures for ideal gas behavior for a gas with a temperature of 300 K if the gas is (a) hydrogen, (b) oxygen, and (c) carbon dioxide.

3.53 Repeat Problem 3.52 for each gas having a temperature of (a) 200 K and (b) 500 K.

3.54 Repeat Problem 3.52, but consider the acceptable range of compressibility factors for approximating a substance as an ideal gas to be between 0.80 and 1.10.

3.55 If you consider the ideal gas law to be an acceptable approximation for substances that have compressibility factors within 10% of 1 (i.e., between 0.90 and 1.10), determine the range of acceptable temperatures for ideal gas behavior for a gas with a pressure of 500 kPa if the gas is (a) hydrogen, (b) oxygen, and (c) carbon dioxide.

3.56 Repeat Problem 3.55 for each gas having a pressure of (a) 2500 kPa and (b) 25 MPa.

3.57 Using both the ideal gas law and van der Waals equation, determine the pressure of N_2 that has a specific volume of 0.95 m^3/kg and a temperature of 350 K.

3.58 Using both the ideal gas law and van der Waals equation, determine the temperature of O_2 having a pressure of 1000 kPa and specific volumes of (a) 0.5 m³/kg, (b) 0.1 m³/kg, and (c) 0.05 m³/kg.

3.59 Using the ideal gas law, the compressibility factor, and the van der Waals equation, determine the temperature of CO_2 having a pressure of 2500 kPa and specific volumes of (a) 0.4 m³/kg, (b) 0.1 m³/kg, and (c) 0.04 m³/kg.

3.60 Nitrogen has a specific volume of 0.25 m³/kg. Using both the ideal gas law and van der Waals equation, tabulate the temperature of the nitrogen for pressures varying between 250 kPa and 2500 kPa, and plot the difference between the two values as a function of pressure.

3.61 Carbon dioxide has a pressure of 1500 kPa. Using both the ideal gas law and van der Waals equation, determine the specific volume of the CO_2 for temperatures of (a) 250 K, (b) 500 K, and (c) 2000 K.

3.62 Oxygen has a pressure of 3000 kPa. Using the ideal gas law, the compressibility factor, and the van der Waals equation, determine the specific volume of the O_2 for temperatures of (a) 200 K, (b) 500 K, and (c) 1500 K.

3.63 Considering lead to be an incompressible substance with c = 0.129 kJ/kg · K, determine the change in internal energy of a 10-kg block of lead as it is cooled from 150°C to 50°C.

3.64 Considering iron to be an incompressible substance with c = 0.42 kJ/kg · K, determine the change in internal energy of 5 kg of iron as it is heated from 25°C to 200°C.

3.65 A block of tin with a volume of 0.002 m³ is cooled from 150°C to 80°C. If the density of tin is 7304 kg/m³ and its specific heat is 0.22 kJ/kg · K, determine the change in total internal energy of the tin during the process.

3.66 Under certain circumstances, both water ice and liquid water can be considered to be incompressible substances with constant specific heats. Determine the change in specific internal energy for (a) ice as it is cooled from –5°C to –25°C at 101.325 kPa, and (b) liquid water as it is heated from 20°C to 40°C at 101.325 kPa. For ice, c = 2.04 kJ/kg · K, and for liquid water, c = 4.18 kJ/kg · K.

3.67 Copper has a density of 8890 kg/m³ and a specific heat of 0.387 kJ/kg · K. Consider that a block of copper undergoes a process that increases its pressure and temperature from 101 kPa and 300 K to 2 MPa and 720 K. Determine the change in specific internal energy and specific enthalpy for the copper in this process. Do not neglect the pressure terms in the enthalpy calculation.

3.68 Oil, with a density of 910 kg/m³ and a specific heat of 1.8 kJ/kg · K, is in a pump. The pump increases the pressure of the oil from 100 kPa to 5000 kPa, whereas the oil temperature increases from 40°C to 90°C. Determine the changes in specific internal energy and specific enthalpy of the oil. At what final pressure would you consider it acceptable to disregard the pressure component of the specific enthalpy change?

3.69 Each of the following incompressible substances is heated by increasing its internal energy by 1500 kJ, with the purpose of increasing the temperature from 300 K to 500 K. Determine the mass of each substance that will result in this temperature increase: (a) aluminum (c = 0.90 kJ/kg · K), (b) gold (c = 0.13 kJ/kg · K), (c) zinc (c = 0.39 kJ/kg · K), (d) gasoline (c = 2.08 kJ/kg · K).

3.70 Each of the following incompressible substances has a mass of 5 kg and is to be cooled from an initial temperature of 450 K by removing 100 kJ of internal energy. Determine the final temperature of each substance: (a) copper (c = 0.395 kJ/kg · K), (b) limestone (c = 0.909 kJ/kg · K), (c) iron (c = 0.45 kJ/kg · K), (d) mercury (c = 0.139 kJ/kg · K).

3.71 Using the steam tables in the appendix, complete the following chart of thermodynamic properties for water. Use computer-based programs or apps to check your results.

T (°C)	*P* (kPa)	*v* (m³/kg)	*h* (kJ/kg)	Phase (Quality If Applicable)
200				Saturated ($x = 0.45$)
150			850	
	500		3483.9	
		1.091		Saturated vapor ($x = 1.0$)

3.72 Using the steam tables in the appendix, complete the following chart of thermodynamic properties for water. Use computer-based programs or apps to check your results.

T (°C)	*P* (kPa)	*v* (m³/kg)	*h* (kJ/kg)	Phase (Quality If Applicable)
150				Saturated ($x = 0.25$)
	500		2150	
400	1000			
	300	0.820		

3.73 Using the steam tables in the appendix, complete the following chart of thermodynamic properties for water. Use computer-based programs or apps to check your results.

T (°C)	*P* (kPa)	*v* (m³/kg)	*h* (kJ/kg)	Phase (Quality If Applicable)
	2000			Saturated vapor ($x = 1.0$)
500			3445	
100	10,000			
20				Saturated ($x = 0.75$)

3.74 Using the steam tables in the appendix, complete the following chart of thermodynamic properties for water. Use computer-based programs or apps to check your results.

T (°C)	*P* (kPa)	*v* (m³/kg)	*h* (kJ/kg)	Phase (Quality If Applicable)
350	200			
	400	0.320		
420	5000			
510	3300			

3.75 Using the steam tables in the appendix, complete the following chart of thermodynamic properties for water. Use computer-based programs or apps to check your results.

T (°C)	*P* (kPa)	*v* (m³/kg)	*h* (kJ/kg)	Phase (Quality If Applicable)
55				Saturated ($x = 0.67$)
	1400	0.0774		
315			3066	
	6900	0.001001		

3.76 Using the steam tables in the appendix, complete the following chart of thermodynamic properties for water. Use computer-based programs or apps to check your results.

T (°C)	*P* (kPa)	*v* (m³/kg)	*u* (kJ/kg)	Phase (Quality If Applicable)
250				Saturated Liquid ($x = 0.0$)
50	15,000			
	10,000		2832	
	600			Saturated ($x = 0.40$)

3.77 Using the steam tables in the appendix, complete the following chart of thermodynamic properties for water. Use computer-based programs or apps to check your results.

T (°C)	*P* (kPa)	*v* (m³/kg)	*u* (kJ/kg)	Phase (Quality If Applicable)
385	10,000			
150				Saturated ($x = 0.82$)
	585	0.250		
	1500			Saturated vapor ($x = 1.0$)

3.78 Find an Internet-based computer program or app that will find the thermodynamic properties of the refrigerant R-134a. Using the program or app, complete the following table of properties for R-134a.

T (°C)	*P* (kPa)	*v* (m³/kg)	*h* (kJ/kg)	Phase (Quality If Applicable)
−20				Saturated liquid ($x = 0.0$)
	400			Saturated vapor ($x = 1.0$)
35		0.015		
20	190			
	700	0.0347		

3.79 Find an Internet-based computer program or app that will find the thermodynamic properties of the refrigerant R-134a. Using the program or app, complete the following table of properties for R-134a.

T (°C)	P (kPa)	v (m³/kg)	h (kJ/kg)	Phase (Quality If Applicable)
−20				Saturated liquid ($x = 0.0$)
	860			Saturated ($x = 0.75$)
	515	0.0094		
−7	175			
44		0.042		

3.80 Find an Internet-based computer program or app that will find the thermodynamic properties of ammonia (which is used as a refrigerant). Using the program or app, complete the following table of properties for ammonia.

T (°C)	P (kPa)	v (m³/kg)	h (kJ/kg)	Phase (Quality If Applicable)
−25				Saturated ($x = 0.20$)
	310			Saturated vapor ($x = 1.0$)
5	120			
20		0.05		
	620	0.250		

3.81 Find an Internet-based computer program or app that will find the thermodynamic properties of ammonia (which is used as a refrigerant). Using the program or app, complete the following table of properties for ammonia.

T (°C)	P (kPa)	v (m³/kg)	h (kJ/kg)	Phase (Quality If Applicable)
	500			Saturated ($x = 0.60$)
0				Saturated vapor ($x = 1.0$)
15	300			
	200	0.20		
−20		0.95		

CHAPTER 4

The First Law of Thermodynamics

Learning Objectives

Upon completion of Chapter 4, you will be able to

4.1 Apply the conservation of mass to thermodynamic systems;

4.2 Identify the concepts that result in the formation of the First Law of Thermodynamics;

4.3 Solve applications of the first law to many types of open systems;

4.4 Use the first law to compute energy balances in common types of closed systems;

4.5 Explain the thermal efficiency of heat engines and the coefficient of performance of refrigerators and heat pumps.

4.1 INTRODUCTION

Where's the energy? That simple question gets to the heart of thermodynamics for most engineers. An engineer might ask that question when they are working with an engine. Fuel, which contains chemical energy waiting to be released, is delivered to the engine. What happens to that energy in the engine? Consider other applications. Electricity flows into an air compressor. Where does the energy in the electricity go? What happens to the energy in a high-speed jet of steam entering a turbine? To follow the energy, we will need to perform energy balances. From those balances, we will learn what happens to energy, and what we might need to use to help move energy.

Energy balances can be performed using the First Law of Thermodynamics, which is fundamental to everyday engineering applications. The First Law of Thermodynamics, also known as the Conservation of Energy, can be phrased simply as follows:

Energy is neither created nor destroyed.

This brief statement leads to a powerful tool used by engineers to analyze the overall energy flows in a system or process. For example, the first law allows engineers to determine heat transfers or work interactions in a process and to calculate expected temperatures in systems.

Although energy is neither created nor destroyed, it can change form inside a system (as we discussed in Chapter 2). Energy can change between internal energy, kinetic energy, and potential energy. Similarly, energy can also be transferred into and out of a system by heat transfer, work, and mass flow. As energy is transferred across the system boundary, it increases or decreases the total energy in the system. The first law can be written to reflect this as

Net Energy Transfer In (or Out) of a System = Energy Change in the System

The first law will be derived by using the concepts developed in Chapter 2. However, because mass transfer can be an integral part of a first law analysis, we first consider the behavior of mass in a system.

4.2 CONSERVATION OF MASS

Just as energy is conserved, a fundamental classical physics concept is the law of the conservation of mass. This law states that *mass is neither created nor destroyed.* (At this point, it should be noted that with Einstein's discovery of the equivalency of mass and energy through $E = mc^2$, where c is the speed of light, the First Law of Thermodynamics and the conservation of mass are more correctly termed as one law of the conservation of energy and mass. However, for non-nuclear processes, it is appropriate in engineering to consider the conservation of mass and the conservation of energy separately.) Like energy, mass can flow into or out of an open system, but unlike energy we would not consider mass to be changing forms. So the conservation of mass can be written as

Net Mass Transfer In (or Out) of a System = Mass Change in the System

Suppose you have a pitcher that contains 2 kg of water, as shown in **Figure 4.1**. You pour 0.5 kg of water into the pitcher, which results in there being 2.5 kg of water in the pitcher. If you then pour 1 kg of water out of the pitcher, the pitcher would be left with 1.5 kg of water. The amount of mass change in the system in each case is equal to the amount added or subtracted.

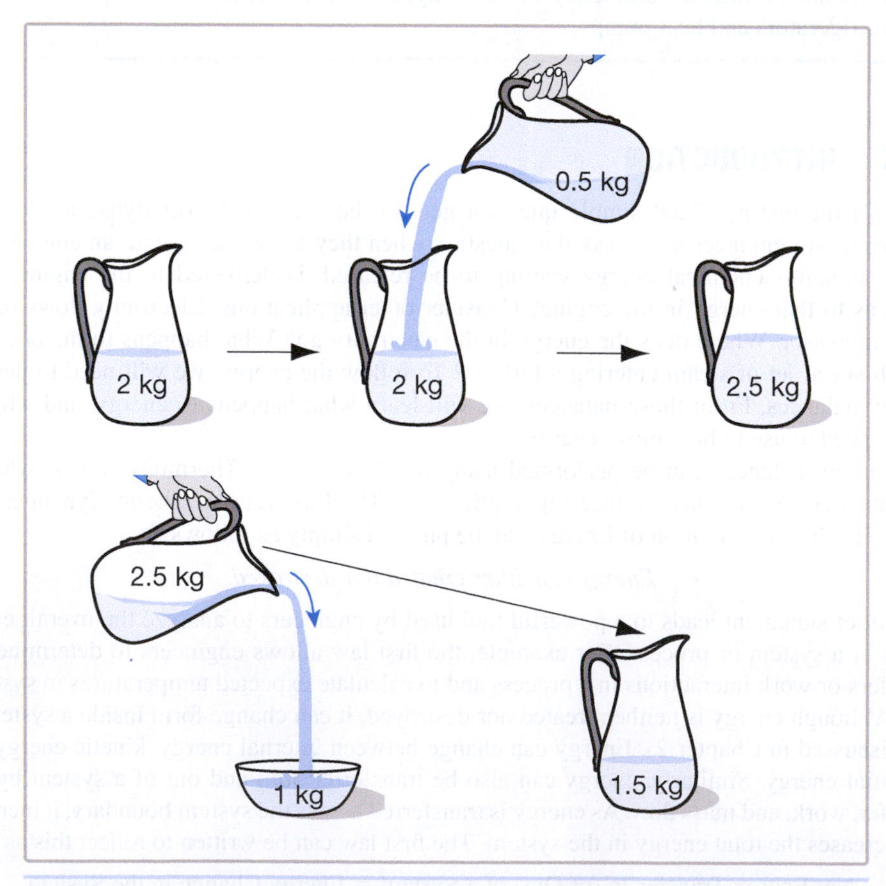

FIGURE 4.1 As mass enters or exits a system (in this case, water being added to or removed from a pitcher), the total mass in the system changes.

We can formally write the conservation mass on a rate basis for a system with multiple inlets and multiple outlets as

$$\sum_{i=1}^{j} \dot{m}_i - \sum_{e=1}^{k} \dot{m}_e = \left(\frac{dm}{dt}\right)_{\text{system}} \tag{4.1}$$

where j is the number of inlets, k is the number of outlets, $\dot{m}_i$ is the mass flow rate through inlet i, and $\dot{m}_e$ is the mass flow rate through outlet e. Equation (4.1) states that the rate of change of the mass of a system is equal to the net rate of mass flow into/out of the system.

A special type of process is a *steady-state, steady-flow* process. In such a process the system is at steady state, meaning that the properties of the system are constant in time. The flow into or out of the system during the process is also constant in time, indicated by the "steady-flow" designation. For example, if a system is undergoing a steady-state, steady-flow process, and the temperature in the system at a point in time is measured to be 40°C, the temperature would still be at 40°C in 1 minute, or 10 minutes, or an hour—however long the process lasts. Similarly, if the process involves an inlet flow of 0.1 kg/s, that flow rate would be the same as long as the process lasts. We will see later in this chapter that many processes can be adequately modeled as steady-state, steady-flow processes.

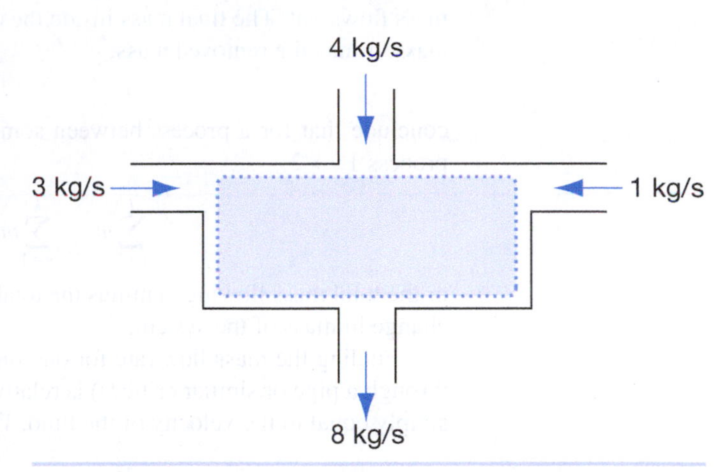

FIGURE 4.2 Open systems can have multiple inlet and outlet streams. For a steady-state process, the net mass flow in equals the net mass flow out.

For a steady-state, steady-flow process, the mass of the system, like any property of the system, will be constant in time. Therefore, for a steady-state, steady-flow process such as in **Figure 4.2**, Eq. (4.1) reduces to

$$\sum_{i=1}^{j} \dot{m}_i - \sum_{e=1}^{k} \dot{m}_e = 0 \tag{4.2}$$

Furthermore, many open systems have only a single inlet and a single outlet, such as in **Figure 4.3**. For a steady-state, steady-flow process in a system with one inlet and one outlet, the conservation of mass reduces to

$$\dot{m}_i - \dot{m}_e = 0 \tag{4.3}$$

From Eq. (4.3), it is clear that the mass flow rate entering the system is equal to the mass flow rate exiting the system, and so we will often drop the subscripts and refer to the mass flow rate simply as $\dot{m}$. Therefore for a steady-state, steady-flow, single-inlet, single-outlet system, we can state that

$$\dot{m}_i = \dot{m}_e = \dot{m} \tag{4.4}$$

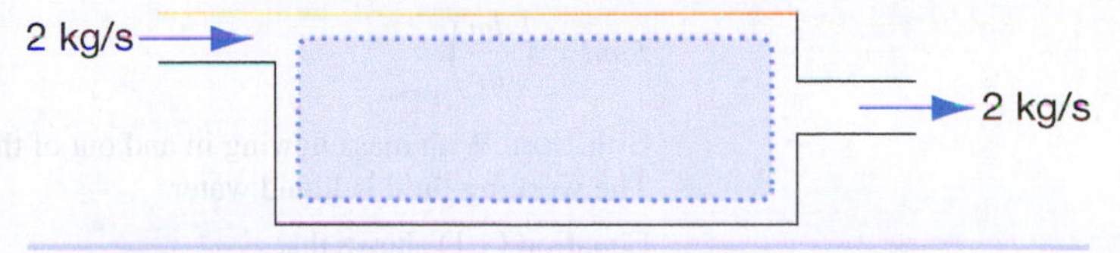

FIGURE 4.3 If an open system has a single inlet and a single outlet, the mass flow rates of the streams will be identical for a steady-state process.

Sometimes, particularly for unsteady processes such as in **Figure 4.4**, we are more concerned with the total flow into or out of a system rather than the mass flow rates. In such a case, we can integrate Eq. (4.1) with respect to time and

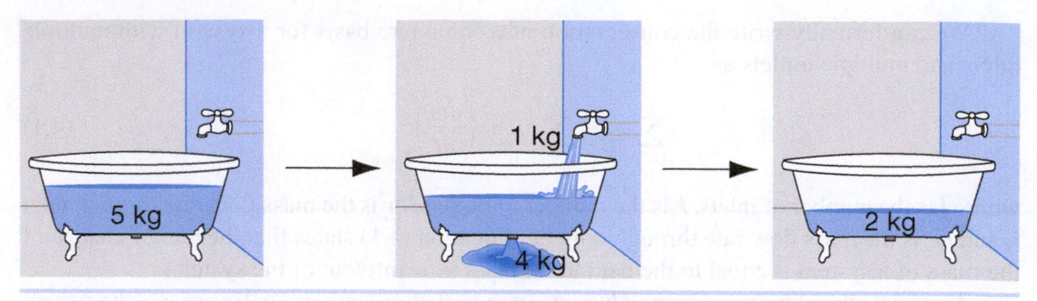

FIGURE 4.4 A container holding a mass of a liquid has some mass added, whereas other mass flows out. The final mass inside the container equals the original mass plus the added mass, minus the removed mass.

conclude that for a process between some initial state 1 and a final state 2 (also written as process 1 → 2),

$$\sum_{i=1}^{j} m_i - \sum_{e=1}^{k} m_e = m_2 - m_1 = \Delta m_{\text{system}} \tag{4.5}$$

or the total mass flowing in minus the total mass flowing out during the process is equal to the change in mass of the system.

Finding the mass flow rate for one-dimensional flow (a common approximation for flow through a pipe or similar orifices) is relatively straightforward. The volumetric flow rate, $\dot{V}$, is simply equal to the velocity of the fluid, V, multiplied by the cross-sectional area, A:

$$\dot{V} = VA \tag{4.6}$$

The mass flow rate of the fluid is then the density, ρ, multiplied by the volumetric flow rate, or, equivalently, the volumetric flow rate divided by the specific volume, v:

$$\dot{m} = \rho\dot{V} = \frac{\dot{V}}{v} = \rho VA = \frac{VA}{v} \tag{4.7}$$

Note that for steady-state, steady-flow problems, the mass flow rate is constant, but the volumetric flow rate may change as the fluid density changes.

▶ EXAMPLE 4.1

Consider the device shown. The device contains three inlets and two outlets. The mass flow rates of water through the three inlets at one instant in time are 2 kg/s, 3 kg/s, and 5 kg/s. The mass flow rates of water through the two outlets at that instant are 7 kg/s and 6 kg/s. Determine the time rate of change of the mass of water in the system at this instant in time.

Given: Inlets: $\dot{m}_1 = 2$ kg/s, $\dot{m}_2 = 3$ kg/s, $\dot{m}_3 = 5$ kg/s

Outlets: $\dot{m}_4 = 7$ kg/s, $\dot{m}_5 = 6$ kg/s

Find: $\left(\frac{dm}{dt}\right)_{\text{system}}$

Solution: With mass flowing in and out of the device, this should be represented by an open system. The working fluid is liquid water.

Equation (4.1) shows that

$$\sum_{i=1}^{j} \dot{m}_i - \sum_{e=1}^{k} \dot{m}_e = \left(\frac{dm}{dt}\right)_{\text{system}}$$

For this problem, $j = 3$ and $k = 2$. There are no simplifying assumptions. When expanded out, Equation (4.1) becomes

$$\left(\frac{dm}{dt}\right)_{\text{system}} = (\dot{m}_1 + \dot{m}_2 + \dot{m}_3) - (\dot{m}_4 + \dot{m}_5)$$
$$= (2\text{ kg/s} + 3\text{ kg/s} + 5\text{ kg/s}) - (7\text{ kg/s} + 6\text{ kg/s})$$
$$= \mathbf{-3\ kg/s}$$

Analysis: Therefore, the device is losing 3 kg/s of water. This is not a steady-state system, and this rate of water loss is not sustainable over time.

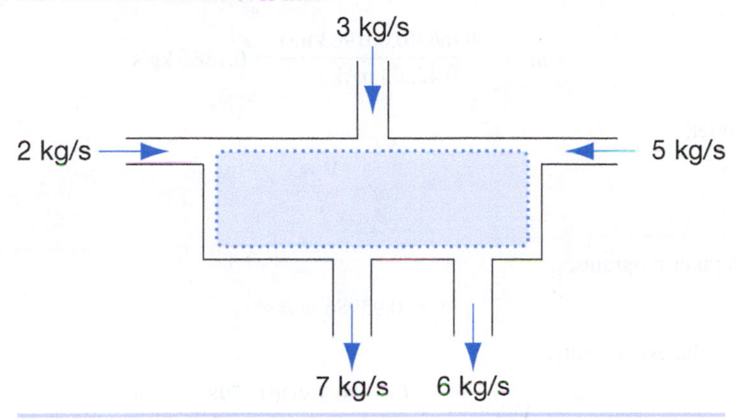

The container and problem described in Example 4.1.

▶ EXAMPLE 4.2

A steady-state, steady-flow device receives steam at a pressure of 500 kPa, a temperature of 200°C, and a velocity of 30 m/s, through a circular inlet with a diameter of 0.05 m. Steam exits the device at 200 kPa, 150°C, through an outlet with a cross-sectional area of 0.015 m^2. Determine the exit velocity of the steam.

Given: $P_1 = 500$ kPa, $T_1 = 200$°C, $V_1 = 30$ m/s, $d_1 = 0.05$ m

$P_2 = 200$ kPa, $T_2 = 150$°C, $A_2 = 0.015$ m^2

Find: V_2

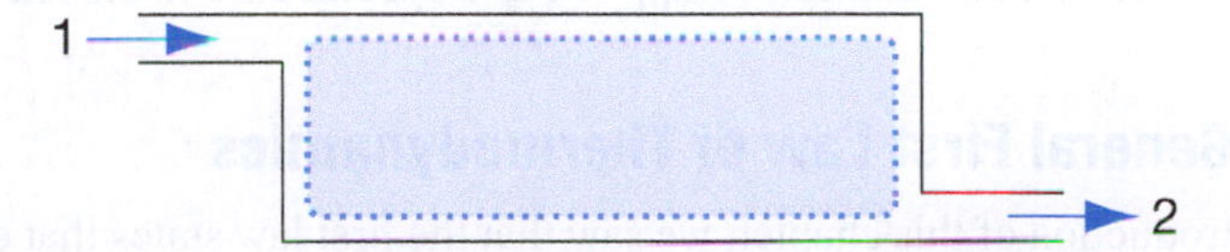

Solution: This is an open system, and the working fluid is steam (water vapor). Thermodynamic properties for the steam can be found from computer programs or superheated steam tables.

Because this is a steady-state, steady-flow device, the mass flow rate entering the system will be identical to the mass flow rate exiting the system. From the conservation of mass, with Eq. (4.1) being set equal to 0 for steady-state:

$$\sum_{i=1}^{j} \dot{m}_i - \sum_{e=1}^{k} \dot{m}_e = \left(\frac{dm}{dt}\right)_{\text{system}}$$

With one inlet and one outlet, this reduces to

$$\dot{m}_1 = \dot{m}_2 = \dot{m}$$

The mass flow rate can be found from the information given about the inlet, using Eq. (4.7):

$$\dot{m} = \frac{V_1 A_1}{v_1}$$

For a circular inlet,

$$A_1 = \frac{\pi d_1^2}{4} = 0.001963 \text{ m}^2$$

The specific volume of steam can be found from computer programs: $v_1 = 0.42503$ m³/kg,

so

$$\dot{m} = \frac{(30 \text{ m/s})(0.001963 \text{ m}^2)}{0.42503 \text{ m}^3\text{/kg}} = 0.1386 \text{ kg/s}$$

For the outlet,

$$\dot{m} = \frac{V_2 A_2}{v_2}$$

From computer programs,

$$v_2 = 0.95986 \text{ m}^3\text{/kg}$$

Solving for the exit velocity,

$$V_2 = \frac{\dot{m} v_2}{A_2} = \frac{(0.1386 \text{ kg/s})(0.95986 \text{ m}^3\text{/kg})}{0.015 \text{ m}^2}$$

$$\mathbf{V_2 = 8.87 \text{ m/s}}$$

Analysis: This is certainly a reasonable exit velocity, considering the inlet velocity and the fact that the exit area is larger than the inlet area. In addition, the specific volume at the exit is roughly twice as large as at the inlet, which will impact the velocity but not dramatically alter the value.

4.3 FIRST LAW OF THERMODYNAMICS IN OPEN SYSTEMS

We are now ready to develop a general equation for the first law, and then apply the general equation to open systems. After covering numerous common open systems, we will develop a simplified form of the first law that can be applied as a special case to closed systems.

4.3.1 The General First Law of Thermodynamics

Recalling the introduction of this chapter, we saw that the first law states that energy is neither created nor destroyed. This can be further interpreted as saying that the net energy transfer into/out of a system is equal to the change in energy of a system. This statement can be put into symbolic form, and we will first develop this on a rate basis:

$$\dot{E}_{\text{transfer}} = \left(\frac{dE}{dt}\right)_{\text{system}} \tag{4.8}$$

where we will consider the rate of energy transfer into the system to be positive. Figure 4.5 is a generic open system that can be used to develop this equation further. Energy is transferred through three mechanisms, and so on a rate basis

$$\dot{E}_{\text{transfer}} = \dot{Q} - \dot{W} + \dot{E}_{\text{mass flow}} \tag{4.9}$$

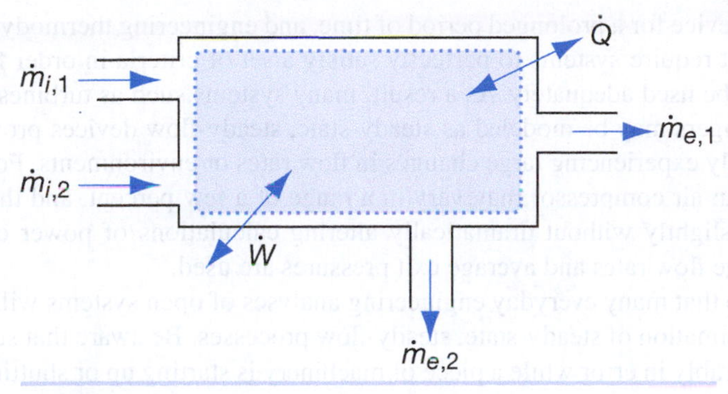

FIGURE 4.5 A representation of a generic open system, showing multiple inlet and outlet mass streams and the possibility of heat transfer and work interactions.

In Eq. (4.9), $\dot{Q}$ is the net rate of heat transfer into the system—this net rate can incorporate multiple sources or sinks. Similarly, the net power, $\dot{W}$, can incorporate multiple work transfer interactions. Considering Eq. (2.33) for the rate of energy transfer via mass flow, Eq. (4.9) can be rewritten as

$$\dot{E}_{\text{transfer}} = \dot{Q} - \dot{W} + \sum_{i=1}^{j} \dot{m}_i \left(h + \frac{V^2}{2} + gz \right)_i - \sum_{e=1}^{k} \dot{m}_e \left(h + \frac{V^2}{2} + gz \right)_e \tag{4.10}$$

The time rate of change of the energy of the system can be set equal to

$$\left(\frac{dE}{dt} \right)_{\text{system}} = \frac{d\left(mu + m\frac{V^2}{2} + mgz \right)_{\text{system}}}{dt} \tag{4.11}$$

Note that in Eq. (4.11), there can be a change in energy of the system if (a) the mass of the system is changing; (b) the internal energy, velocity, or height of the system is changing; or (c) both the mass and the specific energy of the system are changing. Substituting Eqs. (4.10) and (4.11) into Eq. (4.8) leads to a general statement of the first law on a rate basis:

$$\dot{Q} - \dot{W} + \sum_{i=1}^{j} \dot{m}_i \left(h + \frac{V^2}{2} + gz \right)_i - \sum_{e=1}^{k} \dot{m}_e \left(h + \frac{V^2}{2} + gz \right)_e = \frac{d\left(mu + m\frac{V^2}{2} + mgz \right)_{\text{system}}}{dt} \tag{4.12}$$

Remember that the heat transfer rate and power terms in Eq. (4.12) can incorporate multiple contributions and need not be a single heat transfer or power interaction. Equation (4.12) may appear to be a very complicated equation that would be difficult to employ. However, most engineers will rarely use the first law in this complete form; rather, appropriate simplifications will be made to Eq. (4.12) to produce forms of the first law that are adequate for most engineering analyses. Much of the remainder of this chapter is devoted to learning how to make those simplifications and applying the resulting equations to common engineering environments.

4.3.2 Application of the First Law to Steady-State, Steady-Flow, Single-Inlet, Single-Outlet Open Systems

As discussed, a steady-state, steady-flow process is one where the properties of the thermodynamic system are unchanging in time, and the mass flows into and out of the system are also constant. Technically it may be very difficult to achieve pure steady-state, steady-flow operation of a device for a prolonged period of time, and engineering thermodynamic analysis usually does not require systems to perfectly satisfy a set of criteria in order for simplifying assumptions to be used adequately. As a result, many systems such as turbines, compressors, and heat exchangers may be modeled as steady-state, steady-flow devices provided that they are not frequently experiencing large changes in flow rates or environments. For example, the inlet flow into an air compressor may vary in a range of a few percent, and the exit pressure may also vary slightly without dramatically altering calculations of power consumption—provided average flow rates and average exit pressures are used.

This means that many everyday engineering analyses of open systems will be performed using an approximation of steady-state, steady-flow processes. Be aware that such an analysis may be considerably in error while a piece of machinery is starting up or shutting down; however, once a condition close to steady state is achieved, this analysis will usually be adequate. The assumption of steady-state, steady-flow conditions means that the time rate of change of the energy of the system is 0; so the right-hand side of Eq. (4.12) is equal to 0. The resulting equation for the first law on a rate basis for steady-state, steady-flow processes is

$$\dot{Q} - \dot{W} + \sum_{i=1}^{j} \dot{m}_i \left(h + \frac{V^2}{2} + gz \right)_i - \sum_{e=1}^{k} \dot{m}_e \left(h + \frac{V^2}{2} + gz \right)_e = 0 \tag{4.13}$$

Furthermore, many processes involve devices with a single inlet and a single outlet. Equation (4.13) can then be reduced even further by invoking Eq. (4.4):

$$\dot{Q} - \dot{W} + \dot{m}\left(h_{in} - h_{out} + \frac{V_{in}^2 - V_{out}^2}{2} + g(z_{in} - z_{out}) \right) = 0 \tag{4.14}$$

This expression for the first law on a rate basis for a steady-state, steady-flow, single-inlet, single-outlet process can be rewritten into a form somewhat easier to use in calculations. For this form, we will call the outlet state 2 and the inlet state 1 as shown in **Figure 4.6**, and we will move the energy transfer via mass flow component of the expression onto the right-hand side of the equation:

$$\dot{Q} - \dot{W} = \dot{m}\left(h_2 - h_1 + \frac{V_2^2 - V_1^2}{2} + g(z_2 - z_1) \right) \tag{4.15a}$$

or

$$\dot{Q} - \dot{W} = \dot{m}(h_2 - h_1 + \Delta ke + \Delta pe) \tag{4.15b}$$

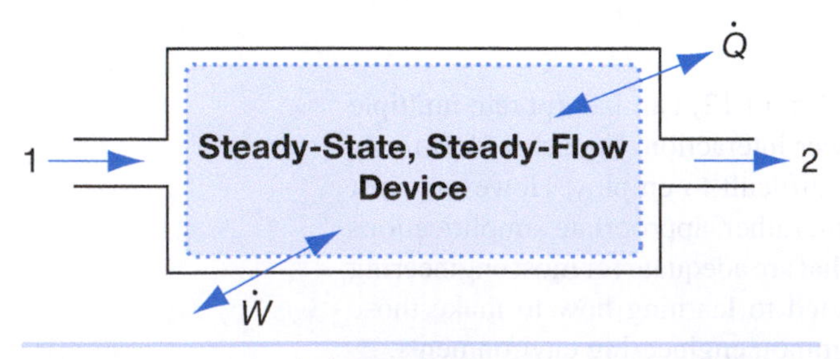

FIGURE 4.6 A single-inlet, single-outlet, steady-state, steady-flow open system.

The form presented in Eq. (4.15b) is a useful shorthand notation when deriving appropriate expressions for a particular situation, because, as discussed in Chapter 2, the changes in kinetic and potential energy are often overwhelmed by the change in internal energy (as included in the enthalpy term).

Careful inspection of industrial facilities, power plants, automobiles, or other pieces of machinery will show that there are many devices in everyday use that can be modeled as steady-state, steady-flow systems with one inlet and one outlet for fluids passing through the system. For example, it is easy to find sections of hose or piping in use with fluids steadily flowing through. Or in a manufacturing plant, we might find an air compressor running steadily to keep pneumatic manufacturing equipment operating smoothly. In an automobile, a fuel pump may be delivering the fuel nearly constantly to the engine as the vehicle drives down the highway. In many cases, the device may not be operating in a strictly steady sense, but as discussed the variations are often small and average out over the operating time of the equipment, which means that they can be safely ignored in most instances.

Some of the more common devices that are seen, and that commonly need thermodynamic analysis, are nozzles, diffusers, turbines, compressors, pumps, and throttling devices. In the following sections, we will consider these devices and demonstrate how to simplify the first law with common assumptions for the devices. Keep in mind, however, that any particular common assumption may not always hold true for that device and so you should concentrate on learning the process of simplifying the first law rather than relying on the derived equations to work in any situation. For example, if you insulate a large turbine, the resulting heat loss will usually be negligible in comparison to the power produced—therefore, turbines are commonly assumed to be adiabatic. However, such an assumption may cause significant errors in predictions of the operation of a small, uninsulated turbine using the first law. So it is important to think through the particular device you are analyzing before automatically making an assumption to simplify the first law for that device.

Nozzles and Diffusers

Nozzles and diffusers are often treated together because the analyses of the devices are very similar even though the devices have opposite purposes. A *nozzle* is a device that accelerates a moving fluid (while simultaneously decreasing its pressure), whereas a *diffuser* is a device that decelerates a moving fluid (while simultaneously increasing its pressure). These devices are shown schematically in **Figure 4.7**. At least one end of these devices will typically be connected to additional piping, and the other end may be attached to piping or may be exposed to the environment. You will find nozzles in many places, ranging from the entrance to a turbine to the end of a garden hose. Diffusers are not always as obvious, but may be found on the exhaust of a device with high fluid speeds so that the outside environment is not disturbed by the high-speed flow. Both nozzles and diffusers are valuable tools for changing the velocity of a flow stream at relatively little cost.

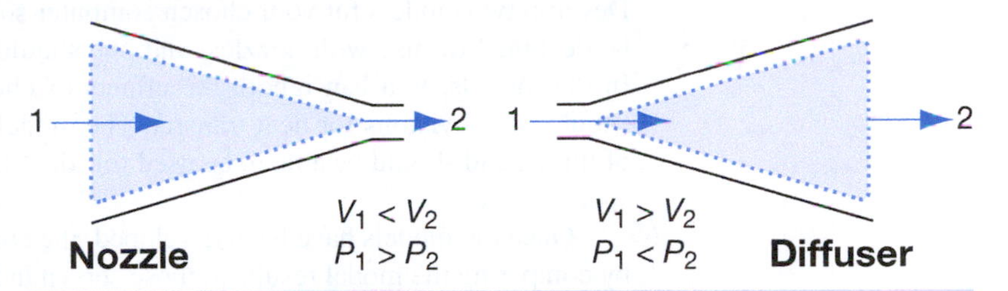

FIGURE 4.7 On the left is a nozzle, with an indication of the velocity of the flow increasing, and the fluid pressure decreasing, between the inlet and the outlet. On the right is a diffuser, with an indication of the velocity decreasing, and the pressure increasing, between the inlet and the outlet.

These devices are often considered to be operating at steady-state, steady-flow conditions, and will have one inlet and one outlet. Therefore, the appropriate form of the first law to use is Eq. (4.15a):

$$\dot{Q} - \dot{W} = \dot{m}\left(h_2 - h_1 + \frac{V_2^2 - V_1^2}{2} + g(z_2 - z_1)\right) \tag{4.15a}$$

For many systems, as mentioned, we will ignore the changes in potential and kinetic energy; however, because the primary purpose of a nozzle or diffuser is to change the velocity of the fluid, we will not neglect the changes in kinetic energy. The changes in potential energy can usually be assumed to be 0, because nozzles and diffusers tend to not significantly impact the height of the fluid. Nozzles and diffusers typically experience no work interactions. There may be an electrical resistance heater on some nozzles to increase the temperature of the fluid passing through (the fluid temperature decreases as the velocity increases), but such an input can be modeled as a heat input as well as a work interaction. So, assuming no changes in potential energy and no work interactions, the first law reduces to

$$\dot{Q} = \dot{m}\left(h_2 - h_1 + \frac{V_2^2 - V_1^2}{2}\right) \tag{4.16}$$

Equation (4.16) can often be simplified further by assuming the system to be adiabatic. Such a condition is a good assumption if the system is insulated, if the fluid is moving very rapidly through the nozzle or diffuser, or if the temperature of the fluid inside the nozzle or diffuser is close to the environmental temperature. With the further assumption of no heat transfer, the first law reduces to

$$h_2 - h_1 + \frac{V_2^2 - V_1^2}{2} = 0 \tag{4.17}$$

Note in Eq. (4.17) that the system becomes independent of the mass flow rate if the heat transfer rate and power are both equal to 0. At times, the inlet velocity to the nozzle or the outlet velocity from the diffuser may be unknown or small in comparison to the other velocity. If it is unknown or much smaller, that velocity term can be set to 0 in Eq. (4.16) or (4.17).

STUDENT EXERCISE

Develop two models for your chosen computer solution platform. One of these should be designed for use with nozzles, and one should be designed for use with diffusers. In the models, you can ignore the effects of changes in potential energy, but there should be provisions for heat transfer. The models should be usable for a wide range of fluids, and should be able to be used for ideal gases with both constant and variable specific heats.

Once the models have been developed, the correctness of the models can be tested by comparing the model results to those shown in Examples 4.3 and 4.4.

▶ EXAMPLE 4.3

An insulated diffuser receives CO_2 gas with a velocity of 150 m/s, at 20°C and 80 kPa. The diffuser inlet has a cross-sectional area of 5 cm². The CO_2 exits at a velocity of 10 m/s, through a cross-sectional area of 20 cm². Assuming constant specific heats for the CO_2, determine (a) the exit temperature of the CO_2 and (b) the exit pressure of the CO_2.

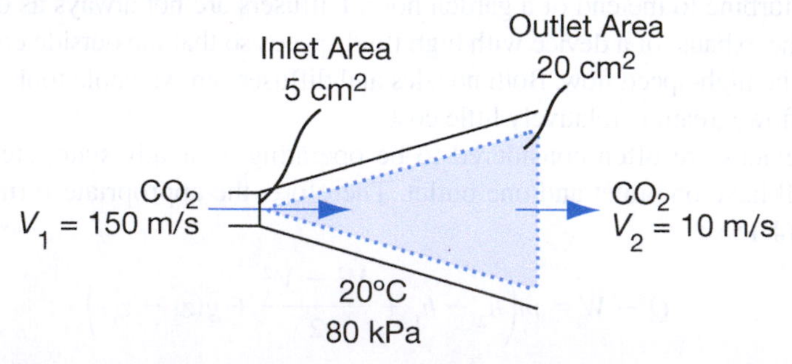

Given: $V_1 = 150$ m/s, $T_1 = 20°C = 293$ K, $P_1 = 80$ kPa, $A_1 = 5$ cm² $= 0.0005$ m²

$V_2 = 10$ m/s, $A_2 = 20$ cm² $= 0.002$ m²

Find: (a) T_2, (b) P_2

Solution: As a diffuser with mass flowing in and out, the system is an open system. The working fluid is CO_2, which will be treated as an ideal gas. Furthermore, the problem specifies assuming constant specific heats.

Assume: $\dot{Q} = \dot{W} = \Delta PE = 0$
Constant specific heats
Steady state, steady flow, single inlet, single outlet
CO_2 behaves as an ideal gas

Necessary equations of state for an ideal gas with constant specific heats:

$$h_2 - h_1 = c_p(T_2 - T_1)$$

$$\rho = \frac{P}{RT} \quad \text{(ideal gas law)}$$

(a) From Eq. (4.15a), the first law for this diffuser is

$$\dot{Q} - \dot{W} = \dot{m}\left(h_2 - h_1 + \frac{V_2^2 - V_1^2}{2} + g(z_2 - z_1)\right)$$

Applying the assumptions, including $h_2 - h_1 = c_p(T_2 - T_1)$ (for an ideal gas):

$$c_p(T_2 - T_1) + \frac{V_2^2 - V_1^2}{2} = 0$$

For CO_2, $c_p = 0.842$ kJ/kg · K.

$$T_2 = \frac{V_1^2 - V_2^2}{2c_p} + T_1 = \frac{(150 \text{ m/s})^2 - (10 \text{ m/s})^2}{2(0.842 \text{ kJ/kg} \cdot \text{K})(1000 \text{ J/kJ})} + 20°\text{C}$$

$T_2 =$ **33.3°C** $= 306.3$ K

(b) From Eq. (4.7), with a constant mass flow rate:

$$\dot{m} = \rho_1 V_1 A_1 = \rho_2 V_2 A_2$$

The density at the inlet state can be found from the ideal gas law, with $R = 0.1889$ kJ/kg · K for CO_2:

$$\rho_1 = \frac{P_1}{RT_1} = \frac{80 \text{ kPa}}{(0.1889 \text{ kJ/kg} \cdot \text{K})(293 \text{ K})} = 1.445 \frac{\text{kg}}{\text{m}^3}$$

Then

$$\rho_2 = \frac{\rho_1 V_1 A_1}{V_2 A_2} = \frac{(1.445 \text{ kg/m}^3)(150 \text{ m/s})(5 \text{ cm}^2)}{(10 \text{ m/s})(20 \text{ cm}^2)} = 5.419 \text{ kg/m}^3.$$

From the ideal gas law for the exit state:

$P_2 = \rho_2 R T_2 = (5.419 \text{ kg/m}^3)(0.1889 \text{ kJ/kg} \cdot \text{K})(306.3 \text{ K})$

$\mathbf{P_2 = 314\ kPa}$

Analysis: With no heat transfer, we would expect the temperature of the CO_2 to increase by a small amount as it passes through a diffuser, with kinetic energy being converted into thermal energy. In addition, as the velocity decreases, the pressure should increase. These two expectations are met by the solution.

Once you have created your model for a diffuser, a more detailed analysis of a diffuser is possible. As an example of possible results, consider what happens to the outlet temperature as the exit velocity for this diffuser is varied between 1.0 m/s and 100 m/s. The following graph illustrates the results:

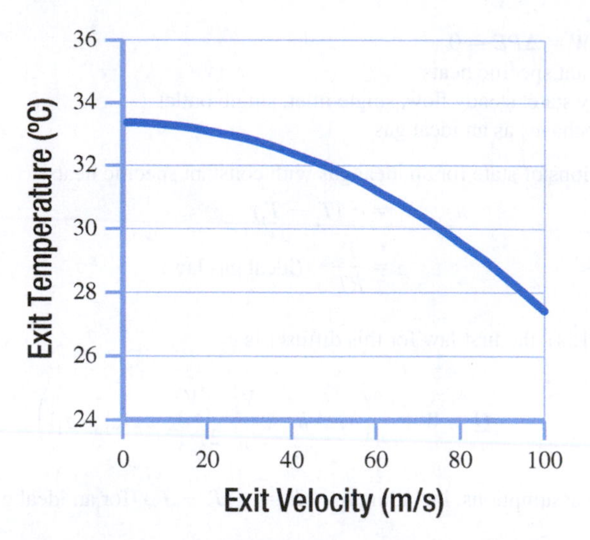

As you can see, the exit temperature becomes hotter the slower the exit velocity becomes, as more kinetic energy is transformed into internal energy.

▶ EXAMPLE 4.4

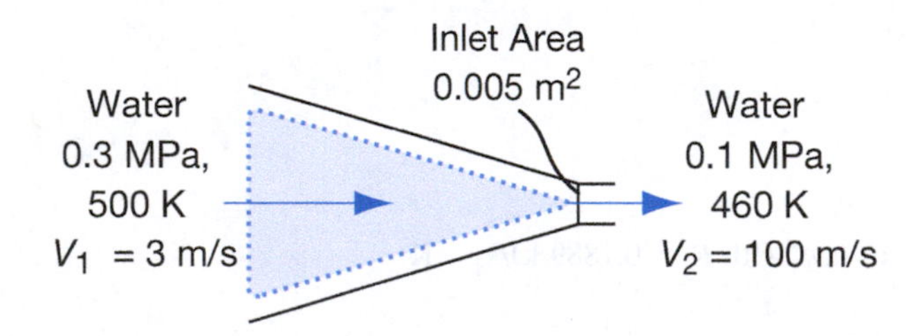

Water vapor is to pass through a nozzle. The water enters the nozzle at 0.3 MPa, 500 K, at a velocity of 3 m/s, and exits the nozzle at 0.1 MPa, 460 K, and 100 m/s. The cross-sectional area of the inlet is 0.005 m². Determine (a) the mass flow rate of the water vapor, and (b) the rate of heat transfer that occurs in the nozzle.

Given: $P_1 = 0.3$ MPa, $T_1 = 500$ K, $V_1 = 3$ m/s, $A_1 = 0.005$ m²
$P_2 = 0.1$ MPa, $T_2 = 460$ K, $V_2 = 100$ m/s

Find: (a) $\dot{m}$, (b) $\dot{Q}$

Solution: As a nozzle with mass flowing in and out, the system is an open system. The working fluid is steam (water vapor), whose properties can be found from computer programs or superheated vapor tables for steam.

Assume: Steady state, steady flow, single inlet, single outlet

$$\dot{W} = \Delta PE = 0$$

From Eq. (4.15a), the first law for this nozzle is

$$\dot{Q} - \dot{W} = \dot{m}\left(h_2 - h_1 + \frac{V_2^2 - V_1^2}{2} + g(z_2 - z_1)\right)$$

Applying the assumptions yields

$$\dot{Q} = \dot{m}h\left(h_2 - h_1 + \frac{V_2^2 - V_1^2}{2}\right)$$

The mass flow rate can be found from Eq. (4.7) applied at the inlet: $\dot{m} = \dfrac{V_1 A_1}{v_1}$

From our computerized properties for steam, $v_1 = 0.75958$ m³/kg, and so

$$\dot{m} = \frac{(3 \text{ m/s})(0.005 \text{ m}^2)}{0.75958 \text{ m}^3/\text{kg}} = \mathbf{0.0197\ kg/s} \quad \textbf{(a)}$$

From the computerized properties for steam, h_1 = 2920.8 kJ/kg and h_2 = 2849.5 kJ/kg.

So the heat transfer rate is

$$\dot{Q} = (0.0197\ \text{kg/s})\left(2849.5\ \text{kJ/kg} - 2920.8\ \text{kJ/kg} + \frac{(100\ \text{m/s})^2 - (3\ \text{m/s})^2}{2(1000\ \text{J/kJ})}\right) = \mathbf{-1.31\ kW} \qquad \text{(b)}$$

Analysis: As the densities of gases (in this case water vapor) tend to be small, we would not expect to have a particularly large mass flow rate for a low-velocity flow passing through a moderately sized opening, and that is the case here. Furthermore, we do not expect large heat transfer rates for nozzles in general, even when the nozzle is uninsulated, as the flow has little time to reside in the nozzle and either lose or gain energy via heat transfer.

Once you have created a model for a nozzle, you can use the model to vary the properties in the nozzle and gain a better understanding of what occurs in the nozzle. For this problem, if you vary the outlet temperature between 400 K and 500 K, you get the following relationship for the heat transfer rate:

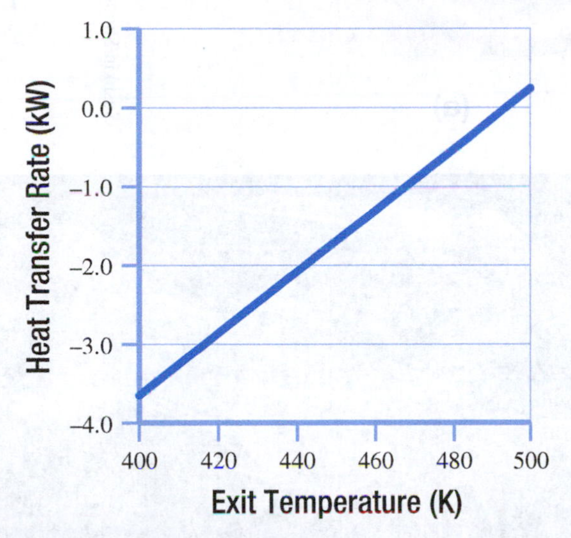

Not surprisingly, to get a cooler exit stream, more heat must be removed from the steam as it flows through the nozzle. Also, at exit temperatures above ~494 K, that heat must be added to the steam to get the desired exit temperature. The 494 K represents the exit temperature that would be achieved by an adiabatic nozzle.

QUESTION FOR THOUGHT/DISCUSSION

Where are nozzles used in common (or uncommon) applications?

Shaft Work Machines (Turbines, Compressors, and Pumps)

Some common devices have the primary purpose of using changes in the energy of a working fluid to either produce power or to use power to accomplish some change in the properties of the working fluid. These devices are, in general, referred to as shaft work machines, and such devices share a common first law analysis. A *turbine* is a shaft work machine, shown in **Figure 4.8**, that takes energy from a working fluid and produces power while expelling the working fluid possessing a reduced energy. Turbines are a fundamental machine used in producing electricity, either in steam power plants or gas power plants. Wind turbines, although in a different form than the more common steam or gas turbines, are becoming commonplace for generating electricity from wind. Turbines are also used as the primary mover on most aircraft engines today, and they provide the power necessary to spin the fans or propellers.

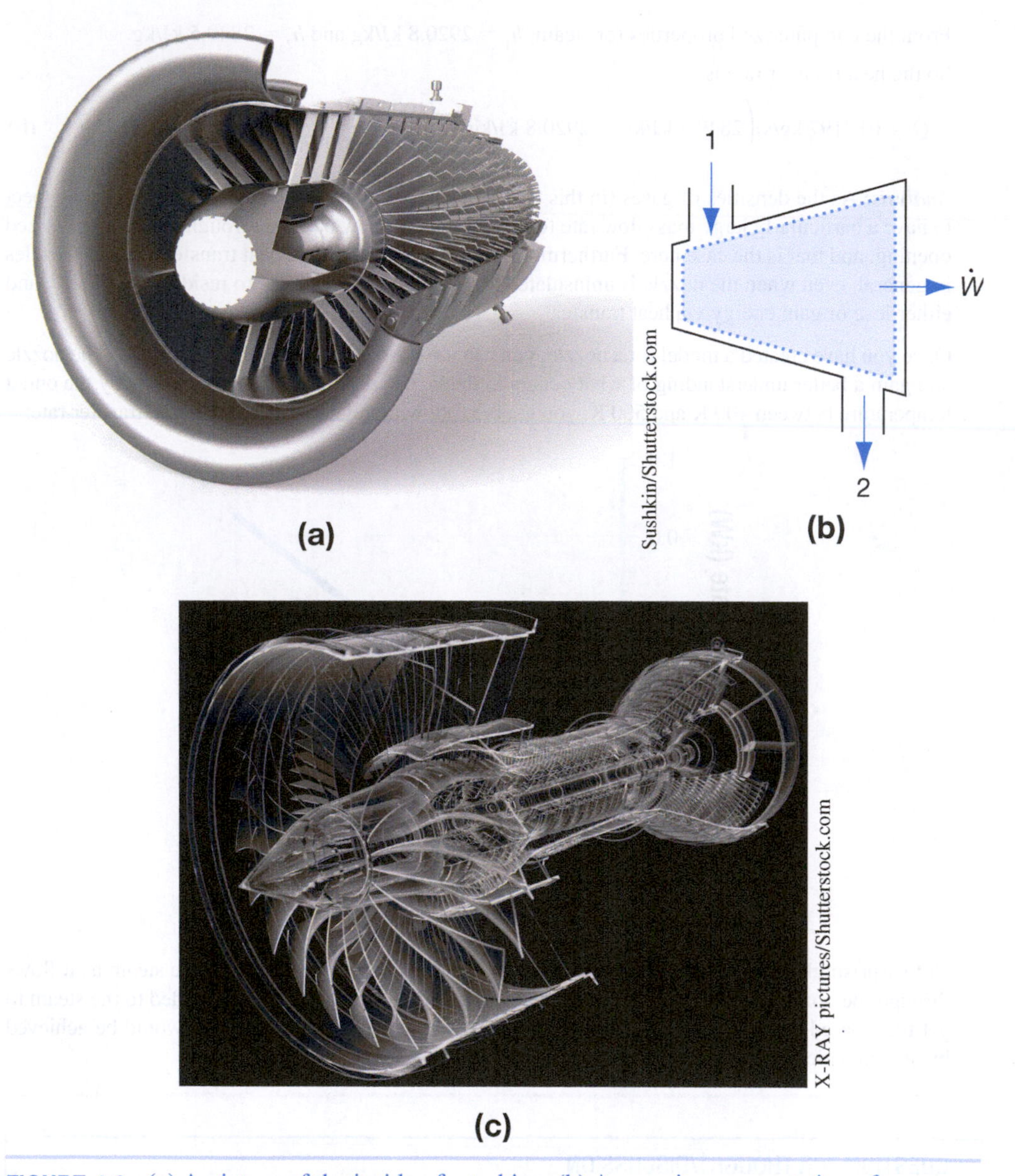

FIGURE 4.8 (a) A picture of the inside of a turbine. (b) A generic representation of a turbine. (c) A cut-away drawing of a turbine-powered turbofan engine.

Most turbines consist of one or more rows of blades attached to a central rotating shaft (the rotor). The working fluid flows into a row of blades, causing the rotor to rotate. There are two factors causing the rotation. The first factor is the force of the fluid impacting the blades and pushing the blades. The second factor is a pressure difference from one side of the blade to the other caused by the flow moving at different speeds over the two sides of the blades. This pressure difference forces the blades to move in the direction of the side with lower pressure. This is much like the flow over the wing of an airplane resulting in a lower pressure on the top of the wing, which then gives lift to the airplane. The fluid can then either be exhausted from the system or collected and directed toward another row of turbine blades. Most turbines contain the system in an outer shell, or stator, but some turbines, such as wind turbines, do not attempt to isolate the flow from the surroundings.

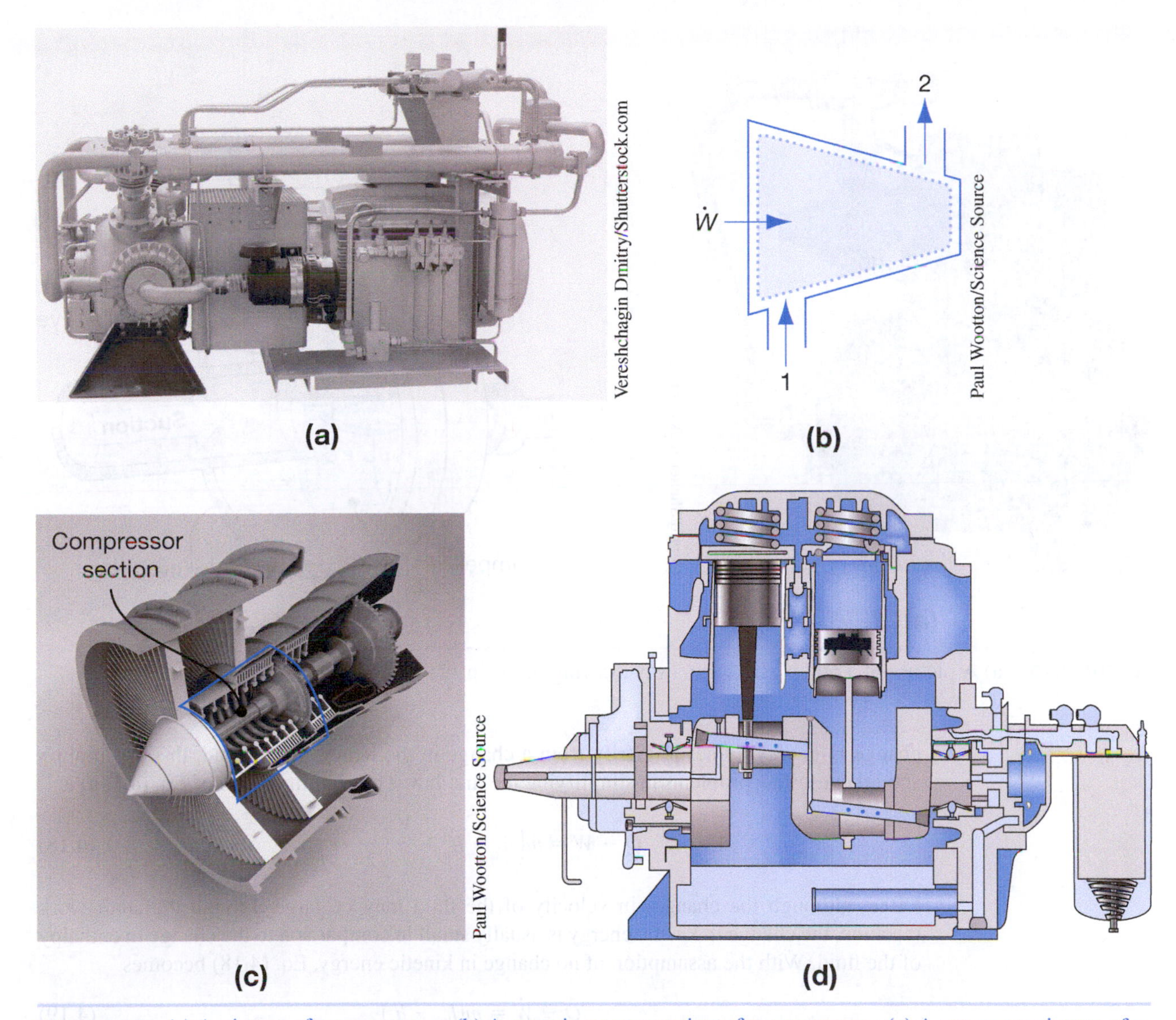

FIGURE 4.9 (a) A picture of a compressor. (b) A generic representation of a compressor. (c) A cut-away picture of a gas turbine engine containing an axial compressor. (d) A cut-away drawing of a reciprocating compressor.

A *compressor* is a shaft work machine whose purpose is to use power to increase the pressure of a gas or vapor flowing through it. As shown in **Figure 4.9**, compressors can either appear to be similar to a turbine, but with the blades positioned in the opposite direction (axial compressors or in some cases centrifugal compressors), or they can be reciprocating, with the air being compressed in a piston–cylinder device. Although axial and centrifugal compressors are more clearly modeled as a steady-state, steady-flow device, reciprocating compressors can also be adequately modeled in such a way if they operate at a high speed.

A pump is a shaft work machine whose purpose is to increase the pressure of a liquid flowing through it through the use of input power. These devices typically use a rotating impeller to increase the fluid pressure, as shown in **Figure 4.10**.

For turbines, compressors, and pumps, a typical assumption is to assume that the change in potential energy of the fluid is negligible. Except in the case of a hydroelectric turbine, whose

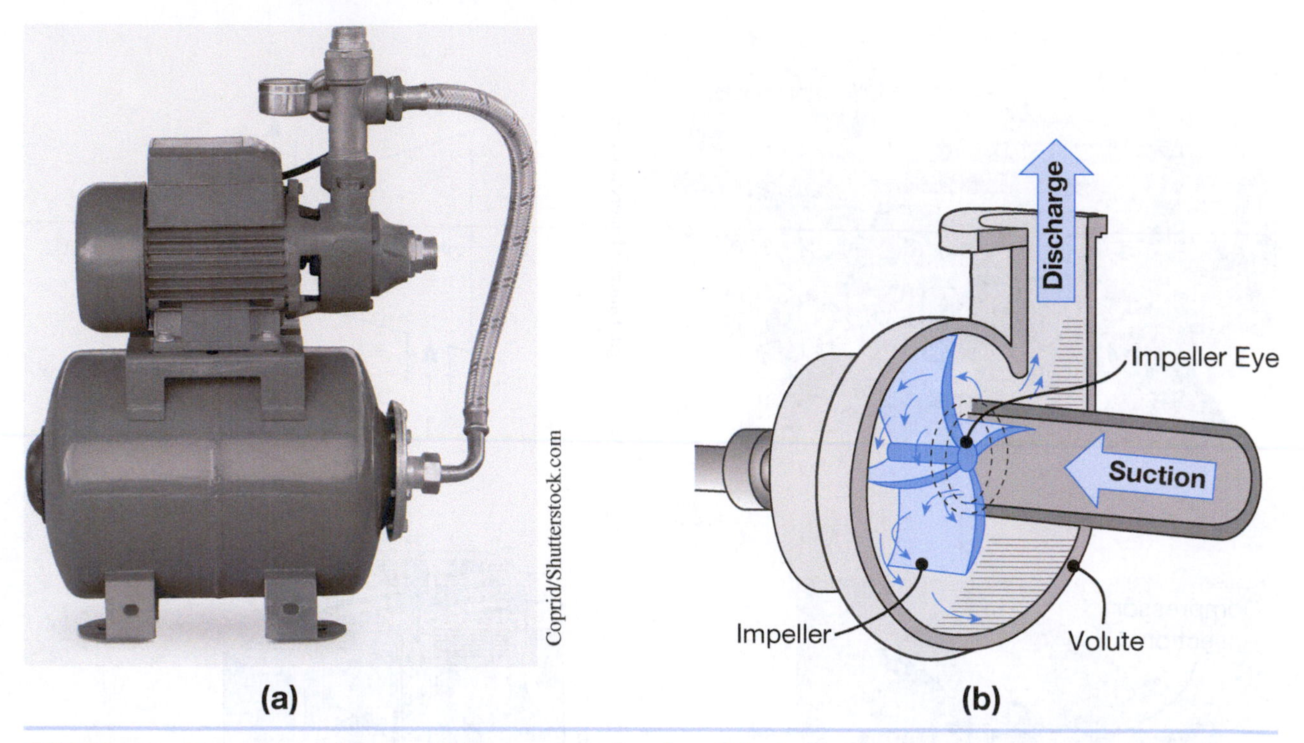

FIGURE 4.10 (a) A picture of a pump. (b) A cut-away drawing of the inside of a pump.

change in energy comes primarily from a change in the height of the fluid, this assumption should cause few problems in simplifying the first law. The resulting form of Eq. (4.15a) is

$$\dot{Q} - \dot{W} = \dot{m}\left(h_2 - h_1 + \frac{V_2^2 - V_1^2}{2}\right) \tag{4.18}$$

Often, although the changes in velocity of the fluid may be large through the shaft work machine, the change in kinetic energy is usually small in comparison to the change in enthalpy of the fluid. With the assumption of no change in kinetic energy, Eq. (4.18) becomes

$$\dot{Q} - \dot{W} = \dot{m}(h_2 - h_1) \tag{4.19}$$

Finally, these devices are often insulated to improve the performance of the device. For example, if heat is lost through a turbine to the surroundings, the amount of power that can be produced by the working fluid decreases by the same amount as the heat loss rate. Similarly, heat loss in a compressor may result in a reduced pressure at the compressor outlet, or will require additional power input to achieve the same pressure. If the shaft work machine is modeled as adiabatic (either due to insulation or due to measurements indicating that the heat transfer rate is very small in comparison to the power produced/used), and if the changes in kinetic and potential energy are assumed to be small, the first law reduces to

$$\dot{W} = \dot{m}(h_1 - h_2) \tag{4.20}$$

Note in Eq. (4.20) that dividing through by the negative sign in front of the power in Eq. (4.15) results in a reversal of the order of the enthalpy terms. Even if all the assumptions are not correct, Eq. (4.20) represents a good starting point to estimate the power produced or required by a shaft work machine, or the impact of the power produced/required on the working fluid.

▶ EXAMPLE 4.5

An insulated air compressor receives 0.55 m³/s of air from the surroundings at 101 kPa and 25°C. The air exits the compressor at 520 kPa and 215°C. Assuming that there are no significant changes in potential or kinetic energy, determine the power consumed by the compressor using an assumption of constant specific heats.

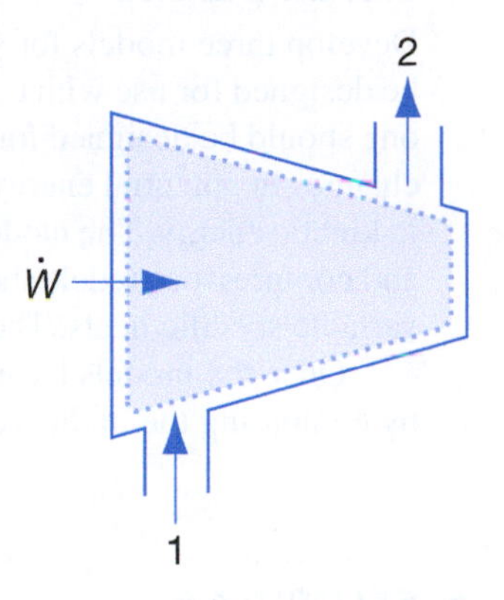

Given: $\dot{V}_1 = 0.055$ m³/s, $T_1 = 25°C = 298$ K, $T_2 = 215°C = 488$ K, $P_1 = 101$ kPa, $P_2 = 520$ kPa

Find: $\dot{W}$

Solution: A compressor is an open system, and the working fluid in this case is air. The air will be treated as an ideal gas, and we are asked to assume constant specific heats in the analysis.

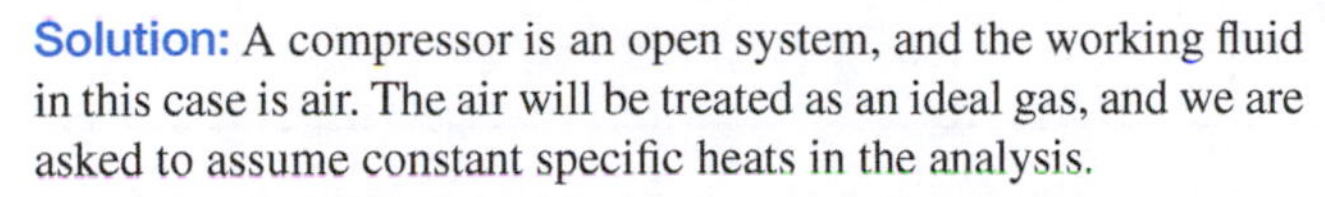

Assume: $\dot{Q} = 0$ (insulated)

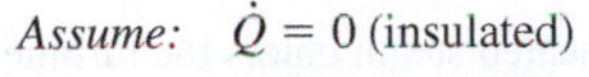

$\Delta KE = \Delta PE = 0$

Ideal gas with constant specific heats

$\dot{m}$ = constant (steady state, steady flow)

Single-inlet, single-outlet device

Necessary equations of state for an ideal gas with constant specific heats:

$$h_2 - h_1 = c_p(T_2 - T_1)$$

$$P\dot{V} = \dot{m}RT \quad \text{(ideal gas law)}$$

First, the first law for single inlet, single outlet, steady-state, steady-flow devices is

$$\dot{Q} - \dot{W} = \dot{m}\left(h_2 - h_1 + \frac{V_2^2 - V_1^2}{2} + g(z_2 - z_1)\right)$$

After the appropriate assumptions are made, the first law reduces to

$$\dot{W} = \dot{m}(h_1 - h_2)$$

Furthermore, for an ideal gas with constant specific heats, we can substitute for the change in enthalpy:

$$\dot{W} = \dot{m}c_p(T_1 - T_2)$$

The mass flow rate is found from the density and the volumetric flow rate (with $R = 0.287$ kJ/kg · K for air):

$$\dot{m} = \frac{P_1\dot{V}_1}{RT_1} = \frac{\left(101\frac{\text{kN}}{\text{m}^2}\right)\left(0.055\frac{\text{m}^3}{\text{s}}\right)}{\left(0.287\frac{\text{kJ}}{\text{kg}\cdot\text{K}}\right)(298\text{ K})} = 0.0650\frac{\text{kg}}{\text{s}}$$

Then, with $c_p = 1.005$ kJ/kg · K for air,

$$\dot{W} = \left(0.0650\frac{\text{kg}}{\text{s}}\right)\left(1.005\frac{\text{kJ}}{\text{kg}\cdot\text{K}}\right)(298\text{ K} - 488\text{ K}) = \mathbf{-12.4\text{ kW}}$$

Analysis: A search of air compressor powers for similar flow rates generally shows that actual air compressors may use a little more power for this flow rate. However, as we will be able to determine later in the book, this compressor is fairly efficient, and there likely would be some heat loss from the compressor in practice. So this power consumption value is reasonable from an engineering perspective.

STUDENT EXERCISE

Develop three models for your chosen computer solution platform. One of these should be designed for use with turbines, one should be designed for use with compressors, and one should be designed for use with pumps. In the models, you can ignore the effects of changes in potential energy, but there should be provisions for heat transfer and changes in kinetic energy. The models should be usable for a wide range of fluids, and the turbine and compressor models should be able to be used for ideal gases with both constant and variable specific heats. The pump model need only be developed for use with liquids.

Once the models have been developed, the correctness of the models can be tested by comparing the model results to those shown in Examples 4.6 through 4.8.

▶ EXAMPLE 4.6

An insulated steam turbine is to produce 100 MW of power. Superheated steam enters the turbine at 10 MPa and 600°C, and saturated water exits the turbine at a pressure of 100 kPa and a quality of 0.93. Determine the necessary mass flow rate of the steam.

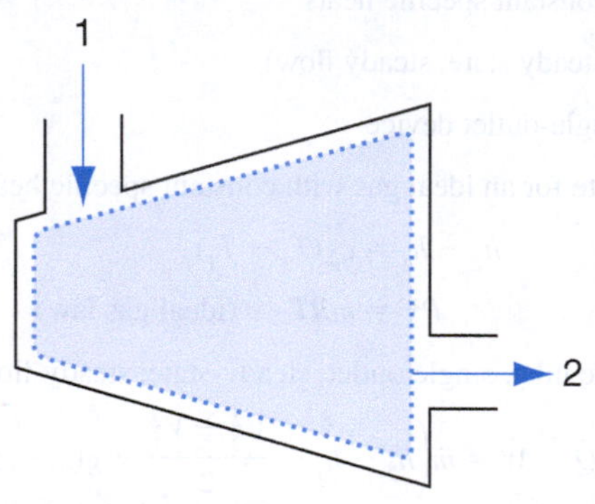

Given: $P_1 = 10$ MPa, $T_1 = 600$°C, $P_2 = 100$ kPa, $x_2 = 0.93$, $\dot{W} = 100$ MW $= 100{,}000$ kW

Find: $\dot{m}$

Solution: A turbine is an open system, and the working fluid is water. The water is superheated steam at the inlet, and becomes a saturated mixture at the exit. Properties can be found from computer programs or steam tables.

Assume: $\dot{Q} = 0$ (insulated)

$\Delta KE = \Delta PE = 0$

$\dot{m} =$ constant (steady state, steady flow)

Single-inlet, single-outlet device

Starting with the first law in Eq. (4.15a),

$$\dot{Q} - \dot{W} = \dot{m}\left(h_2 - h_1 + \frac{V_2^2 - V_1^2}{2} + g(z_2 - z_1)\right)$$

we can apply the assumptions to give

$$\dot{W} = \dot{m}(h_1 - h_2)$$

Using the computerized properties, we find $h_1 = 3625.3$ kJ/kg, $h_2 = 2516.9$ kJ/kg.

Substituting:

$$\dot{m} = \frac{\dot{W}}{h_1 - h_2} = \frac{(100{,}000 \text{ kW})}{(3625.3 - 2516.9) \text{ kJ/kg}} = \mathbf{90.2 \text{ kg/s}}$$

Analysis: Initially, many students are surprised by the large mass flow rates that are encountered in steam turbines used in electricity production. But we must consider that this amount of electricity is used to power cities, and so, in retrospect, the amount of steam being used in these very large turbines is reasonable.

You can use your model of a turbine to analyze turbines thoroughly. Consider the impact of the exit quality on the mass flow rate. In the following graph, the mass flow rate is shown as a function of exit quality for qualities ranging between 0.925 and 1.0.

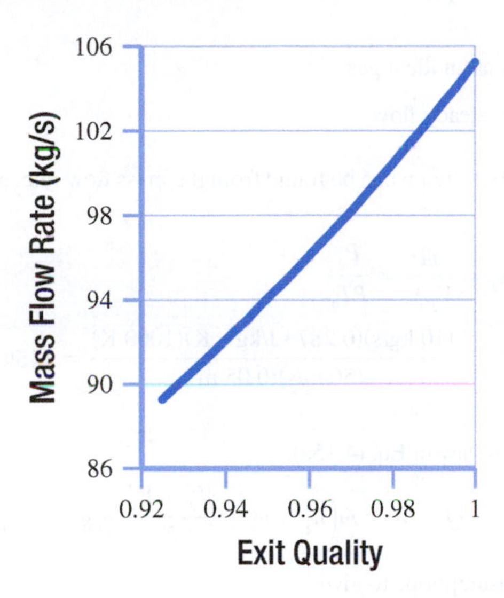

As the exit quality is increased, more energy remains in the exit stream (as represented by the exit enthalpy). As a result, in order to get the 100 MW of power output, additional mass must be sent through the turbine. In future chapters, we will discuss how a higher exit quality relates to more losses in the turbine, resulting in a turbine with a lower operating efficiency. A practical result of this less-than-optimal operation is the need for higher mass flow rates to achieve the same power output. In turn, this requires additional energy to create more high-temperature, high-pressure steam.

▸ **EXAMPLE 4.7**

Air enters an insulated compressor at 400 K and 10 m/s, and exits at 1000 K and 50 m/s. The mass flow rate of the air is 10 kg/s, and the cross-sectional area of the outlet is 0.05 m^2. Considering that the specific heats of the air are variable, determine (a) the pressure of the air at the compressor outlet, and (b) the power used by the compressor.

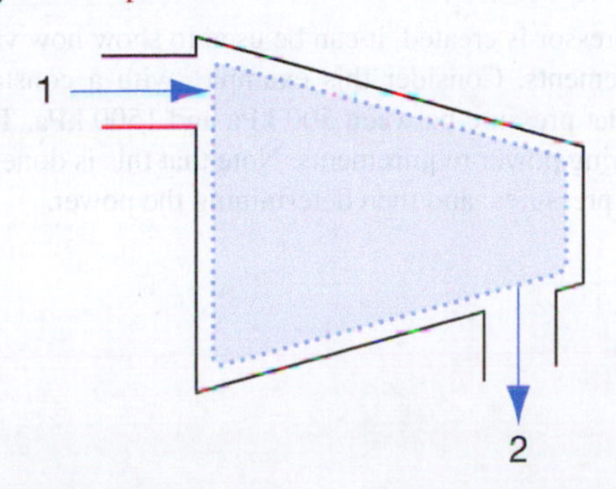

Given: $T_1 = 400$ K, $V_1 = 10$ m/s, $T_2 = 1000$ K, $V_2 = 50$ m/s, $A_2 = 0.05$ m^2, $\dot{m} = 10$ kg/s

Find: (a) P_2, (b) $\dot{W}$

Solution: A compressor is an open system, and the working fluid is air. The air behaves as an ideal gas, and we are told to consider the variability of the specific heats in the solution. Therefore, properties can be found either via computer programs or air tables (such as Table A.3).

Assume: $\dot{Q} = 0$ (insulated)

$\Delta PE = 0$

Air behaves as an ideal gas

Steady state, steady flow

(a) The density of the exit stream can be found from the mass flow rate, and the pressure can be found from the density:

$$\rho_2 = \frac{\dot{m}}{V_2 A_2} = \frac{P_2}{RT_2}$$

$$P_2 = \frac{(10\text{ kg/s})(0.287\text{ kJ/kg}\cdot\text{K})(1000\text{ K})}{(50\text{ m/s})(0.05\text{ m}^2)} = \mathbf{1150\ kPa}$$

(b) Starting with the first law in Eq. (4.15a),

$$\dot{Q} - \dot{W} = \dot{m}\left(h_2 - h_1 + \frac{V_2^2 - V_1^2}{2} + g(z_2 - z_1)\right)$$

we can apply the assumptions to give

$$\dot{W} = \dot{m}\left(h_1 - h_2 + \frac{V_1^2 - V_2^2}{2}\right)$$

Our computerized properties for air with variable specific heats give

$h_1 = 527.02$ kJ/kg

$h_2 = 1172.4$ kJ/kg

Substituting:

$$\dot{W} = (10\text{ kg/s})\left(527.02\text{ kJ/kg} - 1172.4\text{ kJ/kg} + \frac{(10\text{ m/s})^2 - (50\text{ m/s})^2}{2(1000\text{ J/kJ})}\right) = \mathbf{-6470\ kW}$$

The negative sign indicates that the power is into the compressor.

Analysis: Notice that it requires a large amount of power to compress what appears to be a relatively small amount of air. But if we consider that the volume of 1 kg of air at standard temperature and pressure is approximately 1.2 m^3, we can see that 10 kg/s of air is really not a particularly small flow rate.

Once a model for a compressor is created, it can be used to show how variations in outlet conditions impact the power requirements. Consider this example, with a constant mass flow rate and exit velocity, but vary the outlet pressure between 500 kPa and 1500 kPa. The compressor model in the program yields the following power requirements. Note that this is done by determining the new exit temperatures for the new pressures, and then determining the power.

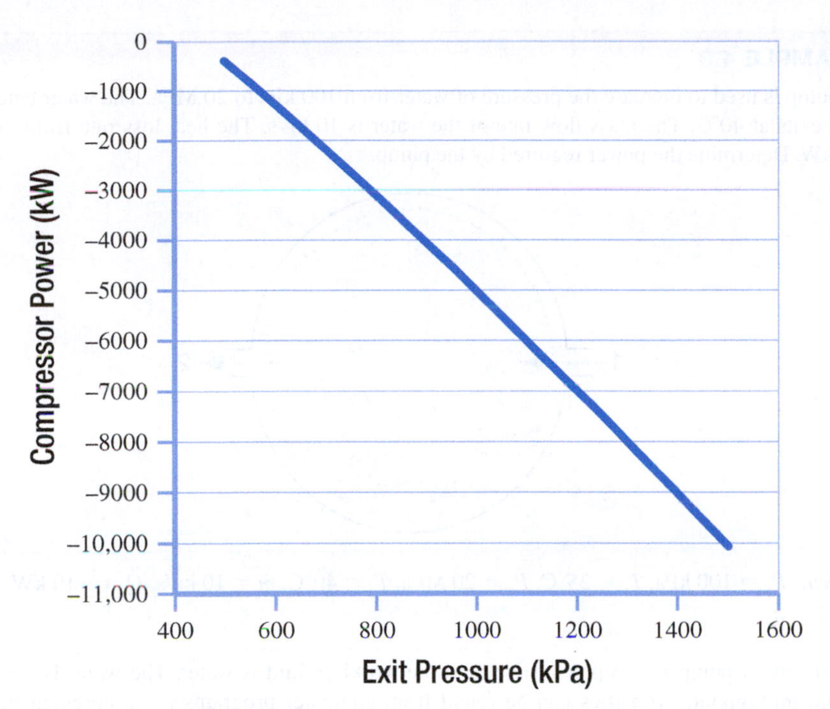

Remember that the compressor power is negative, indicating that power is used by the compressor. As you can see, the power requirement increases dramatically with exit pressure, and the amount of power required to compress even this modest flow of air can become quite large. The following graph shows the exit temperature of the air as a function of exit pressure.

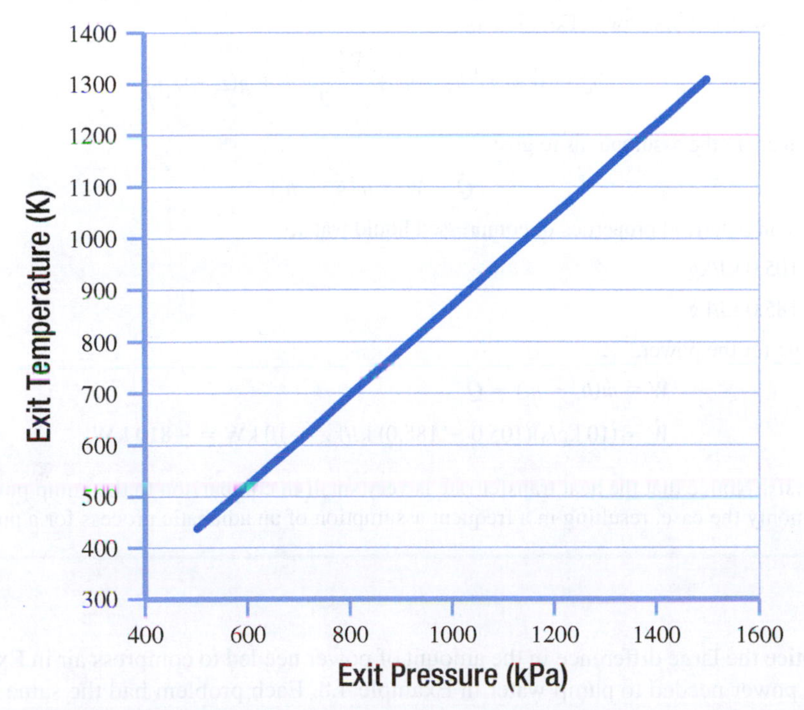

You can see that the temperature of the air exiting the compressor gets very hot at even moderate exit pressures. As a result, many compressors use heat exchangers to cool the air somewhat before it exits the compressor unit.

▶ EXAMPLE 4.8

A pump is used to increase the pressure of water from 100 kPa to 20 MPa. The water enters at 25°C and exits at 40°C. The mass flow rate of the water is 10 kg/s. The heat loss rate from the pump is 10 kW. Determine the power required by the pump.

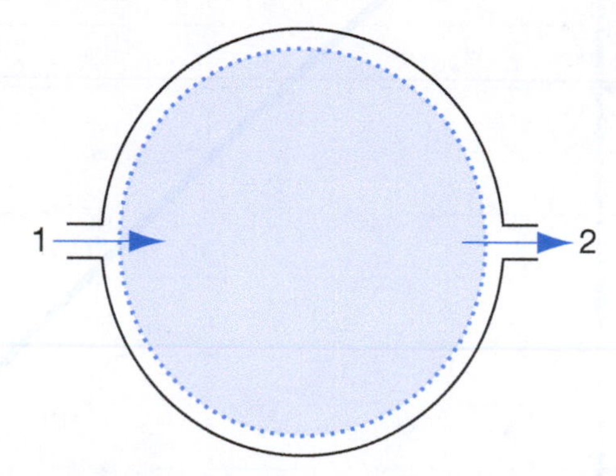

Given: $P_1 = 100$ kPa, $T_1 = 25°C$, $P_2 = 20$ MPa, $T_2 = 40°C$, $\dot{m} = 10$ kg/s, $\dot{Q} = -10$ kW

Find: $\dot{W}$

Solution: A pump is an open system, and the working fluid is water. The water is a compressed liquid throughout. Properties can be found from computer programs or compressed liquid water tables.

Assume: Steady state, steady flow

$\Delta KE = \Delta PE = 0$

Starting with the first law in Eq. (4.15a),

$$\dot{Q} - \dot{W} = \dot{m}\left(h_2 - h_1 + \frac{V_2^2 - V_1^2}{2} + g(z_2 - z_1)\right)$$

we can apply the assumptions to give

$$\dot{Q} - \dot{W} = \dot{m}(h_2 - h_1)$$

From computerized properties of compressed liquid water,

$h_1 = 105.0$ kJ/kg

$h_2 = 185.0$ kJ/kg

Solving for the power,

$$\dot{W} = \dot{m}(h_1 - h_2) + \dot{Q}$$

$$\dot{W} = (10 \text{ kg/s})(105.0 - 185.0) \text{ kJ/kg} - 10 \text{ kW} = \mathbf{-810 \text{ kW}}$$

Analysis: Notice that the heat transfer rate is very small in comparison to the pump power. This is commonly the case, resulting in a frequent assumption of an adiabatic process for a pump.

Notice the large difference in the amount of power needed to compress air in Example 4.7 and the power needed to pump water in Example 4.8. Each problem had the same mass flow rate. But the power needed by the pump was an order of magnitude less than the power needed by the compressor, despite the water being raised to a much higher pressure and despite the presence of a heat loss in the pump. Compressing a gas is a much more power-intensive process than pumping a liquid.

QUESTION FOR THOUGHT/DISCUSSION

Most electricity in the world is produced from turbine-powered generators. What are some advantages of using turbines to power a generator? If turbines had never been invented, what other devices might be used to power electrical generators?

Why is compressing a gas much more power-intensive than pumping a liquid?

Throttling Devices

Although perhaps not as obviously seen as the other devices discussed, throttling devices, or throttles, can play important roles in the performance of a system. Throttles can take on various forms, including orifice plates, partially opened valves, and filters, as shown in **Figure 4.11**. The primary purpose of a throttle is to decrease the pressure of the working fluid with minimal effort and cost. Sometimes the purpose of the throttle is to decrease the flow rate, such as is the case in an engine's throttle, but we will be considering throttles' main purpose as decreasing the fluid's pressure. An example of a common location for such a throttling device is in a refrigerator.

A refrigerator typically works by boiling a refrigerant at a low pressure (and low temperature) by absorbing heat from the space to be cooled. The refrigerant temperature must be lower than the cold space's temperature for this heat transfer to occur. The refrigerant vapor is then pressurized by flowing through a compressor. The refrigerant exits the compressor at a temperature higher than that of the surroundings. By transferring heat to the surroundings, the refrigerant condenses back into a liquid. At this point, the temperature of the refrigerant must be further reduced so it can again absorb heat from the cold space. This is done by sending the refrigerant through a throttling valve, which lowers the pressure (and correspondingly the temperature) of the refrigerant.

Fluids passing through throttling devices will rarely experience any significant changes in potential energy, and these devices are designed to use no power. Because the devices are typically small, we generally do not expect there to be heat gains or losses occurring in them.

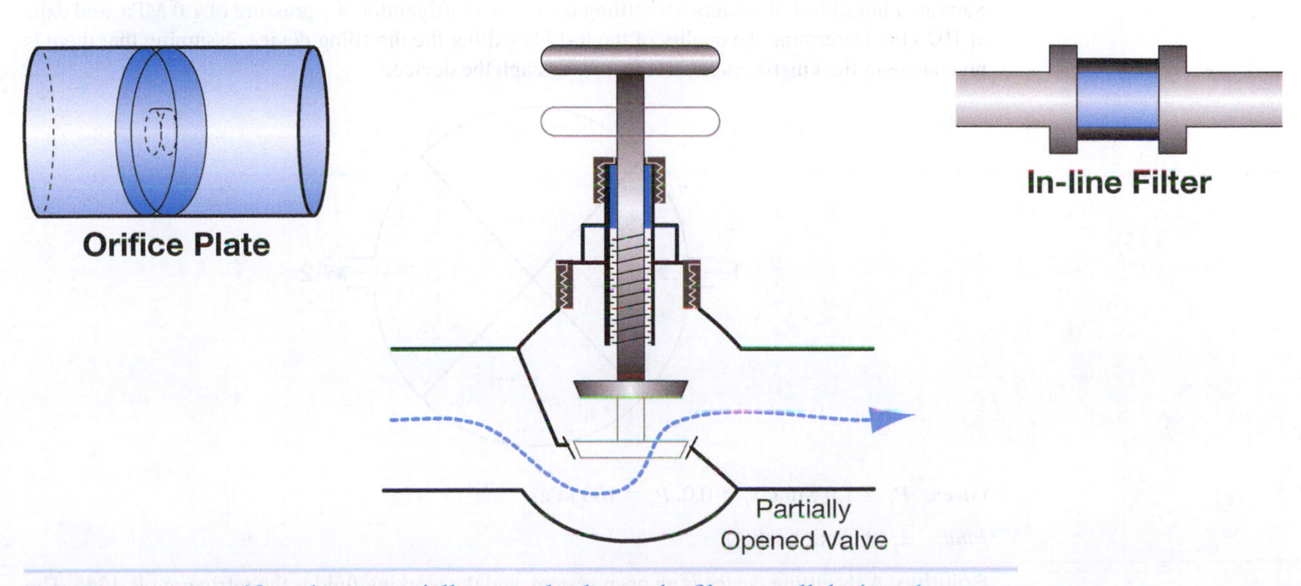

FIGURE 4.11 An orifice plate, a partially opened valve, and a filter are all devices that can be modeled as throttling devices.

For throttling devices, with the assumptions of no heat transfer, no power, and no changes in potential energy, Eq. (4.15a) reduces to

$$h_2 - h_1 + \frac{V_2^2 - V_1^2}{2} = 0 \quad (4.21)$$

Often, the change in the velocity of the working fluid before and after the throttling device is relatively small. As such, for throttling devices, the change in kinetic energy is assumed to be 0, causing Eq. (4.21) to reduce to

$$h_2 = h_1 \quad (4.22)$$

Although Eq. (4.22) may appear to be trivial, it should be remembered that the primary purpose of many throttling devices is to reduce the pressure in a fluid without much economic cost and with a minimal loss in energy. A common application of throttling devices can be seen in Example 4.9.

STUDENT EXERCISE

Develop a model for your chosen computer solution platform that can be used to analyze throttling devices. In the model, you can ignore the effects of changes in potential energy and power, but there should be provisions for heat transfer and kinetic energy changes (although these will often both be 0). The model should be usable for a wide range of fluids, and should be able to be used for ideal gases with both constant and variable specific heats.

Once the model has been developed, the correctness of the model can be tested by comparing the model results to those shown in Example 4.9.

▶ EXAMPLE 4.9

Saturated liquid R-134a enters a throttling device in a refrigerator at a pressure of 1.0 MPa, and exits at 100 kPa. Determine the quality of the R-134a exiting the throttling device, assuming that there is no change in the kinetic energy of the flow through the device.

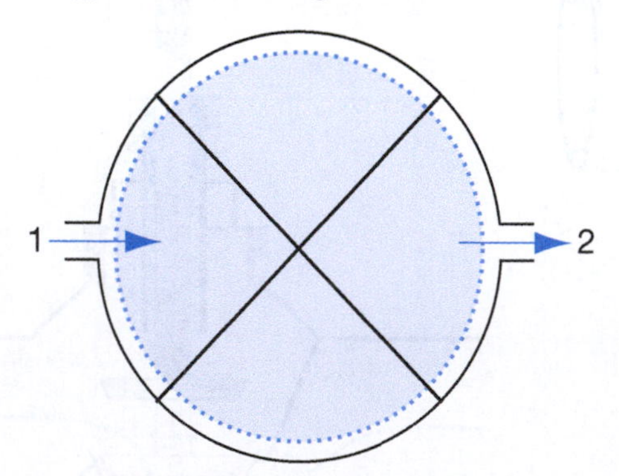

Given: $P_1 = 1.0$ MPa, $x_1 = 0.0$, $P_2 = 100$ kPa

Find: x_2 (the exit quality)

Solution: A throttling device is an open system, and the working fluid is the refrigerant R-134a. The R-134a is initially a saturated liquid, and exits the throttling device as a saturated mixture. Properties can be found from computer tables for R-134a.

Assume: Steady state, steady flow
$\Delta KE = 0$
$\dot{Q} = \dot{W} = \Delta PE = 0$ (standard throttling device assumptions when no other information is known)

Starting with the first law in Eq. (4.15),

$$\dot{Q} - \dot{W} = \dot{m}\left(h_2 - h_1 + \frac{V_2^2 - V_1^2}{2} + g(z_2 - z_1)\right)$$

the assumptions yield that

$$h_1 = h_2$$

From the computerized properties for R-134a, $h_1 = 105.29$ kJ/kg.

Setting $h_2 = 105.29$ kJ/kg, with a $P_2 = 100$ kPa, we find that $\mathbf{x_2 = 0.414}$.

So, 41.4% of the liquid evaporates during the throttling process.

Analysis: While it is difficult to know for certain that this answer is correct, the answer is reasonable in that some of the saturated liquid is expected to vaporize as its pressure is lowered at constant enthalpy through the throttling valve. This results in a saturated mixture, with which the solution is consistent.

QUESTION FOR THOUGHT/DISCUSSION

Throttling devices are often used as an inexpensive way to decrease the pressure of a working fluid. What are other methods of decreasing the pressure of a fluid?

In this section, we have seen several examples of common steady-state, steady-flow, single-inlet, single-outlet devices. At first, it may appear that new equations were developed for each device, but keep in mind that all the equations were derived from Eq. (4.15a) after appropriate assumptions were made. Be aware that for any particular device, you will be called on as an engineer to make appropriate assumptions as to which terms in general equations can be neglected and which are significant, so it is important that you understand how to reduce general equations based upon the assumptions made and what appropriate assumptions may be made.

4.3.3 Application of the First Law to Steady-State, Steady-Flow, Multiple Inlet or Outlet Open Systems

All around us are devices that operate in a steady fashion but that have multiple inlets and outlets. Often these devices take the form of heat exchangers where heat is transferred from one working fluid to another. The radiator of a car, a home's furnace, and the cooling coils on a refrigerator are all examples of common heat exchangers, as shown in **Figure 4.12**.

safakcakir/Shutterstock.com

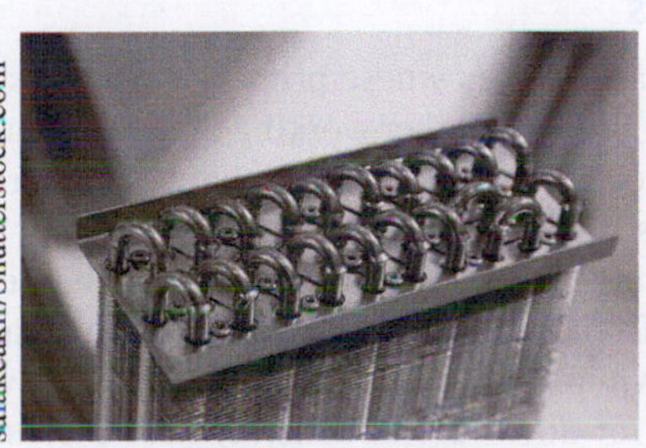
sspopov/iStock/Thinkstock

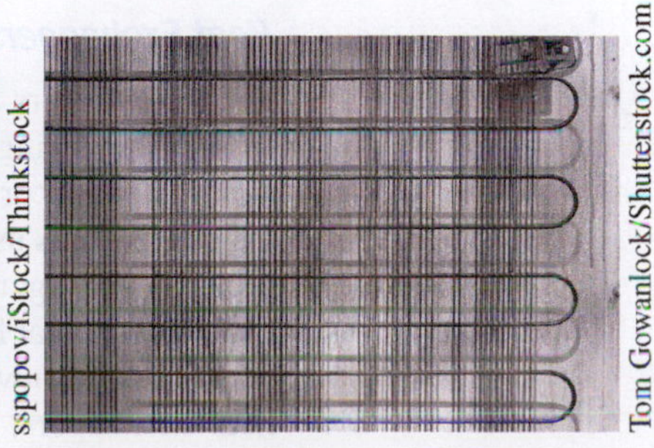
Tom Gowanlock/Shutterstock.com

FIGURE 4.12 Images of several types of heat exchangers.

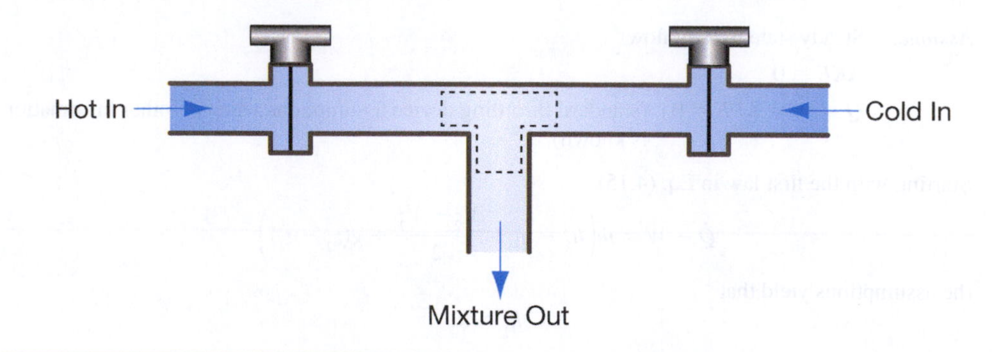

FIGURE 4.13 A schematic diagram of a mixing chamber—in this case, a faucet with two separate inlet streams for hot and cold water, and one outlet stream for the mixed stream.

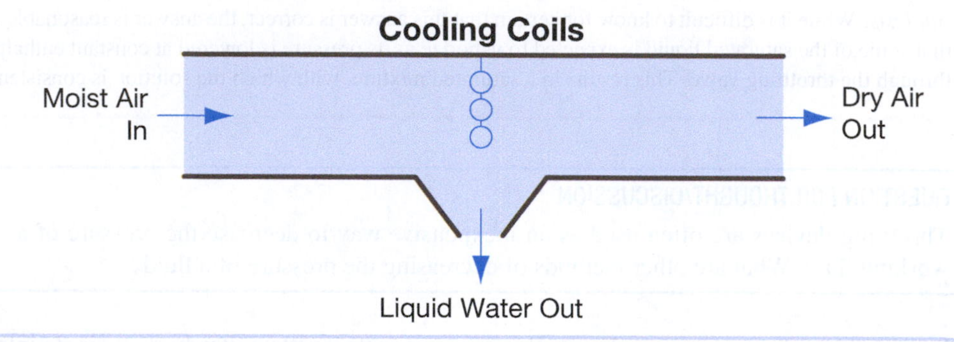

FIGURE 4.14 A dehumidifier works by passing moist air over a cooling coil and removing enough heat to cause water to condense out of the air stream. The liquid water can then be drained away.

Other systems will take two or more fluids and mix them together into a single outlet stream. One example of such a mixing chamber would be a faucet with separate controls for hot and cold water, as shown in **Figure 4.13**. In such a system, there are two inlet streams (the hot water stream and the cold water stream) and a single outlet stream. You can also encounter devices that take a single inlet stream and separate it into multiple outlet streams, such as in **Figure 4.14**. Separation chambers might be used to split the liquid and vapor portions of a boiling system, separate a mixture of gases into individual components, or simply to split one large flow into two or more smaller flows. In this section, we will apply the first law to systems with multiple inlets and/or multiple outlets, concentrating first on heat exchangers, and then considering mixing chambers.

Heat Exchangers

Heat exchangers are devices whose primary purpose is to transfer energy in the form of heat from one working fluid to another working fluid. They are often used either to heat a fluid being sent to another process or to cool a fluid that has been returned from a device. For example, a car's radiator is used to cool the engine coolant coming from the engine by transferring heat to the surrounding air. Conversely, a home's furnace is used to heat air to be sent to the home by transferring heat from combustion products. Heat exchangers can take on many designs, some of which are shown in **Figure 4.15**. Although systems where fluids can mix can be viewed as heat exchangers, for now we will consider heat exchangers to consist of open systems with two separate working fluids that do not mix together. Clearly, such a system will have two

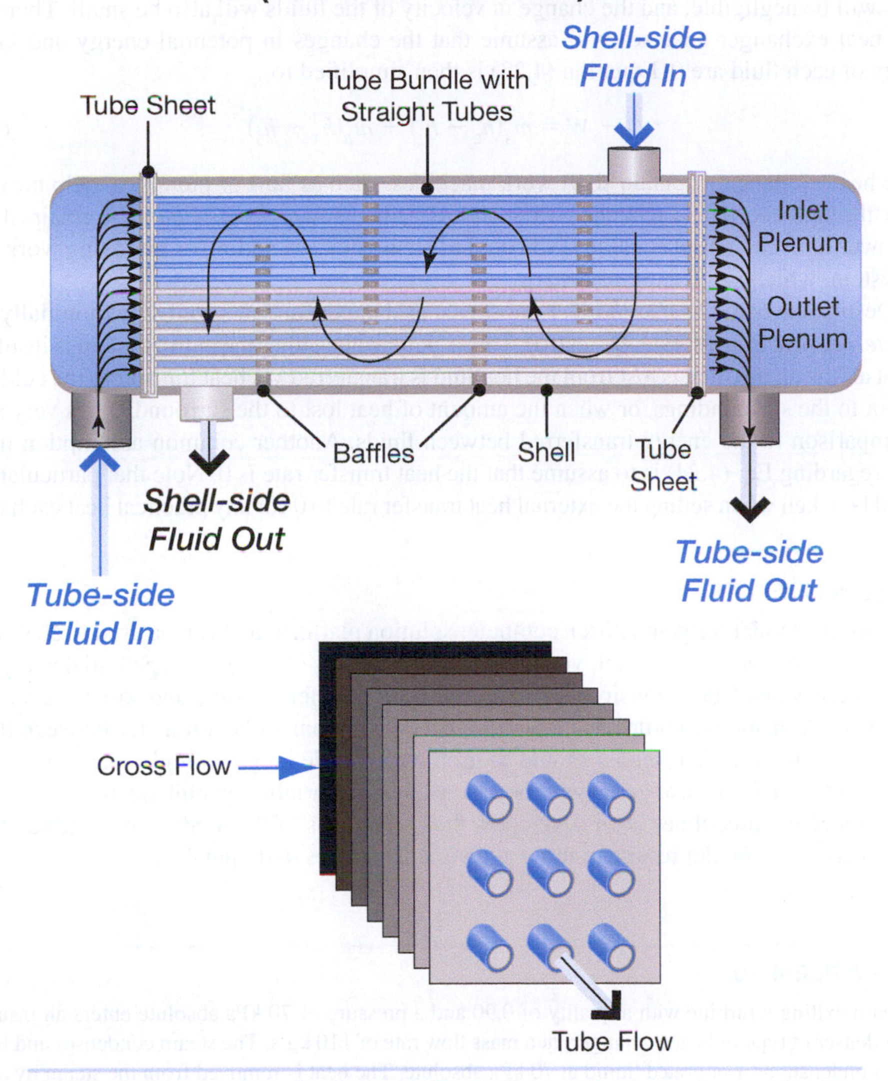

FIGURE 4.15 A schematic diagram of a shell-and-tube heat exhanger (top) and a cross-flow heat exchanger (bottom).

inlets and two outlets, and if we analyze the system once it achieves steady-state, steady-flow conditions, the appropriate form of the first law to be used is Eq. (4.13):

$$\dot{Q} - \dot{W} + \sum_{i=1}^{j} \dot{m}_i \left(h + \frac{V^2}{2} + gz \right)_i - \sum_{e=1}^{k} \dot{m}_e \left(h + \frac{V^2}{2} + gz \right)_e = 0 \tag{4.13}$$

For a two-fluid heat exchanger operating at steady-state, steady-flow conditions, the mass flow rate of each fluid must be the same for that fluid's inlet and outlet. Equation (4.13) can be rewritten for this system, considering two fluids, fluid A and fluid B:

$$\dot{Q} - \dot{W} = \dot{m}_A \left(h_2 - h_1 + \frac{V_2^2 - V_1^2}{2} + g(z_2 - z_1) \right) + \dot{m}_B \left(h_4 - h_3 + \frac{V_4^2 - V_3^2}{2} + g(z_4 - z_3) \right) \tag{4.23}$$

where state 1 is the inlet for fluid A, state 2 is the outlet for fluid A, and states 3 and 4 are the inlet and outlet for fluid B, respectively. For most heat exchangers, the change in height of the fluids will be negligible, and the change in velocity of the fluids will also be small. Therefore, most heat exchanger analyses will assume that the changes in potential energy and kinetic energy of each fluid are 0. Equation (4.23) is then simplified to

$$\dot{Q} - \dot{W} = \dot{m}_A(h_2 - h_1) + \dot{m}_B(h_4 - h_3) \tag{4.24}$$

Some heat exchangers contain shaft work machines, such as fans or pumps, to help move the fluids through the heat exchanger. In such cases, the power term should be retained. But the power term can be set equal to 0 when there are no devices adding or removing work from the system, which is a common scenario.

Despite the name *heat exchanger*, most heat exchangers are analyzed at least initially with the *external* heat transfer rate, $\dot{Q}$, set to 0. This is done when either the entire system is insulated, so that all the energy that is lost from the hot fluid is transferred via heat transfer to the cold fluid and not to the surroundings, or when the amount of heat lost to the surroundings is very small in comparison to the energy transferred between fluids. Another common assumption that is made regarding Eq. (4.24) is to assume that the heat transfer rate is 0. Note that particular care should be taken when setting the external heat transfer rate to 0 for any practical heat exchanger.

STUDENT EXERCISE

Develop a model for your chosen computer solution platform that can be used to analyze heat exchangers. In the model, you can ignore the effects of changes in potential energy, but there should be provisions for external heat transfer, power, and kinetic energy changes. You should set up the models so that calculations of heat transfer between the fluids are possible. The model should be usable for a wide range of fluids, and should be able to be used for ideal gases with both constant and variable specific heats.

Once the model has been developed, the correctness of the model can be tested by comparing the model results to those shown in Examples 4.10 and 4.11.

▶ EXAMPLE 4.10

Steam exiting a turbine with a quality of 0.90 and a pressure of 70 kPa absolute enters an insulated condenser (a type of heat exchanger) at a mass flow rate of 110 kg/s. The steam condenses and leaves the condenser as a saturated liquid at 70 kPa absolute. The heat is removed from the steam by a flow of external cooling water, which enters the condenser at 10°C and exits at 40°C. Determine the mass flow rate of the external cooling water.

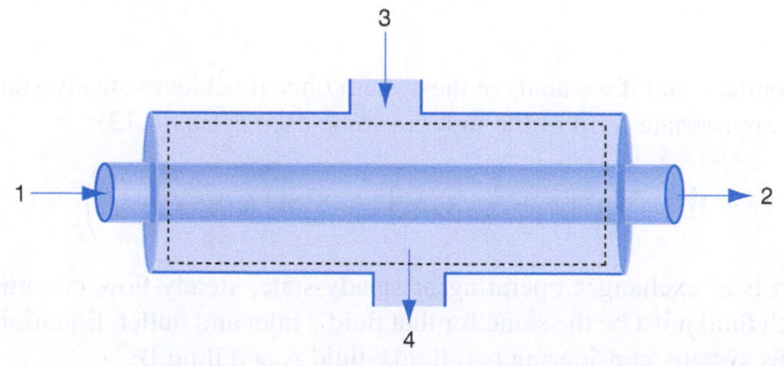

Given: Steam: $P_3 = P_4 = 70$ kPa, $x_3 = 0.90$, $x_4 = 0.0$ (saturated liquid), $\dot{m}_s = 110$ kg/s

Cooling water: $T_1 = 10$°C, $T_2 = 40$°C

Find: $\dot{m}_{cw}$ (the mass flow rate of the cooling water)

Solution: The heat exchanger is an open system, with two inlets and two outlets, and no mixing of the fluids. Water as a saturated mixture enters the heat exchanger and condenses into a saturated liquid at the outlet. Properties for this flow can be found from computer programs or saturated steam tables. Liquid cooling water flows through a pipe, removing heat from the steam. As the cooling water's temperature and pressure (we are given no information about its pressure, but we would not expect a dramatic pressure drop in this system) do not change greatly, it can be treated as an incompressible liquid with constant specific heats.

Assume: Steady state, steady flow, two inlets, two outlets, no mixing of fluids

Cooling water acts as an incompressible liquid with constant specific heats

$\dot{Q} = 0$ (externally insulated)

$\dot{W} = \Delta KE = \Delta PE = 0$ (common heat exchanger assumptions, no information provided to expect other values)

Starting with Eq. (4.23),

$$\dot{Q} - \dot{W} = \dot{m}_{cw}\left(h_2 - h_1 + \frac{V_2^2 - V_1^2}{2} + g(z_2 - z_1)\right) + \dot{m}_s\left(h_4 - h_3 + \frac{V_4^2 - V_3^2}{2} + g(z_4 - z_3)\right)$$

we can simplify the equation using the assumptions to

$$\dot{m}_{cw}(h_2 - h_1) + \dot{m}_s(h_4 - h_3) = 0$$

Using computer programs for steam/water, $h_4 = 375$ kJ/kg, $h_3 = 2430$ kJ/kg

For the liquid cooling water, as an incompressible fluid with constant specific heats: $h_2 - h_1 = c_p(T_2 - T_1)$.

Solving for the mass flow rate of the cooling water,

$$\dot{m}_{cw} = \frac{\dot{m}_s(h_4 - h_3)}{c_p(T_1 - T_2)} = \frac{(110 \text{ kg/s})(375 \text{ kJ/kg} - 2430 \text{ kJ/kg})}{(4.184 \text{ kJ/kg}\cdot\text{K})(10°\text{C} - 40°\text{C})} = \mathbf{1800 \text{ kg/s}}$$

Analysis: This is the equivalent of approximately 1.8 m^3/s. Although this sounds like a large amount of cooling water, it is only what might be expected from a fairly moderately sized coal-fired power plant (~100 MW power output). This also provides us with an idea of the physical size of such systems.

▶ EXAMPLE 4.11

Oil enters the outer tube of an insulated concentric tube heat exchanger at a temperature of 500 K and exits at a temperature of 400 K. The mass flow rate of the oil is 5 kg/s, and the oil has a specific heat of $c_p = 1.91$ kJ/kg · K. Saturated liquid water at 100°C enters the inner tube, and exits at 100°C with a quality of 0.20. Determine the mass flow rate of the water.

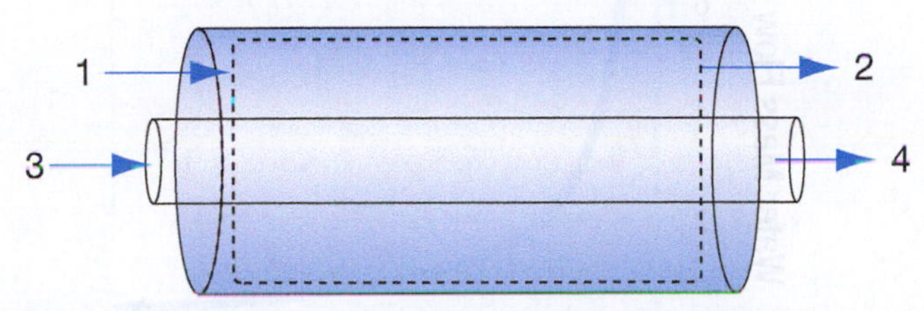

Given: Oil: $T_1 = 500$ K, $T_2 = 400$ K, $\dot{m}_{oil} = 5$ kg/s, $c_{p,oil} = 1.91$ kJ/kg · K

Water: $T_3 = T_4 = 100°$C, $x_3 = 0$ (saturated liquid), $x_4 = 0.20$

Find: $\dot{m}_w$ (the mass flow rate of the water)

Solution: The heat exchanger is an open system, with two inlets and two outlets, and no mixing of the fluids. Water as a saturated liquid enters the heat exchanger, and it begins to boil, exiting as a saturated mixture. Properties for this fluid can be found from computer programs or saturated steam tables. Liquid oil flows through a pipe, cooling as it adds heat to the water. As we have no details about the oil, we will treat it as an incompressible liquid with constant specific heats.

Assume: Steady state, steady flow, two inlets, two outlets, no mixing of fluids

Oil acts as an incompressible liquid with constant specific heats

$\dot{Q} = 0$ (externally insulated)

$\dot{W} = \Delta KE = \Delta PE = 0$ (common heat exchanger assumptions, no information provided to expect other values)

Starting with Eq. (4.23),

$$\dot{Q} - \dot{W} = \dot{m}_{oil}\left(h_2 - h_1 + \frac{V_2^2 - V_1^2}{2} + g(z_2 - z_1)\right) + \dot{m}_w\left(h_4 - h_3 + \frac{V_4^2 - V_3^2}{2} + g(z_4 - z_3)\right)$$

we can simplify the equation using the assumptions:

$$\dot{m}_{oil}(h_2 - h_1) + \dot{m}_w(h_4 - h_3) = 0$$

Using computer programs for water, $h_4 = 869.0$ kJ/kg, $h_3 = 417.5$ kJ/kg

For oil, as an incompressible fluid with constant specific heats, $h_2 - h_1 = c_p(T_2 - T_1)$

Solving for the mass flow rate of the water,

$$\dot{m}_w = \frac{\dot{m}_{oil}c_{p,oil}(T_1 - T_2)}{h_4 - h_3} = \frac{(5 \text{ kg/s})(1.91 \text{ kJ/kg}\cdot\text{K})(500 - 400)\text{ K}}{(869.0 - 417.5)\text{ kJ/kg}\cdot\text{K}} = \mathbf{2.12\ kg/s}$$

Analysis: Typically, we would not expect dramatically different mass flow rates through such a simple heat exchanger, and as the mass flow rate of the water is of the same order of magnitude as the oil, the result is reasonable.

You can use the program model you develop for a heat exchanger to show how operating parameters in the heat exchanger affect each other. For example, you can use the model to show how the mass flow rate of the water will vary as the exit quality of the water changes. For example, the exit quality may be determined by either a requirement downstream of the heat exchanger, or by a desire to meet some heat transfer characteristic in the heat exchanger design (heat will be transferred at different rates to liquids and vapors). In the following graph, the mass flow rate of the water is plotted as a function of exit quality of the water.

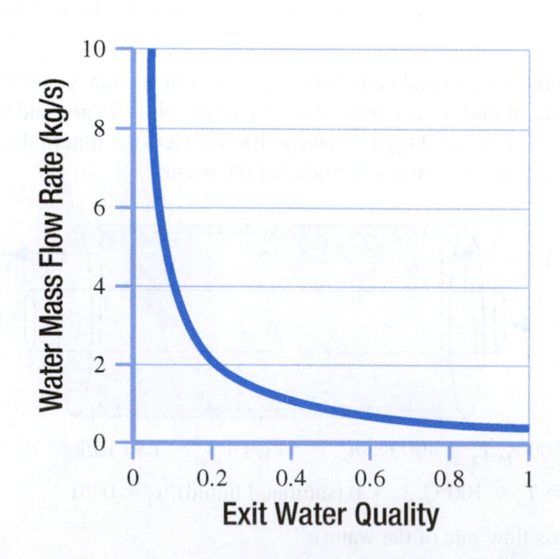

As you can see, there is a large dependency on the mass flow rate for low quality values, but this dependency becomes less significant as the water becomes primarily vapor. So, if you need to keep most of the water in liquid form to maintain the integrity of the heat exchanger design, you must be certain that the mass flow rate of the water remains relatively high.

Before leaving our discussion of heat exchangers, you may notice that a two-fluid, insulated heat exchanger can be adequately modeled as two single-inlet, single-outlet, steady-state, steady-flow devices. In this situation, you would find the heat transfer rate to or from one fluid, reverse the sign on that heat transfer rate, and set that new quantity equal to the heat transfer rate for the other fluid. This is possible because the heat transferred out of one fluid will be transferred into the other fluid (thus leading to the reversal in the sign of the heat transfer but no change in the magnitude of the heat transfer). Either approach, which is treating the whole system as a two-inlet, two-outlet device or treating it as two separate single-inlet, single-outlet devices, is acceptable. Although neither is necessarily preferred, the latter approach is useful if you need to determine how much heat is transferred between fluids.

QUESTION FOR THOUGHT/DISCUSSION

How are heat exchangers used in your home? What are some other common uses for heat exchangers?

Mixing Chambers

It is very common to have to mix two or more streams together to form one outlet stream. One such application, shown in **Figure 4.16**, is an "open feedwater heater" in a steam power plant, where steam that has been extracted from a turbine is mixed with liquid water being directed toward the steam generator to preheat the water and reduce the amount of heat that is needed to create steam. Another application is in air conditioning systems used in buildings when air that has been removed from a space is mixed with some fresh outside air before the mixture is cooled and dehumidified in order to improve the overall air quality in the building, as seen in **Figure 4.17**. Sometimes these mixing processes occur in a formal device to aid in the mixing, and other times these processes occur by two pipes joining together and the flow exiting through a third pipe. In either type of device, the system can be modeled as a mixing chamber.

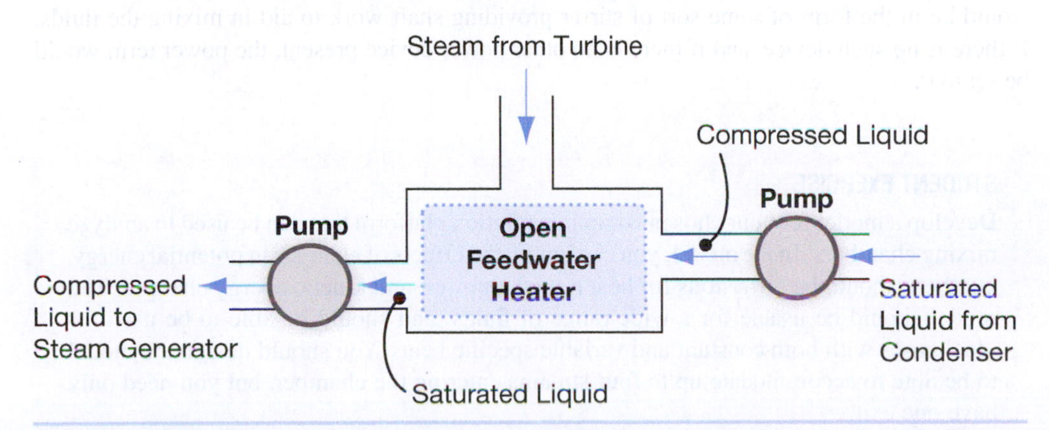

FIGURE 4.16 An open feedwater heater is used in some steam power plants to preheat liquid water going to the steam generator by mixing it with steam extracted from the turbine. This can be modeled as a mixing chamber.

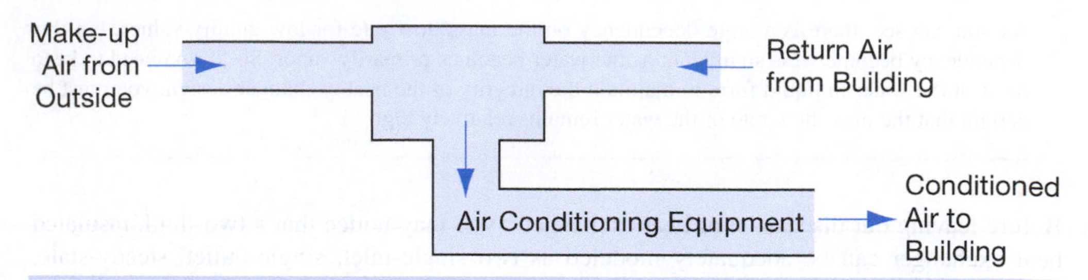

FIGURE 4.17 Many air conditioning systems will mix fresh air from outside a building with interior return air before delivering it to the air conditioning system. This part of the system can be modeled as a mixing chamber.

To analyze a steady-state, steady-flow mixing chamber, we will need Eq. (4.13) for the first law, and will also need to consider the conservation of mass in Eq. (4.2) to determine the mass flow rates. For a common mixing chamber configuration of two inlets (states 1 and 2) and one outlet (state 3), as shown in Figure 4.15, Eq. (4.2) will reduce to

$$\dot{m}_1 + \dot{m}_2 = \dot{m}_3 \quad (4.25)$$

With regard to the first law, common assumptions for mixing chambers include that the changes in height of the fluid and the changes in the velocities of the fluids are small with respect to any changes in enthalpy, so the potential and kinetic energy terms will be neglected. With these assumptions, Eq. (4.13) becomes

$$\dot{Q} - \dot{W} = \dot{m}_3 h_3 - (\dot{m}_1 h_1 + \dot{m}_2 h_2) \quad (4.26)$$

Combining Eqs. (4.25 and 4.26) yields

$$\dot{Q} - \dot{W} = (\dot{m}_1 + \dot{m}_2)h_3 - (\dot{m}_1 h_1 + \dot{m}_2 h_2) \quad (4.27)$$

Just as with heat exchangers, how the heat transfer rate and power terms are handled will depend on the particular situation. Often the heat transfer rate is set to 0, with little or no heat transfer expected out of the system as the fluid rapidly flows through the chamber—however, we must carefully consider the particular system before making this assumption. Whether or not the power term should be set to 0 is usually obvious. Most commonly, any power device would be in the form of some sort of stirrer providing shaft work to aid in mixing the fluids. If there is no such device, and if there is no other power device present, the power term would be set to 0.

STUDENT EXERCISE

Develop a model for your chosen computer solution platform that can be used to analyze mixing chambers. In the model, you can ignore the effects of changes in potential energy, but there should be provisions for heat transfer, power, and kinetic energy changes. The model should be usable for a wide range of fluids, and should be able to be used for ideal gases with both constant and variable specific heats. You should design the model to be able to accommodate up to four streams entering the chamber, but you need only have one exit.

Once the model has been developed, the correctness of the model can be tested by comparing the model results to those shown in Example 4.12.

▶ EXAMPLE 4.12

An insulated mixing chamber receives 5.0 kg/s of superheated steam at 3 MPa and 280°C through one inlet, and saturated liquid water at 3 MPa through a second inlet. The combined stream exits with a quality of 0.80 and a pressure of 3 MPa. A fan, using 10 kW of power, is used to aid the mixing process. Determine the mass flow of the saturated liquid entering the system through the second inlet.

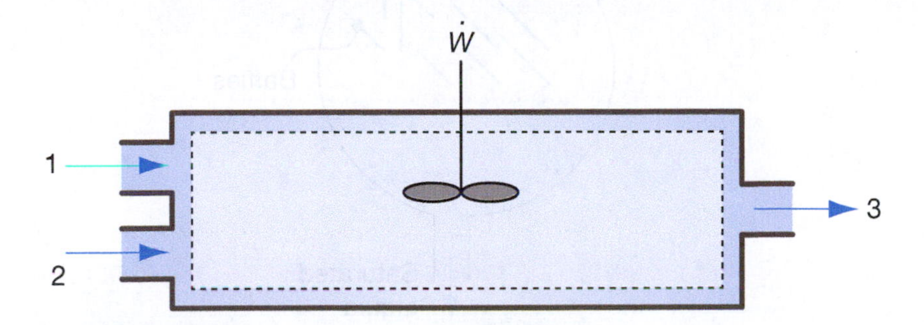

Given: $\dot{m}_1 = 5.0$ kg/s, $P_1 = P_2 = P_3 = 3$ MPa, $T_1 = 280°C$, $x_2 = 0.0$, $x_3 = 0.80$, $\dot{W} = -10$ kW

Find: $\dot{m}_2$ (the mass flow rate of the second inlet stream)

Solution: The mixing chamber is an open system, with two inlets and one outlet. Water as a superheated vapor enters through one inlet, and water as a saturated liquid enters through the second inlet. Water as a saturated mixture exits. Properties for the water can be found from computer programs or saturated steam tables.

Assume: Steady state, steady flow, two inlets, one outlet

$\dot{Q} = 0$ (insulated)

$\Delta KE = \Delta PE = 0$

Beginning with Eq. (4.13),

$$\dot{Q} - \dot{W} + \sum_{i=1}^{j} \dot{m}_i \left(h + \frac{V^2}{2} + gz\right)_i - \sum_{e=1}^{k} \dot{m}_e \left(h + \frac{V^2}{2} + gz\right)_e = 0$$

application of the assumptions yields

$$-\dot{W} = \dot{m}_3 h_3 - (\dot{m}_1 h_1 + \dot{m}_2 h_2)$$

From the conservation of mass, $\dot{m}_1 + \dot{m}_2 = \dot{m}_3$, and so the first law can be further rewritten as

$$-\dot{W} = (\dot{m}_1 + \dot{m}_2) h_3 - (\dot{m}_1 h_1 + \dot{m}_2 h_2)$$

From computerized properties for water, $h_1 = 2941.3$ kJ/kg, $h_2 = 1008.4$ kJ/kg, $h_3 = 2444.96$ kJ/kg.

Substituting:

$$10 \text{ kW} = (5 \text{ kg/s} + \dot{m}_2)(2444.96 \text{ kJ/kg}) - ((5 \text{ kg/s})(2941.3 \text{ kJ/kg}) + \dot{m}_2(1008.4 \text{ kJ/kg}))$$

Solving yields $\dot{m}_2 =$ **1.73 kg/s.**

Analysis: Again, until we gain more experience, we cannot be certain that this value is correct. However, as we are mixing a saturated liquid with slightly superheated steam and getting a saturated mixture exiting, we would at least expect the mass flow rates to be a comparable magnitude, which they are in this case.

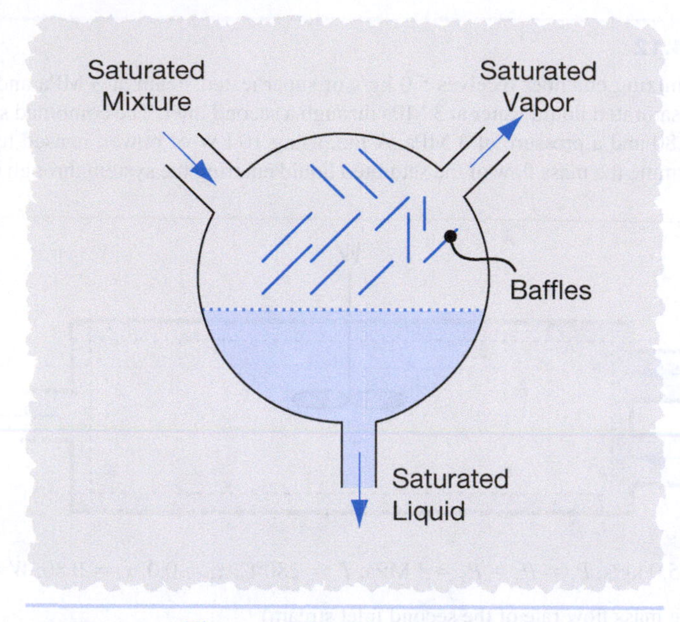

FIGURE 4.18 The steam drum in a steam power plant is a type of separation chamber. Here, a saturated mixture enters and is agitated by flowing over baffles, saturated vapor exits through one port, and saturated liquid exits through another port.

The previous discussion focused on mixing chambers, but the same basic thought process would be used in reducing Eqs. (4.2) and (4.13) for a separation chamber, such as in **Figure 4.18**, or any other steady-state, steady-flow device that contains multiple inlets and multiple outlets and combining or separating streams.

QUESTION FOR THOUGHT/DISCUSSION

What are some everyday processes that can be modeled as mixing chambers?

STUDENT EXERCISE

Develop a model for your chosen computer solution platform that can be used to analyze separation chambers. In the model, you can ignore the effects of changes in potential energy, but there should be provisions for heat transfer, power, and kinetic energy changes. The model should be usable for a wide range of fluids, and should be able to be used for ideal gases with both constant and variable specific heats. You should design the model to have the ability for up to three streams exiting the chamber, but you need only have one entering stream.

4.3.4 Introduction to Unsteady (Transient) Open Systems

So far, we have concentrated on applying the first law to steady-state open systems. It should be clear that many practical systems can be modeled assuming steady-state behavior, even if the system is not perfectly at steady state. However, there are still many engineering applications that cannot be modeled as steady state with any reasonable level of accuracy, because the system's properties vary significantly with time. Some situations are obvious, such as if you

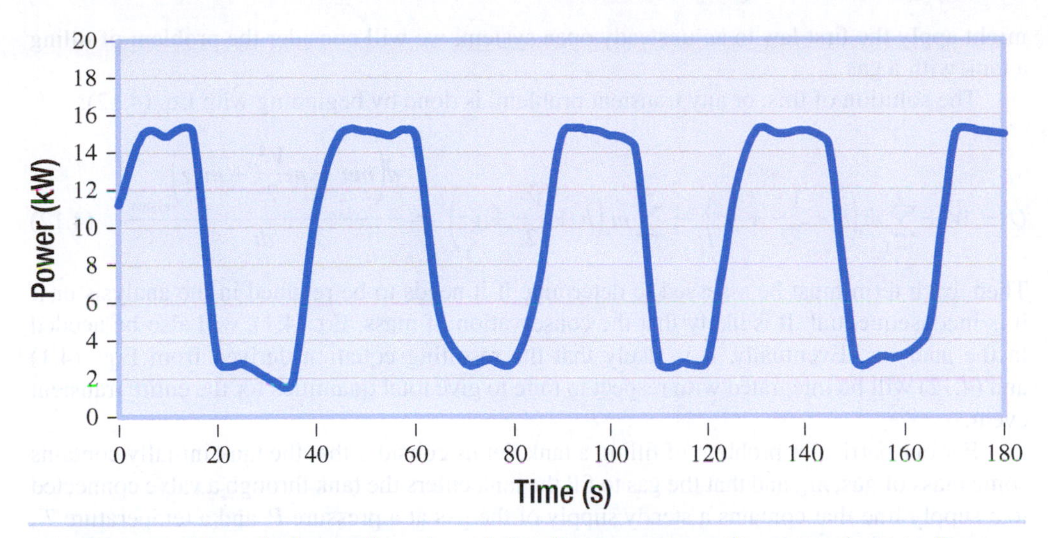

FIGURE 4.19 Many devices, when viewed closely, are not operating at steady state. For example, here the power consumption by an air compressor varies as it operates in filling a tank, and then pauses while the pressure in the tank drops.

are filling a container with a fluid—clearly the mass of the fluid in the container is increasing with time. Or you may have a situation with pulsing flow, such as an air compressor that is used to fill up a storage tank with air, idles while the tank slowly empties for a minute or two, and then again fills up the tank. It is also common to have steady-state, steady-flow devices begin and end their operation in a transient state. For example, after steam begins to flow into a steam turbine, the turbine will heat up for a period of time, after which steady-state operation will be achieved. If you are concerned with the start-up period, a transient analysis will be necessary. **Figure 4.19** shows how power may vary with time for a transiently operating air compressor. **Figure 4.20** presents a graph of how the mass of water in a tank will change as it drains water over time.

Because of the wide variety of problems that could be modeled as unsteady systems, it is not possible to develop a set of standard assumptions that can be typically used for analyzing a device. So in this section, we just introduce you to transient analysis, and we will leave further exploration of the topic to more advanced thermodynamics courses. To illustrate how we

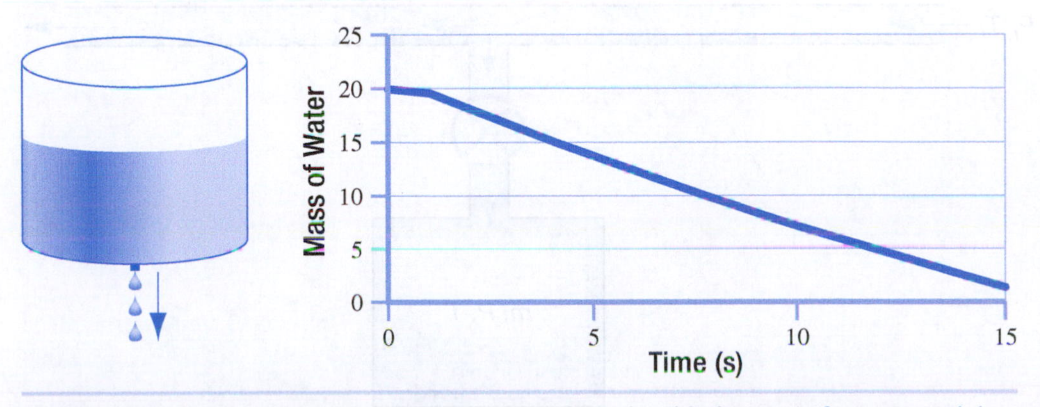

FIGURE 4.20 An example of water draining out of a tank, with the mass of water remaining in the tank as a function of time.

might apply the first law to an unsteady open system, we will consider the problem of filling a tank with a gas.

The solution of this, or any transient problem, is done by beginning with Eq. (4.12):

$$\dot{Q} - \dot{W} + \sum_{i=1}^{j} \dot{m}_i\left(h + \frac{V^2}{2} + gz\right)_i - \sum_{e=1}^{k} \dot{m}_e\left(h + \frac{V^2}{2} + gz\right)_e = \frac{d\left(mu + m\frac{V^2}{2} + mgz\right)_{\text{system}}}{dt} \tag{4.12}$$

Then, each term must be assessed to determine if it needs to be retained in the analysis or if it is inconsequential. It is likely that the conservation of mass, Eq. (4.1), will also be needed in the analysis. Eventually, it is likely that the resulting equation derived from Eqs. (4.1) and (4.12) will be integrated with respect to time to give total quantities for the entire transient event.

For our particular problem of filling a tank, let us consider that the tank initially contains some mass of gas, m_1, and that the gas to fill the tank enters the tank through a valve connected to a supply line that contains a steady supply of the gas at a pressure P_i and a temperature T_i, as shown in **Figure 4.21**. We will assume that we know the volume of the tank and the initial pressure and temperature of the gas in the tank. Gas will flow into the tank until the tank has a new pressure, P_2, with $P_2 < P_i$. Considering this scenario, we can now evaluate whether or not a particular term in Eq. (4.12) can be ignored. Although some work is necessary to open and close the valve, it is likely that this is a trivial amount of work, and so the power term can be set to 0. Because mass only flows into the tank, there is no exiting mass flow and so $\dot{m}_e = 0$, which causes the entire second summation to disappear. Although there may be a high velocity for the inlet flow, the kinetic energy of the incoming flow is likely small in comparison to the inlet flow enthalpy—therefore, the inlet kinetic energy will be set to 0. Similarly, it is unlikely that the inlet potential energy will be significant, and so that will also be set to 0. Finally, the system (i.e., the tank) is stationary, and so its kinetic energy is 0 and its change in potential energy with time is 0. These changes cause the initially complicated Eq. (4.12) to reduce to

$$\dot{Q} + \dot{m}_i h_i = \frac{d(mu)_{\text{system}}}{dt} \tag{4.28}$$

This equation can now be integrated with respect to time. Note that, for such an integration, you will need to know how the inlet enthalpy (or any inlet property) varies with respect to time. In this case, the gas is coming from a steady source, and so the temperature of the

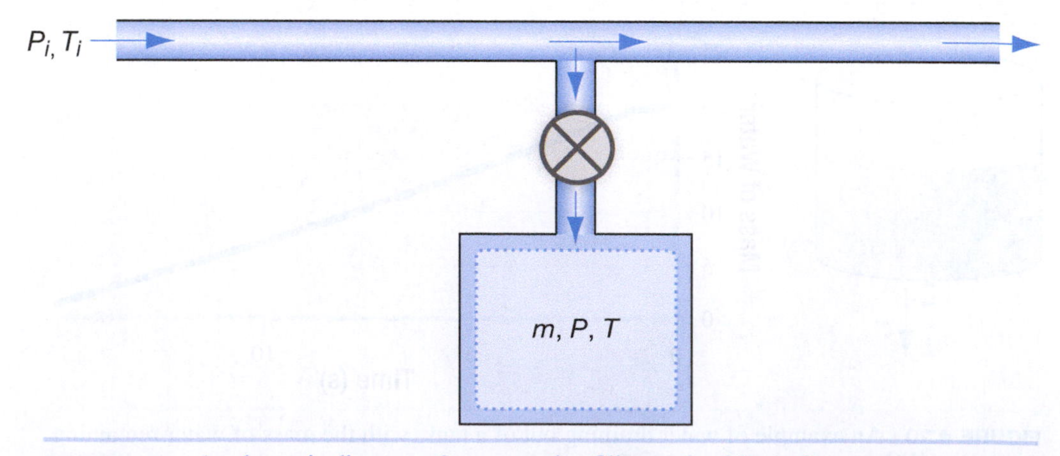

FIGURE 4.21 A schematic diagram of a system that fills a tank with gas from a supply pipeline.

incoming gas is constant and therefore the incoming enthalpy is constant with respect to time, assuming it is modeled as an ideal gas. After integration, the first law becomes

$$Q + m_i h_i = m_2 u_2 - m_1 u_1 \tag{4.29}$$

You may also notice that we can integrate Eq. (4.1) with respect to time, yielding

$$m_i = m_2 - m_1 \tag{4.30}$$

The following example numerically demonstrates this solution procedure for a tank-filling process.

▶ EXAMPLE 4.13

A tank of air, with a volume of 0.25 m³, is to be filled through a valve connected to a supply line. The tank initially contains air at 100 kPa and 20°C. The supply line contains air at 1200 kPa and 25°C. The valve is opened, and air is allowed to flow into the tank until the tank's pressure is 750 kPa. Assume that the tank is insulated for this process. Determine the final temperature of the air inside the tank.

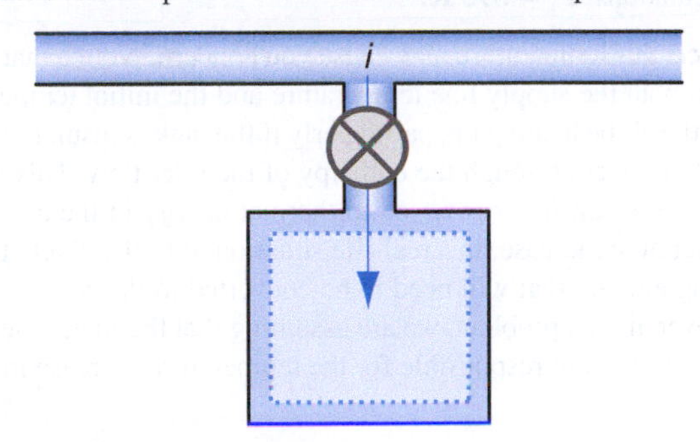

Given: $\forall = 0.25\ \text{m}^3$, $T_1 = 20°\text{C} = 293\ \text{K}$, $P_1 = 100\ \text{kPa}$, $P_2 = 750\ \text{kPa}$, $P_i = 1200\ \text{kPa}$, $T_i = 25°\text{C} = 298\ \text{K}$.

Find: T_2 (the final temperature)

Solution: This is an open system, but it involves an unsteady-state analysis as the contents of the system are changing with time. The working fluid is air, which can be assumed to be an ideal gas. Do not assume the specific heats to be constant; variable specific heats analysis for air can either use Table A.3 for properties, or computer programs. Consider the properties of the air in the supply line (and therefore at the inlet of the tank) to be constant.

Assume: $W = \Delta KE = \Delta PE = 0$ (no information given to suggest otherwise)

$Q = 0$ (tank is insulated)

Air behaves as an ideal gas with variable specific heats

Mass flows in only—no mass flow out of the system ($m_e = 0$)

Start with the general first law as given in Eq. (4.12):

$$\dot{Q} - \dot{W} + \sum_{i=1}^{j} \dot{m}_i \left(h + \frac{V^2}{2} + gz\right)_i - \sum_{e=1}^{k} \dot{m}_e \left(h + \frac{V^2}{2} + gz\right)_e = \frac{d\left(mu + m\frac{V^2}{2} + mgz\right)_{\text{system}}}{dt}$$

Applying the assumptions and integrating with respect to time yields

$$m_i h_i = m_2 u_2 - m_1 u_1$$

From the conservation of mass, $m_i = m_2 - m_1$, so

$$(m_2 - m_1) h_i = m_2 u_2 - m_1 u_1$$

Using the ideal gas law, the masses at the initial and final states can be found:

$$m_1 = \frac{P_1 V}{RT_1} = \frac{(100 \text{ kPa})(0.25 \text{ m}^3)}{(0.287 \text{ kJ/kg} \cdot \text{K})(293 \text{ K})} = 0.297 \text{ kg}$$

$$m_2 = \frac{P_2 V}{RT_2} = \frac{(750 \text{ kPa})(0.25 \text{ m}^3)}{(0.287 \text{ kJ/kg} \cdot \text{K})(T_2)} = 653.3/T_2 \text{ kg}$$

From computer programs for air with variable specific heats, $h_i = 298.18$ kJ/kg, $u_1 = 209.06$ kJ/kg.
Substituting: $(653.3/T_2 - 0.297)$ kg $(298.18$ kJ/kg$) = (653.3/T_2$ kg$)u_2 - 62.09$ kJ

$$194{,}801/T_2 - 653.3u_2/T_2 = 26.469$$

Clearly, there are two unknowns in this equation: T_2 and u_2. Fortunately, u_2 is a function of T_2, and so this problem can be solved either iteratively, or by using the relationship among c_v, T_2, and u_2 to solve for T_2. The first method is preferred, because the tables consider variable specific heats and it is best not to introduce a constant specific heat element to the solution—however, the second approach could yield a reasonable approximation.

Through iterative calculations, $\mathbf{T_2 = 395}$ **K.**

Analysis: At first glance, this temperature may cause you concern. Notice that the final temperature in the tank is higher than both the supply line temperature and the initial temperature in the tank. How can this be? How can the air be heating up, particularly if the tank is insulated? This is a result of flow work being added to the system through the enthalpy of the inlet flow. This additional energy needs to be conserved, and as a result it is converted to thermal energy in the air in the tank. So the flow work results in a temperature increase. In a real-life situation, it is also likely that there will be kinetic energy of the incoming air, and that will need to be converted to thermal energy as the air comes to rest in the tank. However, in this problem, we are assuming that the inlet kinetic energy is negligible, and so that particular effect is not responsible for the temperature increase in the problem.

Notice how quickly the complexity of even this simple scenario could rapidly increase. For example, if the flow into the tank had a time-varying temperature, the problem would immediately increase in difficulty. If there is a leak in the tank, and if the flow properties exiting the tank vary with time, the problem would become very difficult. So, when solving a transient problem, it is very important to make as many simplifying assumptions for the first law as possible while still retaining the important features of the problem.

QUESTION FOR THOUGHT/DISCUSSION

We have modeled many devices, such as turbines and compressors, as steady-state, steady-flow devices. Consider one or two of these devices, and think about the processes involved with starting up the device. What features of the start-up process would move the modeling of the device into an unsteady process, and how should those be considered when applying the First Law of Thermodynamics?

4.4 FIRST LAW OF THERMODYNAMICS IN CLOSED SYSTEMS

So far, we have considered open systems where flow can cross the system boundary. There are many systems, though, that have no mass entering or leaving the system, such as a piston–cylinder device, as shown in **Figure 4.22**. As previously discussed, solids and closed containers often are considered closed systems. Closed systems can be considered special forms of open

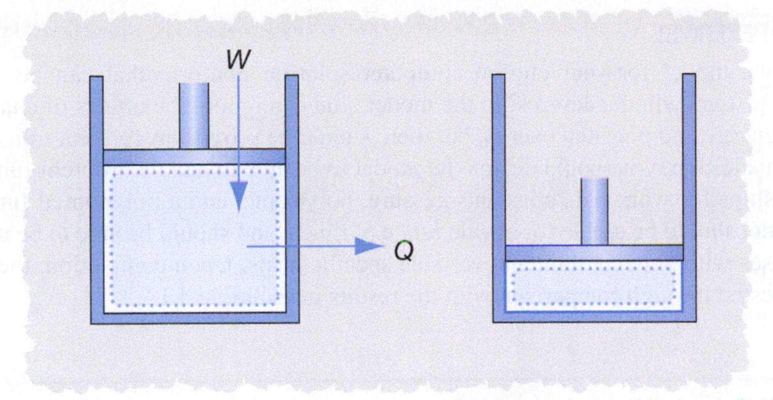

FIGURE 4.22 On the left is a diagram of a piston–cylinder device at the start of a compression process, and on the right is a diagram after the compression process. Work adds energy to the system, whereas energy leaves the system in the form of heat transfer from the gas as it cools.

systems—albeit ones without mass flowing in or out of the system. Therefore, we can derive out an appropriate form of the first law to use with closed systems from Eq. (4.12):

$$\dot{Q} - \dot{W} + \sum_{i=1}^{j} \dot{m}_i \left(h + \frac{V^2}{2} + gz \right)_i - \sum_{e=1}^{k} \dot{m}_e \left(h + \frac{V^2}{2} + gz \right)_e = \frac{d\left(mu + m\frac{V^2}{2} + mgz \right)_{\text{system}}}{dt} \tag{4.12}$$

Because no mass flows in or out of the system, the mass flow rate terms are equal to 0. In addition, the mass of the system does not change with respect to time, so it can be pulled out of the derivative on the right-hand side of Eq. (4.12). The resulting equation for the first law is

$$\dot{Q} - \dot{W} = m \frac{d\left(u + \frac{V^2}{2} + gz \right)_{\text{system}}}{dt} \tag{4.31}$$

Equation (4.31) is the rate balance version of the first law as applied to closed systems. Although this form is used at times, it is more common to integrate Eq. (4.31) with respect to time to consider an entire process taking place in a closed system. When Eq. (4.31) is integrated with respect to time, the resulting form of the first law is

$$Q - W = m\left(u_2 - u_1 + \frac{V_2^2 - V_1^2}{2} + g(z_2 - z_1) \right) \tag{4.32}$$

where state 1 is the initial state and state 2 is the final state. To apply Eq. (4.32) to a process or series of processes acting in a closed system, we either need to know the end states and use Eq. (4.32) to determine either the heat transfer or work done, or we need to know the heat transfer and work and determine either the final or initial state.

As with open systems, the first law can be simplified further for closed systems through making assumptions. Again, these are made through inspection of the particular situation to which the first law is applied. In many engineering applications, the changes in kinetic and potential energy of a closed system are negligible and can be ignored. Also, some closed systems have no work interactions, and some are insulated and have no heat transfer occurring. Several examples of the first law being applied to closed systems can be found next.

STUDENT EXERCISE

Develop a model for your chosen computer solution platform that can be used to analyze piston–cylinder devices. In the model, you can ignore the effects of changes in kinetic energy and potential energy, but there should be provisions for heat transfer and work. In addition, you should design the model to be able to handle different functional relationships for work (i.e., constant-pressure, polytropic, and a polynomial function). The model should be usable for a wide range of fluids, and should be able to be used for ideal gases with both constant and variable specific heats. Upon completion, the model can be tested through comparison with the results in Example 4.14.

▶ EXAMPLE 4.14

2 kg of air, initially at 295 K and 325 kPa, are heated at constant pressure in a piston–cylinder device. The final temperature of the air is 405 K. Assuming that the air behaves as an ideal gas with constant specific heat, determine the amount of heat added during the process.

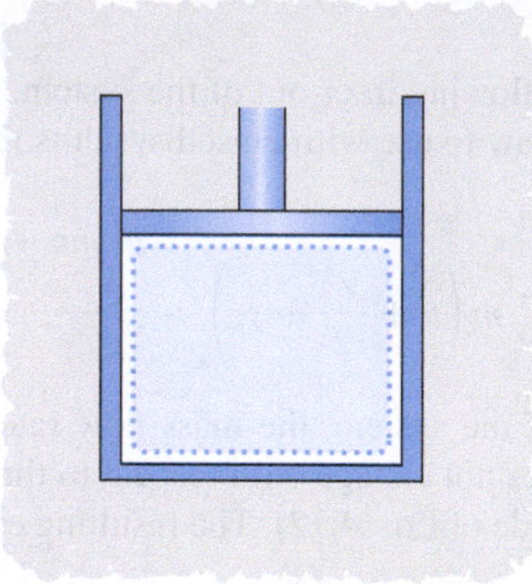

Given: $m = 2$ kg, $T_1 = 295$ K, $P_1 = P_2 = 325$ kPa (constant pressure), $T_2 = 405$ K

Find: Q (the heat added during the process)

Solution: The piston–cylinder device is a closed system, as no mass enters or exits the system during the process. The working fluid is air, which is to be treated as an ideal gas with constant specific heats.

Assume: Air behaves as an ideal gas with constant specific heats

This is a constant-pressure process in a closed system

$\Delta KE = \Delta PE = 0$

From Eq. (4.32), the first law for closed systems is

$$Q - W = m\left(u_2 - u_1 + \frac{V_2^2 - V_1^2}{2} + g(z_2 - z_1)\right)$$

Applying the assumptions yields

$$Q - W = m(u_2 - u_1) = mc_v(T_2 - T_1)$$

The work is moving boundary work for a constant-pressure process:

$$W = \int P\,d\forall = P(\forall_2 - \forall_1)$$

The volumes are found from the ideal gas law:

$$\forall_1 = \frac{mRT_1}{P_1} = \frac{(2\text{ kg})(0.287\text{ kJ/kg}\cdot\text{K})(295\text{ K})}{325\text{ kPa}} = 0.5210\text{ m}^3$$

$$V_2 = \frac{mRT_2}{P_2} = \frac{(2\text{ kg})(0.287\text{ kJ/kg}\cdot\text{K})(405\text{ K})}{325\text{ kPa}} = 0.7153\text{ m}^3$$

$$W = (325\text{ kPa})(0.7153 - 0.5210)\text{ m}^3 = 63.15\text{ kJ}$$

So $$Q = (2\text{ kg})(0.718\text{ kJ/kg}\cdot\text{K})(405 - 295)\text{ K} + 63.15\text{ kJ} = \mathbf{221\text{ kJ}}$$

Analysis: Because the air increased in temperature as the pressure stayed constant, it is expected that the final volume would be larger than the initial volume—as was calculated. This indicates that moving boundary work is done, giving a positive value for the work. As work was done, and the air temperature increased, we would also expect the heat transfer for the process to be positive, and larger than the work. This is what is found, making the answer at least logical.

A computer model of a piston–cylinder device can easily perform these calculations, as well as demonstrate trends that could be expected. For the constant-pressure process described in this example, the following plot shows the amount of work and heat transfer to be expected for various final temperatures. As the final temperature increases at constant pressure, the final volume increases; this increases the amount of work that is done. As expected, the amount of heat transfer added to the air also increases with increasing final temperature, but note that it increases both due to the higher temperature and the contribution of additional work done by the expanding gas.

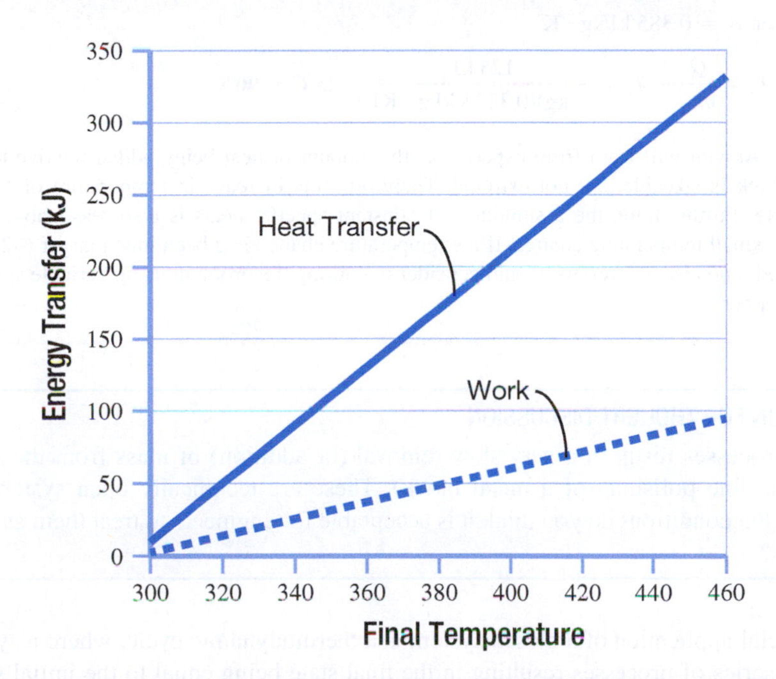

▶ EXAMPLE 4.15

A 5-kg block of copper receives 125 kJ of heat in a furnace. Initially, the copper is at 25°C. Determine the final temperature of the copper.

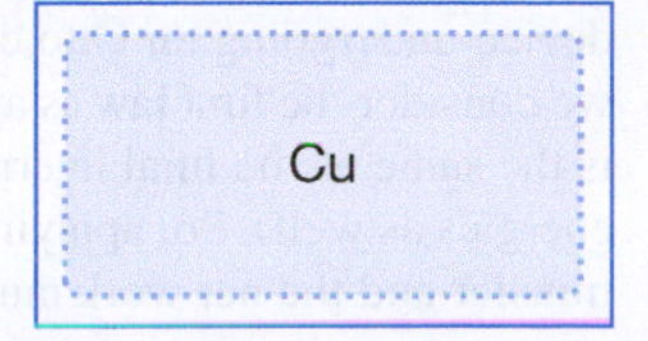

Given: $m = 5$ kg

$Q = 125$ kJ

$T_1 = 25$°C

Find: T_2 (the final temperature)

Solution: A solid object of constant mass is a closed system. As it does not appear to be undergoing any large change in pressure, the copper can be treated as an incompressible substance. To a first approximation, assume constant specific heats in this problem.

Assume: Closed system

Copper acts as an incompressible substance with constant specific heats

$W = \Delta KE = \Delta PE = 0$ (No work because the copper block has a constant volume and so there is no moving boundary work. There is no indication of any other work modes being present.)

From Eq. (4.32), the first law for closed systems is

$$Q - W = m\left(u_2 - u_1 + \frac{V_2^2 - V_1^2}{2} + g(z_2 - z_1)\right)$$

Applying the assumptions yields

$$Q = m(u_2 - u_1) = mc_p(T_2 - T_1)$$

For copper, $c_p = 0.385$ kJ/kg · K

Solving: $T_2 = \dfrac{Q}{mc_p} + T_1 = \dfrac{125 \text{ kJ}}{(5 \text{ kg})(0.385 \text{ kJ/kg} \cdot \text{K})} + 25°\text{C} = \mathbf{90°C}$

Analysis: As you will learn from experience, the amount of heat being added relative to the mass of the block is sizeable, but not extreme. Therefore, this increase in temperature of 65°C seems reasonable. Furthermore, the assumption of constant specific heats is also reasonable due to the relatively small temperature change. If the temperature change had been much larger (~200°C) or if we needed a precise answer, we would consider reworking the problem using variable specific heats for the copper.

QUESTION FOR THOUGHT/DISCUSSION

Some processes result in a very slow removal (or addition) of mass from the system (e.g., the fine polishing of a metal block). These are technically open systems, but under what conditions do you think it is acceptable for engineers to treat them as closed systems?

A special application of a closed system is a thermodynamic cycle, where a system will undergo a series of processes resulting in the final state being equal to the initial state. One type of cycle involves a fluid proceeding through several devices, eventually returning to its initial state so it can repeat the process of flowing through the devices. An example of this is the Rankine cycle, shown in **Figure 4.23**, which is used to model simple steam power plants. A second type of cycle consists of the fluid remaining in one device while undergoing several processes in that device. **Figure 4.24** shows an example of this for a gas in a piston–cylinder device undergoing an Otto cycle; the Otto cycle is used to model a spark-ignition engine. If we consider the first law as applied over the entire cycle, we see that the initial internal energy is the same as the final internal energy ($u_1 = u_2$) (the same is true of the kinetic and potential energies as well). So, applying Eq. (4.32), we see that for a thermodynamic cycle, the net heat transfer and the net work must be equal:

$$Q_{\text{cycle}} = W_{\text{cycle}} \tag{4.33}$$

where these quantities are the summations of the individual process's heat transfers and works.

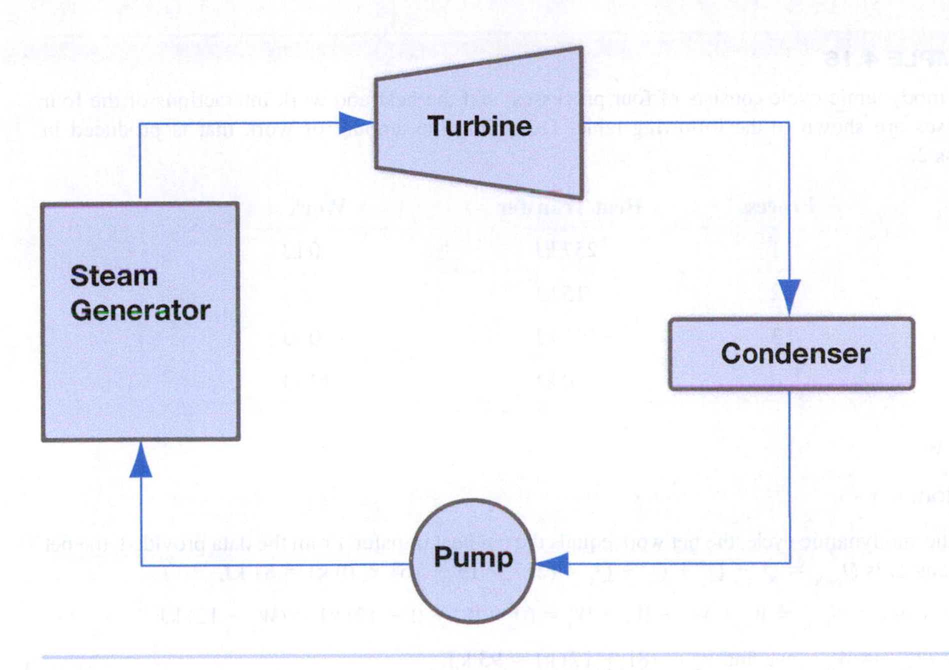

FIGURE 4.23 The four components of a simple Rankine cycle.

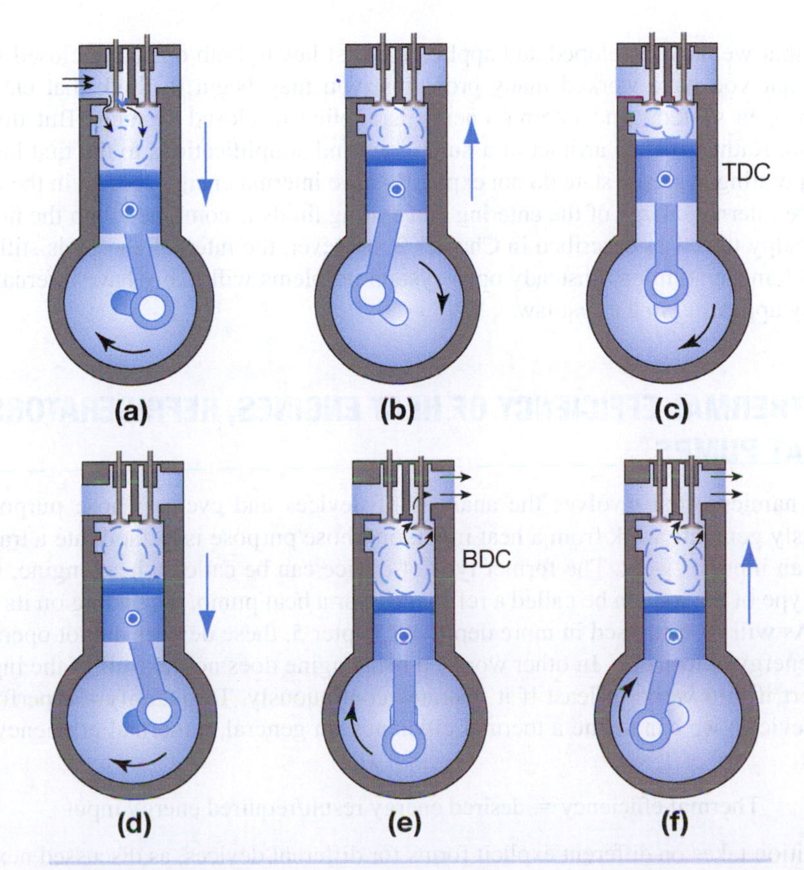

FIGURE 4.24 A series of diagrams showing a piston–cylinder device at various stages of operation in a spark-ignition engine cycle.

▶ EXAMPLE 4.16

A thermodynamic cycle consists of four processes, and the heat and work interactions of the four processes are shown in the following table. Determine the amount of work that is produced in process 2.

Process	Heat Transfer	Work
1	257 kJ	0 kJ
2	−15 kJ	?
3	−161 kJ	0 kJ
4	0 kJ	−12 kJ

Find: W_2

Solution:

For a thermodynamic cycle, the net work equals the net heat transfer. From the data provided, the net heat transfer is $Q_{cycle} = Q_1 + Q_2 + Q_3 + Q_4 = (257 - 15 - 161 + 0)$ kJ = 81 kJ.

The net work is $W_{cycle} = W_1 + W_2 + W_3 + W_4 = (0 + W_2 + 0 - 12)$ kJ $= (W_2 - 12)$ kJ.

Setting $Q_{cycle} = W_{cycle}$, we find $W_2 = (81 + 12)$ kJ = **93 kJ.**

Now that we have developed and applied the first law to both open and closed systems, and now that you have worked many problems, you may begin to think that enthalpy is applied to open systems and internal energy is applied to closed systems. But this is not strictly true. Rather, it is an artifact of assumptions and simplifications to the first law. Open systems operating at steady state do not explicitly have internal energy present in the first law, because the internal energy of the entering and exiting fluids is combined with the flow work in the enthalpy terms, as described in Chapter 2; however, the internal energy is still present in the problem. In addition, unsteady open system problems will likely have internal energy specifically appearing in the first law.

4.5 THERMAL EFFICIENCY OF HEAT ENGINES, REFRIGERATORS, AND HEAT PUMPS

Thermodynamics often involves the analysis of devices and cycles whose purpose is to continuously generate work from a heat input, or whose purpose is to facilitate a transfer of heat with an input of work. The former type of device can be called a heat engine, whereas the latter type of device can be called a refrigerator or a heat pump, depending on its specific purpose. As will be discussed in more depth in Chapter 5, these devices do not operate with a perfect energy conversion. In other words, a heat engine does not take all of the input heat and convert it into work, at least if it operates continuously. To quantify the performance of these devices, we can define a thermal efficiency. In general, a thermal efficiency can be defined as

$$\text{Thermal efficiency} = \text{desired energy result/required energy input}$$

This definition takes on different explicit forms for different devices, as discussed next.

4.5.1 Heat Engines

A heat engine is a device that continuously produces work while receiving heat from a hot thermal reservoir. A thermal reservoir is simply a heat source or sink that is so large that it does not change temperature even if heat is added or removed. For example, if you poured a small amount of hot water on the ground and measured the temperature of the ground 10 m away, you would not notice a change in temperature—the ground would be considered a thermal reservoir. Because these devices operate in a cycle, and because not all heat can be converted to work (to be discussed more in Chapter 5), heat must be rejected to some cool thermal reservoir. A generic diagram of a heat engine is shown in Figure 4.25. A steam or gas power plant can be viewed as a heat engine, as can an internal-combustion engine.

The purpose of a heat engine is to produce work or power, and so its desired energy result is work, W. To get this work, the heat engine requires an energy input of heat from the hot thermal reservoir, Q_H. Therefore, the thermal efficiency of the heat engine, η, is

$$\eta = \frac{W}{Q_H} \tag{4.34a}$$

Note that Eq. (4.34a) can also be written with the work and heat transfer quantities being on a rate basis:

$$\eta = \frac{\dot{W}}{\dot{Q}_H} \tag{4.34b}$$

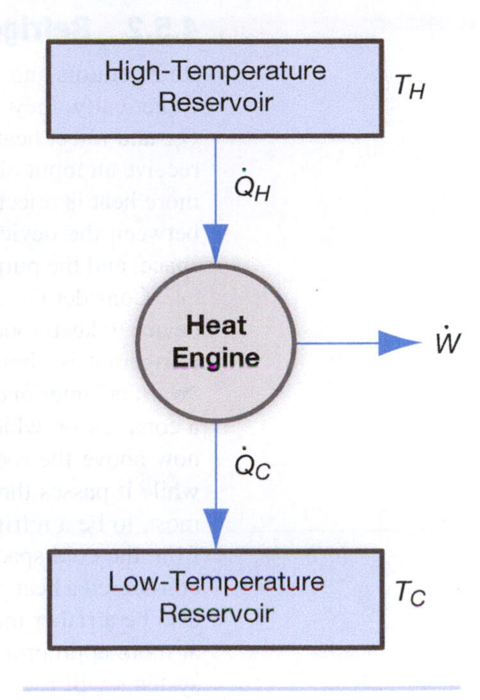

FIGURE 4.25 A heat engine can be modeled as a prime mover (such as a turbine) operating between two thermal reservoirs.

EXAMPLE 4.17

The cycle described in Example 4.16 can be modeled as a heat engine, because work is produced while heat is received from a heat source and rejected to a heat sink. Determine the thermal efficiency of the heat engine thermodynamic cycle described in Example 4.16.

Find: η

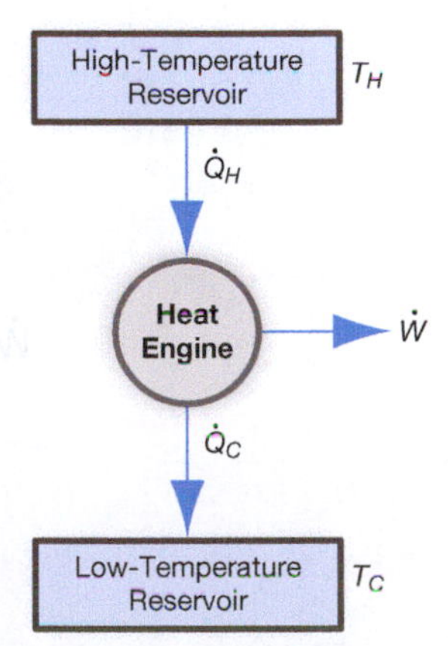

Solution:

The net work generated in the cycle in Example 4.16 is $W = 81$ kJ.

The heat input occurs in process 1, so $Q_H = 257$ kJ.

The thermal efficiency is $\eta = W/Q_H = 81\text{ kJ}/257\text{ kJ} = \mathbf{0.315}$.

Analysis: The thermal efficiency is less than one, and therefore in this case is possible.

QUESTION FOR THOUGHT/DISCUSSION

A power plant may have a thermal efficiency of 40%. This means that 60% of the input heat is lost to the environment and does not produce power. How may this rejected heat alter the local environment around the power plant? What are some ways of transferring the waste heat to the environment, and which are less impactful on the local environment? Do you think that this is a problem on a global scale?

4.5.2 Refrigerators and Heat Pumps

Refrigerators and heat pumps can be viewed as the same device but with a different purpose. Generically, they are shown in Figure 4.26. Both devices remove heat from a cold space, Q_C, and reject heat to a warm space, Q_H, in a cyclic fashion. To do this, both devices need to receive an input of work, W. If we consider the first law as applied to cycles, we will find that more heat is rejected to the warm space than is removed from the cold space. The difference between the devices is that the purpose of the refrigerator is to remove heat from the cold space, and the purpose of the heat pump is to deliver heat to the warm space.

Consider the common household refrigerator shown in Figure 4.27. Most people use this device to keep food cold inside the refrigerator. As described previously, to achieve this, a cold refrigerant is circulated inside the walls of the refrigerator and heat is transferred from the "warmer" interior of the refrigerator to the colder refrigerant. The refrigerant then passes through a compressor, which heats the fluid as it increases its pressure. The refrigerant temperature is now above the room temperature, and so the refrigerant rejects the heat to the cooler room while it passes through coils on the outside of the refrigerator. This device is considered, by most, to be a refrigerator because it is used to keep food cold—its purpose is to remove heat from the cold space. But if we wanted to use this same device to heat the room, it would be considered a heat pump because its primary purpose would be to heat the warm space. It would also be a rather inefficient way to heat the room. However, systems can be designed to work at more appropriate temperatures and operate as heat pumps. For example, some geothermal systems will use the area underground as an evaporator and the area inside a building as the condenser of the system, thus delivering heating to the building rather efficiently.

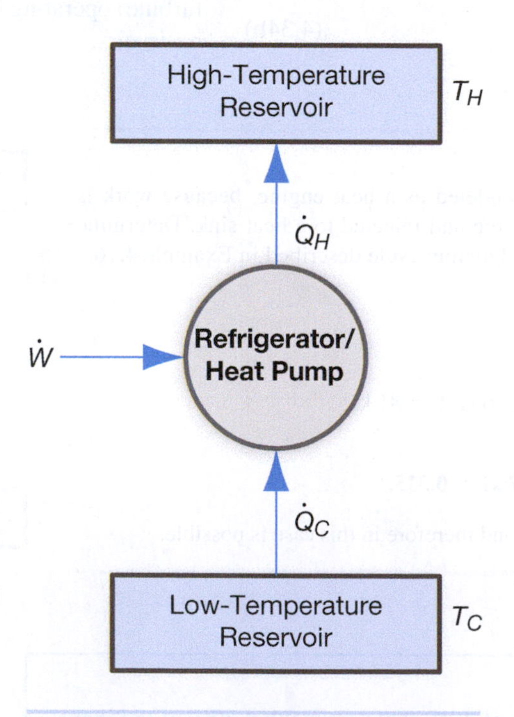

FIGURE 4.26 Both a refrigerator and a heat pump can be modeled as a device operating between two thermal reservoirs, receiving heat from the cold reservoir and rejecting heat to the warm reservoir.

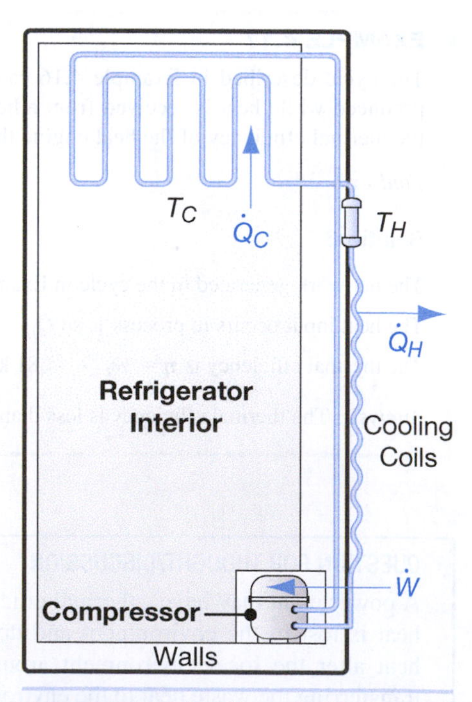

FIGURE 4.27 A schematic diagram of a basic household refrigerator.

You may note that air conditioners also have the purpose of removing heat from a cold space, such as in **Figure 4.28**. Air conditioners, when used for this purpose, are just refrigerators. The different terms are generally used when referring to either the temperatures under consideration or to the type of object that occupies a cold space. So the devices are called refrigerators more commonly when lower temperatures are sought or when inanimate objects are in the cool space, whereas air conditioners generally would operate at a somewhat higher temperature and when animate objects (people, animals) are present. The term *air conditioning* can refer to any change in the state of the air, and so a heat pump, such as that shown in **Figure 4.29**, also performs an air-conditioning operation.

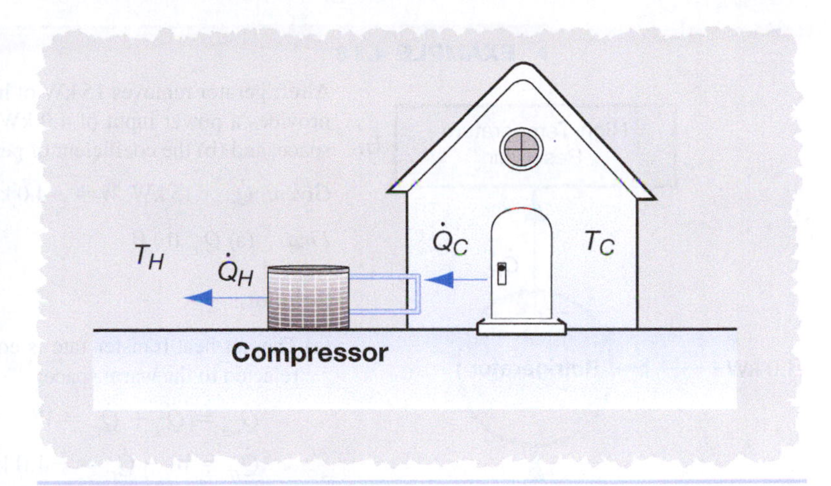

FIGURE 4.28 A schematic diagram of an air-conditioning system, indicating the locations of the thermal reservoirs.

When considering most refrigerators, more heat is removed from the cool space than work is added to the device. If we consider the desired energy result to be the removal of heat, Q_C, and the required energy input to be the work, W, we would see that the thermal efficiency of the refrigerator is usually greater than 1. Because most people do not like to consider efficiencies to be greater than 100%, for refrigerators and heat pumps the thermal efficiency is renamed the coefficient of performance, which is the same thing as the thermal efficiency, but with a different name.

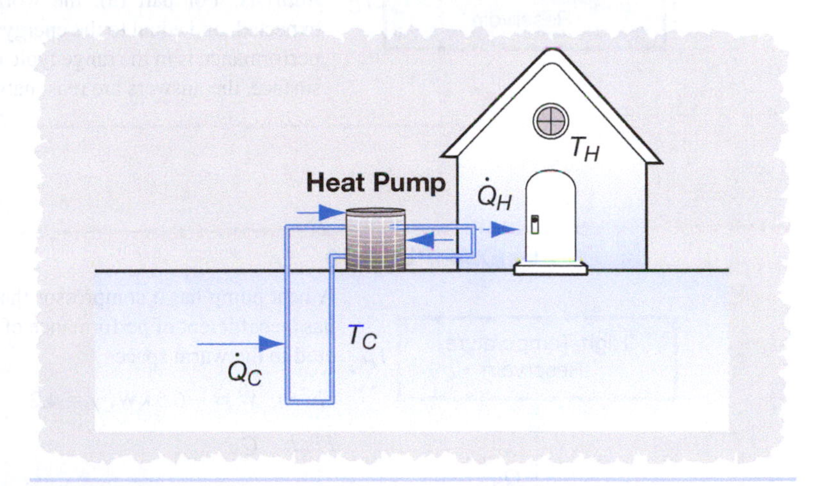

FIGURE 4.29 A schematic diagram of a heat pump system, showing the locations of the thermal reservoirs and appropriate heat transfers.

For a refrigerator, the coefficient of performance, β, is

$$\beta = \frac{Q_C}{|W|} = \frac{\dot{Q}_C}{|\dot{W}|} \tag{4.35}$$

Note that we use the absolute value of the work to keep the coefficient of performance a positive number because the work is added to the system and is therefore a negative value. For a heat pump, the desired energy result is the rejection of heat to the warm space, with the required energy input being the work. So the coefficient of performance of the heat pump, γ, is

$$\gamma = \frac{|Q_H|}{|W|} = \frac{|\dot{Q}_H|}{|\dot{W}|} \tag{4.36}$$

Again, absolute values are used to maintain positive quantities in the calculation.

▶ EXAMPLE 4.18

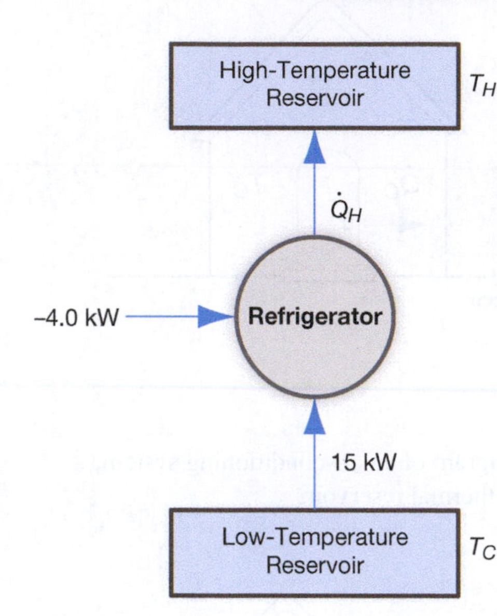

A refrigerator removes 15 kW of heat from a cold space, using a compressor that provides a power input of 4.0 kW. Determine (a) the heat rejected to the warm space, and (b) the coefficient of performance of the refrigerator.

Given: $\dot{Q}_C = 15$ kW, $\dot{W} = -4.0$ kW

Find: (a) $\dot{Q}_H$, (b) β

Solution:

(a) The net heat transfer rate is equal to the net power, so we can find the heat rejected to the warm space:

$$\dot{Q}_{\text{net}} = \dot{Q}_H + \dot{Q}_C = \dot{W}$$

$$\dot{Q}_H = \dot{W} - \dot{Q}_C = -4.0\text{ kW} - 15\text{ kW} = \mathbf{-19\text{ kW}}$$

(b) $$\beta = \frac{\dot{Q}_C}{|\dot{W}|} = \frac{15\text{ kW}}{4.0\text{ kW}} = \mathbf{3.75}$$

Analysis: For part (a), the work rejected to the warm space is negative, as expected, and equal to the energy that was available. In part (b), the coefficient of performance is in the range typically expected for a refrigerator. Therefore, on the surface, the answers are reasonable.

▶ EXAMPLE 4.19

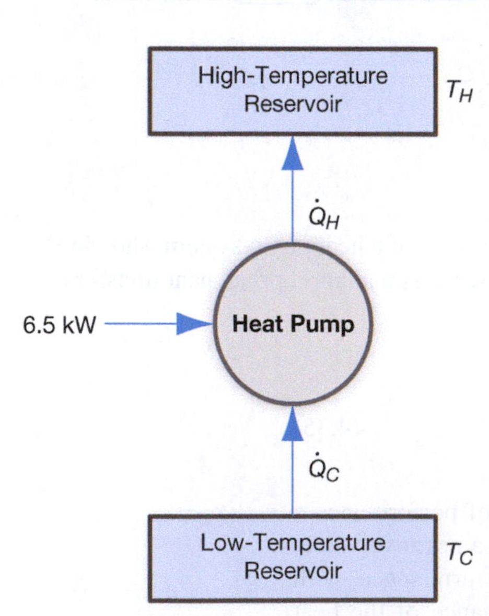

A heat pump has a compressor that operates on 6.5 kW of power. The heat pump has a coefficient of performance of 4.2. Determine the amount of heat that is delivered to the warm space.

Given: $\dot{W} = -6.5$ kW, $\gamma = 4.2$

Find: $\dot{Q}_H$

Solution:

$$\gamma = \frac{|\dot{Q}_H|}{|\dot{W}|}$$

so

$$|\dot{Q}_H| = \gamma|\dot{W}| = (4.2)(6.5\text{ kW}) = 27.3\text{ kW}$$

Because this is heat that is rejected to the warm space, it is heat out of the heat pump:

$$\dot{Q}_H = \mathbf{-27.3\text{ kW}}$$

Analysis: The result is a reasonable value to expect, as it should be larger than the power input by a factor of 4.2.

> **QUESTION FOR THOUGHT/DISCUSSION**
>
> Before mechanical refrigerators were invented, a fairly common method of keeping things such as food cool was to use an icehouse, where blocks of ice were cut from rivers or lakes in the winter and stored in the icehouse until they melted. What advantages do mechanical refrigerators have, and what advantages might icehouses have?
>
> Heat pumps are seen by many as being more energy efficient than typical furnaces or other heating systems. Why aren't heat pumps in more widespread use?

Summary

In this chapter, we developed the First Law of Thermodynamics in equation form from a simple statement on the conservation of energy. We learned how to make appropriate assumptions to simplify the general form of the first law. We applied the first law to open systems, both at steady state and at unsteady state, and to closed systems. To analyze open system problems, we often need to use the conservation of mass, and so we developed the necessary equations to describe this physical law. Finally, we explored the concept of thermal efficiency, and we applied this concept as the thermal efficiency to heat engines and as the coefficient of performance for refrigerators and heat pumps.

Although the first law often is used alone in analyzing engineering devices, it does have its limits. Namely, the first law can suggest that certain processes that we know can't work are possible. To help us determine whether or not a process can work in nature, we can apply the Second Law of Thermodynamics. The principles learned in this chapter will be used in the following chapters to help us analyze processes and devices via the Second Law of Thermodynamics.

KEY EQUATIONS

Conservation of Mass:

General:

$$\sum_{i=1}^{j} \dot{m}_i - \sum_{e=1}^{k} \dot{m}_e = \left(\frac{dm}{dt}\right)_{\text{system}} \tag{4.1}$$

Mass Flow Rate, 1-D Flow:

$$\dot{m} = \rho VA = \frac{VA}{v} \tag{4.7}$$

First Law of Thermodynamics:

General:

$$\dot{Q} - \dot{W} + \sum_{i=1}^{j} \dot{m}_i\left(h + \frac{V^2}{2} + gz\right)_i - \sum_{e=1}^{k} \dot{m}_i\left(h + \frac{V^2}{2} + gz\right)_e = \frac{d\left(mu + m\frac{V^2}{2} + mgz\right)_{\text{system}}}{dt} \tag{4.12}$$

Steady-State, Steady-Flow, Multiple-Inlet, Multiple-Outlet Open Systems:

$$\dot{Q} - \dot{W} + \sum_{i=1}^{j} \dot{m}_i \left(h + \frac{V^2}{2} + gz \right)_i - \sum_{e=1}^{k} \dot{m}_e \left(h + \frac{V^2}{2} + gz \right)_e = 0 \tag{4.13}$$

Steady-State, Steady-Flow, Single-Inlet, Single-Outlet Open Systems:

$$\dot{Q} - \dot{W} = \dot{m} \left(h_2 - h_1 + \frac{V_2^2 - V_1^2}{2} + g(z_2 - z_1) \right) \tag{4.15}$$

Closed Systems (Total Basis):

$$Q - W = m \left(u_2 - u_1 + \frac{V_2^2 - V_1^2}{2} + g\,(z_2 - z_1) \right) \tag{4.32}$$

Thermal Efficiency—Heat Engine:

$$\eta = \frac{W}{Q_H} = \frac{\dot{W}}{\dot{Q}_H} \tag{4.34}$$

Coefficient of Performance—Refrigerator:

$$\beta = \frac{Q_c}{|W|} = \frac{\dot{Q}_c}{|\dot{W}|} \tag{4.35}$$

Coefficient of Performance—Heat Pump:

$$\gamma = \frac{Q_H}{|W|} = \frac{\dot{Q}_H}{|\dot{W}|} \tag{4.36}$$

PROBLEMS

Note: Some of the following problems expect you to use the component models you developed in this chapter, but you can also use your models for many of the other problems as well.

4.1 A tank has two inlets and one outlet. At one instant in time, the tank receives 3.0 kg/s of water through one inlet, and 4.6 kg/s of water through the other inlet, whereas 2.5 kg/s of water flows through the outlet. Determine the rate of change of the mass of the water in the tank at this instant.

4.2 A mixing chamber takes flows from three inlet pipes and combines them into one outlet flow. The chamber operates at steady-state conditions with steady flows. The outlet flow is measured as 3.1 kg/s of water, and two of the inlet flows are known to be 1.2 kg/s and 0.9 kg/s of water. What is the mass flow rate in the third inlet?

4.3 A fire hydrant is fed by one pipe, but then distributes the flow into two hoses. Suppose the fire hydrant receives 0.1 m^3/s of liquid water whose density is 1000 kg/m^3. One of the exit hoses is using 38 kg/s of water. What is the mass flow rate through the second hose?

4.4 Engine oil, with a density of 890 kg/m^3, flows through a circular duct in an engine whose diameter is 5 mm with a volumetric flow rate of 30 cm^3/s. Determine the velocity and mass flow rate of the engine oil.

4.5 Liquid water, with a density of 1000 kg/m^3, flows at a velocity of 8.5 m/s through a circular pipe with a diameter of 42.0 mm. Determine the volumetric flow rate and the mass flow rate of the water.

4.6 Liquid water, with a density of 1000 kg/m^3, flows with a volumetric flow rate of 0.035 m^3/s through a circular pipe with a diameter of 7.5 cm. Determine the velocity and mass flow rate of the water.

4.7 Air, with a pressure of 98.5 kPa and a temperature of 22.0°C, flows through a square duct at 15.0 m/s. Each side of the duct has a length of 15.0 cm. (a) Determine the volumetric flow rate and the mass flow rate of the air. For parts (b) and (c), you should develop a model for the duct. (b) Plot the volumetric flow rate and the mass flow rate of the air for air pressures ranging from 98.5 kPa to 500 kPa while the temperature is maintained at 22.0°C. (c) Plot the volumetric flow and the mass flow rate of the air as the pressure is maintained at 98.5 kPa but as the temperature varies from 22.0°C to 150.°C.

4.8 Steam at 500°C and 10.0 MPa flows into a turbine through a circular duct of 5.0 cm diameter. The volumetric flow rate of the steam is 8.0 m^3/s. Determine the velocity and the mass flow rate of the steam entering the turbine.

4.9 Liquid water with a density of 1000 kg/m^3 flows through a nozzle. At the inlet of the nozzle, the water has a velocity of 3.0 m/s, and the velocity of the water at the outlet is 40.0 m/s. The cross-sectional area of the inlet is 0.005 m^2. (a) Determine the cross-sectional area of the outlet, assuming that the nozzle operates at steady-state, steady-flow conditions. (b) Using your equations, for a constant inlet condition, plot the cross-sectional area of the nozzle outlet as the outlet velocity varies from 10.0 m/s to 100.0 m/s.

4.10 Liquid ethanol with a density of 790 kg/m^3 flows through a nozzle. At the inlet of the nozzle, the ethanol has a velocity of 3 m/s, and the velocity of the ethanol at the outlet is 35 m/s. The cross-sectional area of the outlet is 0.005 m^2. (a) Determine the cross-sectional area of the outlet, assuming that the nozzle operates at steady-state, steady-flow conditions. (b) Using your equations, for a constant outlet condition, plot the cross-sectional area of the nozzle inlet as the inlet velocity varies from 10.0 m/s to 100 m/s.

4.11 Air exits a compressor operating at steady-state, steady-flow conditions of 150°C, 825 kPa, with a velocity of 10 m/s through a circular pipe of 5.0 cm diameter. (a) Determine the outlet volumetric flow rate and the mass flow rate of the air. (b) If the air enters the compressor through an inlet at 20.0°C and 100 kPa and a velocity of 1.0 m/s, determine the volumetric flow rate of the air entering the compressor and the required cross-sectional area of the inlet. (c) Using your equations, plot the inlet volumetric flow rate and the inlet cross-sectional area as the inlet air velocity varies between 0.25 m/s and 10.0 m/s.

4.12 Superheated steam at 2 MPa and 400°C enters a turbine through 4 circular ducts, each with a diameter of 8 cm, at a velocity of 250 m/s. The steam exits the turbine through a single duct with a diameter of 120 cm, as a saturated substance at 10 kPa. The turbine operates at steady-state, steady-flow conditions. (a) Determine the mass flow rate of the steam through the turbine. (b) Plot the exit volumetric flow rate and exit velocity of the steam from the turbine as a function of exit quality, with qualities ranging from 0.88 to 1.0.

4.13 A tank receives water through two inlets and empties through a single outlet. The flow rates through the two inlets are 2.5 kg/s and 0.8 kg/s. The cross-sectional area of the outlet is 0.00015 m^3. Considering the density of the water to be 1000 kg/m^3, determine the exit velocity of the water if the tank is operating under steady-state, steady-flow conditions.

4.14 You are attempting to water some flowers in a garden and only have a leaky watering can to do so. After consulting your local agronomist, you know that the flowers need 2.1 kg of water to flourish. The watering can holds, when filled, 5.0 L of water. The water leaks from the can at a rate of 0.12 kg/s. It takes 30 seconds to walk from the spigot where you fill the watering can to the brim to the flowers. Any water that leaks from the can while you are watering the flowers falls onto the flowers. How many trips between the spigot and the flower bed will you need to make to give the flowers the proper amount of water?

4.15 The mass of water in a container varies by the following relationship:

$$dm/dt = 6.0 - 0.5t + 0.003t^2$$

where t is in seconds and m is in kg. Determine the change in mass of the water in the container during a 5-s period of time.

4.16 The mass of water in a tank varies by the following relationship:

$$dm/dt = -0.1\text{ m}$$

where t is in seconds and m is in kg, with the initial condition of $m = 10$ kg. Plot the mass of water in the tank for a time from 0 to 30 s. At what time does the mass of water in the tank equal 1 kg?

4.17 During a complex process, a system experiences a total increase of energy of 15 kJ. The process involves an energy loss via heat transfer of 10 kJ, a gain of 50 kJ of energy through mass flowing through an inlet, and a loss of 12 kJ of energy from mass flowing through an outlet. How much energy was transferred via work in the process?

4.18 During a process, a system receives 83.5 kJ of energy via heat transfer, loses 72.1 kJ of energy through work, and receives 7.2 kJ of energy through mass flowing through an inlet while losing 8.5 kJ of energy through mass flowing through an outlet. What is the change in energy of the system during this process?

4.19 During a process, a system loses 35 kJ of energy via heat transfer, gains 51 kJ of energy through work, receives 10.5 kJ of energy through mass flowing through the inlet, and experiences a net increase of energy of 15.2 kJ. What is the amount of energy lost from flow through the outlet during the process?

4.20 A device operates under steady-state, steady-flow conditions. The device receives 45.2 kW of heat, the mass flowing in contains 165 kW of energy, and the device produces 65.6 kW of power. Determine the rate of energy flowing out of the system with the mass flowing through the outlet.

4.21 A closed system receives 15.0 kW of power while losing 8.2 kW of heat at one instant in time. Determine the time rate of change of the energy of the system at that instant.

4.22 Liquid water flows through a rough pipe, entering at 20°C and exiting at 22°C, as it warms due to friction with the walls. The mass flow rate of the water is 15.0 kg/s. Assuming that the water is at atmospheric pressure, the pipe is insulated, and there is no work done, what is the rate of energy change in the pipe?

4.23 Steam enters an insulated system pipe with a mass flow rate of 6.0 kg/s, a temperature of 200°C, and a pressure of 150 kPa. The steam exits with the same mass flow rate, but at a temperature of 160°C and a pressure of 140 kPa. Assuming that no work is done, what is the rate of energy change in the pipe?

4.24 A pipe operates at steady-state, steady-flow conditions. Steam enters the pipe at 300°C and 250 kPa, and exits at 250°C and 200 kPa. The volumetric flow rate of the steam entering the pipe is 1.2 m^3/s. Assume that there are no work interactions. (a) Determine the heat transfer rate to or from the pipe. (b) Using your equations, plot the rate of heat transfer for the same inlet conditions and outlet pressure, but for outlet temperatures varying from 150°C to 350°C.

4.25 A pipe operates at steady-state, steady-flow conditions. R-134a enters the pipe at 50°C and 410 kPa absolute, and exits at 345 kPa absolute and a quality of 0.90. The volumetric flow rate of the R-134a entering the pipe is 0.0500 m^3/s. Assume that there are no work interactions.

(a) Determine the heat transfer rate to or from the pipe. (b) Using your equations, plot the rate of heat transfer for the same inlet conditions and outlet pressure, but for outlet qualities varying between 0.0 and 1.0.

4.26 You need to prepare to perform a first law analysis on a device. Upon inspection, you see that the device is small, uninsulated, has one inlet and one outlet, and has no obvious work inputs or outputs. The operator tells you that steam enters through the inlet and liquid water exits through the outlet. The operator says that he turns on the device at the beginning of the day, and turns it off at the end, and never has to pay any attention to its operation. Beginning with the general equation for the First Law of Thermodynamics, develop a simplified expression that should be suitable for use with this device.

4.27 You need to prepare to perform a first law analysis on a device. The device is covered in insulation and appears to have a plunger on the top that can move up and down during operation. There are also electrical wires entering and exiting the walls of the device. The device appears to be sealed. Beginning with the general equation for the first law, develop a simplified expression that should be suitable for use with this device to calculate the overall behavior of this device for a period of time.

4.28 You need to prepare to perform a first law analysis on a device. The device contains two inlets and one outlet. The device is very large, with the two inlets located ~30 m above the outlet. There is a mechanical shaft entering the system, which the operator informs you is to aid in the mixing of the two fluids. The operator also informs you that the flow rates used in this system are low, despite the large size of the system, and that the device operates in a steady manner. Beginning with the general equation for the first law, develop a simplified expression that should be suitable for use with this device.

4.29 A large device contains two inlets at the top and one drain outlet at the bottom. Steam enters one of the inlets and liquid water enters through the other inlet. Liquid water drains from the outlet. The inlets are ~2 m above the drain. You observe that the device has bare walls, which appear to be thin and made of steel. The device operator tells you that the flows vary cyclically over a five-minute period of time, but that the cycle then repeats itself. There is an electric motor with a drive shaft mounted on the outside of the device, and the operator tells you that this controls a stirring device inside the device. Beginning with the general equation for the first law, develop a simplified expression that should be suitable for use with this device.

4.30 Liquid water at 120 kPa and 30°C enters an insulated nozzle at 5 m/s. The water exits at 115 kPa and 26°C. Determine the exit velocity of the water.

4.31 Steam at 400 kPa and 300°C enters an insulated nozzle at 8 m/s. The steam exits the nozzle at 350 kPa and 280°C. Determine the exit velocity of the steam.

4.32 Steam enters an insulated nozzle at 415 kPa and 175°C. The steam exits the nozzle at 275 kPa and 165°C, at a velocity of 130 m/s. Determine the inlet velocity of the steam.

4.33 Air enters an uninsulated nozzle at 5 m/s, 500 kPa, and 50°C. The cross-sectional area of the inlet is 0.0002 m^2. The air exits at a velocity of 120 m/s and a temperature of 48°C. Determine the heat transfer rate for the nozzle.

4.34 Steam enters an insulated nozzle at 1.0 MPa and 400°C, at a velocity of 3 m/s. For an exit pressure of 400 kPa, using your model for a nozzle, plot the exit velocity for exit temperatures ranging from 280°C to 380°C.

4.35 R-134a enters an insulated diffuser at a velocity of 200 m/s at a pressure and temperature of 800 kPa and 60°C, respectively. The R-134a exits at 1.60 MPa and 85°C. Determine the exit velocity.

4.36 Saturated water with a quality of 0.55 and a temperature of 120°C enters an uninsulated diffuser at a velocity of 180 m/s. The mass flow rate is 1.5 kg/s. (a) If the water is to exit as saturated water vapor at a temperature of 120°C, with negligible velocity, determine the necessary heat transfer rate. (b) Using your model of a diffuser, determine and plot the necessary heat transfer rates for the flow to exit as a saturated substance with qualities ranging from 0.65 to 1.0.

4.37 Saturated water with a quality of 0.25 and a temperature of 150°C enters an uninsulated diffuser at a velocity of 155 m/s. The mass flow rate is 2.5 kg/s. (a) If the water is to exit as saturated water vapor at a temperature of 150°C, with a velocity of 7.5 m/s, determine the necessary heat transfer rate. (b) Using your model of a diffuser, determine the necessary heat transfer rates for the flow to exit as a saturated substance with qualities ranging from 0.40 to 1.0. (c) Using your model of a diffuser, determine the necessary heat transfer rates for the flow to exit as a saturated vapor at 150°C, with exit velocities ranging from 0 m/s to 90 m/s.

4.38 Air enters a diffuser with a pressure of 200 kPa, a temperature of 50°C, and a velocity of 250 m/s. The inlet of the diffuser is circular, with a diameter of 2.5 cm. The exit velocity is 20 m/s. Plot the exit temperature for heat transfer rates ranging from −20.0 kW to 20.0 kW. Repeat the calculations for an outlet velocity of 100 m/s.

4.39 Oxygen gas enters an insulated diffuser at 500 kPa, 30°C, and 285 m/s. The diffuser is insulated. Plot the exit velocity of the oxygen for exit temperatures ranging from 32°C to 70°C.

4.40 Steam at a mass flow rate of 75 kg/s enters an insulated turbine at 15.0 MPa and 500°C. The steam exits at 185°C and 800 kPa. The inlet velocity of the steam is 250 m/s, and the exit velocity is 15 m/s. Determine the power produced by the turbine.

4.41 An insulated steam turbine receives steam at 5.0 MPa and 400°C, and the steam exits the turbine at 10 kPa with a quality of 0.95. The volumetric flow rate of the steam entering the turbine is 2.5 m^3/s. Determine the power produced by the turbine.

4.42 An insulated steam turbine produces 25.0 MW of power. Steam enters the turbine at 10.0 MPa, 500°C, and with a velocity of 150 m/s, through a duct with a cross-sectional area of 0.01 m^2. The steam exits with a velocity of 10 m/s, at a pressure of 100 kPa. Determine the exit temperature and, if applicable, the exit quality of the steam.

4.43 An insulated steam turbine produces 50 MW of power. Steam enters the turbine at 14 MPa and 550°C, with a velocity of 125 m/s, through a duct with a cross-sectional area of 185 cm^2. The steam exits with a velocity of 10 m/s, at a temperature of 215°C. Determine the exit pressure and, if applicable, the exit quality of the steam.

4.44 An uninsulated gas turbine receives air at 1000 kPa, 800°C, at a mass flow rate of 12.0 kg/s. The air exits at 100 kPa, 420°C. The turbine produces 4.10 MW of power. (a) Determine the heat transfer rate for the turbine. (b) Using your turbine model, plot the required heat transfer rate for exit temperatures varying between 400°C and 500°C, with the other conditions remaining the same. (c) Using your turbine model and the original conditions, plot the variation in power produced as the heat transfer rate varies between 0.0 MW and −3.0 MW.

4.45 An insulated gas turbine receives air at 1200 kPa, 1000 K, and exhausts the air at 150 kPa and 600 K. The air enters the turbine at 120 m/s and exits at 20 m/s. The turbine produces 9.52 MW of power. Determine the mass flow rate of the air through the turbine. Using your model of a turbine, plot the mass flow rate of the air through the turbine as the power varies between 5.0 MW and 20.0 MW.

4.46 An insulated steam turbine receives steam at 8.0 MPa, 400°C, and a mass flow rate of 35 kg/s. For a steam exit pressure of 20 kPa, using your turbine model, plot the power produced for (a) exit qualities ranging from 0.85 to 1.0, and (b) exit steam temperatures ranging from 65°C to 200°C.

4.47 An insulated steam turbine is to produce 100 MW of power. Superheated steam enters the turbine at 12 MPa and 600°C, and saturated water exits the turbine at a pressure of 80 kPa. There are two design constraints to be considered:

(1) The turbine blade design requires that the quality of the steam to be no less than 0.90 to maintain blade durability.

(2) The maximum mass flow rate of steam that the system can utilize is 100 kg/s.

Plot the mass flow rate of steam necessary to produce 100 MW of power for exit qualities ranging from 0.9 to 1.0, and then determine the range of exit qualities that satisfy both design constraints. Determine the necessary mass flow rate of the steam.

4.48 A steam turbine needs to produce 200 MW of power. Superheated steam enters the turbine at 10 MPa and 500°C, and saturated water exits the turbine at a pressure of 50 kPa. There is a heat loss from the turbine of 10 MW. There are two design constraints:

(1) The quality of the steam can be no less than 0.85 to maintain the turbine blade durability.

(2) The maximum mass flow rate allowed in the system is 250 kg/s.

Plot the mass flow rate of the steam that is necessary to provide the given power with the given heat loss for exit qualities ranging from 0.85 to 1.0. Which exit qualities satisfy the design constraints? What is the minimum mass flow rate that satisfies the constraints?

4.49 An insulated turbine receives superheated steam at 10.0 MPa, and the steam exhausts to a condition of a saturated vapor at 50 kPa. The mass flow rate of the steam is 25 kg/s. Using your turbine model, plot the power produced by the turbine for inlet temperatures ranging from 350°C to 800°C.

4.50 An uninsulated turbine receives superheated steam at 7 MPa, and the steam exits as a saturated vapor at 410 kPa. There is a heat loss of 950 kW from the turbine. The mass flow rate of the steam is 25 kg/s. Using your turbine model, plot the power produced by the turbine for inlet temperatures ranging from 300°C to 500°C.

4.51 An uninsulated steam turbine receives steam at 10.0 MPa, 600°C, and exhausts steam at 10 kPa, 100°C. The mass flow rate of the steam is 45 kg/s. Using your turbine model, plot the power produced by the turbine for heat transfer rates for the turbine ranging from 0 MW to −10.0 MW.

4.52 An uninsulated steam turbine receives 70 kg/s of steam at 12.0 MPa and 500°C, with steam exiting at 20 kPa and 75°C. Using your turbine model, plot the heat transfer rates for turbine powers ranging from 40 MW to 48 MW.

4.53 Air enters an insulated compressor at 100 kPa and 10°C, at a velocity of 1 m/s. The compressor inlet is a rectangle with dimensions of 10 cm × 20 cm. The air exits the compressor at 800 kPa and 260°C. The compressor exit is a circular duct with a diameter of 4 cm. Determine the power required to operate the compressor, and the exit velocity of the air.

4.54 An insulated compressor is used to pressurize N_2 gas from 100 kPa to 1200 kPa. The nitrogen enters at 20°C and exits at 380°C. The mass flow rate of the nitrogen is 0.25 kg/s. Determine the power required to operate the compressor.

4.55 An insulated compressor uses 35.0 kW of power to compress air from 100 kPa to 1400 kPa. The air enters at 25°C, and the mass flow rate is 0.092 kg/s. Determine the temperature of the air exiting the compressor.

4.56 An insulated compressor uses 20 kW of power to compress air from 101 kPa to 550 kPa. The air enters at 25°C, and the volumetric flow rate at the inlet is 70 L/s. Determine the temperature of the air exiting the compressor.

4.57 An uninsulated compressor receives 0.5 m^3/s of saturated water vapor at 100 kPa through a duct with a cross-sectional area of 0.005 m^2. Superheated steam exits at 800 kPa and 350°C, at a velocity of 15 m/s. The compressor uses 175 kW of power. Determine the rate of heat loss from the compressor.

4.58 An insulated air compressor is to pressurize air from 100 kPa to 700 kPa. The air enters at 15°C and exits at 250°C. (a) Using your model of a compressor, plot the power required for this compressor for mass flow rates varying from 0.005 kg/s to 5 kg/s. (b) If the maximum power that can be supplied to the compressor via the current electrical system in use is 300 kW, determine the range of mass flow rates of compressed air that this compressor can produce.

4.59 An insulated air compressor receives 0.25 kg/s of air at 100 kPa and 20°C. The air is compressed to 900 kPa. For exit temperatures ranging from 290°C to 400°C, use your compressor model to plot the power required by the compressor. (The higher the exit temperature, the less efficient the compressor is.)

4.60 An insulated air compressor receives 0.300 kg/s of air at 101 kPa and 20°C. The air is compressed to 850 kPa. For exit temperatures ranging from 300°C to 400°C, use your compressor model to plot the power required by the compressor. (The higher the exit temperature, the less efficient the compressor is.)

4.61 An uninsulated air compressor is to compress 0.10 kg/s of air from 100 kPa and 20°C to 900 kPa and 330°C. Using your compressor model, determine the power requirement for this process for heat transfer rates ranging from −2.0 kW to −10.0 kW.

4.62 Superheated steam enters an insulated compressor at 150°C and 120 kPa. The mass flow rate of the steam is 0.75 kg/s. The steam is to exit the compressor at 1.0 MPa. Plot the exit temperature of the steam for compressor power inputs ranging from 400 kW to 600 kW. Compressors with lower efficiency will use more power. If you define the efficiency of the compressor as the actual power used divided by the minimum possible power (in this case, 400 kW), what compressor efficiency corresponds to an outlet steam temperature of 480°C?

4.63 An insulated pump receives 0.05 m^3/s of liquid water at 15°C and 100 kPa. The water exits at 15.0 MPa and 19°C. Determine the power required by the pump.

4.64 An insulated pump receives liquid water at 20°C and 100 kPa, and exhausts the water at 20.0 MPa and 40°C. The mass flow rate of the water is 15 kg/s. Determine the power required by the pump.

4.65 An insulated pump receives liquid engine oil ($\rho = 880$ kg/m^3, $c = 1.85$ kJ/kg · K) at 25°C and 105 kPa, and exhausts the oil at 20 MPa and 60°C. The mass flow rate of the oil is 0.500 kg/s. Determine the power required by the pump.

4.66 An adiabatic pump receives liquid water at 25°C and 100 kPa. The water exits at 30.0 MPa and 40.0°C. Using your pump model, plot the power required for this pumping operation for entering volumetric flow rates ranging from 0.001 m^3/s to 1.0 m^3/s.

4.67 A throttling valve in a refrigerator receives saturated liquid R-134a at 800 kPa, and exhausts the R-134a as a saturated mixture at 180 kPa. (a) Assuming no changes in kinetic energy of the R-134a, determine the quality of the exiting mixture. (b) For the same inlet conditions, use your throttling device model and plot the exit quality of the R-134a as the exit pressure is varied between 100 kPa and 700 kPa.

4.68 Saturated water vapor at 1000 kPa is throttled to 100 kPa. The velocity of the steam remains constant throughout the process. Determine the exit temperature of the water vapor.

4.69 Air passes through a round pipe containing a dirty filter. The filter can be thought of as a throttling device. Consider the filter to be insulated. The velocity of the air entering the filter is 30 m/s, and the velocity of the air exiting the filter is 15 m/s. The air enters the filter at 400 kPa and 20°C, and exits the filter at 350 kPa. Assuming ideal gas behavior for the air, what is the exit temperature?

4.70 Saturated liquid ammonia at 800 kPa is throttled to a lower pressure. Assume no changes in kinetic or potential energy. Plot the quality of the saturated ammonia mixture for exit pressures ranging from 200 kPa to 600 kPa.

4.71 Saturated liquid water at 2000 kPa enters a throttling valve at 5 m/s. The stream exits the throttling valve at 400 kPa, at a velocity of 25 m/s. (a) Determine the quality of the exiting saturated mixture. (b) Using your model of a throttling device, plot the exit quality for the same inlet conditions and same outlet pressure, but for exit velocities ranging from 1 m/s to 100 m/s.

4.72 Saturated liquid R-134a at 830 kPa enters a throttling valve. The R-134a exits the throttling valve at a pressure of 200 kPa. Assume that there is no change in velocity. (a) Determine the exit quality of the R-134a. (b) Using your model of a throttling device, determine the exit quality of the R-134a for exit pressures ranging from 100 kPa to 700 kPa.

4.73 An insulated throttling valve receives saturated liquid R-134a. The exiting flow must have a pressure of 200 kPa. (a) Using your model of a throttling device, plot the exit quality of the R-134a as a function of inlet pressure, for pressures ranging from 250 kPa to 1000 kPa. (b) Upon exiting the throttling valve, the refrigerant is to be used in a heat exchanger. To assure proper operation of the heat exchanger, the quality exiting the throttling valve must be no greater than 0.25. What range of inlet pressures to the throttling valve will satisfy this design constraint?

4.74 A simple insulated concentric tube heat exchanger is used to cool hot oil by transferring heat to water. Hot oil enters the center tube at 120°C and a mass flow rate of 5.0 kg/s. The oil exits the heat exchanger at 80°C. Liquid water flows through the annulus surrounding the central tube. The water enters at 10°C. (a) If the water exits at 50°C, determine the mass flow rate of the water. (b) Using your heat exchanger model to acquire the data, plot the required mass flow rate of the water for water exit temperatures ranging from 15°C to 75°C. Consider the specific heat of the oil to be 1.90 kJ/kg · K, and the specific heat of the liquid water to be 4.18 kJ/kg · K.

4.75 A simple insulated concentric tube heat exchanger is used to cool hot water by transferring heat to cool air. Hot water enters the center tube at 100°C and a mass flow rate of 3 kg/s. The water exits the heat exchanger at 40°C. Air flows through the annulus surrounding the central tube. The air enters at 10°C. (a) If the air exits at 35°C, determine the mass flow rate of the air. (b) Using your heat exchanger model to acquire the data, plot the required mass flow rate of the air for air exit temperatures ranging from 12.5°C to 40°C. (c) Consider that the maximum allowed exit temperature of the air is 30°C. Plot the maximum exit water temperature as a function of the mass flow rate of the air for air mass flow rates varying between 2 kg/s and 30 kg/s. Consider the specific heat of the air to be 1.00 kJ/kg · K, and the specific heat of the liquid water to be 4.184 kJ/kg · K.

4.76 Water is to be condensed from a saturated vapor at 200 kPa to a saturated liquid at 200 kPa in a simple insulated concentric tube heat exchanger. The mass flow rate of the water is 2.5 kg/s. The cooling fluid flows through the other tube, entering at 10°C and exiting at a temperature of no higher than 100°C. Plot the mass flow rate of the coolant over an exit temperature range of 15°C to 100°C if the coolant is (a) engine oil (c = 1.91 kJ/kg · K), (b) helium gas (c_p = 5.193 kJ/kg · K), and (c) carbon dioxide gas (c_p = 0.846 kJ/kg · K).

4.77 A tankless water heater can be modeled as a heat exchanger where heat from hot combustion products are rapidly transferred to the water. Suppose you require a tankless water heater to provide 0.001 m^3/s of water at 50°C. Assuming the system is externally insulated, plot the rate at which heat must be delivered to the water from the combustion process for inlet water temperatures ranging from 2°C to 25°C.

4.78 Water is to be cooled by passing through the tube in an insulated cross-flow heat exchanger while air flows over the outside surface of the tube. The water enters the 4.0-cm-diameter pipe with a velocity of 15 m/s, a temperature of 150°C, and a quality of 0.10. The water exits as a liquid at 100°C. (This can be approximated as a saturated liquid at 100°C.) The air enters the heat exchanger at 15°C. (a) For a mass flow rate of the air of 5.0 kg/s, determine the air exit temperature. (b) Using your heat exchanger model to acquire the data, plot the exit temperature of the air as the air mass flow rate varies from 4.0 kg/s to 25 kg/s.

4.79 Steam exiting the turbine of a power plant passes through the insulated shell-and-tube heat exchanger condenser. The mass flow rate of the steam is 145 kg/s, and the steam enters the heat exchanger as a saturated mixture with a quality of 0.95 at 20 kPa. The steam exits the condenser as a saturated liquid at 20 kPa. Liquid cooling water is used to condense the steam. The cooling water enters the condenser at 10°C. (a) Determine the exit temperature of the liquid cooling water if the mass flow rate is 4000 kg/s. (b) Using your heat exchanger model, plot the cooling water mass flow rate and volumetric flow rate (entering the condenser) for exit temperatures ranging from 15°C to 55°C. (c) If the exit temperature of the cooling water is regulated such that it must be no greater than 40°C, and if the cooling water flow rate through the condenser can be adjusted in a range from 500 kg/s to 2000 kg/s, what is the range of maximum possible flow rates of steam through the condenser?

4.80 You need to cool 8.0 kg/s of superheated steam at 1000 kPa from 325°C to 200°C. You can either use R-134a that is available as a saturated liquid at 0°C, or air that is available at 15°C and 100 kPa. Due to other operational factors, the R-134a must not have an exit quality greater than 0.60, and the air cannot have an exit temperature greater than 75°C. Design a system that will meet the constraint for each substance, and comment on the relative practicality of each system.

4.81 Superheated steam, at a mass flow rate of 15.0 kg/s, enters an insulated mixing chamber at 500 kPa and 200°C. In the chamber, the steam is mixed with liquid water at 500 kPa and 40°C. Determine the mass flow rate of the entering liquid water required for the combined flow to exit as a saturated liquid at 500 kPa.

4.82 Superheated ammonia enters a mixing chamber at a pressure of 350 kPa absolute at a mass flow rate of 2.5 kg/s. Saturated liquid ammonia enters the chamber at a rate of 250 g/s at 350 kPa absolute. Heat is added to the chamber at a rate of 130 kW. If the combined exit flow is saturated vapor ammonia at 350 kPa absolute, determine the temperature of the superheated ammonia that entered the chamber.

4.83 Two streams of liquid water at 500 kPa are to be mixed together and heated in an uninsulated mixing chamber. Stream A enters at 10°C, with a mass flow rate of 5 kg/s. Stream B enters at 30°C. The exit stream is to exit at 45°C. Plot the amount of heat that needs to be added as the mass flow rate of stream B varies between 1 kg/s and 20 kg/s.

4.84 Liquid water at 10°C enters an insulated mixing chamber at a rate of 180 kg/s. In the chamber, the water is mixed with a second stream of liquid water, which enters at 90°C. Using your mixing chamber model to calculate the data, plot the exit temperature of the combined stream for mass flow rates of the second stream ranging from 20 kg/s to 500 kg/s.

4.85 Compressed liquid water at 1000 kPa and 30°C enters an insulated mixing chamber at 50 kg/s. The water is mixed in the chamber with superheated steam at 1000 kPa. The combined flow exits as a saturated liquid at 1000 kPa. Using your mixing chamber model to calculate the data, plot the mass flow rate of the superheated steam for entering steam temperatures ranging from 200°C to 500°C.

4.86 Saturated liquid water at 200 kPa enters an insulated mixing chamber at a rate of 15 kg/s. The water is mixed with superheated steam at 200 kPa and 125°C. The combined flow will exit as a saturated mixture at 200 kPa. Using your mixing chamber model to calculate the data, plot the quality of the combined exit flow for superheated steam entrance flow rates ranging from 5 kg/s to 200 kg/s.

4.87 An insulated centrifugal separator is to be used to separate the liquid and vapor portions of a saturated mixture flow. A saturated mixture of water with a quality of 0.30 and a temperature of 120°C enters the system with a mass flow rate of 4.50 kg/s. Saturated liquid water exits through one exhaust pipe at a temperature of 120°C at a rate of 3.00 kg/s, and 1.50 kg/s of saturated water vapor exits at 120°C through the other exhaust pipe. (a) Determine the amount of power that was added to the system to generate these outlet flows. (b) Using your separation chamber model, plot the power required to produce flows of saturated vapor at 120°C ranging from 1.40 kg/s to 4.0 kg/s.

4.88 A distillation unit is to be used to generate pure water vapor. 2.0 kg/s of liquid water at 25°C enters the system. 0.2 kg/s of saturated water vapor leaves through one port, and 1.8 kg/s of saturated liquid leaves through a second port (along with impurities in the water that can be ignored in this analysis). The pressure throughout the whole system is 100 kPa. There is no work done in the process, and all energy is added through heat transfer. (a) Determine the rate of heat transfer needed to produce this flow of saturated water vapor. (b) Using your separation chamber model, plot the heat transfer rates needed to produce saturated water vapor flow rates between 0.1 kg/s and 1.0 kg/s (along with corresponding changes to the saturated liquid water exit flow rates).

4.89 A distillation unit is to be used to generate pure water vapor. 2.5 kg/s of liquid water at 40°C enters the system, and 0.5 kg/s of saturated water vapor leaves through one port and 2 kg/s of saturated liquid leaves through a second port (along with impurities in the water that can be neglected in the analysis). The pressure throughout the whole system is 140 kPa absolute. A stirring system is used, which is powered by a 300 W motor—assume that all of the motor's work enters the system. All other energy is added through heat transfer. (a) Determine the rate of heat transfer needed to produce this flow of saturated water vapor. (b) Using your separation chamber model, plot the heat transfer rates needed to produce saturated water vapor flow rates between 0.250 kg/s and 2 kg/s (along with corresponding changes to the saturated liquid water exit flow rates).

4.90 An initially evacuated tank is to be filled with N_2 gas from a supply line. The N_2 gas in the supply line is at 20°C and 800 kPa. The volume of the tank is 0.5 m^3, and the tank is to be filled until the pressure in the tank is 750 kPa. Determine the amount of N_2 that flows into the tank, and the final temperature of the N_2 in the tank, assuming that the tank is insulated.

4.91 An initially evacuated tank is to be filled with O_2 gas from a supply line that contains O_2 at 1000 kPa and 25°C. The tank volume is 0.25 m^3, and the tank is to be filled until the pressure inside the tank is 900 kPa. Determine the amount of heat transfer required if the temperature of the O_2 in the tank is to be 40°C.

4.92 An initially evacuated tank is to be filled with CO_2 gas from a supply line that contains CO_2 at 1 MPa and 20°C. The tank volume is 40 liters, and the tank is to be filled until the pressure inside the tank is 800 kPa. Determine the amount of heat transfer required if the temperature of the CO_2 in the tank is to be 25°C.

4.93 A tank contains CO_2 gas at 25°C and 500 kPa. A valve on the tank is opened slightly so that the CO_2 slowly exits the tank until the pressure inside the tank drops to 100 kPa. The volume of the tank is 0.25 m^3. Determine the amount of heat transfer necessary so that the CO_2 in the tank remains at 25°C throughout the process.

4.94 An insulated tank initially contains 1.25 kg N_2 gas at 150 kPa and 25°C. The tank is to receive 3.0 kg of N_2 gas from a supply line with N_2 at 20°C. Determine the final temperature and pressure of the N_2 in the tank.

4.95 Consider a tank water heater which contains 0.18 m^3 of water. Several users of the water simultaneously need water, such that water is removed from the heater at a rate of 2 kg/s. Initially, the water exits at 47°C. Water at 10°C is added at a rate of 2 kg/s to maintain the total volume of water in the tank. The burner provides 20 kW of heat input to the tank of water. Assuming that the water is well mixed in the tank, plot the exit temperature of the water as a function of time for 5 min of operation.

4.96 A 150-liter tank (insulated) water heater is initially filled with water at 40°C. A shower is turned on, drawing 30 liters of water from the tank. At that point, water at 10°C begins to enter the tank at a rate of 6 liters/min. However, water continues to be emptied from the tank at a rate of 9 liters/min. When the fresh water flow starts, a burner providing 10 kW of heat begins operation. Assume that the water in the tank and the fresh water are well mixed. Determine the temperature of the water exiting the tank at the following times after the fresh water flow begins: 1 min, 5 min, 10 min, and 20 min.

4.97 A pressure cooker is filled with 2.0 kg of liquid water at 30°C. Heat is added to the system, and the relief valve releases saturated water vapor when the pressure reaches 150 kPa. Model the system, and then use the model to plot the amount of heat transfer that occurs as the mass of the water present in the pressure cooker ranges from 1.95 kg to 1.00 kg.

4.98 A computer chip with a mass of 50 g, made of copper, experiences an electrical work input of 10 W. The chip is insulated. Initially, the chip has a temperature of 20°C. What is the temperature of the chip after one minute of operation?

4.99 On a warm summer day, you decide to leave a fan running in your room. The room is 4 m × 5 m × 2.5 m. The room is full of dry air. When you leave the room, the air temperature is 25°C. The fan has a 75 W motor. Assuming that the air pressure stays constant at 101 kPa and that the room is perfectly insulated, determine the air temperature in the room when you return in 8 hours.

4.100 Someone leaves a 100-W light bulb on before leaving and sealing up a 2 m × 4 m × 4 m room full of air. Initially the air temperature is 20°C, and the air pressure is 100 kPa. Assuming that the mass of the air in the room stays constant and ignoring any pressure changes, (a) determine the temperature of the air in the room after 8 hours. (b) What would the temperature of the room be after 8 hours if someone had also left a 50-W fan operating in the room at the same time?

4.101 Consider the room from Problem 4.100, without the fan operating. What would be the temperature in the room after 8 hours if the 100-W light bulb was replaced with (a) a 23-W CFL bulb, and (b) a 13-W LED bulb.

4.102 A balloon is filled with air, initially at 35°C and 200 kPa at a volume of 0.005 m³. The air is then compressed at constant temperature until the volume is 0.002 m³. Determine the final pressure of the air and the amount of heat transfer that occurs.

4.103 A balloon initially contains air at 25°C and 450 kPa, with a volume of 0.10 m³. The air inside the balloon is heated and the balloon expands until the volume is 0.25 m³. The pressure and volume follow the relationship $PV^{1.3}$ = constant during the process. Determine the final temperature of the air and the amount of heat added.

4.104 Air occupies a balloon that has an initial volume of 0.05 m³. Initially, the air has a temperature of 20°C and a pressure of 1000 kPa. The air is heated, and the balloon expands until the volume is 0.25 m³. The pressure and volume are related through PV^n = constant. Develop a computer program model for analyzing such a process/device. Plot the final temperature of the air, the work done, and the heat transfer for values of n = 1.1, 1.2, 1.3, 1.4, and 1.5.

4.105 N_2 gas fills a flexible container with an initial volume of 0.05 m³, a temperature of 30°C, and a pressure of 800 kPa. The N_2 gas is to expand in an isothermal manner until the pressure is 100 kPa. (a) Determine the amount of heat transfer required for this process to maintain an isothermal nature. (b) Develop a computer model to simulate this process, and then plot the amount of heat transfer required for an isothermal process with final pressures ranging between 100 kPa and 600 kPa.

4.106 Air is located in a piston–cylinder device. Initially, the air is 20°C and the pressure is 200 kPa, and the volume is 0.1 m³. (a) The air is heated at constant pressure until the volume is 0.3 m³. Determine the heat transfer for the process. (b) For the same initial conditions, use your piston–cylinder device model to generate data, and plot the heat transferred for an isobaric expansion to volumes ranging from 0.15 m³ to 0.50 m³.

4.107 Steam occupies a piston–cylinder device at 600 kPa and 250°C, at a volume of 0.5 m³. The steam is cooled at constant pressure until the volume is 0.1 m³. (a) Determine the work done and the heat transfer for the process. (b) Using your computer model for a piston–cylinder assembly, plot the work and heat transfer for the process for final volumes ranging between 0.05 m³ and 0.40 m³.

4.108 A piston–cylinder device, initially at a volume of 0.025 m³, contains steam. The steam is initially at 500 kPa and 300°C. The steam is to be cooled at constant pressure. Plot the work done, the heat transfer for the process, and the final volume of the cylinder for final temperatures between 151.86°C (with the steam being a saturated vapor) and 275°C.

4.109 2 kg of saturated liquid water is to be boiled at a constant pressure of 100 kPa in a piston–cylinder device until the water is all saturated vapor. Determine the work and heat transfer for the process.

4.110 2.5 kg of saturated water vapor is to be condensed at a constant pressure of 160 kPa in a piston–cylinder device until all the water is a saturated liquid. Determine the work and heat transfer for the process.

4.111 A saturated mixture with a quality of 0.10 of R-134a is to expand in an insulated flexible container from a pressure of 1.00 MPa and a volume of 0.02 m³ until the pressure is

500 kPa and the quality is 0.5. Determine the final temperature and volume of the R-134a, and the work done in the process.

4.112 A 2-kg block of iron is to be dropped from the roof of a tall building. The initial temperature of the iron is 25°C, and the height of the building is 150 m. (a) Assuming that the block is insulated and ignoring the effects of the atmosphere, determine the final temperature of the iron when it comes to rest on the ground. (b) Develop a computer model for this problem, and plot the final temperature of the iron that would be expected for building heights ranging from 10 m to 500 m.

4.113 An aluminum ball is shot from a cannon. As it leaves the muzzle of the cannon, the ball has a temperature of 100°C and a velocity of 250 m/s. Experiencing no air resistance, the ball hits a target and drops to a velocity of 0. Assuming that there is no heat lost in the process, what is the temperature of the aluminum ball after the ball comes to rest?

4.114 A copper roof is being installed on a 100-m-tall building. A copper tile is dropped and falls to the ground. Assume that there is no air resistance and that the temperature of the copper tile stays constant (at 25°C) during its fall to the ground. When the copper tile hits the ground, all of its kinetic energy is converted into thermal energy, with no heat lost in the process. What is the temperature of the copper tile after it comes to rest on the ground?

4.115 You are given a block of an unknown metal, with the metal being either aluminum (c = 0.903 kJ/kg · K), copper (c = 0.385 kJ/kg · K), iron (c = 0.447 kJ/kg · K), or lead (c = 0.129 kJ/kg · K). You wish to decide what metal it is made of by dropping the block from the roof of a 100-m-tall building at sea level and measuring its temperature increase. Before you drop the block, you measure its temperature to be 20.3°C. Immediately after the block comes to rest on the ground, its temperature is measured to be 22.8°C. Assuming that there was no heat loss from the block during the process, which metal is the block most likely made of?

4.116 Water is placed on a stove to boil in a container that has a flexible lid so that it can expand with increasing amounts of water vapor while not losing any water to the surroundings. The mass of the water is 2 kg. Initially, the water is a saturated liquid at 100°C, and the boiling process is at constant pressure. 10 kW of heat is added to the water, and the process continues until three-fourths of the water has become water vapor. (a) Determine the time elapsed for the process. (b) Develop a computer model to simulate this problem, and plot the net heat transfer, the net work done, and the quality of the water all as a function of time. Consider the time to start at 25 s, and to end when the boiling process is complete (i.e., when the water is a saturated vapor).

4.117 1.5 kg of air is initially at 200 kPa and 30°C in an expandable container. The air is heated so that it expands following the relationship $PV^{1.2}$ = constant. The container also holds a fan that adds a total of 50 kJ of work to the system during the process. The pressure at the end of the process is 100 kPa. (a) Determine the final volume of the air, the final temperature, and the amount of heat transfer during the process. (b) Develop a computer model for this problem, and plot the final volume, the final temperature, and the amount of heat transfer during the process for final pressures ranging between 50 kPa and 190 kPa.

4.118 A thermodynamic cycle consists of four processes, and the heat and work interactions of the four processes are shown in the following table. Determine the amount of work that is produced in process 3.

Process	Heat Transfer	Work
1	192 kJ	0 kJ
2	−18 kJ	50 kJ
3	−110 kJ	?
4	0 kJ	−32 kJ

4.119 A thermodynamic cycle consists of three processes, and the heat and work interactions of the three processes are shown in the following table. Determine the amount of heat transfer that is produced in process 3.

Process	Heat Transfer	Work
1	92 kJ	10 kJ
2	11 kJ	−32 kJ
3	? kJ	17 kJ

4.120 A thermodynamic cycle consists of four processes, and the heat and work interactions of the four processes are shown in the following table. Determine the amount of power that is produced in process 4.

Process	Heat Transfer Rate	Power
1	0 kW	−15 kW
2	110 kW	37 kW
3	−26 kW	0 kW
4	−21 kW	?

4.121 A thermodynamic cycle consists of four processes, and the heat and work interactions of the four processes are shown in the following table. Determine the amount of work that is produced in process 2.

Process	Heat Transfer	Work
1	25 kW	30 kW
2	0 kW	?
3	−40 kW	0 kW
4	5 kW	−20 kW

4.122 A thermodynamic cycle consists of five processes, and the heat and work interactions of the processes are shown in the following table. Determine the heat transfer rate in process 3.

Process	Heat Transfer Rate	Power
1	5 kW	–10 kW
2	12 kW	62 kW
3	?	–21 kW
4	– 15 kW	0 kW
5	9 kW	8 kW

4.123 A cyclic heat engine produces 10 kW of power while receiving 25 kW of heat as an input. Determine the heat rejection rate to the cool space, and the thermal efficiency of the heat engine.

4.124 An automobile engine produces 85 kW of power and has an efficiency of 0.32. Determine the heat input rate from the combustion process as well as the rate of heat lost to the environment.

4.125 A motorcycle engine produces 18.5 kW of power at a thermal efficiency of 0.25. Determine the heat input rate from the combustion process as well as the rate of heat lost to the environment.

4.126 A coal-fired power plant receives heat at a rate of 625 MW from the combustion of coal, and rejects 430 MW of heat to cooling water received from a lake. Determine the thermal efficiency of the power plant.

4.127 A steam power plant has an efficiency of 30%. The plant produces 250 MW of power. The heat that is input to the plant is used to produce steam in the boiler. The water enters the boiler as a saturated liquid at 8 MPa and exits as a saturated vapor at 8 MPa. Determine the mass flow rate of the water flowing through the boiler.

4.128 Plot the power produced and heat rejection rate to a cool reservoir for a cyclic heat engine that receives 100 kW of heat from a warm reservoir, for thermal efficiencies ranging from 0.05 to 0.70.

4.129 A household refrigerator has a coefficient of performance of 3.25 and is able to remove heat from its interior at a rate of 1.82 kW. Determine the power used in the compressor to operate the refrigerator.

4.130 A refrigerator using a 375 W motor has a coefficient of performance of 3.05. 7.5 liters of tea (modeled as water) are placed into the refrigerator at 25°C. How long will it take for the refrigerator to cool the tea to 5°C, assuming that this is the only heat load to be considered? Does this seem realistic? If not, what assumptions were made that make the calculation appear unrealistic?

4.131 An air-conditioning unit provides 125 kW of cooling while using power at a rate of 58 kW. Determine the coefficient of performance of this air conditioner (refrigerator).

4.132 A refrigerator removes 25 kW of heat from a cold space while rejecting 35 kW of heat to a warm space. Determine the power used and the coefficient of performance of the refrigerator.

4.133 A refrigerator is to remove 15 kW of heat from a cool space. Plot the power required to do so for coefficients of performance ranging between 1.5 and 10.

4.134 A refrigerator is to cool 2 kg of water (at atmospheric pressure) from 40°C to 5°C. The coefficient of performance of the refrigerator is 3.10. Determine the work needed to operate the refrigerator during this cooling process.

4.135 A heat pump is to deliver 59 kW of heat to a warm space while using 15 kW of power. Determine the coefficient of performance of the heat pump.

4.136 A heat pump uses a 4.5 kW motor to deliver 17.1 kJ/s of heat to a building. Determine the coefficient of performance of the heat pump.

4.137 A heat pump receives 15 kW of heat from a cool space and uses 5 kW of power. Determine the coefficient of performance of the heat pump, and the heat delivered to the warm space.

4.138 A heat pump operates continuously with 0.5 kg/s of R-134a condensing from a saturated vapor to a saturated liquid at 1200 kPa to provide the heat rejected to the warm space. If the coefficient of performance of the heat pump is 4.50, determine the power required to drive the compressor operating in the heat pump.

4.139 A heat engine is set up to supply power to operate a refrigerator. The refrigerator is to remove 1.5 kW of heat from a cold space and has a coefficient of performance of 2.75. What is the necessary heat input to the heat engine used to power the refrigerator if the thermal efficiency of the heat engine is (a) 0.25, (b) 0.50, (c) 0.75, and (d) 0.90?

DESIGN/OPEN-ENDED PROBLEMS

4.140 A factory requires 0.85 m^3/s of compressed air (on average) (at standard temperature and pressure) delivered at a system pressure exiting the compressor of 720 kPa to operate all the compressed-air equipment in the facility. In addition, it is expected that on average, an additional 5% of the compressed air is lost to leaks in the system. Occasionally, demand on the compressed air system will reach 1.0 m^3/s. If the primary compressor system fails, it is desired to have a back-up system that will be able to provide 0.70 m^3/s of compressed air. The interior temperature of the factory is 35°C in the vicinity of the air compressors, and the average outdoor temperature of the factory is 10°C. Design a system of air compressors that will satisfy the needs of the facility, and specify the source of the inlet air. Your system may or may not include a storage tank for compressed air. You may search air compressor manufacturer sources for details on specific air compressors.

4.141 A steam power plant has a thermal efficiency of 35%, and the electric generator of the plant has an efficiency of 95%. The power plant is to generate an average of 300 MW of electricity over the course of a year (8760 hours). The heat source of the plant is the combustion of a fuel. The combustor is able to convert 80% of the available energy in the fuel (as represented by the "lower heating value" of the fuel) into heat used to convert the liquid water into steam. Using the Internet as a source of data on lower heating values of possible fuels and current fuel prices, design and justify a fuel source (with an annual fuel cost) for this power plant.

4.142 Oil, with a mass flow rate of 2 kg/s, is to pass through a heat exchanger to be cooled from 90°C to 25°C. The specific heat of the oil is 1.88 kJ/kg · K. Choose the parameters of a cooling fluid (the fluid itself, the inlet and outlet temperatures, and the flow rate of the fluid) that can be used in the heat exchanger to accomplish the desired cooling (assuming that the heat transfer between fluids is ideal).

4.143 A home currently is using an outdated, inefficient lighting system. Typically in the evening, there are two 40-W incandescent bulbs, four 75-W incandescent bulbs, and eight 100-W incandescent bulbs on in the house (for an average of 6 hours a day), and outside the house there are four 90-W halogen flood lights that are on for an average of 12 hours a day. The light outputs from each source are 450 lumens (40-W incandescent), 1100 lumens (75-W incandescent), 1600 lumens (100-W incandescent), and 1350 lumens (90-W halogen). Design a lighting system that will produce the same illumination, and determine the annual energy savings with the new system as well as the money saved from the reduced electricity bill.

4.144 A building's heating system requires that air be delivered at a rate of 0.20 m^3/s at a temperature of 30°C. The inside air can be returned to the furnace at any temperature between 18°C and 22°C. It is required that at least 10% of the air that is delivered back to the building is fresh air from the outside. During the heating season, the outside air temperature will range between −5°C and 18°C. Design a system that will be able to accommodate the varying outdoor conditions, and specify the amounts of air that will be taken from inside and outside the building for a variety of conditions. In addition, specify the requirements (or identify an actual piece of equipment) that will describe the furnace.

4.145 The owner of a house wishes to install a system on his driveway to melt the snow in winter (the driveway is sloped so that any liquid water would run off the driveway). The driveway is 15 meters long and 3 meters wide, and is made of concrete. Assume that any heat removed from the driveway is coming directly from the snow. The homeowner desires that the system be able to melt up to 2.5 cm of snow per hour. Assume that the amount of energy needed to heat the snow to the melting point and then melt the snow is 350 kJ/kg, and the density of the snow is 70 kg/m^3. Design a heating system that will accomplish this melting process.

4.146 A condenser in a steam power plant is used to take steam exiting the turbine and condense it into a saturated liquid at the same pressure. This is typically done by using a flow of external cooling water that then transfers the heat to the environment. Consider a power plant that has steam exiting the turbine at a quality of 0.95 and a pressure of 30 kPa, at a mass flow rate of 200 kg/s. Design a condenser that will use external cooling water to remove heat from the steam to convert the steam into a saturated liquid. The external cooling water will be taken from a lake with a water temperature of 10°C and cannot return to the lake at a temperature higher than 30°C. Choose the number of tubes and the diameter of tubes that will be used for the cooling water.

4.147 Design a 0.20-m^3 tank water heater that will supply water for 1 hour within a temperature range of 38–45°C at a rate of 10 L per minute. Fresh water is to be added to maintain a constant volume in the tank from a source at 8°C. Design both the burner size and the control scheme for operating the burner to maintain the water in the desired exit temperature range.

CHAPTER 5

Introduction to the Second Law of Thermodynamics

Learning Objectives

Upon completion of Chapter 5, you will be able to:

5.1 Describe the nature of the Second Law of Thermodynamics, recognize its uses and implications, and state the classical statements of the second law;

5.2 Explain the difference between a reversible and an irreversible process;

5.3 Calculate the maximum possible efficiency (Carnot efficiency) of a heat engine, and the maximum possible coefficient of performance of a refrigerator and a heat pump; and

5.4 Recognize the concept of a perpetual motion machine.

5.1 THE NATURE OF THE SECOND LAW OF THERMODYNAMICS

In Chapter 4, we learned about the First Law of Thermodynamics, and from this it should be apparent that the first law is a simple accounting of all of the energy in the system and processes under study. In fact, the first law is stated succinctly in three words: "Energy is conserved." The first law is obeyed in every process observed in nature. Unfortunately, the first law also allows for some processes and results that have never been observed, and ones that we have no reason to believe will ever occur. For example, what is expected if an ice cube is dropped into a glass of boiling water at 100°C at atmospheric pressure, as illustrated in **Figure 5.1**? We expect that the boiling water will cool, that the ice cube will melt, and that the water originally in the ice will increase in temperature—eventually all the water will be the same temperature as thermal equilibrium is reached. However, if you were to perform a simple energy balance, an alternative result is that some of the hot water in the glass would freeze, making the ice cube larger, and the remaining water will become vapor—this leaves a larger ice cube surrounded by superheated water vapor at an elevated temperature. We know that this result does not happen in nature. But this unusual scenario is allowed by the first law, because the energy in the system is conserved.

As another example, consider a heat exchanger in which air is used to cool oil, as shown in **Figure 5.2**. Depending on the flow rates, if the oil enters the heat exchanger at 80°C, and the air enters at 10°C, we might expect the air to exit at 35°C and the oil to exit at 45°C. However, the first law also allows a solution where the air cools to –15°C and the oil heats to 115°C. Again, we know that the former result is possible, whereas the latter result does not occur in nature. However, we cannot eliminate the second option using only the first law, because energy is conserved in each scenario.

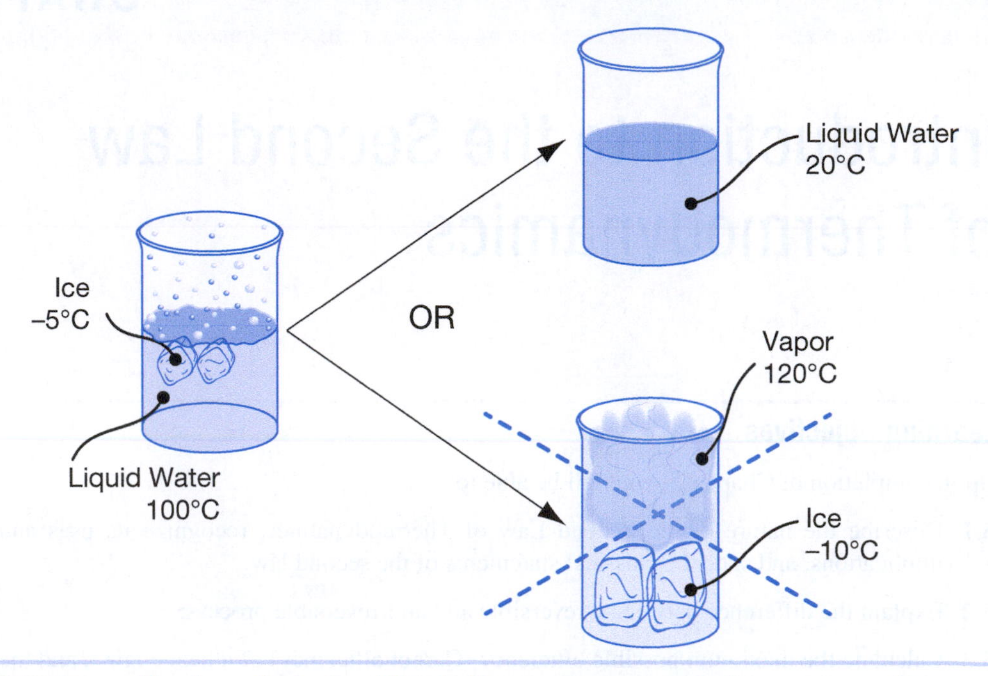

FIGURE 5.1 If an ice cube is placed into boiling water at atmospheric pressure, we know that the ice cube will melt and the boiling water will cool, such that all the water will reach some intermediate temperature. We know that the ice cube will not grow colder and larger.

If the first law describes processes that may or may not occur, how can we determine if a proposed process can actually occur in nature? To make such a determination, we need to call on the Second Law of Thermodynamics. The second law describes conditions that

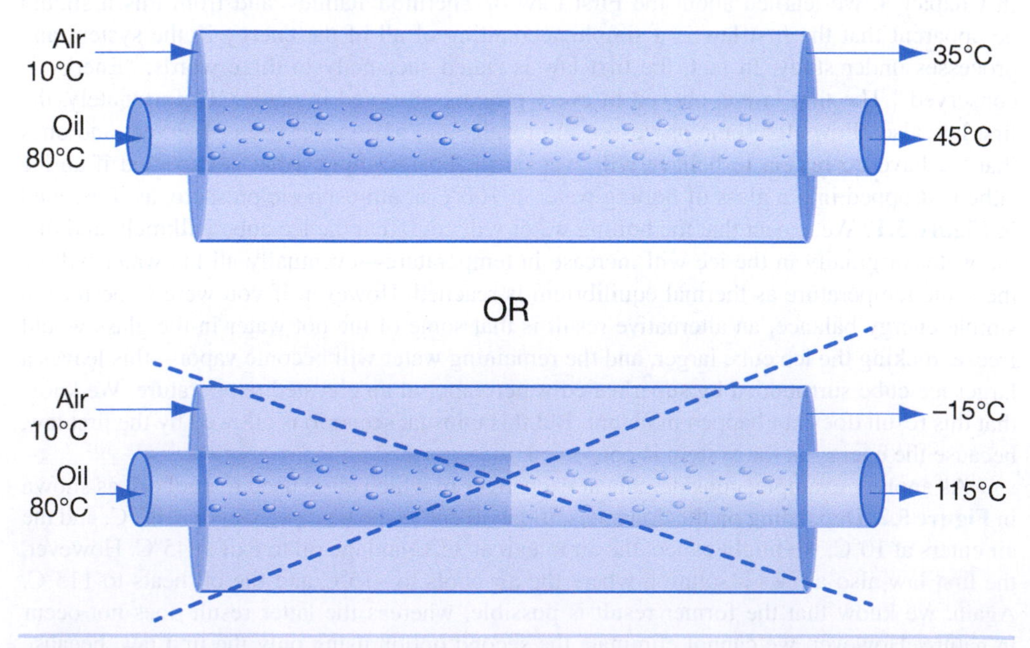

FIGURE 5.2 As two fluids flow through a heat exchanger, we know that the hotter fluid will cool whereas the colder fluid will warm. We know that the hotter fluid will not get hotter while the colder fluid gets cooler.

dictate how processes actually proceed in nature. A second law analysis of the two options involving the boiling water and the ice cube would indicate that the result with the water all reaching an equilibrium temperature is possible, whereas the result with the expanding ice cube and additional water boiling is not possible. A second law analysis of the heat exchanger would indicate that the cooling of the oil and heating of the air is possible, but not the reverse.

How easy is it to describe how every process in the world proceeds? Or how easy is it to describe all factors that influence whether or not a process can occur? As you can probably guess, it is difficult to succinctly summarize these. As a result, there is no simple statement of the second law such as there is for the first law. There is no single equation that covers all possible situations in nature, although we will eventually develop an equation that is useful in many situations. There are many aspects of the second law that must all be satisfied for a process to be possible. These aspects will lead to a number of related statements that combined constitute the Second Law of Thermodynamics. Because there are multiple aspects of the second law, it is important to remember that a process that satisfies one aspect of the second law may not satisfy other aspects of the second law, and in that case the process would not be possible.

QUESTION FOR THOUGHT/DISCUSSION

Suppose you have a gas cylinder filled with air at 500 kPa. What happens to the air when the valve is opened? Why doesn't air enter the cylinder?

5.2 SUMMARY OF SOME USES OF THE SECOND LAW

We could assume that if all the second law does is tell us what is possible in nature, then we do not need to concern ourselves with it because we are only interested in what does occur. We might think that if we do not see that a process works, it must not satisfy the second law. But as engineers, we would like to have some idea if a process is even possible before proceeding with the completion of a design and construction of a device. Furthermore, a process may actually work but the device designed to complete that process may be poorly constructed. Some situations where the second law would be violated may be obvious, such as those described in the previous section, but others may not be so clear. Could a proposed electric power plant have a thermal efficiency of 75%? Most power plants may have thermal efficiencies below 50%, but does that mean that a new facility can't have a revolutionary design? Or, could an insulated turbine have steam enter at 5 MPa and 500°C and exit at 1 MPa and 200°C? At first glance, we really wouldn't know if this is possible. The second law will give us a suite of tools that can be used to help us analyze situations that are not intuitively obvious.

The following are some uses of the second law:

1. Determining the maximum thermal efficiency of a heat engine and the maximum coefficient of performance of a refrigerator or heat pump
2. Determining the efficiency of devices such as turbines, pumps, compressors, nozzles, and diffusers through comparison of their actual performance to their ideal performance
3. Determining the feasibility of a proposed process
4. Determining the effectiveness utilization of available energy
5. Comparing different processes to determine which is most efficient thermodynamically
6. Determining the direction of a chemical reaction
7. Establishing a thermodynamic or absolute temperature scale, independent of the properties of any substance

As this list illustrates, aspects of the second law are present in many disparate areas. Many of these uses are somewhat related; others, such as the establishment of a thermodynamic temperature scale, may seem out of place. However, application of the second law in the area of temperature scales was critical to the development of thermodynamics over time and influences thermodynamic analysis to this day.

So, not only does the second law tell us whether or not a process is possible in nature, it also allows us to determine the quality of the process—whether it is wasteful of energy or whether it makes good use of the available energy to achieve a desired purpose. For instance, suppose you wish to boil water at atmospheric pressure. To do this, you know that you have to raise the temperature of the water to 100°C. But how should you heat the water? You could heat the water by building a fire that has a temperature of 200°C or one that has a temperature of 1500°C. Both will boil the water, and the fire with a temperature of 1500°C may very well boil the water more quickly. But there will also be much more energy released in the hotter fire than needed, which results in a waste of that energy and a less efficient use of energy. The second law will give us the tools to show that is indeed the case. In the past, energy conservation was not on the forefront of public consciousness, but today green living and energy conservation are widely covered in the media, and most people are aware of the importance of conserving energy. Energy conservation will gain increasing prominence in the future as energy resources become less abundant and therefore more expensive.

Now that we have established the purpose of the second law, we begin our in-depth consideration of the second law by discussing two historic statements from early in the development of the Second Law of Thermodynamics.

5.3 CLASSICAL STATEMENTS OF THE SECOND LAW

As mentioned previously, it is very difficult, if not impossible, to develop a single statement that will encompass all aspects of the second law. Instead, there are many statements that embody parts of the second law. The earliest of these statements date to the 19th century, when scientists and engineers attempted to formulate explanations and laws for what they were seeing involving energy as it was utilized in practice. Two such classical statements of the second law, neither of which fully describes the second law but both of which describe key elements based on which thermodynamics has developed, are presented here.

The first classical statement of the second law is the Kelvin-Planck statement regarding heat engines, illustrated in **Figure 5.3**. Recall that heat engines produce work after receiving an input of heat, and that a thermal reservoir is a large body that does not change temperature as energy is added or removed through heat transfer. If we observe a heat engine that is operating continuously, we can see that more energy is input as heat than is output as work. With this observation, the Kelvin-Planck statement of the second law is as follows:

It is impossible for a continuously operating heat engine to produce work while exchanging heat with a single thermal reservoir.

This means that some of the energy that was initially received by the heat engine as heat from a high-temperature thermal reservoir must be rejected to a low-temperature thermal reservoir, and that the efficiency of the heat engine must be less than 100%. Why? To illustrate this, consider a situation where we will take liquid water and boil it at atmospheric pressure in a piston–cylinder device, such as in **Figure 5.4**. The water receives heat at 150°C, and as the water boils, it expands. Work is done as the water expands. To operate continuously, the volume must undergo a contraction, and this means water must return to liquid form by condensing. To condense, the water will reject heat to some other thermal reservoir, at

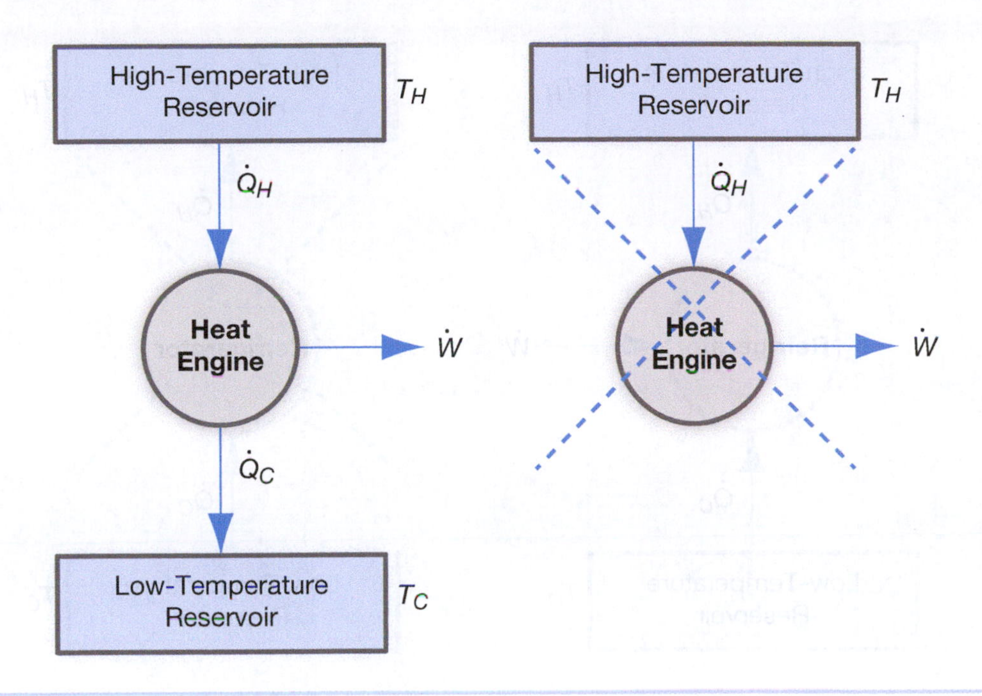

FIGURE 5.3 A heat engine that is continuously operating must reject heat to some low-temperature reservoir. It cannot convert all the input heat into power.

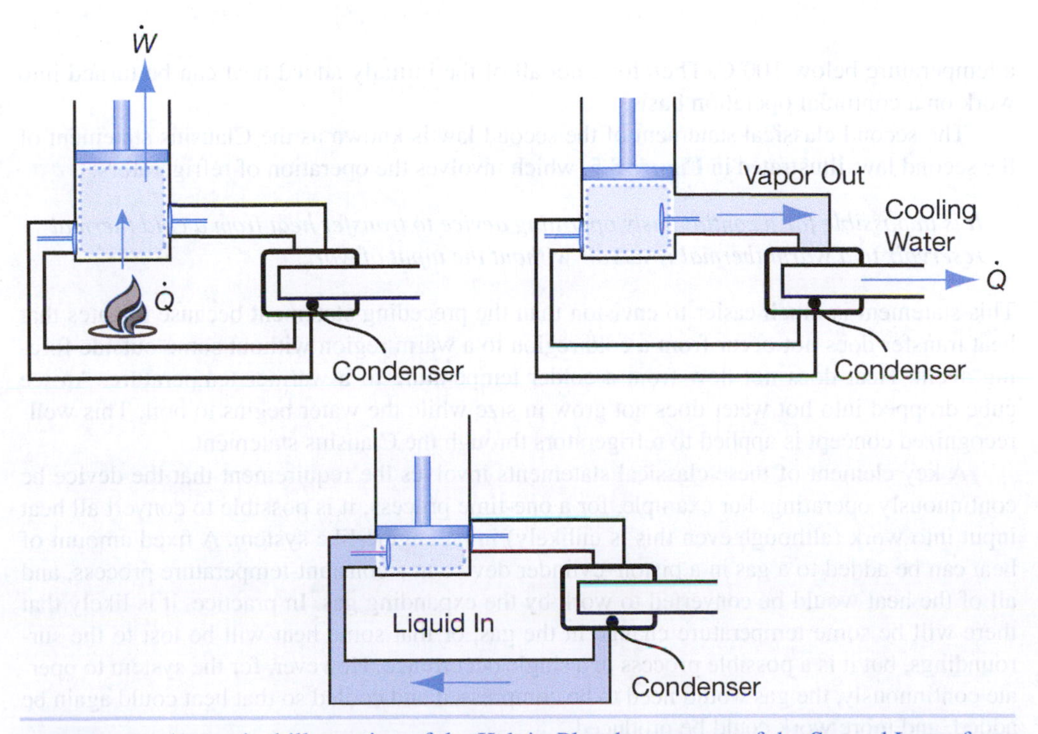

FIGURE 5.4 A practical illustration of the Kelvin-Planck statement of the Second Law of Thermodynamics. Water vapor expands when heated in a piston–cylinder device. An exhaust port opens, and the vapor exits the piston–cylinder, allowing the piston to return to its original position. However, before reentering the piston as liquid, the water vapor must be condensed, which requires a rejection of heat to a low-temperature reservoir through a condenser.

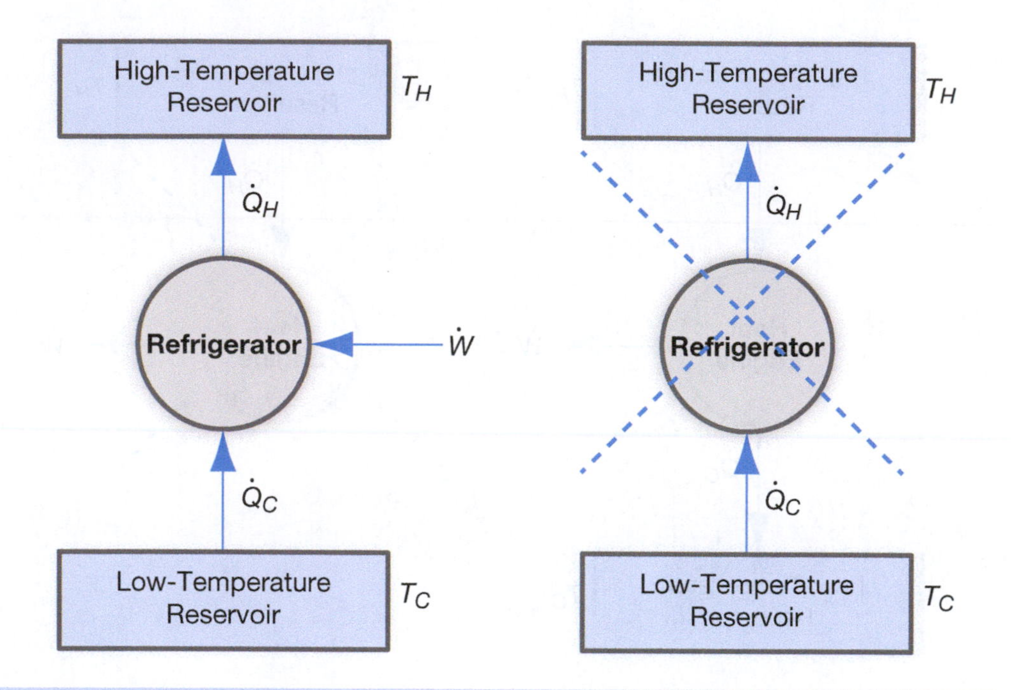

FIGURE 5.5 According to the Clausius statement of the Second Law of Thermodynamics, in order for a refrigerator to transfer heat from a low-temperature reservoir to a high-temperature reservoir, it must receive an input of work.

a temperature below 100°C. Therefore, not all of the initially added heat can be turned into work on a continual operation basis.

The second classical statement of the second law is known as the Clausius statement of the second law, illustrated in **Figure 5.5**, which involves the operation of refrigerators:

> *It is impossible for a continuously operating device to transfer heat from a cold thermal reservoir to a warm thermal reservoir without the input of work.*

This statement is much easier to envision than the preceding statement because it states that heat transfer does not occur from a cold region to a warm region without some outside forcing agent. Heat does not flow from a colder temperature to a warmer temperature. An ice cube dropped into hot water does not grow in size while the water begins to boil. This well-recognized concept is applied to refrigerators through the Clausius statement.

A key element of these classical statements involves the requirement that the device be continuously operating. For example, for a one-time process, it is possible to convert all heat input into work (although even this is unlikely) in an engine-like system. A fixed amount of heat can be added to a gas in a piston–cylinder device in a constant-temperature process, and all of the heat would be converted to work by the expanding gas. In practice, it is likely that there will be some temperature change in the gas, or that some heat will be lost to the surroundings, but it is a possible process in a single occurrence. However, for the system to operate continuously, the gas would need to be compressed and cooled so that heat could again be added, and more work could be produced.

From these statements of the second law, further development of our understanding how things happen in nature became possible. The next step is the application of these ideas, along with development of the thermodynamic temperature scale, to determine how efficient heat engines and refrigerators can become.

QUESTION FOR THOUGHT/DISCUSSION

While nothing is truly a perfect thermal reservoir, what are some examples of objects or volumes that can be acceptably modeled as thermal reservoirs for particular processes? (For example, the walls of a building probably can be modeled as having a constant temperature if someone leans up against the building for a minute, while the walls probably shouldn't be modeled as a thermal reservoir if the building is sitting in the sun for a day.)

5.4 REVERSIBLE AND IRREVERSIBLE PROCESSES

We have mentioned that the second law can be used to compare the actual performance of a device to its ideal performance. The ideal performance of a device is obtained if the device is operating in a *reversible* manner. Such operation is described as a reversible process. A reversible process is one that, once it is completed, can be performed in the opposite direction (reversed) after which both the system and the surroundings will be returned to states identical to those which existed before the original process took place. So, if a system undergoes a reversible process from state 1 to 2, and then the process is reversed and performed from state 2 to 1, both the system and the surroundings are identical to their original conditions at state 1.

Conversely, an *irreversible process* is one that cannot be reversed such that the system and the surroundings return to their same initial states from before the process began. All real processes are irreversible; no process is truly reversible, although some may be very close to being reversible and can be adequately modeled as such.

Sometimes a process may at first look reversible, only to be revealed as irreversible upon further inspection. For example, if we put a kettle of water on a gas stove, as shown in Figure 5.6, heat the water from 20°C to 50°C, and then cool the water back to 20°C after

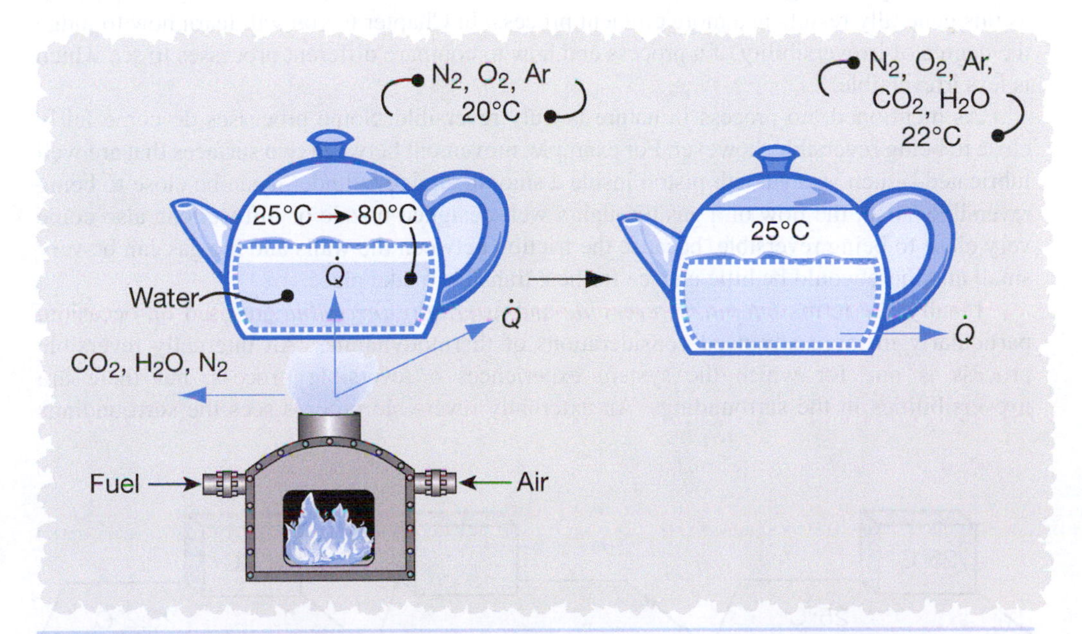

FIGURE 5.6 As an illustration of irreversibility, consider this situation of heating water through a combustion process. Through heat addition, the water is heated from 25°C to 80°C. Although the system and the surroundings may initially appear unchanged after the water cools back to 25°C, we see that the fuel and air (part of the surroundings) in the combustion process have been converted to combustion products, and the surroundings have warmed slightly.

turning off the flame, we may think that the process was reversible because the water is back to its initial condition. However, the surroundings have been changed substantially. First, some natural gas and air have undergone chemical reactions and been converted to other gases such as carbon dioxide and water vapor. Second, the surrounding air has received a heat input from the combustion as well as from the cooling water. Although the temperature of the air may not have increased dramatically, it did increase. Just because the water temperature returned to the initial state, and just because the air temperature did not change noticeably, the process was irreversible because the molecular composition and the temperature of the immediate surroundings did change.

There are many causes of irreversibility. Five of the most noteworthy and common causes of irreversibility are (1) friction, (2) heat transfer across a finite temperature difference, (3) unrestrained expansion of a gas, (4) mixing processes, and (5) chemical reactions. Friction occurs in any process involving motion in nature, and so any such process is irreversible. Consider a situation where a block of wood is pushed across a surface from point A to point B, and then the block is pushed back to A, as shown in **Figure 5.7**. The first process may appear to be reversible, because the block is in its initial state, until we consider that friction occurred as the block was moved. Careful measurement of the temperatures of the block and the surface would indicate that the temperature of each was increased. Futhermore, the air was disturbed and heated (slightly) through the process. These changes persist after the block is returned to A, and so the process is irreversible.

Unrestrained expansion of a gas is another simple example of irreversibility. If a gas is held in one half of a container by a membrane, such as in **Figure 5.8**, and the membrane is punctured, the gas will flow to fill up the entire container. However, the gas will not spontaneously all flow back to occupy just half of the container, and so the initial process is irreversible—work must be put into the system to push the gas to one side of the container, thus altering the surroundings.

One thing that engineers attempt to do is reduce the amount of irreversibility in a process, as this generally results in a more efficient process. In Chapter 6, you will learn how to judge the degree of irreversibility of a process and how to compare different processes to see which is less irreversible.

As mentioned, no process in nature is truly reversible. Some processes do come fairly close to being reversible, however. For example, movement between two surfaces that are well lubricated—such as a smooth piston inside a smooth engine cylinder—can be close to being reversible. Also, the flow of a gas through a well-designed nozzle or diffuser can also come very close to being reversible, because the friction between the walls and the gas can be very small and there would be little chance for heat transfer to take place.

Finally, the terms *internally reversible* and *externally reversible* are used on occasion, particularly in more advanced considerations of thermodynamics. An internally reversible process is one for which the system experiences a reversible process, but there are irreversibilities in the surroundings. An externally reversible process sees the surroundings

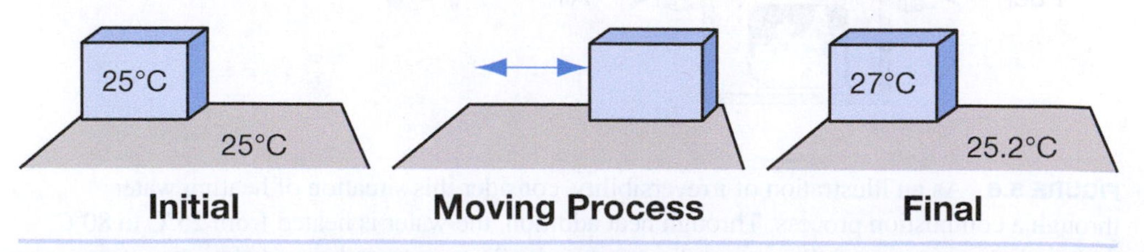

FIGURE 5.7 Pushing a block along a surface and returning it to its initial position may look like a reversible process, but careful investigation shows that frictional heating has increased the temperature of the block and surface—thus the process is irreversible.

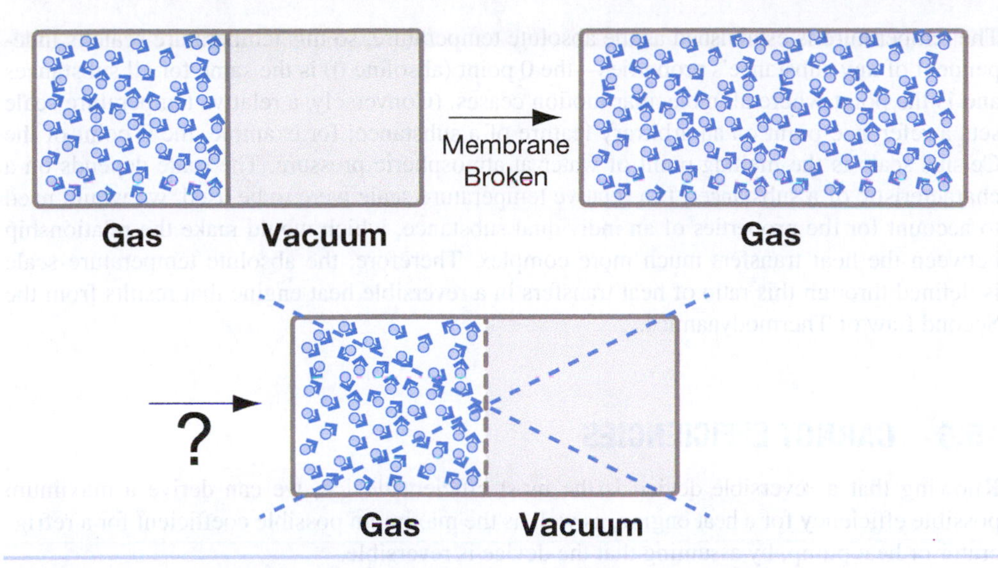

FIGURE 5.8 Unrestrained gas expansion is irreversible. Gas will expand into a vacuum when it is allowed to do so, but the resulting system will not spontaneously return to being half gas and half vacuum.

experiencing a reversible process, whereas irrerversibilities occur inside the system. For most basic thermodynamic purposes, this distinction is not particularly important, because in reality both the system and the surroundings are likely to be experiencing irreversibilities.

> **QUESTION FOR THOUGHT/DISCUSSION**
>
> Engineers often try to reduce the factors (such as friction) that lead to irreversibilities in a process. But such reductions often cost money. How would you decide whether the added costs of reducing the irreversibilities of a process are justified by the advantages gained by doing so?

5.5 A THERMODYNAMIC TEMPERATURE SCALE

We have noted that the second law can be used to create a temperature scale that is independent of the properties of a given material. Details on how this is formulated can be found elsewhere; although the final result of the formulation is of great use to engineers, the detailed development is of less use until higher level studies of thermodynamics are taken. Sadi Carnot, a 19th-century French engineer, found that all reversible heat engines operating between thermal reservoirs of the same two temperatures would have the same thermal efficiency. Using this concept, we can demonstrate that the ratio of the heat transfers between the two thermal reservoirs is a function of the temperatures of the two reservoirs: $\frac{|Q_H|}{|Q_C|} = f(T_H, T_C)$. We can show that this function is a ratio of a function of each temperature: $f(T_H, T_C) = g(T_H)/g(T_C)$. This function can take various forms, but the simplest is to state that $g(T) = T$, with the temperature being based off of absolute 0—an absolute temperature scale:

$$\left(\frac{|Q_H|}{|Q_C|}\right)_{\text{rev}} = \frac{T_H}{T_C} \tag{5.1}$$

The temperature is established as the absolute temperature, so this temperature scale is independent of any substance's properties—the 0 point (absolute 0) is the same for all substances and is the point where all molecular motion ceases. (Conversely, a relative temperature scale sets a reference point on an arbitrary feature of a substance; for example, the 0 point of the Celsius scale is the melting point of water at atmospheric pressure. The scale depends on a characteristic of a substance.) If a relative temperature scale were to be used, we would need to account for the properties of an individual substance, which would make the relationship between the heat transfers much more complex. Therefore, the absolute temperature scale is defined through this ratio of heat transfers in a reversible heat engine that results from the Second Law of Thermodynamics.

5.6 CARNOT EFFICIENCIES

Knowing that a reversible device is the most efficient device, we can derive a maximum possible efficiency for a heat engine, as well as the maximum possible coefficient for a refrigerator or heat pump, by assuming that the device is reversible.

Consider the heat engine shown in Figure 5.9. Recall from Eq. (4.34a) that the thermal efficiency of a heat engine is

$$\eta = W/Q_H \tag{4.34a}$$

Furthermore, recall that for a cycle, such as that of a heat engine, the net heat transfer equals the net work:

$$Q_{\text{cycle}} = W_{\text{cycle}} \tag{4.33}$$

For the heat engine, $W = W_{\text{cycle}}$, and

$$Q_{\text{cycle}} = Q_H - |Q_C|$$

Therefore, $W = Q_H - |Q_C|$, because Q_H is a heat input and is already a positive number. Substituting into Eq. (4.34a),

$$\eta = \frac{Q_H - |Q_C|}{Q_H} = 1 - \frac{|Q_C|}{Q_H} \tag{5.2}$$

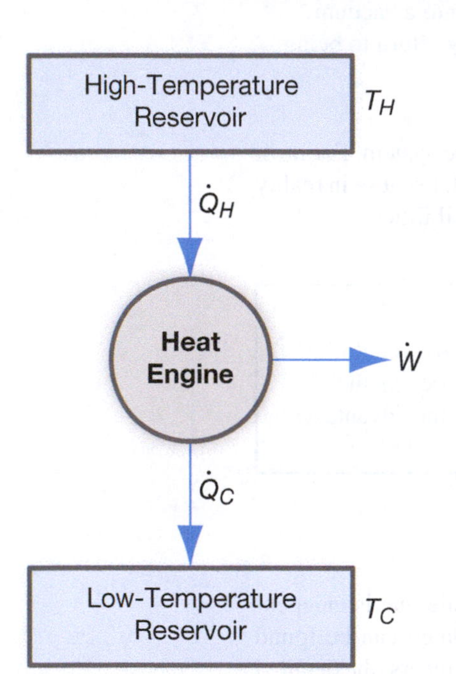

FIGURE 5.9 A generic model of a heat engine.

Now, if the heat engine is reversible, the efficiency will be equal to the maximum efficiency and can be found by substituting Eq. (5.1), which applies for reversible heat engines:

$$\eta_{\text{max}} = 1 - \frac{T_C}{T_H} \tag{5.3}$$

This maximum efficiency for a heat engine is sometimes referred to as the Carnot efficiency, and sometimes as the reversible efficiency. Note that it is only dependent on the absolute temperatures of the thermal reservoirs.

Let us now consider refrigerators and heat pumps, as shown in Figure 5.10. The maximum coefficient of a refrigerator or a heat pump is found in the same manner. Applying the same equivalency between the cycle's net work and net heat transfer, Eqs. (4.35) and (4.36) can be rewritten for any refrigerator or heat pump as

$$\beta = \frac{Q_C}{|Q_H| - Q_C} \tag{5.4}$$

and

$$\gamma = \frac{|Q_H|}{|Q_H| - Q_C} \tag{5.5}$$

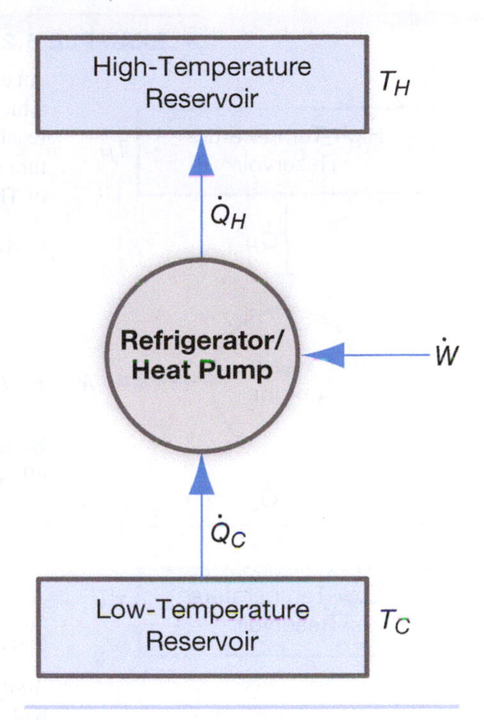

FIGURE 5.10 A generic model of a refrigerator or heat pump.

Like heat engines, maximum values for the coefficients of performance exist for reversible refrigerators and heat pumps. So, for the reversible form of these devices, Eq. (5.1) can be substituted into Eqs. (5.4) and (5.5):

$$\beta_{\max} = \frac{T_C}{T_H - T_C} \tag{5.6}$$

and

$$\gamma_{\max} = \frac{T_H}{T_H - T_C} \tag{5.7}$$

Naturally, for any heat engine, refrigerator, or heat pump, the thermal efficiency of the coefficient of performance must be less than or equal to the maximum possible value for the device operating between the two given thermal reservoirs:

Heat Engine: $\eta \leq \eta_{\max}$
Refrigerator: $\beta \leq \beta_{\max}$
Heat Pump: $\gamma \leq \gamma_{\max}$

This is another statement of an aspect of the second law, and if a proposed device violates this restriction, the device violates the second law and cannot exist.

▸ EXAMPLE 5.1

A reversible heat engine operates between a hot thermal reservoir at 950 K and a low-temperature thermal reservoir at 300 K. The heat engine delivers 525 kW of power. Determine the rate at which heat is received from the high-temperature reservoir.

Given: $T_H = 950$ K, $T_C = 300$ K, $\dot{W} = 525$ kW

Find: $\dot{Q}_H$

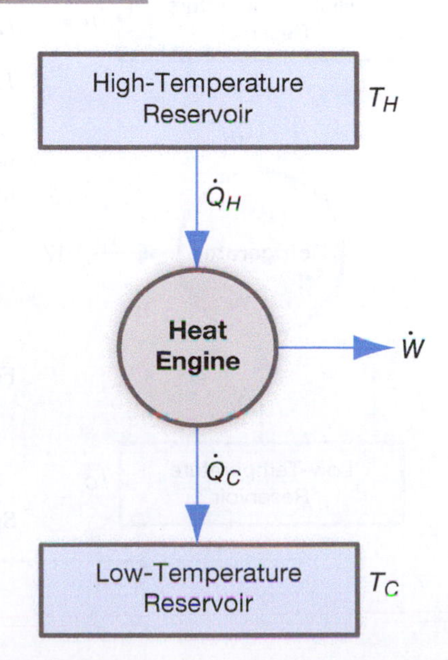

Solution: The reversible heat engine has a thermal efficiency equal to the Carnot thermal efficiency. This will lead to the actual thermal efficiency (Eq. (4.34b)) being set equal to the maximum possible thermal efficiency (Eq. (5.3)):

$$\eta = \eta_{\max} = 1 - \frac{T_C}{T_H} = 0.684$$

Inserting this value into Eq. (4.34b) yields the heat input from the high-temperature reservoir.

$$\dot{Q}_H = \frac{\dot{W}}{\eta} = \frac{525 \text{ kW}}{0.684} = \mathbf{767\ kW}$$

▶ EXAMPLE 5.2

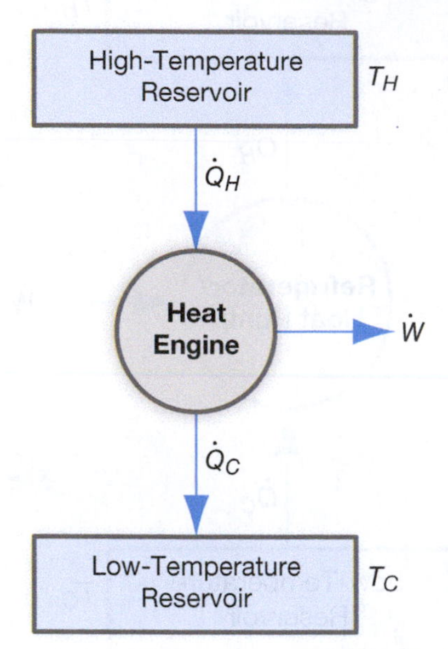

An engineer proposes to develop a power plant (which can be modeled as a heat engine) which will operate at an efficiency of 65%. The steam flowing through the power plant reaches a peak temperature of 660 K, and rejects heat to the environment at a temperature of 290 K. Is this proposed power plant feasible, or does it violate the Second Law of Thermodynamics?

Given: $T_H = 660$ K

$T_C = 290$ K

$\eta_{\text{claim}} = 0.65$

Find: Does this plant violate the Second Law of Thermodynamics?

Solution: To answer this question, we need to find the maximum possible thermal efficiency for a heat engine operating between these two temperatures. From Eq. (5.3):

$$\eta_{\max} = 1 - \frac{T_C}{T_H} = 1 - \frac{290 \text{ K}}{660 \text{ K}} = 0.561$$

As $\eta_{\text{claim}} > \eta_{\max}$, the proposed system **violates the Second Law of Thermodynamics** and is not possible.

Analysis: This type of analysis considers one aspect of the second law. If the power plant had not violated this aspect, it still may not have met all aspects of the second law. As long as one aspect of the second law is violated, a proposed system or device cannot work.

▶ EXAMPLE 5.3

A Carnot refrigerator provides 25.0 kW of cooling to a space at 5°C while using 8.20 kW of power. What is the temperature of the warm space to which the heat is rejected?

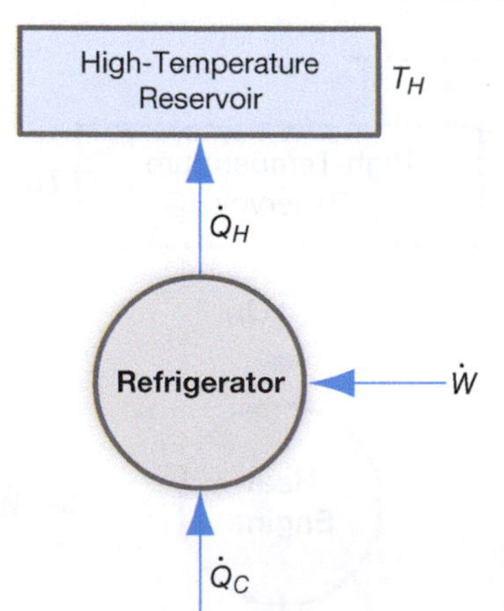

Given: $T_C = 5°\text{C} = 278$ K, $\dot{Q}_C = 25.0$ kW, $|\dot{W}| = 8.20$ kW

Find: T_H

Solution: The Carnot refrigerator operates at the maximum possible coefficient of performance for a refrigerator operating between two thermal reservoirs. We will interpret the "warm space" to be the high-temperature thermal reservoir. From Eq. (4.35), the coefficient of performance of the refrigerator is

$$\beta = \frac{\dot{Q}_C}{|\dot{W}|} = 3.049$$

For a Carnot refrigerator, Eq. (5.6) leads to

$$\beta = \beta_{\max} = \frac{T_C}{T_H - T_C} = \frac{278 \text{ K}}{T_H - 278 \text{ K}} = 3.049$$

Solving yields $\boldsymbol{T_H = 369}$ **K.**

▶ EXAMPLE 5.4

A young engineer proposes a new heat pump design for your company. The heat pump operates between temperatures of 8.0°C and 22.0°C and claims to be able to deliver 15.0 kW of heat to the warm space while using only 0.60 kW of power. Should you have your company develop this design, or does it violate the Second Law of Thermodynamics?

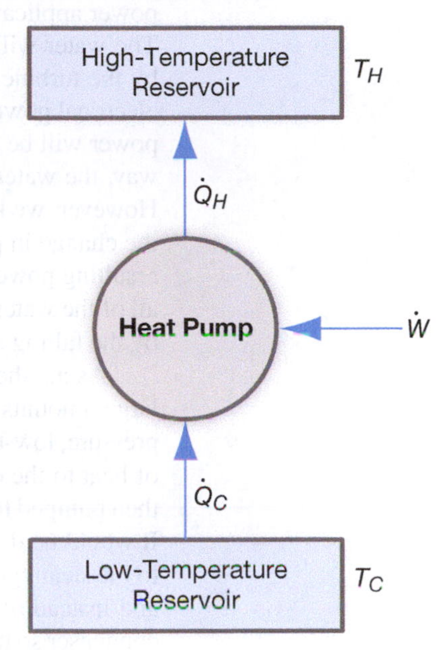

Given: $T_H = 22.0°C = 295$ K; $T_C = 8.0°C = 281$ K; $\dot{Q}_H = -15.0$ kW; $\dot{W} = -0.60$ kW

Find: Does the proposed system violate the Second Law of Thermodynamics?

Solution:

We will compare the claimed coefficient of performance of the heat pump to that which would be possible with an ideal heat pump operating between the given thermal reservoir temperatures. First, use Eq. (5.7) to find the Carnot coefficient of performance:

$$\gamma_{max} = \frac{T_H}{T_H - T_C} = \frac{295 \text{ K}}{295 \text{ K} - 281 \text{ K}} = 21.07$$

Next, calculate the claimed coefficient of performance from Eq. (4.36):

$$\gamma = \frac{|\dot{Q}_H|}{|\dot{W}|} = \frac{15.0 \text{ kW}}{0.60 \text{ kW}} = 25.0$$

Comparing the claimed and maximum coefficients of performance, we see that $\gamma > \gamma_{max}$. Therefore, the proposed heat pump **violates the second law** and your company should not pursue development.

Analysis: While this proposed design violates the second law, you do not want to completely discourage your young engineer. You could suggest that he recheck his measurements because minor changes in the parameters could make the design at least feasible in terms of the second law.

QUESTION FOR THOUGHT/DISCUSSION

The maximum coefficient of performance for a refrigerator can go to infinity. What combination of environmental and refrigeration conditions may lead to a refrigerator having a high maximum possible coefficient of performance but for which it would be unreasonable to try to operate the refrigerator?

5.7 PERPETUAL MOTION MACHINES

Throughout history, there have been many claims of devices that can run forever (or at least a very long time) with no new input of energy. These devices are given the name perpetual motion machines, and in one way or another, these machines have violated either the first or second law of thermodynamics. A machine does not need to run forever to fall into this category—rather, any machine that violates a law of thermodynamics is labeled a perpetual motion machine. A machine that violates the First Law of Thermodynamics (by violating the energy balance, usually through the creation of energy) is known as a perpetual motion machine—Type 1 (PMM1), and a machine that violates the Second Law of Thermodynamics is a perpetual motion machine—Type 2 (PMM2). Sometimes it is easy to see how a perpetual motion machine violates the laws of thermodynamics, but in other cases the devices are clever and require substantial analysis to determine how they fail.

For example, consider the following device proposed to be used in a pumped–stored hydropower application. A reservoir of water will be located at a height above a hydropower turbine. The water will be released, and the potential energy of the water will be converted into power by the turbine. The mechanical power will power an electrical generator, and some amount of electrical power (say 100 kW) will be sent to the outside electrical grid. The remaining electrical power will be used to operate a pump to pump the water back into the elevated reservoir. In this way, the water will be able to again fall through the turbine, and this process can last forever. However, we know from the first law that the amount of work produced will be no greater than the change in potential energy of the water as it falls from the elevated reservoir. If some of the resulting power is sent to the outside world, there will be insufficient power remaining to pump all of the water back into the reservoir (doing so will require at least as much power as produced by the falling water). Therefore, this device is creating energy and is a PMM1.

As another example, consider the following proposed device. A steam power plant rejects large amounts of heat to the environment through the condenser. The condenser receives low-pressure, low-temperature steam from the turbine exhaust. The condenser facilitates the rejection of heat to the environment, causing the steam to become a saturated liquid. The liquid water is then pumped to a high pressure and sent through a steam generator, where it receives a heat input. It would be desirable to improve the efficiency of the power plant by decreasing the heat input to the steam generator. It is proposed to take the heat that is rejected from the condensing steam, and instead of directing that to the environment, direct it toward the liquid water that exits the condenser so that it becomes heated prior to entering the pump. However, this violates the second law in at least two ways and is a PMM2. First, the proposed system removes one of the thermal reservoirs with which the cycle exchanges heat. This violates the Kelvin-Planck statement of the second law by operating in a cyclic mode, producing work, while exchanging heat with only one thermal reservoir. Second, and perhaps more obviously, to heat the liquid water with the rejected heat, the heat would have to flow from a colder substance (the condensing steam) to a hotter substance (the heating liquid water). We cannot transfer heat from a colder system to a hotter system without adding work, so this too is a violation of the second law.

Frequently, a proposed device is a PMM2 in that it will be frictionless and expect processes to be reversible. Because the ability of energy to do work continually decreases as it is used, these features render the operation of a device impossible and relegate what often appears to be a clever device into just another failed perpetual motion machine.

QUESTION FOR THOUGHT/DISCUSSION

People continually try to create and patent perpetual motion machines. What is so tantalizing about the concept of a perpetual motion machine?

Summary

In this chapter, we have surveyed the basic elements of the Second Law of Thermodynamics. We have explored the wide variety of uses of the second law, and have seen how the second law describes irreversibilities in processes. We also considered some of the historical elements of the law with regard to the performance of heat engines, refrigerators, and heat pumps. In Chapter 6, we will discuss more of the numerical applications of the second law that are commonly employed in engineering practice.

KEY EQUATIONS

Maximum Thermal Efficiency—Heat Engine:

$$\eta_{max} = 1 - \frac{T_C}{T_H} \tag{5.3}$$

Maximum Coefficient of Performance—Refrigerator:

$$\beta_{max} = \frac{T_C}{T_H - T_C} \tag{5.6}$$

Maximum Coefficient of Performance—Heat Pump:

$$\gamma_{max} = \frac{T_H}{T_H - T_C} \tag{5.7}$$

PROBLEMS

5.1 Describe some of the characteristics that make the following processes irreversible:

- **(a)** liquid water flowing through a pipe
- **(b)** a block of lead being melted from the heat of a furnace
- **(c)** air being compressed isothermally in a piston–cylinder assembly
- **(d)** hydrogen and oxygen reacting to form water

5.2 Describe some of the characteristics that make the following processes irreversible:

- **(a)** liquid water at 101 kPa being heated and vaporized to steam at 101 kPa and 200°C
- **(b)** carbon dioxide gas being mixed with nitrogen gas
- **(c)** a ball rolling down an inclined plane
- **(d)** air escaping a popped balloon

5.3 Describe some characteristics that make the following processes irreversible:

- **(a)** natural gas burning with air in a furnace to produce warm air for a home
- **(b)** a football being kicked through the air
- **(c)** sodium and chloride atoms combining to form table salt
- **(d)** water and soap being used to clean oil from a surface

5.4 Provide some suggestions as to how the following processes could have lower amounts of irreversibility:

- **(a)** a lamp providing a certain amount of light to a room
- **(b)** compressed air exiting a cylinder through a fully opened valve
- **(c)** cooking food to a temperature of 75°C

5.5 Provide some suggestions as to how the following processes could have lower degrees of irreversibility:

- **(a)** a piston sliding against a cylinder wall
- **(b)** condensing water vapor at 101 kPa
- **(c)** heating a metal to its melting point in a furnace

5.6 Provide some suggestions as to how the following processes could have lower amounts of irreversibility:

- **(a)** liquid water flowing through a pipe
- **(b)** steam expanding through a turbine
- **(c)** the tire of an automobile rolling down a road

5.7 A reversible heat engine operates between temperatures of 1200 K and 350 K. The heat engine produces 150 kW of power. Determine the rate of heat input from the high-temperature reservoir and the rate of heat rejected to the low-temperature reservoir.

5.8 A reversible heat engine receives 250 kW of heat from a high-temperature reservoir at 600 K and produces 140 kW of power. What is the temperature of the low-temperature reservoir?

5.9 A reversible heat engine receives 7.4 kJ/s of heat from a high-temperature reservoir at 610 K and produces 3.1 kW of power. What is the temperature of the low-temperature reservoir?

5.10 A reversible heat engine receives 500 kW of heat from a reservoir at 750 K and rejects 200 kW of heat to the low-temperature reservoir. What is the temperature of the low-temperature reservoir?

5.11 A heat engine operates between reservoirs at 700 K and 400 K. The heat engine receives 1500 kW of heat from the high-temperature reservoir. What is the maximum power that the heat engine can produce?

5.12 A heat engine produces 200 kW of power after receiving 550 kW of heat. The temperature of the low-temperature reservoir is 350 K. What is the minimum allowable temperature for the high-temperature reservoir?

5.13 A heat engine produces 110 kW of power after receiving 215 kJ/s of heat input. The temperature of the low-temperature reservoir is 290 K. What is the minimum allowable temperature for the high-temperature reservoir?

5.14 A heat engine receives 650 kW of heat from a reservoir at 500 K and produces 200 kW of power. What is the maximum possible temperature of the low-temperature reservoir?

5.15 A heat engine receives steam at 300°C and releases liquid water at 30°C. The heat engine produces 1500 kW of power. What is the minimum amount of heat that must be added to the heat engine through the steam?

5.16 Consider Problem 5.15. The steam is produced from a thermal reservoir at 500°C, and the water is released to a thermal reservoir at 10°C. Assuming that the temperatures of the steam and liquid water can be adjusted accordingly, what is the minimum amount of heat that must be added to the heat engine through the steam under these conditions to produce 1500 kW of power?

5.17 A Carnot heat engine is to be used to produce 250 kW of power using a heat input of 525 kW. For high-temperature reservoirs ranging in temperature between 500 K and 1000 K, plot the corresponding low-temperature reservoir temperatures.

5.18 A Carnot heat engine receives 1000 kW of heat from a reservoir at 750 K. Plot the power produced for temperatures of the low-temperature reservoir varying between 300 K and 600 K.

5.19 A reversible heat engine receives 19 kJ/s of heat from a reservoir at 700 K. Plot the power produced for temperatures of the low-temperature reservoir varying between 200 K and 600 K.

5.20 A proposed heat engine draws its input heat from a space at 1000 K and rejects the heat to a space at 400 K. The engine is to produce 300 kW of power from a heat input of 900 kW. Is this process possible?

5.21 A proposed heat engine receives heat at 500 K and rejects heat at 350 K. The engine is to produce 100 kW of power from a heat input of 150 kW. Is this process possible?

5.22 A proposed heat engine is to produce 95 kW of power while receiving 170 kJ/s of heat. The engine receives heat at 780 K and rejects heat at 360 K. Is this process possible?

5.23 An engineer brings you a basic design for a steam power plant. The heat is input to the plant from a source at 500°C, and heat is rejected to cooling water at 20°C. The power plant is to produce 250 MW of power while receiving heat at a rate of 700 MW. Is this process possible?

5.24 Following up on the work of the engineer in Problem 5.23, another engineer proposes adding several modifications to the steam power plant. He proposes the following in his design: heat will be added to the steam at 550°C, heat will be rejected to cooling water at 20°C, and 300 MW of power will be produced from a heat input of 400 MW. Is this process possible?

5.25 A proposed gas power plant is to receive heat from a source at 1200 K. It is to produce 10 MW of power while receiving a heat input of 20 MW. The gas power plant is to deliver its exhaust gases to a manufacturing process that requires a temperature of 700 K. Is this process possible?

5.26 A steam power plant, which can be modeled as a heat engine, is to receive steam at 550°C. Plot the maximum possible thermal efficiency of the power plant for heat rejection temperatures ranging from 0°C to 100°C.

5.27 A steam power plant, which can be modeled as a heat engine, is to reject heat to cooling water at 295 K. Plot the maximum possible thermal efficiency of the power plant for heat input temperatures ranging between 500 K and 800 K.

5.28 A steam power plant is to reject heat to the environment. Two possible heat sinks are available: a lake with a steady water temperature of 280 K, and air with a temperature that can rise to 300 K during the day. The plant receives a heat input of 115 MW. Plot the maximum possible power production for heat input temperatures ranging between 500 K and 800 K for both possible heat rejection temperatures.

5.29 You are advising an investor who is looking for new products to support. The investor asks you to review a plan she has received from an inventor who claims he will provide a new type of engine for an automobile. The investor is skeptical of the inventor's claims. The inventor claims that the engine will produce 150 kW of power while receiving heat at a rate of 210 kW. The engine is to receive heat at a temperature of 600°C and reject the heat at a temperature of 80°C. Should you advise the investor to consider the inventor's proposal or to steer clear of it? Why?

5.30 A high-school teacher asks you for assistance in evaluating a science project. The students working on the project have built a small engine which they measure as producing 5.1 kW of power while rejecting heat at a rate of 5.8 kW. (The students could not figure out how to measure the heat input rate.) The engine receives the heat from a source at 315°C, and the heat is rejected at 30°C. Are the results reported by the students possible?

5.31 A Carnot refrigerator removes 15 kW of heat from a low-temperature reservoir at 5°C while drawing a power input of 5 kW. What is the temperature of the high-temperature reservoir to which the heat is rejected?

5.32 A Carnot refrigerator operates between thermal reservoirs of −5°C and 25°C. The refrigerator uses 25 kW of power. Determine the rate of heat removal from the low-temperature reservoir.

5.33 A reversible air conditioner (modeled as a refrigerator) operates between thermal reservoirs of −2°C and 25°C. The refrigerator uses 25 kW of power. Determine the rate of heat removal from the low-temperature reservoir.

5.34 A Carnot refrigerator removes 35 kW of heat from a cool space using 10 kW of power. The heat is rejected to a space at 310 K. Determine the temperature of the cold space.

5.35 A Carnot refrigerator operates in a kitchen whose temperature is 22°C. The refrigerator uses 1.5 kW of power when operating. If the refrigerator is to remove 3 MJ of heat from the interior of the refrigerator, which is maintained at 2°C, for how long must the refrigerator operate?

5.36 Consider the refrigerator in Problem 5.35. Plot the amount of time the refrigerator must operate for interior temperatures ranging from 1°C to 20°C.

5.37 Plot the coefficient of performance versus the temperature of a high-temperature reservoir for a Carnot refrigerator whose low-temperature reservoir is 5°C and whose high-temperature reservoir varies between 7°C and 50°C.

5.38 A certain household refrigerator rejects heat at a temperature of 30°C so that it will work even when the inside house temperature becomes uncomfortably hot. Plot the maximum possible coefficient of performance for this refrigerator as a function of low-temperature reservoir temperatures ranging between −20°C and 5°C.

5.39 A Carnot air conditioner is to remove 100 kW of heat from a building at a temperature of 10°C. Plot the power used for the process for heat rejection temperatures ranging between 25°C and 60°C.

5.40 A refrigerator uses 1.5 kW of power to maintain a cool space at 275 K while rejecting heat at 315 K. Determine the maximum rate of heat removal that can be achieved by this refrigerator.

5.41 A continuously operating refrigerator needs to remove 500 kJ of heat from an object in a certain period of time. The heat is removed at 280 K, and the heat is rejected to a space at 305 K. Determine the minimum amount of work that is needed to accomplish this process.

5.42 A refrigerator removes 280 kW of heat from a cold space at 15°C while using 125 kW of power. What is the maximum temperature possible for the high-temperature space to which the heat is rejected?

5.43 Refrigerators are often rated by specifying their refrigeration capacity in "tons of refrigeration," where one ton of refrigeration is equal to 211 kJ/min. An air conditioning system has a capacity of 8 tons of refrigeration when removing heat from a space at 25°C. Plot the minimum power requirement for the compressor in the refrigerator (in kW) for high-temperature heat rejection spaces ranging in temperature from 25°C to 50°C.

5.44 A proposed air conditioner is to remove 250 kW of heat from air at 15°C, reject the heat at 35°C, and use 50 kW of power. Is this process possible?

5.45 A proposed refrigeration system is to remove 50 kW of heat from a space at −3°C and reject the heat to a space at 25°C. The system is to use 20 kW of power. Is this process possible?

5.46 A proposed refrigeration system is to remove 11 kJ/s of heat from a space at 0°C, reject the heat at 40°C, and use 1.2 kW of power. Is this process possible?

5.47 An inventor claims to have invented a power-saving refrigerator that can keep objects at −5°C while the refrigerator is located in an environment that is maintained at 25°C. To prove his system works as claimed, he places a 2-kg block of heated iron (c = 0.447 kJ/kg · K) at 75°C into the refrigerator. The power meter on the refrigerator reads that it is using 100 W of electrical work. The refrigerator is operated for 1 min, and the temperature of the iron is read to be −5°C. Is this possible?

5.48 You assign a young engineer to design an air conditioning system for a building. She proposes a system which will remove heat at 5°C and reject heat at 40°C. 850 kW of heat is to be removed from the cooled air, and the system will use compressors providing 140 kW of power to accomplish the task. Does her proposed system violate the Second Law of Thermodynamics?

5.49 An inventor wishes to create a new refrigeration system to cool systems at very low temperatures. He designs a system that, if it works as planned, will remove 5 kW of heat from a space at 125 K, reject the heat at 300 K, and use 4 kW of power. Does this system violate the Second Law of Thermodynamics?

5.50 One difficulty in cooling systems to very low temperatures is the large amounts of power that are required. Suppose a refrigerator is to remove 2 kW of heat from a cold space and reject the heat to a space at 300 K. Plot the minimum power required and the maximum coefficient of performance possible for such a refrigerator for cold-space temperatures ranging from 50 K to 290 K.

5.51 A salesman at a store wants to sell you a refrigerator. He tells you that the refrigerant removes heat from the interior of the refrigerator at a temperature of 0°C and rejects heat to the environment at 35°C. He also tells you that the refrigerator uses 200 W of power to remove 5 kW of heat from the refrigerator. Should you believe his claims about the refrigerator?

5.52 You overhear a couple comparing window air conditioning units in a store. The man is skeptical of the claims on the sign describing the performance of the refrigerator, but the woman seems confident they are reasonable. Looking at the sign, you notice that the air conditioner is to remove heat with refrigerant at 10°C, reject heat with refrigerant at 40°C, use 1 kW of power, and remove 7 kW of heat. Is the man or woman more likely to be correct? Why?

5.53 A Carnot heat engine is to supply power to run a Carnot refrigerator. The heat engine receives 100 kW of heat from a source at 500 K and rejects heat to a thermal sink at 320 K. The refrigerator removes heat from a cool space at 280 K and rejects the heat to a space at 320 K. Determine the rate at which heat is removed from the cool space.

5.54 A Carnot heat engine is to be used to supply power to run a Carnot refrigerator. The heat engine receives heat at 600 K and rejects heat at 300 K. The refrigerator removes 50 kW of heat from a space at 276 K and rejects heat to a space at 300 K. Determine the rate at which heat must be added to the Carnot heat engine.

5.55 A Carnot heat engine is to be used to supply power to operate a Carnot refrigerator. The heat engine receives 90 kW of heat from a source at 500 K and rejects heat to a sink at 300 K. The refrigerator removes heat from a cool space at 280 K and rejects heat to a space at 320 K. Determine the rate at which heat is removed from the cool space.

5.56 A Carnot heat engine is to be used to supply the power to run a Carnot refrigerator. The heat engine receives 1000 kW of heat and rejects heat to a space at 350 K. The refrigerator removes 900 kW of heat from a space at 270 K and rejects the heat to a space at 350 K. What is the temperature of the heat engine's heat source?

5.57 A reversible heat pump delivers 250 kW of heat to a space at 40°C while removing heat from a space at 10°C. Determine the power used by the heat pump.

5.58 A reversible heat pump is set up to operate between a cold space at −5°C and a warm space at 23°C. The heat pump uses 1.2 kW of power. Determine the rate at which heat is delivered to the warm space.

5.59 A Carnot heat pump uses 5 kW of power to deliver heat to a space at 35°C after removing heat from a space at 10°C. Determine the rates at which heat is delivered to the warm space and removed from the cold space.

5.60 A Carnot heat pump delivers 315 kW of heat to a space at 30°C while removing heat from a space at 5°C. Determine the power used by the heat pump, in kW.

5.61 A reversible heat pump uses 8 kW of power to remove 20 kW of heat from a cold space at 275 K. Determine the rate at which heat is delivered to the warm space and the temperature of the warm space.

5.62 A heat pump is to deliver heat to a space at 310 K after removing heat from a space at 280 K. The heat is delivered to the warm space at a rate of 150 kW. What is the minimum power needed to provide this heating rate?

5.63 A heat pump uses 12 kW of power to deliver 70 kW of heat. The heat is delivered to a space that is 315 K. What is the minimum temperature allowed for the cold-space side of the heat pump?

5.64 A heat pump circulates its refrigerant between reservoirs at 5°C and 30°C. The pump can use a maximum of 15 kW of power to move the refrigerant. Determine the maximum amount of heat that can be delivered to the warm space.

5.65 A heat pump circulates its refrigerant between reservoirs at 2°C and 25°C. The pump can use a maximum of 3.75 kW to move the refrigerant. Determine the maximum amount of heat that can be delivered to the warm space.

5.66 A heat pump uses 10 kW of power to deliver 40 kW of heat. The pump operates by circulating refrigerant underground, where the temperature is 10°C, and then delivering the refrigerant to a warmer space. Determine the maximum possible temperature for the warm space.

5.67 It is planned to use a heat pump to provide heat to a house in the colder months of the year. The heat is to be delivered to the inside of the house at 22°C. The cold heat source is the outside air. The heat pump operates at a power of 1.5 kW. Plot the maximum possible coefficient of performance and the maximum heat transfer rate at which heat can be supplied to the interior of the house for outdoor air temperatures ranging from −15°C to 10°C. If the outside air temperature is 0°C, what is the minimum amount of operation time that it will take for the heat pump to deliver 10 MJ of energy to the house?

5.68 A heat pump operates with a low-temperature reservoir of 8°C. The pump delivers 20 kW of heat to the high-temperature reservoir. For high-temperature reservoir temperatures ranging between 15°C and 50°C, plot the maximum possible coefficient of performance and the minimum power needed to provide the heat rate.

5.69 A deep lake filled with water is to be used as the low-temperature reservoir for a Carnot heat pump. The water temperature drops with depth, beginning at 298 K on the surface and falling to 3°C at the lake bed. Consider the high-temperature reservoir to be at 310 K. You can control the performance of the heat pump by selecting the temperature of the water to which the piping should be installed. If 10 kW of heat is to be removed from the cold space, plot the power needed by the Carnot heat pump and the rate at which heat is delivered to the warm space for the temperature range in the lake.

5.70 A heat pump salesman comes to your door and describes the performance of the heat pump he is trying to sell. The salesman notes that the cold space (located underground at your home) is a steady 10°C and that you wish to deliver heat to your home at a temperature of 25°C. He says that the heat pump will use 500 W of power to deliver 10 kW of heat to your home. Does this claim violate the Second Law of Thermodynamics?

5.71 In the future, as an engineering manager at an HVAC company, you need to choose between two promising engineering students for an entry-level position. To each of them, you pose the following question: "A heat pump is to supply 25 kW of heat to a space at 300 K while receiving heat from a space at 280 K. You must choose between two suggested systems, one of which uses 1000 W of power and the other of which uses 3 kW of power. Which do you choose?" Candidate A says that you should use the 1000-W system and Candidate B says that you should use the 3-kW system. Which candidate should you choose to hire, and why?

5.72 A hopeful inventor who has not studied thermodynamics (nor kept up to date with current technology) believes he has created a wonderful device for heating a space. He circulates refrigerant between a cold space and a warm space, using a compressor to move the refrigerant. The cold space is kept at –3°C while the warm space is at 25°C. He believes his system can deliver 50 kW of heat while the system's compressor uses 1 kW of power. What do you tell the inventor about his proposed system, which he has called a "pump of heat"?

5.73 A Carnot heat engine is set up to operate a Carnot heat pump. The Carnot heat engine receives 20 kW of heat from a source at 500°C and rejects heat to a cold space at 30°C. The Carnot heat pump receives heat from a space at 0°C and rejects heat to a space at 30°C (i.e., the heat engine and the heat pump have the same heat sink). What is the rate at which the heat pump delivers heat to the 30°C space, and what is the total rate at which both the heat pump and the heat engine deliver heat to that 30°C space?

DESIGN/OPEN-ENDED PROBLEMS

5.74 Propose a device that could operate indefinitely but that violates the First Law of Thermodynamics and therefore is a perpetual motion machine—Type 1. Thoroughly describe this device so that it may appear possible to someone else. Share your device idea with nonengineers and observe their reactions. Describe these reactions, and also describe why the device violates the First Law of Thermodynamics.

5.75 Propose a device that could operate indefinitely but that violates the Second Law of Thermodynamics and therefore is a perpetual motion machine—Type 2. Thoroughly describe this device so that it may appear possible to someone else. Share your device idea with nonengineers and observe their reactions. Describe these reactions, and also describe why the device violates the Second Law of Thermodynamics.

5.76 Research historical records to find a description of a proposal for a perpetual motion machine. Describe the proposed device, and discuss how it violates the laws of thermodynamics.

5.77 Devise a layout for a system that will use a reversible heat engine to power a reversible refrigerator in your home. The refrigerator must cool 4 L of water from 30°C to 4°C in 15 minutes. Determine (or estimate) the temperature of the room containing the refrigerator (to give you the heat sink for the refrigerator). The heat engine must use a high-temperature thermal reservoir that is readily available in the home in the summer months (so that a furnace is not likely to be in use). Your layout should include the temperatures of the thermal reservoir, the available heat transfer rate from the hot reservoir of the heat engine, and your determinations of power for the devices.

CHAPTER 6

Entropy

Learning Objectives

Upon completion of Chapter 6, you will be able to

6.1 Explain the concept of entropy and the principle of entropy generation;

6.2 Calculate changes in entropy during a process for various types of substances;

6.3 Apply the entropy balance to open and closed system applications;

6.4 Compute the isentropic efficiency for devices such as turbines and compressors;

6.5 Describe the concept of irreversibility of a process; and

6.6 Recognize the consistent and complementary nature of methods of quantifying the relative irreversibility of different processes.

In Chapter 5, we described some of the fundamentals of the Second Law of Thermodynamics. We also discussed a few calculation tools that can help in determining if a thermodynamic cycle is possible, and how close to ideal a particular cycle may be. However, these tools do not extend to the level of individual devices, which means that we still need to develop concepts and associated analysis methods to determine if a given device can operate as proposed, or whether the proposed operation violates the second law. To assist us in performing this type of analysis, as well as to provide us with a tool that can be used in more advanced applications for tracking the energy lost in processes, we will introduce the thermodynamic concept of entropy.

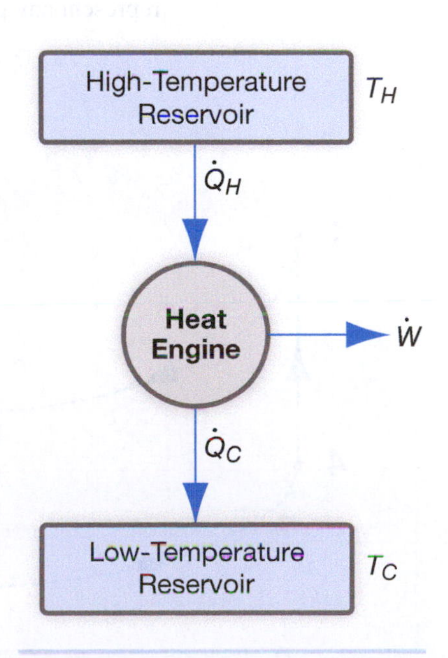

FIGURE 6.1 A standard model of a heat engine with thermal reservoirs.

6.1 ENTROPY AND THE CLAUSIUS INEQUALITY

To begin this study, we must develop a definition of entropy. For this development, we will follow the path outlined by Rudolph Clausius in the 19th century. Consider a standard reversible cyclic heat engine, as shown in Figure 6.1. Note that if we take the cyclic integral of the heat transfers (where all the heat transfers that occur in the cycle are added together), we can write

$$\oint \delta Q = Q_H - \left|Q_C\right| > 0 \quad (6.1)$$

The value from the cyclic integral of the heat transfers is greater than 0 because the net heat transfer is equal to the net work, and the purpose of a heat engine is to produce work: a positive quantity. It can also be noted that if the heat rejected from the cycle and the work produced by the cycle must add to the heat input, then the magnitude of the heat rejection is less than the magnitude of the heat input. If we divide each of the heat transfers by the absolute temperature of the thermal reservoir with which the heat transfer occurs, and if we recall Eq. (5.1), we find that

$$\oint \frac{\delta Q}{T} = \frac{Q_H}{T_H} - \frac{|Q_C|}{T_C} = 0 \tag{6.2}$$

Also recall that heat transfer over a finite temperature difference is one cause of irreversibility. Suppose that the temperature difference between the two thermal reservoirs begins to decrease. This should result in smaller amounts of irreversibility. As the difference between T_H and T_C becomes infinitesimally small, the processes become reversible, and we would now have a reversible heat engine. However, we need a temperature difference to drive a heat transfer, because the amount of heat transfer would also become infinitesimally small, approaching 0. For any heat engine, we can conclude

$$\oint \delta Q \geq 0 \tag{6.3a}$$

with the value equal to 0 for a reversible heat engine, and

$$\oint \left(\frac{\delta Q}{T}\right)_{\text{rev}} = 0 \tag{6.3b}$$

for a reversible heat engine.

Recall that, by definition, a thermodynamic cycle, such as in **Figure 6.2**, is a series of processes for which the beginning and end states are identical. If you integrate a thermodynamic property over a thermodynamic cycle, you will find that the value of the integral is 0, because a thermodynamic property's value depends only on the local thermodynamic equilibrium state and not on how that state was achieved. For a thermodynamic cycle, considering B to represent any property,

$$\oint dB = B_{\text{final}} - B_{\text{initial}} = 0 \tag{6.4}$$

This implies that the quantity $(\delta Q/T)_{\text{rev}}$, as shown in Eq. (6.3b), is a differential of a thermodynamic property. This property is named entropy, and is designated by S:

$$dS = \left(\frac{\delta Q}{T}\right)_{\text{rev}} \tag{6.5}$$

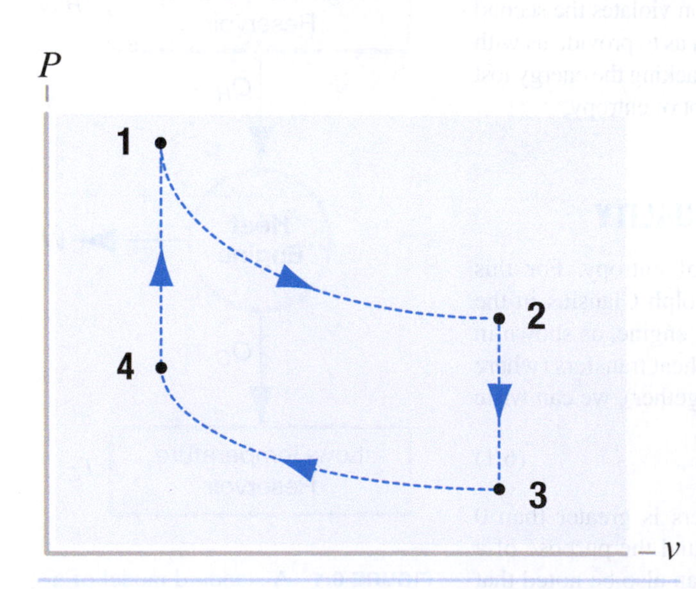

FIGURE 6.2 A cycle consisting of four processes represented on a P-v diagram. The beginning and the end states of the cycle are the same.

Although this is the formal mathematical definition of entropy, it does little to provide us with an intuitive understanding of what entropy actually is. A "reversible heat transfer divided by the temperature at which the heat is transferred" does not clarify the nature of entropy. One way to think of entropy is as a measurement of the molecular disorder in a system, as illustrated in **Figure 6.3**. A crystalline solid with a well-ordered lattice pattern of atoms at a low temperature has very little molecular disorder, and correspondingly has very low entropy. Conversely, a high-temperature gas, with free atoms and molecules traveling at high speeds in all directions, has a large amount of molecular disorder, and accordingly has high entropy. As heat is added to a system, it will cause the molecules to move at higher speeds and perhaps break free from surrounding molecules—this increases the disorder, and entropy, of the system. If heat is removed from the system, the molecules will slow down and the system will become more orderly, which causes the entropy of the system to decrease.

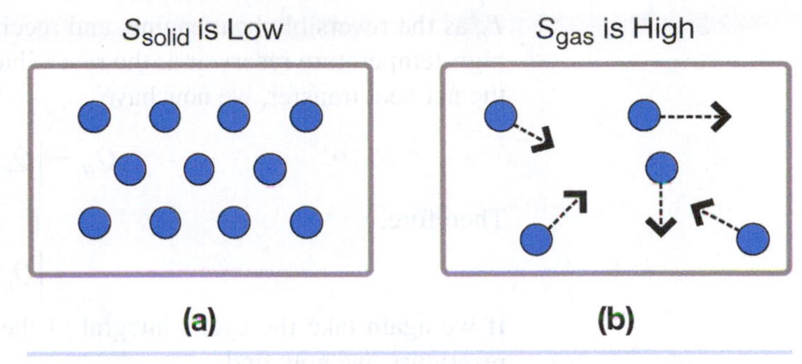

FIGURE 6.3 (a) The entropy (and disorder) of a crystalline solid is low, with only slight oscillations of the atoms and slight disorder of the atoms in the crystal structure. (b) The random, fast movement of atoms or molecules in a gas results in a system with high entropy (and disorder).

This interpretation of entropy is formalized with the Third Law of Thermodynamics. Remember that scientific laws are statements based on observations that have never been shown to be incorrect—yet they have not been absolutely proven. The third law states that the entropy of a pure crystalline substance at a temperature of absolute 0 is 0:

$$\lim_{T \to 0\text{K}} S_{\text{pure crystal}} = 0$$

The reason for this is that at absolute 0, all motion ceases, and for a pure crystal, there is only one possible arrangement of atoms. Therefore, there is no uncertainty in the location of each atom, and there is no disorder in the system. Impure crystals will have an entropy greater than 0 even at a temperature of absolute 0, as there is disorder in the atomic arrangement and uncertainty in the particular placement of the atoms. Any substance at temperatures above absolute 0 will have an entropy greater than 0 because there will at least be disorder resulting from motions of electrons, vibrations and rotations of molecules, and translational motion of atoms and molecules. Although the third law is very useful in setting up an absolute entropy scale, it generally is not used beyond this in engineering calculations.

A second way to think of entropy is as a measure of the quality, or usefulness, of the energy available in a system. The lower the entropy, the more easily used the energy in the system is. For example, consider a fuel such as gasoline in an automobile's fuel tank. The entropy of this system is rather low, and we can easily release the chemical energy bound up in the gasoline by burning it with air. In an engine, this will cause work to be produced, which will propel the automobile. It will also cause hot exhaust gases to be expelled from the engine through the tailpipe, and heat will also be lost to the air through the engine coolant and from the heating of the engine itself. If we wait for the engine to cool after the gasoline has been burned, we will find that the energy in all of the world has not changed (according to the first law), but the energy is now dispersed in the form of very slightly warmer air and ground temperatures. This larger system has a higher entropy than the initial gasoline, and it is very difficult to use the energy of the slightly heated air and ground to do anything. So, the increased entropy represents a decreased ability for the available energy to be used to do work. Both of these interpretations will be of additional use to us after we complete our study of the developments attributed to Clausius.

Returning to the heat engine, consider a situation where the temperatures of the reservoir differ by a finite amount, such that irreversibilities are now present. In this case, Eq. (6.1) still describes the cyclic integral of the heat transfers. But the irreversibility reduces the amount of work that is produced:

$$W_{\text{irr}} < W_{\text{rev}} \tag{6.6}$$

where W_{irr} represents the work produced in an irreversible process and W_{rev} represents the work produced in a reversible process. Consider an irreversible heat engine that has the same T_H and

T_C as the reversible heat engine, and receives the same amount of heat equal to Q_H from the high-temperature reservoir as the reversible heat engine. Considering that the net work equals the net heat transfer, we now have

$$Q_H - |Q_C|_{irr} < Q_H - |Q_C|_{rev} \tag{6.7}$$

Therefore,

$$|Q_C|_{irr} > |Q_C|_{rev} \tag{6.8}$$

If we again take the cyclic integral of the heat transfers divided by the temperatures of the reservoirs, we now find

$$\oint \frac{\delta Q}{T} = \frac{Q_H}{T_H} - \frac{|Q_C|_{irr}}{T_C} < 0 \tag{6.9}$$

recalling that the Q_C in Eq. (6.2) was actually $Q_{C,rev}$ because the heat engine under consideration was reversible. Therefore, for irreversible heat engine cycles, we have

$$\oint \delta Q \geq 0 \tag{6.10a}$$

and

$$\oint \frac{\delta Q}{T} < 0 \tag{6.10b}$$

It should be noted that the same analysis can be performed for refrigeration cycles as well, and the results will be the same, except that the cyclic integral of the heat transfers is less than or equal to 0 (i.e., Eq. (6.10a) would be $\oint \delta Q \leq 0$). Combining the results from the reversible and irreversible cycles, we are left with Clausius' inequality:

$$\oint \frac{\delta Q}{T} \leq 0 \tag{6.11}$$

where the value is equal to 0 for reversible cycles, and less than 0 for irreversible cycles.

> **QUESTION FOR THOUGHT/DISCUSSION**
>
> What is an example of a device or system that undergoes multiple heat transfers during its cyclic operation?

6.2 ENTROPY GENERATION

Let us consider an irreversible thermodynamic cycle consisting of two processes. Process 1-2 contains all of the irreversibilities of the cycle, whereas process 2-1 is a reversible process. The cycle is shown in Figure 6.4. The Clausius inequality can now be expanded into

$$\oint \frac{\delta Q}{T} = \int_1^2 \left(\frac{\delta Q}{T}\right) + \int_2^1 \left(\frac{\delta Q}{T}\right)_{rev} \leq 0 \tag{6.12}$$

When the two processes are both reversible, Equation (6.12) is equal to 0 (i.e., there are no irreversibilities in process 1-2). Substituting with Eq. (6.5),

$$\int_1^2 \left(\frac{\delta Q}{T}\right) + \int_2^1 dS \leq 0$$

and

$$\int_1^2 \left(\frac{\delta Q}{T}\right) + S_1 - S_2 \leq 0$$

Rearranging,

$$S_2 - S_1 \geq \int_1^2 \left(\frac{\delta Q}{T}\right) \tag{6.13}$$

where the equality exists for a reversible process, and the inequality exists for an irreversible process. We can consider the right-hand side of Eq. (6.13) to be the entropy transferred due to heat transfer. It can be noted that the change in entropy for a system during a process is greater than or equal to the amount of entropy transferred during heat transfer. Considering that the change in entropy is equal to the entropy transferred by heat transfer for a reversible process, we can note that the difference between the actual entropy change and that experienced from the heat transfer in a reversible process is a measure of the irreversibility of a process. This difference can be called the entropy that is generated during a process, or simply the entropy generation: S_{gen}. Incorporating this concept into Eq. (6.13) and rewriting yields

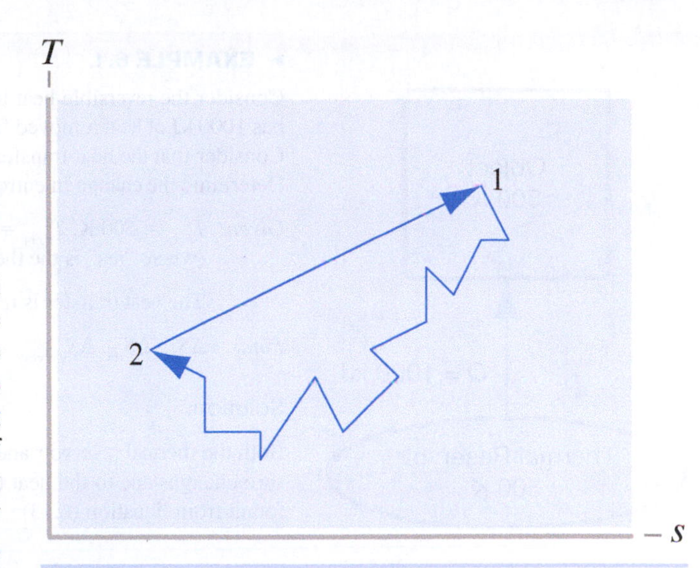

FIGURE 6.4 A two-process cycle represented on a *T-s* diagram. Process 1-2 is irreversible, whereas process 2-1 is considered reversible.

$$S_{gen} = S_2 - S_1 - \int_1^2 \left(\frac{\delta Q}{T}\right) \tag{6.14}$$

Using Eq. (6.14), the amount of entropy that is generated during a process can be calculated. Note that for irreversible processes, which all real processes should be considered, entropy is always generated. If we have an ideal, or reversible, process, the entropy generation is 0. But at no time does the overall entropy of the universe decrease during a process. If we consider the changes in entropy of a system and all of its surroundings during a process, we will find that, at best, the total entropy of the system and the surroundings (i.e., the universe) remains the same, and for any real process the total entropy increases. This does not mean that the entropy of a system cannot decrease during a process, nor that the entropy of the surroundings cannot decrease. However, if the entropy of one of these parts of the universe decreases, the entropy of the other must increase by at least the same amount. For example, if a system loses heat to the surroundings, the entropy of the system will decrease, but the entropy of the surroundings will increase by an equal or greater amount, as will be illustrated in Examples 6.1 and 6.2.

As a result of the requirement of the second law that the entropy of the universe will always increase or stay the same during a process, the entropy generation for a process can be used to determine if the process is impossible by violating the second law. According to an entropy generation perspective, the following rules apply with respect to a process:

$S_{gen} > 0$	Process is possible and irreversible
$S_{gen} = 0$	Process is possible and reversible
$S_{gen} < 0$	Process is impossible

Keep in mind that just because the analysis of the entropy generation of a proposed process is positive, the process is not guaranteed to be possible. If faulty assumptions are made or poor approximations are used, an entropy generation calculation could indicate that a process is possible, but further analysis may show that the process violates some other aspect of the second law. Great care must be taken when making assumptions in entropy generation calculations.

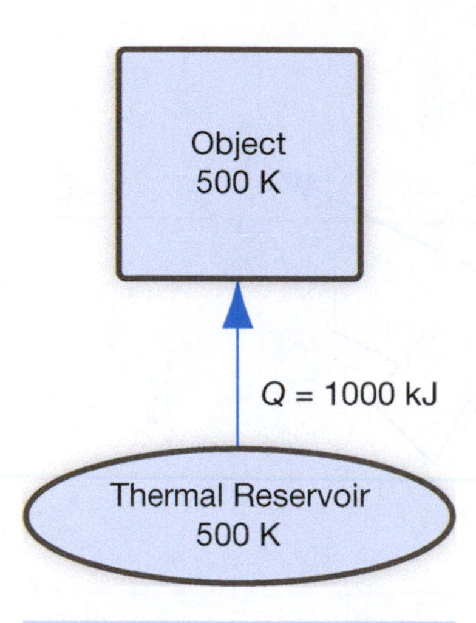

FIGURE 6.5 Scenario described in Example 6.1. Heat is being transferred reversibly between two regions of equal temperature. This is a reversible process.

▶ EXAMPLE 6.1

Consider the reversible heat transfer shown in Figure 6.5. A thermal reservoir at 500 K has 1000 kJ of heat removed from it, and that heat is added to a large object also at 500 K. Consider that the heat transfer has an insignificant effect on the properties of the object. Determine the change in entropy of the thermal reservoir, the object, and the universe.

Given: $T_{b,\text{res}} = 500$ K, $T_{b,\text{obj}} = 500$ K, $Q_{\text{res}} = -1000$ kJ, $Q_{\text{obj}} = +1000$ kJ (where "res" is the thermal reservoir, and "obj" is the object)

The heat transfer is reversible.

Find: $\Delta S_{\text{res}}, \Delta S_{\text{obj}}, \Delta S_{\text{universe}}$

Solution:

Both the thermal reservoir and the object are too large to experience temperature or pressure changes due to the heat transfer. So, the change in entropy for each will simply be found from Equation (6.13):

$$S_2 - S_1 = \int_1^2 \left(\frac{\delta Q}{T}\right)$$

because this is a reversible heat transfer. Because there is only one temperature for the reservoir, and only one temperature for the object, each can be considered to have isothermal boundaries, so that the integral will reduce to

$$\Delta S = S_2 - S_1 = Q/T_b$$

where T_b is the boundary temperature over which the heat transfer occurs.

For the reservoir, $Q_{\text{res}} = -1000$ kJ, and $T_{b,\text{res}} = 500$ K, so

$$\Delta S_{\text{res}} = Q_{\text{res}}/T_{b,\text{res}} = (-1000\text{ kJ})/500\text{ K} = \mathbf{-2\ kJ/K}$$

For the object, $Q_{\text{obj}} = +1000$ kJ, and $T_{b,\text{obj}} = 500$ K, so

$$\Delta S_{\text{obj}} = Q_{\text{obj}}/T_{b,\text{obj}} = (+1000\text{ kJ})/500\text{ K} = \mathbf{+2\ kJ/K}$$

Because these are the only two items in consideration for the "universe,"

$$\Delta S_{\text{universe}} = \Delta S_{\text{res}} + \Delta S_{\text{obj}} = \mathbf{0\ kJ/K}$$

Analysis: For a reversible process, the change in entropy of the universe during a process is 0, yet one portion of the universe saw a decrease in entropy and one portion saw an increase in entropy.

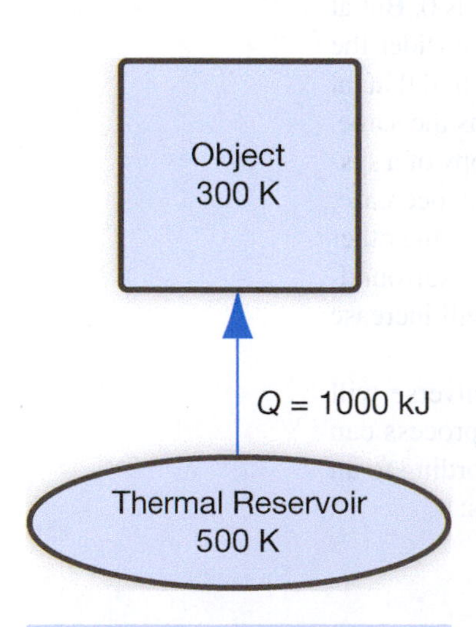

FIGURE 6.6 Scenario described in Example 6.2. Heat is being transferred across a finite temperature difference between two regions of different temperature. This is an irreversible process.

▶ EXAMPLE 6.2

Now, let's consider the previous example, but make the heat transfer irreversible by having the object have a temperature of 300 K, as shown in Figure 6.6. Determine the change in entropy of the thermal reservoir, the object, and the universe.

Given: $T_{b,\text{res}} = 500$ K, $T_{b,\text{obj}} = 300$ K, $Q_{\text{res}} = -1000$ kJ, $Q_{\text{obj}} = +1000$ kJ (where "res" is the thermal reservoir, and "obj" is the object)

The heat transfer is irreversible.

Find: $\Delta S_{\text{res}}, \Delta S_{\text{obj}}, \Delta S_{\text{universe}}$

Solution:

In this case, the analysis for the reservoir is the same as in Example 6.1, but the analysis for the object changes.

For the reservoir, $Q_{res} = -1000$ kJ, and $T_{b,res} = 500$ K, so

$$\Delta S_{res} = Q_{res}/T_{b,res} = (-1000 \text{ kJ})/500 \text{ K} = \mathbf{-2\ kJ/K}$$

For the object, $Q_{obj} = +1000$ kJ, and $T_{b,obj} = 300$ K, so

$$\Delta S_{obj} = Q_{obj}/T_{b,obj} = (+1000 \text{ kJ})/300 \text{ K} = \mathbf{+3.33\ kJ/K}$$

Because these are the only two items in consideration for the "universe,"

$$\Delta S_{universe} = \Delta S_{res} + \Delta S_{obj} = \mathbf{1.33\ kJ/K}$$

Analysis: For an irreversible process, there is a net increase in entropy in the universe. This can be related to the idea that the generation of entropy indicates that the energy has lost some ability to do work or be transferred in the form of heat. When there was no change in the entropy of the universe (the reversible process in Example 6.1), the 1000 kJ of energy was still in a system with a temperature of 500 K—it had the same ability to be used to do work or to warm some other system as it had before it was transferred. But when the energy was placed into a lower temperature system, it lost some ability to be used. For example, at 500 K the energy could have been used to heat a fluid at 400 K; but at 300 K, the energy cannot be used to heat a fluid at 400 K—it can only heat a fluid at a lower temperature.

To fully utilize Eq. (6.14), we need to be able to evaluate the changes in entropy of a system as well as the amount of entropy transfer due to heat transfer. We will return to the latter, but first let us consider how to evaluate the change in entropy for a system as it undergoes changes during a process.

QUESTION FOR THOUGHT/DISCUSSION

A freezer will turn liquid water into ice, which reduces the entropy of the water. How can this process occur?

6.3 EVALUATING CHANGES IN THE ENTROPY OF A SYSTEM

As discussed, it is rather difficult to measure exactly what entropy is from an engineering standpoint. Can we "measure" the quality of the energy of a system? We could make some attempt to measure the molecular disorder of a system, but it would be rather difficult to quantify the result. We can say that one set of molecules is more disordered than another, but by how much? The result of this is that we do not have an "entropy meter" that can be used to measure the entropy of a system. In fact, it is even difficult to understand what the units of entropy physically represent: kJ/K, or energy unit/temperature unit in general. (You might want to consider this with respect to the usefulness of energy to be transferred to another system in the form of heat. If you have a fixed amount of energy, an increase in the entropy associated with that energy would mean that the temperature decreased. As such, the energy is only able to be used to warm even colder objects, making the energy less useful.) Rather, we will have to derive values of the entropy of a system using other properties, and develop equations of state relating changes in those properties to changes in entropy of a system.

Consider the first law for a closed system with no changes in kinetic or potential energy, on a differential basis. This can be written as

$$\delta Q - \delta W = dU \tag{6.15}$$

For a simple compressible process,

$$\delta W = P dV \tag{6.16}$$

For a reversible process, from Eq. (6.5),

$$\delta Q = T dS \tag{6.17}$$

Combining Eqs. (6.16) and (6.17) into (6.15) yields

$$T dS = dU + P d\forall \tag{6.18}$$

Considering that $H = U + PV$,

$$dH = dU + P d\forall + \forall dP \tag{6.19}$$

Substituting Eq. (6.19) into Eq. (6.18),

$$T dS = dH - \forall dP \tag{6.20}$$

Equations (6.18) and (6.20) are known as the Gibbs equations and are named after Josiah Gibbs, who made great contributions to the development of thermodynamics in the 19th century. These equations can be divided by the system's mass, yielding relationships based on intensive properties:

$$T ds = du + P dv \tag{6.21}$$

$$T ds = dh - v dP \tag{6.22}$$

These relationships, combined with the definitions of specific heats, give us the tools we need to find changes in the entropy of a system based on easily measured properties. However, you might question whether these only apply to reversible processes, because we used that restriction in developing equations from the first law (i.e., with Eq. (6.17)). But recall that the values of a property depend only on the local thermodynamic equilibrium state, and not on how that state was attained. Equations (6.21) and (6.22) contain only properties. Therefore, changes in the properties depend only on the thermodynamic states at the beginning and at the end of the process, and not on the type of process used to go between the states. So the equations work for all processes, both reversible and irreversible.

For some substances, such as water or refrigerants, the relationships for the specific heats as a function of temperature are too complex to allow for easy hand calculation of the changes in entropy. For such substances, it is best to use the computer software that you've chosen for finding the properties of such substances to calculate the value of the entropy of the substance when given two other thermodynamic properties, such as the temperature and pressure, or the temperature and specific volume. This is done in the same manner as was used for finding values for the specific enthalpy or specific internal energy. The relationship for the quality applies as well:

$$x = \frac{s - s_f}{s_g - s_f} \tag{6.23}$$

For some idealized classes of substances, the Gibbs equations can be used to develop equations for quick hand calculations of the change in entropy for these substances if needed.

6.3.1 Change in Entropy for Ideal Gases

For ideal gases, we first need to consider whether to include the variability of the specific heats in the analysis. As discussed in Chapter 3, although it is relatively easy to perform calculations considering the variability of specific heats, quick estimates and calculations that are sufficiently accurate for many engineering purposes can be achieved with an assumption of constant specific heats for ideal gases.

Ideal Gases with Constant Specific Heats

As you can see in Eqs. (6.21) and (6.22), the change in specific entropy will depend on a change in either the specific internal energy or the specific enthalpy. These changes, for an ideal gas, are related to the change in temperature through

$$du = c_v dT \tag{3.14}$$

and

$$dh = c_p dT \tag{3.13}$$

These expressions can be substituted into the Gibbs equations, and the resulting equations are easily integrated if the specific heats are assumed to be constant. Incorporating in the ideal gas law (Eq. (3.12b), $P = RT/v$ or $v = RT/P$) allows Eqs. (6.21) and (6.22) to become

$$ds = c_v \frac{dT}{T} + R\frac{dv}{v}$$

and

$$ds = c_p \frac{dT}{T} - R\frac{dP}{P}$$

With the assumption of constant specific heats, integration yields

$$s_2 - s_1 = c_v \ln\frac{T_2}{T_1} + R\ln\frac{v_2}{v_1} \tag{6.24}$$

and

$$s_2 - s_1 = c_p \ln\frac{T_2}{T_1} - R\ln\frac{P_2}{P_1} \tag{6.25}$$

Calculations of the change in entropy using Eqs. (6.24) and (6.25) provide the same result, and the choice of which equation to use depends on which properties are readily known. In both cases, the change in entropy is found through knowledge of the values of easily measured properties at the end states of the process.

Ideal Gases Considering Variable Specific Heats

In the case where the variability of the specific heats for the ideal gas is included, it is necessary to consider the impact of the changing specific heat on the change in specific enthalpy in Eq. (6.22), or the change in specific internal energy in Eq. (6.21). Typically, this only involves use of Eq. (6.22). Inclusion of Eq. (3.13), the ideal gas law, and subsequent integration of Eq. (6.22) yields

$$s_2 - s_1 = \int_1^2 c_p \frac{dT}{T} - R\ln\frac{P_2}{P_1} \tag{6.26}$$

Because the enthalpy change portion of Eq. (6.26) is only a function of temperature for ideal gases, an entropy function based on the temperature of the ideal gas, s_T^o, is often used to aid in completing the calculation:

$$s_T^o = \int_{T_{\text{ref}}}^{T} c_p \frac{dT}{T} \tag{6.27}$$

where T_{ref} is a reference state temperature. Although the choice of this temperature impacts the actual value of s_T^o, the choice will cancel out when the difference between two entropies is determined. The values for s_T^o are typically evaluated at a pressure of 101.325 kPa. Using Eq. (6.27),

Eq. (6.26) can be rewritten to provide an expression that can be used to determine the change in specific entropy of an ideal gas with the variability of the specific heats taken into account:

$$s_2 - s_1 = s^o_{T_2} - s^o_{T_1} - R \ln \frac{P_2}{P_1} \quad (6.28)$$

Modern computer calculations handle this calculation differently depending on the program and the substance. Some programs are available that calculate the entropy directly for the gas, considering both the temperature and pressure immediately. For such programs, we would just determine the entropy immediately, and the difference between the two values could be found without the additional pressure correction. In other cases, only the temperature-dependent portion is available, and when that is the case, the pressure correction in Eq. (6.28) is needed. It can also be difficult to find computer programs for determining the properties for less common gases—for such gases a table of values may be needed. On such tables, the values for s^o_T are given as a function of temperature alone, and are considered to be at 101.325 kPa.

Table A.3 in the appendix has been previously introduced for finding the properties of air while considering the variability of the specific heats. It should be noted that this table contains values of the temperature-dependent portion of the entropy, s^o_T, and may be used for the values in Eq. (6.28) for problems involving air with consideration of variable specific heats.

▶ EXAMPLE 6.3

Consider oxygen gas to behave as an ideal gas. The O_2 begins a process at 300 K and 100 kPa, and ends the process at 700 K and 250 kPa. Determine the change in specific entropy of the O_2 (a) considering the specific heats to be constant, and (b) considering the specific heats to be variable.

Given: $T_1 = 300$ K, $T_2 = 700$ K, $P_1 = 100$ kPa, and $P_2 = 250$ kPa

Find: Δs for (a) constant specific heats, and (b) variable specific heats

Solution: The working fluid for this problem is O_2.

Assume: The O_2 is an ideal gas.

(a) Considering the specific heats to be constant, for O_2, we will take the value of $c_p = 0.918$ kJ/kg · K, and $R = 0.2598$ kJ/kg · K.

Using Eq. (6.25),

$$s_2 - s_1 = c_p \ln \frac{T_2}{T_1} - R \ln \frac{P_2}{P_1}$$

$$s_2 - s_1 = (0.918 \text{ kJ/kg} \cdot \text{K}) \ln \frac{700 \text{ K}}{300 \text{ K}} - (0.2598 \text{ kJ/kg} \cdot \text{K}) \ln \frac{250 \text{ kPa}}{100 \text{ kPa}}$$

$$\mathbf{s_2 - s_1 = 0.540 \text{ kJ/kg} \cdot \text{K}}$$

(b) Employing variable specific heats, the temperature-dependent portions of the entropy at 101.325 kPa can be found using computer programs for O_2 properties:

$$s^o_{T_2} = 7.230 \text{ kJ/kg} \cdot \text{K} \qquad s^o_{T_1} = 6.413 \text{ kJ/kg} \cdot \text{K}$$

Using Eq. (6.28), $s_2 - s_1 = s^o_{T_2} - s^o_{T_1} - R \ln \frac{P_2}{P_1}$.

$$s_2 - s_1 = (7.230 - 6.413) \text{ kJ/kg} \cdot \text{K} - (0.2598 \text{ kJ/kg} \cdot \text{K}) \ln(250 \text{ kPa}/100 \text{ kPa}) = \mathbf{0.579 \text{ kJ/kg} \cdot \text{K}}$$

Analysis: Comparing the results from (a) and (b), it is clear that some error is introduced by using constant specific heats. The size of the error can be reduced by using a value for the specific heat at an average temperature for the process. In this case, if we had used the value of the specific heat for O_2 at a temperature of 500 K (0.972 kJ/kg · K), the value for the change in entropy for constant specific heats would be 0.586 kJ/kg · K, which is in error by only 1.1%.

6.3.2 Change in Entropy for Incompressible Substances

A simple expression for the entropy change for incompressible substances can also be developed. Recall from Chapter 3 that an incompressible substance is one whose specific volume (or density) is constant. We normally assume solids and liquids to be incompressible, unless the substance experiences a very large pressure change during a process. If v = constant, $dv = 0$, and for incompressible substances, Eq. (6.21) reduces to

$$T\,ds = c_v\,dT \tag{6.29}$$

Considering that $c_p = c_v = c$ for an incompressible substance, Eq. (6.29) becomes

$$ds = c\frac{dT}{T} \tag{6.30}$$

If the specific heat of the incompressible substance is assumed to be constant, Eq. (6.30) can be integrated to yield

$$s_2 - s_1 = c\ln\frac{T_2}{T_1} \tag{6.31}$$

▶ **EXAMPLE 6.4**

A 2.0-kg block of iron at 20°C is dropped into a large bucket of boiling water at 100°C. The iron is heated until its temperature is 100°C. Determine the increase in entropy of the iron during this process.

Given: $m = 2.0$ kg, $T_1 = 20°C = 293$ K, $T_2 = 100°C = 373$ K

Find: ΔS (the change in total entropy for the iron)

Solution: The substance of interest is iron.

Assume: The iron behaves as an incompressible substance with constant specific heats.

For the change in total entropy:

$$S_2 - S_1 = m(s_2 - s_1)$$

Assuming that the iron behaves as an incompressible substance with constant specific heats, Eq. (6.31) can be used for the change in specific entropy. For iron, $c = 0.45$ kJ/kg · K.

$$S_2 - S_1 = m\left(c\ln\frac{T_2}{T_1}\right) = (2.0\text{ kg})\left((0.45\text{ kJ/kg}\cdot\text{K})\ln\frac{373\text{ K}}{293\text{ K}}\right)$$

$$\mathbf{S_2 - S_1 = 0.217\ kJ/K}$$

Analysis: As the temperature of the iron increases, the entropy of the iron increases as expected.

6.4 THE ENTROPY BALANCE

As discussed in Section 6.2, entropy is generated in any real (irreversible) process. This means that entropy is not a conserved quantity, but rather that the entropy present in the universe is constantly increasing. As seen from Eq. (6.14), the change in the entropy of a system during a process is equal to the entropy transported in or out of the system plus the entropy generated.

Entropy Change = Entropy Transported + Entropy Generated

You may note that this is somewhat different from our energy balance, as represented by the first law in Chapter 4. Energy is a conserved quantity and is not created or destroyed. As such, the energy balance states that the energy change is equal to the net energy transport in the system.

Entropy is transported in two ways: (1) via heat transfer, and (2) via mass flow. The entropy transport via heat transfer has already been discussed. Recalling the development of the energy balance in Chapter 4, it is logical that if mass flows into or out of a system, it will carry the entropy present in the mass along with it. Therefore, entropy transport via mass flow is an expected feature of an open system that has mass flowing in or out of the system.

Just as for energy in Chapter 4, we will develop expressions for the entropy balance for both open and closed systems. We will begin with a rate form of the balance applied to general open systems, and then simplify this for the specific case of closed systems.

6.4.1 Entropy Balance in Open Systems

Considering the entropy transport methods and the entropy generation, the entropy rate balance can be written for an open system, such as in **Figure 6.7**, as

$$\frac{dS_{\text{system}}}{dt} = \sum_{\text{inlets}} \dot{m}_i s_i - \sum_{\text{outlets}} \dot{m}_e s_e + \int \frac{\dot{Q}}{T} + \dot{S}_{\text{gen}} \tag{6.32}$$

The $\dot{m}$ terms represent the mass flow rate into or out of the system through an inlet or outlet, and $\dot{S}_{\text{gen}}$ is the rate of entropy generation during the process. This general equation can be applied to both steady- and unsteady-state systems, can accommodate any number of inlets and outlets for the mass flow, and can be used for any type of heat transfer environment. These possibilities could result in a very complicated environment, and the complications that can be introduced through the consideration of all of these aspects can move the calculations beyond the scope of this text. However, there are some logical simplifications that can ease the application of Eq. (6.32).

First, the entropy transport via heat transfer can become difficult to evaluate if the heat transfer rate is unsteady and if the amount of heat transfer varies with location on the system boundary. Furthermore, if the temperature of the boundary is changing, we need to know the functional relationship between the heat transfer rate and the temperature over which that heat transfer occurs in order to evaluate the integral. The evaluation of this integral can be simplified in two common situations. First, if the system is adiabatic ($\dot{Q} = 0$), the entropy transport via heat transfer is 0. The second case is for a situation that has constant values of the heat transfer rate over discrete sections of the system boundary that are at constant temperatures. This second case is commonly referred to as the system having isothermal boundaries. If we can relate the amounts of heat transfer to specific areas of the system boundary that have constant temperatures, we can use this simplification. This leads to the integral for the entropy transport via heat transfer being changed to a summation over discrete locations:

$$\int \frac{\dot{Q}}{T} = \sum_{j=1}^{n} \frac{\dot{Q}_j}{T_{b,j}} \tag{6.33}$$

where the system boundary is divided into n discrete isothermal sections. .

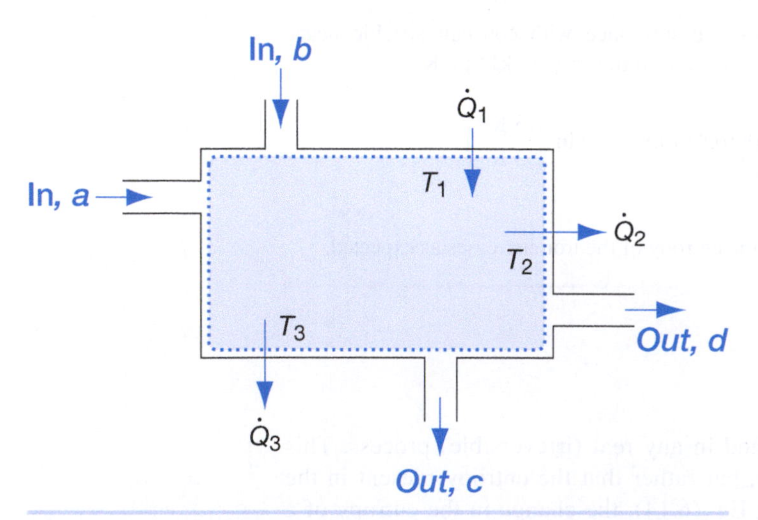

FIGURE 6.7 A generic open system, with multiple inlets and outlets, and several heat transfers across surfaces at potentially different temperatures.

An example of how a situation with multiple sections of the system boundary with different heat transfer rates can occur is shown in **Figure 6.8**. Here, an object is floating on the surface of a container of water. The object also rests against the edge of the container, and part of the object is in contact with the air. As you can see, all three contact points have not reached thermal equilibrium: the air is at 20°C, the water is at 60°C, and the wall is at 40°C. There are different heat transfers associated with each section.

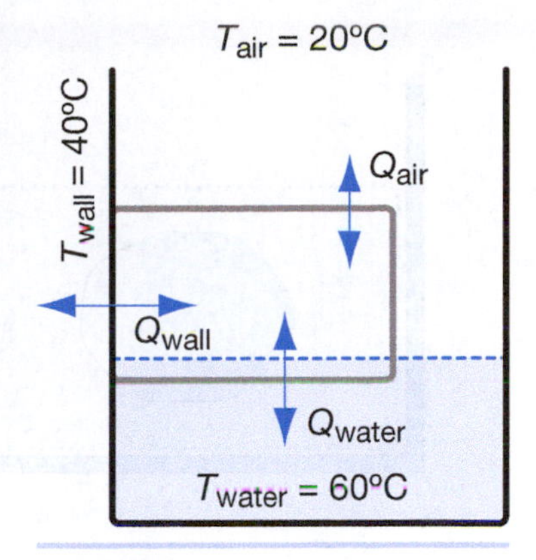

FIGURE 6.8 An object floating on the surface of water and resting against a wall may experience three different heat transfers with regions at three different temperatures.

▶ EXAMPLE 6.5

Determine the entropy transport via heat transfer for the system shown in **Figure 6.9**, with a system in contact with three isothermal boundaries and one insulated side.

Given: $\dot{Q}_1 = -5$ kW, $T_{b,1} = 20°C = 293$ K; $\dot{Q}_2 = -1$ kW, $T_{b,2} = 40°C = 313$ K;

$\dot{Q}_3 = 8$ kW, $T_{b,3} = 60°C = 333$ K, $\dot{Q}_4 = 0$ (insulated)

Find: $\int \frac{\dot{Q}}{T}$ (The entropy transport via heat transfer)

Solution: With the heat transfers in the system shown being discrete and each over a particular isothermal boundary, the entropy transport via heat transfer can be found using Eq. (6.33):

$$\int \frac{\dot{Q}}{T} = \sum_{j=1}^{3} \frac{\dot{Q}_j}{T_{b,j}} = \frac{\dot{Q}_1}{T_{b,1}} + \frac{\dot{Q}_2}{T_{b,2}} + \frac{\dot{Q}_3}{T_{b,3}} + \frac{\dot{Q}_4}{T_{b,4}} = \mathbf{0.00377\ kW/K}$$

Analysis: As there is a net heat transfer to the system, it may be natural to expect that the entropy transport is into the system. However, note that there is a dependence on the temperatures over which the heat transfers occur. If $T_{b,3}$ had been 400 K, the result would have been a net entropy transport out of the system if the heat transfer rates had stayed the same. Entropy transport can be positive or negative, and it is not always safe to assume the direction without performing the calculation.

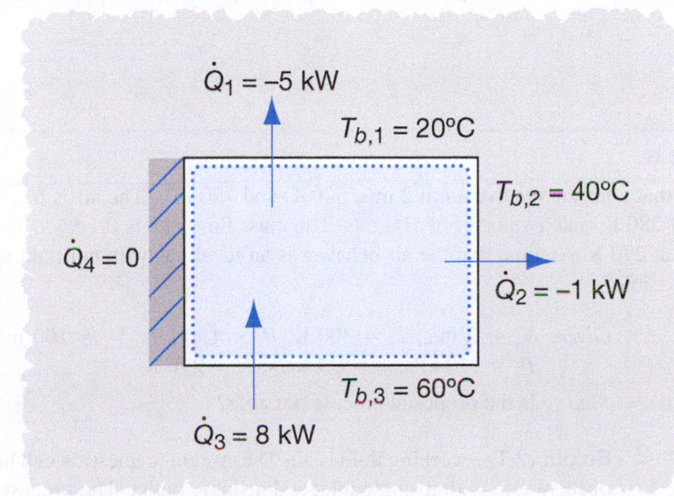

FIGURE 6.9 A box that is experiencing three heat transfers across surfaces that have three different boundary temperatures. The fourth side of the box is insulated and experiences no heat transfer.

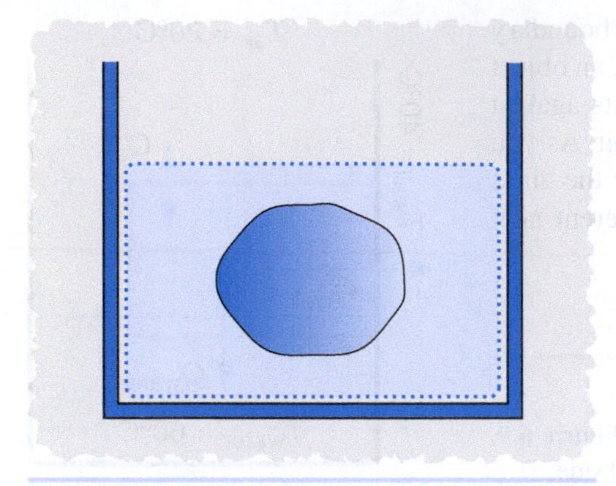

FIGURE 6.10 An object submerged in a liquid will usually have a single isothermal boundary.

Keep in mind that this summation can become even simpler if there is only one boundary temperature. This situation may happen if the system is surrounded by one substance, such as air or a liquid, and that fluid is at a constant temperature. In this case, the summation disappears, because there is only one term to add. Figure 6.10 illustrates one example of a single isothermal boundary: an object submerged in a tank of a liquid, with the liquid at a uniform temperature. Objects in an oven will also often have a uniform surface temperature. In fact, most objects exposed to the environment will experience a uniform surface temperature provided that any parts of the surface not exposed to the environment (such as the contact area between a device and its mounting brackets) have minimal heat transfer through them.

Returning to Eq. (6.32), a second common assumption is that the system is at steady state. In this case, the entropy of the system is constant with time: $dS_{system}/dt = 0$. With isothermal boundaries, for steady-state operation Eq. (6.32) reduces (with rearrangement) to

$$\dot{S}_{gen} = \sum_{outlets} \dot{m}_e s_e - \sum_{inlets} \dot{m}_i s_i - \sum_{j=1}^{n} \frac{\dot{Q}_j}{T_{b,j}} \tag{6.34}$$

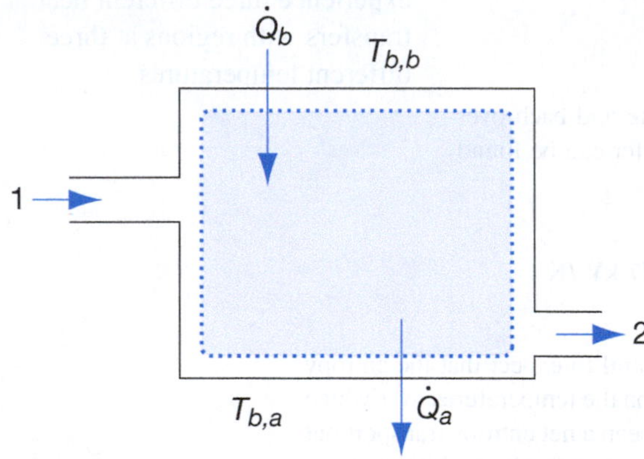

FIGURE 6.11 A single-inlet, single-outlet open system experiences two heat transfers across isothermal boundaries.

Furthermore, if the steady-state system with isothermal boundaries has only one inlet and one outlet, such as in Figure 6.11, the entropy balance reduces to

$$\dot{S}_{gen} = \dot{m}(s_2 - s_1) - \sum_{j=1}^{n} \frac{\dot{Q}_j}{T_{b,j}} \tag{6.35}$$

As discussed in Chapter 4, there are many possible applications for which Eq. (6.35) would be appropriate.

▶ EXAMPLE 6.6

It is proposed that a nozzle receive air at 2 m/s, 300 K, and 150 kPa. The air is to exit at 110 kPa, a temperature of 280 K, and a velocity of 100 m/s. The mass flow rate is 0.1 kg/s. The surface of the nozzle is kept at 270 K. Assume that the air behaves as an ideal gas with constant specific heats. Is this process possible?

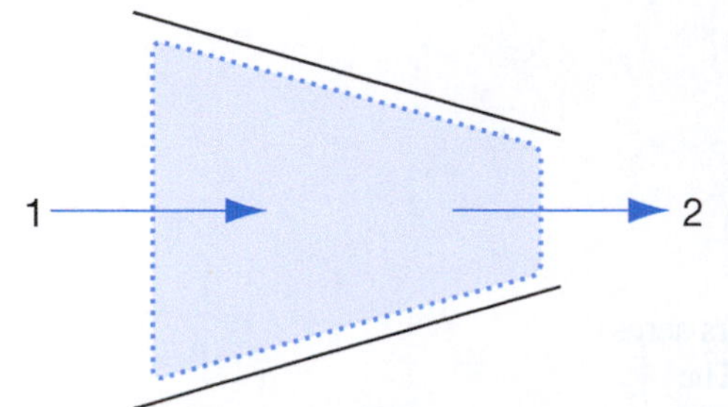

Given: $V_1 = 2$ m/s, $T_1 = 300$ K, $P_1 = 150$ kPa, $V_2 = 100$ m/s, $T_2 = 280$ K, $P_2 = 110$ kPa, $\dot{m} = 0.1$ kg/s, $T_b = 270$ K

Find: Is the proposed process possible?

Solution: The working fluid is air. The system in question can be modeled as an open system with a single inlet and a single outlet. The process is steady-state, steady-flow.

Assume: The air behaves as an ideal gas with constant specific heats. Any heat transfer occurs over a single isothermal boundary. $\Delta PE = 0$

For air, take $c_p = 1.005$ kJ/kg · K, $R = 0.287$ kJ/kg · K.

To test this process, we will find the entropy generation rate. First, we need to find if there is any heat transfer. Assuming that there is no work and no change in potential energy, the first law reduces to

$$\dot{Q} = \dot{m}\left(h_2 - h_1 + \left(\frac{V_2^2 - V_1^2}{2(1000\ \text{J/kJ})}\right)\right) = \dot{m}\left(c_p(T_2 - T_1) + \left(\frac{V_2^2 - V_1^2}{2(1000\ \text{J/kJ})}\right)\right) = -1.51\ \text{kW}$$

This heat transfer rate is now used in the entropy generation rate calculation. The entropy generation rate is then found from Eq. (6.35), with only one isothermal surface over which heat transfer occurs:

$$\dot{S}_{gen} = \dot{m}(s_2 - s_1) - \frac{\dot{Q}}{T_b} = \dot{m}\left(c_p \ln\frac{T_2}{T_1} - R\ln\frac{P_2}{P_1}\right) - \frac{\dot{Q}}{T_b} = \mathbf{0.00756\ kW/K}$$

With the assumptions made, this indicates that the process is **possible** as the entropy generation rate is greater than 0.

Analysis: The small entropy generation rate indicates that this process has only a small amount of irreversibility.

▶ EXAMPLE 6.7

An uninsulated pump receives saturated liquid water at 15°C and exhausts the water at 13.4 MPa absolute and 25°C. The water flows through the pump at a rate of 7.0 kg/s. The pump uses 375 kW for the pumping process. The surface temperature of the pump is 40°C. Determine the entropy generation rate of the pump.

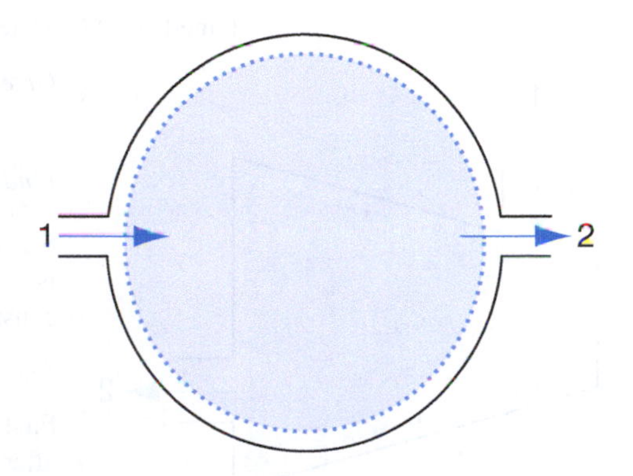

Given: $x_1 = 0.0$, $T_1 = 15°\text{C}$, $P_2 = 13.4\ \text{MPa}$, $T_2 = 25°\text{C}$, $\dot{m} = 7.0\ \text{kg/s}$, $T_b = 40°\text{C} = 313\ \text{K}$

$\dot{W} = -375\ \text{kW}$

Find: $\dot{S}_{gen}$

Solution: The working fluid is water. The system is an open system with a single inlet and a single outlet, and the process can be considered steady-state, steady-flow.

Assume: $\Delta KE = \Delta PE = 0$. Any heat transfer is through an isothermal boundary.

As the water pressure becomes high, do not assume that the water is incompressible. Rather, use computer programs or tables for the thermodynamic properties.

With the assumptions, the first law (Eq. (4.15a)) reduces to

$$\dot{Q} = \dot{m}(h_2 - h_1) + \dot{W}$$

Using computer-based properties:

$$h_1 = 63.0\ \text{kJ/kg},\ h_2 = 179.4\ \text{kJ/kg}$$

Solving for the heat transfer,

$$\dot{Q} = 566\ \text{kW}$$

With this, and with a single isothermal boundary, Eq. (6.35) can be modified and used to find the entropy generation rate:

$$\dot{S}_{gen} = \dot{m}(s_2 - s_1) - \frac{\dot{Q}}{T_b}$$

From a computer-based property program:

$$s_1 = 0.2245\ \text{kJ/kg}\cdot\text{K},\ s_2 = 0.5672\ \text{kJ/kg}\cdot\text{K}$$

Solving,

$$\dot{S}_{gen} = 0.591 \text{ kJ/kg} \cdot \text{K}$$

Analysis: The positive entropy generation rate indicates that the process is possible and irreversible. This is expected, as there was nothing particularly suspicious in the given information: water is being pumped to a higher pressure, and the temperature of the water is staying below that of the surface of the pump.

STUDENT EXERCISE

Revise your open system component computer models (i.e., turbine, compressor, heat exchanger, etc.) so that the entropy of the entrance and exit states can be calculated or input from other sources, and so that the entropy generation rate can be calculated for each component.

▶ EXAMPLE 6.8

A steam turbine receives 5 kg/s of superheated steam at 6.0 MPa and 400°C, and it exhausts at 500 kPa and 200°C. The turbine produces 1500 kW of power, and the surface of the turbine is maintained at 50°C. Determine the rate of entropy generation for the turbine.

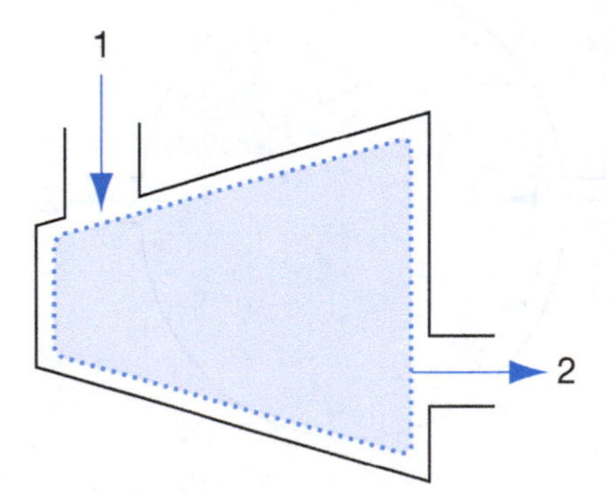

Given: $\dot{m} = 5$ kg/s, $T_1 = 400°C$, $P_1 = 6.0$ MPa; $T_2 = 200°C$, $P_2 = 500$ kPa, $\dot{W} = 1500$ kW, $T_b = 50°C = 323$ K

Find: $\dot{S}_{gen}$

Solution: The working fluid is water, in the form of superheated steam. The system is an open system with a single inlet and a single outlet, and the process can be considered steady-state, steady-flow.

Assume: $\Delta KE = \Delta PE = 0$. Any heat transfer is through an isothermal boundary.

First, we need to determine if there is any heat transfer for this process. In this case, after applying the assumptions, the first law reduces to

$$\dot{Q} = \dot{m}(h_2 - h_1) + \dot{W}$$

Using a computer program, $h_1 = 3177.2$ kJ/kg and $h_2 = 2855.4$ kJ/kg.

So $\dot{Q} = -109$ kW.

With this, and with a single isothermal boundary, Eq. (6.35) can be modified and used to find the entropy generation rate:

$$\dot{S}_{gen} = \dot{m}(s_2 - s_1) - \frac{\dot{Q}}{T_b}$$

From a computer program, $s_1 = 6.5408$ kJ/kg · K, $s_2 = 7.0592$ kJ/kg · K

Substituting, $\dot{S}_{gen}$ = **2.93 kW/K.**

Analysis: This indicates that the process is possible. The calculation of entropy generation rate is most useful at this point for determining which of several scenarios may be the best in terms of having the least irreversibility. Such a calculation is facilitated through a computer program. You are encouraged to modify your model of a turbine (and other components) to allow for the calculation of entropy generation rates. If you have done so, you can test the model for entropy generation rates for different levels of power production with the same inlet and outlet states. Such a calculation will yield results similar to the following:

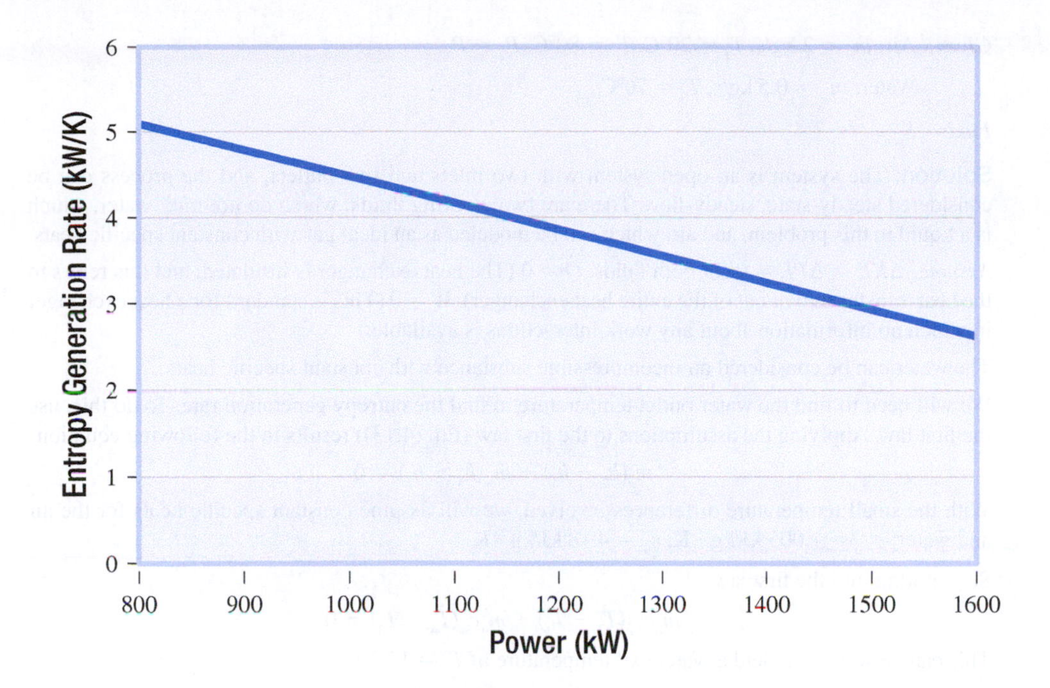

So, the more power that is produced, the less irreversible the process. Another way to interpret this is that when more energy is lost to heat transfer, the turbine is less efficient at producing work. This is one reason why turbines are often insulated: so that less energy is lost to heat transfer to the environment. Insulating a turbine would also have the effect of changing the outlet state, so this simple analysis is not quite sufficient. But it is adequate enough to provide some of the insight needed to produce better performance from a turbine.

▸ EXAMPLE 6.9

A proposed insulated heat exchanger has air entering at a mass flow rate of 2 kg/s and a temperature of 20°C. The air exits at the same pressure and at a temperature of 80°C. Water, at a mass flow rate of 0.5 kg/s, enters at a temperature of 70°C. Determine the rate of entropy generation for the proposed heat exchanger.

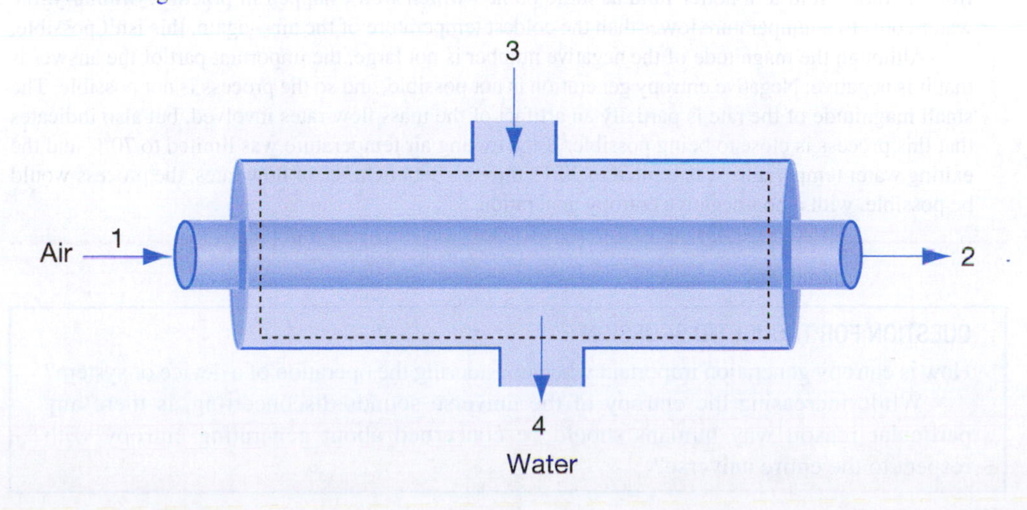

Given: Air: $\dot{m}_a = 2$ kg/s, $T_1 = 20°C$, $T_2 = 80°C$, $P_2 = P_1$

Water: $\dot{m}_w = 0.5$ kg/s, $T_3 = 70°C$

Find: $\dot{S}_{gen}$

Solution: The system is an open system with two inlets and two outlets, and the process can be considered steady-state, steady-flow. There are two working fluids, which do not mix: water, which is a liquid in this problem, and air, which can be modeled as an ideal gas with constant specific heats.

Assume: $\Delta KE = \Delta PE = 0$ for both fluids. $\dot{Q} = 0$ (The heat exchanger is insulated, and this refers to the heat transfer into or out of the entire heat exchanger). $\dot{W} = 0$ (This is standard for a heat exchanger in which no information about any work interactions is available.)

The water can be considered an incompressible substance with constant specific heats.

We will need to find the water outlet temperature to find the entropy generation rate. To do this, use the first law. Applying the assumptions to the first law (Eq. (4.13)) results in the following equation:

$$\dot{m}_a(h_2 - h_1) + \dot{m}_w(h_4 - h_3) = 0$$

With the small temperature differences involved, we will assume constant specific heats for the air and water: $c_{p,a} = 1.005$ kJ/kg · K, $c_w = 4.18$ kJ/kg · K

Substituting into the first law:

$$\dot{m}_a c_{p,a}(T_2 - T_1) + \dot{m}_w c_w(T_4 - T_3) = 0$$

This can be solved to yield a water exit temperature of $T_4 = 12.3°C$.

The entropy generation rate can now be found from Eq. (6.34), with $\dot{Q} = 0$:

$$\dot{S}_{gen} = \dot{m}_a(s_2 - s_1) + \dot{m}_w(s_4 - s_3)$$

With constant specific heats,

$$s_2 - s_1 = c_{p,a} \ln \frac{T_2}{T_1} - R \ln \frac{P_2}{P_1} = (1.005 \text{ kJ/kg} \cdot \text{K}) \ln \frac{353 \text{ K}}{293 \text{ K}} = 0.187 \text{ kJ/kg} \cdot \text{K}$$

$$s_4 - s_3 = c_w \ln \frac{T_4}{T_3} = (4.18 \text{ kJ/kg} \cdot \text{K}) \ln \frac{285.3 \text{ K}}{343 \text{ K}} = -0.770 \text{ kJ/kg} \cdot \text{K}$$

Solving for the entropy generation rate yields

$$\dot{S}_{gen} = \mathbf{-0.011\ kW/K}$$

Analysis: Note that this number is negative, indicating that this process is impossible. If we inspect the temperatures, it should be apparent that this process is not possible as designed. The air enters as the cooler fluid but is heated above the hottest temperature of the water. This requires that heat goes from a colder fluid to a hotter fluid at some point—which won't happen in practice. Similarly, the water cools to a temperature lower than the coldest temperature of the air—again, this isn't possible.

Although the magnitude of the negative number is not large, the important part of the answer is that it is negative. Negative entropy generation is not possible, and so the process is not possible. The small magnitude of the rate is partially an artifact of the mass flow rates involved, but also indicates that this process is close to being possible. If the exiting air temperature was limited to 70°C and the exiting water temperature was limited to 20°C, through better choice of flow rates, the process would be possible, with a non-negative entropy generation.

QUESTION FOR THOUGHT/DISCUSSION

How is entropy generation important when considering the operation of a device or system?

While increasing the entropy of the universe sounds disconcerting, is there any particular reason why humans should be concerned about generating entropy with respect to the entire universe?

6.4.2 Second Law Implications for Heat Exchangers

As discussed in Chapter 5, the second law requires that heat be transferred from a hot substance to a cold substance unless there is an input of work. At first glance, when applied to heat exchangers we may think that this would require that the exit temperature of the hotter fluid would always be higher than the exit temperature of the colder fluid. How could the fluid that was originally colder become warmer than the fluid that was originally hotter?

For concentric-tube heat exchangers and shell-and-tube heat exchangers, there are two general classifications that describe the flow pattern: parallel flow and counter flow. These are shown in Figures 6.12 and 6.13. For simplicity, we will just consider the concentric-tube design. For parallel-flow heat exchangers, the two fluids enter the heat exchanger on the same side and flow in the same general direction. For counter-flow heat exchangers, the two fluids enter on opposite ends of the heat exchanger and flow in the opposite direction through the heat exchanger. The second law requires that at any point in the heat exchanger, the heat must flow from the hotter fluid to the colder fluid. For parallel-flow heat exchangers, the implication of this is shown in Figure 6.12. As you can see, in order for the second law to be maintained, the colder fluid must always be at a temperature lower than (or equal to) the temperature of the hotter fluid. As a result, the colder fluid must exit at a temperature equal to or lower than the hotter fluid temperature:

$$T_{\text{cold, exit}} \leq T_{\text{hot, exit}} \quad \text{(Parallel-flow heat exchangers)}$$

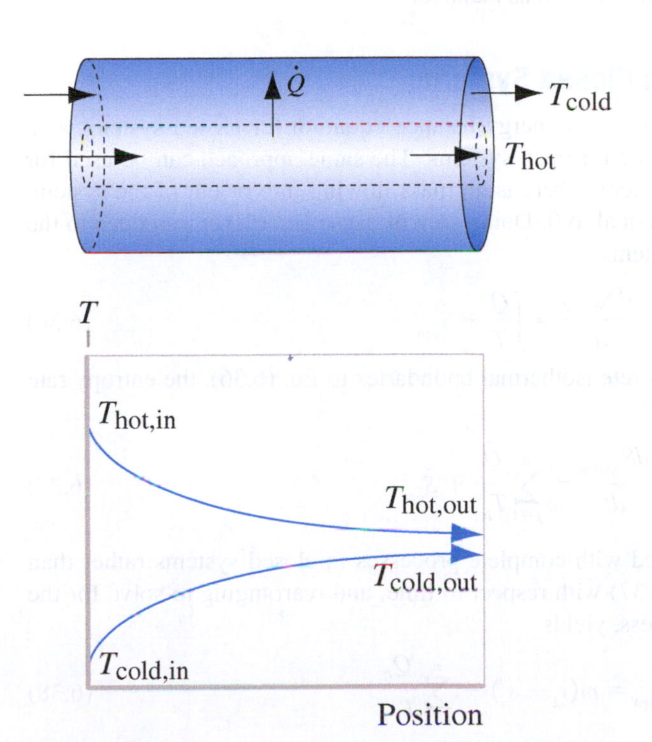

FIGURE 6.12 A concentric-tube heat exchanger, with both fluids flowing in the same direction (parallel flow). The hot fluid cools while the cold fluid's temperature increases, but the hot fluid will exit at a temperature no colder than the exit temperature of the cold fluid.

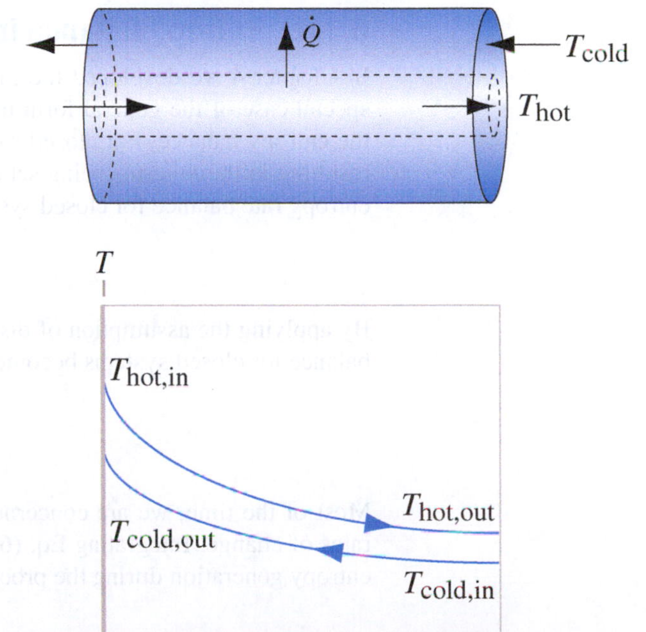

FIGURE 6.13 A concentric-tube heat exchanger, with the fluids flowing in opposite directions (counter-flow). At any location in the heat exchanger, the hot fluid is warmer than the cold fluid. However, due to the opposite flow direction, it is possible to have the hot fluid exit at a colder temperature than that at which the cold fluid exits.

However, a different conclusion can result when the second law is applied to the counter-flow heat exchanger, as shown in Figure 6.13. In this case, the second law only requires that the colder fluid exit at a temperature colder than the hot fluid inlet temperature, and that the hot fluid exit at a temperature hotter than the cold fluid inlet:

$$T_{\text{cold, exit}} \leq T_{\text{hot, enter}} \qquad \text{(Counter-flow heat exchanger)}$$

$$T_{\text{cold, enter}} \leq T_{\text{hot, exit}}$$

As a result, the cold fluid can exit a counter-flow heat exchanger at a higher temperature than that of the hot fluid exit temperature. (Depending on the flow rates and fluid temperatures, the cold fluid may still exit at a lower temperature than the hot fluid, but it is not required to do so.) Therefore, counter-flow heat exchangers are often used when a large amount of temperature change is needed for one or both of the fluids.

One issue to be concerned with when analyzing heat exchangers is that the direction of the flow is not normally incorporated into an entropy generation rate analysis. Therefore, we could calculate that a given set of conditions leads to a positive entropy generation rate, but because the system would require temperatures to violate the second law as described in the preceding discussion, the system is not possible. For example, a parallel-flow heat exchanger could have a positive entropy generation rate, but also have the fluid that initially was colder exit at a temperature higher than the initially warmer fluid exit temperature. Although allowed by an entropy generation analysis, subsequent second law analysis would show that the specific proposed condition will not work as planned.

6.4.3 Entropy Balance in Closed Systems

In Chapter 4 we developed the first law's energy balance equation for closed systems as a special case of the general form used for open systems. The same approach can be used for the entropy balance. For closed systems, there is no mass flowing into or out of the system, resulting in the $\dot{m}$ terms being set equal to 0. Doing so will cause Eq. (6.32) to reduce to the entropy rate balance for closed systems:

$$\frac{dS_{\text{system}}}{dt} = \int \frac{\dot{Q}}{T} + \dot{S}_{\text{gen}} \tag{6.36}$$

By applying the assumption of discrete isothermal boundaries to Eq. (6.36), the entropy rate balance for closed systems becomes

$$\frac{dS_{\text{system}}}{dt} = \sum_{j=1}^{n} \frac{\dot{Q}_j}{T_{b,j}} + \dot{S}_{\text{gen}} \tag{6.37}$$

Most of the time, we are concerned with complete processes in closed systems rather than rates of change. Integrating Eq. (6.37) with respect to time, and rearranging to solve for the entropy generation during the process, yields

$$S_{\text{gen}} = m(s_2 - s_1) - \sum_{j=1}^{n} \frac{Q_j}{T_{b,j}} \tag{6.38}$$

with state 1 being the initial state and state 2 being the final state. Equation (6.38) further simplifies if there is only one isothermal boundary for the entire process, or if the process is adiabatic.

▶ EXAMPLE 6.10

Pressurized water, with a mass of 1.0 kg, is to be boiled from a saturated liquid to a saturated vapor at a pressure of 200 kPa. The heat is being added over a surface that is maintained at 500 K. Determine the amount of entropy generated during this process.

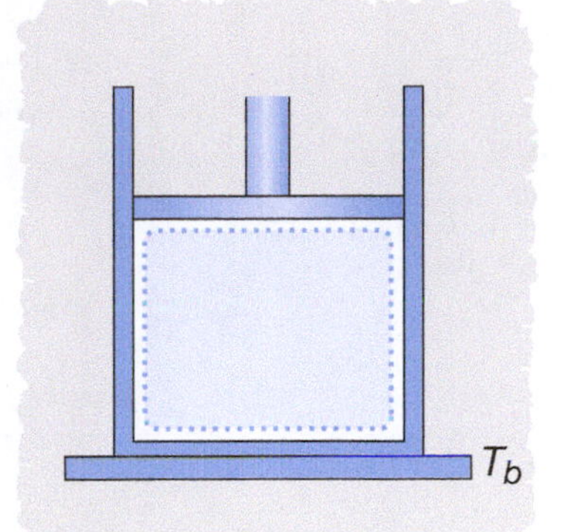

Given: $P_1 = P_2 = 200$ kPa, $m = 1$ kg, $x_1 = 0.0$, $x_2 = 1.0$, $T_b = 500$ K

Find: $\dot{S}_{gen}$

Solution: This is a closed system, with water (as a saturated substance) as the working fluid.

Assume: $\Delta KE = \Delta PE = 0$. The heat transfer occurs over an isothermal boundary.

First, the amount of heat transfer that is needed is found from the first law for closed systems. If the pressure is held constant, the volume of the container holding the water must be expanding. The moving boundary work is

$$W = \int_1^2 P\,d\forall = mP(v_2 - v_1)$$

From a computer property program, $v_1 = 0.0010605$ m³/kg and $v_2 = 0.8857$ m³/kg.

So

$$W = 176.9 \text{ kJ}$$

Assuming that there is no change in the kinetic or potential energy of the system, the first law reduces to

$$Q = m(u_2 - u_1) + W$$

From the software, $u_1 = 504.49$ kJ/kg and $u_2 = 2529.5$ kJ/kg.

Solving,

$$Q = 2202 \text{ kJ}$$

Equation (6.38) is now used to calculate the entropy generation, with one isothermal boundary.

$$S_{gen} = m(s_2 - s_1) - \frac{Q}{T_b}$$

Again, from a computer property program, $s_1 = 1.5301$ kJ/kg · K and $s_2 = 7.1271$ kJ/kg · K.

Solving,

$$\mathbf{S_{gen} = 1.19 \text{ kJ/K}}$$

Analysis: The positive result indicates that the process is possible and irreversible. This is not surprising, as the temperature of the area where the heat is being transferred from is above the temperature of the steam.

Having completed this calculation, what would the entropy generation be if the heat had been added over a boundary equal in temperature to that of the boiling water (i.e., $T_b = 120.2$°C $= 393.2$ K)? This can be found to be 0; the process would be reversible because there is no irreversibility caused by heat transfer over a finite temperature difference (because there is no temperature difference).

You may want to use the computer software to build a model of this system and test the entropy generation for different boundary temperatures. (Alternatively, you can easily use a spreadsheet to calculate out the results, because the only quantity changing is the boundary temperature.) Plotting the results yields the following:

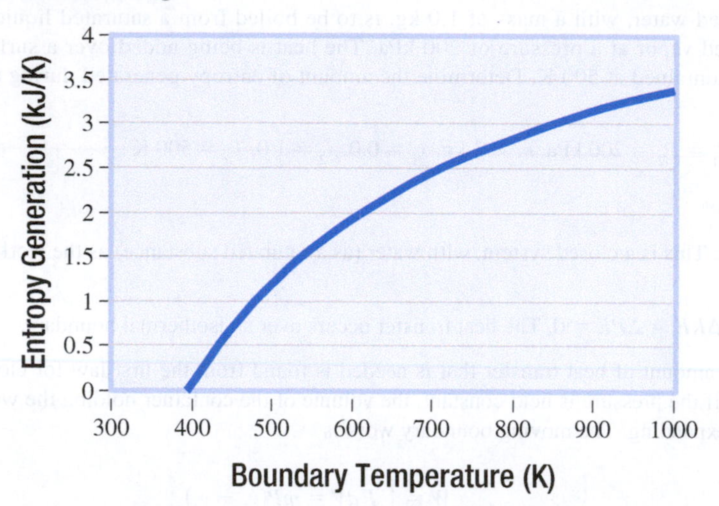

Note that temperatures with negative entropy generation results are not presented, because those processes are impossible. Heat would need to go from a colder source to a hotter sink to boil the water at 200 kPa.

▶ EXAMPLE 6.11

A 2.5 kg thin sheet of iron is placed in an oven to be heated. The iron begins the process at 10°C, and is removed when its temperature is 150°C. The oven temperature is maintained at 300°C. Determine the entropy generation in the process.

Iron

Given: $m = 2.5$ kg, $T_1 = 10°\text{C} = 283$ K, $T_2 = 150°\text{C} = 423$ K, $T_b = 300°\text{C} = 573$ K

Find: $\dot{S}_{\text{gen}}$

Solution: This is a closed system, and the working substance is iron. The iron can be treated as an incompressible substance, with constant specific heats for this problem.

Assume: $\Delta KE = \Delta PE = 0$. $W = 0$ (as there is no indication that any work is taking place in the process). The heat transfer occurs over an isothermal boundary.

First, the heat transfer needs to be found from the first law, which for a closed system takes the following form after applying the assumptions:

$$Q = m(u_2 - u_1)$$

Treating the specific heat of iron as a constant ($c = 0.443$ kJ/kg · K), the first law can be rewritten and solved for the heat transfer:

$$Q = mc(T_2 - T_1) = 155 \text{ kJ}$$

The entropy generation is then found from Eq. (6.38), with one isothermal boundary temperature:

$$S_{\text{gen}} = m(s_2 - s_1) - \frac{Q}{T_b} = mc \ln\frac{T_2}{T_1} - \frac{Q}{T_b}$$

Solving,

$$S_{\text{gen}} = 0.182 \text{ kJ/K}$$

Analysis: The positive value of the entropy generation indicates the process is possible and irreversible.

► **EXAMPLE 6.12**

Someone who is not an engineer at the company using the process in Example 6.11 suggests that energy could be saved by heating the iron with the oven maintained at 120°C. Is such a process possible?

Solution: Here, the analysis used in Example 6.11 is the same, except T_b = 120°C = 393 K. Performing the revised calculation yields: S_{gen} = 0.052 kJ/K.

This suggests that the process is possible, but we know that we can't heat a substance up to 150°C by transferring heat from a source at 120°C. So the process is not possible. But what is the problem with the analysis? Why would the entropy balance approach say the process is possible when it clearly is not?

Analysis: The problem is that the boundary temperature is not isothermal. The system under consideration is the iron sheet. This sheet is heating up, and therefore the boundary temperature is heating up over time, and it is not isothermal. The outer surface of the iron is being heated quickly by the oven, and then heat is conducting through the iron and the iron is being removed when its temperature is 150°C. For a thin sheet, the temperature is likely to quickly become uniform, but nothing in the problem relied on this being a thin sheet. A block of iron would have a hotter surface temperature than interior temperature, and yet nothing would be different in the analysis. In Example 6.11, the boundary temperature was taken at 300°C. Although this is probably appropriate for a block of iron, it is likely too high for a thin sheet; if the surface temperature is 300°C, the center temperature is likely close to that. A more appropriate temperature to take would be 150°C, although even then the process is not isothermal. However, because we are only using entropy generation at this point to compare different possible processes, taking a standard assumption on the isothermal nature of the boundary temperature and applying it consistently should not lead to faulty conclusions on which process may be better if we are comparing processes: the actual value of the entropy generation may be incorrect with the isothermal boundary assumption, but the general trends should be correct.

However, that conclusion breaks down in Example 6.12, when the second law is clearly being violated. At that point, with the oven temperature being close to (or in this case below) the desired iron temperature, the effects of a non-isothermal boundary become much more critical to the analysis. This is where it becomes very important to be certain that your assumptions are valid when using entropy generation to analyze whether or not a process is possible.

6.4.4 Use of Entropy Generation

As discussed with the results of Example 6.12, you may be wondering what use the entropy generation has if it is susceptible to erroneous calculations when incorrect simplifying assumptions are applied. In particular, if a system is assumed to have isothermal boundaries in order to make the calculation more feasible but then if that assumption leads to incorrect conclusions, particularly for closed systems that are not at steady state, what use is the entropy generation result? Its first use, as discussed in Section 6.2, is to tell us whether or not a process is potentially possible. It is not the only factor to be considered, and other aspects of the second law should be considered (such as the direction of heat transfer based on temperature). However, if an entropy generation calculation indicates that a process has a negative entropy generation, it is probable that the process is impossible. Poor assumptions tend to increase the likelihood of a process being considered possible through an entropy generation calculation, and so it is very unlikely that you would analyze a process and find it to have a negative entropy generation when the process is in fact possible. The entropy generation calculation may not be the final word on whether a process is possible, but it is a good initial step in this determination.

But what is of much more value is that the entropy generation calculation can allow us to compare different processes that we know are possible to see which is better thermodynamically (i.e., which have a lower degree of irreversibility). Consider Example 6.11. Suppose we wanted to see if an alternative process with an oven temperature of 250°C would be more or less irreversible than the 300°C oven temperature. We know that both could heat the iron to a temperature of 150°C. Even though the assumption of an isothermal boundary set at the oven temperature is questionable, if it is applied consistently we should get a conclusion that is valid, even though the quantitative entropy generation calculation may be inaccurate (i.e., if one possible process has a lower entropy generation than a second possible process, the one with the lower entropy generation has less irreversibility). Such a calculation would show that the entropy generation for a boundary temperature of 250°C is smaller, and so that process has less irreversibility. As the temperature difference between the iron and the oven is smaller, it is reasonable to expect that the process is less irreversible—a conclusion confirmed through the entropy generation calculation.

Another example of how processes can be compared is with the steady-state turbine and insulation considered in Example 6.8. There we found that the better insulated the turbine is, resulting in less heat transfer, the lower the entropy generation rate and the lower the irreversibility. This resulted in greater power output, which would indicate better performance for the turbine.

However, just because a process is "better thermodynamically" by being closer to reversible does not mean that the process is more sensible to use. Suppose you have a choice between two processes, one of which costs $100,000 more per year to implement. The more expensive process is "less irreversible" because its entropy generation is closer to 0. However, you calculate that your savings achieved with this process amounts to $1000 per year, indicating that a simple payback time period of 100 years would be necessary to pay for the "better" process. Obviously, the thermodynamically better process does not make sense economically. Another example is that you may have a process that requires a material to be heated to 200°C very quickly. Although the material could be heated by a heat source whose temperature is 205°C, you may find that the process occurs more rapidly with a heat source at 500°C. Here, the need for rapid heating encourages you to use the higher temperature heat source, even though more entropy would be generated and the process is more irreversible.

We can at times be caught up in trying to minimize entropy generation, but it is important to note that there is no fundamental problem with generating entropy. Although entropy generation normally indicates that a process is less efficient, there are times when increasing the efficiency is not economically sensible, or where other factors negate the importance of the energy efficiency of a process; there may be productivity efficiencies or economic considerations that supersede energy efficiency factors.

QUESTION FOR THOUGHT/DISCUSSION

What implications does entropy generation have with regard to fossil fuel resources and their use on earth?

6.5 ISENTROPIC EFFICIENCIES

It is often of great interest to either determine how close a device is to operating at its ideal performance, or conversely how to use a known efficiency of a device to help determine the outlet conditions of the fluid flowing through it and how much power is produced or consumed by the device. Both of these activities can be done through using what is known as the *isentropic efficiency*. The isentropic efficiency is most commonly applied to turbines, compressors, and pumps. Occasionally it is also used for nozzles and diffusers.

The term *isentropic efficiency* is somewhat of a misnomer, but it is understandable that the term is used. We have already established that ideal processes are reversible. Furthermore, we have discussed how we often seek to have adiabatic turbines, compressors, and pumps in order to either maximize the power output (for turbines) or minimize power consumption (for compressors and pumps). Therefore, for these devices we are seeking to compare the actual performance to that which would be seen for a reversible and adiabatic device. Consider Eq. (6.35):

$$\dot{S}_{gen} = \dot{m}(s_2 - s_1) - \sum_{j=1}^{n} \frac{\dot{Q}_j}{T_{b,j}} \tag{6.35}$$

For a reversible process, $\dot{S}_{gen} = 0$, and for an adiabatic process, $\dot{Q} = 0$. Substituting these into Eq. (6.35) yields

$$0 = \dot{m}(s_2 - s_1) - 0$$

which logically reduces to

$$s_2 = s_1$$

The entropy is constant for a reversible and adiabatic process, and a constant entropy process is also called an isentropic process. Therefore, because a reversible and adiabatic process implies that the process is isentropic, we call the type of comparison reflecting this an isentropic efficiency although it is more correctly termed a reversible and adiabatic efficiency. Be aware, though, that an isentropic process does not need to be reversible and adiabatic. We could have a constant entropy process where heat is removed from a system and entropy is generated but whose contributions from each balance out, which causes the process to be isentropic.

6.5.1 Isentropic Processes in Ideal Gases

To solve problems involving isentropic processes for fluids such as water and refrigerants, we generally find the appropriate property values using the initial entropy and the final pressure, temperature, or specific volume—depending on what is known. This will provide an isentropic end state. However, with ideal gases, certain relationships can be used to describe the processes in more detail.

First, consider the case of an *ideal gas with constant specific heats* assumed. For isentropic processes, Eqs. (6.24) and (6.25) can be written as

$$c_v \ln \frac{T_2}{T_1} + R \ln \frac{v_2}{v_1} = 0$$

and

$$c_p \ln \frac{T_2}{T_1} - R \ln \frac{P_2}{P_1} = 0$$

For ideal gases,

$$c_p = \frac{kR}{k-1} \quad \text{and} \quad c_v = \frac{R}{k-1} \tag{6.39}$$

where k is the specific heat ratio: $k = c_p/c_v$. Substitution of these relationships into the Gibbs equations for isentropic processes yields

$$\frac{T_2}{T_1} = \left(\frac{P_2}{P_1}\right)^{\frac{k-1}{k}} \tag{6.40}$$

and

$$\frac{T_2}{T_1} = \left(\frac{v_1}{v_2}\right)^{k-1} \tag{6.41}$$

Equating Eqs. (6.40) and (6.41) further results in

$$\frac{P_2}{P_1} = \left(\frac{v_1}{v_2}\right)^k \tag{6.42}$$

which can also be written as Pv^k = constant. This is a polytropic process, with $n = k$. Therefore, an isentropic process for an ideal gas with constant specific heats is represented as a polytropic process with $n = k$, and the moving boundary work for this can be found from Eq. (2.16c), with $n = k$. Furthermore, Eqs. (6.40)–(6.42) can be used to relate the common properties of temperature, pressure, and specific volume for these processes.

If the fluid being used is air, and if the variability of the specific heats is being considered, Table A.3 in the appendix offers an opportunity to quickly find states for an isentropic process. Table A.3 contains a column labeled "P_r" and a column labeled "v_r". The values presented in these tables are relative pressures and relative specific volumes, respectively, for air when it undergoes an isentropic process. Individually, the numbers in these columns have little utility, but for isentropic processes, the ratio of the values in the column equals the ratio of the respective actual properties. Therefore, for air with variable specific heats undergoing an isentropic process:

$$\frac{P_2}{P_1} = \frac{P_{r,2}}{P_{r,1}} \quad \text{and} \quad \frac{v_2}{v_1} = \frac{v_{r,2}}{v_{r,1}} \tag{6.43a, b}$$

If we know the actual pressure ratio or the actual specific volume ratio in an isentropic process, and the initial state, we can use either Eq. (6.43a) or (6.43b) to calculate a final relative pressure or specific volume value and then use Table A.3 to find the corresponding final temperature, specific enthalpy, or specific internal energy value, as needed. Most computer programs that find thermodynamic properties do not include these relative values, so it is most likely that you will need to use a table such as Table A.3 to take advantage of this calculation. In addition, such tables are not common for gases other than air. Example 6.13 demonstrates this calculation procedure.

▶ EXAMPLE 6.13

Air undergoes an isentropic compression from 300 K and 100 kPa to a final pressure of 1500 kPa. Considering air to be an ideal gas with variable specific heats, determine the temperature of the air at the end of the process and the change in specific enthalpy for the process.

Given: $T_1 = 300$ K, $P_1 = 100$ kPa, $P_2 = 1500$ kPa

Find: T_2 (the final temperature) and Δh (the change in specific enthalpy)

Solution: The working fluid is air.

Assume: Air as an ideal gas with variable specific heats, isentropic process ($s_2 = s_1$)

For this example, we will use Table A.3. From this table, at a temperature of 300 K:

$$h_1 = 300.19 \text{ kJ/kg} \qquad P_{r,1} = 1.70203$$

From Eq. (6.43a),

$$P_{r,2} = (P_2/P_1)\,P_{r,1} = (1500 \text{ kPa}/100 \text{ kPa})(1.70203) = 25.53$$

Because $P_{r,2}$ does not appear directly on the table, linear interpolation will be used to find $\boldsymbol{T_2}$ **= 678 K** (which is the final temperature) and $h_2 = 689.36$ kJ/kg.

Therefore, the change in enthalpy is $\boldsymbol{\Delta h = h_2 - h_1}$ **= 389.2 kJ/kg.**

Remember that the P_r and v_r columns in Table A.3 **only** apply to isentropic processes. Furthermore, note that these values do not have a unit associated with them.

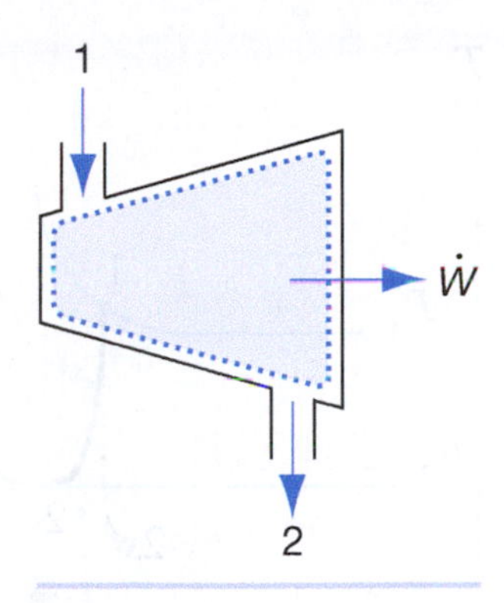

FIGURE 6.14 A turbine, with state 1 being the inlet flow and state 2 being the exit flow.

6.5.2 Isentropic Turbine Efficiency

For work-producing devices, such as turbines, irreversible devices will produce less work than their reversible counterparts operating between the same pressures. To allow for the efficiency to be less than 1, we define the isentropic efficiency, η_s, for a work-producing device to be

$$\eta_{s,\text{ work-producing device}} = \frac{\text{Actual Work}}{\text{Isentropic Work}}$$

Specifically, for the adiabatic turbine shown in **Figure 6.14**, with state 1 as the inlet state and state 2 as the outlet state,

$$\eta_{s,t} = \frac{h_1 - h_2}{h_1 - h_{2s}} \tag{6.44}$$

This equation is developed by recognizing that the mass flow rate terms in the numerator and denominator will cancel, assuming there are no changes in kinetic or potential energy and that the heat transfer is 0. The quantity h_1 represents the inlet state enthalpy, h_2 represents the actual outlet state enthalpy, and h_{2s} represents the isentropic process outlet state enthalpy. These values, as well as a comparison between the isentropic process and the actual process, can be shown graphically, as in **Figure 6.15**. This figure contains *T-s* (temperature vs. specific entropy) diagrams for potential situations involving a steam turbine with the turbine receiving superheated steam.

Note that $\eta_{s,t} \leq 1$, and if we calculated an isentropic efficiency above 1, the process would violate the second law.

Although states 1 and 2 are straightforward to understand, state 2*s* requires more description. For state 2*s*, we assume that the pressure is identical to the actual outlet pressure, and the entropy is equal to the inlet state entropy:

$$P_{2s} = P_2$$

$$s_{2s} = s_1$$

Knowing these two properties, we can find the enthalpy of the outlet isentropic outlet state from a computer program for steam or refrigerants. Alternatively, if we are dealing with an ideal gas, we can use the ideal gas relationships described in Section 6.5.1 to find isentropic outlet state properties.

Three possible scenarios are shown for a process through a steam turbine in Figure 6.15: Figure 6.15a shows the result if the hypothetical isentropic outlet state and the actual outlet state are both saturated mixtures; Figure 6.15b shows the diagram for an isentropic outlet state that is a saturated mixture, but the actual outlet state is superheated; and Figure 6.15c shows the diagram for both the isentropic outlet state and the actual outlet states being superheated. All three of these scenarios can exist in practice, depending on the particular conditions under which the turbine operates. Notice that depending on the scenario, the isentropic and actual outlet states may have different temperatures, but they have the same pressure by definition. In case (a), because both states 2 and 2*s* are saturated at the same pressure, they have the same temperature. However, in cases (b) and (c), $T_2 > T_{2s}$ because the superheated vapor in state 2 has a higher temperature than the ideal outlet state temperature.

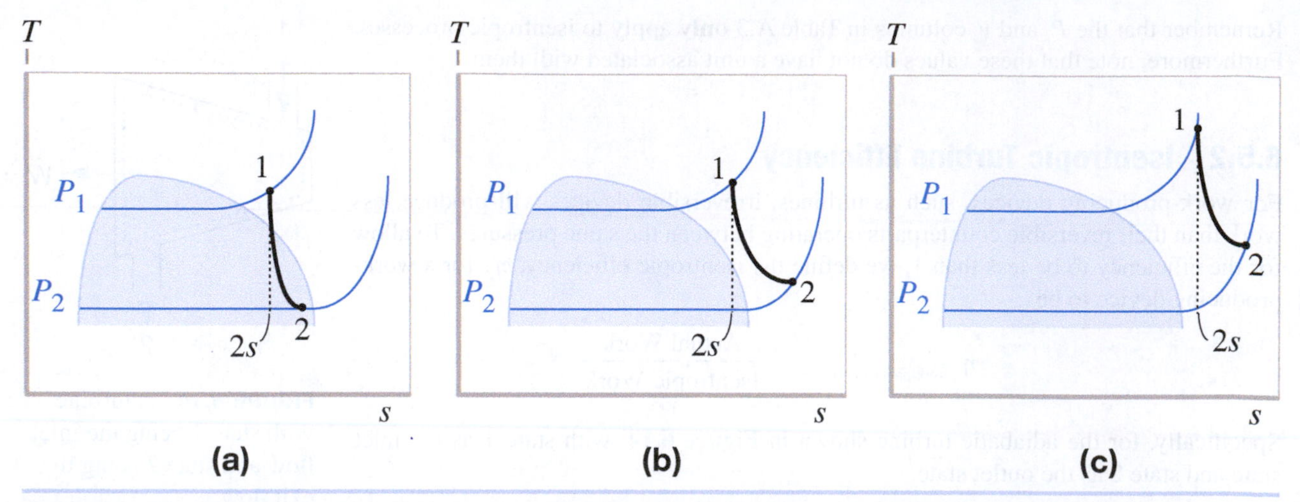

FIGURE 6.15 Three possible scenarios for the behavior of a steam turbine (with superheated steam entering) on a *T-s* diagram. (a) Both the hypothetical isentropic outlet state (2*s*) and the actual outlet state (2) are saturated mixtures—in this case, $T_{2s} = T_2$. (b) The isentropic outlet state is a saturated mixture, but the actual outlet state is a superheated vapor. (c) Both the isentropic outlet state and the actual outlet state are superheated vapors. In both (b) and (c), $T_{2s} \neq T_2$.

STUDENT EXERCISE

Modify your computer model of a turbine so that you can incorporate the use of an isentropic efficiency in an analysis of a turbine. The model should either allow for direct calculation of the actual outlet temperature or should provide the actual outlet specific enthalpy such that that value could be used with another source to determine the outlet temperature.

▶ EXAMPLE 6.14

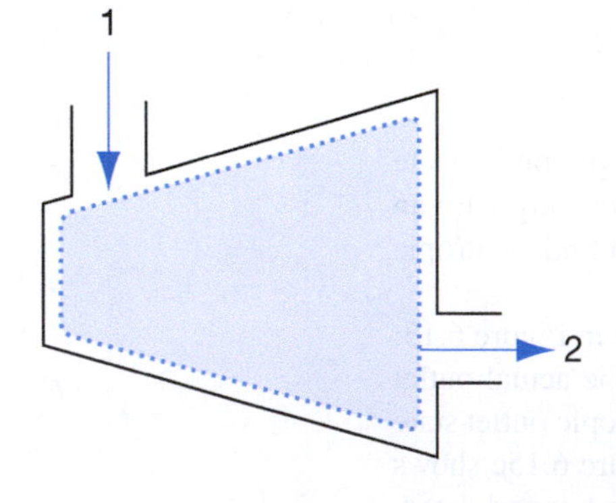

Steam enters a turbine at a flow rate of 5 kg/s, at a temperature of 600°C and a pressure of 10 MPa. The steam exits at a pressure of 100 kPa. The isentropic efficiency of the turbine is 0.75. Determine the power output from the turbine, assuming it is insulated.

Given: $P_1 = 10$ MPa, $T_1 = 600°C$, $\dot{m} = 5$ kg/s, $P_2 = 100$ kPa, $\eta_{s,t} = 0.75$

Find: $\dot{W}$

Solution: The system is an open system with a single inlet and a single outlet. The working fluid is water, in the form of steam. The steam is initially superheated, and it is unknown whether the steam is superheated or a saturated mixture at the exit.

Assume: $\Delta KE = \Delta PE = 0$

$\dot{Q} = 0$ (insulated)

For a turbine with no heat transfer or changes in kinetic or potential energy, the first law reduces to

$$\dot{W} = \dot{m}(h_1 - h_2)$$

From computer software, $h_1 = 3625.3$ kJ/kg. To find h_2, we need to use the isentropic efficiency, which first requires us to find state 2*s*.

From software, $s_1 = 6.9029$ kJ/kg · K.

For state 2*s*, $s_{2s} = s_1 = 6.9029$ kJ/kg · K, and $P_{2s} = P_2 = 100$ kPa. This is a saturated mixture, with $x_{2s} = 0.925$, and $h_{2s} = 2506.1$ kJ/kg.

From

$$\eta_{s,t} = \frac{h_1 - h_2}{h_1 - h_{2s}}$$

we find $h_2 = 2785.9$ kJ/kg. This state turns out to be a superheated vapor, with $T_2 = 154.8°C$.

Solving for the power produced yields

$$\dot{W} = \mathbf{4200\ kW}$$

Analysis: This is a good example of how the ideal outlet state from a steam turbine might be a saturated mixture, but the actual outlet state of the nonideal turbine is superheated. Always remember that the ideal outlet state is just that—it is not the actual state and only would exist in practice if the turbine has any losses. Use the actual outlet enthalpy to find the actual outlet temperature.

▶ EXAMPLE 6.15

Air, behaving as an ideal gas with constant specific heats, enters a turbine at 1000 kPa and 700 K, and exits at 100 kPa and 425 K. Determine the isentropic efficiency of the turbine.

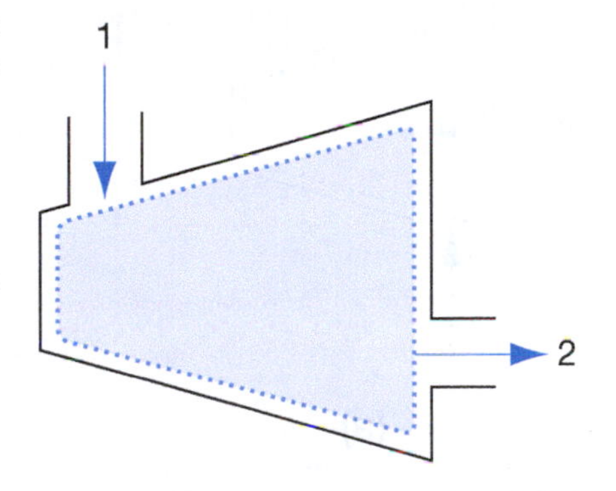

Given: $P_1 = 1000$ kPa, $T_1 = 700$ K, $P_2 = 100$ kPa, $T_2 = 425$ K

Find: $\eta_{s,t}$

Solution: The system is a turbine with the working fluid being air (which can be modeled as an ideal gas).

Assume: The air behaves as an ideal gas with constant specific heats.

For the specific heats, in this example we will use the values at 300 K:

$$k = 1.4,\ c_p = 1.005\ \text{kJ/kg} \cdot \text{K}$$

We need to find the isentropic outlet state. Recall that $P_{2s} = P_2$. For an ideal gas with constant specific heats, we can use Eq. (6.40) to relate the isentropic outlet state temperature to the pressure ratio:

$$\frac{T_{2s}}{T_1} = \left(\frac{P_{2s}}{P_1}\right)^{\frac{k-1}{k}}$$

where state 2 is now designated 2*s* to emphasize it is an isentropic outlet state. This yields

$$T_{2s} = 362.6\ \text{K}$$

Now, Eq. (6.44) can be used to find the isentropic efficiency of the turbine. Using Eq. (6.44), we will incorporate the changes in specific enthalpy for an ideal gas with constant specific heats:

$$\eta_{s,t} = \frac{h_1 - h_2}{h_1 - h_{2s}} = \frac{c_p(T_1 - T_2)}{c_p(T_1 - T_{2s})} = \frac{(T_1 - T_2)}{(T_1 - T_{2s})} = \mathbf{0.815}$$

Analysis: Note that this is a possible result for an actual turbine. If the isentropic efficiency had been greater than one, it would be known that the given information was incorrect (or impossible if it represented a proposed turbine system).

Also, note that the form of the equation used based on temperatures will only work for an ideal gas with constant specific heats (or an incompressible substance with constant specific heats). When the change in specific enthalpy cannot be reduced to a constant multiplied by a change in temperatures (such as for a substance near or undergoing a phase change or an ideal gas with variable specific heats), the ratio of the actual temperature difference to the isentropic temperature difference will not result in the correct answer for the isentropic efficiency.

In general, if you have a well-designed turbine operating at design conditions, you would expect the efficiency to be somewhere between 75 and 90%. These isentropic efficiencies can get quite low when the turbine is being operated well away from design conditions.

QUESTION FOR THOUGHT/DISCUSSION

For a turbine, how are the isentropic efficiency, entropy generation rate, and power produced related?

6.5.3 Isentropic Compressor and Pump Efficiencies

For work-absorbing devices, such as compressors and pumps, the definition of the isentropic efficiency is changed so that its value remains less than or equal to 1. A nonideal compressor or pump will use more power than an ideal device, so that we define the isentropic efficiency of a work-absorbing device to be

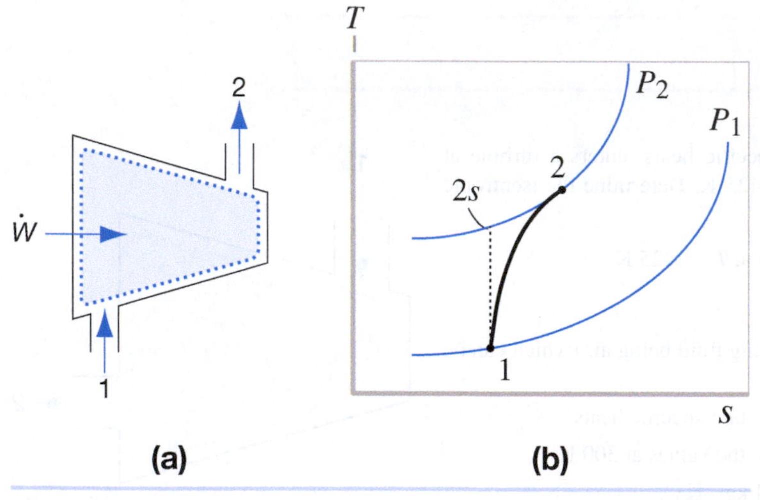

FIGURE 6.16 (a) A schematic diagram of a gas compressor, with state 1 being the inlet state and state 2 being the outlet state. (b) A representation of the non-isentropic process of flow through a compressor on a *T-s* diagram.

$$\eta_{s,\ \text{work-absorbing device}} = \frac{\text{Isentropic Work}}{\text{Actual Work}}$$

For the compressor shown in **Figure 6.16**, the isentropic efficiency is given by

$$\eta_{s,c} = \frac{h_1 - h_{2s}}{h_1 - h_2} \tag{6.45}$$

The same expression is used for the isentropic efficiency of the pump shown in **Figure 6.17**:

$$\eta_{s,p} = \frac{h_1 - h_{2s}}{h_1 - h_2} \tag{6.46}$$

Figure 6.16 also contains a *T-s* diagram for an isentropic and an actual process involving

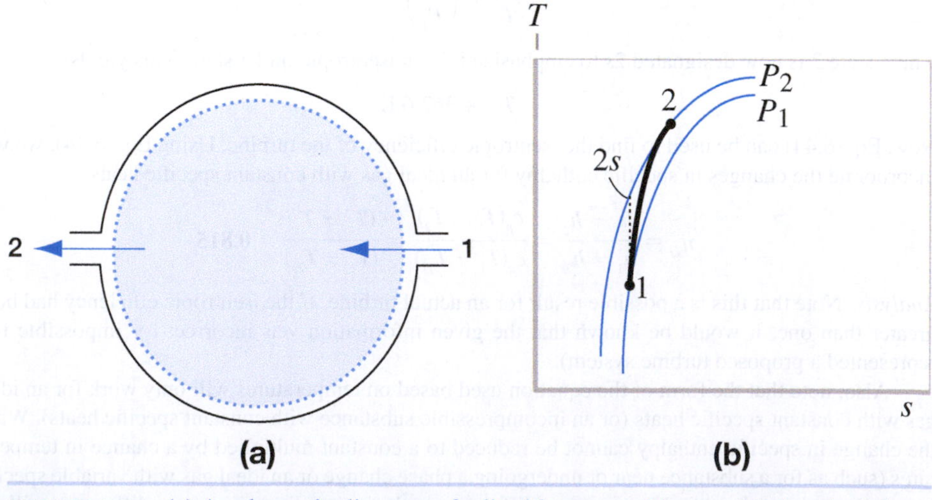

FIGURE 6.17 (a) A schematic diagram of a liquid pump, with state 1 being the inlet state and state 2 being the outlet state. (b) A representation of the non-isentropic process of flow through a pump on a *T-s* diagram.

the compression of a gas, whereas Figure 6.17 contains a *T-s* diagram for the isentropic and actual process of pumping a liquid. Note that the temperature change experienced by a pumping process is typically small. In Eqs. (6.45) and (6.46), state 1 represents the inlet state, state 2 represents the actual outlet state, and state 2*s* represents the isentropic outlet state. The isentropic outlet state is defined in the same way as for turbines: $P_{2s} = P_2$ and $s_{2s} = s_1$. Again, keep in mind that the isentropic efficiency of a pump or compressor must be less than or equal to 1, and if one of these devices has an isentropic efficiency greater than 1, it would violate the second law. Typically, compressors that are well designed and operating near design conditions will have isentropic efficiencies between 0.70 and 0.85, whereas pumps often have somewhat lower isentropic efficiencies, between 0.60 and 0.80.

STUDENT EXERCISE

Modify your computer model of a compressor so that you can incorporate the use of an isentropic efficiency in an analysis of a compressor. The model should allow for either the direct calculation of the actual outlet temperature, or provide the actual outlet specific enthalpy such that that value could be used with another source to determine the outlet temperature.

▶ EXAMPLE 6.16

A compressor receives air at 100 kPa and 300 K and compresses the air to 500 kPa. The isentropic efficiency of the compressor is 0.7. Determine the actual outlet temperature of the air.

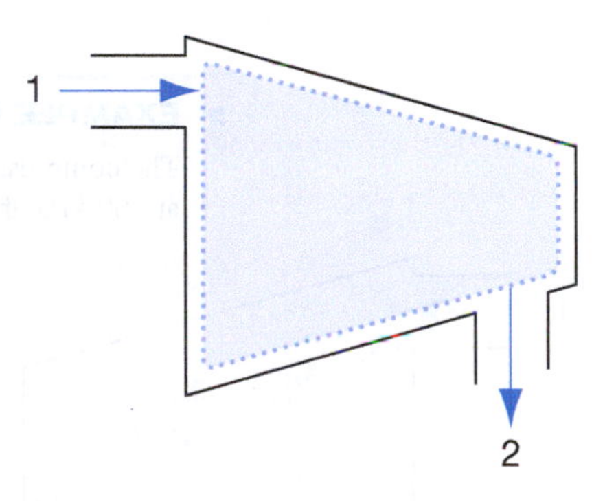

Given: $P_1 = 100$ kPa, $T_1 = 300$ K, $P_2 = 500$ kPa, $\eta_{s,c} = 0.70$

Find: T_2

Solution: This is a basic, nonideal compressor. The working fluid is air, which can be treated as an ideal gas.

Assume: Constant specific heats for the air.

Take the specific heat ratio of air to be the specific heat ratio at 300 K: $k = 1.4$.

The first step is to use the isentropic relations for an ideal gas with constant specific heat to determine the isentropic outlet temperature (Eq. (6.40)):

$$\frac{T_{2s}}{T_1} = \left(\frac{P_{2s}}{P_1}\right)^{\frac{k-1}{k}}$$

Recalling that P_{2s} is defined to equal P_2, $T_{2s} = 475.1$ K

Then, applying the isentropic efficiency equation (Eq. 6.45) for a compressor,

$$\eta_{s,c} = \frac{h_1 - h_{2s}}{h_1 - h_2} = \frac{c_p(T_1 - T_{2s})}{c_p(T_1 - T_2)} = \frac{(T_1 - T_{2s})}{(T_1 - T_2)} = 0.70$$

Solving for T_2 yields $\mathbf{T_2 = 550\ K}$

Analysis: It should be apparent that the actual outlet temperature will increase by a greater amount as the isentropic efficiency of the compressor decreases. To check this, you may wish to experiment by using different values of the isentropic efficiency of the turbine. Such a calculation will lead to results similar to the following:

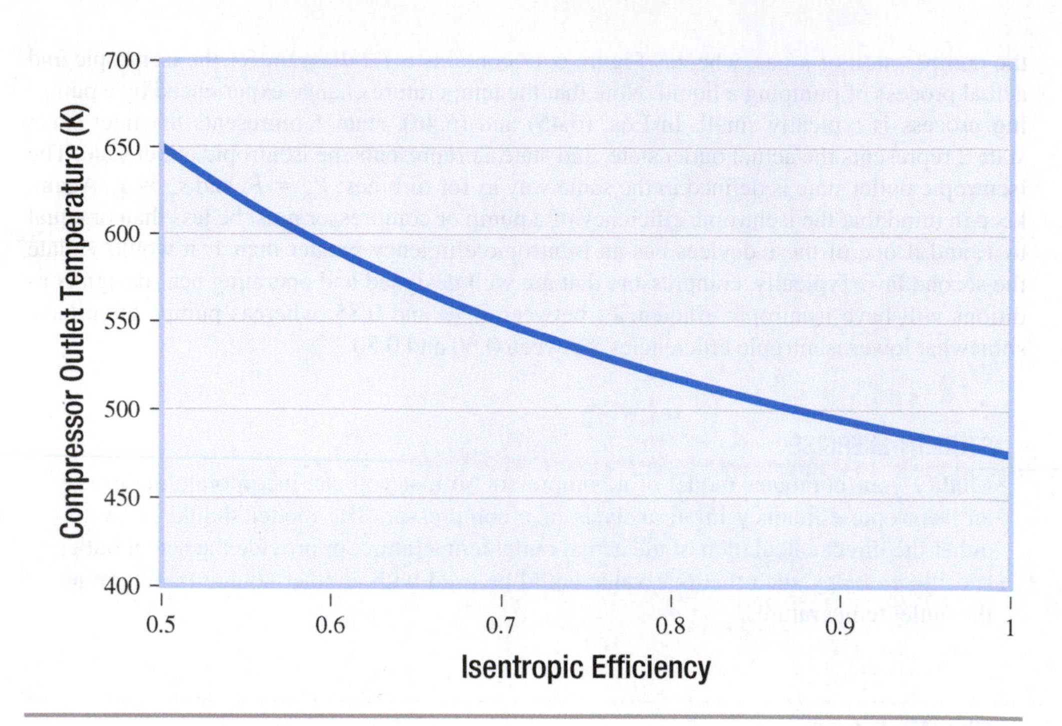

▶ EXAMPLE 6.17

The compressor in a refrigerator must compress 14 g/s of R-134a refrigerant from a saturated vapor at 200 kPa absolute to a pressure of 825 kPa absolute. You measure the temperature of the R-134a leaving the compressor as 45°C. What is the isentropic efficiency of the compressor, and what is the power consumed by the compressor?

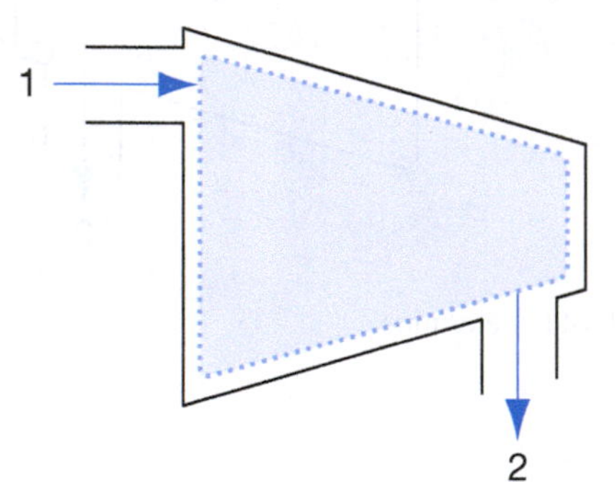

Given: $P_1 = 200$ kPa, $x_1 = 1.0$, $P_2 = 825$ kPa, $T_2 = 45°C$, $\dot{m} = 14$ g/s $= 0.014$ kg/s

Find: $\eta_{s,c}$, $\dot{W}$

Solution: The compressor is an open system with a single inlet and a single outlet, operating under steady-state, steady-flow conditions. The working fluid is R-134a, for which properties must be found from either a computer program or tables.

Assume: $\Delta KE = \Delta PE = 0$, $\dot{Q} = 0$ (insulated)

Assuming the compressor is adiabatic and has no changes in kinetic or potential energy, the power consumed by the compressor can be found from the reduced first law:

$$\dot{W} = \dot{m}(h_1 - h_2)$$

From a computerized properties program, $h_1 = 393$ kJ/kg, and $h_2 = 429$ kJ/kg.

Substitution yields

$$\dot{W} = -0.504 \text{ kW}$$

To find the isentropic efficiency, we first need the inlet entropy. From the same properties program,

$$s_1 = 0.9503 \text{ kJ/kg} \cdot \text{K}$$

Setting $s_{2s} = s_1$ and $P_{2s} = P_2$, h_{2s} can be found: $h_{2s} = 419$ kJ/kg.

Solving for the isentropic efficiency with Eq. (6.45), $\eta_{s,c} = \dfrac{h_1 - h_{2s}}{h_1 - h_2} = \mathbf{0.722.}$

Analysis: The isentropic efficiency is consistent with what can be expected from a refrigerator compressor.

Although the isentropic efficiency of pumps and compressors is an important consideration in a design, the use of a nonideal pump often does not affect the outlet temperature of the fluid and the power consumption of the device nearly as dramatically as the use of a nonideal compressor. The reason is that in comparison to the amount of power used by a compressor, the amount of power used by a pump is small. In addition, the specific heats of liquids are often larger than the specific heats of gases. Therefore, a smaller amount of power impacting a substance whose temperature changes more slowly with an energy input leads to a much smaller temperature change. Inefficiencies will increase the power used and the enthalpy exiting, but the difference in enthalpy will only marginally alter the temperature and increase the power consumption.

It can be difficult to determine a change in temperature of the liquid exiting the nonideal pump, so we normally concern ourselves with the change in enthalpy instead. However, the use of an isentropic process to find the isentropic outlet state enthalpy can be problematic in a liquid. However, we can easily overcome this difficulty by assuming an incompressible substance. Consider Eq. (6.22):

$$T ds = dh - v dP \tag{6.22}$$

Integration of Eq. (6.22) yields

$$\int T ds = h_2 - h_1 - \int v dP \tag{6.47}$$

Considering an isentropic process ($ds = 0$) for an incompressible substance (v = constant), Eq. (6.47) can be reduced to

$$h_2 - h_1 = v(P_2 - P_1) \tag{6.48}$$

Equation (6.48) can be used to approximate the change in enthalpy for an incompressible liquid undergoing an isentropic process (in which case $h_2 = h_{2s}$).

▶ EXAMPLE 6.18

Liquid water enters an adiabatic pump at 20°C and 100 kPa, at a flow rate of 2.5 kg/s. The water exits at 2000 kPa. The isentropic efficiency of the pump is 0.65. Determine power consumed by the pump.

Given: $P_1 = 100$ kPa, $T_1 = 20$°C, $P_2 = 2000$ kPa, $\dot{m} = 2.5$ kg/s, $\eta_{s,p} = 0.65$

Find: $\dot{W}$

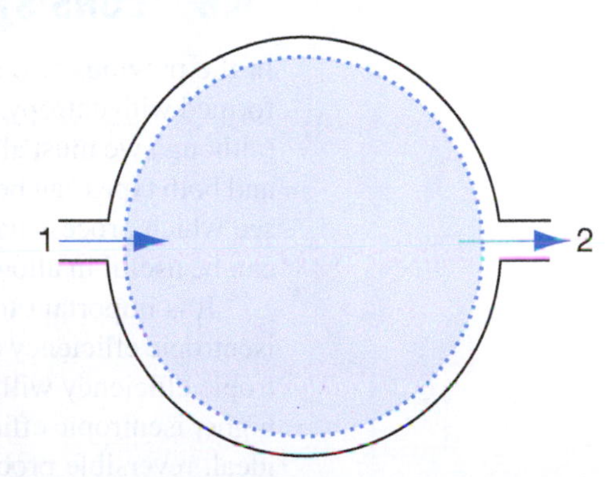

Solution: The pump is an open system with a single inlet and a single outlet, operating at steady-state, steady-flow conditions. The working fluid is liquid water.

Assume: $\Delta KE = \Delta PE = 0$, $\dot{Q} = 0$ (insulated). The water is an incompressible substance. (However, as the pressure change is large, we will not assume the specific heats are constant, and we will instead use computer programs for the enthalpy values.)

With an adiabatic pump, and assuming that there are no changes in kinetic or potential energy, the power consumed will be found from the first law as

$$\dot{W} = \dot{m}(h_1 - h_2)$$

From software, we find $h_1 = 83.96$ kJ/kg.

The isentropic outlet state will be found from Eq. (6.48):

$$h_{2s} - h_1 = v(P_{2s} - P_1)$$

keeping in mind that $P_{2s} = P_2$. From the software, $v = v_1 = 0.0010018$ m³/kg.

Solving,

$$h_{2s} = 85.86 \text{ kJ/kg}.$$

Using the isentropic efficiency of the pump (Eq. (6.46)),

$$\eta_{s,p} = \frac{h_1 - h_{2s}}{h_1 - h_2}$$

and solving for h_2 gives

$$h_2 = 86.89 \text{ kJ/kg}.$$

Substituting and solving for the power gives

$$\dot{W} = \dot{m}(h_1 - h_2) = \mathbf{-7.32\ kW}$$

Analysis: If we were to find the outlet temperature of the water, we would find $T_2 \approx 21°C$. Because of the small change in the temperature, it is often better to use the approach of Eq. (6.48) to find isentropic outlet states of the water leaving a pump.

QUESTION FOR THOUGHT/DISCUSSION

Which of these processes should more effort be devoted to: improving the isentropic efficiencies of gas compressors or improving the isentropic efficiencies of liquid pumps?

STUDENT EXERCISE

Modify your computer model of a pump to incorporate the isentropic efficiency of the pump. The liquid passing through the pump experiences little change in temperature, so it is optional whether you add the capability to calculate the actual outlet temperature.

6.6 CONSISTENCY OF ENTROPY ANALYSES

In the previous two sections, we considered two different types of analysis that can be performed with entropy. Both types can provide insights as to whether or not a process is possible (although we must always consider if appropriate assumptions are being made in the analysis), and both types can be used to compare different processes from a thermodynamic viewpoint to see which process has the least irreversibility. Performing an analysis of reversible processes can be useful in allowing us to understand the ideal limits that can be achieved with a process.

It is important to remember that the results of the entropy generation analysis and of the isentropic efficiency analysis are consistent. A turbine, compressor, or pump with a lower isentropic efficiency will also see a higher entropy generation rate than a competing device with a higher isentropic efficiency. Both types of analysis are comparing an irreversible process to an ideal, reversible process. The less efficient a device is, the more entropy that is generated by the device. Similarly, a less efficient device will result in a turbine producing less power, or a compressor or pump using more power. This is illustrated in Example 6.19.

▶ EXAMPLE 6.19

A steam turbine receives 25 kg/s of steam at 5.0 MPa and 500°C, and the steam exits at 150 kPa. The turbine is adiabatic. Using your computer model of a turbine, plot the power produced, the entropy generation rate, and the outlet temperature for isentropic efficiencies varying between 0.40 and 1.0.

Given: $\dot{m} = 25$ kg/s, $P_1 = 5.0$ MPa, $T_1 = 500°C$, $P_2 = 150$ kPa

Find: $\dot{W}$, $\dot{S}_{gen}$, T_2 for varying isentropic efficiencies

Solution: The system is an open system with a single inlet and a single outlet. The working fluid is water, in the form of steam. The steam is initially superheated, and it is unknown whether the steam is superheated or a saturated mixture at the exit.

Assume: $\dot{Q} = \Delta KE = \Delta PE = 0$, steady-state, steady-flow process

Applying the assumptions, from the first law,

$$\dot{W} = \dot{m}(h_1 - h_2)$$

In addition, from Eq. (6.35),

$$\dot{S}_{gen} = \dot{m}(s_2 - s_1)$$

and from Eq. (6.44),

$$h_2 = h_1 - \eta_{s,t}(h_1 - h_{2s})$$

Using the computer model, the following plots can be obtained:

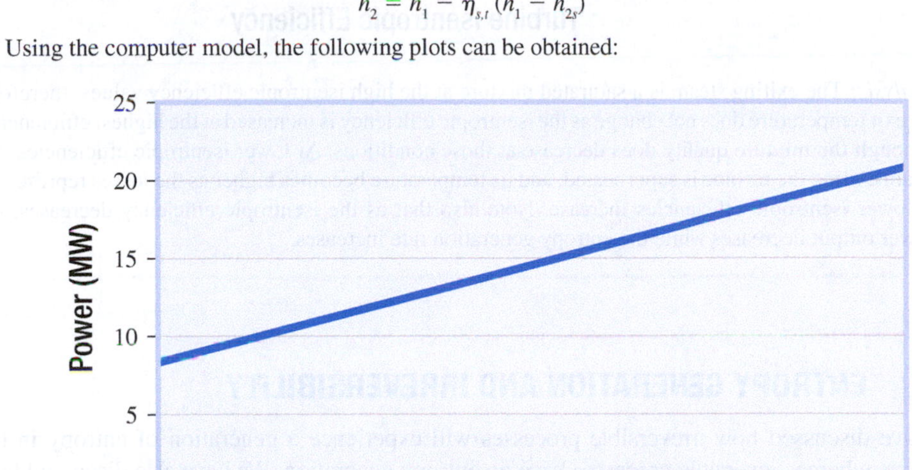

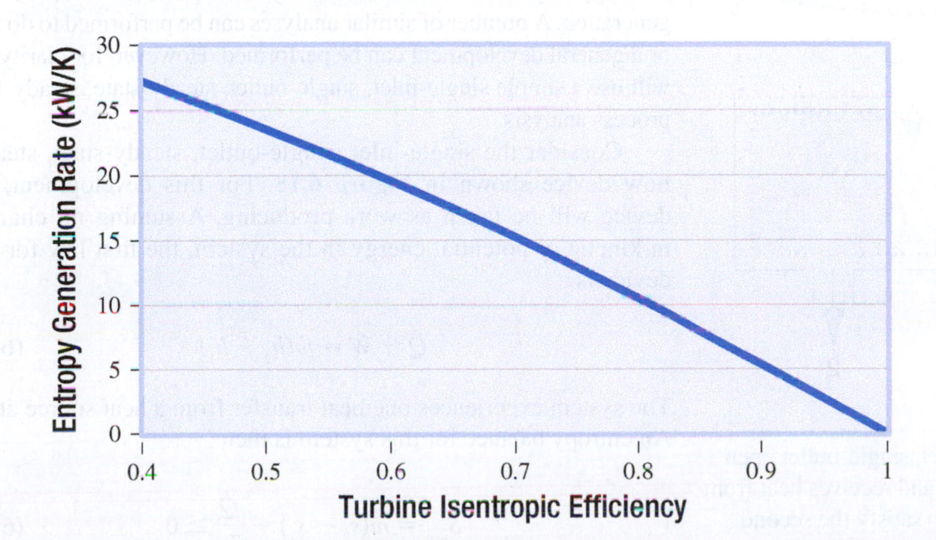

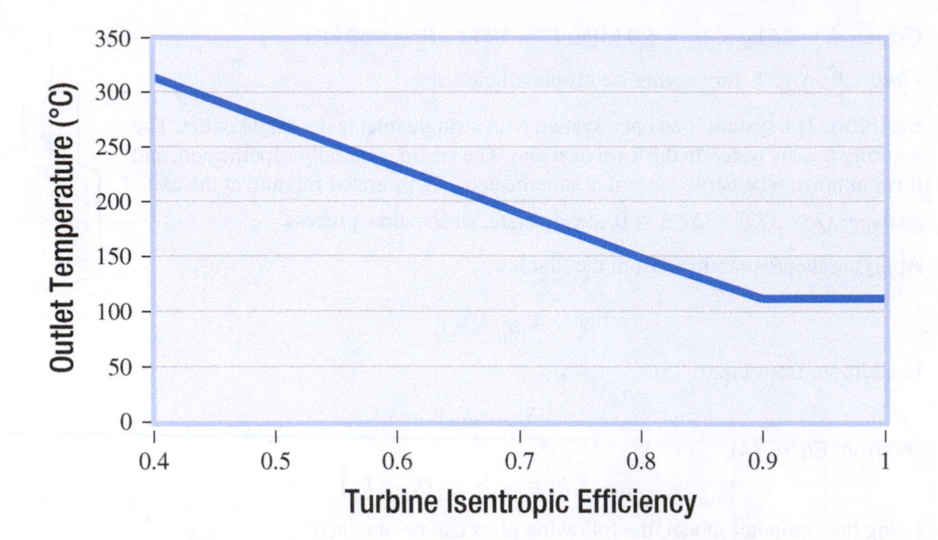

Analysis: The exiting steam is a saturated mixture at the high isentropic efficiency values; therefore, the exit temperature does not change as the isentropic efficiency is increased at the highest efficiencies, although the mixture quality does decrease at those conditions. At lower isentropic efficiencies, the steam exiting the turbine is superheated, and its temperature becomes higher as the losses represented in lower isentropic efficiencies increase. Note also that as the isentropic efficiency decreases, the power output decreases while the entropy generation rate increases.

6.7 ENTROPY GENERATION AND IRREVERSIBILITY

We have discussed how irreversible processes will experience a generation of entropy in the universe, whereas reversible processes have no entropy generation. We have also discussed how we can use entropy generation to compare processes to determine which is "less irreversible." Along these lines, irreversibility is a measure of the potential work lost during a process. In this section, we explicitly relate the irreversibility of a process to the entropy generation. A number of similar analyses can be performed to do this, or a general development can be performed. However, for clarity, we will use a simple single-inlet, single-outlet, steady-state, steady-flow process analysis.

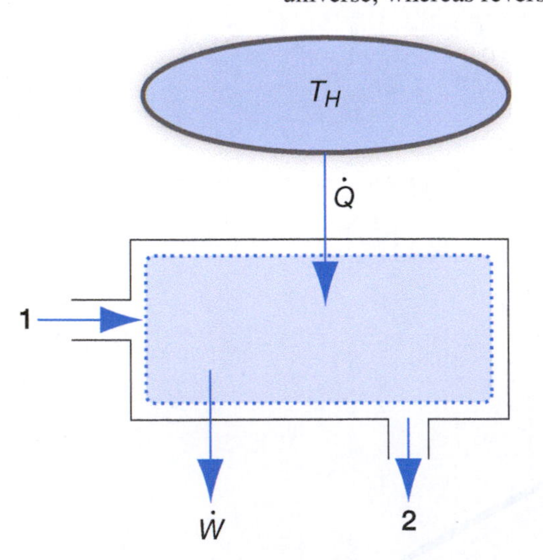

FIGURE 6.18 A single-inlet, single-outlet open system that produces work and receives heat from a thermal reservoir at T_H. To satisfy the second law, heat will also be rejected to the surroundings.

Consider the single-inlet, single-outlet, steady-state, steady-flow device shown in **Figure 6.18**. For this development, the device will be taken as work-producing. Assuming no changes in kinetic or potential energy in the system, the first law for this device is

$$\dot{Q} - \dot{W} = \dot{m}(h_2 - h_1) \tag{6.49}$$

The system experiences one heat transfer from a heat source at T_H. An entropy balance for this system is then

$$\dot{S}_{\text{gen}} = \dot{m}(s_2 - s_1) - \frac{\dot{Q}}{T_H} \geq 0 \tag{6.50}$$

Let us define the rate of irreversibility, $\dot{I}$, as

$$\dot{I} = \dot{W}_{rev} - \dot{W}_{irr} \tag{6.51}$$

where $\dot{W}_{rev}$ is the power produced in a reversible process, and $\dot{W}_{irr}$ is the power produced in an irreversible process. (Note that the definition could be reversed for a work-absorbing device in order to keep the irreversibility as a positive number.)

Let us consider that both an irreversible process and the reversible process have the same inlet and outlet states (h_1, s_1, h_2, and s_2) and have the same heat transfer rate ($\dot{Q}$) from the same thermal reservoir (T_H). For the reversible process to have an entropy generation rate of 0, there must be an additional entropy transport term in Eq. (6.50), because the existing terms produce a positive entropy generation rate for an irreversible process. Clearly, there can be no additional entropy transport via mass flow in Eq. (6.50), but if we consider the early development of the entropy rate balance concept, there is the possibility of having an additional entropy transport via heat transfer (see Eq. (6.34)). The natural place for this heat transfer to occur is with the surroundings, which will be at T_0. Calling this heat transfer rate $\dot{Q}_0$, the entropy rate balance for a reversible process can be written as

$$\dot{S}_{gen} = \dot{m}(s_2 - s_1) - \frac{\dot{Q}}{T_H} - \frac{\dot{Q}_0}{T_0} = 0 \tag{6.52}$$

Solving Eq. (6.52) for the heat transfer rate with the surroundings yields

$$\dot{Q}_0 = \dot{m}T_0(s_2 - s_1) - \frac{\dot{Q}T_0}{T_H} \tag{6.53}$$

Considering that there are now two heat transfers for the reversible process, the first law for the reversible power is

$$\dot{W}_{rev} = -\dot{m}(h_2 - h_1) + \dot{Q} + \dot{Q}_0 \tag{6.54}$$

Substituting Eq. (6.53) into Eq. (6.54) yields

$$\dot{W}_{rev} = -\dot{m}(h_2 - h_1) + T_0\dot{m}(s_2 - s_1) + \dot{Q} - \dot{Q}\frac{T_0}{T_H} \tag{6.55}$$

The irreversible process power is found from Eq. (6.49); the scenario presented there works for an irreversible process:

$$\dot{W}_{irr} = -\dot{m}(h_2 - h_1) + \dot{Q} \tag{6.56}$$

Substituting Eqs. (6.55) and (6.56) into Eq. (6.51) yields

$$\dot{I} = T_0\dot{m}(s_2 - s_1) - T_0\frac{\dot{Q}}{T_H} \tag{6.57}$$

Recalling the expression for the entropy generation rate in this scenario from Eq. (6.51), the rate of irreversibility is

$$\dot{I} = T_0\dot{S}_{gen} \tag{6.58}$$

Although we performed this development for a specific situation, similar expressions can be developed using devices with multiple inlets and outlets, or in transient situations. Closed systems can also be used, which will produce a nonrate expression of

$$I = T_0 S_{gen} \tag{6.59}$$

Clearly, the entropy generation during a process can be directly related to the amount of irreversibility. In turn, the irreversibility can be seen as the lost potential work for a process as a result of the process being nonreversible. For work-absorbing devices, the irreversibility represents the amount of additional work needed from the universe for a process to be performed as a result of the process not being reversible—this work cannot be used for other purposes.

To this point, we have been concerned with two types of efficiencies: the thermal efficiency and the isentropic efficiency. The thermal efficiency is also referred to as a first law efficiency. We can also consider a second law efficiency, which relates the actual performance of a device to its best possible performance. The isentropic efficiencies are specific applications of this general concept: the isentropic efficiencies previously developed are second law efficiencies for adiabatic devices. But other devices are clearly nonadiabatic, and the general second law efficiency accounts for this. For a heat engine, the second law efficiency can be written as

$$\eta_{2nd,\,HE} = \frac{\eta}{\eta_{max}} \tag{6.60}$$

For a refrigerator and heat pump, the second law efficiencies can be written as

$$\eta_{2nd\,ref} = \frac{\beta}{\beta_{max}} \tag{6.61}$$

and

$$\eta_{2nd,\,HP} = \frac{\gamma}{\gamma_{max}} \tag{6.62}$$

respectively. These definitions can provide us with a better picture of how well these devices are performing than the performance data gleaned using only a thermal efficiency. For example, a power plant with a thermal efficiency of 40% would sound like it is performing poorly. But if it only has a maximum possible efficiency of 55%, it does not appear to be performing as poorly.

For general work-producing devices, the second law efficiency is written as

$$\eta_{2nd,\,work\text{-}producing} = \frac{\dot{W}}{\dot{W}_{rev}}, \tag{6.63}$$

whereas work-absorbing devices use

$$\eta_{2nd,\,work\text{-}absorbing} = \frac{\dot{W}_{rev}}{\dot{W}} \tag{6.64}$$

for their second law efficiency, such as already done for isentropic efficiencies. Note that all of these forms of the second law efficiency will have values less than or equal to 1 for real processes.

▶ EXAMPLE 6.20

Suppose a coal-fired power plant is operating with a thermal efficiency of 35%. The steam in the power plant is produced at 780 K, and heat is rejected at a temperature of 310 K. Determine the plant's second law efficiency.

Given: $\eta = 0.35$, $T_H = 780$ K, $T_C = 310$ K

Find: η_{2nd} (the second law efficiency)

Solution: The system under consideration is the entire power plant, and we need only know the given general characteristics to determine the second law efficiency.

First, we can find η_{max} from Eq. (5.3):

$$\eta_{max} = 1 - \frac{T_C}{T_H} = 0.60$$

As a power plant is a heat engine, Eq. (6.60) is used to find the second law efficiency:

$$\eta_{2nd} = \frac{\eta}{\eta_{max}} = \mathbf{0.583}$$

Analysis: While still indicating a loss of potential work in the system, this calculation recognizes that the facility was unable to reach a 100% efficiency, so this comparison is useful.

▶ EXAMPLE 6.21

For a temperature of the surroundings of 298 K, determine the second law efficiency of the turbine in Example 6.8.

Given: $T_0 = 298$ K, $\dot{S}_{gen} = 2.93$ kW/K, $\dot{W} = 1500$ kW

Find: $\eta_{2nd,\ turbine}$ (the second law efficiency of the turbine)

Solution: The system under consideration is a single-inlet, single-outlet open system that will be considered to be operating under steady-state, steady-flow conditions.

The rate of irreversibility of the process is found from Eq. (6.58):

$$\dot{I} = T_0\dot{S}_{gen} = 873.1 \text{ kW}$$

From the definition of the irreversibility,

$$\dot{W}_{rev} = \dot{W} + \dot{I} = 2373 \text{ kW}$$

From Eq. (6.61),

$$\eta_{2nd,\ turbine} = \frac{\dot{W}}{\dot{W}_{rev}} = 0.632$$

QUESTION FOR THOUGHT/DISCUSSION

How does reducing the irreversibility of a device improve its performance?

A further, more advanced thermodynamic concept that can be used in second law analysis and consideration of irreversibility is *availability* (also known as *exergy*). Availability analysis merges together first law analysis and entropy balances to give consideration to how much energy can actually be used. For example, suppose you develop a device which will produce work as it removes energy from a heated block of metal. The metal is heated to 600 K, and work is produced as it cools. From a strictly first law perspective, you should be able to produce work until the metal cools to a temperature of 0 K. However, if we assume that the device is in an area where the temperature is 298 K, we know from the second law that the block of metal will only cool to 298 K. Therefore, there is only energy in the block available to produce work in this device as it cools from 600 K to 298 K: this is representative of the availability of the block of metal.

A second example of how not all energy can be used to perform useful work is in a piston–cylinder device, with gases expanding inside the device. The device is likely located in the atmosphere, so part of the work done by the gases is only pushing the piston against the atmospheric pressure: not all the work being done by the expanding gases can be transferred to some other process. Only the extra work done after providing energy to move the piston against the atmosphere can be employed to perform a useful task.

The second law tells us that we cannot recover all energy in a substance to do work, and an availability analysis can be used to assist with this analysis. Although you have acquired many of the tools to perform an availability analysis, the concept is not as commonly employed by engineers as the tools we have already developed. Therefore, more detailed discussion of availability analysis will be left for more advanced courses in thermodynamics.

Summary

In this chapter, we have explored the basics of the thermodynamic property entropy. We have learned how to find properties for isentropic processes, and we have learned how to perform entropy generation analyses and isentropic efficiency and other second law efficiency analyses or processes. At this point, you may still be unsure what entropy actually is. This is a fairly typical reaction from students just beginning to learn thermodynamics. The concept of entropy becomes clearer as you advance in your studies of thermodynamics. For example, in Chapter 1 we noted that the material we discuss in this text falls under the heading of classical thermodynamics, and we noted that there is another branch of thermodynamics known as statistical thermodynamics. Although the concept of entropy is one of the more difficult topics to understand in classical thermodynamics, it is straightforward in statistical thermodynamics. The most important thing you should glean from your study of this chapter is an understanding of how to use entropy and perform entropy analyses. It is natural to want a firm understanding of every concept you learn, but we urge you to not be too frustrated if you are struggling with understanding the complete nature of entropy. As we continue our discussions in this text, you will do well as long as you can properly apply the analysis tools that a study of entropy provides.

KEY EQUATIONS

Gibbs Equations:

$$TdS = dU + P d\forall \quad (6.18)$$

$$TdS = dH - \forall dP \quad (6.20)$$

Entropy Change—Ideal Gas with Constant Specific Heats:

$$s_2 - s_1 = c_v \ln \frac{T_2}{T_1} + R \ln \frac{v_2}{v_1} \tag{6.24}$$

$$s_2 - s_1 = c_p \ln \frac{T_2}{T_1} - R \ln \frac{P_2}{P_1} \tag{6.25}$$

Entropy Change—Ideal Gas with Variable Specific Heats:

$$s_2 - s_1 = s^o_{T_2} - s^o_{T_1} - R \ln \frac{P_2}{P_1} \tag{6.28}$$

Entropy Change—Incompressible Substances with Constant Specific Heats:

$$s_2 - s_1 = c \ln \frac{T_2}{T_1} \tag{6.31}$$

Entropy Rate Balance—Steady-State, Steady-Flow Open Systems with Multiple Inlets and Outlets and Isothermal Boundaries:

$$\dot{S}_{\text{gen}} = \sum_{\text{outlets}} \dot{m}_e s_e - \sum_{\text{inlets}} \dot{m}_i s_i - \sum_{j=1}^{n} \frac{\dot{Q}_j}{T_{b,j}} \tag{6.34}$$

Entropy Rate Balance—Steady-State, Steady-Flow Open Systems with a Single Inlet and a Single Outlet and Isothermal Boundaries:

$$\dot{S}_{\text{gen}} = \dot{m}(s_2 - s_1) - \sum_{j=1}^{n} \frac{\dot{Q}_j}{T_{b,j}} \tag{6.35}$$

Entropy Rate Balance—Closed System with Isothermal Boundaries:

$$S_{\text{gen}} = m(s_2 - s_1) - \sum_{j=1}^{n} \frac{Q_j}{T_{b,j}} \tag{6.38}$$

Isentropic Processes—Ideal Gas with Constant Specific Heats:

$$\frac{T_2}{T_1} = \left(\frac{P_2}{P_1}\right)^{\frac{k-1}{k}} \tag{6.40}$$

$$\frac{T_2}{T_1} = \left(\frac{v_1}{v_2}\right)^{k-1} \tag{6.41}$$

$$\frac{P_2}{P_1} = \left(\frac{v_1}{v_2}\right)^{k} \tag{6.42}$$

Isentropic Efficiency:

State 1 is the inlet, and State 2 is the outlet:

Turbine: $$\eta_{s,t} = \frac{h_1 - h_2}{h_1 - h_{2s}} \tag{6.44}$$

Compressor or Pump: $$\eta_{s,c} = \eta_{s,p} = \frac{h_1 - h_{2s}}{h_1 - h_2} \tag{6.45, 6.46}$$

Isentropic Compression of an Incompressible Substance with Constant Specific Heats:

$$h_2 - h_1 = v(P_2 - P_1) \tag{6.48}$$

PROBLEMS

If you have not already done so, you should modify your computer models of different devices and processes to accommodate entropy generation calculations. In addition, you should modify your models of turbines, compressors, and pumps to incorporate an isentropic efficiency.

6.1 3150 kJ of heat is transferred into a system which has an isothermal boundary temperature of 400 K. Determine the change in entropy of the system as a result of this heat transfer.

6.2 65 kJ of heat is added to a block of ice as it begins to melt at 0°C. Determine the change in entropy of the system resulting from this heat transfer.

6.3 2 kg of saturated water at 100°C initially has a quality of 0.75. The water is in a compressible container and is surrounded by cooler air. Heat is removed from the water until the quality is 0.50. Determine the amount of heat transfer that occurs and the change in entropy of the water.

6.4 125 kJ of heat is removed from steam as it condenses at 120°C. Determine the change in entropy of the system resulting from this heat transfer.

6.5 A large block of material whose temperature is 175°C floats on water. The bottom part of the material is immersed in water, which is 10°C, while the top of the material is in contact with the air, which is 25°C. The block is removed after transferring 50 kJ of heat to the water and 40 kJ of heat to the air. Determine the changes in entropy of the water, the air, and the block, assuming that all of the temperatures are maintained during the process.

6.6 A container of water is placed on an outdoor table in the early evening. The table has been sitting in the sun all day and has a temperature of 35°C. 8 kJ of heat is transferred from the table to the water. A cool breeze of air at 15°C blows over the other sides of the container, transferring 15 kJ of heat from the water to the air. Consider the water temperature to be 25°C. Determine the changes in entropy of the container of water, the air, and the table, assuming the water, air, and table temperatures are constant during the process.

6.7 Water at 5°C flows over the outside of a pump whose surface temperature is steady at 30°C. 5 kW of heat escapes from the pump to the water. Determine the rates of entropy change for the water and the pump as a result of this heat transfer.

6.8 Air flows through an uninsulated turbine whose surface temperature is steady at 60°C. 25 kW of heat escapes the turbine through the wall. Determine the rate of entropy loss of the air as a result of this heat transfer.

6.9 Cold water flows through a nozzle whose surface temperature is heated and held steady at 40°C. 1.6 kW of heat is added to the water as it flows through the nozzle. Determine the rate of entropy gain of the water as a result of the heat transfer.

6.10 An object undergoes a process involving two reversible heat transfers. During this process, 50 kJ of heat is added to the object across a surface temperature of 60°C, and 35 kJ of heat is removed across a surface temperature of 10°C. Determine the entropy change of the object during this process.

6.11 A reversible heat transfer of 150 kJ takes place to a steel chain whose surface temperature is constant. By measuring other properties, you determine that the increase in entropy of the chain is 0.460 kJ/K. What is the surface temperature of the chain?

6.12 Water is boiled in a steam power plant by transferring heat from combustion gases. The water is to boil (change from a saturated liquid to a saturated vapor) at a temperature of 300°C. The combustion gases are supplied at a temperature of 1400 K. If the rate at which the water is to boil is 15 kg/s, what is the rate of entropy change of the water, and what is the rate of entropy change of the combustion gases? Is the process reversible?

6.13 In an old boiler, water is heated by flowing through a heat exchanger through which pass combustion products at a temperature of 610 K. The water boils at a temperature of 120°C. The water flows through the boiler at a rate of 0.7 kg/s. Assume that the temperature of the combustion gases does not drop through the heat exchanger. What is the rate of entropy change of the water, and what is the rate of entropy change of the combustion gases? Is the process reversible?

6.14 Water is boiled in a nuclear steam power plant by transferring heat from the reactor core. The water is to boil (change from a saturated liquid to a saturated vapor) at a temperature of 300°C. The reactor core is maintained at 900 K. If the rate at which the water is to boil is 15 kg/s, what is the rate of entropy change of the water, and what is the rate of entropy change of the reactor core? Is the process reversible? How does this compare to the process in Problem 6.12?

6.15 Consider the boiler in Problem 6.13. Suppose the heat exchanger is upgraded so that the combustion gases are at 500 K. What is the rate of entropy change of the water, and what is the rate of entropy change of the combustion gases? Is this process "better" than the process described in Problem 6.13?

6.16 Determine the change in specific entropy for air as it is cooled from 200°C and 300 kPa to 50°C and 200 kPa (a) assuming constant specific heats, and (b) incorporating the variability of the specific heats.

6.17 Determine the change in specific entropy for air as it is heated from 25°C and 100 kPa to 300°C and 400 kPa (a) assuming constant specific heats, and (b) incorporating the variability of the specific heats.

6.18 Determine the change in specific entropy for air as it expands from 500 K and 500 kPa to 325 K and 100 kPa (a) assuming constant specific heats, and (b) incorporating the variability of the specific heats.

6.19 Determine the change in specific entropy for air as it is cooled from 150°C and 350 kPa absolute to 50°C and 140 kPa absolute (a) assuming constant specific heats, and (b) incorporating the variability of the specific heats.

6.20 Mercury is heated from 30°C to 150°C. Determine the change in specific entropy for the mercury in this process.

6.21 Iron is cooled from 300°C to 50°C. Determine the change in specific entropy for the iron in this process.

6.22 A copper electrical wire is heated from −10°C to 50°C as a current flows through it on a winter day. Determine the change in specific entropy for the copper wire in the process.

6.23 Carbon dioxide gas is isentropically expanded from 100°C and 500 kPa to 200 kPa. Determine the final temperature of the CO_2 and the amount of work per kg of CO_2 produced in the process if the process is adiabatic.

6.24 Nitrogen undergoes an isentropic compression from 25°C and 100 kPa to 800 kPa. Determine the final temperature of the nitrogen and the amount of work per kg of N_2 used in the process if it is adiabatic.

6.25 Oxygen undergoes an isentropic expansion from 120°C and 280 kPa absolute to 140 kPa absolute. Determine the final temperature of the oxygen and the amount of work per kg of O_2 produced in the process if it is adiabatic.

6.26 The compression ratio of an engine is the ratio of the maximum volume to the minimum volume of the cylinder as it goes through a cycle. Air enters the cylinder of an engine whose compression ratio is 9.5, while the cylinder is at its maximum volume. Initially, the air is at 40°C and 90 kPa. Compression occurs until the cylinder is at its minimum volume. Assuming that the compression is isentropic and that the air behaves as an ideal gas with constant specific heats, (a) determine the final temperature and pressure of the air after the compression process. (b) With the same assumptions, plot the final temperature and pressure for compression ratios ranging from 7 to 12. (c) If the air temperature cannot exceed 530°C, what is the maximum allowed compression ratio for the engine?

6.27 Consider Problem 6.26, but the engine is operating in cold weather. The air begins the compression process at 0°C and 92 kPa. Assuming that the compression is isentropic and that the air behaves as an ideal gas with constant specific heats, (a) determine the final temperature and pressure of the air after the compression process. (b) With the same assumptions, plot the final temperature and pressure for compression ratios ranging from 7 to 12. (c) If the air temperature cannot exceed 530°C, what is the maximum allowed compression ratio for the engine?

6.28 Steam flows through an isentropic turbine, exiting at 50 kPa with a quality of 0.90. What is the inlet temperature of the steam if the inlet pressure is (a) 1.0 MPa, (b) 5.0 MPa, and (c) 10.0 MPa?

6.29 Steam at 1000 kPa and 500°C is to expand through an isentropic turbine to a pressure of 100 kPa. Determine the final temperature and quality (if applicable) of the steam.

6.30 Steam enters an isentropic turbine at 400°C. The steam expands to a pressure of 50 kPa. Calculate and plot the exit temperature and quality (if applicable) for inlet steam pressures ranging between 500 kPa and 5000 MPa. What is the qualitative relationship between the exit steam quality and the inlet steam pressure?

6.31 Steam enters an isentropic turbine at 480°C and a pressure of 11 MPa absolute. Calculate and plot the exit quality for outlet steam pressures of 40 kPa absolute to 1.2 MPa absolute. What is the qualitative relationship between the exit steam pressure and the exit steam quality?

6.32 Steam enters an isentropic turbine at 5 MPa. The steam expands to a pressure of 50 kPa. Calculate and plot the exit temperature and quality (if applicable) for inlet steam temperatures ranging between 264°C (a saturated vapor) and 600°C. What is the qualitative relationship between the exit steam quality and the inlet steam temperature?

6.33 Steam enters an isentropic turbine at 5 MPa and 500°C. The steam expands through the turbine. Calculate and plot the exit temperature and quality (if applicable) for outlet steam pressures ranging between 50 kPa and 3 MPa. What is the qualitative relationship between the exit steam quality or exit temperature and the outlet steam pressure?

6.34 Suppose you are able to supply an isentropic turbine with steam at 500°C. The exit pressure is 25 kPa. What is the inlet pressure that would correspond to an outlet quality of (a) 0.8, (b) 0.9, and (c) 1.0?

6.35 Air expands isentropically in a piston–cylinder device from 800 K and 800 kPa to 400 kPa. Assume the air behaves as an ideal gas with constant specific heats. Determine the exit temperature and the work per unit mass used in the process.

6.36 Air is to be compressed isentropically in a piston–cylinder device from 300 K and 100 kPa to 1000 kPa. Assume the air behaves as an ideal gas with constant specific heats. Determine the exit temperature and the work per unit mass used in the process.

6.37 Air expands isentropically in a turbine from 800 K and 1000 kPa to a pressure of 150 kPa. The mass flow rate of the air is 2.0 kg/s. Assume the air behaves as an ideal gas with constant specific heats and that the turbine is adiabatic. Determine the exit temperature of the air and the power produced in the turbine.

6.38 Air is compressed isentropically in a compressor from 15°C and 101.3 kPa absolute to 800 kPa. Determine the exit temperature of the air and the work per unit mass used in the process (a) assuming the specific heats are constant, and (b) considering the variability of the specific heats.

6.39 (a) Steam expands in an adiabatic turbine from 6 MPa and 500°C to 100 kPa and 150°C. The mass flow rate of the steam is 8 kg/s. Determine the power produced and the rate of entropy generation for the turbine. (b) Plot the power produced and the rate of entropy generation from the turbine for the same inlet conditions and outlet pressure, but with outlet temperatures ranging from 100°C to 300°C.

6.40 (a) R-134a flows into an adiabatic compressor of a refrigerator at a rate of 0.5 kg/s. Inlet conditions are a saturated vapor at −4°C. The R-134a leaves the compressor at 1.0 MPa and 60°C. Determine the power used by the compressor and the rate of entropy generation for the compressor. (b) Plot the power consumed and the rate of entropy generation for the compressor for the same inlet conditions and the same outlet temperature, but for outlet pressures varying between 500 kPa and 1400 kPa. What does this indicate about the effect of irreversibilities on the outlet conditions for a compressor?

6.41 An adiabatic nozzle receives air at 20°C, 150 kPa, and 2 m/s. The air exits at 200 m/s and 105 kPa. (a) Determine the entropy generation per kg of air for the nozzle. (b) Plot the entropy generation per kg of air for exit pressures ranging from 80 kPa to 140 kPa. At what exit pressures is the operation of the nozzle impossible as described?

6.42 An adiabatic diffuser receives oxygen at −10°C, 101 kPa absolute, and 275 m/s. The oxygen exits at 3 m/s and 140 kPa absolute. Assuming constant specific heats, (a) determine the entropy generation per kg of oxygen for the diffuser. (b) Plot the entropy generation per kg of oxygen for exit pressures ranging from 125 kPa absolute to 350 kPa absolute, with the same exit velocity. At what exit pressure is the operation of the diffuser impossible as described?

6.43 R-134a enters an adiabatic throttling valve as a saturated liquid at 1200 kPa. The R-134a expands to 150 kPa. (a) What is the exit quality and the entropy generation per kg of R-134a for the process? (b) Plot the exit quality and entropy generation per kg of R-134a for exit pressures ranging between 100 kPa and 1000 kPa.

6.44 Air enters an adiabatic throttling valve at 425 K and 500 kPa. The mass flow rate of the air is 0.015 kg/s. Assume that the change in kinetic energy is negligible. Plot the entropy generation rate for exit pressures ranging from 100 kPa to 450 kPa.

6.45 An uninsulated gas turbine receives 10 kg/s of air at 2 MPa and 1000 K, and the air exits at 600 K and 150 kPa. 4 MW of power is produced by the turbine, whose surface is maintained at 110°C. Determine the entropy generation rate of the turbine. Is this process possible?

6.46 An uninsulated steam turbine receives 25 kg/s of steam at 10 MPa and 450°C and exhausts the steam at 500 kPa and a quality of 0.98. 12 MW of power is produced by the turbine. The surface temperature of the turbine is 80°C. (a) Determine the entropy generation rate for the given conditions. (b) Plot the entropy generation rate for the turbine for values of the power produced ranging between 9 MW and 13 MW.

6.47 An uninsulated steam turbine is to produce 6 MW of power from steam entering at a flow rate of 20 kg/s, a temperature of 400°C, and a pressure of 8 MPa while exiting as a saturated vapor at 3 MPa. The surface temperature of the turbine is maintained at 50°C. Calculate the entropy generation rate. Is this process possible? If not, what are some possible problems with this steam turbine concept?

6.48 An uninsulated steam turbine produces 25 MW of power from steam entering at 8.25 MPa absolute and 480°C, and exiting as a saturated vapor at 550 kPa absolute, at a flow rate of 30 kg/s. The surface temperature is maintained at 50°C. Calculate the entropy generation rate. Is this process possible? If not, what are some possible problems with this steam turbine concept?

6.49 An uninsulated air compressor is to receive 0.5 kg/s of air at 100 kPa and 300 K and compress the air to 620 K and 1000 kPa. The compressor loses 20 kW of heat to the surroundings, across a surface that is maintained at 300 K. (a) Determine the power consumed by the compressor and the entropy generation rate. Is the process possible? (b) Maintaining the same inlet conditions, exit pressure, surface temperature, and heat transfer rate, plot the power consumed and the entropy generation rate for exit temperatures ranging from 450 K to 700 K. At what exit temperatures is operation not possible? (c) Maintaining the same inlet and outlet conditions and surface temperature as in part (a), plot the power consumed and the entropy generation rate as the heat transfer varies from −100 kW to +50 kW. At what heat transfer rates is operation not possible?

6.50 You are to evaluate two possible steam turbines that are needed to produce 10 MW of power for a small factory.

Option A: An insulated turbine that receives steam at 1000 kPa and 300°C and has an exit pressure of 30 kPa and an exit quality of 0.98.

Option B: An uninsulated turbine that receives steam at 4000 kPa and 500°C, exhausts the steam at 50 kPa and a quality of 0.95, and has a mass flow rate of 11.2 kg/s and a surface temperature of 50°C.

By calculating the entropy generation for each turbine, determine which option has a lower degree of irreversibility.

6.51 You have two choices for an air compressor which will produce 2 kg/s of compressed air at 750 kPa, and you wish to evaluate which is the better choice thermodynamically (i.e., which has the lowest entropy generation rate). In both cases, the air to be compressed originally is at 20°C and 100 kPa. The two options are as follows.

Option A: An uninsulated compressor which has a surface temperature of 30°C and exhausts the air at 270°C, using 530 kW of power.

Option B: An insulated compressor which exhausts the air at 320°C.

By calculating the entropy generation for each compressor, determine which option has a lower degree of irreversibility.

6.52 You are to evaluate two possible steam turbines that are needed to produce 150 MW of power for a small power plant.

Option A: An insulated turbine that receives steam at 10,000 kPa and 400°C and has an exit pressure of 50 kPa and an exit quality of 0.84.

Option B: An uninsulated turbine that receives steam at 12,000 kPa and 500°C, exhausts the steam at 25 kPa and a quality of 0.88, and has a mass flow rate of 152 kg/s and a surface temperature of 75°C.

By calculating the entropy generation for each turbine, determine which option has a lower degree of irreversibility.

6.53 Liquid water enters a nozzle at 20°C, 250 kPa, and 2 m/s. The water exits at 18°C, 200 kPa, and 30 m/s. The mass flow rate of the water is 0.25 kg/s. The surface of the nozzle is maintained at 25°C. Determine the heat transfer rate and the rate of entropy generation for the nozzle. Is this process possible?

6.54 Air enters a nozzle at 80°C, 200 kPa, and 5 m/s, and exits at 78°C, 160 kPa, and 150 m/s. The mass flow rate of the air is 0.75 kg/s. The surface of the nozzle is maintained at 80°C. Determine the heat transfer rate and the rate of entropy generation for the nozzle. Is this process possible?

6.55 Steam enters an insulated nozzle at 1000 kPa and 300°C, at a velocity of 5 m/s. The mass flow rate of the steam is 2.5 kg/s. The steam exits at 250 m/s. Determine the exit temperature if (a) the nozzle is reversible and (b) the exit pressure is 300 kPa. (c) Plot the entropy generation rate for exit pressures ranging from 200 kPa to 900 kPa.

6.56 Air enters an adiabatic nozzle at 80°C, 300 kPa absolute, and 1.2 m/s. The air exits at 120 m/s. The mass flow rate of the air is 0.5 kg/s. The surface of the nozzle is maintained at 90°C. (a) Determine the exit pressure if the nozzle is reversible. (b) Plot the entropy generation rate for exit pressures ranging from 100 kPa absolute to 299 kPa absolute.

6.57 A diffuser receives air at 80 kPa and 100°C at a velocity of 150 m/s and a mass flow rate of 2.1 kg/s. The air exits the diffuser at 120 kPa and 135°C, at a velocity of 15 m/s. The surface of the diffuser is maintained at 25°C. Determine the entropy generation rate for the process. Is the process possible? If not, suggest some possible explanations.

6.58 An insulated diffuser receives steam at 300 kPa and 200°C at a velocity of 200 m/s. The steam is to leave the diffuser at a velocity of 30 m/s at a pressure of 315 kPa. Determine the exit temperature and the entropy generation per kg of steam for the process.

6.59 It is proposed to have a nozzle accelerate a flow of air from 3 m/s to 100 m/s. The air is to enter the nozzle at 500 kPa and 20°C and is to exit the nozzle at 300 kPa and 10°C. The surface of the nozzle is maintained at 0°C. By calculating the entropy generation rate (per kg of air), determine if this process is possible.

6.60 A nozzle is needed to accelerate steam flowing into a steam turbine. The steam turbine needs to receive the steam at 10 MPa, 500°C, and 250 m/s. An engineer proposes a nozzle to accomplish this by taking steam at 10.1 MPa, 520°C, and entering at a velocity of 2 m/s. The nozzle is to be uninsulated, with a surface temperature of 200°C. Will this nozzle provide the necessary inlet conditions for the turbine?

6.61 Air exiting a wind tunnel needs to be slowed down by flowing through a diffuser. The exit conditions for the diffuser need to be 100 kPa, 20°C, and 2 m/s. The air will enter the diffuser from conditions of 80 kPa, 15°C, and 100 m/s. An uninsulated diffuser with a surface temperature of 15°C will be employed. Will this diffuser be able to produce the required outlet conditions?

6.62 An insulated counter-flow heat exchanger is used to cool oil coming from a machine. The oil, which has a specific heat of 1.95 kJ/kg · K, enters at 150°C and exits at 50°C. The mass flow rate of the oil is 0.5 kg/s. It is cooled by air, which enters at 10°C and exits at 60°C. The pressure change of the air is negligible. Determine the required mass flow rate of the air and the entropy generation rate of the heat exchanger.

6.63 An insulated counter-flow heat exchanger is used to cool water coming from a manufacturing process. The water enters at 55°C and exits at 27°C, with a mass flow rate of 0.70 kg/s. The water is cooled by air, which enters at 24°C and flows with a rate of 4.1 kg/s. The pressure change of the air is negligible. Determine the exit temperature of the air and the entropy generation rate of the heat exchanger.

6.64 An insulated parallel-flow heat exchanger is used to transfer heat from steam to air. The steam enters at 500 kPa and 300°C and exits at 500 kPa and 160°C. The mass flow rate of the steam is 2 kg/s. The air enters at 100 kPa and 40°C and exits at 100 kPa and 200°C. Determine the required mass flow rate of the air and the entropy generation rate of the heat exchanger. Is this arrangement possible for a parallel-flow heat exchanger?

6.65 An uninsulated heat exchanger, whose surface is at 40°C, is used to cool engine coolant with air. The coolant, whose specific heat is 3.20 kJ/kg · K, enters the heat exchanger at 120°C and exits at 50°C. The mass flow rate of the coolant is 1.2 kg/s. The air enters at 100 kPa and 15°C and exits at 35°C at the same pressure. The mass flow rate of the air is 10 kg/s. Determine the entropy generation rate for this heat exchanger.

6.66 An insulated heat exchanger is proposed to be used to cool hot air with a flow of cool water. The air enters at 200 kPa and 100°C and is to exit at 195 kPa and 0°C. The mass flow rate of the air is 2.0 kg/s. The water enters the heat exchanger at 30°C and exits at 95°C. Calculate the entropy generation rate of the proposed heat exchanger. Is this heat exchange process possible?

6.67 Suppose the heat exchanger in Problem 6.66 is uninsulated and has a surface maintained at 5°C. Determine a rate of heat transfer to the surroundings and a mass flow rate of water for which the heat exchanger would work.

6.68 An insulated heat exchanger is used to cool oil with air. The oil has a mass flow rate of 0.5 kg/s, enters with a temperature of 60°C, and exits at 30°C. The specific heat of the oil is 1.82 kJ/kg · K. The air enters at 10°C and can have a flow rate up to 0.75 kg/s. Determine the maximum temperature that the air can exit at if the heat exchanger is (a) parallel-flow, and (b) counter-flow if the process is possible.

6.69 An insulated heat exchanger is to be used to condense steam exiting a turbine. The steam enters the heat exchanger at a mass flow rate of 30 kg/s, at a pressure of 60 kPa and a quality of 0.90, and exits the heat exchanger as a saturated liquid at 60 kPa. Liquid water is used to remove the heat. The liquid water enters at 10°C. Plot the mass flow rate of the liquid water and the entropy generation rate of the heat exchanger for cooling water exit temperatures ranging from 15°C to 85°C.

6.70 An insulated heat exchanger is used to condense steam coming from a turbine. The steam enters the heat exchanger at a mass flow rate of 68 kg/s, at a pressure of 35 kPa absolute and a quality of 0.85, and exits the heat exchanger as a saturated liquid at 35 kPa absolute. The heat is removed by liquid water, which enters at 20°C. Plot the exit temperature of the liquid water and the entropy generation rate of the heat exchanger for cooling water flow rates ranging from 700 kg/s to 2300 kg/s.

6.71 An insulated heat exchanger is to be used to heat air for a building with heat from combustion products. The air enters at 10°C and exits at 25°C and has a mass flow rate of 1.5 kg/s. The air enters the heat exchanger at 130 kPa and exits at 120 kPa. The combustion products, which can be modeled as air, enter at 250°C and 105 kPa and exit at 100 kPa.

Plot the required mass flow rate of the combustion products and the entropy generation rate for the heat exchanger for combustion product exit temperatures ranging from 240°C to 60°C.

6.72 Your manager has asked you to select a heat exchanger, based solely on thermodynamic considerations, to be used to condense steam from an industrial process. The steam has a mass flow rate of 4.5 kg/s and enters the heat exchanger as a superheated vapor at 200 kPa and 140°C. The steam is to exit at a pressure of 200 kPa, as a saturated liquid. Your two options for the heat exchanger are as follows.

Option A: An insulated heat exchanger which uses air. The air enters at 15°C and 120 kPa and exits at 110°C and 100 kPa.

Option B: An insulated heat exchanger which uses liquid water. The water enters at 20°C and exits at 50°C.

Calculate the entropy generation rate for each heat exchanger and choose the one which has the lower amount of irreversibility.

6.73 You need to select a heat exchanger, based solely on thermodynamic considerations, to be used to cool oil from an industrial process. The oil has a mass flow rate of 3.0 kg/s and enters the heat exchanger at 150°C. The oil needs to exit at 50°C and has a specific heat of 1.75 kJ/kg · K. Your two options for the heat exchanger are as follows.

Option A: An insulated cross-flow heat exchanger which uses N_2. The N_2 enters at 25°C and 140 kPa and exits at 45°C and 105 kPa.

Option B: An insulated counter-flow heat exchanger which uses liquid water. The water enters at 15°C and exits at 80°C, both at atmospheric pressure.

Calculate the entropy generation rate for each heat exchanger and choose the one which has the lower amount of irreversibility.

6.74 An insulated mixing chamber is used to heat liquid water to a saturated liquid by adding superheated steam. The liquid water enters at 1000 kPa, a temperature of 150°C, and a mass flow rate of 20 kg/s. The superheated steam enters at 1000 kPa and 340°C. The combined stream exits as a saturated liquid at 1000 kPa. (a) Determine the entropy generation rate for the mixing chamber. (b) Plot the entropy generation rate for the mixing chamber as a function of steam inlet temperature, for steam inlet temperatures ranging from 320°C to 500°C.

6.75 When supplying heated air for a building, one often chooses to mix in some fresh outside air with air that has been heated from the building as it passes through the furnace. An insulated mixing chamber is used to combine two streams of air to be used in a building. One stream of air, brought from the outside, enters at 2 kg/s, at a pressure of 120 kPa and a temperature of 5°C. The second stream of air, coming from the building's furnace, has a mass flow rate of 8 kg/s, a pressure of 120 kPa, and a temperature of 35°C. The combined stream is then delivered to the warm space at 120 kPa. Determine the rate of entropy generation for this mixing chamber.

6.76 1.5 kg of air is to be compressed in a piston–cylinder device from 100 kPa and 20°C to 500 kPa and 100°C. The process consumes 280 kJ of work. The surface of the cylinder is maintained at 20°C throughout the process. Determine the entropy generated during this process.

6.77 35 g of air is to be compressed in a piston–cylinder device from 105 kPa absolute and 10°C to 550 kPa absolute and 105°C, using 60 kJ of work. The surface of the cylinder is maintained at 10°C throughout the process. Determine the entropy generated during this process.

6.78 Steam, with a mass of 0.5 kg, is located in a piston–cylinder device whose surface is maintained at 70°C. The steam is initially at 500 kPa and 400°C. If the steam undergoes a constant-pressure process to a volume half of the initial volume, plot the final temperature and entropy generation for processes involving heat transfers ranging from −200 kJ to +200 kJ.

6.79 An insulated container contains nitrogen at a pressure of 200 kPa and a temperature of 20°C. Initially, a membrane divides the container into two halves, each with a volume of 0.25 m^3. The membrane is removed, and the nitrogen expands to fill the whole container. Determine the entropy generation for this process. Is this process possible?

6.80 An insulated container contains nitrogen at a pressure of 100 kPa and a temperature of 20°C. The volume of the container is 0.50 m^3. The nitrogen spontaneously flows into half of the container, and a membrane is inserted to hold it in a volume of 0.25 m^3, with the other 0.25 m^3 being evacuated. Determine the entropy generation for this process. Is this process possible?

6.81 An insulated aluminum bar is subjected to a 120-V electric current of 1.5 A for a period of 10 minutes. The mass of the aluminum bar is 5 kg, and the initial temperature is 5°C. Determine the entropy generated in the aluminum during this process.

6.82 An insulated copper bar, with a mass of 2.5 kg, is subjected to a 120-V electric current of 5 A for a period of 2 minutes. The copper is initially at 10°C. Determine the entropy generated in the copper during this process.

6.83 An iron spring with a mass of 1.5 kg and a spring constant of 1.5 kN · m stretches from a length of 0.25 m to a length of 1.0 m. The spring is initially at 15°C. Assuming no heat transfer, determine the entropy generated in the iron during the process. Explain the result.

6.84 An insulated balloon initially contains 0.5 kg of air at 550 kPa and 75°C. The air expands until the pressure is 400 kPa and the volume is 0.12 m^3. Determine the entropy generation for this process. Is the process possible?

6.85 An insulated rigid container is filled with 0.5 kg of liquid water at 20°C and 100 kPa. A stirring device inside the container adds 20 kJ of work to the water. Modeling the water as an incompressible substance, determine the entropy generated during the process.

6.86 A 5-kg block of lead is to be cooled from 200°C to 25°C by immersing it into a stream of flowing water which has a temperature of 20°C. Determine the entropy generation for this process.

6.87 It is desired to heat copper from 20°C to 150°C. Determine the minimum isothermal boundary temperature required for this process.

6.88 A 12-kg iron bolt is to be heat-treated by heating it from 10°C to 120°C. Determine the minimum isothermal boundary temperature required for this process.

6.89 An insulated gas turbine has air flowing through it at a rate of 5.0 kg/s. The air enters at 900 K and 1000 kPa and exits at 120 kPa. The isentropic efficiency of the turbine is 0.82. Determine the exit temperature and the power produced by the turbine (a) considering the specific heats of air to be constant (evaluated at 900 K), and (b) considering the specific heats of air to be variable.

6.90 Steam is to flow through an insulated turbine whose isentropic efficiency is 0.80. The steam enters at 2000 kPa and 400°C and has a mass flow rate of 15 kg/s. The exit pressure is 200 kPa. Determine the exit temperature (and quality, if applicable) and power produced by the turbine during this process.

6.91 Steam flows through an insulated turbine whose isentropic efficiency is 0.70. The steam enters as a saturated vapor with a pressure of 14 MPa and a mass flow rate of 70 kg/s. The exit pressure is 43 kPa absolute. Determine the exit temperature (and quality, if applicable) and power produced by the turbine during this process.

6.92 Steam enters an insulated turbine at 1000 kPa and 250°C and exits the turbine at 100 kPa. The turbine has an isentropic efficiency of 0.70. The mass flow rate of the steam is 25 kg/s. Determine the exit temperature (and quality, if applicable) and power produced by the turbine.

6.93 Steam enters an insulated turbine at 5000 kPa and 400°C and exits as a saturated vapor at 200 kPa. Determine the isentropic efficiency of the turbine.

6.94 An engineer proposes a turbine design which will have steam entering at 2000 kPa and 400°C and exiting at 200 kPa with a quality of 0.90. The turbine is to be insulated. Calculate the isentropic efficiency of this proposed steam turbine. Is the process possible?

6.95 Steam enters an insulated turbine at 15 MPa and 400°C at a mass flow rate of 30 kg/s. The steam exits at 50 kPa. Plot the exit quality of the steam, the power produced, and the entropy generation rate for isentropic turbine efficiencies ranging between 0.75 and 1.0.

6.96 It is desired to have steam exit an insulated turbine at 75 kPa with a quality of 0.95. The steam is to be supplied to the turbine at a temperature of 400°C. (a) Determine the minimum pressure at which the steam can be supplied for these conditions to be met. (b) Plot the isentropic efficiency of the turbine for inlet pressures ranging from the minimum pressure found in part (a) to a pressure of 5 MPa needed to obtain the desired outlet conditions.

6.97 An insulated steam turbine has an exit pressure of 20 kPa and must have an exit steam quality of 0.92. The steam is supplied to the turbine at 5 MPa. (a) Determine the maximum allowable inlet temperature for the steam. (b) Plot the isentropic efficiency of the turbine for inlet temperatures ranging from 400°C to the maximum allowable temperature in part (a).

6.98 A steam turbine with an isentropic efficiency of 0.85 exhausts steam at 100 kPa. (a) For an inlet steam temperature of 500°C, plot the exit quality (and temperature, if the outlet condition is superheated) for inlet pressures ranging from 1000 kPa to 10 MPa. (b) For an inlet steam pressure of 5 MPa, plot the exit quality (and temperature, if the outlet condition is superheated) for inlet temperatures ranging between 275°C and 600°C.

6.99 An insulated steam turbine receives steam at 425°C and 14 MPa and exhausts steam to 70 kPa absolute. The flow rate of the steam is 70 kg/s. Plot the power produced and the exit temperature and quality (if applicable) for turbine isentropic efficiencies ranging between 0.25 and 1.0.

6.100 Combustion products (modeled as air) enter an insulated turbine at 1000 kPa and 800 K. The mass flow rate of the gases is 12 kg/s. The gases exit the turbine at 100 kPa. Plot the exit temperature of the gases, the power produced by the turbine, and the entropy generation rate for isentropic turbine efficiencies between 0.70 and 1.0 (a) assuming constant specific heats for the gases, and (b) considering variable specific heats for the gases.

6.101 Air enters an insulated gas turbine at 700 kPa and 300°C. The air exits at 100 kPa and 200°C. Determine the isentropic efficiency of this gas turbine. Is this process possible?

6.102 Air enters an insulated compressor at 100 kPa and 20°C, at a mass flow rate of 1.5 kg/s. The air exits at 700 kPa and 290°C. Determine the isentropic efficiency of the compressor and the power consumed in the process.

6.103 R-134a enters a compressor of an insulated refrigeration unit as a saturated vapor at 150 kPa. The isentropic efficiency of the compressor is 0.75, and the R-134a exits at 500 kPa. Determine the exit temperature, the entropy generation rate, and the power consumed for a mass flow rate of 1.5 kg/s.

6.104 Ammonia enters a compressor of an insulated refrigerator as a saturated vapor at 310 kPa absolute. The ammonia exits at 66°C and 690 kPa absolute. The mass flow rate of the ammonia is 0.55 kg/s. Determine the isentropic efficiency of the compressor, the entropy generation rate, and the power consumed.

6.105 A young inventor wishes to design a compressor for a refrigerator which will take saturated vapor R-134a at 100 kPa and discharge it as a saturated vapor at 600 kPa. He wants to insulate the system as well. Calculate the isentropic efficiency of this proposed device. Is the device possible?

6.106 A heat pump is to use R-134a. It is proposed to use an insulated compressor that receives saturated vapor at 4°C and exhausts the R-134a at 500 kPa and 30°C. Calculate the isentropic efficiency of the proposed compressor. Is the device possible?

6.107 You are asked to consider replacing an old air compressor in a factory with a new, more efficient compressor, but you need to determine the energy savings that will result from switching. Currently, the factory needs 2 kg/s of compressed air. The air intake to the compressor is 100 kPa and 15°C. The current compressor produces compressed air at 400 kPa and a temperature of 240°C. You find a new compressor which will produce the compressed air at 400 kPa with an isentropic efficiency of 80%. How much less power does the new compressor use than the old compressor?

6.108 R-134a enters an insulated compressor as a saturated vapor at 200 kPa. The R-134a exits the compressor at 1000 kPa. The mass flow rate is 0.5 kg/s. Plot the exit temperature, the power consumed, and the entropy generation rate for isentropic compressor efficiencies ranging between 0.50 and 1.0.

6.109 R-134a enters an insulated compressor as a saturated vapor at 150 kPa absolute. The R-134a exits at 410 kPa absolute. The mass flow rate is 0.45 kg/s. Plot the isentropic compressor efficiency, the power consumed, and the entropy generation rate for exit temperatures ranging between 15°C and 65°C.

6.110 Air enters an insulated compressor at 100 kPa and 290 K. The air exits at 600 kPa, and the mass flow rate of the air is 1.25 kg/s. Plot the isentropic efficiency, the power consumed, and the entropy generation rate for exit temperatures ranging between 485 K and 650 K.

6.111 Oxygen gas is directed through an insulated compressor. The O_2 enters at 100 kPa and 30°C and exits at 800 kPa. Plot the isentropic efficiency and the exit temperature for entropy generation rates per kg of O_2 ranging between 0 kJ/kg · K and 1 kJ/kg · K.

6.112 Nitrogen gas flows through an insulated compressor at a rate of 0.25 kg/s. The N_2 enters at 100 kPa and −10°C and exits at 700 kPa. Plot the exit temperature and entropy generation rate for isentropic efficiencies ranging between 0.40 and 1.0.

6.113 Liquid water is directed through an isentropic pump. The water enters the pump at 20°C and 100 kPa and exits at 2 MPa. The mass flow rate is 20 kg/s. Determine the power consumed and the approximate exit temperature for the water.

6.114 An isentropic pump is used to pressurize liquid water. The water enters at 25°C and 100 kPa and has a mass flow rate of 15 kg/s. Plot the power consumed by the pump for exit pressures ranging between 500 kPa and 20 MPa.

6.115 An isentropic pump is used to pressurize liquid water. The water enters at 10°C and 105 kPa absolute, at a mass flow rate of 4.5 kg/s. Plot the power consumed by the pump for exit pressures ranging between 700 kPa absolute and 14 MPa absolute.

6.116 Liquid water is pumped at a rate of 5 kg/s. The water enters at 20°C and 100 kPa and exits at 5000 kPa. The pump uses 30 kW of work to accomplish this task. What is the isentropic efficiency of the pump?

6.117 Engine oil, with c = 1.97 kJ/kg · K and v = 0.00118 m³/kg, flows through a pump with an isentropic efficiency of 0.75. The mass flow rate of the oil is 0.50 kg/s. The oil enters at 30°C and 150 kPa and exits at 3000 kPa. Determine the power used by the pump and the approximate exit temperature of the oil.

6.118 Water enters an isentropic pump at 10°C and 120 kPa and exits at 2 MPa. The flow rate of the water is 8 kg/s. (a) If the isentropic efficiency of the pump is 0.65, determine the power consumed by the pump. (b) Plot the power consumed and the entropy generation rate for isentropic efficiencies ranging from 0.40 to 1.0.

6.119 You have an insulated pump which receives water at 10°C and 100 kPa and exhausts the water at 3000 kPa. You also know that the isentropic efficiency of the pump is 0.75. If the pump uses 15 kW of power, what is the mass flow rate of the water through the pump?

6.120 An insulated pump receives water at 10°C and 100 kPa absolute and exhausts the water at 3.5 MPa. The mass flow rate of the water is 2.7 kg/s, and the pump uses 12 kW of power. What is the isentropic efficiency of the pump?

6.121 An insulated pump is used to increase the pressure of liquid water at 15°C and 100 kPa to 5000 kPa. The flow rate of the water through the pump is 2 kg/s. Plot the isentropic efficiency, the exit temperature, and the entropy generation rate of the pump for pump powers ranging between −10 kW and −20 kW.

6.122 It is desired to use an insulated pump to increase the pressure of 3 kg/s of liquid water from 100 kPa to 4000 kPa. The water enters at a temperature of 10°C, and the pump uses 10 kW of power. Calculate the isentropic efficiency of the pump and the entropy generation rate of the pump. Is the process possible?

6.123 Consider a system that uses both an insulated compressor and an insulated turbine, with air as the working fluid. The air enters the compressor at 100 kPa and 10°C and is compressed to 1000 kPa. The air is then further heated, entering the turbine at 1000 kPa and 1000 K. The air expands through the turbine to 100 kPa. Consider the net power from the system to be equal to the sum of the compressor power and the turbine power. The mass flow rate of the air through the system is 10 kg/s. Determine the net power from the system, the compressor entropy generation rate, and the turbine entropy generation rate if (a) the turbine and the compressor are both isentropic, and (b) the isentropic efficiencies of the turbine and the compressor are both 0.80.

6.124 Consider a system that uses both an insulated pump and an insulated turbine, with water as the working fluid. The pump is used to increase the pressure of saturated liquid water at 50 kPa to a pressure of 5 MPa. The water is then boiled, and the turbine is used to produce power from steam expanding from a saturated vapor at 5 MPa to a pressure of 50 kPa. Consider the net power from the system to be equal to the sum of the pump and turbine powers. The mass flow rate of the water through the system is 25 kg/s. Determine the net power from the system and the entropy generation rates from the pump and turbine if (a) the pump and turbine are both isentropic, and (b) the isentropic efficiencies of the turbine and pump are both 0.80.

6.125 For the steam turbine in Problem 6.39(a), determine the rate of irreversibility for an ambient temperature of 298 K.

6.126 Plot the irreversibility per kg of R-134a for the conditions of the throttling valve outlined in Problem 6.43(b), with an ambient temperature of 25°C.

6.127 Calculate the rate of irreversibility for an ambient temperature of 25°C for the turbine in Problem 6.48.

6.128 Calculate the rate of irreversibility for the heat exchanger in Problem 6.66 and an ambient temperature of 298 K.

6.129 Calculate the amount of irreversibility for the process in Problem 6.79 and an ambient temperature of 298 K.

6.130 Calculate the amount of irreversibility for the lead cooling process described in Problem 6.86 and an ambient temperature of 25°C.

6.131 The steam flowing through a power plant receives heat at 600°C and rejects heat to the environment at 40°C. The thermal efficiency of the power plant cycle is 0.360. What is the second law efficiency of the power plant cycle?

6.132 The steam in a power plant receives heat at 666 K and rejects heat at 300 K. The thermal efficiency of the power plant cycle is 0.290. What is the second law efficiency of the power plant cycle?

6.133 A gas-turbine power plant receives heat at 1200 K and rejects heat at 700 K. The thermal efficiency of the power plant cycle is 0.240. What is the second law efficiency of the power plant cycle?

6.134 A household freezer (a refrigerator acting at colder temperatures) removes heat from a space at −15°C and rejects heat at 35°C. The coefficient of performance of the freezer is 2.75. What is the second law efficiency of the freezer?

6.135 A household refrigerator removes heat at a temperature of 2°C and rejects the heat at 40°C. The coefficient of performance of the refrigerator is 3.10. What is the second law efficiency of the refrigerator?

6.136 An air conditioning cycle removes heat at 4°C and rejects heat at 50°C. The coefficient of performance of the air conditioning cycle is 3.25. What is the second law efficiency of the air conditioner (which can be thought of as a refrigerator)?

6.137 A heat pump cycle receives heat at 10°C and rejects heat at 30°C. The coefficient of performance of the cycle is 7.50. What is the second law efficiency of the heat pump?

DESIGN/OPEN-ENDED PROBLEMS

6.138 You are to design an air compressor system that may contain a storage tank. The factory in which it is installed requires 0.025 m³/s (at standard atmospheric conditions) of air delivered at a gage pressure of 620 kPa. The air is to be cooled after compression so that its temperature leaving the cooler does not exceed 65°C. The isentropic efficiency of the compressor decreases as it is operated at reduced loads; assume that this relationship follows $\eta_c = 0.85 - (100\text{-}\%\ \text{load})/100 \times 0.85$ (i.e., the compressor operating at 80% load would have an isentropic efficiency of 0.68). The % load is defined as the actual power consumed by the air compressor divided by the maximum power consumed by the compressor. The air compressor receives air at 26°C and 100 kPa absolute to be compressed. Design a system that will satisfy these demands (locating a commercially available compressor).

6.139 You need to design a steam turbine that will deliver 150 MW of power. The turbine can receive steam that has a pressure no greater than 10,000 kPa, at a temperature no higher than 400°C. The outlet from the turbine can be no cooler than 30°C. If superheated steam exits the turbine, the turbine has an isentropic efficiency of 0.80; however, if a saturated mixture exits the turbine, the isentropic efficiency decreases with decreasing quality following the relationship $\eta_t = 0.80 - (1 - x_{out}) \times 0.5$. In addition, the cost of the steam turbine increases as the mass flow rate required to flow through the turbine increases: Cost = \$15,000,000 + \$40,000$\dot{m}$. Design and price a system that will meet the demands for the steam turbine.

6.140 Oil must be cooled in a heat exchanger from a temperature of 110°C to 30°C at constant pressure (101.3 kPa). The oil has a specific heat of 1.82 kJ/kg · K, and its density is 915 kg/m^3. The mass flow rate of the oil is 2.5 kg/s. Design an insulated concentric-tube heat exchanger that will accomplish this task. The velocity of either fluid may not exceed 8 m/s. You may choose the fluid to be used, and its inlet and outlet properties, and you should size the tubes (although the length can be ignored in this design). Consider several possibilities, and choose the design with the lowest entropy generation rate for the heat exchanger.

6.141 A condenser in a steam power plant is used to convert the steam exiting the turbine into a saturated liquid at the same pressure. Typically, this is done in a shell-and-tube configuration, where the steam passes through a shell, which contains many tubes carrying a cooling fluid. Design a condenser for a power plant in which steam is condensed from a quality of 0.90 to a saturated liquid at a pressure of 15 kPa. The cooling fluid flowing through the pipes is liquid water at 125 kPa. The water enters the condenser at 5°C. The mass flow rate of the steam is 150 kg/s. The cooling water cannot exceed a velocity of 10 m/s, and the diameter of any tube can be no larger than 8 cm. Your design should include the number of tubes used.

6.142 In a manufacturing process, it is necessary to supply a stream of water at a rate of 5.0 kg/s at a temperature of 70°C. To create this stream, you have several sources of water that can be mixed together: Source 1: Liquid water at 20°C; Source 2: Liquid water at 50°C; Source 3: Liquid water at 95°C; Source 4: Superheated steam at 120°C. The pressure of all the sources and the resulting mixed stream is 101.3 kPa. Design a system to mix appropriate streams in an insulated chamber to accomplish this task, and try to minimize the entropy generation rate of the mixing process.

6.143 You need to develop a system that will supply water to a shower at 40°C at a rate of 1.5 kg/s. You have available to you an insulated water heater that heats water to up to 60°C from a source that is maintained at 125°C, a water heater that provides water steadily at 35°C that is heated to that temperature from a source maintained at 45°C, and a cold water stream at 15°C that is also used as a supply line for the two water heaters. Design a system to mix appropriate streams in an insulated chamber to give the required output, and try to minimize the entropy generation of the water heating and mixing processes.

6.144 An inventor wishes to create a system that will utilize pumped hydropower to provide a continuous amount of power to a building. He plans to put water in a storage tank 30 m above the building and use a turbine to create power from the water as it falls through a pipe to the ground from the tower. He will use the power created to pump the water back to the tank, as well as to provide enough power to light the building. There will be no other power inputs. Explain to the inventor why his proposed system will not run forever.

6.145 Discuss why each of these processes cannot happen, and reinforce your discussion with appropriate calculations: (a) a substance being throttled from a lower pressure to a higher pressure, (b) a gas spontaneously contracting into a volume half its initial size (leaving a vacuum in the other half), and (c) boiling water poured into liquid water at the freezing point and producing superheated steam and solid ice.

6.146 An inventor comes to you with a proposal. He proposes that he will create an engine that runs forever using only a small tank of hydrogen as a fuel source. The hydrogen will burn with the oxygen in air and produce water. The engine's power will operate a generator which will produce electricity, and the electricity will be used in an electrolysis process to convert the water to hydrogen and oxygen. The hydrogen will be captured and returned to the engine to be burned. Explain to the inventor why his proposed system will not run forever.

CHAPTER 7

Power Cycles

Learning Objectives

Upon completion of Chapter 7, you will be able to

7.1 Explain the role of power cycles in the modern world;

7.2 Apply cycle analysis techniques to the Rankine vapor power cycle and variations of the Rankine cycle, including using additional components;

7.3 Analyze the Brayton power cycle, along with possible modifications to the Brayton cycle;

7.4 Examine the Otto, Diesel, and Dual cycles, and compare the predictions from these cycles to the actual behavior of practical internal combustion engines; and

7.5 Model how variations in properties within the cycles affect the performance of the cycles.

7.1 INTRODUCTION

Many thermodynamic applications involve a series of processes resulting in a thermodynamic cycle. Recall from Chapter 1 that a thermodynamic cycle is defined as a series of processes for which the beginning of the first process and the end of the last process are identical, so that the working fluid can continually proceed through the series of processes. One subset of thermodynamic cycles is *power cycles*. The purpose of a power cycle is to generate work or power after the working fluid receives energy in the form of heat. The power is produced by the working fluid flowing through a prime mover. Examples of prime movers are turbines and piston-cylinder devices (such as in reciprocating internal combustion engines).

In many ways, thermodynamic power cycles are at the heart of modern society's way of life. Most of the electricity in use in the world is produced in thermal power plants that use a power cycle to produce the energy that will spin the shaft of the electric generator, which, in turn, produces the electricity. Vehicles that use internal-combustion engines, such as automobiles, trucks, and buses, rely on a power cycle incorporated into the engine to produce the power that moves the people and goods inside the vehicles. Airplanes are propelled through the air either using gas turbines or internal-combustion engines. Some electricity, such as that produced via wind turbines, photovoltaic solar panels, and hydroelectric power, does not come from power cycles; however, such means currently produce less than 20% of the world's electricity. If most of the electricity and most of the engine-based transportation modes in the world were eliminated because power cycles did not exist, modern society would have a much different appearance. The world would likely look much like it did in the 18th century rather than how it does in the 21st century. The contrast is illustrated in **Figures 7.1** and **7.2**.

Power cycles use working fluids that either remain as a gas throughout the entire process or undergo a phase change between a liquid and a vapor. The former power cycles are considered "gas power cycles" and the latter are considered "vapor power cycles." Although any gas can be used in a gas power cycle, the gas used is typically air, or perhaps

stockelements/Shutterstock.com

FIGURE 7.1 Modern society relies on electricity and motorized vehicles. Power cycles are fundamental to electricity production and combustion engine operation.

Ksenia Ragozina/Shutterstock.com

FIGURE 7.2 Life before modernization would have used technologies that did not depend on power cycles.

combustion products (frequently modeled as air). Meanwhile, water is the dominant working fluid used in vapor power cycles, although certain systems can use other fluids with different boiling/condensation properties for specialized applications. For example, a lower temperature heat source (such as some geothermal applications) may be better suited for power generation if a refrigerant is used as the working fluid.

In this chapter, we will explore the means of analyzing many contemporary thermodynamic power cycles. With the tools that you will learn, you will be able to apply thermodynamic concepts to other cycles not explicitly covered in this chapter, and perhaps even yet to be developed. You should realize by the end of the chapter that the basics of cycle analysis are straightforward, but complexities arise as cycles become more elaborate; however, the principles used in basic cycle analysis are the same as those used in more complicated cycles.

QUESTION FOR THOUGHT/DISCUSSION

Many power cycles rely on the combustion of fossil fuels. What are some ongoing efforts to produce electricity and facilitate transportation that do not use power cycles fueled by fossil fuel combustion, and how likely is it that these will become the dominant form of electricity production or transportation power in the near future?

7.2 THE IDEAL CARNOT POWER CYCLE

Based on the information presented in the previous section, you should understand that power cycles are a form of a heat engine, and, as such, the maximum possible efficiency of the power cycle can be found from modeling it as a Carnot heat engine. The specific application of this concept to power cycles is called the Carnot power cycle. The Carnot power cycle consists of four reversible processes:

Process 1-2: Isothermal heat addition
Process 2-3: Isentropic expansion
Process 3-4: Isothermal heat rejection
Process 4-1: Isentropic compression

The resulting *T-s* diagram for the Carnot power cycle is a rectangle. For a vapor Carnot power cycle, the *T-s* diagram is taken typically as falling below the vapor dome, as shown in **Figure 7.3a**. For a gas Carnot power cycle, the rectangle is completely in the gaseous phase, as shown in **Figure 7.3b**.

Recall that for a thermodynamic cycle,

$$W_{net} = Q_{net} \tag{7.1}$$

Heat is transferred in processes 1-2 and 3-4. Upon integrating Eq. (6.5) for the processes in the Carnot power cycle,

$$Q_{12} = mT_H(s_2 - s_1) \quad \text{and} \quad Q_{34} = mT_C(s_4 - s_3) \tag{7.2a, b}$$

where T_H is the temperature of process 1-2, and T_C is the temperature of process 3-4. However, because of the two isentropic processes, Eq. (7.2b) can be rewritten as

$$Q_{34} = mT_C(s_1 - s_2) \tag{7.2c}$$

The net heat transfer is equal to the sum of Q_{12} and Q_{34}:

$$Q_{net} = Q_{12} + Q_{34} = mT_H(s_2 - s_1) + mT_C(s_1 - s_2) = m(T_H - T_C)(s_2 - s_1) \tag{7.3}$$

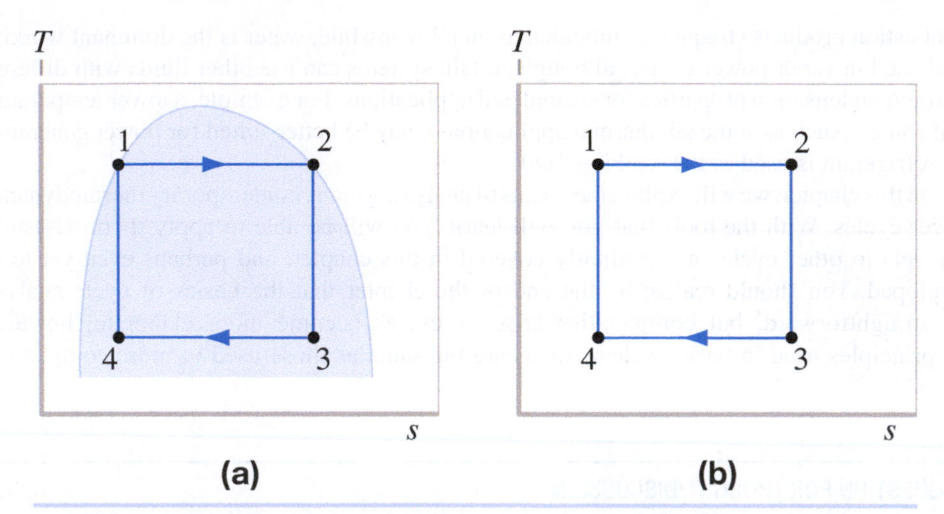

FIGURE 7.3 *T-s* diagrams illustrating the processes in (a) a vapor Carnot power cycle, and (b) a gas Carnot power cycle.

Setting Eq. (7.3) equal to the net work, as described in Eq. (7.1), the thermal efficiency of a Carnot power cycle can be found from Eq. (4.34) as

$$\eta = \frac{W_{net}}{Q_H} = \frac{m(T_H - T_C)(s_2 - s_1)}{mT_H(s_2 - s_1)} = \frac{T_H - T_C}{T_H} \tag{7.4}$$

as $Q_H = Q_{12}$. Equation (7.4) quickly reduces to the expression found for the maximum thermal efficiency possible through a Carnot heat engine in Eq. (5.3) as

$$\eta = 1 - \frac{T_C}{T_H} \quad \text{(Thermal efficiency of the Carnot power cycle)} \tag{7.5}$$

Because the Carnot power cycle represents the reversible (ideal) form of any power cycle, the maximum possible efficiency of any power cycle operating between a maximum temperature and a minimum temperature is

$$\eta_{max} = 1 - \frac{T_C}{T_H} \tag{7.6}$$

where T_C is the coldest temperature in the cycle and T_H is the hottest temperature. This expression works for both vapor power cycles and gas power cycles.

▶ EXAMPLE 7.1

Air flows through a gas Carnot power cycle. The air is heated to a temperature of 600°C in the heat addition process, and is cooled to 200°C in the heat rejection process. What is the thermal efficiency of this cycle?

Given: $T_H = 600°C = 873$ K $\quad T_C = 200°C = 473$ K

Find: η

Solution: The cycle under consideration is a Carnot power cycle, and the working fluid is a generic gas.

The thermal efficiency of any Carnot power cycle is found using Eq. (7.5):

$$\eta = 1 - \frac{T_C}{T_H}$$

as a Carnot power cycle's actual thermal efficiency is equal to the maximum possible efficiency for a power cycle operating between two temperatures.

So,

$$\eta = 1 - \frac{T_C}{T_H} = 1 - \frac{473\ \text{K}}{873\ \text{K}} = \mathbf{0.458 = 45.8\%}$$

Analysis: Remember to use absolute temperatures in this calculation, as one temperature is being divided by another temperature.

In the preceding discussion, the thermal efficiency expression used Q_H to represent the heat input. Because there was only one heat input, this was appropriate. However, some of the power cycles to be considered will have multiple heat inputs. Therefore, for the rest of this discussion, we will use Q_{in} to represent the heat input to a cycle, such that the thermal efficiency expression becomes

$$\eta = \frac{W_{net}}{Q_{in}} = \frac{\dot{W}_{net}}{\dot{Q}_{in}}$$

It can also be noted that the expressions are appropriate for either a set mass flowing through the cycle once or for a mass flow rate.

Although conventionally we associate the maximum and minimum temperatures of the cycle with T_H and T_C, for a Carnot heat engine we are actually concerned with the temperatures of the thermal reservoir. For a power cycle, the high-temperature reservoir must have a temperature equal to or greater than the maximum temperature in the cycle, and the low-temperature reservoir temperature must have a temperature equal to or less than the minimum temperature in the cycle. We could find an alternative maximum possible efficiency for a cycle using these reservoir temperatures, and this efficiency would be greater than that found using the maximum and minimum temperatures in the system. However, material or heat transfer device limitations may make it impossible for the working fluid to reach these reservoir temperatures, and so using the maximum and minimum temperatures of the fluid in the cycle is still an appropriate tool for determining the maximum possible efficiency of a particular cycle.

7.3 THE RANKINE CYCLE

The primary vapor power cycle in thermodynamics is the Rankine cycle, named for William Rankine, a 19th-century Scottish engineer who developed a thermodynamic cycle that described a steam engine containing an external condenser. This cycle, often in modified form, is the basis for the majority of electricity generation in the United States today. Most coal-fired power plants, some natural gas-fired power plants, and most nuclear power plants (as well as some facilities using other fuels) use a form of the Rankine cycle to generate the mechanical power transmitted to an electrical generator to produce electricity. In 2019, approximately 60% of the electricity generated in the United States came from these steam power plants. The Rankine cycle can also be used in a cogeneration system, where both electricity and process steam (for manufacturing or heating purposes) are simultaneously developed.

Water (steam) is the most common working fluid used in the Rankine cycle. Other substances that undergo phase changes at appropriate temperatures and pressures can also be used, but a fundamental component of the Rankine cycle is the phase change between a liquid and a vapor. Because it is the dominant fluid used in Rankine cycle systems, we will often use the terms *water* or *steam* in reference to the working fluid of the Rankine cycle; remember, though, that other fluids undergoing liquid–vapor phase changes are acceptable for use. To study the Rankine cycle, we will begin with the ideal, basic Rankine cycle. Then we will introduce nonideal processes into the analysis. We will then look at three common modifications of the basic Rankine cycle that will improve the cycle's thermal efficiency.

7.3.1 Ideal, Basic Rankine Cycle

The ideal, basic Rankine cycle consists of four processes, which are performed in four pieces of equipment: a pump, a steam generator, a turbine, and a condenser. The layout of the cycle hardware is shown in **Figure 7.4**. The steam generator in the basic Rankine cycle consists of an economizer section (which raises the temperature of the liquid water entering the steam generator to that of a saturated liquid at the pressure of the steam generator) and a boiler section (which causes the water to evaporate from a saturated liquid to a saturated vapor). A schematic diagram of a simple steam generator is shown in **Figure 7.5**. The four processes in the ideal, basic Rankine cycle are as follows:

Process 1-2: Isentropic compression
Process 2-3: Constant-pressure heat addition
Process 3-4: Isentropic expansion
Process 4-1: Constant-pressure heat removal.

These four processes are shown on a *T-s* diagram in **Figure 7.6**. For the basic Rankine cycle, two additional requirements are employed. First, the steam exiting the steam generator (state 3) is a saturated vapor at a subcritical pressure ($x_3 = 1$). The design of the steam generator is such that the water will exit the device as a saturated vapor in the basic Rankine cycle. Second, the water exiting the condenser is assumed to be a saturated liquid ($x_1 = 0$). The condenser is essentially a shell-and-tube heat exchanger, as shown in **Figure 7.7**. Although the condenser

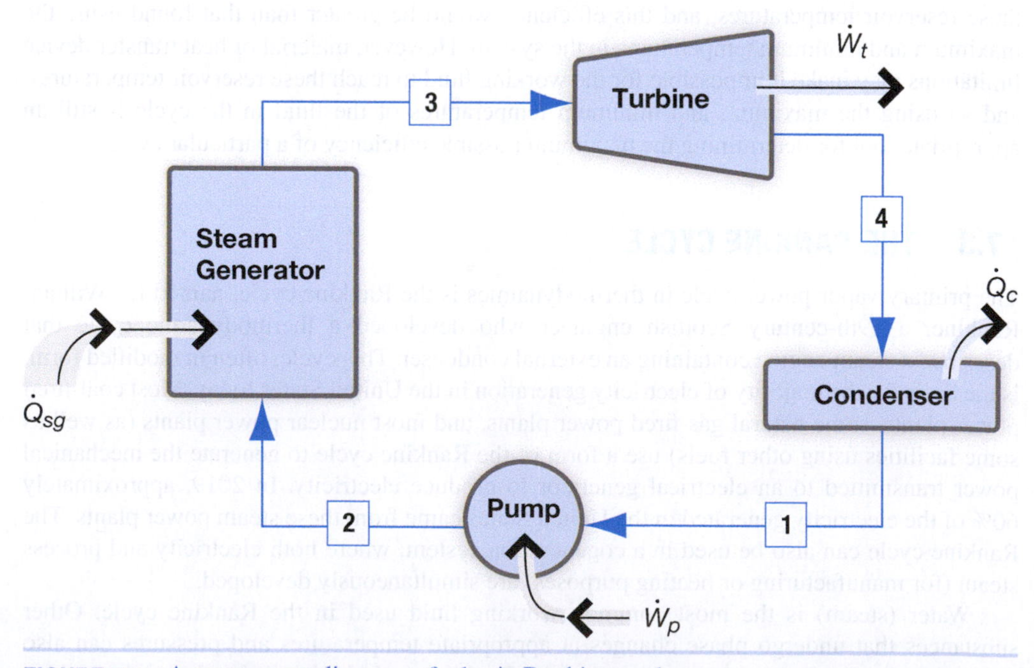

FIGURE 7.4 A component diagram of a basic Rankine cycle.

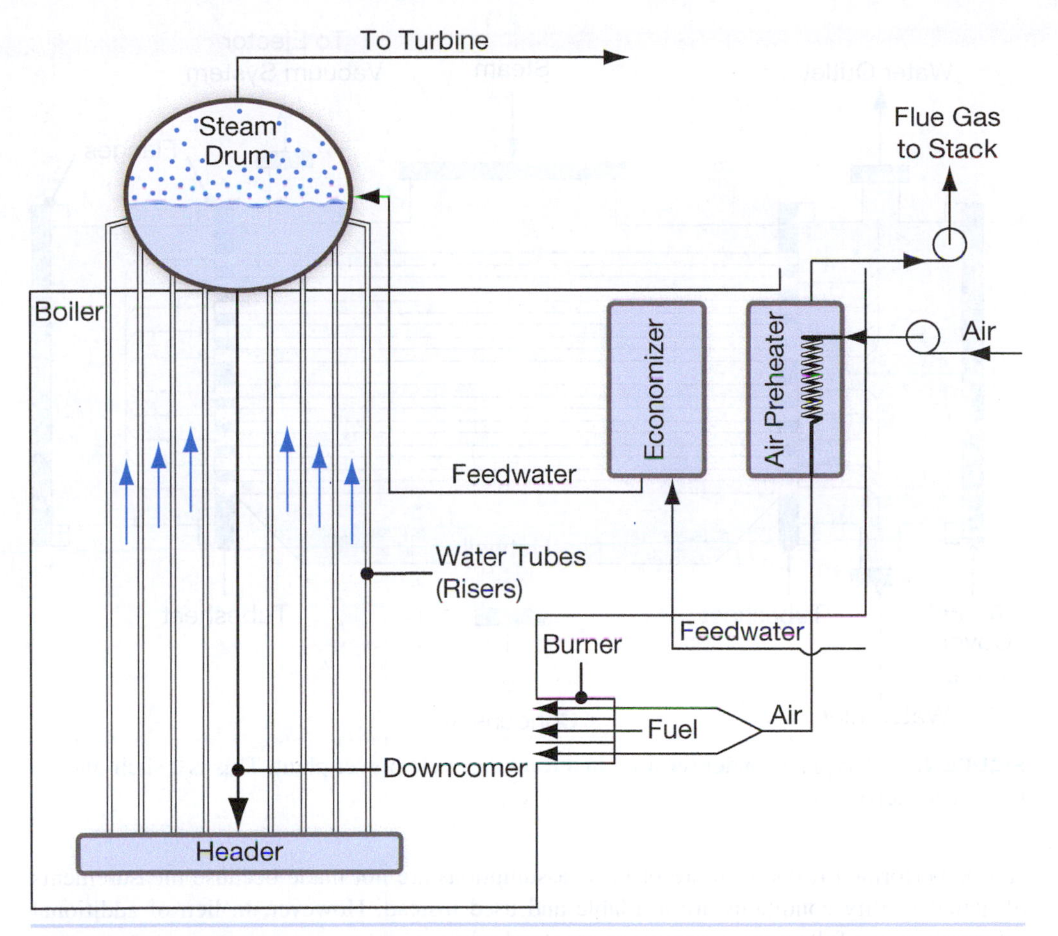

FIGURE 7.5 An illustration of a steam drum-type steam generator used in many combustion-based Rankine cycle power plants.

design could allow additional cooling of the liquid water below the saturation temperature, the water exiting the condenser is typically assumed to be a saturated liquid at the condenser pressure unless additional information is known.

In the Rankine cycle, the turbine and pump can each be treated as shaft work machines, as discussed in Chapter 4. The steam generator and the condenser are both heat exchangers. However, at this time, we are less concerned with analyzing the entire device and more concerned with considering how much heat is added to the cycle in the steam generator, and how much heat is removed from the cycle during the steam condensation in the condenser. This information can later be used to determine how much fuel may be needed to provide the heat in the steam generator, or how much external cooling water (or cooling air, or whatever external fluid is used to remove heat from the system) is needed in the condenser. Figure 7.8 contains a diagram of how the Rankine cycle components fit into an entire steam power plant scheme, with fuel being added for combustion, with an electric generator, and with external cooling water being used to remove heat to the environment through a cooling tower (in many power plants, a natural body of water is used to remove the heat from the cycle).

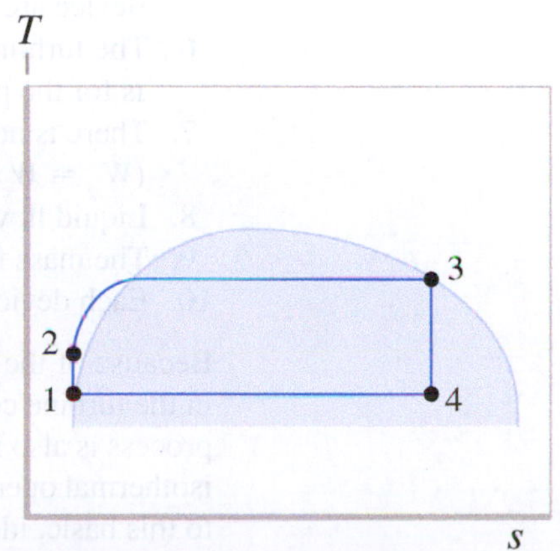

FIGURE 7.6 The *T-s* diagram for the ideal, basic Rankine cycle.

In conducting a thermodynamic analysis of the ideal, basic Rankine cycle, several assumptions are commonly made. The same sort of analysis

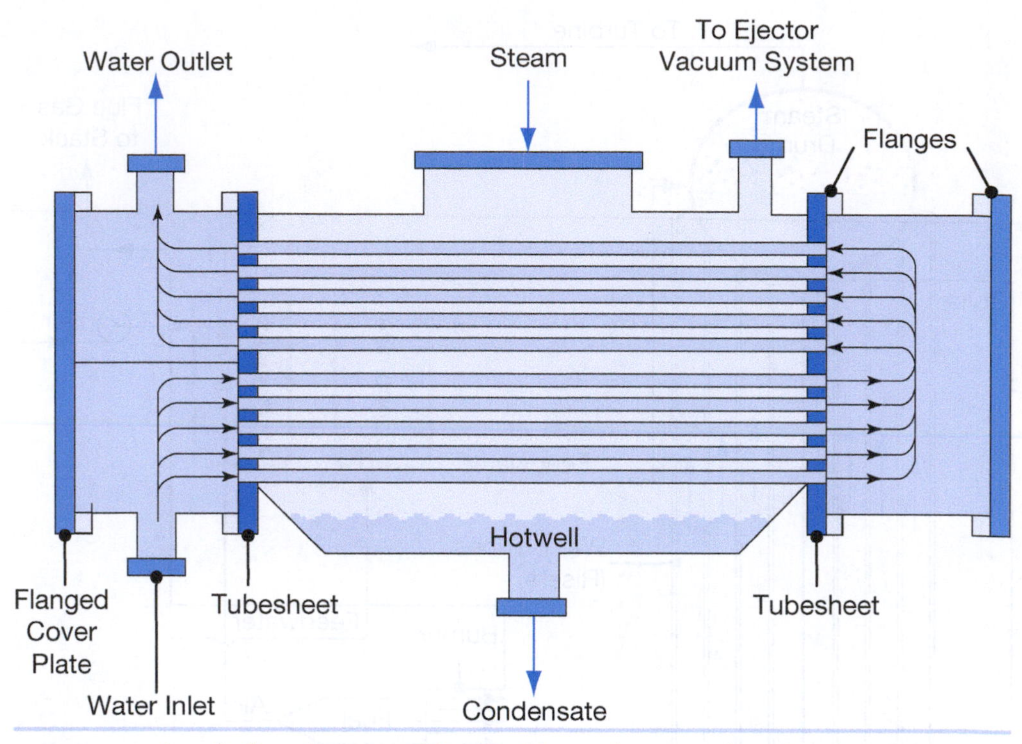

FIGURE 7.7 A typical condenser used in a Rankine cycle power plant. This is a shell-and-tube heat exchanger.

can be performed if one or more of these assumptions are not made because measurements of actual facility conditions are available and used instead. However, in lieu of additional information, the following assumptions are standard:

1. $s_1 = s_2$ (By definition, the pump is isentropic in the ideal cycle.)
2. $P_2 = P_3$ (The heat addition is considered to be at constant pressure.)
3. $s_3 = s_4$ (By definition, the turbine is isentropic in the ideal cycle.)
4. $P_4 = P_1$ (The heat removal is considered to be at constant pressure.)
5. The changes in kinetic energy and potential energy of the working fluid through each device are negligible.
6. The turbine and pump are both adiabatic. ($\dot{Q}_t = \dot{Q}_p = 0$, where t is for the turbine and p is for the pump.)
7. There is no power used or produced in the steam generator or condenser. ($\dot{W}_{sg} = \dot{W}_c = 0$, where sg is for the steam generator and c is for the condenser.)
8. Liquid flowing through the pump is incompressible.
9. The mass flow rate is constant through each device.
10. Each device operates as a steady-state, steady-flow device.

Because of the constant-pressure assumption in the condenser and the isentropic assumption in the turbine combined with the inlet to the turbine being a saturated vapor, the condensation process is also isothermal for the ideal, basic Rankine cycle, as can be seen in Figure 7.6. This isothermal operation of the heat removal process may not be true once modifications are made to this basic, ideal Rankine cycle, so you should become more accustomed to thinking of it as a constant-pressure process rather than a constant-temperature process.

We will model each device as a steady-state, steady-flow, single-inlet, single-outlet open system. In Chapter 4, such an analysis was common for turbines and pumps. However, it was

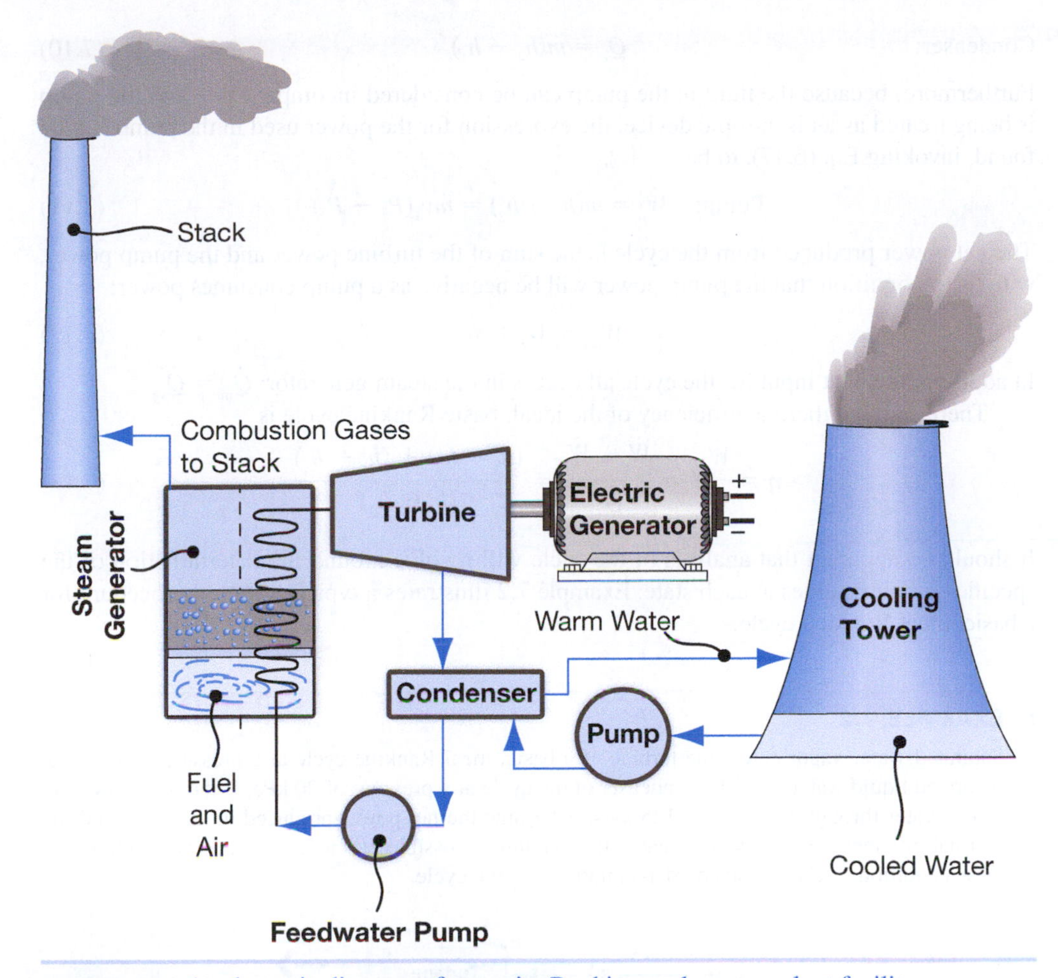

FIGURE 7.8 A schematic diagram of an entire Rankine cycle power plant facility, including the cycle itself, a combustion process, and a cooling tower for rejecting heat to the environment.

not what was used for heat exchangers, such as the steam generator and the condenser. In this application, we are not concerned with the overall analysis of the steam generator and condenser; we are only concerned with the heat added to or removed from the water. Therefore, it is more appropriate to treat the analysis of these devices as one inlet and one outlet for the water/steam, and with heat either coming from some source or being rejected to some sink.

As developed in Chapter 4, the first law of thermodynamics for a steady-state, steady-flow, single-inlet, single-outlet device is

$$\dot{Q} - \dot{W} = \dot{m}\left(h_{out} - h_{in} + \frac{V_{out}^2 - V_{in}^2}{2} + g(z_{out} - z_{in})\right) \tag{4.15a}$$

The assumptions made previously for the Rankine cycle can be applied to Eq. (4.15a) to provide a reduced form of the first law for each device.

Pump: $$\dot{W}_p = \dot{m}(h_1 - h_2) \tag{7.7}$$

Steam generator: $$\dot{Q}_{sg} = \dot{m}(h_3 - h_2) \tag{7.8}$$

Turbine: $$\dot{W}_t = \dot{m}(h_3 - h_4) \tag{7.9}$$

Condenser: $$\dot{Q}_c = \dot{m}(h_4 - h_1) \tag{7.10}$$

Furthermore, because the fluid in the pump can be considered incompressible and the pump is being treated as an isentropic device, the expression for the power used in the pump can be found, invoking Eq. (6.47), to be

$$\text{Pump:} \quad \dot{W}_p = \dot{m}(h_1 - h_2) = \dot{m}v_1(P_1 - P_2) \tag{7.11}$$

The net power produced from the cycle is the sum of the turbine power and the pump power, with the recognition that the pump power will be negative as a pump consumes power:

$$\dot{W}_{\text{net}} = \dot{W}_t + \dot{W}_p \tag{7.12}$$

In addition, the heat input for the cycle all occurs in the steam generator: $\dot{Q}_{\text{in}} = \dot{Q}_{sg}$.

Therefore, the thermal efficiency of the ideal, basic Rankine cycle is

$$\eta = \frac{\dot{W}_{\text{net}}}{\dot{Q}_{\text{in}}} = \frac{\dot{W}_t + \dot{W}_p}{\dot{Q}_{sg}} = \frac{(h_3 - h_4) + (h_1 - h_2)}{(h_3 - h_2)} \tag{7.13}$$

It should be apparent that analysis of the cycle will revolve around the determination of the specific enthalpy values at each state. Example 7.2 illustrates a typical solution procedure for a basic, ideal Rankine cycle.

▶ EXAMPLE 7.2

Saturated water vapor enters the turbine of a basic, ideal Rankine cycle at a pressure of 10 MPa. Saturated liquid water exits the condenser of the cycle at a pressure of 20 kPa. If the mass flow rate of the water through the cycle is 125 kg/s, determine the net power produced in the cycle and the thermal efficiency of the cycle. What is the maximum possible efficiency for a power cycle under these conditions? Use the standard assumptions for the cycle.

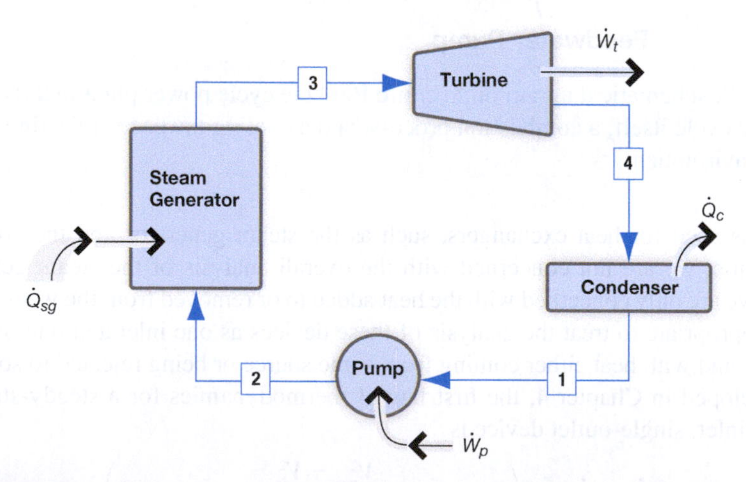

Given: $x_1 = 0$, $P_1 = 20$ kPa, $x_3 = 1$, $P_3 = 10$ MPa, $\dot{m} = 125$ kg/s

Find: $\dot{W}$, η, η_{max}

Solution: The cycle under consideration is a basic, ideal Rankine cycle. The working fluid is water, which exists in both saturated and compressed liquid form in the cycle.

Assume: $P_2 = P_3$, $P_1 = P_4$, $s_1 = s_2$, $s_3 = s_4$, $\dot{W}_{sg} = \dot{W}_c = \dot{Q}_t = \dot{Q}_p = 0$, $\Delta KE = \Delta PE = 0$ for each device. Steady-state, steady-flow systems. Liquid water in pump is incompressible.

The first step is to find the specific enthalpy values of each state, which will be done here by using a property calculator program.

For state 3, use the given conditions.

State 3: $P_3 = 10$ MPa, $x_3 = 1.0$ $h_3 = 2724.7$ kJ/kg

$s_3 = 5.6141$ kJ/kg · K

For state 4, use the pressure from state 1 and the entropy from state 3, because those values correspond to state 4's values for pressure and entropy, from the assumptions.

State 4: $P_4 = P_1 = 20$ kPa $s_4 = s_3 = 5.6141$ kJ/kg · K

$x_4 = (s_4 - s_f)/(s_g - s_f) = 0.6758$

$h_4 = x_4 h_{fg} + h_f = 1845.1$ kJ/kg

For state 1, use the given conditions. However, we will find the specific volume at state 1, rather than the specific entropy, because the specific volume will be used in the calculation of the specific enthalpy of state 2.

State 1: $P_1 = 20$ kPa, $x_1 = 0.0$ $h_1 = 251.40$ kJ/kg

$v_1 = 0.0010172$ m³/kg

For state 2, use the relationship presented in Eq. (6.47) for isentropic flow of an incompressible substance.

State 2: $P_2 = 10$ MPa $= 10{,}000$ kPa $h_2 = h_1 + v_1(P_2 - P_1) = 261.6$ kJ/kg

With the specific enthalpy values known, Eqs. (7.7)–(7.10) can be used to find any needed power values or heat transfer rate values.

For the turbine power: $\dot{W}_t = \dot{m}(h_3 - h_4) = 109{,}950$ kW

For the pump power: $\dot{W}_p = \dot{m}(h_1 - h_2) = -1275$ kW

For the heat transfer rate of the steam generator: $\dot{Q}_{sg} = \dot{m}(h_3 - h_2) = 307{,}900$ kW

Therefore, the net power is $\dot{W}_{net} = \dot{W}_t + \dot{W}_p = 108{,}700$ kW $=$ **109 MW**

and the thermal efficiency is $\eta = \dfrac{\dot{W}_{net}}{\dot{Q}_{in}} = \dfrac{\dot{W}_t + \dot{W}_p}{\dot{Q}_{sg}} = \mathbf{0.353.}$

The maximum possible efficiency can be found by using the maximum and minimum temperatures in the system. These are the temperatures at states 3 and 1, respectively, which are equal to

$$T_H = T_3 = T_{sat}(P_3) = 311.1°\text{C} = 584.1 \text{ K}$$

$$T_C = T_1 = T_{sat}(P_1) = 60.06°\text{C} = 333 \text{ K}$$

Then the maximum or Carnot efficiency is found from Eq. (7.6):

$$\eta_{max} = 1 - \frac{T_C}{T_H} = \mathbf{0.430}$$

Analysis: Several observations can be made regarding these results.

(1) The flow rates through large-scale power plants, and therefore the equipment size in the plants, can be enormous. This cycle had a net power of about 100 MW. Many power plants are on the order of 1000 MW, or 1 GW. So, for such a plant, we would expect the water flow rate to be higher by about a factor of 10: ~1200 kg/s. Although this may sound unbelievable, after considering the density of liquid water, we realize that this is on the order of 1 m³/s of a volumetric flow rate of liquid water.

(2) The amount of power needed to compress the liquid is very small in comparison to the power produced in the turbine. In this case, only 1.16% of the power produced in the turbine was needed to pump the liquid water. In actual power plants with inefficient devices, this is normally slightly higher, on the order of 2 to 3%. But this characteristic is a major advantage to using a Rankine cycle system in comparison to other cycles that we will study later in this chapter.

(3) The thermal efficiency of steam power plants is not high. Typically, much more heat is rejected to the environment than power is produced. In this case, with a thermal efficiency of ~35%, ~65% of the input heat is rejected to the environment through the condenser: ~199 MW in comparison to the 108 MW of power produced. Management of this heat is a primary concern for plant designers and operators.

(4) The basic, ideal Rankine cycle has a thermal efficiency less than that of a corresponding Carnot cycle, but it is not far below. Although the raw thermal efficiency is fairly low, the maximum possible thermal efficiency is also fairly low. In this case, the second law efficiency of the cycle is 0.821 (found from 0.353/0.430), which indicates that the cycle is actually performing nearly as well as could be expected from a second law perspective.

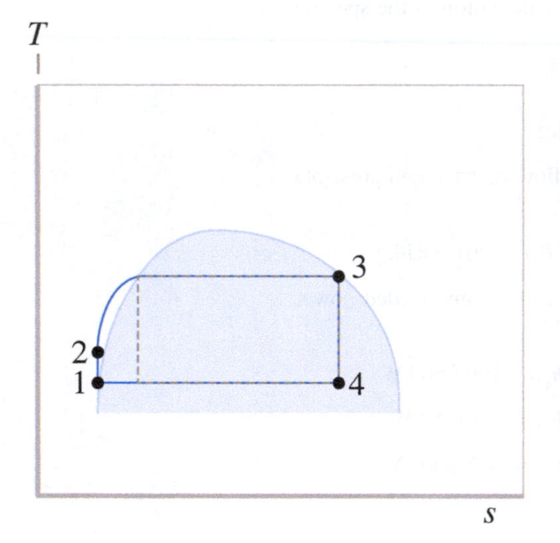

FIGURE 7.9 A comparison of the *T-s* diagrams of the vapor Carnot power cycle (shown in a dashed line) and the ideal, basic Rankine cycle.

As noted in Example 7.2, the basic, ideal Rankine cycle is less efficient than the Carnot cycle, but not by a great amount. If we compare the *T-s* diagrams of these two cycles, as seen in **Figure 7.9**, we can see that there are considerable similarities between the cycles. The basic, ideal Rankine cycle's *T-s* diagram is nearly a rectangle, with the difference being on the left-hand side. This is a result of practicality. We know that we will not in practice have isentropic turbines or pumps, but attempting to create an isentropic device that will take a saturated mixture and compress it to a saturated liquid is unrealistic. Such a process results in substantial irreversibilities, and it has been found to be more practical to use a standard pump and compress a liquid. As mentioned in Example 7.2, pumps do not require much power for what they accomplish, and so the use of a pump is seen as a good trade-off to develop an efficient working system as opposed to trying to develop a system that is theoretically better, but much less efficient in practice.

The use of the rectangle of a Carnot power cycle's *T-s* diagram can be helpful to quickly determine if a cycle is reasonably close to the theoretical maximum efficiency of a power cycle. To be closer, thermodynamically, to an ideal efficiency, the actual cycle's *T-s* diagram should be similar to a rectangle. The more rectangular the cycle's *T-s* diagram, the closer the cycle is to ideal and the better the second law efficiency of the cycle.

STUDENT EXERCISE

As seen in Example 7.2, the solution of a Rankine cycle problem involves sequentially determining enthalpy values and then using simplified first law equations that have been developed using standard assumptions for each component. Previously, you have developed computer models for the turbine and the pump. Now, develop computer models for your chosen solution platform for the steam generator and the condenser. Then combine the four models into one model for a basic, ideal Rankine cycle. Alternatively, you may choose to develop a new model for the basic, ideal Rankine cycle with all common assumptions already included in the programmed equations. With either approach, your final model should be capable of determining the net power output (or the required mass flow rate to give a desired power output) and thermal efficiency if given the steam generator and condenser pressures. For future flexibility in exploring the Rankine cycle, you may also develop the model so that it can provide property information for the working fluid if given net work and heat input information.

7.3.2 Nonideal, Basic Rankine Cycle

As discussed in Chapter 6, no real processes are truly reversible. As such, it is unrealistic to model actual steam power plant cycles with the ideal, basic Rankine cycle. The assumption of an isentropic turbine and an isentropic pump will lead to an overprediction of the thermal efficiency of an actual Rankine cycle. Fortunately, it is a simple process to incorporate a nonisentropic turbine and a nonisentropic pump into the analysis of the basic Rankine cycle. To do so, the calculations needed in an ideal Rankine cycle are still performed, and the isentropic efficiencies of the turbine and pump are then applied to provide the actual outlet states from these devices.

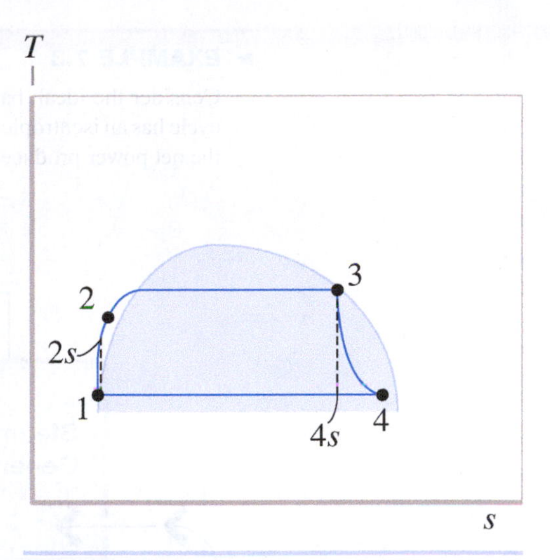

FIGURE 7.10 A comparison between *T-s* diagrams of the ideal and nonideal basic Rankine cycles.

Figure 7.10 illustrates on a *T-s* diagram the differences that can be expected between the ideal and nonideal basic Rankine cycles. The isentropic outlet states from the pump and turbine, which are obtained through an analysis of the ideal, basic Rankine cycle, are denoted with an *s* (2*s* and 4*s*, respectively), whereas the actual outlet states from nonisentropic devices are denoted with the appropriate state number. It should be noted that the differences between the outlet state of an isentropic pump (state 2*s*) and the outlet state of a nonisentropic pump are in reality very small, due to the small change in enthalpy experienced when pumping a liquid, but are exaggerated in Figure 7.10 for clarity.

Incorporating the nonisentropic processes into the basic Rankine cycle analysis first involves the solution procedure for an ideal, basic Rankine cycle as outlined in Section 7.3.1. However, the outlet state of the pump in the ideal cycle should now be labeled 2*s*, and the outlet state in the turbine should be labeled 4*s*. Equations (6.44) and (6.46), with state numbering appropriate to the cycle components, are then used to calculate the actual outlet states from the pump and turbine:

$$\eta_{s,p} = \frac{h_1 - h_{2s}}{h_1 - h_2} \tag{7.14}$$

and

$$\eta_{s,t} = \frac{h_3 - h_4}{h_3 - h_{4s}} \tag{7.15}$$

Remember, states 2 and 4 are the true outlet states from the pump and turbine, respectively, and the isentropic outlet states (2*s* and 4*s*) are theoretical tools used only to find these actual outlet states—or alternatively to find the isentropic efficiencies of the devices if the actual outlet conditions are known. Example 7.3 illustrates incorporating the nonideal turbine and pump into the basic Rankine cycle analysis.

STUDENT EXERCISE

If your computer model of the basic Rankine cycle is not able to incorporate a nonisentropic turbine or pump, modify the model to accommodate the isentropic efficiency of a turbine and a pump.

▶ EXAMPLE 7.3

Consider the ideal, basic Rankine cycle described in Example 7.2. Consider that the turbine of the cycle has an isentropic efficiency of 0.80, and the pump has an isentropic efficiency of 0.70. Determine the net power produced in the cycle and the thermal efficiency of the cycle.

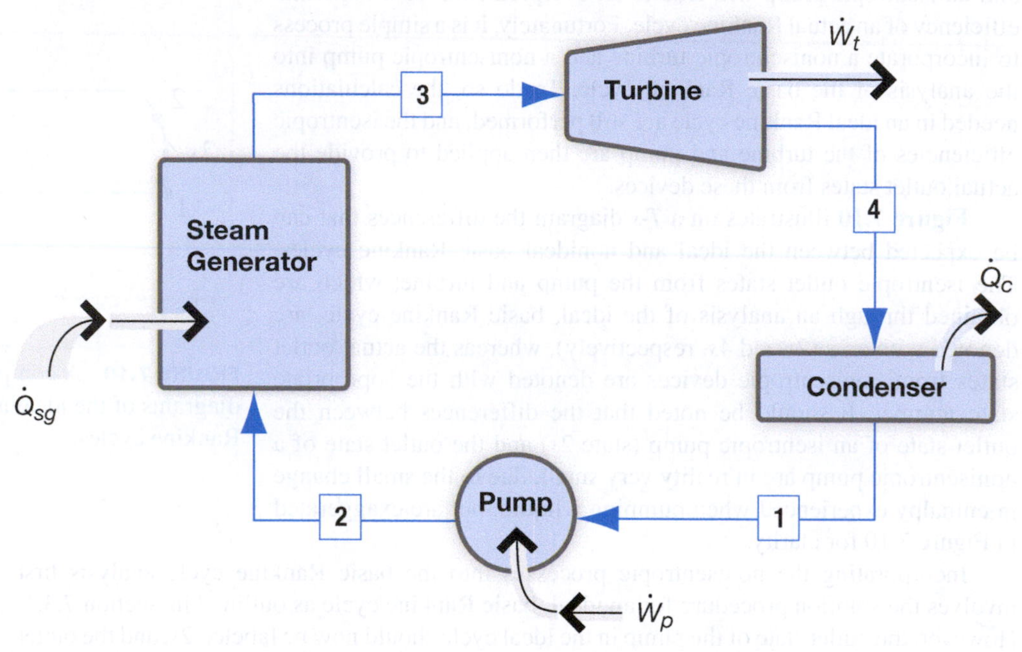

Given: Results from Example 7.2, $\eta_{s,t} = 0.80$, $\eta_{s,p} = 0.70$

Find: $\dot{W}$, η

Solution: The cycle under consideration here is a nonideal, basic Rankine cycle. The nonideal nature is seen by having a turbine and a pump that are not isentropic. To perform this analysis, we will first take the results of the enthalpy calculations from Example 7.2. Recall that the outlet states from the turbine and pump in an isentropic analysis are renamed with a subscript *s*. Therefore, for use in this example, the results from Example 7.2 are

$$h_1 = 251.40 \text{ kJ/kg} \qquad h_{2s} = 261.6 \text{ kJ/kg}$$

$$h_3 = 2724.7 \text{ kJ/kg} \qquad h_{4s} = 1845.1 \text{ kJ/kg}$$

Equations (7.14) and (7.15) are now used to find the actual outlet states from the pump and turbine:

$$h_2 = h_1 + \frac{h_{2s} - h_1}{\eta_{s,p}} = 265.97 \text{ kJ/kg}$$

and

$$h_4 = h_3 - \eta_{s,t}(h_3 - h_{4s}) = 2021 \text{ kJ/kg}$$

These values can also be easily found with computer property software and using the appropriate isentropic efficiencies in your turbine and pump computer models of the components.

With these new specific enthalpy values known, Eqs. (7.7)–(7.10) can be used to find any needed power values or heat transfer rate values.

For the turbine power: $\dot{W}_t = \dot{m}(h_3 - h_4) = 87{,}963 \text{ kW}$

For the pump power: $\dot{W}_p = \dot{m}(h_1 - h_2) = -1821 \text{ kW}$

For the heat transfer rate of the steam generator: $\dot{Q}_{sg} = \dot{m}(h_3 - h_2) = 307{,}341 \text{ kW}$

Therefore, the net power is $\dot{W}_{\text{net}} = \dot{W}_t + \dot{W}_p = 86{,}140 \text{ kW} = \mathbf{86.1\ MW}$

and the thermal efficiency is $\eta = \frac{\dot{W}_{net}}{\dot{Q}_{in}} = \frac{\dot{W}_t + \dot{W}_p}{\dot{Q}_{sg}} = \mathbf{0.280.}$

Analysis: The turbine (and net) power and thermal efficiency decrease significantly whereas the pump power increases with the incorporation of realistic turbine and pump isentropic efficiencies. You may also note that in comparison to Example 7.2, the heat input into the steam generator decreased slightly. Although decreasing the heat input to the steam generator is one method to increase the thermal efficiency, the choice of doing this by having a less efficient pump is not an effective means of increasing the thermal efficiency of the cycle.

The maximum possible thermal efficiency (i.e., the Carnot efficiency) will not change between an ideal and a nonideal, basic Rankine cycle. Therefore, the reduction in thermal efficiency seen in a nonideal cycle pushes the thermal efficiency further from the optimal situation. Reasons for this can be seen in **Figure 7.11**, which contains *T-s* diagrams of both the nonideal, basic Rankine cycle and the Carnot cycle. It can be seen in Figure 7.11 that the right-hand side of the *T-s* diagram is less rectangular than in the ideal Rankine cycle, which indicates that the increase in entropy seen in the turbine is further lowering the thermal efficiency from what would be seen in the Carnot cycle. Similarly, the increase in entropy in the pump makes the cycle less ideal; however, the left-hand side of the *T-s* diagram of the cycle was already considerably different from the rectangle in the Carnot cycle, and the small difference introduced by a nonisentropic pump is fairly insignificant.

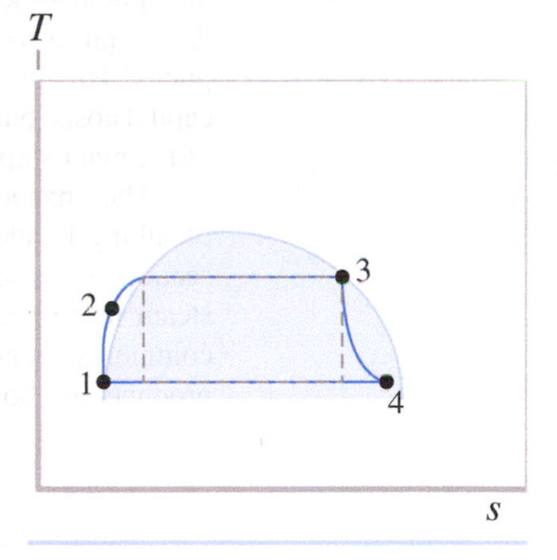

FIGURE 7.11 A comparison between the *T-s* diagrams of the vapor Carnot power cycle (shown in a dashed box) and the nonideal, basic Rankine cycle.

It is desirable to increase the thermal efficiency of steam power plants. As seen in Example 7.3, it is reasonable to expect that plants following a basic Rankine cycle with realistic components will have thermal efficiencies below 30%. The lower the thermal efficiency of a power plant, the more fuel that needs to be consumed in order to get the same net power output. Fuel costs money, and so it is desirable to have a higher thermal efficiency to save plant operators money on fuel costs. However, if we consider ourselves limited to the basic Rankine cycle using water, the highest temperature that could be achieved for the steam entering the turbine is the critical temperature of 374.14°C (647 K). On the heat rejection side of the cycle, it is reasonable to expect that the heat will be rejected to the environment, and it is reasonable to consider a standard environmental temperature at 25°C (298 K); the steam exiting the turbine and condensing in the condenser cannot be colder than the environment temperature if heat is to be transferred from the steam to the environment. A Carnot cycle operating between these two temperatures will have an efficiency of 0.539. This is a low maximum possible efficiency, and when we consider that the isentropic efficiency of a realistic turbine under these conditions will not be extremely high (due to the high moisture content of the flow of the steam through the turbine), it is reasonable to conclude that the best possible thermal efficiency that could be achieved with a realistic basic Rankine cycle in practice would be less than 40%, and likely even lower.

Later, we will discuss some options for increasing the thermal efficiency of the Rankine cycle. In general, these options require additional capital costs in constructing the cycle, and so they are not universally used. Smaller Rankine cycle facilities will often rely on the less expensive option of a basic Rankine cycle. Operators of systems that use co-generation, during which both process steam and electricity from a Rankine cycle are produced, may not be particularly concerned with increasing the cycle's thermal efficiency significantly because they may view any electricity produced as a bonus. Furthermore, one large-scale power plant design also relies on the basic Rankine cycle—a boiling water reactor version of a nuclear power plant. A boiling water reactor nuclear power plant uses saturated steam to enter the turbine, following the basic Rankine cycle design. This is done because in the reactor only a portion of the water in the nuclear reactor is allowed to vaporize, whereas most remains a liquid. The resulting saturated steam (both liquid and vapor are present, so it is by necessity a saturated mixture) is extracted and sent directly toward the turbine. Additional heating is not provided because there is no temperature source available and it is desirable to keep the amount of equipment subjected to radiation from the reactor core to a minimum.

7.3.3 Rankine Cycle with Superheat

The simplest method to increase the maximum possible thermal efficiency, which should in principle increase the actual thermal efficiency as well, is to increase the highest temperature in the system—the temperature of the steam entering the turbine. This will require superheating the steam in the steam generator, and it is a practice done in most large-scale steam power plants. This process requires modifying the design of the steam generator, and so increases capital costs, but these increased costs are paid back through fuel cost savings over the lifetime of the plant's operation.

The change in the design of the steam generator is shown in Figure 7.12. A section of tubing is added after the steam drum boiler, and the saturated steam exiting the boiler section is passed through this tubing. Although there is some potential for superheating steam in a pressurized-water reactor-type nuclear power plant, superheating steam is most commonly found in facilities using fossil-fuel combustion as the heat source. The combustion products are hot enough after exiting the boiler section to transfer heat to the steam in the

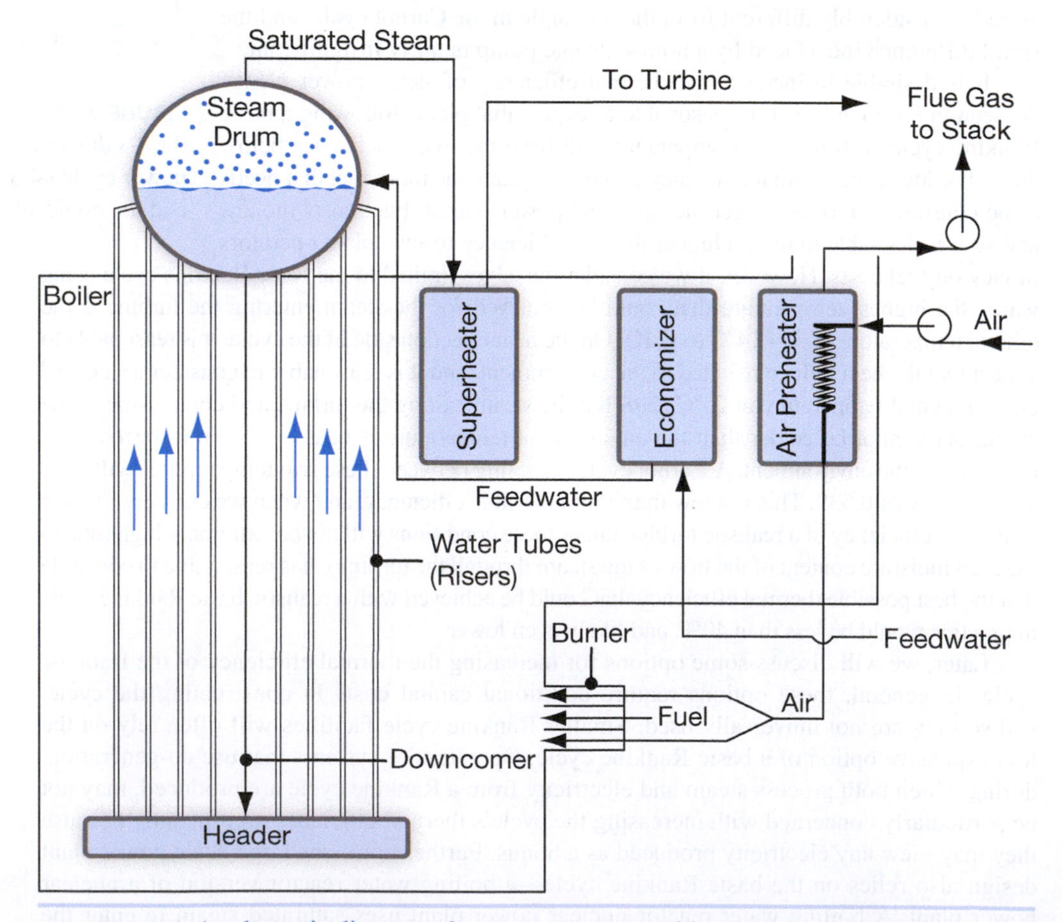

FIGURE 7.12 A schematic diagram of a steam generator with a superheater section added.

superheater, thereby causing the steam to become superheated. The pressure loss of the steam in the superheater section in an actual system is low. Theoretically, the steam could be heated to the maximum temperature of the combustion products entering the steam generator, but the temperature of the superheated steam is usually kept below approximately 600°C due to considerations of material properties and material durability.

The resulting cycle is the Rankine cycle with superheat. The only change in the hardware of the cycle is an additional heating stage in the steam generator, and because the steam generator can be considered as a single component in the Rankine cycle analysis, the schematic diagram of the cycle shown in Figure 7.4 is applicable to the Rankine cycle with superheat. The effects of adding superheat to the Rankine cycle are shown in the *T-s* diagrams of ideal Rankine cycles with superheat presented in **Figure 7.13**. As you can see, state 3 is now superheated. Two possible scenarios for state 4 are also shown in Figure 7.13. In Figure 7.13a, the isentropic expansion through the turbine results in a saturated mixture exiting at state 4—leading to a constant-pressure and constant-temperature heat rejection process through the condenser. In Figure 7.13b, the expansion through the turbine does not cause the steam to begin to condense in the turbine, and so superheated steam exits the turbine at state 4. As you can see, the resulting flow through the condenser is still at constant pressure but is no longer at constant temperature because the steam must first cool to a saturated vapor before it can begin to condense.

The nonideal Rankine cycle with superheat, which is what would be seen in practice, offers an additional possible complexity in the analysis of the cycle. Three potential scenarios are shown in **Figure 7.14** for this nonideal cycle. Figure 7.14a presents the possibility that both the theoretical isentropic expansion through the turbine and the non-isentropic expansion result in substances that are saturated (states 4*s* and 4, respectively.) Figure 7.14b presents the possibility of the isentropic expansion resulting in a saturated mixture but the nonisentropic actual expansion being a superheated vapor. Figure 7.14c has both the isentropic and nonisentropic processes resulting in superheated vapors. Although these different scenarios do not alter the solution procedure, it is useful to be aware of the different possibilities when developing and interpreting your analysis.

The analysis of the Rankine cycle with superheat does not change from the basic Rankine cycle, with the exception that the steam entering the turbine (state 3) is no longer a saturated vapor, but rather is a superheated vapor. Therefore, the specific enthalpy and specific entropy of state 3 are found as a superheated vapor. The rest of the analysis is identical to the basic Rankine cycle.

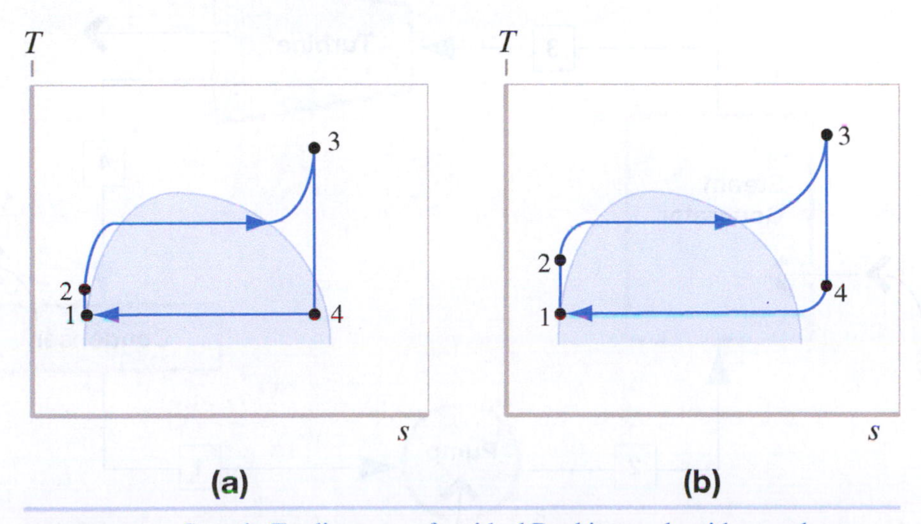

FIGURE 7.13 Sample *T-s* diagrams of an ideal Rankine cycle with superheat. The steam leaving the turbine (state 4) can be either in the form of (a) a saturated mixture, or (b) a superheated vapor.

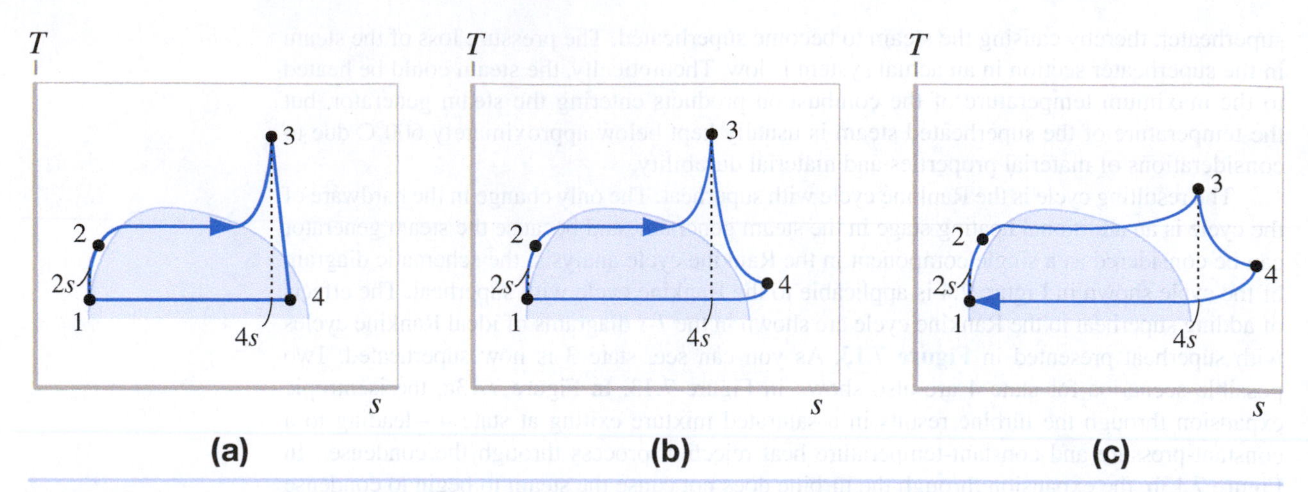

FIGURE 7.14 Sample *T-s* diagrams of a nonideal Rankine cycle with superheat. (a) Both the ideal isentropic turbine outlet state (state 4*s*) and the actual outlet state (state 4) are saturated mixtures. (b) State 4*s* is saturated, whereas state 4 is superheated vapor. (c) Both state 4*s* and 4 are superheated vapors.

STUDENT EXERCISE

If your computer model of the Rankine cycle is not designed to use superheated steam entering the turbine, modify the model to accommodate the possibility of superheated steam.

▶ EXAMPLE 7.4

The ideal, basic Rankine cycle of Example 7.2 is to be modified so that a superheat section is added to the steam generator. The steam generator now produces superheated steam at 10 MPa and 600°C. Determine the net power output for this Rankine cycle with superheat, and determine the thermal efficiency of the cycle.

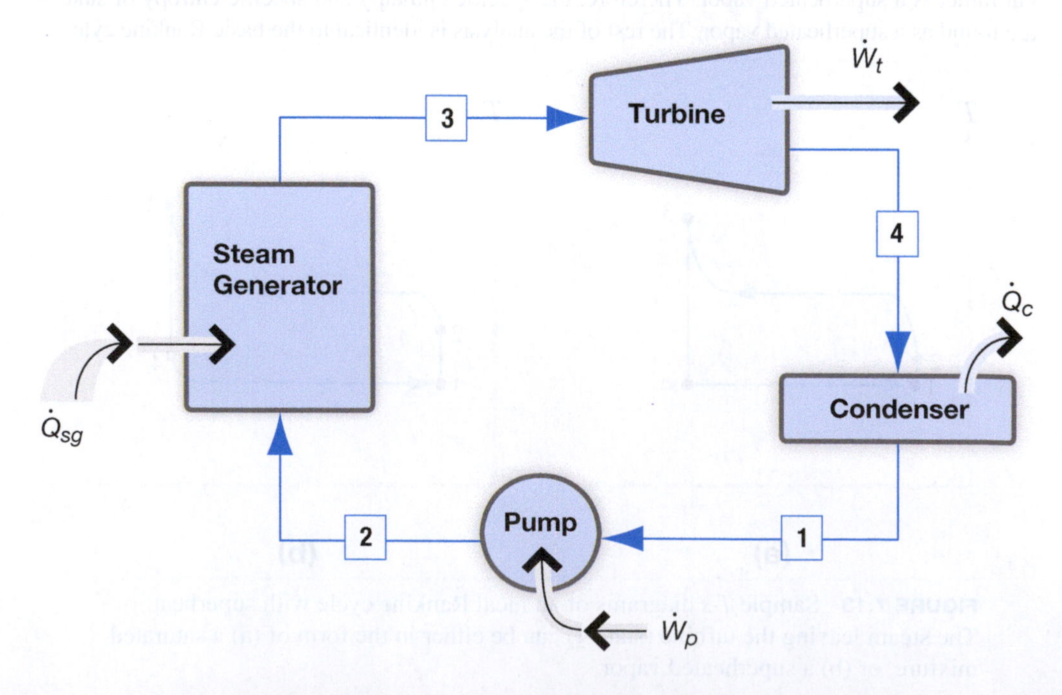

Given: $x_1 = 0$, $P_1 = 20$ kPa, $T_3 = 600°C$, $P_3 = 10$ MPa, $\dot{m} = 125$ kg/s

Find: $\dot{W}$, η

Solution: The cycle under consideration in this problem is an ideal Rankine cycle with superheat. The working fluid is water, in the form of a compressed liquid, a saturated substance, and a superheated vapor at various points in the cycle.

Assume: $P_2 = P_3$, $P_1 = P_4$, $s_1 = s_2$, $s_3 = s_4$, $\dot{W}_{sg} = \dot{W}_c = \dot{Q}_t = \dot{Q}_p = 0$, $\Delta KE = \Delta PE = 0$ for each device. Steady-state, steady-flow systems. Liquid water in the pump is incompressible.

The first step is to find the specific enthalpy values of each state, which will be done here using a property calculator program.

For state 3, use the given conditions.

State 3: $P_3 = 10$ MPa, $T_3 = 600°C$ $\quad h_3 = 3625.3$ kJ/kg

$s_3 = 6.9029$ kJ/kg · K

For state 4, use the pressure from state 1 and the entropy from state 3, because those values correspond to state 4's values for pressure and entropy, based on the assumptions given.

State 4: $P_4 = P_1 = 20$ kPa $\quad s_4 = s_3 = 6.9029$ kJ/kg · K

$x_4 = (s_4 - s_f)/(s_g - s_f) = 0.8579$

$h_4 = 2274.6$ kJ/kg

For state 1, use the given conditions. However, we will find the specific volume at state 1, rather than the specific entropy, because the specific volume will be used in calculating the specific enthalpy of state 2.

State 1: $P_1 = 20$ kPa, $x_1 = 0.0$ $\quad h_1 = 251.40$ kJ/kg

$v_1 = 0.0010172$ m³/kg

For state 2, use the relationship presented in Eq. (6.48) for isentropic flow of an incompressible substance.

State 2: $P_2 = 10$ MPa $= 10{,}000$ kPa $\quad h_2 = h_1 + v_1(P_2 - P_1) = 261.6$ kJ/kg

With the specific enthalpy values known, Eqs. (7.7)–(7.10) can be used to find any needed power values or heat transfer rate values.

For the turbine power: $\dot{W}_t = \dot{m}(h_3 - h_4) = 168{,}838$ kW

For the pump power: $\dot{W}_p = \dot{m}(h_1 - h_2) = -1275$ kW

For the heat transfer rate of the steam generator: $\dot{Q}_{sg} = \dot{m}(h_3 - h_2) = 420{,}463$ kW

Therefore, the net power is $\dot{W}_{net} = \dot{W}_t + \dot{W}_p = 167{,}560$ kW $=$ **167.6 MW**

and the thermal efficiency is $\eta = \dfrac{\dot{W}_{net}}{\dot{Q}_{in}} = \dfrac{\dot{W}_t + \dot{W}_p}{\dot{Q}_{sg}} =$ **0.399.**

Analysis:

(1) The maximum possible efficiency can still be found by using the maximum and minimum temperatures in the system. These are the temperatures at states 3 and 1, respectively, which are equal to

$$T_H = T_3 = T_{sat}(P_3) = 600°C = 873 \text{ K}$$

$$T_C = T_1 = T_{sat}(P_1) = 60.06°C = 333 \text{ K}$$

Then, the maximum or Carnot efficiency is found from Eq. (7.6):

$$\eta_{max} = 1 - \frac{T_C}{T_H} = \mathbf{0.619}$$

This is a substantial increase from the Carnot efficiency for the basic Rankine cycle as shown in Example 7.2.

(2) Although the thermal efficiency increased from the basic Rankine cycle by 4.6 percentage points, the cycle's second law efficiency (found from the ratio of the cycle's thermal efficiency to the corresponding maximum possible efficiency) dropped considerably (from 82% to 64.5%). This indicates that the Rankine cycle with superheat is not as close to being reversible (i.e., it is worse "thermodynamically") and that there is substantial room to attempt to further improve the thermal efficiency of the Rankine cycle with superheat.

(3) Remember that, in practice, the thermal efficiency is more useful than the second law efficiency. A power plant operator is likely to choose to use the more thermally efficient cycle (i.e., with superheat) than the cycle that is closer to being reversible because the thermally efficient cycle will result in fuel cost savings by requiring less fuel to produce the same amount of power.

As discussed in Example 7.4, the Rankine cycle with superheat is farther from ideal than the basic Rankine cycle. This is clearly obvious in the *T-s* diagrams in Figures 7.13 and 7.14, where the Rankine cycle with superheat's *T-s* diagram no longer even remotely resembles a rectangle. As a result, although adding superheat increases the thermal efficiency of a Rankine cycle, it also increases the maximum possible efficiency by a larger amount. The Rankine cycle with superheat has more irreversibility than a Rankine cycle without superheat, but this illustrates that it is important to keep in mind the big picture. We are seeking higher thermal efficiencies, and we are most concerned with achieving these efficiencies, even at the possibility of having a more irreversible process. The larger difference between the thermal efficiency and the maximum possible efficiency for a Rankine cycle with superheat in comparison to a Rankine cycle without superheat indicates that there should be opportunities for further improvement in the efficiency of the Rankine cycle with superheat.

QUESTION FOR THOUGHT/DISCUSSION

As the Rankine cycle with superheat will almost certainly have a higher thermal efficiency than a basic Rankine cycle, why would someone choose to employ a basic Rankine cycle power plant?

7.3.4 Rankine Cycle with Reheat

One device that can plague Rankine cycle power plants is an inefficient turbine, and one factor that can cause the efficiency of a turbine to decrease is the presence of liquid in a saturated mixture steam flow. Larger amounts of liquid suspended in the steam will result in larger liquid droplet sizes, which will provide more impedance to the rotation of the turbine blades around the central rotor. Larger amounts of liquid can also result in liquid accumulating on the blades, which further resists the motion. These impediments to the motion reduce the isentropic efficiency of a turbine, and so we would expect that a turbine with lower quality steam flowing through portions of the turbine will have a lower isentropic efficiency than a similar steam turbine with dry steam, or at least saturated steam with a higher quality. (On a related note, the presence of larger water droplets and films of water on the blades also increases the likelihood of damage being done to the blades during operation, and subsequently a decreased lifetime of the blades before they need replacement.)

As you can see, adding superheat to the Rankine cycle should have an auxiliary benefit of increasing the exit quality of the steam from the turbine, which should lead to higher isentropic efficiencies of the turbine. In some cases, however, the high pressure used in the steam generator leads to an undesirably low steam quality by the time the steam exits the turbine. This is illustrated in **Figure 7.15**, which shows an isentropic expansion on a *T-s* diagram from two pressures to a common turbine exhaust pressure. As you can see, the higher the turbine inlet pressure, the lower the exit quality for the same outlet pressure.

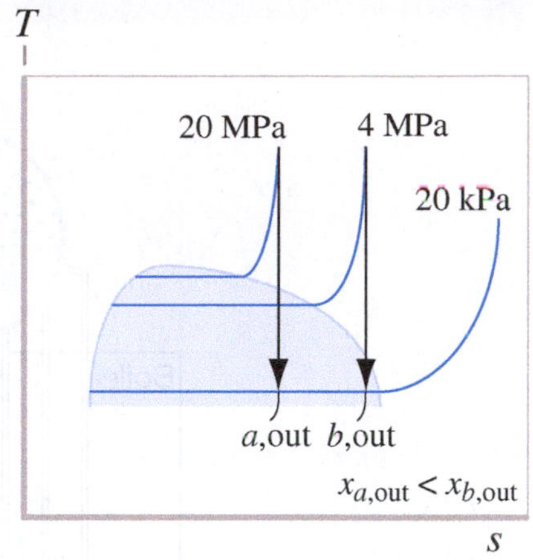

FIGURE 7.15 If superheated steam entering a turbine at a particular temperature undergoes isentropic expansion in a turbine to a saturated mixture, a higher inlet pressure will result in a lower exit quality.

To help avoid the undesirably low exit quality from the turbine, a reheat stage can be added. In the Rankine cycle with reheat, the steam is extracted partway through the turbine system and returned to another section of the steam generator—the reheat section. In practice, the reheat section is physically similar to the superheater section, but it is located just downstream of the superheater with respect to the flow of the combustion products. The combustion products are usually still hot enough to reheat the steam to a temperature nearly that of the steam exiting the superheater. The steam is then directed back toward the turbine system, where it continues its expansion. **Figure 7.16** shows a schematic diagram of the steam generator system containing a reheater section. **Figure 7.17** presents a schematic diagram of a Rankine cycle with one stage of reheat, and **Figure 7.18a** shows a *T-s* diagram of an ideal Rankine cycle with reheat. A nonisentropic turbine will alter the *T-s* diagram, which is shown in **Figure 7.18b**. More than one stage of reheat can be used, although it is usually not advantageous to use more than one or two stages. Furthermore, we generally want to direct superheated vapor to the reheater section, although some reheat systems can manage to receive a saturated mixture with a very high quality.

A secondary benefit to using reheat is that higher temperature fluids have a higher specific heat. As a result, using reheat can lead to a greater change in enthalpy through the turbine, resulting in a greater turbine power output. For example, consider two isentropic expansions of steam from 10 MPa to 1 MPa, with one beginning at a temperature of 700°C and the other beginning at a temperature of 500°C. The former will result in an exit temperature of 312°C and a specific enthalpy change of −792.3 kJ/kg. The latter, beginning at 500°C, results in an exit temperature of 181.5°C and a specific enthalpy change of −591.1 kJ/kg. So, the steam entering the turbine at a higher temperature can provide a greater power output from the turbine—although there is also a greater heat input requirement to increase the enthalpy of the steam at the higher temperatures. The net effect is often a small additional increase in the thermal efficiency of the cycle, in comparison to one without reheat. Greater increases in the thermal efficiency are seen if a higher isentropic efficiency of the turbine is experienced when reheat is used.

The analysis of a Rankine cycle with reheat is very similar to the analyses of previous versions of the Rankine cycle. The changes, other than in the numbering of the states for some of the components, involve using an additional heat input calculation and expanding the turbine analysis into multiple sections of turbine. The reheat process is considered to be a constant-pressure heat addition process. As such, referring to Figure 7.17, $P_5 = P_4$, and from the processes of the basic Rankine cycle, we also have $P_1 = P_6$ and $P_2 = P_3$. It is assumed that there is no change in kinetic energy or potential energy through the reheat section, and that there is no power consumed in the reheat process. The resulting heat transfer rate for the reheater is found from the first law as

$$\dot{Q}_r = \dot{m}(h_5 - h_4) \tag{7.16}$$

and the resulting heat input rate for the cycle is

$$\dot{Q}_{\text{in}} = \dot{Q}_{sg} + \dot{Q}_r = \dot{m}[(h_3 - h_2) + (h_5 - h_4)] \tag{7.17}$$

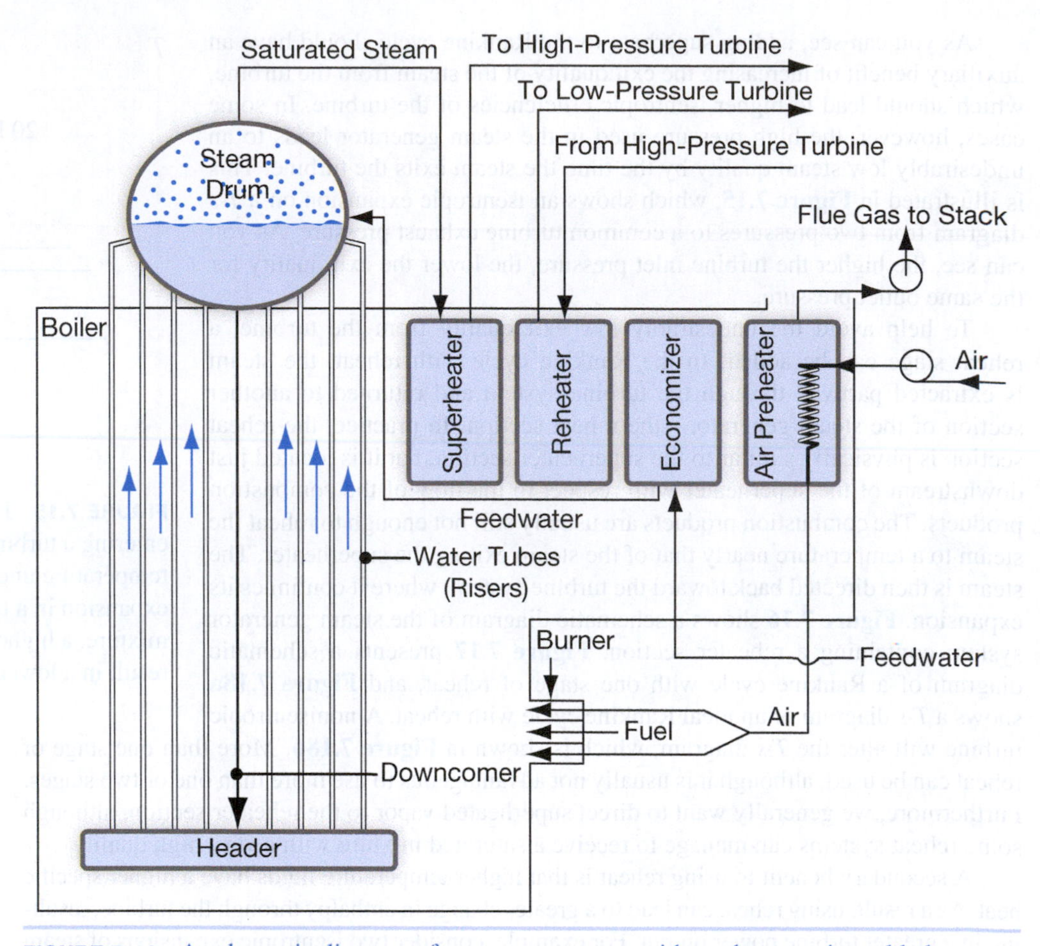

FIGURE 7.16 A schematic diagram of a steam generator with a reheat section added.

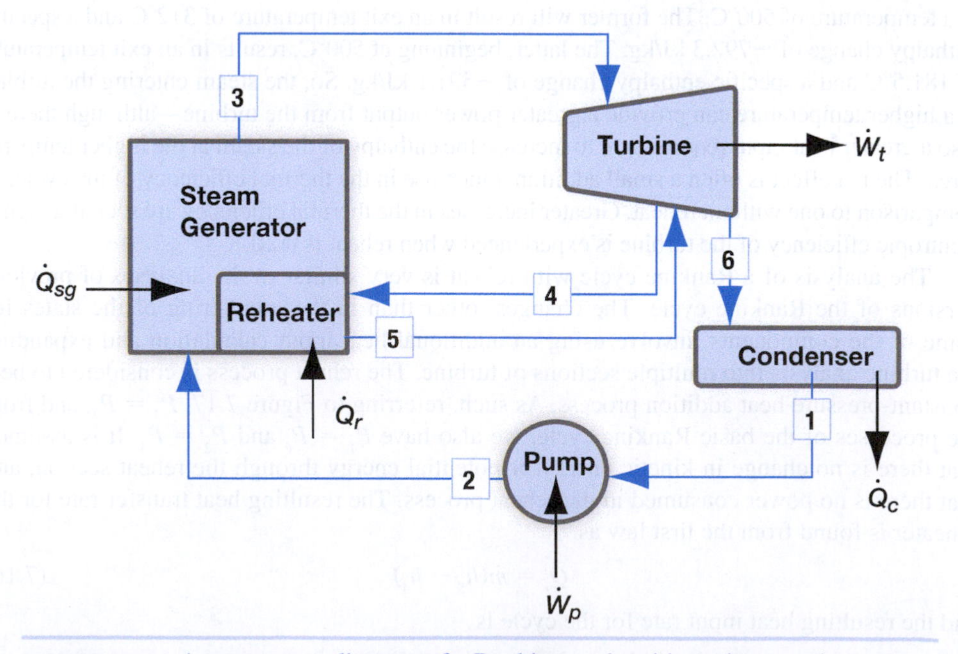

FIGURE 7.17 A component diagram of a Rankine cycle with a reheat section.

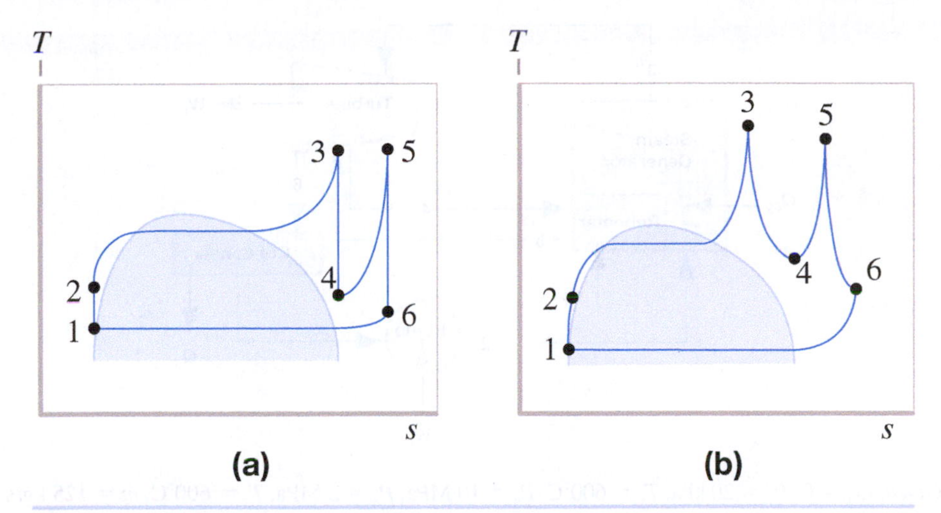

FIGURE 7.18 *T-s* diagrams of a Rankine cycle with reheat (and superheat): (a) isentropic turbine and pump, (b) nonisentropic turbine and pump.

Although reheaters are part of the steam generator, they are treated separately in the equations for heat input because they involve receiving steam at a different inlet condition—essentially, the addition of a reheater section turns the steam generator into a multiple-inlet, multiple-outlet device. Rather than viewing it specifically as such, we model it as two (or more, depending on the number of reheat stages) single-inlet, single-outlet devices in series.

It can also be seen that there are two turbine sections, and the power from these two sections must be combined to calculate the power from the turbine:

$$\dot{W}_t = \dot{W}_{t1} + \dot{W}_{t2} = \dot{m}[(h_3 - h_4) + (h_5 - h_6)] \tag{7.18}$$

The pump power expression is the same as the basic Rankine cycle, and the heat transfer rate through the condenser is similar, with only the state numbers changing ($\dot{Q}_c = \dot{m}(h_6 - h_1)$). The thermal efficiency is still found as the net power divided by the rate of heat input:

$$\eta = \frac{\dot{W}_{net}}{\dot{Q}_{in}} = \frac{\dot{W}_t + \dot{W}_p}{\dot{Q}_{sg} + \dot{Q}_r} = \frac{(h_3 - h_4 + h_5 - h_6) + (h_1 - h_2)}{(h_3 - h_2) + (h_5 - h_4)} \tag{7.19}$$

Note that as additional components are added to the Rankine cycle, there are more expressions to be incorporated into the analysis. But if we follow the working fluid flow through the devices and determine the appropriate first law for each device, the analysis remains straightforward and you will be able to correctly analyze more complex cycles.

STUDENT EXERCISE

Modify your computer model for the Rankine cycle to incorporate one or more reheat stages.

▶ EXAMPLE 7.5

The ideal Rankine cycle with superheat of Example 7.4 is to be modified with the addition of a reheat section. For the reheat process, the steam is removed from the turbine at a pressure of 2 MPa and reheated to a temperature of 600°C. Determine the net power produced by the cycle and the thermal efficiency of this Rankine cycle with reheat.

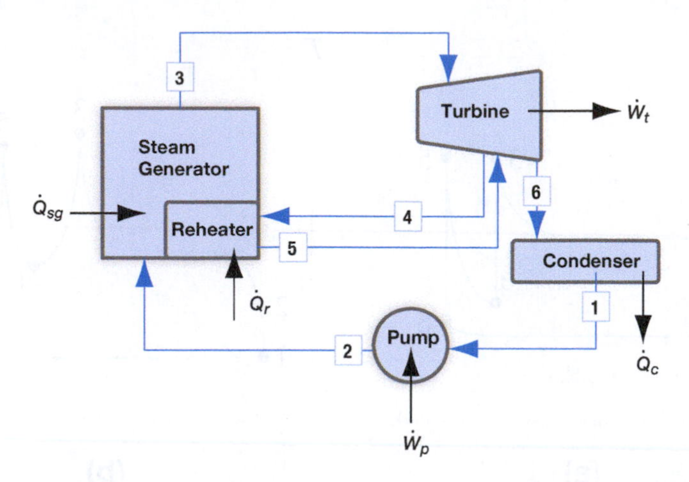

Given: $x_1 = 0$, $P_1 = 20$ kPa, $T_3 = 600°C$, $P_3 = 10$ MPa, $P_4 = 2$ MPa, $T_5 = 600°C$, $\dot{m} = 125$ kg/s

Find: $\dot{W}$, η

Solution: The cycle under consideration is an ideal Rankine cycle with reheat. The working fluid is water, and the water exists as a compressed liquid, a saturated substance, and a superheated vapor at various points in the cycle.

Assume: $P_2 = P_3$, $P_1 = P_6$, $P_4 = P_5$, $s_1 = s_2$, $s_3 = s_4$, $s_5 = s_6$, $\dot{W}_{sg} = \dot{W}_c = \dot{W}_r = \dot{Q}_t = \dot{Q}_p = 0$, $\Delta KE = \Delta PE = 0$ for each device. Steady-state, steady-flow systems. Liquid water in the pump is incompressible.

The first step is to find the specific enthalpy values of each state, using your property calculator program of choice.

For state 3, use the given conditions.

State 3: $P_3 = 10$ MPa, $T_3 = 600°C$ $\quad h_3 = 3625.3$ kJ/kg

$s_3 = 6.9029$ kJ/kg · K

For state 4, use the pressure from state 4 and the entropy from state 3, because those values correspond to state 4's values for pressure and entropy, based on the assumptions given.

State 4: $P_4 = 2.0$ MPa $\quad s_4 = s_3 = 6.9029$ kJ/kg · K (this is superheated vapor)

$h_4 = 3104.9$ kJ/kg

States 5 and 6 are found in a similar fashion:

State 5: $P_5 = 2$ MPa, $T_5 = 600°C$ $\quad h_5 = 3690.1$ kJ/kg

$s_5 = 7.7024$ kJ/kg · K

State 6: $P_6 = P_1 = 20$ kPa $\quad s_6 = s_5 = 7.7024$ kJ/kg · K (this is a saturated mixture)

$x_6 = 0.9709$

$h_6 = 2541.0$ kJ/kg

States 1 and 2 are identical to those found in Example 7.6:

$h_1 = 251.40$ kJ/kg $\quad h_2 = 261.6$ kJ/kg

With the specific enthalpy values known, Eqs. (7.7) and (7.17)–(7.19) can be used to find any needed power values or heat transfer rate values.

For the turbine power: $\dot{W}_t = \dot{W}_{t1} + \dot{W}_{t2} = \dot{m}[(h_3 - h_4) + (h_5 - h_6)] = 208{,}688$ kW

For the pump power: $\dot{W}_p = \dot{m}(h_1 - h_2) = -1275$ kW

The heat transfer rate into the system is the sum of the primary steam generator and the reheater:

$$\dot{Q}_{\text{in}} = \dot{Q}_{sg} + \dot{Q}_r = \dot{m}[(h_3 - h_2) + (h_5 - h_4)] = 493{,}613 \text{ kW}$$

Therefore, the net power is

$$\dot{W}_{\text{net}} = \dot{W}_t + \dot{W}_p = 207{,}413 \text{ kW} = \mathbf{207.4\ MW}$$

and the thermal efficiency is

$$\eta = \frac{\dot{W}_{\text{net}}}{\dot{Q}_{\text{in}}} = \frac{\dot{W}_t + \dot{W}_p}{\dot{Q}_{sg}} = \mathbf{0.420}$$

Analysis:

(1) This is a significant increase in the cycle efficiency with the addition of reheat, and it does not even include the added benefit in a practical system of improving the turbine's isentropic efficiency. The increase in thermal efficiency seen with reheat is often not this large, but is present here because the pressure of the reheat process was chosen well.

(2) Although it is still nowhere near rectangular in shape, the *T-s* diagram of the Rankine cycle with reheat can be argued to be slightly more rectangular than the diagram for the Rankine cycle with superheat alone. As such, the thermal efficiency is slightly closer to the ideal Carnot cycle's efficiency, and this version of the Rankine cycle is slightly closer to reversible.

When incorporating nonisentropic turbine sections into the analysis, we must remember to apply the isentropic efficiency to each turbine section. This is left as an exercise.

7.3.5 Rankine Cycle with Regeneration

With thermal efficiencies in the range of 30–40% for the forms of the Rankine cycle discussed previously, it should be obvious that much of the input heat is ultimately rejected to the environment (60–70%). If some of this heat could be captured and used as heat input to the cycle, the heat input load could be reduced and the thermal efficiency could be increased. This is the concept behind the various forms of the Rankine cycle with regeneration, with "regeneration" being the internal transfer of heat within the cycle.

A fundamental problem with directly using the heat removed through the condenser to heat the water elsewhere in the cycle is that the working fluid passing through the condenser is at the lowest temperature in the cycle. From the Second Law of Thermodynamics, heat is transferred from a hotter substance to a colder substance, and so although the heat can be sent to the environment from the condensing working fluid, the heat transfer only occurs because the environment is at an even lower temperature. It is not possible to use this energy to heat the working fluid elsewhere in the cycle, because the temperatures of the fluid elsewhere in the cycle are hotter than in the condenser. Therefore, to achieve regeneration, it is necessary to extract energy from the working fluid before the working fluid reaches the condenser.

If we look at a *T-s* diagram of a Rankine cycle, the most logical use for the regeneration is to preheat the working fluid heading toward the steam generator. If the compressed liquid enters the steam generator at a hotter temperature, less energy is needed in the steam generator to heat the compressed liquid to the saturated liquid state; this results in an overall reduction in heat input. This heat transfer process can be achieved through relatively simple heat exchangers.

One such regeneration scheme is shown in **Figure 7.19**. In this scheme, compressed liquid exiting the pump is sent through tubing wrapped around the turbine and flows in a direction opposite that of the flow through the turbine. In this way, heat is steadily transferred from the vapor flowing through the turbine to the liquid flowing toward the steam generator.

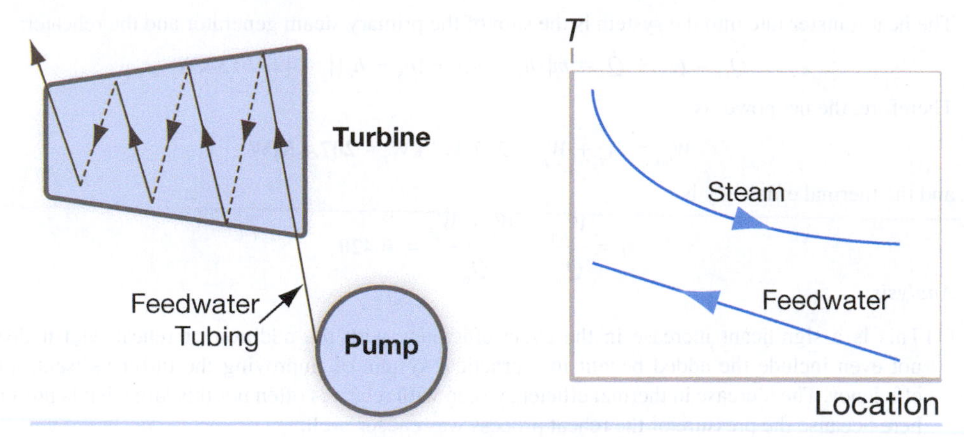

FIGURE 7.19 A potential method of utilizing regenerative heat transfer would be to have the boiler feedwater continuously being heated from heat lost through the turbine. Although this process is possible, it is not practical.

Thermodynamically, this is a very good process from a second law perspective because a rather small temperature difference between the fluids drives the heat transfer. However, from a practical, engineering standpoint, this is a rather unwieldy solution due to the complexity of the hardware required.

In practice, two techniques are used often to achieve regeneration: (1) the open feedwater heater and (2) the closed feedwater heater, where "feedwater" refers to the water being sent to the steam generator from the condenser. Both of these techniques result in some loss of total flow through the turbine, and so the power produced per unit total mass flow rate is reduced for the cycle. However, with proper choice of parameters, the heat input per unit total mass flow rate is reduced by a greater percentage than the power reduction, leading to an improved thermal efficiency for the cycle.

Rankine Cycle with an Open Feedwater Heater

An open feedwater heater is a mixing chamber, where flow extracted from the turbine is mixed with flow coming from a pump, and the combined exit stream is then directed toward the steam generator. All three streams should be at the same pressure to allow for proper flow without backflow toward the turbine or pump. The relative mass flow rates of the two inlet streams are typically set so that the exit stream is a saturated liquid. Because the steam from the turbine has expanded to a lower pressure than that of the steam generator, a second pump between the open feedwater heater and the steam generator is needed to raise the working fluid pressure to that of the steam generator. Figure 7.20 shows a schematic component diagram of a Rankine cycle with one open feedwater heater, and Figure 7.21 shows *T-s* diagrams for both the ideal version of the cycle and a cycle with a nonisentropic turbine and pumps. Open feedwater systems may or may not contain superheat, but the cycles shown in Figures 7.20 and 7.21 do contain superheat.

When analyzing this version of the Rankine cycle, we must keep track of how much flow is flowing through each component. We find the turbine power by taking a summation of the product of the mass flow rate through and the enthalpy change for each turbine section, and we find the pump power by adding the products of the mass flow rate and enthalpy change for each pump. The entire mass of working fluid will flow through the steam generator; however, if a reheat section is used, a reduced mass flow rate may flow through the reheater if the feedwater heater is positioned at a higher pressure than the reheater. Also, when analyzing the

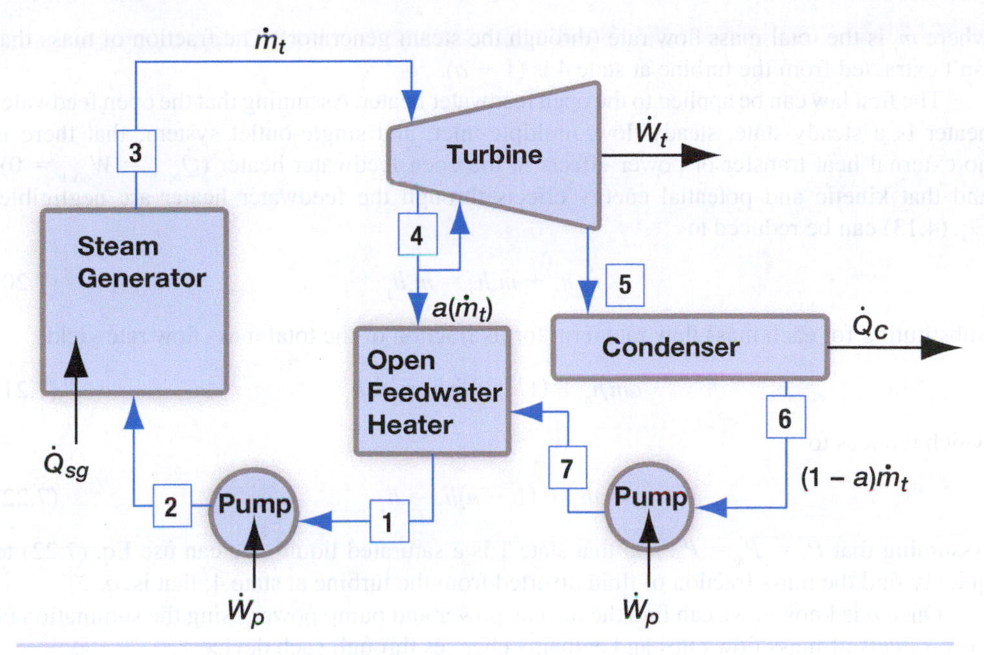

FIGURE 7.20 A component diagram of a Rankine cycle with one open feedwater heater.

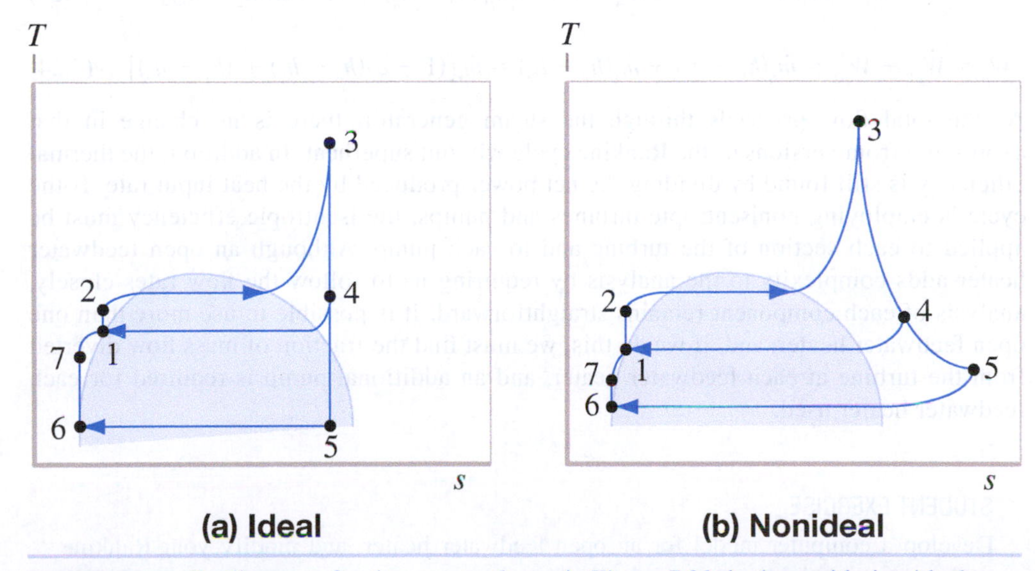

FIGURE 7.21 *T-s* diagrams for the system shown in Figure 7.20, both considering ideal turbine and pumps, and nonideal turbine and pumps.

Rankine cycles of increased complexity, remember that the pressure continues to drop through the turbine sections and that the pumps are used to increase pressure.

It is important to determine the fraction of the total mass flow extracted from the turbine to be directed toward the feedwater heater. We will call this fraction a. Referring to Figure 7.20, you can see that the following mass flow equivalencies hold:

$$\dot{m}_1 = \dot{m}_2 = \dot{m}_3 = \dot{m}_t$$

$$\dot{m}_4 = a\dot{m}_t$$

$$\dot{m}_5 = \dot{m}_6 = \dot{m}_7 = (1 - a)\dot{m}_t$$

where $\dot{m}_t$ is the total mass flow rate (through the steam generator). The fraction of mass that isn't extracted from the turbine at state 4 is $(1 - a)$.

The first law can be applied to the open feedwater heater. Assuming that the open feedwater heater is a steady-state, steady-flow, multiple-inlet, and single-outlet system; that there is no external heat transfer or power effects in the open feedwater heater ($\dot{Q}_{\text{ofwh}} = \dot{W}_{\text{ofwh}} = 0$); and that kinetic and potential energy effects through the feedwater heater are negligible, Eq. (4.13) can be reduced to

$$\dot{m}_4 h_4 + \dot{m}_7 h_7 = \dot{m}_1 h_1 \tag{7.20}$$

Substituting for each mass flow rate term for its fraction of the total mass flow rate yields

$$a\dot{m}_t h_4 + (1 - a)\dot{m}_t h_7 = \dot{m}_t h_1 \tag{7.21}$$

which reduces to

$$ah_4 + (1 - a)h_7 = h_1 \tag{7.22}$$

Assuming that $P_1 = P_4 = P_7$, and that state 1 is a saturated liquid, we can use Eq. (7.22) to quickly find the mass fraction of fluid diverted from the turbine at state 4; that is, a.

Once a is known, we can find the turbine power and pump power using the summation of the products of mass flow rates and enthalpy changes through each device:

$$\dot{W}_t = \dot{m}_3(h_3 - h_4) + \dot{m}_5(h_4 - h_5) = \dot{m}_t[(h_3 - h_4) + (1 - a)(h_4 - h_5)] \tag{7.23}$$

$$\dot{W}_p = \dot{W}_{p1} + \dot{W}_{p2} = \dot{m}_6(h_6 - h_7) + \dot{m}_1(h_1 - h_2) = \dot{m}_t[(1 - a)(h_6 - h_7) + (h_1 - h_2)] \tag{7.24}$$

As the total flow proceeds through the steam generator, there is no change in that expression from versions of the Rankine cycle without superheat. In addition, the thermal efficiency is still found by dividing the net power produced by the heat input rate. If the cycle is employing nonisentropic turbines and pumps, the isentropic efficiency must be applied to each section of the turbine and to each pump. Although an open feedwater heater adds complexity to the analysis by requiring us to follow the flow rates closely, analysis of each component remains straightforward. It is possible to use more than one open feedwater heater, and, if we do this, we must find the fraction of mass flow diverted from the turbine at each feedwater heater, and an additional pump is required for each feedwater heater used.

STUDENT EXERCISE

Develop a computer model for an open feedwater heater, and modify your Rankine cycle system model to accommodate one or more open feedwater heaters.

▶ EXAMPLE 7.6

The ideal Rankine cycle with superheat of Example 7.4 is to be modified with the addition of an open feedwater heater. The pressure of the open feedwater heater is at 1 MPa. Saturated liquid water exits the open feedwater heater and enters a pump, where it is pressurized to the steam generator pressure of 10 MPa. Determine the net power produced by the cycle, and determine the thermal efficiency of this Rankine cycle with one open feedwater heater.

Given: $x_1 = 0$, $x_6 = 0$, $P_6 = 20$ kPa, $T_3 = 600°\text{C}$, $P_3 = 10$ MPa, $P_4 = P_7 = P_1 = 1$ MPa, $\dot{m}_t = 125$ kg/s

Find: $\dot{W}$, η

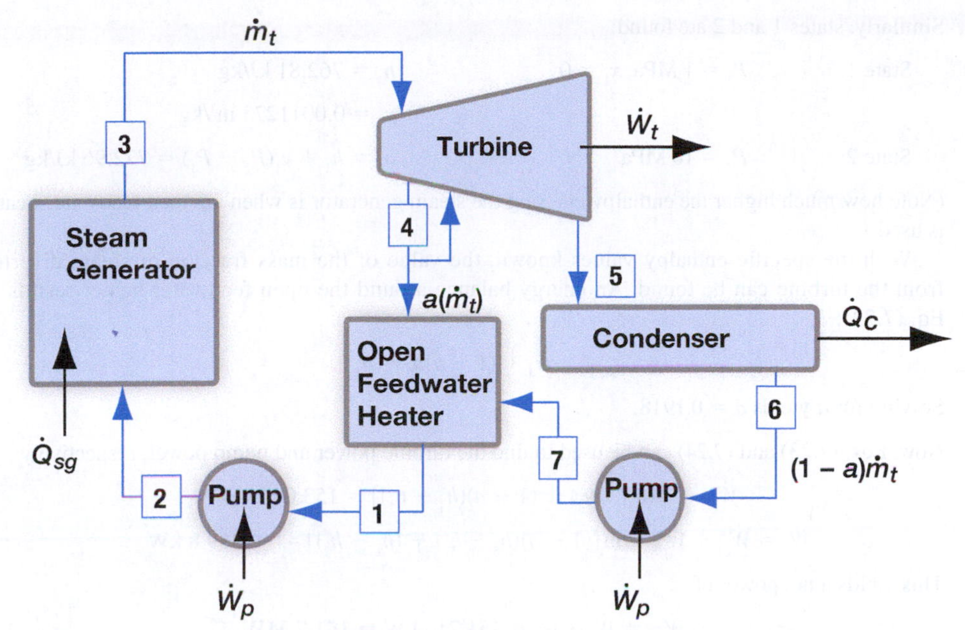

Solution: The cycle under consideration is a Rankine cycle with one open feedwater heater. The working fluid is water, which will be a compressed liquid, a saturated substance, and a superheated vapor at various points in the cycle.

Assume: $P_2 = P_3, P_5 = P_6, P_1 = P_4 = P_7, s_1 = s_2, s_3 = s_4 = s_5, \dot{W}_{sg} = \dot{W}_c = \dot{W}_{\text{ofwh}} = \dot{Q}_t = \dot{Q}_p = \dot{Q}_{\text{ofwh}} = 0$, $\Delta KE = \Delta PE = 0$ for each device. Steady-state, steady-flow systems. Liquid water in the pumps is incompressible.

As is usually the case for Rankine cycle anlaysis, the first step is to find the specific enthalpy values of each state, using a property calculator.

For state 3, use the given conditions.

State 3: $P_3 = 10$ MPa, $T_3 = 600°C$ $\quad h_3 = 3625.3$ kJ/kg

$s_3 = 6.9029$ kJ/kg · K

For state 4, use the pressure from state 1 and the entropy from state 3, because those values correspond to state 4's values for pressure and entropy, based on the assumptions given.

State 4: $P_4 = 1$ MPa $\quad s_4 = s_3 = 6.9029$ kJ/kg · K

$h_4 = 2931.7$ kJ/kg

Similarly,

State 5: $P_5 = P_6 = 20$ kPa $\quad s_5 = s_3 = 6.9029$ kJ/kg · K

$x_5 = (s_5 - s_f)/(s_g - s_f) = 0.8579$

$h_5 = 2274.6$ kJ/kg

For state 6, use the given conditions. However, we will find the specific volume at state 1, rather than the specific entropy, because the specific volume will be used in the calculation of the specific enthalpy of state 7.

State 6: $P_6 = 20$ kPa, $x_6 = 0.0$ $\quad h_6 = 251.40$ kJ/kg

$v_6 = 0.0010172$ m³/kg

For state 7, use the relationship presented in Eq. (6.48) for isentropic flow of an incompressible substance.

State 7: $P_7 = 1$ MPa $= 1000$ kPa $\quad h_7 = h_6 + v_6(P_7 - P_6) = 252.4$ kJ/kg

Similarly, states 1 and 2 are found:

State 1: $P_1 = 1$ MPa, $x_1 = 0$ $h_1 = 762.81$ kJ/kg

$v_1 = 0.0011273$ m^3/kg

State 2: $P_2 = 10$ MPa $h_2 = h_1 + v_1(P_2 - P_1) = 772.96$ kJ/kg

(Note how much higher the enthalpy entering the steam generator is when an open feedwater heater is used.)

With the specific enthalpy values known, the value of the mass fraction of steam diverted from the turbine can be found. An energy balance around the open feedwater heater results in Eq. (7.22):

$$ah_4 + (1 - a)h_7 = h_1$$

Solving for a yields $a = 0.1918$.

Now, Eqs. (7.23) and (7.24) can be used to find the turbine power and pump power, respectively:

$$\dot{W}_t = \dot{m}_t[(h_3 - h_4) + (1 - a)(h_4 - h_5)] = 153{,}084 \text{ kW}$$

$$\dot{W}_p = \dot{W}_{p1} + \dot{W}_{p2} = \dot{m}_t[(1 - a)(h_6 - h_7) + (h_1 - h_2)] = -1369.8 \text{ kW}$$

This yields a net power of

$$\dot{W}_{\text{net}} = \dot{W}_t + \dot{W}_p = 151{,}715 \text{ kW} = \mathbf{151.7\ MW}$$

For the heat transfer rate of the steam generator, $\dot{Q}_{sg} = \dot{m}(h_3 - h_2) = 356{,}540$ kW

and the thermal efficiency is

$$\eta = \frac{\dot{W}_{\text{net}}}{\dot{Q}_{\text{in}}} = \frac{\dot{W}_t + \dot{W}_p}{\dot{Q}_{sg}} = \mathbf{0.426}$$

Analysis: Although the net power decreases while using an open feedwater heater, the heat input drops as well, and by a larger percentage. (The net power decreased by 9.5%, whereas the heat input decreased by 15.2%.) The result is that the thermal efficiency increases with the use of feedwater heating—in this case the thermal efficiency increased by 2.7 percentage points.

Open feedwater heaters are simple devices, but they can be difficult to maintain and operate. As such, most large-scale steam power plants use no more than one open feedwater heater in their system, and this is in place primarily to aid "deaeration," or the removal of noncondensible gases (such as air that may have leaked in through the condenser where the steam pressure is below atmospheric pressure) from the water. A more common type of feedwater heating is to use a closed feedwater heater.

Rankine Cycle with a Closed Feedwater Heater

A closed feedwater heater is essentially a small condenser that has the basic form of a shell-and-tube heat exchanger. Steam is extracted from the turbine system and directed toward the closed feedwater heater. The feedwater flows through the tubes and receives heat from the steam condensing on the outside of the tubes. In such an arrangement, the condensate from the steam typically exits as a saturated liquid ($x_7 = 0$ for the numbering systems used in **Figures 7.22** and **7.23**). This condensed steam then must be combined with the feedwater at another point in the cycle. Figures 7.22 and 7.23 provide two options, as shown in component diagrams for a Rankine cycle with one closed feedwater heater. Figure 7.22 shows the more common option of throttling the condensed steam to a lower pressure and then combining the resulting saturated mixture back with the main flow in the condenser, where additional heat is recovered through condensation

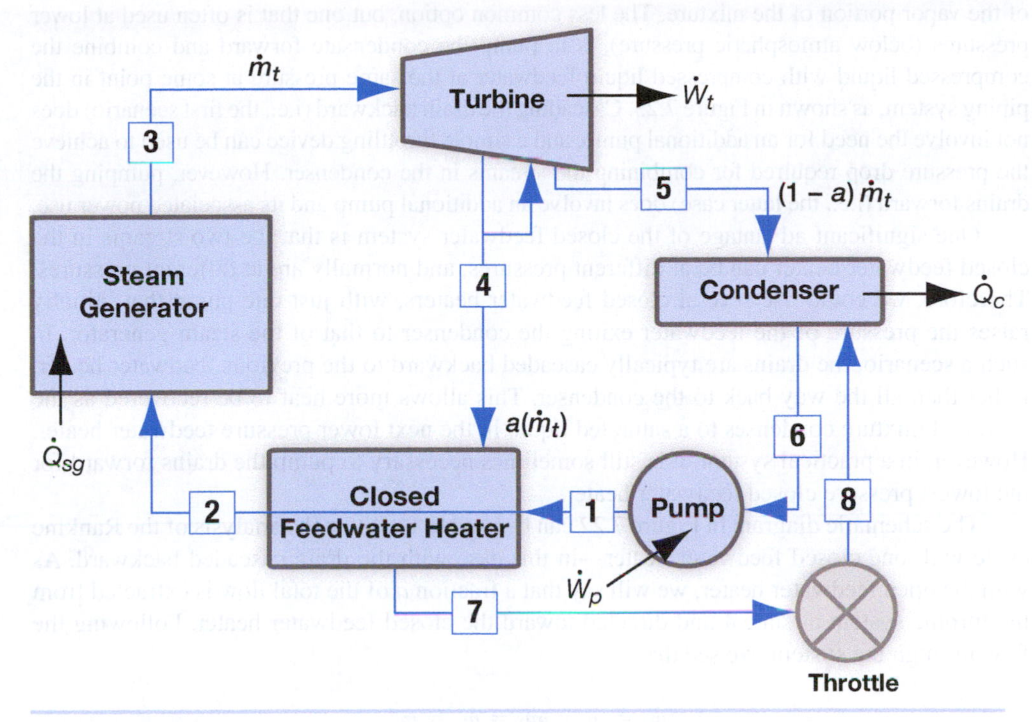

FIGURE 7.22 A component diagram of a Rankine cycle with one closed feedwater heater, with its drain cascaded backward.

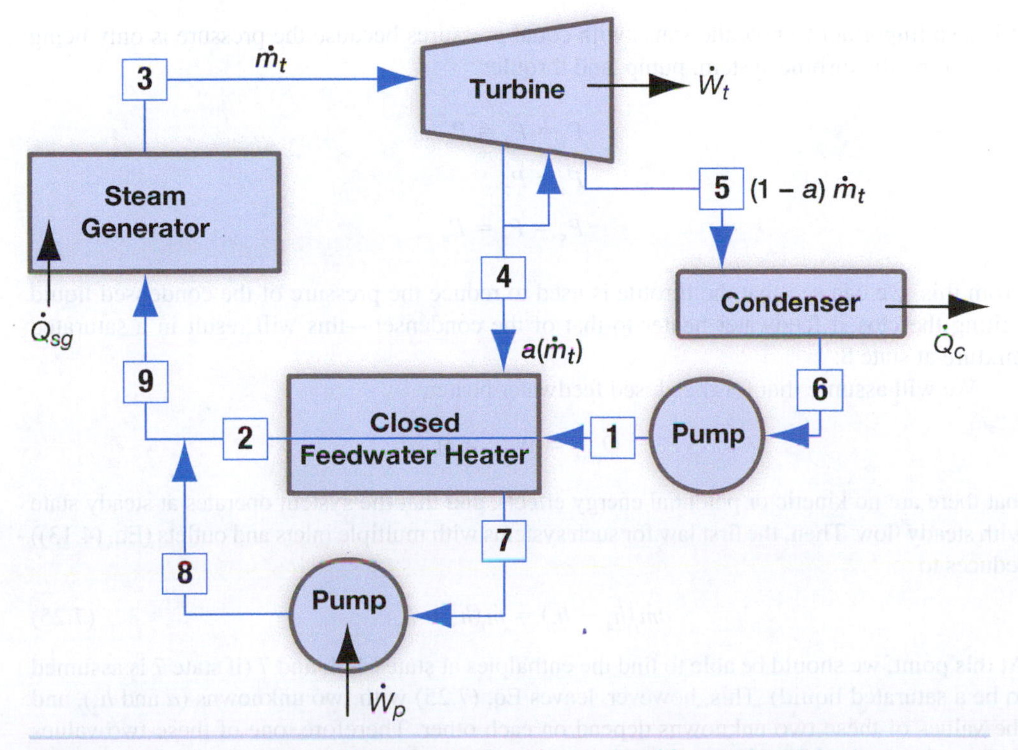

FIGURE 7.23 A component diagram of a Rankine cycle with one closed feedwater heater, with its drain pumped forward.

of the vapor portion of the mixture. The less common option, but one that is often used at lower pressures (below atmospheric pressure), is to pump the condensate forward and combine the compressed liquid with compressed liquid feedwater at the same pressure at some point in the piping system, as shown in Figure 7.23. Cascading the drain backward (i.e., the first scenario) does not involve the need for an additional pump, and a simple throttling device can be used to achieve the pressure drop required for combining the streams in the condenser. However, pumping the drains forward (i.e., the latter case) does involve an additional pump and its associated power use.

One significant advantage of the closed feedwater system is that the two streams in the closed feedwater heater can be at different pressures, and normally are at different pressures. Therefore, we could use several closed feedwater heaters, with just one pump that initially raises the pressure of the feedwater exiting the condenser to that of the steam generator. In such a scenario, the drains are typically cascaded backward to the previous feedwater heater, rather than all the way back to the condenser. This allows more heat to be recovered as the saturated mixture condenses to a saturated liquid in the next lower pressure feedwater heater. However, in a practical system, it is still sometimes necessary to pump the drains forward for the lowest pressure closed feedwater heater.

The schematic diagram in Figure 7.22 can be used to facilitate the analysis of the Rankine cycle with one closed feedwater heater—in this case with the drain cascaded backward. As with the open feedwater heater, we will say that a fraction a of the total flow is extracted from the turbine system at state 4 and directed toward the closed feedwater heater. Following the flow through the system, we see that

$$\dot{m}_1 = \dot{m}_2 = \dot{m}_3 = \dot{m}_6 = \dot{m}_t$$
$$\dot{m}_4 = \dot{m}_7 = \dot{m}_8 = a\dot{m}_t$$
$$\dot{m}_5 = (1 - a)\dot{m}_t$$

It is also important to note the states with equal pressures because the pressure is only being changed by the turbine system, pump, and throttle:

$$P_1 = P_2 = P_3$$
$$P_4 = P_7$$
$$P_5 = P_6 = P_8$$

From this, we can see that the throttle is used to reduce the pressure of the condensed liquid exiting the closed feedwater heater to that of the condenser—this will result in a saturated mixture at state 8.

We will assume that for the closed feedwater heater,

$$\dot{Q}_{cfwh} = \dot{W}_{cfwh} = 0$$

that there are no kinetic or potential energy effects, and that the system operates at steady state with steady flow. Then, the first law for such systems with multiple inlets and outlets (Eq. (4.13)) reduces to

$$a\dot{m}_t(h_4 - h_7) = \dot{m}_t(h_2 - h_1) \tag{7.25}$$

At this point, we should be able to find the enthalpies at states 1, 4, and 7 (if state 7 is assumed to be a saturated liquid). This, however, leaves Eq. (7.25) with two unknowns (a and h_2), and the values of these two unknowns depend on each other. Therefore, one of these two values will need to be specified. Often, rather than directly specifying h_2, a temperature at state 2 is given and the enthalpy at state 2 is taken to be that of a saturated liquid at T_2. When choosing the temperature at state 2, keep in mind that the second law requires that the temperature of the feedwater exiting the closed feedwater heater be less than or equal to the temperature of the steam entering: $T_2 \leq T_4$.

Once we know the enthalpy values and extracted mass fraction, we can calculate the turbine power, pump power, heat input, and thermal efficiency using techniques employed previously:

$$\dot{W}_t = \dot{m}_3(h_3 - h_4) + \dot{m}_5(h_4 - h_5) = \dot{m}_t[(h_3 - h_4) + (1 - a)(h_4 - h_5)] \quad (7.26)$$

$$\dot{W}_p = \dot{m}_6(h_6 - h_1) = \dot{m}_t(h_6 - h_1) \quad (7.27)$$

$$\dot{Q}_{sg} = \dot{m}_t(h_3 - h_2) \quad (7.28)$$

We can also analyze a system with a closed feedwater heater with the drain pumped forward. We leave this as an end-of-chapter exercise.

STUDENT EXERCISE

Develop a computer model for a closed feedwater heater, and modify your Rankine cycle system model to accommodate one or more closed feedwater heaters, with options of pumping the drains forward or cascading the drains backward.

▶ EXAMPLE 7.7

The ideal Rankine cycle with superheat of Example 7.4 is to be modified with the addition of a closed feedwater heater. The steam is extracted from the turbine system at a pressure of 1 MPa to be sent to the closed feedwater heater. Saturated liquid water exits the drain of the closed feedwater heater and is directed through a trap (throttle) and then to the condenser. The feedwater exits the closed feedwater heater at a temperature of 175°C. Determine the net power produced by the cycle, and determine the thermal efficiency of this Rankine cycle with one closed feedwater heater.

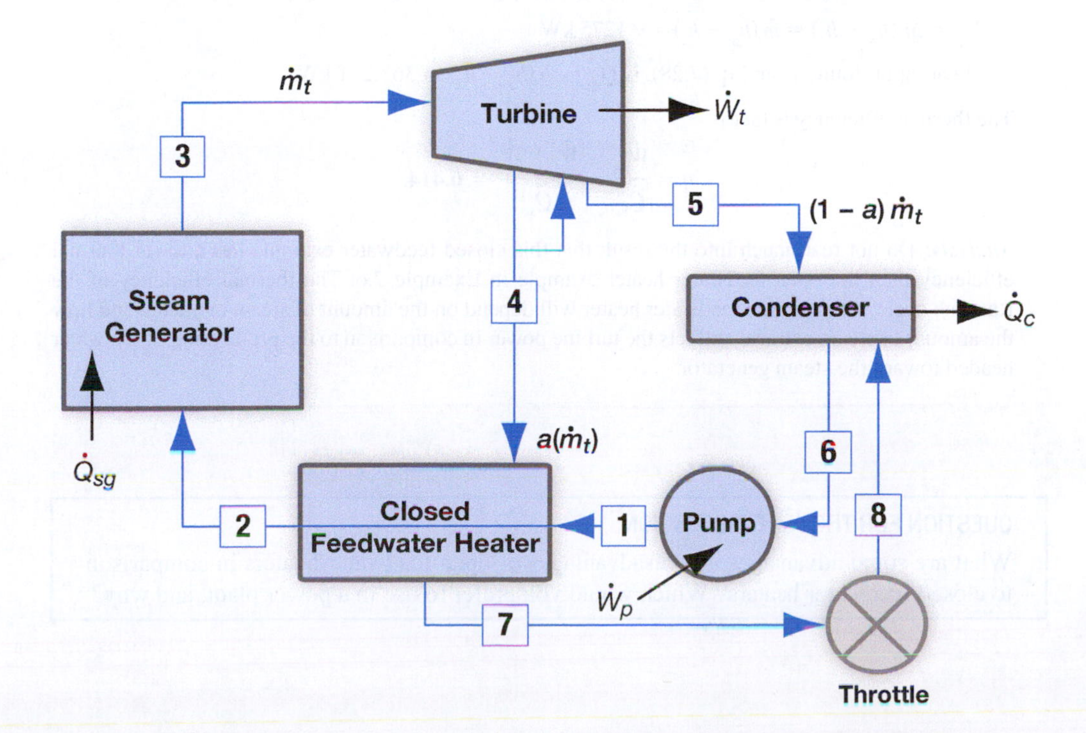

Given: $x_6 = 0, x_7 = 0, P_6 = 20$ kPa, $T_3 = 600°C$, $P_3 = 10$ MPa, $P_4 = P_7 = 1$ MPa, $T_2 = 170°C$, $\dot{m}_t = 125$ kg/s

Find: $\dot{W}, \eta$

Solution: The cycle under consideration is a Rankine cycle with one closed feedwater heater cascaded backward. The working fluid is water, which will be a compressed liquid, a saturated substance, and a superheated vapor at various points in the cycle.

Assume: $P_1 = P_2 = P_3, P_5 = P_6 = P_8, P_4 = P_7, s_1 = s_6, s_3 = s_4 = s_5,$
$\dot{W}_{sg} = \dot{W}_c = \dot{W}_{cfwh} = \dot{Q}_t = \dot{Q}_p = \dot{Q}_{cfwh} = 0, \Delta KE = \Delta PE = 0$ for each device. Steady-state, steady-flow systems. Liquid water in the pump is incompressible.

Many of the enthalpy values have been found in Examples 7.4 and 7.5 already (note that some state numbers have changed):

$h_1 = 261.6$ kJ/kg

$h_3 = 3625.3$ kJ/kg

$h_4 = 2931.7$ kJ/kg

$h_5 = 2274.6$ kJ/kg

$h_6 = 251.40$ kJ/kg

State 2: As a compressed liquid, take $h_2 = h_f(T_2) = 719.21$ kJ/kg.

State 7: As a saturated liquid, $h_7 = h_f = 762.81$ kJ/kg.

State 8: For a throttling process, $h_8 = h_7 = 762.81$ kJ/kg.

Using Eq. (7.25), $a\dot{m}_t(h_4 - h_7) = \dot{m}_t(h_2 - h_1)$, the fraction of mass extracted can be found to be

$$a = 0.211$$

The turbine power and pump power are found from Eqs. (7.26) and (7.27):

$$\dot{W}_t = \dot{m}_3(h_3 - h_4) + \dot{m}_5(h_4 - h_5) = \dot{m}_t[(h_3 - h_4) + (1 - a)(h_4 - h_5)] = 151{,}507 \text{ kW}$$

$$\dot{W}_p = \dot{m}_6(h_6 - h_1) = \dot{m}_t(h_6 - h_1) = -1275 \text{ kW}$$

The heat input, found from Eq. (7.28), is $\dot{Q}_{sg} = \dot{m}_t(h_3 - h_2) = 363{,}261$ kW.

The thermal efficiency is then

$$\eta = \frac{\dot{W}_{\text{net}}}{\dot{Q}_{\text{in}}} = \frac{\dot{W}_t + \dot{W}_p}{\dot{Q}_{sg}} = \mathbf{0.414.}$$

Analysis: Do not read much into the result that this closed feedwater example has a lower thermal efficiency than the open feedwater heater example in Example 7.6. The thermal efficiency of the Rankine cycle with a closed feedwater heater will depend on the amount of steam extracted and how the amount of steam extracted affects the turbine power in comparison to the pre-heating of the water headed toward the steam generator.

QUESTION FOR THOUGHT/DISCUSSION

What are some advantages and disadvantages of open feedwater heaters in comparison to closed feedwater heaters? Which would you prefer to use in a power plant, and why?

Large steam power plants often will use a Rankine cycle with many of these additional components added. For example, you may find a Rankine cycle with superheat, one reheater, one open feedwater heater, and several closed feedwater heaters. (Also note that actual power plants will have pressure losses through piping and components that we do not incorporate into the Rankine cycle analysis here, but we could easily include those losses if we had measurements of the pressure and temperature before and after each component—we would just use the enthalpy corresponding to those specific temperature and pressure combinations.) When analyzing a cycle with numerous additional components, it is important to think through how to develop expressions for turbine power, closed feedwater heaters, heat input, and so on. Always remember to develop expressions based on the first law. Determine which states have the same pressure, recalling that we treat turbines, pumps, and throttles as the devices to impart a pressure change in the system. Follow the mass through, and keep track of how much mass has been diverted to different components. Then multiply the particular mass flow rate at a location by its change in enthalpy to get a heat transfer rate or power for that component. Examples 7.8 and 7.9 will illustrate how to develop equations for particular situations that may arise when dealing with multiple components.

▶ EXAMPLE 7.8

Consider the turbine in the modified Rankine cycle shown. Steam enters the turbine at state 3, and a fraction a is removed and sent to a closed feedwater heater at state 4. The remaining steam expands further and is extracted and directed toward a reheater at state 5, and returns at state 6. After further expansion, a fraction b is extracted and directed toward an open feedwater heater at state 7, and, after more expansion, a fraction c is extracted at state 8 and directed toward a closed feedwater heater. The remaining steam exits at state 9. Determine the expression to be used for the power from the turbine.

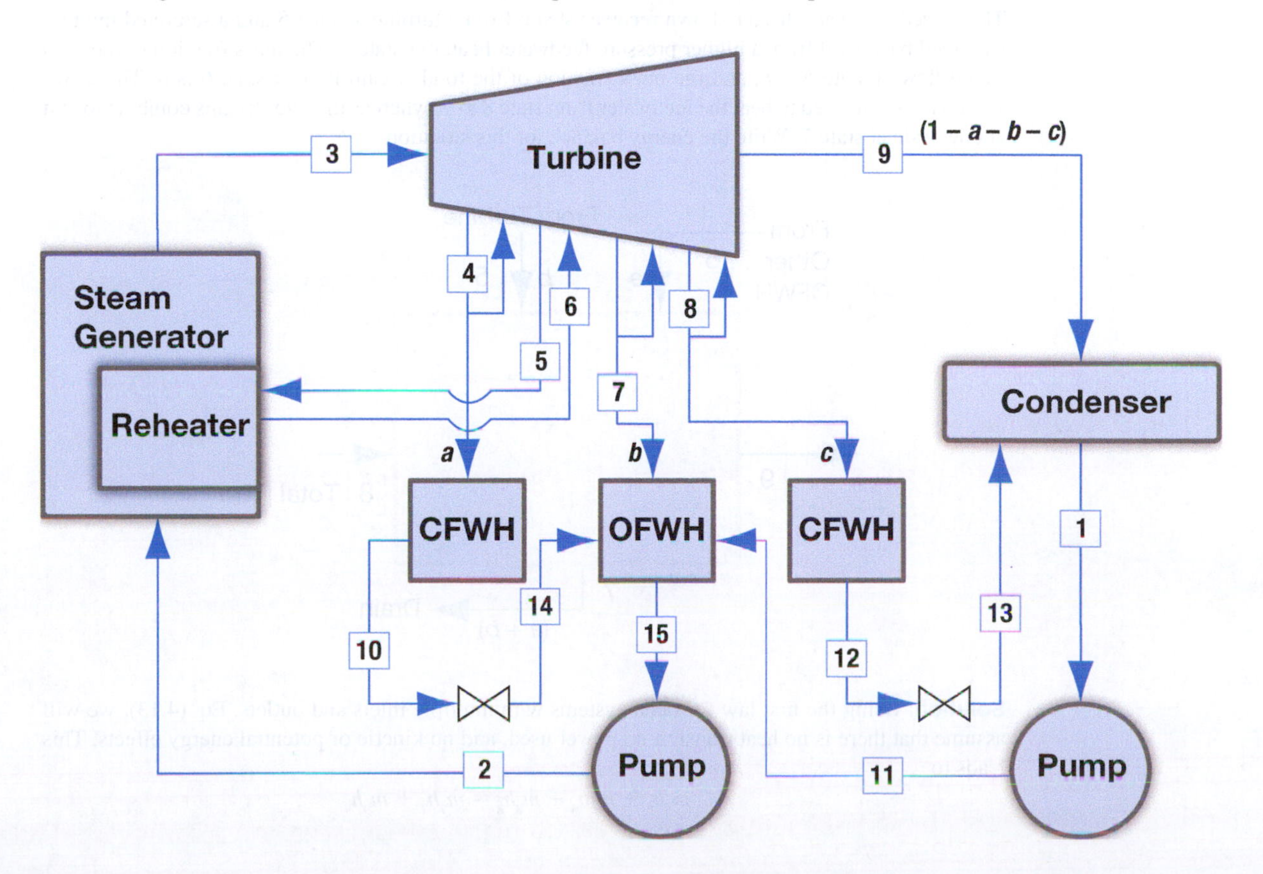

Solution: Considering the mass flow rates,

$$\dot{m}_3 = \dot{m}_t$$
$$\dot{m}_4 = a\dot{m}_t$$
$$\dot{m}_5 = \dot{m}_6 = (1 - a)\dot{m}_t$$
$$\dot{m}_7 = b\dot{m}_t$$
$$\dot{m}_8 = c\dot{m}_t$$
$$\dot{m}_9 = (1 - a - b - c)\dot{m}_t$$

Flow between state 7 and 8 through the turbine is given by

$$(1 - a - b)\dot{m}_t$$

Multiplying the flow through the turbine between states by the change in enthalpy between the states and adding the result will give an expression for the turbine power:

$$\dot{W}_t = \dot{m}_t[(h_3 - h_4) + (1 - a)(h_4 - h_5) + (1 - a)(h_6 - h_7) + (1 - a - b)(h_7 - h_8) + (1 - a - b - c)(h_8 - h_9)]$$

Analysis: The heat added through the reheater will be $\dot{Q}_r = (1 - a)\dot{m}_t(h_6 - h_5)$, because only a fraction of the total flow through the steam generator goes through the reheater.

▶ EXAMPLE 7.9

The closed feedwater heater shown receives steam from a turbine at state 5 and a saturated mixture cascaded backward from a higher pressure feedwater heater at state 6. The mass fraction of the total steam flow at state 5 is b, and the mass fraction of the total stream flow at state 6 is a. These two streams both are used to heat the feedwater from state 8 to 9, whereas the two streams combine to exit at one drain at state 7. Write the energy balance for this situation.

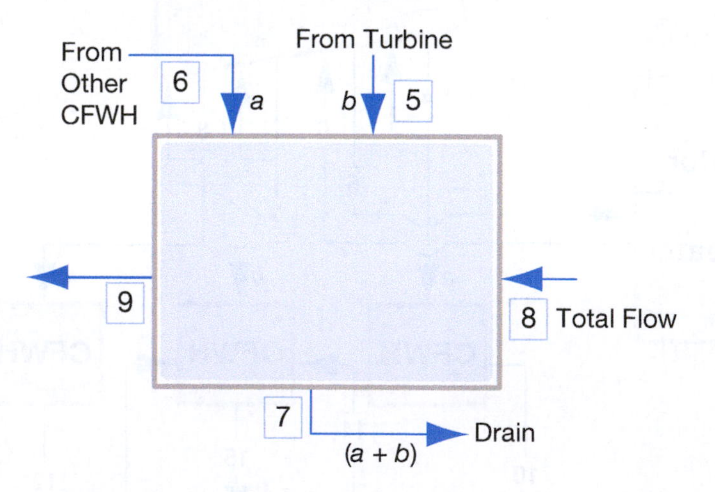

Solution: Using the first law for open systems with multiple inlets and outlets, Eq. (4.13), we will assume that there is no heat transfer, no power used, and no kinetic or potential energy effects. This leads to

$$\dot{m}_5 h_5 + \dot{m}_6 h_6 + \dot{m}_8 h_8 = \dot{m}_7 h_7 + \dot{m}_9 h_9$$

Considering that

$$\dot{m}_8 = \dot{m}_9 = \dot{m}_t$$
$$\dot{m}_6 = a\dot{m}_t$$
$$\dot{m}_5 = b\dot{m}_t$$
$$\dot{m}_7 = \dot{m}_5 + \dot{m}_6 = (a + b)\dot{m}_t$$

we can substitute and see

$$bh_5 + ah_6 - (a + b)h_7 = h_9 - h_8$$

when $\dot{m}_t$ is divided out from each term.

QUESTION FOR THOUGHT/DISCUSSION

In the last two decades, many countries have shown increasing interest in reducing their use of coal in steam power plants. What are some reasons for this interest in changing primary fuel sources?

7.4 GAS (AIR) POWER CYCLES AND AIR STANDARD CYCLE ANALYSIS

The Rankine cycle, with its various modifications, describes a vapor power cycle—one that has the working fluid changing phase. Although they are commonly used, such cycles have a disadvantage in that they often require multiple working fluids: one fluid flows through the power cycle itself, a second fluid provides the energy for a heat source such as high-temperature combustion products, and a third fluid may be needed to remove heat from the cycle. Such requirements usually limit vapor power cycles to stationary applications, and also increase the price of the system using the cycle.

Gas power cycles, on the other hand, typically use a single working fluid that begins as air and is then mixed with fuel. The resulting mixture is burned, resulting in combustion products that then flow through a prime mover to provide power. Finally, the combustion products are exhausted to the atmosphere. The working fluid usually begins as air, so this category of cycles is also known as air power cycles. (Although this description fits the common practical form exhibited by gas power cycles, the cycles can also be made in a closed-loop format with heat being transferred to some gas flowing through the cycle, and heat being rejected from the gas to the surroundings. This is conceptually similar to how a Rankine cycle is used, but without a phase change in the fluid.)

You may observe that this practical operational description does not fit that of a thermodynamic cycle, for which the working fluid begins and ends the cycle in the same thermodynamic state. As such, several assumptions are made to enable these practical systems to be analyzed as thermodynamic cycles. These assumptions permit what is called an air standard cycle (ASC) analysis. The assumptions made for ASC analysis are as follows:

1. The cycle is a closed-loop system with the working fluid being air that flows through the cycle.
2. The air behaves as an ideal gas.
3. All processes in the cycle are internally reversible.
4. The combustion process is replaced with a heat addition process, which receives heat from an external source.
5. The exhaust and intake processes are replaced with a single heat-rejection process which returns the air to its initial state.

If a gas other than air is being used, a corresponding gas cycle analysis can be performed, with *gas* replacing *air* in the preceding assumptions. Sometimes a "cold ASC" analysis is performed, where the specific heats of the air are held constant at values found at 25°C.

There are many practical air power cycles in use, but we will focus on three cycles of particular practical interest. The Brayton cycle is the basic cycle used for gas turbine applications, the Otto cycle is the theoretical cycle describing spark-ignition internal combustion engine operation, and the diesel cycle is the historical theoretical cycle describing compression-ignition engine operation. All of these cycles can be used in both stationary and mobile operations. In addition, although the equipment that uses these cycles generally operates in an open format, where fresh air is regularly ingested and combustion products are exhausted, the cycles can be operated in a closed loop, with heat transfers described in the cycles themselves.

7.5 BRAYTON CYCLE

The Brayton cycle was developed by the American engineer George Brayton in the 19th century, who improved on the work seen in earlier cycles devised by Stirling, Ericsson, and Lenoir. The Brayton cycle is applied in gas turbines and therefore is encountered today in electricity generation, aircraft engines, and other gas turbine applications. Some gas turbines are illustrated in **Figure 7.24**. The cycle itself can be seen as a gas power cycle equivalent of the basic Rankine cycle because it consists of the following four processes (in an ideal cycle):

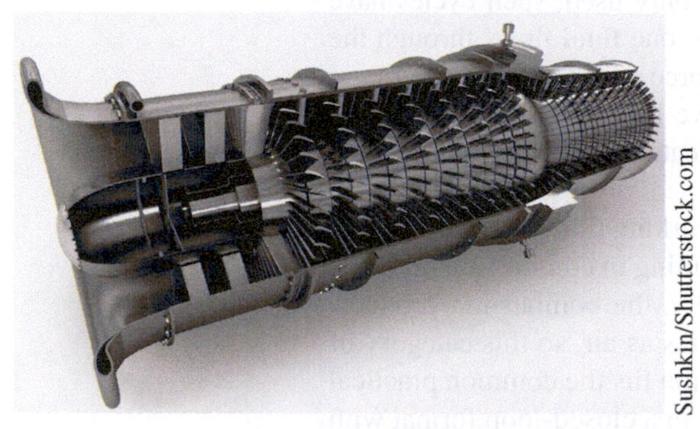

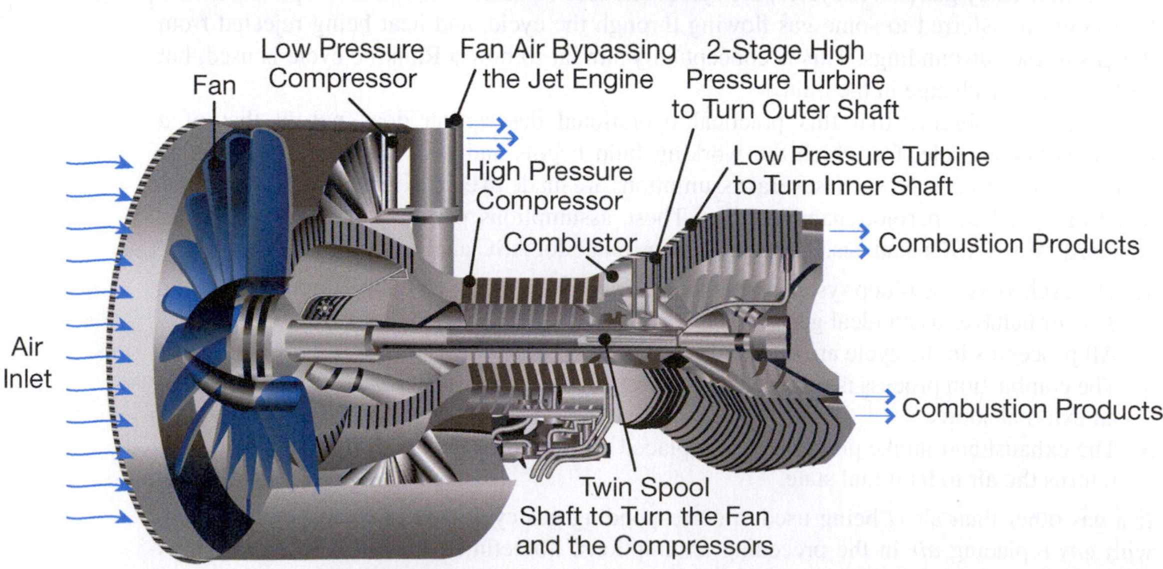

FIGURE 7.24 Cut-away images of gas turbines.

Process 1-2: Isentropic compression
Process 2-3: Constant-pressure heat addition
Process 3-4: Isentropic expansion
Process 4-1: Constant-pressure heat removal

Figure 7.25 shows the four processes on both a *T-s* diagram and a *P-v* diagram. Recall that because the cycle uses gases, the working fluid is considered to be well away from the vapor dome, and so the vapor dome does not appear on these diagrams. The equipment used in practice to achieve these processes is shown schematically in Figure 7.26. The gas first passes through a compressor, then is directed toward a combustor (where fuel is usually added and combustion

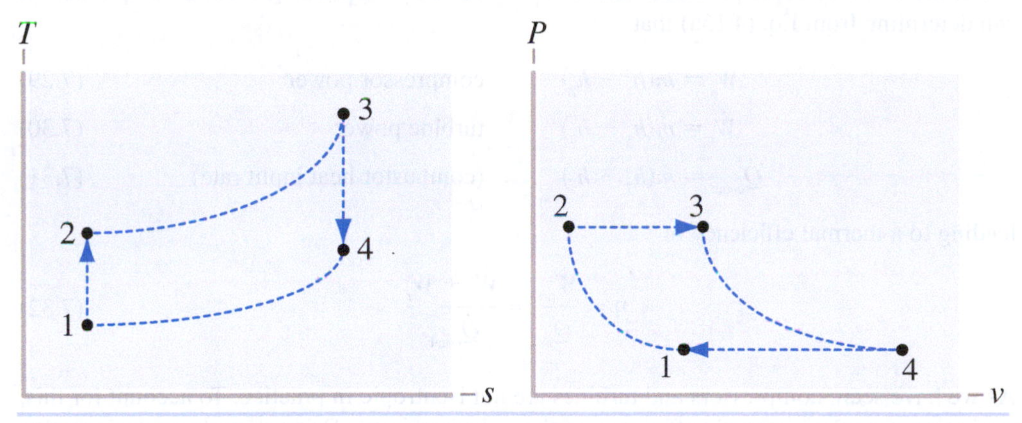

FIGURE 7.25 *T-s* and *P-v* diagrams of the processes in an ideal, basic Brayton cycle.

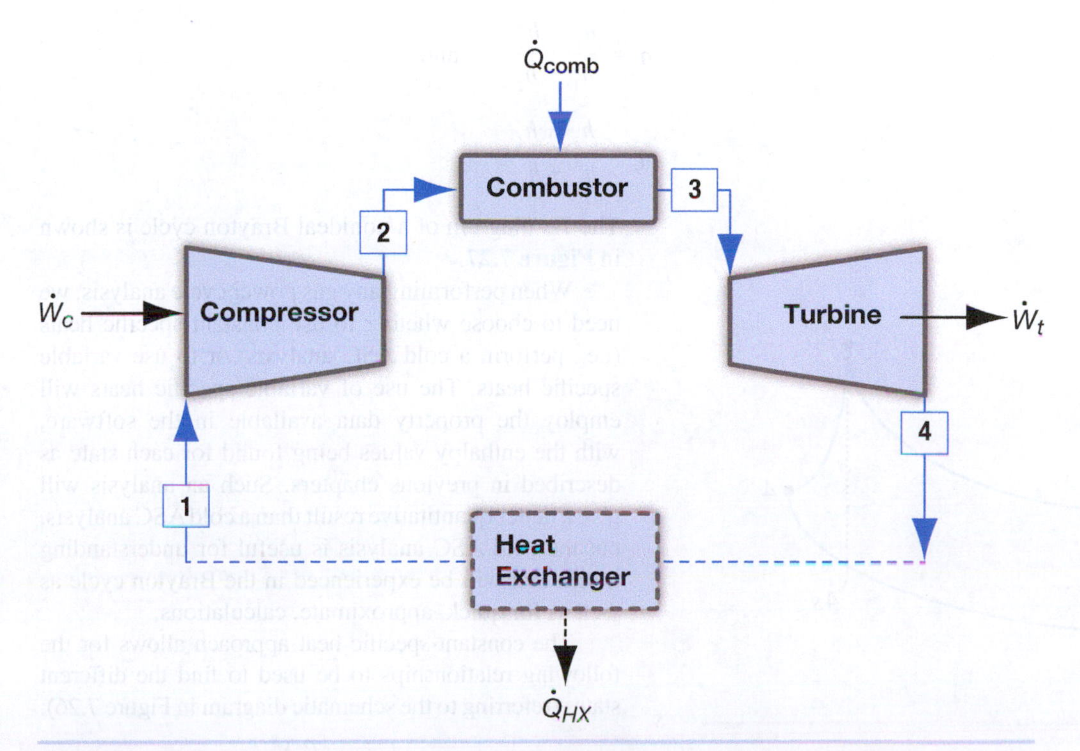

FIGURE 7.26 A component diagram of a basic Brayton cycle. Most working systems do not have the heat exchanger between the turbine exhaust and the compressor inlet.

takes place), and then flows through a turbine. To complete a thermodynamic cycle analysis, a heat exchanger is hypothetically used between states 4 and 1—this is illustrated as a dotted line in Figure 7.26. If the Brayton cycle was operated in a closed loop, this additional heat exchanger would be present in a closed-loop system. Also in such a system, the combustor would be a heat exchanger and there would be no fuel added to the gas acting as the working fluid.

The assumptions that are commonly made with the Brayton cycle are that each device operates as a steady-state, steady-flow, single-inlet, single-outlet device; that the ASC assumptions are in place; that there are no changes in kinetic or potential energy through each device; and that $\dot{Q}_c = \dot{Q}_t = \dot{W}_{comb} = \dot{W}_{HX} = 0$, where c designates the compressor and "comb" designates the combustor. In addition, $P_2 = P_3$ and $P_1 = P_4$. For the ideal Brayton cycle, the two isentropic processes lead to $s_1 = s_2$ and $s_3 = s_4$. Using these assumptions, we can determine from Eq. (4.15a) that

$$\dot{W}_c = \dot{m}(h_1 - h_2) \qquad \text{compressor power} \tag{7.29}$$

$$\dot{W}_t = \dot{m}(h_3 - h_4) \qquad \text{turbine power} \tag{7.30}$$

$$\dot{Q}_{comb} = \dot{m}(h_3 - h_2) \qquad \text{(combustor heat input rate)} \tag{7.31}$$

leading to a thermal efficiency of

$$\eta = \frac{\dot{W}_{net}}{\dot{Q}_{in}} = \frac{\dot{W}_c + \dot{W}_t}{\dot{Q}_{comb}} \tag{7.32}$$

As we have seen, compressors and turbines are not isentropic in practice. To account for this, we incorporate the isentropic efficiency of these devices in the Brayton cycle analysis so that the cycle is now considered a nonideal Brayton cycle:

$$\eta_c = \frac{h_1 - h_{2s}}{h_1 - h_2} \quad \text{and}$$

$$\eta_t = \frac{h_3 - h_4}{h_3 - h_{4s}}$$

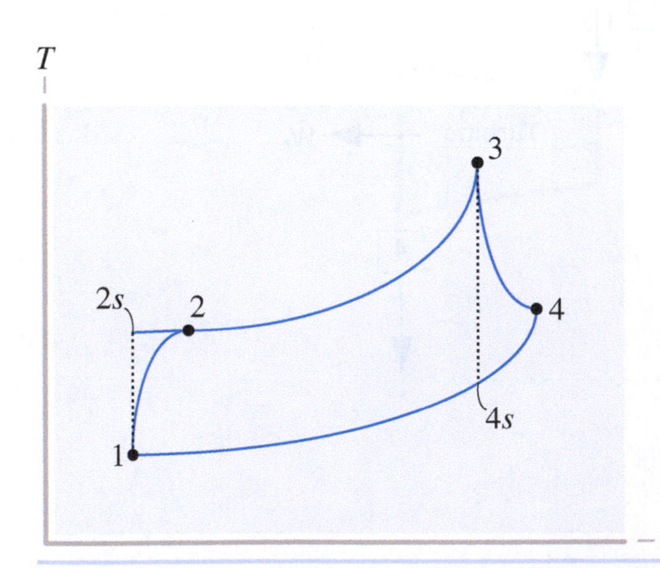

FIGURE 7.27 A *T-s* diagram of a nonideal, basic Brayton cycle.

The *T-s* diagram of a nonideal Brayton cycle is shown in **Figure 7.27**.

When performing any gas power cycle analysis, we need to choose whether to use constant specific heats (i.e., perform a cold ASC analysis) or to use variable specific heats. The use of variable specific heats will employ the property data available in the software, with the enthalpy values being found for each state as described in previous chapters. Such an analysis will give a better quantitative result than a cold ASC analysis, but the cold ASC analysis is useful for understanding trends that will be experienced in the Brayton cycle as well as for quick, approximate, calculations.

The constant-specific heat approach allows for the following relationships to be used to find the different states (referring to the schematic diagram in Figure 7.26):

$$\frac{T_{2s}}{T_1} = \left(\frac{P_2}{P_1}\right)^{\frac{k-1}{k}}$$

$$\frac{T_{4s}}{T_3} = \left(\frac{P_4}{P_3}\right)^{\frac{k-1}{k}}$$

and $\Delta h = c_p \Delta T$ for each change in enthalpy

STUDENT EXERCISE

Build a computer model for the Brayton cycle. You can either use your existing models for the compressor and the turbine or modify the models to automatically incorporate the standard Brayton cycle assumptions. You will also need to develop a model for the combustor. Your model should be able to accommodate a nonisentropic turbine and a nonisentropic compressor. The model should be capable of using either constant or variable specific heats (for this, you may choose to set up two similar system models). The model should determine net power and thermal efficiency values given temperature and pressure data, or be able to determine temperature data if given heat input or work information.

▶ EXAMPLE 7.10

Air enters the compressor of a Brayton cycle at 300 K and 100 kPa, at a rate of 10 kg/s. The air is compressed to a pressure of 1000 kPa, and then the air is heated in the combustor to a temperature of 1200 K. Finally, the air expands in the turbine to a pressure of 100 kPa. Considering variable specific heats for the air, determine the power produced by the cycle and the thermal efficiency of the cycle if (a) the turbine and compressor are isentropic, and (b) the turbine and compressor have isentropic efficiencies of 0.80.

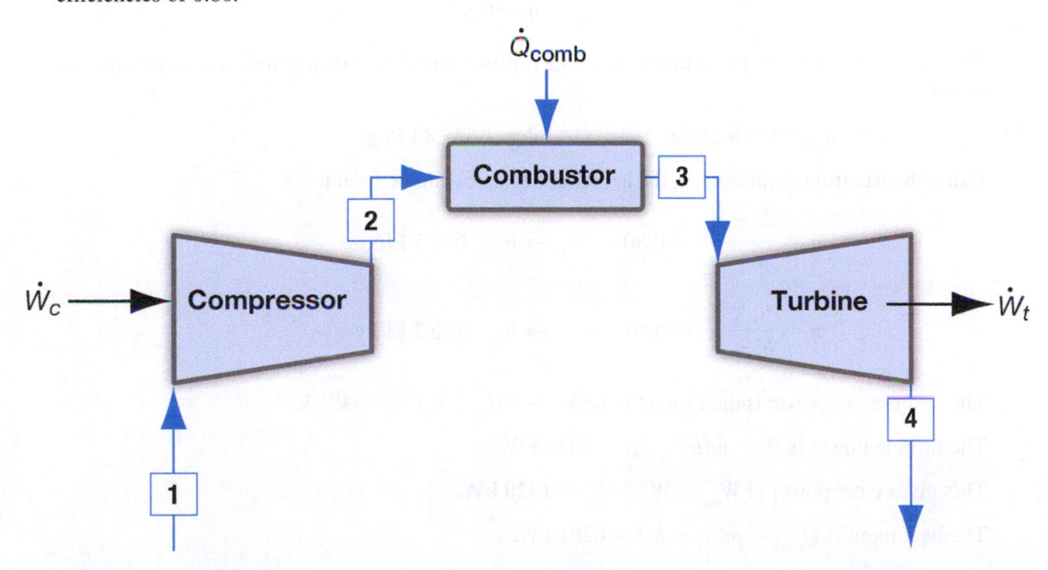

Given: $P_1 = P_4 = 100$ kPa, $P_2 = P_3 = 1000$ kPa, $T_1 = 300$ K, $T_3 = 1200$ K, $\dot{m} = 10$ kg/s

Find: $\dot{W}$, η for the cases of (a) isentropic turbine and compressor, and (b) nonisentropic turbine and compressor with the given efficiency.

Solution: The cycle under study is the basic Brayton cycle. The working fluid is air, and we will consider the specific heats of the air to be variable (using Table A.3 for properties, although computer programs could also be employed).

Assume: $P_1 = P_4$, $P_2 = P_3$, $\dot{Q}_c = \dot{Q}_t = \dot{W}_{comb} = 0$, $\Delta KE = \Delta PE = 0$ for each device. Steady-state, steady-flow processes through each device.

(a) For the isentropic turbine and compressor, the exit states can be found using Table A.3 in the appendix and considering isentropic processes for air:

$$P_{r2} = (P_2/P_1)P_{r1}$$

At 300 K, $P_{r1} = 1.386$, so $P_{r2} = 13.86$.

This value corresponds to $T_2 = 574$ K.

At 1200 K, $P_{r3} = 238.0$. Then, $P_{r4} = P_{r3}(P_4/P_3) = 23.80$, which corresponds to $T_4 = 665$ K.

Each of these temperatures corresponds to the following enthalpy values for air:

$$h_1 = 300.19 \text{ kJ/kg} \qquad h_3 = 1277.79 \text{ kJ/kg}$$
$$h_2 = 579.8 \text{ kJ/kg} \qquad h_4 = 675.8 \text{ kJ/kg}$$

The compressor power is then found to be $\dot{W}_c = \dot{m}(h_1 - h_2) = -2796$ kW.

The turbine power is $\dot{W}_t = \dot{m}(h_3 - h_4) = 6020$ kW.

This gives a net power of $\dot{W}_{\text{net}} = \dot{W}_c + \dot{W}_t =$ **3220 kW.**

The heat input is $\dot{Q}_{\text{comb}} = \dot{m}(h_3 - h_2) = 6980$ kW,

and the thermal efficiency is

$$\eta = \frac{\dot{W}_{\text{net}}}{\dot{Q}_{\text{in}}} = \frac{\dot{W}_c + \dot{W}_t}{\dot{Q}_{\text{comb}}} = \mathbf{0.462}$$

(b) The isentropic efficiencies of the turbine and compressor are now 0.80:

$$\eta_c = \eta_t = 0.80$$

The outlet states from the compressor and turbine found in (a) are now the isentropic outlet states:

$$h_{2s} = 579.8 \text{ kJ/kg} \qquad h_{4s} = 675.8 \text{ kJ/kg}$$

Using the isentropic efficiencies, the actual outlet states can be found:

$$\eta_c = \frac{h_1 - h_{2s}}{h_1 - h_2} = 0.80 \qquad \rightarrow h_2 = 649.7 \text{ kJ/kg}$$

$$\eta_t = \frac{h_3 - h_4}{h_3 - h_{4s}} = 0.80 \qquad \rightarrow h_4 = 796.2 \text{ kJ/kg}$$

The compressor power is then found to be $\dot{W}_c = \dot{m}(h_1 - h_2) = -3495$ kW.

The turbine power is $\dot{W}_t = \dot{m}(h_3 - h_4) = 4816$ kW.

This gives a net power of $\dot{W}_{\text{net}} = \dot{W}_c + \dot{W}_t =$ **1320 kW.**

The heat input is $\dot{Q}_{\text{comb}} = \dot{m}(h_3 - h_2) = 6281$ kW,

and the thermal efficiency is

$$\eta = \frac{\dot{W}_{\text{net}}}{\dot{Q}_{\text{in}}} = \frac{\dot{W}_c + \dot{W}_t}{\dot{Q}_{\text{comb}}} = \mathbf{0.210}$$

Analysis:

(1) The ideal Brayton cycle has a very favorable thermal efficiency in comparison to the Rankine cycle.

(2) However, real systems do not have isentropic turbines and compressors, and when the non-isentropic nature of the turbine and compressor is introduced, the performance of the Brayton cycle deteriorates substantially. The difference between the Rankine and Brayton cycle

performance deterioration is that, as a percentage of the power produced by the turbine, the power consumed by the compressor in a Brayton cycle is much larger than the power consumed by the pump in a Rankine cycle. The pump power is nearly insignificant in the Rankine cycle, whereas the compressor power represents a large loss in the Brayton cycle. So, when both devices in the Brayton cycle are nonisentropic, the impact is multiplied and the net power quickly drops, as does the thermal efficiency. Even though the inefficient compressor does allow for less heat addition in the combustor due to the higher compressor exit temperature, the reduction in heat transfer effect is small in comparison to the power drop.

(3) Because of these operational issues, the Brayton cycle by itself is not used for large-scale, baseline electricity production to the extent that the Rankine cycle is used.

As can be seen in Example 7.10, the thermal efficiency from a practical Brayton cycle is usually small in comparison to the Rankine cycle. As such, for electricity generation, the Brayton cycle is typically used for small, peak-demand type plants that are turned on when electricity demand is high and shut down when the demand is reduced. The Brayton cycle is also incorporated into combined cycle facilities, which take advantage of the hot temperature of the turbine exhaust in a Brayton cycle to heat water in the Rankine cycle; such combined cycle systems can achieve thermal efficiencies above 50%.

Modifications can be made to the Brayton cycle to attempt to improve the thermal efficiency of the cycle. In practice, these are not often done because the Brayton cycle systems are designed to be small and relatively inexpensive to build. The possible modifications have not been found in practice to dramatically increase the overall thermal efficiency of Brayton cycle systems, and so they generally aren't worth the additional cost and size associated with the equipment.

The Brayton cycle can be modified in three ways: (1) intercooling of the gases partway through the compressor, (2) reheating of the gases partway through the turbine, and (3) using regenerative heat transfer. These modifications are described next.

7.5.1 Brayton Cycle with Intercooling

The intercooling modification is designed to take advantage of the fact that less power is required to compress a low-temperature gas than to compress a high-temperature gas. As shown in **Figure 7.28**, a heat exchanger, known as an intercooler, can be placed in the middle of the compression process in the Brayton cycle. The gases are partially compressed and are then sent through the intercooler to reduce the gas temperature. Assuming no power is used in the intercooler, and neglecting changes in the kinetic and potential energy of the gas, the amount of heat that is removed through intercooling is

$$\dot{Q}_{ic} = \dot{m}(h_3 - h_2) \tag{7.33}$$

where state 3 is the outlet from the intercooler and state 2 is the inlet to the intercooler as shown in Figure 7.28. The cooled gas is then directed through another compressor to complete the compression process for the Brayton cycle. Although this process will lead to a reduction in the amount of power needed for the complete compression, the reduction in power is usually not significant enough to warrant the extra complexity of adding an intercooler—the difference in temperatures (and corresponding change in specific heats at the higher temperature) is usually not large enough to dramatically alter the power requirements for compression. Furthermore, to get sufficient flow of a cooling fluid through the intercooler, some power will likely be necessary to operate a pump or fan. Finally, additional heat is needed in the combustor because the gases coming from the compressor are cooler, and this works against improving the thermal efficiency of the cycle.

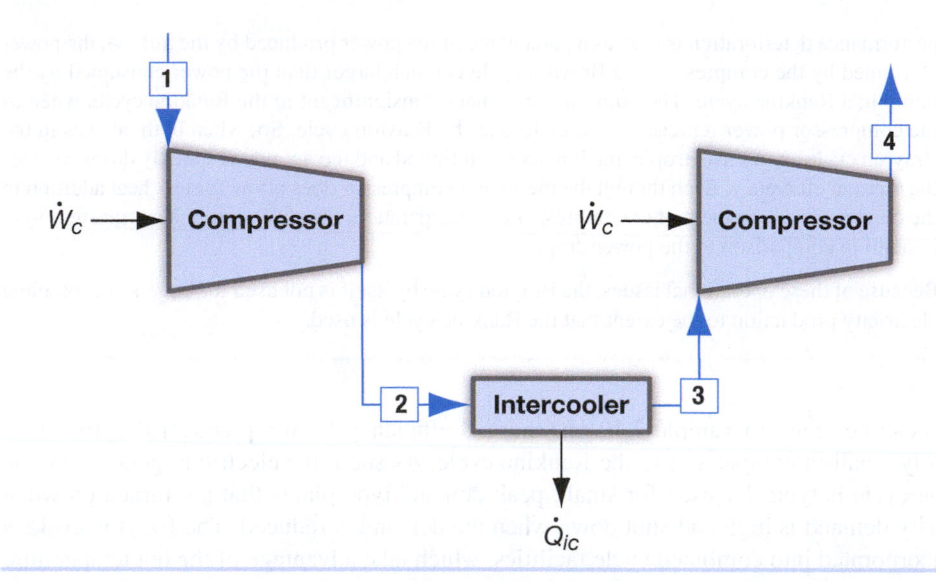

FIGURE 7.28 A component diagram of two compressors with an intercooler located between the compressors.

► EXAMPLE 7.11

It is proposed to use an intercooler during the compression process for a Brayton cycle. Air enters the compressor at 15°C and 100 kPa and exits the compression process at 1.2 MPa. The mass flow rate of the air is 2 kg/s. Assume the compression to be isentropic. Determine the power required for the process to be completed with a single compressor. Next add an intercooler at a pressure of 400 kPa that cools the air back to 15°C, and then complete the compression through a second compressor to 1.2 MPa. Determine the power required for this process with intercooling. Consider the specific heats to be variable for both systems.

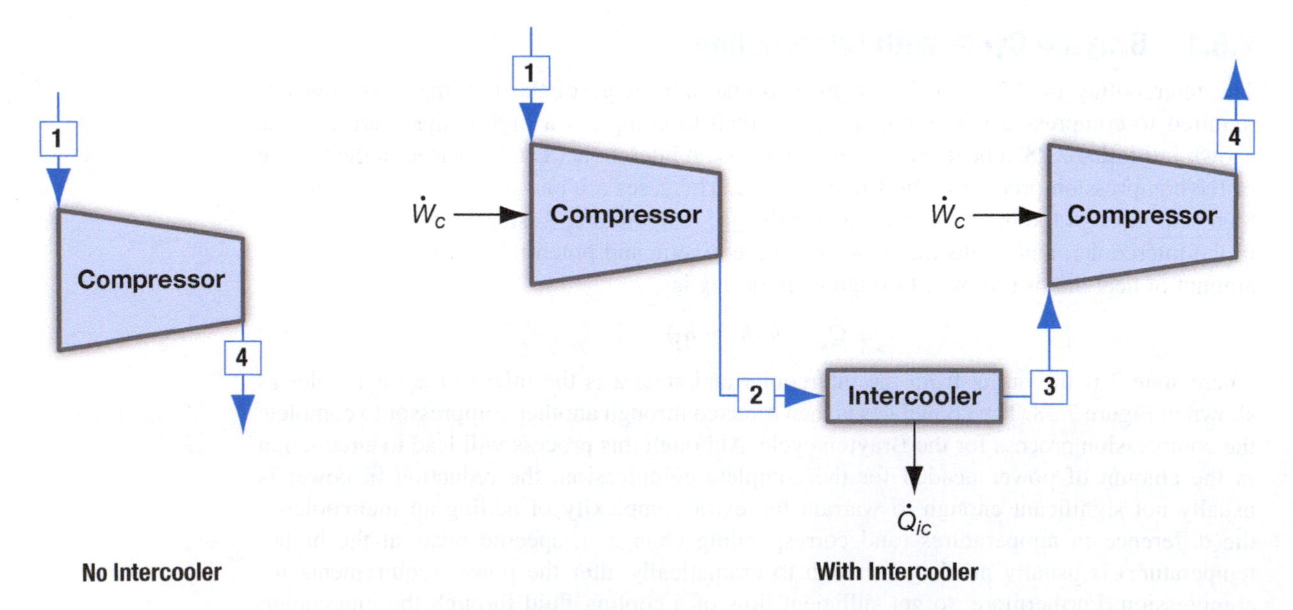

Given: $T_1 = 15°C = 288$ K, $T_3 = 15°C = 288$ K, $P_1 = 100$ kPa, $P_2 = P_3 = 400$ kPa, $P_4 = 1.2$ MPa, $\dot{m} = 2$ kg/s

Find: $\dot{W}_c$, both with and without intercooling

Solution: Here our region of interest is only the compression process in a Brayton cycle with intercooling. We will focus on just the compressors. The working fluid is air, and we will include the variability of the specific heats in our analysis (using Table A.3).

Assume: $P_2 = P_3$. The compressors are considered isentropic, so $s_1 = s_2$ and $s_3 = s_4$. $\dot{Q}_c = 0$, $\Delta KE = \Delta PE = 0$ for all devices. The compressors are acting as steady-state, steady-flow devices.

For no intercooling, from Table A.3, $h_1 = 289$ kJ/kg, $P_{r1} = 123$ kPa.

$P_{r4} = P_{r1}(P_4/P_1) = 1.47$ MPa, which corresponds to $T_4 = 582$ K, $h_4 = 588$ kJ/kg.

The power needed for one compressor (assuming adiabatic operation with no changes in kinetic or potential energy) is

$$\dot{W}_c = \dot{m}(h_1 - h_4) = -598 \text{ kW}$$

For intercooling, $h_1 = 289$ kJ/kg, $P_{r1} = 123$ kPa.

$P_{r2} = P_{r1}(P_2/P_1) = 492$ kPa, which corresponds to $T_2 = 428$ K, $h_2 = 430$ kJ/kg.

After intercooling, $T_3 = 288$ K, $h_3 = 289$ kJ/kg, and $P_{r3} = 123$ kPa.

$P_{r4} = P_{r3}(P_4/P_3) = 369$ kPa, which corresponds to $T_2 = 428$ K, $h_2 = 396$ kJ/kg.

The total compressor power is then

$$\dot{W}_c = \dot{m}[(h_1 - h_2) + (h_3 - h_4)] = -496 \text{ kW}$$

Analysis: This is a significant power reduction (~17%), but it is offset in practice by the added system complexity and the additional heat input required in the combustor. In addition, it is likely that some power will be needed to send the air and cooling fluid through the intercooler, which will further reduce the apparent power reduction benefit.

7.5.2 Brayton Cycle with Reheat

Adding reheat to the Brayton cycle involves partially expanding the gases through the turbine, directing the gas to either an additional combustor where more fuel is added and more heat is generated or directing the gas to another heat exchanger to receive heat from another source, and then completing the expansion through the turbine. This process is shown in Figure 7.29. Reheating is to take advantage of the ability of higher temperature gases to produce more power as they expand. Again, this system adds complexity. Although reheating does lead to

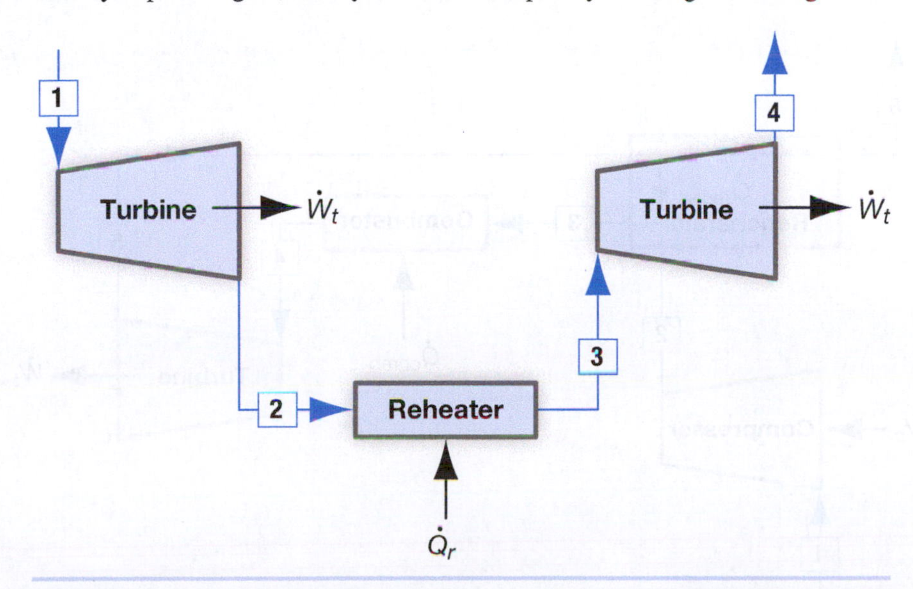

FIGURE 7.29 A component diagram for two turbines with a reheater.

greater power output, the extra power is not usually worth the added complexity. However, some gas turbine combustion schemes designed to reduce some pollutant formation may use this technique. The amount of additional heat produced is found from

$$\dot{Q}_r = \dot{m}(h_3 - h_2) \tag{7.34}$$

where state 3 is the inlet to the second turbine section and state 2 is the outlet from the first turbine section, as shown in Figure 7.29.

7.5.3 Brayton Cycle with Regeneration

The gases exiting the turbine in the Brayton cycle are typically hotter than the gases exiting the compressor. Therefore, it is possible to transfer heat from the turbine exhaust gases to the gases entering the combustor, as shown in **Figure 7.30**. This would be done through a heat exchanger. Ideally, the gases entering the combustor could be heated to a temperature equal to that of the gases leaving the turbine. This theoretical limit leads to a definition of a regenerator effectiveness, $\varepsilon_{\text{regen}}$:

$$\varepsilon_{\text{regen}} = \frac{\dot{Q}_{\text{regen, act}}}{\dot{Q}_{\text{regen, max}}} = \frac{h_3 - h_2}{h_5 - h_2} \tag{7.35}$$

where the states are numbered as shown in Figure 7.30. Although this process will reduce the heat input requirement through the combustor, like intercooling and reheating it adds considerable complexity to a system whose primary advantage is often simplicity. The temperature increase to be gained is often not worth the added cost and complexity of the regenerator.

As mentioned, the Brayton cycle can be combined with the Rankine cycle and considered a combined cycle; one possible configuration is shown in **Figure 7.31**. A combined cycle will produce power from two turbines, will have pump and compressor work requirements, and will have heat input from just the combustor in the Brayton cycle or with additional heat input from supplemental firing for the steam in the Rankine cycle. The transfer of heat between the gas turbine exhaust gases and the steam is achieved directly through a heat exchanger, or, alternatively, as a hotter air temperature entering the combustion process for the Rankine cycle steam generator.

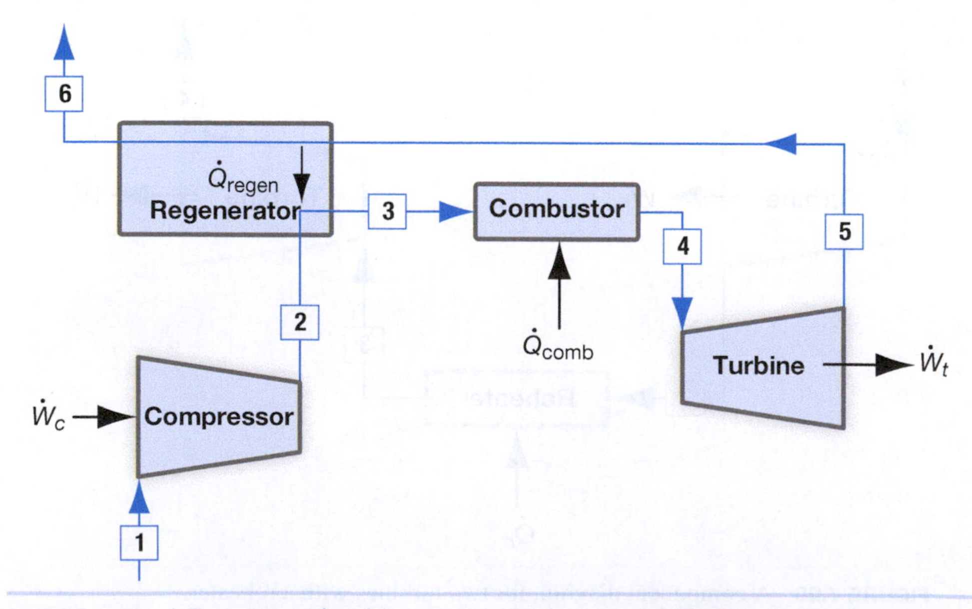

FIGURE 7.30 A Brayton cycle with a regenerator incorporated before the combustor.

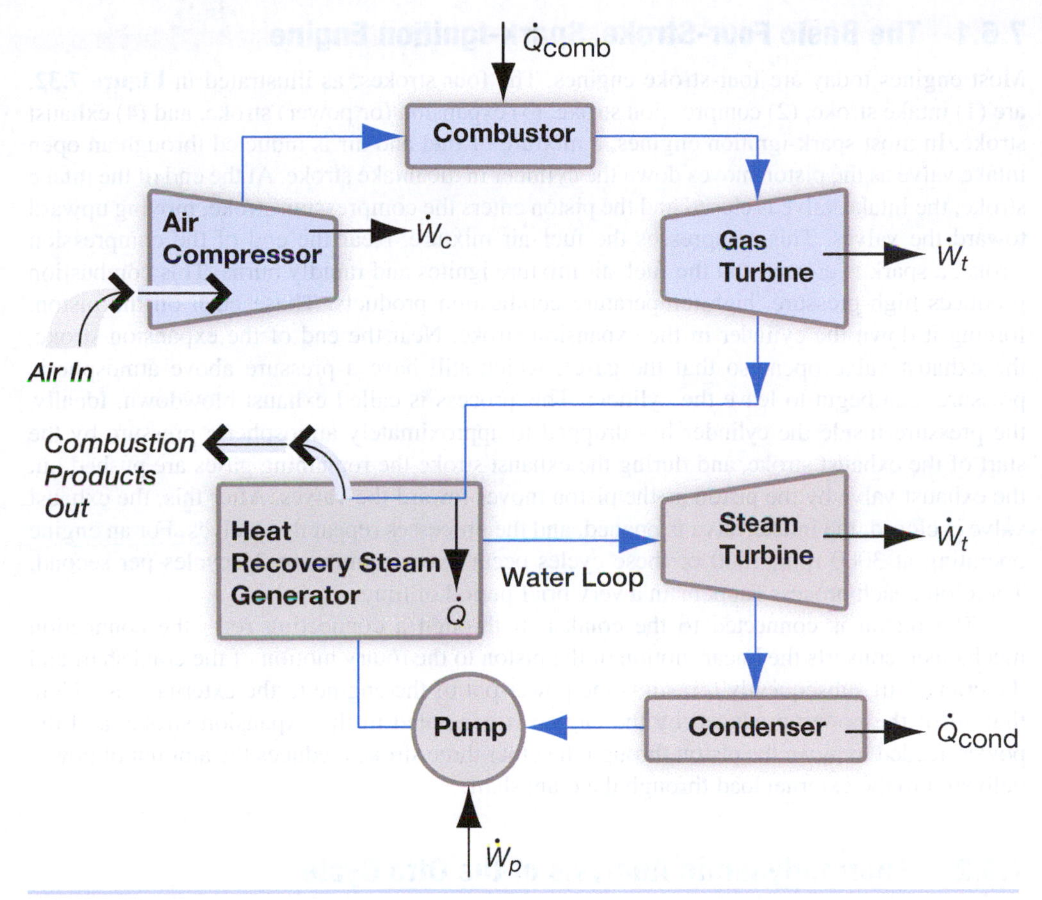

FIGURE 7.31 A component diagram of a combined cycle layout, with the hot gases exiting the turbine of the Brayton cycle and being used as heat input to the Rankine cycle.

STUDENT EXERCISE

You have learned all of the analysis techniques necessary for performing a thermodynamic analysis of a combined cycle. Modify your computer models as necessary to assist in such an analysis.

QUESTION FOR THOUGHT/DISCUSSION

As pure Brayton cycle power plants tend to have lower thermal efficiencies than Rankine cycle power plants, why would a Brayton cycle power plant be desirable? Under what situations does a Brayton cycle power plant make sense?

7.6 OTTO CYCLE

The basic thermodynamic model for a spark-ignition internal-combustion engine (such as a gasoline engine) is the Otto cycle. Unlike the Brayton and Rankine cycles, where a working fluid flows through a variety of components, the Otto cycle consists of four processes occurring in a single piston–cylinder device. Before describing the Otto cycle, it may be helpful to review how a spark-ignition internal-combustion engine operates.

7.6.1 The Basic Four-Stroke, Spark-Ignition Engine

Most engines today are four-stroke engines. The four strokes, as illustrated in Figure 7.32, are (1) intake stroke, (2) compression stroke, (3) expansion (or power) stroke, and (4) exhaust stroke. In most spark-ignition engines, a mixture of fuel and air is inducted through an open intake valve as the piston moves down the cylinder in the intake stroke. At the end of the intake stroke, the intake valve is closed and the piston enters the compression stroke, moving upward toward the valves. This compresses the fuel–air mixture. Near the end of the compression stroke, a spark is created and the fuel–air mixture ignites and rapidly burns. This combustion produces high-pressure, high-temperature combustion products. These push on the piston, forcing it down the cylinder in the expansion stroke. Near the end of the expansion stroke, the exhaust valve opens so that the gases, which still have a pressure above atmospheric pressure, can begin to leave the cylinder. This process is called exhaust blowdown. Ideally, the pressure inside the cylinder has dropped to approximately atmospheric pressure by the start of the exhaust stroke, and during the exhaust stroke the remaining gases are pushed out the exhaust valve by the piston as the piston moves toward the valves. After this, the exhaust valve is closed, the intake valve is opened, and the processes repeat themselves. For an engine operating at 3000 rpm, 1500 of these cycles occur every minute, or 25 cycles per second. Therefore, each process happens in a very brief period of time.

The piston is connected to the crankshaft through a connecting rod—the connection mechanism converts the linear motion of the piston to the rotary motion of the crankshaft and the crankshaft subsequently transmits the power out of the engine to the external load. Note that all of the power produced by the engine is generated in the expansion stroke, and the power needed to move the piston through the other three strokes reduces the amount of power delivered to the external load through the crankshaft.

7.6.2 Thermodynamic Analysis of the Otto Cycle

The thermodynamic cycle for the spark-ignition engine was developed by Nicholas Otto in 1876. As developed, the cycle consists of four processes involving two piston strokes. We can add two additional piston strokes, but the work from those two strokes will cancel out if the pressures of the two strokes are identical. (Conversely, if the pressures are not taken

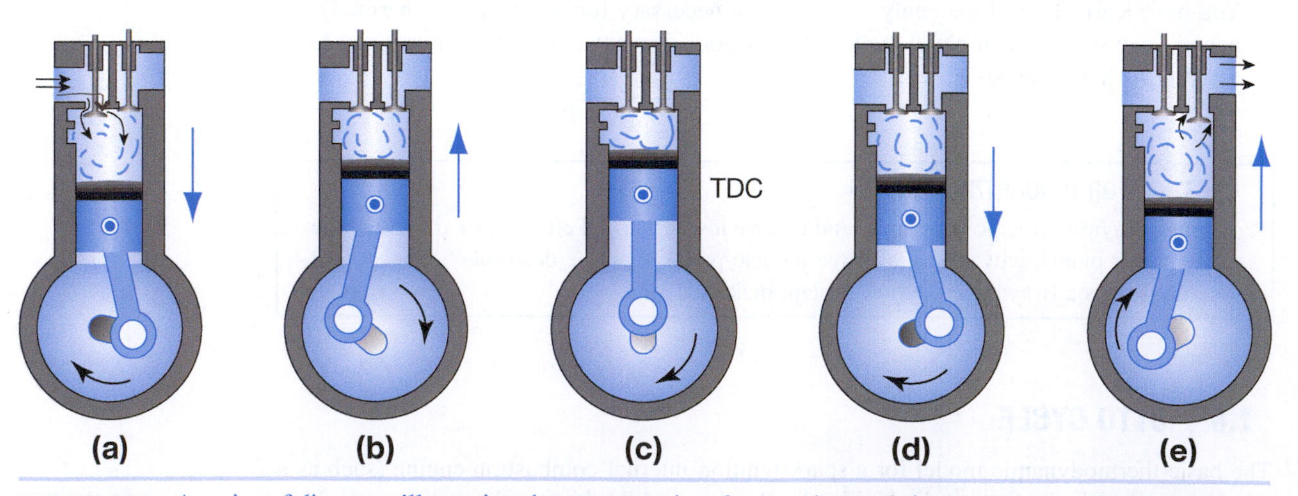

FIGURE 7.32 A series of diagrams illustrating the processes in a four-stroke spark-ignition engine. The illustrated processes are (a) the intake stroke, (b) the compression stroke, (c) combustion, (d) the power stroke, and (e) the exhaust stroke.

to be identical, we can obtain a small change to the net work of the cycle by incorporating the moving boundary work of the two strokes.) The four processes of the Otto cycle are as follows:

Process 1-2: Isentropic compression
Process 2-3: Constant-volume heat addition
Process 3-4: Isentropic expansion
Process 4-1: Constant-volume heat rejection

These four processes are shown in a *P-v* and a *T-s* diagram in **Figure 7.33**. Two additional processes can be added, as also shown in Figure 7.33:

Process 1-1′: Constant-pressure contraction
Process 1′-1: Constant-pressure expansion

If the pressures are taken as being equal for states 1 and 1′, as we will do in this analysis, the moving boundary work for the two processes cancels.

As you can see in Figure 7.33, the Otto cycle is a reasonable representation for an actual engine, with the primary difference being the region near the end of the expansion stroke when the actual engine experiences exhaust blowdown. You may be concerned with the isentropic requirements for processes 1-2 and 3-4, but a well-lubricated engine does not experience much friction loss between the piston and the cylinder wall, and the processes occur over a very short period of time, minimizing the opportunity for heat transfer to occur.

Because the four processes are occurring in one device and there is no flow in or out of the device (with a fixed mass of air), the system to be analyzed is a closed system. The first law for a closed system is given in Eq. (4.32):

$$Q - W = m\left(u_f - u_i + \frac{V_f^2 - V_i^2}{2} + g(z_f - z_i)\right) \tag{4.32}$$

where i represents the initial state for a process and f is the final state for the process. The work that occurs in the system is moving boundary work: $W = \int P d\mathcal{V}$. As such, there will be no work taking place when the volume is constant during a process: $W_{23} = W_{41} = 0$, where 23 refers to process 2-3 and 41 refers to process 4-1. In addition, we will assume that there is no heat transfer in the compression and expansion processes: $Q_{12} = Q_{34} = 0$, where 12 refers to process 1-2

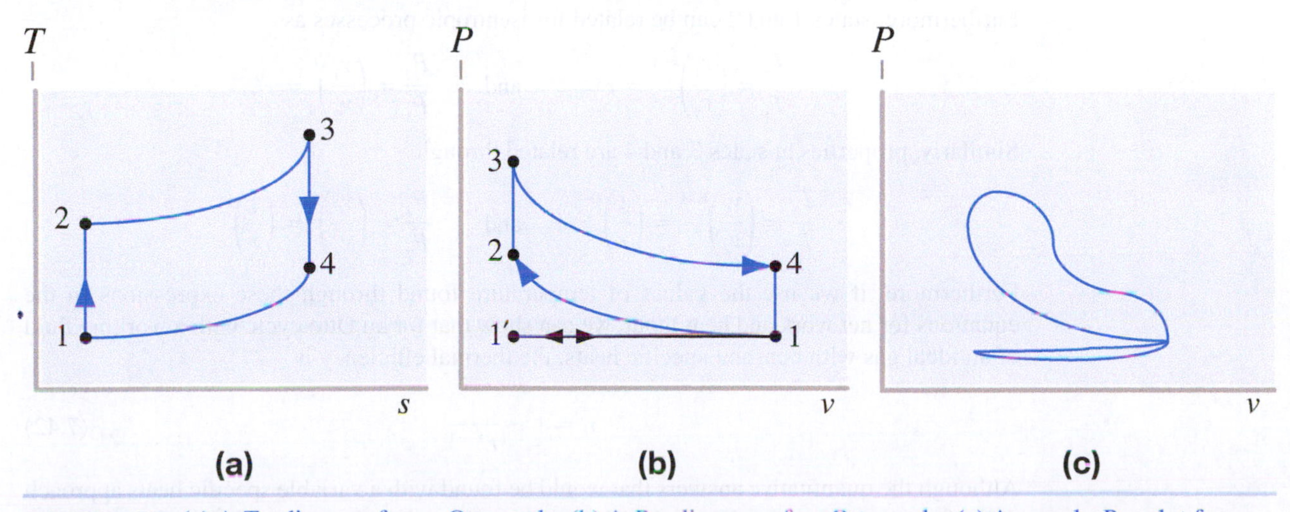

FIGURE 7.33 (a) A *T-s* diagram for an Otto cycle. (b) A *P-v* diagram of an Otto cycle. (c) A sample *P-v* plot for a spark-ignition engine.

and 34 refers to process 3-4. We will also assume that there are no changes in kinetic or potential energy for all of the processes. This leads to the following relationships to analyze each process:

$$W_{12} = m(u_1 - u_2) \tag{7.36}$$

$$Q_{23} = m(u_3 - u_2) \tag{7.37}$$

$$W_{34} = m(u_3 - u_4) \tag{7.38}$$

$$Q_{41} = m(u_1 - u_4) \tag{7.39}$$

The net work for the cycle is $W_{net} = W_{12} + W_{34}$, and the heat input is $Q_{in} = Q_{23}$. Therefore, the thermal efficiency for the Otto cycle is

$$\eta = \frac{W_{net}}{Q_{in}} = \frac{W_{12} + W_{34}}{Q_{23}} \tag{7.40}$$

You will likely find that this calculation of the thermal efficiency is higher than that seen in actual engines. The primary cause of this discrepancy is the replacement of the intake and exhaust strokes with a constant-volume heat removal. Due to this discrepancy, despite the large change in temperatures the Otto cycle is sometimes solved using a constant specific heat approach—the added accuracy of the variable specific heat approach still results in a rather inaccurate prediction of the thermal efficiency of an actual engine. Although the quantitative predictions of the thermal efficiency and work outputs and temperatures may be in error in comparison to the values seen in real engines, the Otto cycle is valuable in elucidating trends as modifications are made to the engine parameters or operating conditions.

The compression ratio of the Otto cycle is often used to assist in moving from one state to another in a thermodynamic analysis of the cycle. The compression ratio, r, is defined as

$$r = \frac{\forall_{max}}{\forall_{min}} = \frac{v_{max}}{v_{min}} = \frac{v_1}{v_2} = \frac{v_4}{v_3} \tag{7.41}$$

because $v_{max} = v_1 = v_4$ and $v_{min} = v_2 = v_3$ in an Otto cycle. Isentropic relationships for an ideal gas can then be used to relate properties at states 1 and 2, and states 3 and 4. If we choose to incorporate variable specific heats in the analysis, Table A.3 in the appendix will be helpful in helping analyze the two isentropic processes.

If we consider the specific heats to be constant, all changes in internal energy become

$$\Delta u = c_v \Delta T$$

Furthermore, states 1 and 2 can be related for isentropic processes as

$$\frac{T_2}{T_1} = \left(\frac{v_1}{v_2}\right)^{k-1} = r^{k-1} \quad \text{and} \quad \frac{P_2}{P_1} = \left(\frac{v_1}{v_2}\right)^{k} = r^{k}$$

Similarly, properties at states 3 and 4 are related through

$$\frac{T_4}{T_3} = \left(\frac{v_3}{v_4}\right)^{k-1} = \left(\frac{1}{r}\right)^{k-1} \quad \text{and} \quad \frac{P_4}{P_3} = \left(\frac{v_3}{v_4}\right)^{k} = \left(\frac{1}{r}\right)^{k}$$

Furthermore, if we use the values of temperature found through these expressions in the equations for net work and heat input, we can show that for an Otto cycle with a working fluid of an ideal gas with constant specific heats, the thermal efficiency is

$$\eta = 1 - \frac{1}{r^{k-1}} \tag{7.42}$$

Although the quantitative answers that would be found with a variable specific heats approach differ from those found in Eq. (7.42), the trend is still correct: as seen in **Figure 7.34**, a higher compression ratio leads to a higher thermal efficiency.

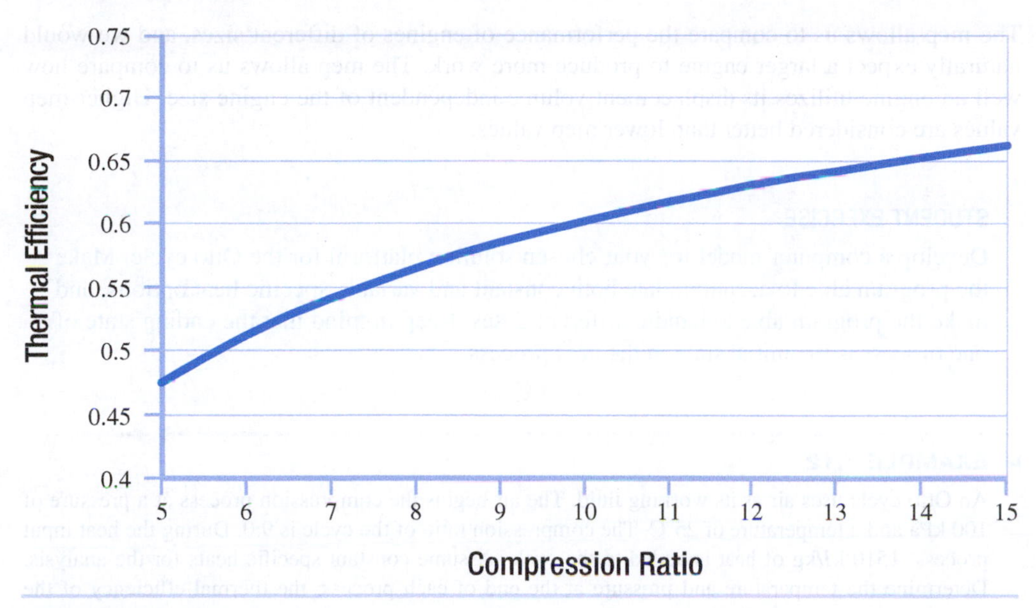

FIGURE 7.34 The thermal efficiency of the Otto cycle as a function of compression ratio, considering the working fluid to be an ideal gas with constant specific heats.

However, spark-ignition engines typically have compression ratios between 8 and 12. Why are higher compression ratios not used if they would lead to a more efficient engine? The primary reason is that the fuel–air mixture that is being compressed in the compression stroke will be more prone to self-igniting during the compression stroke because higher temperatures are achieved in this stroke due to the higher compression ratio. Self-ignition of the fuel leads to an increase in pressure too early in the cycle, and engine work is wasted compressing these higher pressure gases. Self-ignition can also lead to engine knock, and in extreme cases to engine damage. The octane number of a fuel is a measure of the self-ignition characteristics of a fuel, and fuels with higher octane numbers typically require higher temperatures to achieve ignition. High-performance engines typically require fuels with a higher octane number, because these engines generally experience higher cylinder pressures and correspondingly higher temperatures during the compression stroke. So, although higher compression ratios would seem to lead to better engine performance, in fact the mixture of fuel and air used will be more likely to self-ignite and lead to a reduction in the performance of an actual engine.

For either a constant or variable specific heats analysis, the pressures at states 2 and 3 can be related through

$$\frac{P_3}{P_2} = \frac{T_3}{T_2} \tag{7.43}$$

Equation (7.43) is what will result from application of the ideal gas law at each state, with the v and R terms canceling in the resulting ratio.

Another quantity of interest in engines and the Otto cycle is the mean effective pressure: mep. The mep represents the pressure that would produce the same net work output from the cycle or engine using the same change in volume during the cycle (the engine displacement) if the work were produced in a constant-pressure process:

$$\text{mep} = \frac{W_{\text{net}}}{V_{\text{max}} - V_{\text{min}}} = \frac{W_{\text{net}}/m}{v_1 - v_2} \tag{7.44}$$

The mep allows us to compare the performance of engines of different sizes, and we would naturally expect a larger engine to produce more work. The mep allows us to compare how well an engine utilizes its displacement volume independent of the engine size. Higher mep values are considered better than lower mep values.

STUDENT EXERCISE

Develop a computer model for your chosen solution platform for the Otto cycle. Make the program able to accommodate both constant and variable specific heat options, and make the program able to handle different gases. Keep in mind that the ending state of one process is the initial state in the next process.

▶ EXAMPLE 7.12

An Otto cycle uses air as its working fluid. The air begins the compression process at a pressure of 100 kPa and a temperature of 25°C. The compression ratio of the cycle is 9.0. During the heat input process, 1510 kJ/kg of heat is added to the cycle. Assume constant specific heats for the analysis. Determine the temperature and pressure at the end of each process, the thermal efficiency of the process, and the mean effective pressure of the cycle.

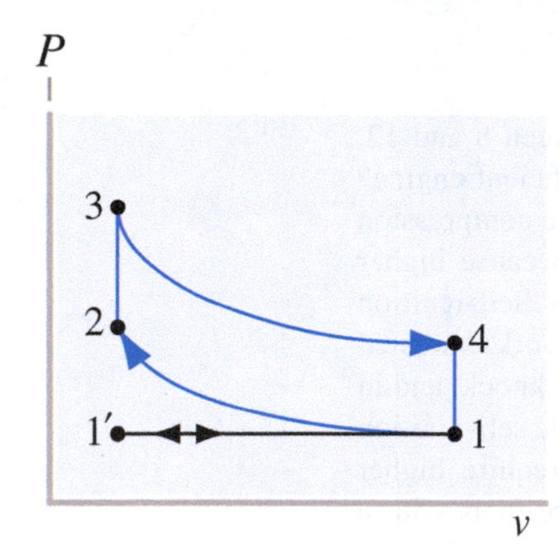

Given: $P_1 = 100$ kPa, $T_1 = 25°C = 298$ K, $r = 9$, $Q_{23}/m = 1510$ kJ/kg

Find: T and P at the end of each process, η, and mep

Solution: The cycle under study is the Otto cycle. The working fluid is air, and for simplicity we will use constant specific heats. Each process is considered to be a closed system process.

Assume: ASC assumptions. $Q_{12} = Q_{34} = W_{23} = W_{41} = 0$. $\Delta KE = \Delta PE = 0$ for each process.

For this problem, we will use specific heats at 25°C, $c_v = 0.718$ kJ/kg · K, and $k = 1.40$. Use $R = 0.287$ kJ/kg · K.

For the isentropic compression of an ideal gas with constant specific heats,

$$T_2 = T_1 r^{k-1} = (298\text{ K})(9)^{0.40} = \mathbf{718\text{ K}}$$

$$P_2 = P_1 r^k = \mathbf{2170\text{ kPa}}$$

The temperature at state 3 is found using the first law for the process (Eq. (7.37)):

$$\frac{Q_{23}}{m} = (u_3 - u_2) = c_v(T_3 - T_2) = 1510\text{ kJ/kg}$$

Solving yields $T_3 = \mathbf{2820\text{ K}}$.

Then using the relationship in Eq. (7.43) yields $P_3 = P_2\,(T_3/T_2) = \mathbf{8520\text{ kPa}}$.

For the isentropic expansion of an ideal gas with constant specific heats,

$$T_4 = T_3\,(1/r)^{k-1} = \mathbf{1170\text{ K}}$$

$$P_4 = P_3\,(1/r)^k = \mathbf{393\text{ kPa}}$$

The net work done in the cycle, per unit mass, is

$$W_{net}/m = W_{12}/m + W_{34}/m = (u_1 - u_2) + (u_3 - u_4) = c_v\,(T_1 - T_2 + T_3 - T_4) = 833\text{ kJ/kg}$$

and the thermal efficiency is

$$\eta = \frac{W_{net}/m}{Q_{in}/m} = \frac{W_{12}/m + W_{34}/m}{Q_{23}/m} = \mathbf{0.585}$$

(Note: For constant specific heats, the efficiency could also have been found using Eq. (7.42).)

To find the mean effective pressure, values are needed for the maximum and minimum specific volumes in the cycle:

$$v_1 = RT_1/P_1 = (0.287 \text{ kJ/kg} \cdot \text{K})(298 \text{ K})/100 \text{ kPa} = 0.855 \text{ m}^3/\text{kg}$$

$$v_2 = v_1/r = 0.0950 \text{ m}^3/\text{kg}$$

Then the mean effective pressure can be found from Eq. (7.44):

$$\text{mep} = \frac{W_{net}/m}{v_1 - v_2} = \frac{883 \text{ kJ/kg}}{(0.855 - 0.0950) \text{ m}^3/\text{kg}} = 1160 \text{ kPa}$$

Analysis: As discussed, these values tend to be high in comparison to an actual engine. This is due to both the assumptions made in the ASC analysis and to the use of constant specific heats, with values taken at the low temperature end of the cycle. Although better predictions could be gained either by using a specific heat value at an average cycle temperature or by using variable specific heats, predictions of the peak temperature and pressure and thermal efficiency will usually still be high.

QUESTION FOR THOUGHT/DISCUSSION

While the basic four-stroke, spark-ignition engine cycle has changed little over the past 100 years, what are some modifications that have been made to such engines that improve engine performance but also push the actual cycle farther from the pure Otto cycle?

7.7 DIESEL CYCLE

The Diesel cycle, developed by Rudolph Diesel in 1893, models early types of compression-ignition (or diesel) engines. Although modern compression-ignition engines often have a modified thermodynamic cycle that describes their performance more accurately, the basics of the Diesel cycle are useful in understanding how the compression-ignition engine differs from the spark-ignition engine.

7.7.1 The Basic Four-Stroke, Compression-Ignition Engine

Figure 7.35 shows the basic processes in a four-stroke, compression-ignition engine. Initially, the air is inducted into the cylinder during the intake stroke. At the end of the intake stroke, the intake valve is closed and the air is compressed in the compression stroke. Near the end of the compression stroke, fuel is injected into the hot, compressed air in the cylinder. Combustion begins and continues into the expansion stroke; fuel will also continue being injected into the expansion stroke if large amounts of power are needed. Near the end of the compression stroke, the exhaust valve opens, allowing for exhaust blowdown to occur. Finally, the exhaust stroke is used to force the combustion products out of the cylinder.

Although the compression-ignition engine cycle is fairly similar to the spark-ignition engine cycle, there are some important differences. First, only air is inducted into the cylinder. As a result, you cannot have pre-ignition of the gases before the desired time in the cycle because no fuel is present during the compressional heating of the intake gases. Second, the fuel is initially ignited by being heated by the hot air whose temperature is above the self-ignition temperature of the fuel. This is the mechanism that leads to pre-ignition of fuel in the spark-ignition engine; however, here the pre-ignition of the fuel is desired because the combustion of the fuel cannot start until the fuel is injected into the cylinder. This timing is chosen to be late enough in the compression stroke to avoid large

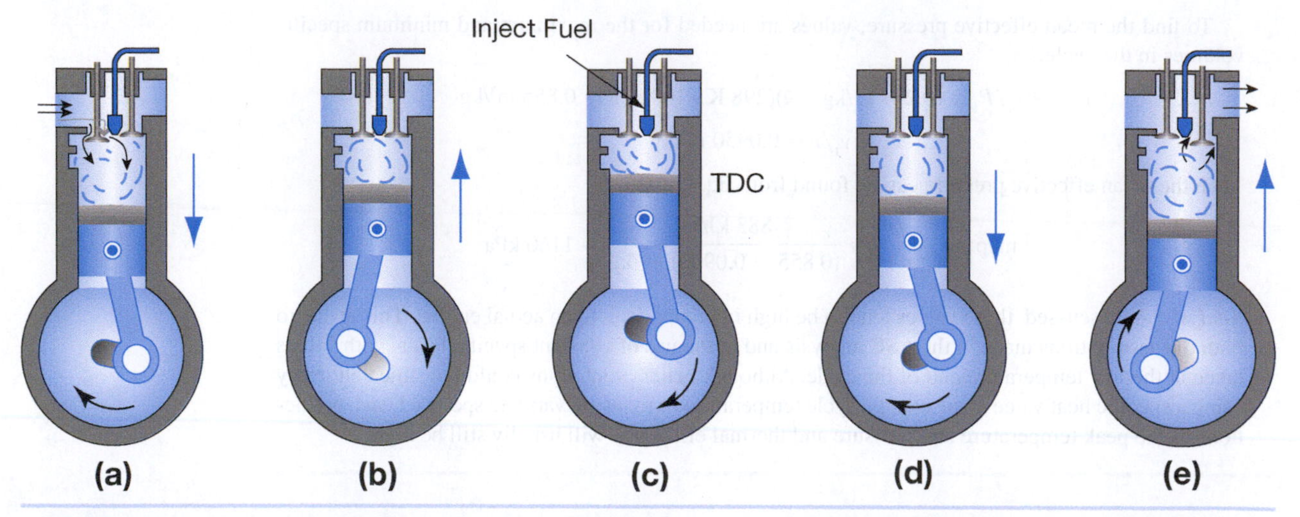

FIGURE 7.35 A series of diagrams illustrating the processes in a four-stroke compression-ignition engine. The illustrated processes are (a) the intake stroke, (b) the compression stroke, (c) fuel injection, (d) combustion and the power stroke, and (e) the exhaust stroke.

increases in the pressure of the gases in the cylinder before the end of the compression stroke. Third, the amount of power is controlled not by throttling the air–fuel mixture inducted into the engine, but rather by the length of time that the fuel is injected into the cylinder.

7.7.2 Thermodynamic Analysis of the Diesel Cycle

The Diesel cycle consists of four processes, with two additional processes that enable the cycle to model a four-stroke cycle. As in the Otto cycle, we will have those two additional strokes cancel out by having the same pressure, but a modified analysis could be performed with the strokes by having different pressures and not canceling. The processes of the Diesel cycle are as follows:

Process 1-2: Isentropic compression
Process 2-3: Constant-pressure heat addition
Process 3-4: Isentropic expansion
Process 4-1: Constant-volume heat rejection

These four processes are shown in a *P-v* and a *T-s* diagram in Figure 7.36. The two additional processes that can be added, as also shown in Figure 7.36, are

Process 1-1′: Constant-pressure contraction
Process 1′-1: Constant-pressure expansion

The analysis of the Diesel cycle will be identical to that of the Otto cycle, with the exception of process 2-3 and a small but important modification to process 3-4. All processes take place in a closed system, and we will assume that $Q_{12} = Q_{34} = W_{41} = 0$ and that there are no changes in kinetic or potential energy for each process.

With process 2-3, there will be moving boundary work, because the heat addition takes place at constant pressure. For moving boundary work at constant pressure,

$$W_{23} = \int_2^3 P\,d\forall = mP_2(v_3 - v_2) \tag{7.45}$$

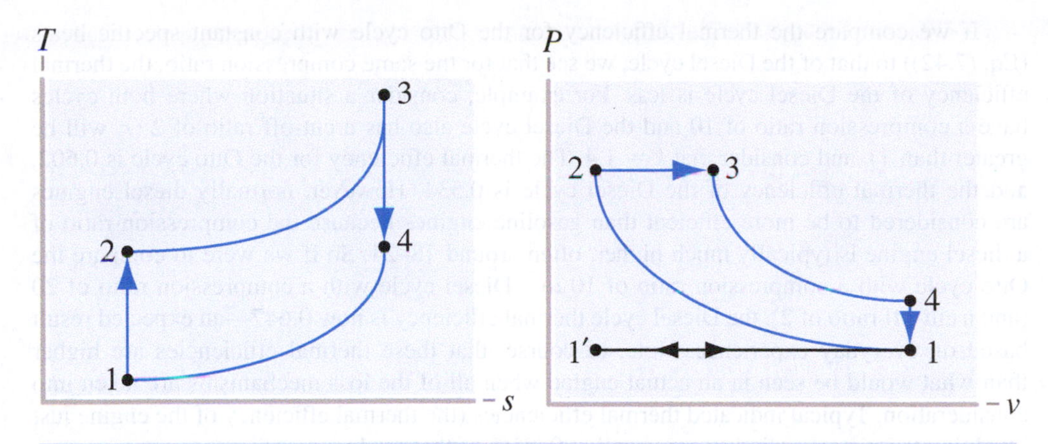

FIGURE 7.36 The *T-s* and *P-v* diagrams of a Diesel cycle.

When we consider Eq. (4.32), we see that the heat transfer is

$$Q_{23} = m(u_3 - u_2) + W_{23} = m(u_3 - u_2 + P_2(v_3 - v_2)) = m(h_3 - h_2) \tag{7.46}$$

because $h = u + Pv$ and $P_2 = P_3$ for this constant-pressure process. The net work for the cycle is the sum of the work for processes 1-2, 2-3, and 3-4:

$$W_{net} = W_{12} + W_{23} + W_{34} = m[(u_1 - u_2) + P_2(v_3 - v_2) + (u_3 - u_4)] \tag{7.47}$$

The thermal efficiency is calculated as

$$\eta = \frac{W_{net}}{Q_{in}} = \frac{W_{12} + W_{23} + W_{34}}{Q_{23}} \tag{7.48}$$

The process of relating the properties between states 3 and 4 changes in a subtle fashion: the volume at the start of the isentropic expansion is no longer equal to the minimum volume. Therefore, for the Diesel cycle,

$$r = \frac{v_{max}}{v_{min}} = \frac{v_1}{v_2} \neq \frac{v_4}{v_3}$$

the compression ratio cannot be used to relate the specific volumes at states 3 and 4. Rather, the isentropic relationships between the two states must explicitly use the specific volumes at the two states.

One parameter that can be used to relate states 2 and 3 is the cut-off ratio, r_c. The cut-off ratio represents the ratio of volumes between the end and the beginning of the heat addition process. Because process 2-3 involves an *ideal gas at constant pressure*, the volume ratio is also equal to the temperature ratio:

$$r_c = \frac{v_3}{v_2} = \frac{T_3}{T_2} \tag{7.49}$$

For an ideal gas with constant specific heats, the thermal efficiency of the Diesel cycle can be reduced to a function of the cycle's compression ratio and cut-off ratio. Using relations between the temperatures for constant specific heats, and finding the appropriate changes in internal energy and enthalpy, the thermal efficiency for a Diesel cycle using an ideal gas with constant specific heats can be shown to be

$$\eta = 1 - \frac{1}{r^{k-1}}\left[\frac{r_c^k - 1}{k(r_c - 1)}\right] \quad \text{(constant specific heats only)} \tag{7.50}$$

If we compare the thermal efficiency for the Otto cycle with constant specific heats (Eq. (7.42)) to that of the Diesel cycle, we see that for the same compression ratio, the thermal efficiency of the Diesel cycle is less. For example, consider a situation where both cycles have a compression ratio of 10 and the Diesel cycle also has a cut-off ratio of 2 (r_c will be greater than 1), and consider that $k = 1.4$. The thermal efficiency for the Otto cycle is 0.602, and the thermal efficiency of the Diesel cycle is 0.534. However, normally diesel engines are considered to be more efficient than gasoline engines because the compression ratio of a diesel engine is typically much higher, often around 18–24. So if we were to compare the Otto cycle with a compression ratio of 10 to a Diesel cycle with a compression ratio of 20 (and a cut-off ratio of 2), the Diesel cycle thermal efficiency is now 0.647—an expected result based on everyday experience. Note, of course, that these thermal efficiencies are higher than what would be seen in an actual engine when all of the loss mechanisms are taken into consideration. Typical indicated thermal efficiencies (the thermal efficiency of the engine just involving the piston–cylinder) are usually 80–85% of these values.

STUDENT EXERCISE

Develop a computer model for your solution platform for the Diesel cycle. Make the program able to accommodate both constant and variable specific heat options, and make your program able to handle different gases.

▶ EXAMPLE 7.13

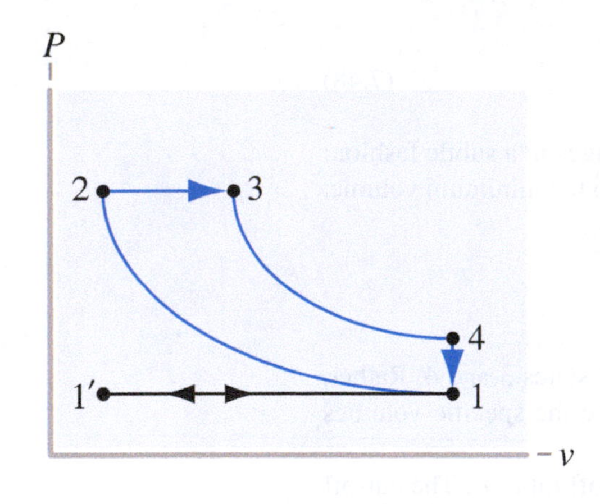

Air flows through a Diesel cycle and begins the compression process at 40°C and 90 kPa. The compression ratio of the cycle is 18. During the heat addition process, 1100 kJ/kg of heat is added. Treating the air as an ideal gas with constant specific heats, determine the temperature and pressure at each state point, the net work per unit mass, and the thermal efficiency of the cycle.

Given: $T_1 = 40°\text{C} = 313\text{ K}$, $P_1 = 90\text{ kPa}$, $Q_{23}/m = 1100\text{ kJ/kg}$, $r = 18$

Find: T and P at the end of each process, W_{net}/m, and η

Solution: The cycle under study is the Diesel cycle. The working fluid is air, and for simplicity we will use constant specific heats. Each process is considered to be a closed system process.

Assume: ASC assumptions. $Q_{12} = Q_{34} = W_{41} = 0$. $\Delta KE = \Delta PE = 0$ for each process.

The values for the constant specific heats will be taken at 300 K: $k = 1.4$, $c_p = 1.005$ kJ/kg · K, $c_v = 0.718$ kJ/kg · K, $R = 0.287$ kJ/kg · K.

Using the typical formulation for the Diesel cycle, with constant specific heats,

$$T_2 = T_1 r^{k-1} = \mathbf{995\ K}$$

$$P_2 = P_1 r^k = \mathbf{5150\ kPa}$$

$$P_3 = P_2 = \mathbf{5150\ kPa} \text{ (by definition for the Diesel cycle)}$$

From Eq. (7.46), $Q_{23}/m = h_3 - h_2 = c_p(T_3 - T_2) \rightarrow \mathbf{T_3 = 2090\ K.}$

To relate states 3 and 4, the specific volumes of states 3 and 4 are needed:

From the ideal gas law: $v_1 = RT_1/P_1 = 0.998\text{ m}^3\text{/kg}$

$$v_2 = v_1/r = 0.05545\text{ m}^3\text{/kg}$$

Considering a constant-pressure process for an ideal gas, $v_3 = v_2(T_3/T_2) = 0.1165\ \text{m}^3/\text{kg}$.

Finally, for the constant-volume heat rejection, $v_4 = v_1 = 0.998\ \text{m}^3/\text{kg}$. So $T_4 = T_3(v_3/v_4)^{k-1} =$ **885 K** and $P_4 = P_3(v_3/v_4)^k =$ **255 kPa.**

The net work per unit mass is found from Eq. (7.47):

$$W_{net}/m = (W_{12} + W_{23} + W_{34})/m = [(u_1 - u_2) + P_2(v_3 - v_2) + (u_3 - u_4)]$$

$$= c_v(T_1 - T_2 + T_3 - T_4) + P_2(v_3 - v_2)$$

$$\mathbf{W_{net}/m = 690\ kJ/kg}$$

and the thermal efficiency is from Eq. (7.48):

$$\eta = \frac{W_{net}/m}{Q_{in}/m} = \frac{(W_{12} + W_{23} + W_{34})/m}{Q_{23}/m} = \mathbf{0.627}$$

Analysis: The use of the constant specific heat approach again leads to values for the temperatures, pressures, and thermal efficiency that are somewhat high in comparison to an actual diesel engine operating under these conditions.

7.8 DUAL CYCLE

From our discussions of the thermal efficiencies of the Otto cycle and the Diesel cycle with the same compression ratio, we see that the constant-volume heat addition process (representing the very rapid burning of the fuel in a spark-ignition engine) is more efficient than the constant-pressure heat addition process (representing the slower burning of fuel as it is injected into the cylinder in a compression-ignition engine). Therefore, engine designers have altered the timing of the fuel injection in compression-ignition engines so that the fuel will begin to be injected slightly earlier in the cycle. This results in a portion of the combustion occurring at nearly constant volume near the end of the compression stroke. The process is better modeled by the Dual cycle, which combines elements of the Otto cycle and the Diesel cycle and is shown in Figure 7.37. The Dual cycle consists of the following:

Process 1-2: Isentropic compression
Process 2-a: Constant-volume heat addition
Process a-3: Constant-pressure heat addition
Process 3-4: Isentropic expansion
Process 4-1: Constant-volume heat removal

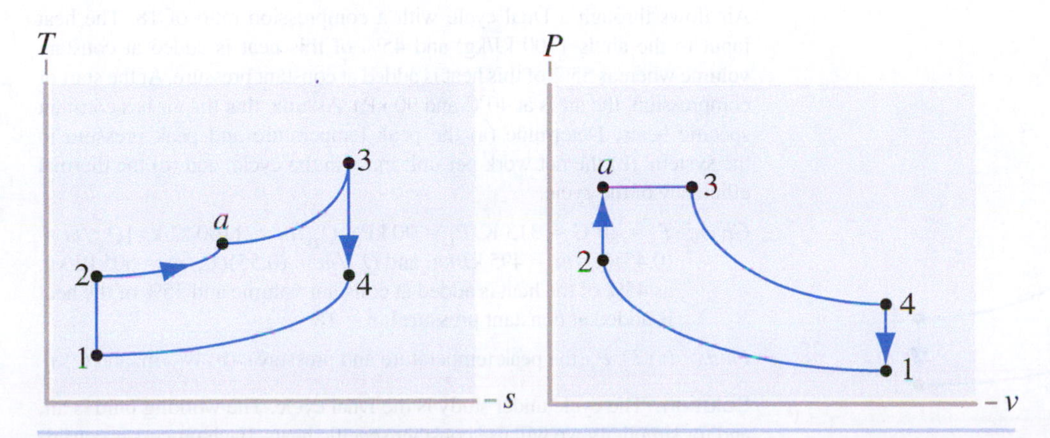

FIGURE 7.37 The *T-s* and *P-v* diagrams of a Dual cycle.

As you can see, the heat addition process is split into two components. After assumptions similar to those in the Otto and Diesel cycles are made, the heat addition can be found from

$$Q_{in} = Q_{2a} + Q_{a3} = m[(u_a - u_2) + (h_3 - h_a)] \tag{7.51}$$

There is also a small change to the net power, but it still is similar to that determined for the Diesel cycle:

$$W_{net} = W_{12} + W_{a3} + W_{34} = m[(u_1 - u_2) + P_a(v_3 - v_a) + (u_3 - u_4)] \tag{7.52}$$

Also note that $v_2 = v_a$, and $P_a = P_3$, by definition. The cut-off ratio is $r_c = v_3/v_a$.

In analyzing a Dual cycle, we need to determine how much heat is added at constant volume, and how much heat is added at constant pressure. This distribution of heat can be given as a percentage of the total heat addition or can be given indirectly through a pressure ratio, α, where

$$\alpha = \frac{P_a}{P_2} = \frac{T_a}{T_2} \tag{7.53}$$

Again, an expression for the thermal efficiency of a Dual cycle using an ideal gas with constant specific heats can be formulated:

$$\eta = 1 - \frac{1}{r^{k-1}}\left[\frac{\alpha r_c^k - 1}{k\alpha(r_c - 1) + \alpha - 1}\right] \quad \text{(constant specific heats only)} \tag{7.54}$$

By comparing cycles with the same compression ratio and cut-off ratio, we find that for the same compression ratio and same cut-off ratio,

$$\eta_{Otto} > \eta_{Dual} > \eta_{Diesel} \tag{7.55}$$

Again note that in practice the compression-ignition engines represented by the Dual and Diesel cycles have higher compression ratios than spark-ignition engines, so actual compression-ignition engines have higher thermal efficiencies than spark-ignition engines.

STUDENT EXERCISE

Develop a computer model for your chosen platform for the Dual cycle. Create the program so it can accommodate both constant and variable specific heat options and can handle different gases.

▶ EXAMPLE 7.14

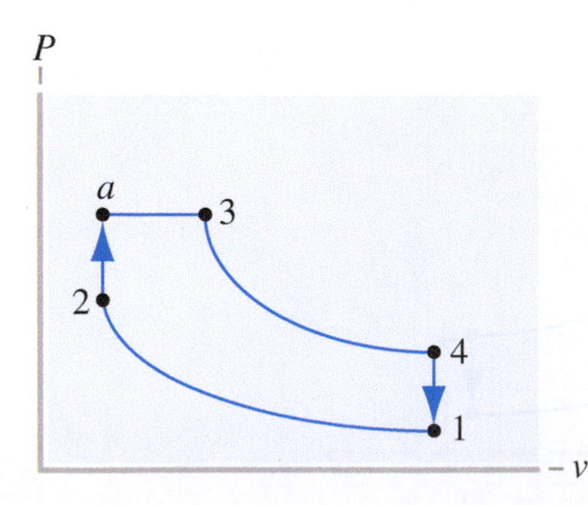

Air flows through a Dual cycle with a compression ratio of 18. The heat input to the air is 1100 kJ/kg, and 45% of this heat is added at constant volume whereas 55% of this heat is added at constant pressure. At the start of compression, the air is at 40°C and 90 kPa. Assume that the air has constant specific heats. Determine (a) the peak temperature and peak pressure in the system, (b) the net work per unit mass in the cycle, and (c) the thermal efficiency of the cycle.

Given: $T_1 = 40°C = 313$ K, $P_1 = 90$ kPa, $Q_{23}/m = 1100$ kJ/kg $[Q_{2a}/m = (0.45)Q_{23}/m = 495$ kJ/kg, and $Q_{a3}/m = (0.55)Q_{23}/m = 605$ kJ/kg, as 45% of the heat is added at constant volume and 55% of the heat is added at constant pressure], $r = 18$

Find: (a) T_3, P_3 (the peak temperature and pressure), (b) W_{net}/m, and (c) η

Solution: The cycle under study is the Dual cycle. The working fluid is air, and for simplicity we will use constant specific heats. Each process is considered to be a closed system process.

Assume: ASC assumptions. $Q_{12} = Q_{34} = W_{2a} = W_{41} = 0$. $\Delta KE = \Delta PE = 0$ for each process.

The values for the constant specific heats will be taken at 300 K: $k = 1.4$, $c_p = 1.005$ kJ/kg · K, $c_v = 0.718$ kJ/kg · K, $R = 0.287$ kJ/kg · K.

Using the typical formulation for the Dual cycle, with constant specific heats,

$$T_2 = T_1 r^{k-1} = 995 \text{ K}$$
$$P_2 = P_1 r^k = 5150 \text{ kPa}$$

The constant-volume heat addition process is treated like the heat addition process in the Otto cycle:

$$Q_{2a}/m = (u_a - u_2) = c_v (T_a - T_2) \rightarrow T_a = 1684.4 \text{ kJ/kg}$$
$$P_a = P_2 (T_a/T_2) = 8720 \text{ kPa}$$

The constant-pressure heat addition process is treated like the heat addition process in the Diesel cycle:

$$Q_{a3}/m = h_3 - h_a = c_p(T_3 - T_a) \rightarrow T_3 = \mathbf{2286\ K}$$
$$P_3 = P_a = \mathbf{8720\ kPa}$$

The temperature and pressure at state 3 represent the peak temperature and pressure.

To find state 4, we first need to find the specific volume at state 3, similar to a Diesel cycle:

From the ideal gas law, $v_1 = RT_1/P_1 = 0.998$ m³/kg,

$v_2 = v_1/r = 0.05545$ m³/kg, and for a constant-volume process, $v_a = v_2$.

Considering a constant-pressure process for an ideal gas, $v_3 = v_a(T_3/T_a) = 0.07527$ m³/kg.

Finally, for the constant-volume heat rejection, $v_4 = v_1 = 0.998$ m³/kg.

So $T_4 = T_3(v_3/v_4)^{k-1} = 813$ K and $P_4 = P_3(v_3/v_4)^k = 234$ kPa.

The net work per unit mass is found from Eq. (7.47):

$$W_{net}/m = (W_{12} + W_{a3} + W_{34})/m = [(u_1 - u_2) + P_a(v_3 - v_a) + (u_3 - u_4)]$$
$$= c_v(T_1 - T_2 + T_3 - T_4) + P_a(v_3 - v_a)$$

$$\mathbf{W_{net}/m = 741\ kJ/kg}$$

and the thermal efficiency is

$$\eta = \frac{W_{net}/m}{Q_{in}/m} = \frac{(W_{12} + W_{a3} + W_{34})/m}{Q_{23}/m} = \mathbf{0.673}$$

Analysis: Keeping in mind that the calculated values are high in comparison to those found in actual engines, we can still observe the qualitative trends. Examples 7.13 and 7.14 were set up in such a way as to allow for two different analyses of a compression-ignition engine—the Diesel cycle represents older forms of the engines, while the Dual cycle is a better representation of many modern compression-ignition engines. As expected, the analysis confirms that the thermal efficiency of the Dual cycle should be higher than that of the Diesel cycle, due to the inclusion of a portion of the heat addition process being at constant volume. The constant-volume approach is a more "efficient" way of adding heat than the constant-pressure approach. Beginning the combustion earlier in the cycle leads to a portion of the heat addition occurring at nearly-constant pressure. To achieve this, the fuel injection process timing was changed to occur earlier in the compression stroke. This results in ignition of the fuel late in the compression stroke, and a portion of the heat addition occurring at near-constant volume conditions and an increase in engine efficiency.

QUESTION FOR THOUGHT/DISCUSSION

If compression-ignition engines are more fuel efficient than spark-ignition engines, why aren't more automobiles powered using compression-ignition engines?

7.9 ATKINSON/MILLER CYCLE

In recent years, engine manufacturers have focused increasingly on the Atkinson cycle, particularly in its use as part of a hybrid system. Essentially, the Atkinson cycle is a modified Otto cycle, with different lengths for the compression stroke and the expansion stroke. Historically, this was achieved through complex mechanical linkages, but in most engines using the cycle today, it is achieved through creative valve timings, which keep the intake valve open for a longer duration than in a standard Otto cycle engine. This causes some of the fresh air–fuel mixture to be pushed back into the intake system, which effectively shortens the compression stroke length because the compression process does not begin until the intake valve closes. The expansion stroke utilizes the full length of the piston run in the cylinder. Pushing some of the fresh mixture into the intake system does reduce the power produced by the engine, as there is less fuel to burn. However, this is not seen as problematic when used in a hybrid drive system because the electric motor can provide an additional power input. The use of modified valve timings to achieve the desired results is often referred to as the Miller cycle. But, because engine manufacturers refer to this as the Atkinson cycle, we will also refer to it as such. (Furthermore, many engines that have used the Miller cycle in the past have also been supercharged to increase the density of the air in the cycle, which in turn helps maintain the power that is otherwise lost as the fresh air–fuel mixture is pushed back into the intake system.)

Summary

There are many more power cycles that have been proposed and/or implemented, such as the Kalina cycle and the Lenoir cycle, that have not been covered in this chapter. However, the cycles discussed in this chapter have the most practical use today. They also provide you with a set of analysis tools that can be applied to other thermodynamic power cycles you may encounter. With the analysis tools learned in this chapter, you should be comfortable with modifying one of the cycles discussed based on the situation encountered in an actual device. For example, there will often be pressure losses in steam power cycles, and such pressure losses can be incorporated into the Rankine cycle analysis by finding the enthalpy of a state in a different way that depends on the actual properties of the steam rather than on a set of standard, idealized assumptions. Such modifications are relatively minor, and with practice you should have little difficulty making them as needed.

PROBLEMS

7.1 A Carnot power cycle is designed to operate between temperatures of 300 K and 650 K. Determine the thermal efficiency of this cycle.

7.2 A Carnot power cycle has an efficiency of 52% and rejects heat to an environment at 10°C. Assuming that the low temperature of the fluid in the cycle is also 10°C, determine the high temperature of the fluid in the cycle.

7.3 A Carnot cycle has a maximum temperature of 400°C and has a thermal efficiency of 58%. What is the maximum temperature to which heat could be rejected through the cycle?

7.4 A certain power cycle has a thermal efficiency of 41%, and its working fluid has a maximum temperature of 570 K and a minimum temperature of 305 K. Would you consider this cycle to be performing reasonably in comparison to its thermodynamic maximum possible thermal efficiency, or should substantial effort be made to improve its efficiency further?

7.5 Suppose a power cycle has a working fluid whose maximum temperature is 500°C and whose minimum temperature is 75°C. The thermal efficiency of this particular cycle is 23%. Would you recommend that efforts be made to improve the thermal efficiency within the given temperature range, or do you believe that this is an adequate actual thermal efficiency?

7.6 An automobile engine is powered by a power cycle which operates between 2700 K and 800 K. The thermal efficiency of the engine is 45%. Would you recommend that efforts be made to improve the thermal efficiency within this temperature range, or is the thermal efficiency adequate?

7.7 A basic, ideal Rankine cycle using water as the working fluid has a saturated vapor exiting the steam generator at 18 MPa and has a saturated liquid exiting the condenser at 25 kPa. The steam generator receives a heat input of 250 MW. Determine the mass flow rate of the water, the net power output from the cycle, and the thermal efficiency of the cycle. How does the actual thermal efficiency compare to the maximum possible efficiency for a power cycle operating over this temperature range?

7.8 A basic, ideal Rankine cycle, using water as the working fluid, has a saturated vapor exiting the steam generator at 15 MPa and saturated liquid exiting the condenser at 20 kPa. The mass flow rate of the water through the cycle is 60 kg/s. Determine the net power produced, the heat input to the steam generator, and the thermal efficiency of the cycle.

7.9 A basic, ideal Rankine cycle, using water as the working fluid, has a saturated vapor exiting the steam generator at 12 MPa and saturated liquid exiting the condenser at 40 kPa. The mass flow rate of the water through the cycle is 150 kg/s. Determine the net power produced, the heat input to the steam generator, and the thermal efficiency of the cycle.

7.10 A basic, ideal Rankine cycle, using water as the working fluid, has a saturated vapor exiting the steam generator at 13.8 MPa absolute and saturated liquid exiting the condenser at 7.0 kPa absolute. The mass flow rate of the water through the cycle is 90 kg/s. Determine the net power produced, the heat input to the steam generator, and the thermal efficiency of the cycle.

7.11 A basic, ideal Rankine cycle, using water as the working fluid, has saturated vapor entering the turbine at 18 MPa. The mass flow rate of the water is 50 kg/s. The water exits the condenser as a saturated liquid. Plot the net power produced, the turbine exit quality, and the thermal efficiency of the cycle for condenser pressures ranging between 10 kPa and 1000 kPa.

7.12 A basic, ideal Rankine cycle, using water as the working fluid, has saturated liquid water exiting the condenser at 40°C. The mass flow rate of the water is 120 kg/s. The water enters the turbine as a saturated vapor. Plot the net power produced, the turbine exit quality, and the thermal efficiency of the cycle for the turbine inlet temperature ranging between 200°C and 370°C.

7.13 A basic, nonideal Rankine cycle, using water as the working fluid, has saturated vapor entering the turbine at 340°C and exiting the condenser as a saturated liquid at 40°C. The isentropic efficiency of the turbine is 0.80 and the isentropic efficiency of the pump is 0.70. For a mass flow rate of 100 kg/s, determine the net power produced and the thermal efficiency of the cycle.

7.14 Repeat Problem 7.8, but as a nonideal cycle with an isentropic turbine efficiency of 78% and a pump isentropic efficiency of 65%.

7.15 Repeat Problem 7.10, but as a nonideal cycle with an isentropic turbine efficiency of 81% and a pump isentropic efficiency of 72%.

7.16 A basic, nonideal Rankine cycle, using water as the working fluid, has saturated vapor entering the turbine at 20 MPa and saturated liquid exiting the condenser at 20 kPa. The isentropic efficiency of the pump is 0.72, and the mass flow rate of the water is 150 kg/s. Plot the net power produced, the exit quality from the turbine, and the thermal efficiency of the cycle for turbine isentropic efficiencies ranging between 0.5 and 1.0.

7.17 A basic, nonideal Rankine cycle, using water as the working fluid, has saturated vapor entering the turbine at 15 MPa and saturated liquid exiting the condenser at 50 kPa. The mass flow rate of the water is 90 kg/s. Plot the net power produced, the exit quality from the turbine, and the thermal efficiency of the cycle for (a) turbine isentropic efficiencies ranging between 0.5 and 1.0, with a constant pump isentropic efficiency of 0.75; and (b) pump isentropic efficiencies ranging between 0.30 and 1.0, with a constant turbine isentropic efficiency of 0.80.

7.18 A basic, nonideal Rankine cycle, using water as the working fluid, has saturated vapor entering the turbine at 13.8 MPa absolute and saturated liquid exiting the condenser at 27 kPa absolute. The mass flow rate of the water is 80 kg/s. Plot the net power produced, the exit quality from the turbine, and the thermal efficiency of the cycle for (a) turbine isentropic efficiencies ranging between 0.5 and 1.0, with a constant pump isentropic efficiency of 0.70; and (b) pump isentropic efficiencies ranging between 0.30 and 1.0, with a constant turbine isentropic efficiency of 0.75.

7.19 A basic, nonideal Rankine cycle, using water as the working fluid, has saturated vapor entering the turbine at 18 MPa and saturated liquid exiting the condenser. The mass flow rate of the water is 200 kg/s. The isentropic efficiency of the turbine is 0.80 and the isentropic efficiency of the pump is 0.70. Plot the net power produced, the exit quality from the turbine, and the thermal efficiency of the cycle for condenser pressures ranging between 10 kPa and 300 kPa.

7.20 An ideal Rankine cycle with superheat has water as the working fluid. Superheated steam, with a pressure of 20 MPa and a temperature of 600°C, enters the turbine, and a saturated liquid at 25 kPa exits the condenser. The mass flow rate of the water is 200 kg/s. Determine the net power produced, the exit state of the water from the turbine, and the thermal efficiency of the cycle.

7.21 Repeat Problem 7.20, but consider the cycle to be nonideal, with a turbine isentropic efficiency of 80% and a pump isentropic efficiency of 68%.

7.22 A Rankine cycle with superheat uses water as its working fluid. Superheated steam exits the steam generator at 12.4 MPa absolute and 590°C and exits the condenser as a saturated liquid at 30 kPa absolute. The mass flow rate of the water is 100 kg/s. Determine the net power produced, the exit state of the water from the turbine, and the thermal efficiency of the cycle if (a) the turbine and pump are isentropic, and (b) the turbine isentropic efficiency is 75% and the pump isentropic efficiency is 67%.

7.23 A Rankine cycle with superheat uses water as its working fluid. Superheated steam enters the turbine at 17 MPa and 550°C, while saturated liquid exits the condenser at 25 kPa. The net power produced in the cycle is 525 MW. External cooling water enters the condenser at a temperature of 10°C and exits at 40°C. Determine the mass flow rate of the steam and the mass flow rate of the external cooling water if (a) the turbine and pump are isentropic, and (b) the isentropic efficiency of the turbine is 82% and the pump isentropic efficiency is 71%.

7.24 A Rankine cycle with superheat uses water as its working fluid. Superheated steam enters the turbine at 12 MPa and 450°C, while saturated liquid exits the condenser at 35 kPa. The net power produced in the cycle is 145 MW. External cooling water enters the condenser at a temperature of 10°C and exits at 30°C. Determine the mass flow rate of the steam and the mass flow rate of the external cooling water if (a) the turbine and pump are isentropic, and (b) the isentropic efficiency of the turbine is 77% and the pump isentropic efficiency is 69%.

7.25 A Rankine cycle with superheat uses water as its working fluid. Steam enters the turbine at 16 MPa, 550°C, and saturated liquid exits the condenser at 15 kPa. The cycle produces 300 MW of power. Consider the isentropic efficiency of the pump to be 0.75. Plot the necessary mass flow rate of the water and the thermal efficiency of the cycle for turbine isentropic efficiencies ranging from 0.60 to 1.0.

7.26 A Rankine cycle with superheat uses water as the working fluid. Superheated steam enters the turbine at 18 MPa, 575°C, while saturated liquid exits the condenser at 20 kPa. The mass flow rate of the water is 220 kg/s. The isentropic efficiency of the pump is 0.70. Plot the net power produced, the exit quality (and temperature) of the water from the turbine, and the thermal efficiency of the cycle for turbine isentropic efficiencies ranging from 0.40 to 1.0.

7.27 A Rankine cycle with superheat has a turbine with an isentropic efficiency of 0.75 and a pump with an isentropic efficiency of 0.65. Superheated water vapor enters the turbine at a pressure of 20 MPa. The condenser pressure is 10 kPa. The net power produced by the cycle is 750 MW. Plot the required mass flow rate of the steam, the exit quality (and temperature) of the water leaving the turbine, and the thermal efficiency of the cycle for turbine inlet temperatures ranging between 400°C and 600°C.

7.28 A Rankine cycle with superheat has a turbine with an isentropic efficiency of 0.75 and a pump with an isentropic efficiency of 0.65. Superheated water vapor enters the turbine at a temperature of 600°C. The condenser pressure is 10 kPa. The net power produced by the cycle is 750 MW. Plot the required mass flow rate of the steam, the exit quality (and temperature) of the water leaving the turbine, and the thermal efficiency of the cycle for turbine inlet pressures ranging between 1 MPa and 20 MPa.

7.29 A Rankine cycle with superheat receives 950 kW of heat input through the steam generator, producing steam which exits the steam generator at 11 MPa absolute and 540°C. The condenser pressure is 10 kPa absolute. External cooling water enters the condenser at 20°C and exits the condenser at 40°C. The isentropic efficiency of the pump is 0.75. Plot the mass flow rate of the external cooling water for turbine isentropic efficiencies ranging between 0.40 and 1.0.

7.30 A Rankine cycle with superheat, using water as the working fluid, has superheated water vapor entering the turbine at 18 MPa and 570°C and saturated liquid exiting the condenser. The mass flow rate of the water is 200 kg/s. The isentropic efficiency of the turbine is 0.80 and the isentropic efficiency of the pump is 0.70. Plot the net power produced, the exit quality (and temperature) of the water leaving the turbine, and the thermal efficiency of the cycle for condenser pressures ranging between 10 kPa and 500 kPa.

7.31 A Rankine cycle with superheat and reheat has steam entering the turbine at 15 MPa and 500°C. The steam is extracted and sent to the reheater at a pressure of 3 MPa, and returns to the turbine at 500°C. The condenser pressure is 20 kPa. The cycle produces 400 MW of power. Determine the mass flow rate of the steam, the exit quality (or temperature) of the water leaving the turbine, and the thermal efficiency of the cycle for (a) an isentropic turbine and pump, and (b) a turbine with an isentropic efficiency of 0.75 and a pump with an isentropic efficiency of 0.65.

7.32 A Rankine cycle with superheat and reheat has steam entering the turbine at 20.7 MPa absolute and 540°C. The steam is extracted and sent to the reheater at a pressure of 5.5 MPa absolute, and returns to the turbine at 525°C. The condenser pressure is 15 kPa absolute. The mass flow rate of the steam is 90 kg/s. Determine the net power produced, the exit quality (or temperature) of the water leaving the turbine, and the thermal efficiency of the cycle for (a) an isentropic turbine and pump, and (b) a turbine with an isentropic efficiency of 0.75 and a pump with an isentropic efficiency of 0.65.

7.33 A Rankine cycle with superheat and reheat has steam entering the turbine at 18 MPa and 600°C. The steam is extracted and sent to the reheater and returns to the turbine at 600°C. The condenser pressure is 20 kPa. The turbine and pump are isentropic. The mass flow rate of the steam is 150 kg/s. Plot the net power produced, the exit quality (or temperature) of the water leaving the turbine, and the thermal efficiency of the cycle for reheat pressures ranging from 500 kPa to 10 MPa.

7.34 Repeat Problem 7.33, but consider the turbine isentropic efficiency to be 0.80 and the pump isentropic efficiency to be 0.70.

7.35 A Rankine cycle with superheat and reheat has steam entering the turbine at 20 MPa and 500°C. The steam is extracted and sent to the reheater and returns to the turbine at 500°C. The condenser pressure is 15 kPa. The mass flow rate of the water is 125 kg/s. The turbine isentropic efficiency is 0.82 and the pump isentropic efficiency is 0.68. Plot the net power produced and the thermal efficiency of the cycle for reheat pressures ranging from 1 MPa to 7.5 MPa.

7.36 A Rankine cycle with superheat and reheat delivers steam to a turbine at 20 MPa. The condenser pressure is 27 kPa. The turbine has an isentropic efficiency of 0.80, and the pump has an isentropic efficiency of 0.72. The cycle produces 600 MW of power. Plot the mass flow rate of the steam, the net heat input, the exit quality (and temperature) of the water leaving the turbine, and the thermal efficiency of the cycle for (a) a turbine inlet and reheat temperature of 550°C and a reheat pressure ranging from 500 kPa to 10 MPa; (b) a reheat pressure of 5 MPa, reheat temperature of 500°C, and turbine inlet temperatures from the steam generator ranging between 500°C and 650°C; and (c) a reheat pressure of 5 MPa, a turbine inlet temperature from the steam generator of 600°C, and reheat return temperatures ranging from 400°C to 600°C.

7.37 A Rankine cycle employs one open feedwater heater. Steam initially enters the turbine at 18 MPa, 600°C, and saturated liquid water exits the condenser at 10 kPa. The open feedwater heater receives steam extracted from the turbine at 800 kPa, and the combined stream exiting the feedwater heater is a saturated liquid at 800 kPa. The mass flow rate of water through the steam generator is 150 kg/s. Determine the net power produced and the thermal efficiency of the cycle if (a) the turbine and pumps are isentropic, and (b) the turbine isentropic efficiency is 0.80 and the isentropic efficiencies of both pumps is 0.70.

7.38 Consider a Rankine cycle using one open feedwater heater. Steam initially enters the turbine at 17 MPa, 550°C, and saturated liquid water exits the condenser at 25 kPa. The open feedwater heater receives steam extracted from the turbine at 1500 kPa, and the combined stream exiting the feedwater heater is a saturated liquid at 1500 kPa. The mass flow rate of water through the steam generator is 250 kg/s. Determine the net power produced and the thermal efficiency of the cycle if (a) the turbine and pumps are isentropic, and (b) the turbine isentropic efficiency is 0.81 and the isentropic efficiency of both pumps is 0.71.

7.39 A Rankine cycle with one open feedwater heater has a water mass flow rate through the steam generator of 140 kg/s. Superheated steam enters the turbine system at 19.3 MPa absolute and 540°C, and saturated liquid water exits the condenser at 30 kPa absolute.

The open feedwater heater receives extracted steam from the turbine at 2.4 MPa, and saturated liquid water at 2.4 MPa exits the open feedwater heater. Determine the net power produced by the cycle and the thermal efficiency of the cycle if (a) the turbine and pumps are isentropic, and (b) the turbine isentropic efficiency is 0.82 and the isentropic efficiencies of both pumps is 0.65.

7.40 Consider the ideal Rankine cycle with superheat described in Problem 7.20. An open feedwater heater (and accompanying pump) is to be added to the cycle, with the pressure of the open feedwater heater being 1 MPa. Determine the net power produced, the fraction of water extracted from the turbine, and the thermal efficiency of this modified cycle.

7.41 Repeat Problem 7.40, but plot the net power produced, the fraction of water extracted from the turbine, and the thermal efficiency of the cycle for open feedwater heater pressures ranging between 100 kPa and 5000 kPa.

7.42 Consider the ideal Rankine cycle with superheat described in Problem 7.20. A closed feedwater heater with its drain cascaded backward is to be added to the cycle. Steam is extracted and sent to the feedwater heater at a pressure of 1 MPa. The fraction of mass extracted from the turbine is 0.12. Determine the net power produced, the temperature of the boiler feedwater after the closed feedwater heater, and the thermal efficiency of the cycle.

7.43 Repeat Problem 7.42, but plot the net power produced, the temperature of the boiler feedwater after the closed feedwater heater, and the thermal efficiency of the cycle for fractions of mass flow extracted from the turbine ranging between 0.05 and 0.25.

7.44 A Rankine cycle with one closed feedwater heater with its drain cascaded backward has a water mass flow rate through the steam generator of 140 kg/s. Superheated steam enters the turbine system at 19.3 MPa and 540°C, and saturated liquid water exits the condenser at 28 kPa absolute. The closed feedwater heater receives extracted steam from the turbine at 2.4 MPa absolute, and the temperature of the feedwater exiting the closed feedwater heater is 210°C. Determine the net power produced by the cycle and the thermal efficiency of the cycle if (a) the turbine and pumps are isentropic, and (b) the turbine isentropic efficiency is 0.82 and the isentropic efficiencies of both pumps is 0.65.

7.45 Reconsider the cycle described in Problem 7.44. Plot the net power produced by the cycle, the fraction of mass extracted from the turbine and delivered to the closed feedwater heater, and the thermal efficiency of the cycle for a system with an isentropic turbine and pumps, for feedwater temperatures exiting the closed feedwater heater ranging from 90°C to 220°C.

7.46 A Rankine cycle with a closed feedwater heater with the drains cascaded backward through a throttle to the condenser has steam entering the turbine initially at 550°C and 15 MPa, and saturated liquid water exiting the condenser at 20 kPa. The closed feedwater heater receives steam extracted from the turbine at 2000 kPa. The isentropic efficiency of the turbine is 0.78, and that of the pumps is 0.70. The mass flow rate of the water through the steam generator is 200 kg/s. External cooling water enters the condenser at 10°C and exits at 35°C. Plot the extracted mass fraction of water, the net power produced in the cycle, the thermal efficiency of the cycle, and the mass flow rate of the external cooling water for temperatures of the feedwater exiting the closed feedwater heater ranging from 100°C to 210°C.

7.47 Develop the equations that are needed to calculate the net power and the thermal efficiency for a Rankine cycle with one closed feedwater heater with the drains pumped forward. Create a computer model which incorporates this scenario.

7.48 Develop the equations that are needed to calculate the net power and the thermal efficiency for a Rankine cycle with two open feedwater heaters.

7.49 Develop the equations that are needed to calculate the net power and the thermal efficiency for a Rankine cycle with one open feedwater heater and one closed feedwater heater, where the closed feedwater heater uses a steam extraction from the turbine at a higher pressure than the open feedwater heater. Furthermore, the drain of the closed feedwater heater is cascaded backwards to the open feedwater heater. The boiler feed pump following the open feedwater heater is placed between the two feedwater heaters.

7.50 Develop the equations that are needed to calculate the net power and the thermal efficiency for a Rankine cycle with two closed feedwater heaters and one reheater. The reheater is located at a pressure between the extraction pressures for the two closed feedwater heaters. The drain of the higher-pressure feedwater heater is cascaded backward to the lower-pressure closed feedwater heater, and the drain of the lower-pressure closed feedwater heater is cascaded backward to the condenser.

7.51 Consider the Rankine cycle described in Problem 7.20, but which will now include a closed feedwater heater with the drains pumped forward. The steam is extracted from the turbine at 1 MPa, and the boiler feedwater exiting the closed feedwater heater has a temperature of 170°C. The extracted water exits the feedwater heater as a saturated liquid at 1 MPa and is pumped to a pressure of 20 MPa, when it is mixed with the remainder of the boiler feedwater. Determine the fraction of mass extracted from the turbine and directed toward the closed feedwater heater, the net power from the cycle, and the thermal efficiency of the cycle.

7.52 A Rankine cycle has both an open feedwater heater and a closed feedwater heater, with its drains cascaded back toward the condenser. Steam enters the turbine at a rate of 200 kg/s, at 20 MPa and 600°C, and is extracted at 2 MPa to be sent to the open feedwater heater. Additional steam is extracted at 500 kPa and sent toward the closed feedwater heater. The water exits the condenser as a saturated liquid at 15 kPa, and the water exits the open feedwater heater as a saturated liquid at 2 MPa. The feedwater exiting the closed feedwater heater is at a temperature of 150°C. The pump following the condenser increases the pressure of the water to 2000 kPa, and the pump after the open feedwater heater increases the pressure to 20 MPa. Determine the fraction of the total mass flow rate extracted from the turbine for each feedwater heater, the net power of the cycle, and the thermal efficiency of the cycle if (a) the turbine and pumps are isentropic, and (b) the turbine and pump isentropic efficiencies are both 0.80.

7.53 Consider the cycle described in Problem 7.52, but reverse the locations of the open feedwater heater and closed feedwater heater (i.e., the closed feedwater heater extraction pressure is 2 MPa and the open feedwater heater pressure is 500 kPa). The drain from the closed feedwater heater is throttled to 500 kPa and sent into the open feedwater heater. The pump after the condenser raises the water pressure to 500 kPa, and the pump after the open feedwater heater raises the pressure to 20 MPa. The feedwater exits the closed feedwater heater at a temperature of 210°C. Determine the fraction of the total mass flow rate extracted from the turbine for each feedwater heater, the net power of the cycle, and the thermal efficiency of the cycle if (a) the turbine and pumps are isentropic, and (b) the turbine and pump isentropic efficiencies are both 0.80.

7.54 You are given various parameters for designing a Rankine cycle with one open feedwater heater and one reheater. The steam enters the turbine at 18 MPa and 550°C, and the condensate leaves the condenser at 12 kPa as a saturated liquid. Steam can be extracted from the turbine at pressures of 4 MPa and 800 kPa. Consider the isentropic efficiency of the turbine to be 0.80 and the isentropic efficiency of the pumps to be 0.70. The steam returns to the turbine from the reheater at 550°C. Determine whether a higher thermal efficiency is obtained by having the reheater at 4 MPa and the feedwater heater at 800 kPa, or if the feedwater heater should be at 4 MPa and the reheater at 800 kPa.

7.55 A Rankine cycle with reheat and a closed feedwater heater (with its drain cascaded back to the condenser) has 150 kg/s of superheated steam at 20 MPa absolute and 600°C entering the turbine. Steam is extracted at a pressure of 6.9 MPa absolute and sent to the reheater, where it returns to the turbine at 570°C. The feedwater temperature leaving the closed feedwater heater is equal to 5°C below the temperature of the steam leaving the turbine and being sent to the closed feedwater heater. The extracted water quality in the drain of the closed feedwater heater is 0. The condenser pressure is 15 kPa absolute. The isentropic efficiency of the turbine is 0.78 and of the pump is 0.68. Plot the net power produced, the mass flow rate of the steam extracted and sent to the closed feedwater heater, and the thermal efficiency of the cycle for pressures of the extracted steam directed toward the closed feedwater heater from the turbine ranging between 150 kPa absolute and 950 kPa absolute.

7.56 Consider a Rankine cycle with superheat. The steam enters the turbine at 20 MPa and 500°C, and the condenser pressure is 15 kPa. Plot the ratio of the work used by the pump to that produced by the turbine (per kg of steam) for (a) isentropic efficiencies of the turbine ranging between 0.30 and 1.0 for a set pump isentropic efficiency of 0.75, and (b) isentropic efficiencies of the pump ranging between 0.30 and 1.0 for a set isentropic efficiency of the turbine of 0.80.

7.57 Consider a Rankine cycle with superheat. The steam enters the turbine at 19 MPa absolute and 540°C, and the condenser pressure is 20 kPa absolute. Plot the ratio of the work used by the pump to that produced by the turbine (per kg of steam) for (a) isentropic efficiencies of the turbine ranging between 0.30 and 1.0 for a set pump isentropic efficiency of 0.70, and (b) isentropic efficiencies of the pump ranging between 0.30 and 1.0 for a set isentropic efficiency of the turbine of 0.75.

7.58 A Brayton cycle has air entering the compressor at 100 kPa and 290 K. The air exits the compressor at a pressure of 1000 kPa, and heat is added through the combustor so that the air enters the turbine at 1100 K. The mass flow rate of the air is 15 kg/s. Consider the air to have variable specific heats. Determine the net power produced by the cycle, the ratio of the compressor power to the turbine power, and the thermal efficiency of the cycle for (a) an isentropic turbine and compressor, and (b) an isentropic turbine efficiency of 0.80 and an isentropic compressor efficiency of 0.70.

7.59 Repeat Problem 7.58, but consider the specific heats of the air to be constant (with values evaluated at 300 K).

7.60 A Brayton cycle has air entering the compressor at 100 kPa and 25°C. The air exits the compressor at a pressure of 1100 kPa. The mass flow rate of the air is 40 kg/s. 25 MW of heat is added through the combustor. Consider the air to have variable specific heats. Determine the net power produced by the cycle, the ratio of the compressor power to the turbine power, and the thermal efficiency of the cycle for (a) an isentropic turbine and compressor, and (b) an isentropic turbine efficiency of 0.75 and an isentropic compressor efficiency of 0.75.

7.61 A Brayton cycle using N_2 gas in a closed loop has the N_2 entering the compressor at 80 kPa and 15°C. The nitrogen exits the compressor at 600 kPa and enters the turbine at 1000 K. The cycle produces a net power of 10 MW. The isentropic efficiency of both the compressor and the turbine is 0.78. Assume constant specific heats for the nitrogen. Determine the required mass flow rate and the thermal efficiency of the cycle. How would these values change if the gas used was CO_2?

7.62 A Brayton cycle has air entering the compressor at 100 kPa and 24°C, and the air exits the compressor at 1.4 MPa. The air exits the combustor at 1200°C. The mass flow rate of the air is 11 kg/s. Consider the air to have variable specific heats. Determine the net power produced by the cycle, the ratio of the compressor power to the turbine power, and the thermal efficiency of the cycle for (a) an isentropic turbine and compressor, and (b) an isentropic turbine efficiency of 0.78 and an isentropic compressor efficiency of 0.72.

7.63 A Brayton power cycle has air with a mass flow rate of 50 kg/s entering a compressor at 100 kPa and 25°C. The air exits the compressor at 1200 kPa and exits the combustor at 1200 K. The isentropic efficiency of the turbine is 0.80. Consider the air to have variable specific heats. Plot the net power produced by the cycle, the ratio of the compressor power to turbine power, and the thermal efficiency for the cycle for isentropic efficiencies of the compressor ranging between 0.50 and 1.0.

7.64 A Brayton power cycle has air with a mass flow rate of 50 kg/s entering a compressor at 100 kPa and 25°C. The air exits the compressor at 1200 kPa and exits the combustor at 1200 K. The isentropic efficiency of the compressor is 0.75. Consider the air to have variable specific heats. Plot the net power produced by the cycle, the ratio of the compressor power to turbine power, and the thermal efficiency for the cycle for isentropic efficiencies of the turbine ranging between 0.50 and 1.0.

7.65 A Brayton power cycle has air with a mass flow rate of 60 kg/s entering the compressor at 100 kPa and 25°C. The air exits the compressor at 1300 kPa. The isentropic efficiency of the turbine is 0.80 and the isentropic efficiency of the compressor is 0.75. Consider the air to have constant specific heats. Plot the net power produced by the cycle, the turbine inlet temperature, and the thermal efficiency for the cycle for combustor heat inputs ranging between 5000 kW and 50,000 kW.

7.66 Repeat Problem 7.65, but use variable specific heats for the air.

7.67 A Brayton cycle using air receives the air into the compressor at 100 kPa and 20°C. The air exits the compressor at 1.1 MPa and has a mass flow rate of 30 kg/s. The isentropic efficiency of the turbine is 0.80, and the isentropic efficiency of the compressor is 0.78. Consider the air to have variable specific heats. Plot the net power produced by the cycle and the thermal efficiency for the cycle for turbine inlet temperatures ranging between 800°C and 1400°C.

7.68 A Brayton cycle using air receives the air into the compressor at 100 kPa and 290 K. The air has a mass flow rate of 50 kg/s. The air enters the turbine at 1250 K. Consider the air to have constant specific heats. Plot the ratio of the compressor power to the turbine power, the net power produced, and the thermal efficiency of the cycle for compressor outlet pressures ranging between 400 kPa and 1500 kPa for (a) an isentropic compressor and turbine, and (b) a turbine with an isentropic efficiency of 0.80 and a compressor with an isentropic efficiency of 0.75.

7.69 A Brayton cycle using air receives air into the compressor at 100 kPa. The air has a volumetric flow rate of 30 m^3/s entering the compressor, and the pressure leaving the compressor is 1000 kPa. The air enters the turbine at 1200 K. Considering the air to have variable specific heats, plot the net power produced and the thermal efficiency of the cycle for compressor inlet air temperatures ranging between –40°C and 40°C.

7.70 It is proposed to use a Brayton cycle with an intercooler added to the compressor. Air enters the compressor at 100 kPa and 25°C and eventually leaves the compressor at a pressure of 1500 kPa. The mass flow rate of the air is 50 kg/s. Consider the specific heats of the air to be constant. The isentropic efficiency of the compressor is 0.80. Determine the power required by the compressor if (a) the compression occurs in one step, and (b) the air is compressed to 400 kPa, directed toward the intercooler where it is cooled to 25°C, and then returned to the compressor where it completes the compression to 1500 kPa.

7.71 A Brayton cycle with an intercooler has a mass flow rate of 30 kg/s. Air enters the first compressor at 100 kPa and 300 K; is compressed to an intermediate pressure and sent through an intercooler, which cools the air to 305 K; and then returns to the compressor to be further compressed to a pressure of 1000 kPa. The air enters the turbine at 1100 K and expands through the turbine to 100 kPa. The isentropic efficiency of both the compressor and the turbine is 0.78. Consider the specific heat of the air to be variable. Plot the net power produced by the cycle, the amount of heat removed by the intercooler, and the thermal efficiency of the cycle for intercooler pressures ranging between 200 kPa and 800 kPa.

7.72 Repeat Problem 7.62, but include a reheater in the turbine. The air expands through the turbine to a pressure of 400 kPa and is then directed through a reheater to be heated to 1175°C, at which point it returns to the turbine to continue expansion to 100 kPa.

7.73 A Brayton cycle with a reheater has a mass flow rate of 30 kg/s. Air enters the first compressor at 100 kPa and 300 K and is compressed to a pressure of 1200 kPa. The air enters the turbine at 1100 K and expands through the turbine to an intermediate reheat pressure. The air returns to the turbine at 1100 K and continues to expand through the turbine to a pressure of 100 kPa. The isentropic efficiency of both the compressor and the turbine is 0.78. Consider the specific heat of the air to be variable. Plot the net power produced by the cycle, the amount of heat added by the reheater, and the thermal efficiency of the cycle for reheater pressures ranging between 200 kPa and 1000 kPa.

7.74 A Brayton cycle uses a regenerator between the compressor and combustor to reduce the heat needed to be added through the combustor. Air enters the compressor at 100 kPa and 10°C and is compressed to 1000 kPa. The air then passes through a regenerator with a regenerator effectiveness of 0.80. The air expands through the turbine to a pressure of 100 kPa. The mass flow rate of the air is 20 kg/s. The isentropic efficiency of the turbine is 0.80, and the isentropic efficiency of the compressor is 0.75. Consider the specific heats of the air to be variable. Plot the net power produced by the cycle, the heat input, and the thermal efficiency of the cycle for turbine inlet temperatures ranging between 1000 K and 1400 K.

7.75 A basic Brayton cycle is to be combined with a basic Rankine cycle with superheat to form a combined cycle. The air enters the compressor of the Brayton cycle at 100 kPa and 25°C, is compressed to 1200 kPa, and is heated in the combustor to 1400 K. The isentropic efficiency of the compressor is 0.78 and of the gas turbine is 0.80. The mass flow rate of the air is 200 kg/s. The exhaust gases from the turbine are sent through a heat exchanger to heat the water in the Rankine cycle. The exhaust gases from the Brayton cycle exit the heat exchanger at 80°C. The steam exits this steam generator at a pressure of 15 MPa and a temperature of 400°C. The condenser pressure of the Rankine cycle is 10 kPa, the isentropic efficiency of the steam turbine is 0.80, and the isentropic efficiency of the pump is 0.70. Considering no additional heat to be added to the steam, determine (a) the mass flow rate of the steam in the Rankine cycle, (b) the total net power produced in the combined cycle, and (c) the thermal efficiency of the combined cycle.

7.76 A combined Brayton-Rankine cycle receives air at 300 K and 100 kPa into its air compressor. The air is compressed to 1.2 MPa and then heated in the combustor to 1250°C. The air expands in the gas turbine to 0.1 MPa and is then sent through a heat exchanger to transfer heat to the steam in the Rankine cycle. The air exits the heat exchanger at 65°C. Steam exits the steam generator at 540°C and 19 MPa absolute. The mass flow rate of the air is 90 kg/s. The condenser pressure is 20 kPa absolute. The air turbine and compressor both have isentropic efficiencies of 0.78, and the steam turbine isentropic efficiency is 0.75 while the pump isentropic efficiency is 0.70. Plot the ratio of the mass flow rate of the steam to the air, the additional amount of supplemental heating required by the Rankine cycle, and the thermal efficiency of the combined cycle for Brayton cycle net power outputs ranging between 100 MW and 400 MW.

7.77 A combined Brayton-Rankine cycle receives air at 20°C and 100 kPa into its air compressor, in which it is compressed to 1300 kPa. The air is then heated in the combustor to 1200 K. The air expands in the gas turbine to 100 kPa and is then sent through a heat exchanger to transfer heat to the steam in the Rankine cycle. The air exits the heat exchanger as cold as 80°C (but possibly warmer if no supplemental heating is required). Steam exits the steam generator at 450°C and 15 MPa. The condenser pressure is 10 kPa. The air turbine and compressor both have isentropic efficiencies of 0.78, and the steam turbine isentropic efficiency is 0.79 while the pump isentropic efficiency is 0.71. The total power to be produced by the combined cycle is 400 MW. Plot the ratio of the mass flow rate of the steam to the air, the additional amount of supplemental heating required by the Rankine cycle, and the thermal efficiency of the combined cycle for Rankine cycle net power outputs ranging between 100 MW and 300 MW.

7.78 An Otto cycle uses air as the working fluid. The air begins the compression process at 95 kPa and 35°C. The temperature of the air after the heat addition process is 2650 K. The compression ratio of the cycle is 9.6. Treating the air as an ideal gas with constant specific heats, determine (a) the temperature and pressure at the end of each process, (b) the net work per kg of air produced in the cycle, (c) the thermal efficiency of the cycle, and (d) the mean effective pressure.

7.79 An Otto cycle has air operating as the working fluid. The air begins the compression process at 90 kPa and 40°C. During the heat addition process, 1400 kJ/kg of heat is added to the air. The compression ratio of the cycle is 9.2. Treating the air as an ideal gas with variable specific heats, determine (a) the temperature and pressure at the end of each process, (b) the net work per kg of air produced in the cycle, (c) the thermal efficiency of the cycle, and (d) the mean effective pressure.

7.80 Repeat Problem 7.79, but assume the specific heats of the air to be constant with values taken at (a) 300 K, and (b) 1000 K.

7.81 An Otto cycle engine has air beginning the compression process at 95 kPa absolute and 40°C. The air begins the expansion process at a temperature of 2600°C. The compression ratio of the engine is 9.8. Treating the air as an ideal gas with constant specific heats, determine (a) the temperature and pressure at the end of each process, (b) the net work per kg of air produced in the cycle, (c) the thermal efficiency of the cycle, and (d) the mean effective pressure.

7.82 Air flows through an Otto cycle, beginning the compression process at 85 kPa and 50°C. During the heat addition process, 1250 kJ/kg of heat is added to the air. Plot the maximum temperature and pressure in the system, the net work per kg of air produced in the cycle, the thermal efficiency of the cycle, and the mean effective pressure for compression ratios ranging between 7 and 12, assuming that the air behaves as an ideal gas with variable specific heats.

7.83 Air flows through an Otto cycle, beginning the compression process at a pressure of 90 kPa. The volume of gas present in the cylinder at the start of compression is 0.001 m^3. During the heat addition process, 1200 kJ/kg of heat is added to the air. The compression ratio of the cycle is 8.5. Plot the work produced during one cycle, the maximum temperature in the cycle, the thermal efficiency of the cycle, and the mean effective pressure for temperatures at the beginning of compression ranging between 0°C and 100°C, assuming that the air behaves as an ideal gas with constant specific heats.

7.84 Air flows through an Otto cycle, beginning the compression process at a temperature of 50°C. The volume of gas present in the cylinder at the start of compression is 0.001 m^3. During the heat addition process, 1200 kJ/kg of heat is added to the air. The compression ratio of the cycle is 8.5. Plot the work produced during one cycle, the maximum pressure in the cycle, the thermal efficiency of the cycle, and the mean effective pressure for pressures at the beginning of compression ranging between 50 kPa and 150 kPa, assuming that the air behaves as an ideal gas with constant specific heats.

7.85 Air flows through an Otto cycle, beginning the compression process at a temperature of 45°C and a pressure of 95 kPa. The volume of gas present in the cylinder at the start of compression is 0.0025 m^3. The compression ratio of the cycle is 9.3. The engine completes 1500 cycles per minute. Plot the power produced by the engine, the rate of heat addition for the engine, the thermal efficiency of the cycle, and the mean effective pressure for peak cycle temperatures varying between 2000 K and 3000 K, assuming that the air behaves as an ideal gas with variable specific heats.

7.86 Air flows through an Otto cycle, beginning the compression process at a temperature of 50°C and a pressure of 90 kPa absolute. The volume of gas present in the cylinder at the start of compression is 2 liters. The compression ratio of the cycle is 9.5. The engine completes 1400 cycles per minute. Plot the power produced by the engine, the maximum temperature of the cycle, the thermal efficiency of the cycle, and the mean effective pressure for levels of heat addition varying between 100 kJ/kg and 1000 kJ/kg of air, assuming that the air behaves as an ideal gas with variable specific heats.

7.87 A Diesel cycle has air operating as the working fluid. The air begins the compression process at 100 kPa and 40°C. During the heat addition process, 1200 kJ/kg of heat is added to the air. The compression ratio of the cycle is 18. Treating the air as an ideal gas with variable specific heats, determine (a) the temperature and pressure at the end of each process, (b) the net work per kg of air produced in the cycle, (c) the thermal efficiency of the cycle, and (d) the mean effective pressure.

7.88 Repeat Problem 7.87, but assume the specific heats of the air to be constant with values taken at (a) 300 K, and (b) 1000 K.

7.89 A Diesel cycle has air operating as the working fluid. The air begins the compression process at 100 kPa and 30°C. The temperature after the heat addition process is 2500 K. The compression ratio of the cycle is 19. Treating the air as an ideal gas with constant specific heats, determine (a) the temperature and pressure at the end of each process, (b) the net work per kg of air produced in the cycle, (c) the thermal efficiency of the cycle, and (d) the mean effective pressure.

7.90 A Diesel cycle engine has air beginning the compression process at 98 kPa absolute and 40°C. The air begins the expansion process at a temperature of 2200°C. The compression ratio of the engine is 20. Treating the air as an ideal gas with constant specific heats, determine (a) the temperature and pressure at the end of each process, (b) the net work per kg of air produced in the cycle, (c) the thermal efficiency of the cycle, and (d) the mean effective pressure.

7.91 Air flows through a Diesel cycle, beginning the compression process at 100 kPa and 50°C. During the heat addition process, 1400 kJ/kg of heat is added to the air. Plot the maximum temperature and pressure in the system, the net work per kg of air produced in the cycle, the thermal efficiency of the cycle, and the mean effective pressure for compression ratios ranging between 16 and 25, assuming that the air behaves as an ideal gas with variable specific heats.

7.92 Air flows through a Diesel cycle, beginning the compression process at 100 kPa and 50°C. The compression ratio of the engine is 20. Plot the maximum temperature and pressure in the system, the net work per kg of air produced in the cycle, the thermal efficiency of the cycle, the cut-off ratio, and the mean effective pressure for heat inputs during the heat addition process varying between 500 kJ/kg and 1500 kJ/kg, assuming that the air behaves as an ideal gas with variable specific heats.

7.93 Air flows through a Diesel cycle, beginning the compression process at a pressure of 100 kPa. The volume of gas present in the cylinder at the start of compression is 0.0035 m^3. During the heat addition process, 1000 kJ/kg of heat is added to the air. The compression ratio of the cycle is 19. Plot the work produced during one cycle, the maximum temperature in the cycle, the thermal efficiency of the cycle, and the mean effective pressure for temperatures at the beginning of compression ranging between 0°C and 100°C, assuming that the air behaves as an ideal gas with constant specific heats.

7.94 Air flows through a Diesel cycle, beginning the compression process at a temperature of 50°C. The volume of gas present in the cylinder at the start of compression is 0.0025 m^3. During the heat addition process, 1000 kJ/kg of heat is added to the air. The compression ratio of the cycle is 19. Plot the work produced during one cycle, the maximum pressure in the cycle, the thermal efficiency of the cycle, and the mean effective pressure for pressures at the beginning of compression ranging between 90 kPa and 250 kPa, assuming that the air behaves as an ideal gas with constant specific heats.

7.95 Air flows through a Diesel cycle, beginning the compression process at a temperature of 45°C and a pressure of 100 kPa. The volume of gas present in the cylinder at the start of compression is 0.0041 m^3. The compression ratio of the cycle is 20. The engine completes 1250 cycles per minute. Plot the power produced by the engine, the rate of heat addition for the engine, the thermal efficiency of the cycle, and the mean effective pressure for peak cycle temperatures varying between 1800 K and 2200 K, assuming that the air behaves as an ideal gas with variable specific heats.

7.96 Air flows through a Diesel cycle, beginning the compression process at a temperature of 50°C and a pressure of 90 kPa absolute. The volume of gas present in the cylinder at the start of compression is 3.4 liters. The compression ratio of the cycle is 22. The engine completes 1100 cycles per minute. Plot the power produced by the engine, the maximum temperature of the cycle, the thermal efficiency of the cycle, and the mean effective pressure for levels of heat addition varying between 500 kJ/kg and 1000 kJ/kg of air, assuming that the air behaves as an ideal gas with variable specific heats.

7.97 A Dual cycle has air operating as the working fluid. The air begins the compression process at 100 kPa and 40°C. During the heat addition process, 1200 kJ/kg of heat is added to the air. The pressure ratio in the cycle is 2, and the compression ratio of the cycle is 18. Treating the air as an ideal gas with variable specific heats, determine (a) the temperature and pressure at the end of each process, (b) the net work per kg of air produced in the cycle, (c) the thermal efficiency of the cycle, and (d) the mean effective pressure.

7.98 Repeat Problem 7.97, but treat the air as an ideal gas with constant specific heats evaluated at (a) 300 K, and (b) 1000 K.

7.99 A Dual cycle has air operating as the working fluid. The air begins the compression process at 100 kPa and 35°C. The temperature after the heat addition process is 2600 K. The pressure ratio in the cycle is 1.4, and the compression ratio of the cycle is 19. Treating the air as an ideal gas with constant specific heats, determine (a) the temperature and pressure at the end of each process, (b) the net work per kg of air produced in the cycle, (c) the thermal efficiency of the cycle, and (d) the mean effective pressure.

7.100 Air flows through a Dual cycle, beginning the compression process at 100 kPa and 50°C. During the heat addition process, 1400 kJ/kg of heat is added to the air. Take the pressure ratio of the cycle to be 1.75. Plot the maximum temperature and pressure in the system, the net work per kg of air produced in the cycle, the thermal efficiency of the cycle, and the mean effective pressure for compression ratios ranging between 16 and 25, assuming that the air behaves as an ideal gas with variable specific heats.

7.101 Air flows through a Dual cycle, beginning the compression process at 100 kPa and 50°C. During the heat addition process, 1200 kJ/kg of heat is added to the air. Take the compression ratio of the cycle to be 18. Plot the maximum temperature and pressure in the system, the net work per kg of air produced in the cycle, the thermal efficiency of the cycle, and the mean effective pressure for pressure ratios ranging between 1 and 3, assuming that the air behaves as an ideal gas with variable specific heats.

7.102 Air flows through a Dual cycle, beginning the compression process at 100 kPa and 50°C. Consider the compression ratio to be 20 and the pressure ratio of the cycle to be 1.75. Plot the maximum temperature and pressure in the system, the net work per kg of air produced in the cycle, the thermal efficiency of the cycle, and the mean effective pressure for cut-off ratios varying between 1.1 and 1.3, assuming that the air behaves as an ideal gas with variable specific heats.

7.103 Air flows through a Dual cycle, beginning the compression process at a temperature of 50°C and a pressure of 90 kPa absolute. The volume of gas present in the cylinder at the start of compression is 2.8 liters. The compression ratio of the cycle is 18 and the pressure ratio is 1.5. The engine completes 1100 cycles per minute. Plot the power produced by the engine, the maximum temperature of the cycle, the thermal efficiency of the cycle, and the mean effective pressure for levels of heat addition varying between 500 kJ/kg and 1500 kJ/kg of air, assuming that the air behaves as an ideal gas with variable specific heats.

7.104 Treating the air as an ideal gas with constant specific heats, plot the following:

(a) the thermal efficiency of an Otto cycle, Diesel cycle, and Dual cycle, with the Diesel cycle and Dual cycle having cut-off ratios of 3 and the Dual cycle having a pressure ratio of 1.5, as the compression ratio of each varies between 5 and 15.
(b) the thermal efficiency of a Diesel cycle and a Dual cycle, with the Dual cycle having a pressure ratio of 1.5, with both cycles having a compression ratio of 18, and with the cut-off ratio of each varying between 1.2 and 4.
(c) the thermal efficiency of a Diesel cycle and a Dual cycle, with the Dual cycle having a pressure ratio of 1.5, with both cycles having a cut-off ratio of 2.5, and with the compression ratio of each varying between 15 and 25.
(d) the thermal efficiency of a Dual cycle, with a compression ratio of 18, and with the pressure ratio varying between 1 and 2.

DESIGN/OPEN-ENDED PROBLEMS

7.105 Design a Rankine cycle with one reheater and one open feedwater heater that has a high thermal efficiency. The maximum pressure and temperature in the system are 20 MPa and 580°C, and the condenser pressure is 20 kPa. The isentropic efficiency of the turbine is 0.82, and the isentropic efficiency of the pumps is 0.69. The net power to be developed from the cycle is 450 MW.

7.106 Design a Rankine cycle with two closed feedwater heaters that has a high thermal efficiency. The maximum pressure in the cycle is 20 MPa and the maximum temperature is 620°C. The condenser pressure is 20 kPa absolute. The drains from the feedwater heaters can be either pumped forward or cascaded backward. The isentropic efficiency of the turbine is 0.77, and the isentropic efficiency of the pumps is 0.65. The net power to be produced by the cycle is 600 MW.

7.107 Design a Rankine cycle with one reheater, one open feedwater heater, and two closed feedwater heaters with drains cascaded backward that has a high thermal efficiency. The maximum temperature and pressure in the cycle are 600°C and 18 MPa. The condenser pressure is 10 kPa. The isentropic efficiency of the turbine is 0.80, and the isentropic efficiency of the pumps is 0.70. The net power to be produced by the cycle is 750 MW. In addition, determine the required external cooling water mass flow rate for a temperature increase through the condenser of 20°C.

7.108 Design a Rankine cycle that has a maximum allowed pressure of 20 MPa, a maximum allowed temperature of 550°C, and a minimum allowed pressure of 10 kPa. The isentropic efficiency of the turbine is 0.84, and the isentropic efficiency of the pumps is 0.60. The thermal efficiency of the cycle needs to be at least 42%. The net power to be produced by the cycle is 400 MW. Include a design for the external cooling water system with a maximum allowed flow rate for the external cooling water of 10,000 kg/s.

7.109 A basic Rankine cycle is to produce 200 MW of power. The inlet steam pressure can be no higher than 15 MPa, and the temperature of the outside environment is 15°C. The isentropic efficiency of the turbine is 0.80, and the isentropic efficiency of the pump is 0.75. Design a basic cycle specifying state properties at each state, the required mass flow rate of the steam, the thermal efficiency of the cycle, and the required mass flow rate of either cooling water or air in the condenser.

7.110 Repeat Problem 7.109, but the cycle can now contain superheated steam. The steam temperature in the cycle can be no higher than 600°C. In addition, the steam quality in the turbine cannot fall below 0.85.

7.111 Repeat Problem 7.110, but the cycle should now contain one reheater and one open feedwater heater.

7.112 Repeat Problem 7.110, but the cycle should now contain one reheater and one closed feedwater heater. If the pressure of the closed feedwater heater is above 100 kPa, the drains should be cascaded backward; otherwise the drains should be pumped forward.

7.113 Repeat Problem 7.110, but the cycle may now contain up to two reheaters and should have one open feedwater heater and one closed feedwater heater. If the pressure of the closed feedwater heater is above 100 kPa, the drains should be cascaded backward; otherwise the drains should be pumped forward.

7.114 Repeat Problem 7.110, but the cycle may contain up to two reheaters, up to three open feedwater heaters, and up to four closed feedwater heaters. If the pressure of the closed feedwater heater is above 100 kPa, the drains should be cascaded backward; otherwise the drains should be pumped forward. Attempt to design a system with the highest possible efficiency.

7.115 Design a Rankine cycle with superheat and one reheater that will produce 400 MW of power. The maximum pressure allowed in the cycle is 20 MPa, and the minimum allowable pressure is 10 kPa. The maximum allowable temperature is 500°C. The isentropic efficiency of the pump is 0.74. The isentropic efficiency of a particular turbine section is dependent on the isentropic outlet state quality:

If exit steam is superheated, $\eta_{s,t} = 0.83$

For exit steam quality between 0.6 and 1.0: $\eta_{s,t} = 0.83 - 0.8(1 - x)$

For exit steam quality below 0.6: $\eta_{s,t} = 0.50$

Your design should specify state properties at each state, the required mass flow rate of the steam, and the thermal efficiency of the cycle.

7.116 A Brayton cycle is to produce 100 MW of power, using a compressor with an isentropic efficiency of 0.78 and a turbine with an isentropic efficiency of 0.83. The air entering the compressor is at 100 kPa and 20°C. The maximum allowable pressure in the system is 2.0 MPa, and the maximum allowable temperature is 1400°C. Design a Brayton cycle that meets these requirements, indicating the temperature and pressure at each state and the mass flow rate for the system, as well as the thermal efficiency of the cycle.

7.117 Design a Brayton cycle (with any additional components you choose) that has a compressor inlet condition of air entering at 100 kPa and 20°C and a maximum allowable pressure and temperature of 1500 kPa and 1400 K, respectively. The turbine and compressor both have isentropic efficiencies of 0.80. The cycle should have a high thermal efficiency. For a mass flow rate of 25 kg/s, determine the net power produced and the thermal efficiency of the cycle, justifying your choices or air properties and components. You may use either constant or variable specific heats in your analysis.

7.118 Repeat Problem 7.116, but now you may include up to two intercoolers, up to two reheaters, and no more than one regenerator in the design. Design a Brayton cycle that meets these requirements, indicating the temperature and pressure at each state and the mass flow rate for the system, as well as the thermal efficiency of the cycle.

7.119 An Otto cycle engine has material properties that limit the maximum pressure allowed in the cycle to 7100 kPa and the maximum allowable temperature to 2800 K. Air enters the compression process at no cooler than 40°C and 95 kPa. The compression ratio of the engine must be at least 9.0. Assume the specific heats of the air are constant, with values taken at 1000 K. Design a system with high thermal efficiency that meets these constraints. You should indicate the temperature and pressure at each state in the cycle, and state the cycle's thermal efficiency.

7.120 Design a Dual cycle that meets the following constraints: the compression ratio must be at least 14 but no more than 22, the maximum allowable pressure in the system is 7.0 MPa, the maximum allowable temperature is 2500 K, the air begins the compression stroke at 30°C and 100 kPa, and the specific heats of the air may be assumed to be constant. As part of your design, indicate the temperature and pressure at each state in the cycle and the thermal efficiency of the cycle.

CHAPTER 8

Refrigeration Cycles

Learning Objectives

Upon completion of Chapter 8, you will be able to

8.1 Interpret the fundamentals of practical refrigeration systems;

8.2 Analyze vapor-compression refrigeration cycles and reversed Brayton refrigeration cycles; and

8.3 Describe additional refrigeration concepts, such as absorption refrigeration and cascade refrigeration.

8.1 INTRODUCTION

In Chapter 7, we described and analyzed many types of power cycles. The purpose of a power cycle is to produce work after receiving an input of heat; during the process, heat is also rejected to a colder area. If a power cycle is reversed, it can successfully move heat from a colder space to a hotter space after experiencing an addition of work. **Figure 8.1** shows a basic schematic diagram of this process, in which heat is received from a cold space, work is added to the device, and heat is rejected to the warm space. The thermodynamic cycles that follow this basic process are known as refrigeration cycles.

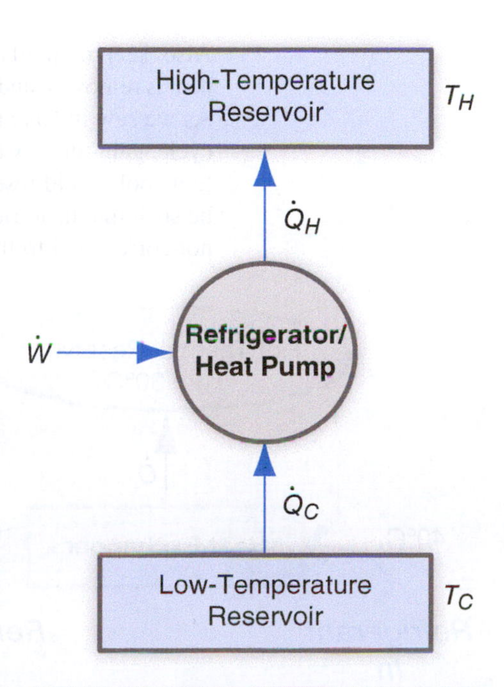

FIGURE 8.1 A schematic diagram of the basic lay-out of a refrigeration cycle.

Refrigeration cycles are used in many applications, and the devices shown in **Figure 8.2** that use them are referred to by different names depending on the application. A device or a cycle whose purpose is to remove heat from a cool space is known as a refrigerator or an air conditioner. The term *refrigerator* is often used for a system that maintains a space at a lower temperature to store objects, such as food. The term *air conditioner* is usually reserved for devices and cycles that remove heat from a space but that maintain that space at a temperature that is somewhat higher than would be in a refrigerator. You might expect to have people or animals present in such a space. (Although walking into a cold refrigerator on a hot day may be temporarily pleasant, it soon becomes too cold for most people.) Alternatively, the purpose of the device may be to transfer heat to a warm space for the purpose of heating the space. Such a device and cycle is called a heat pump.

As discussed in Chapter 5, there are theoretical maximum possible coefficients of performance for refrigerators and heat pumps. These can be applied to the refrigeration cycles, although we must be careful about

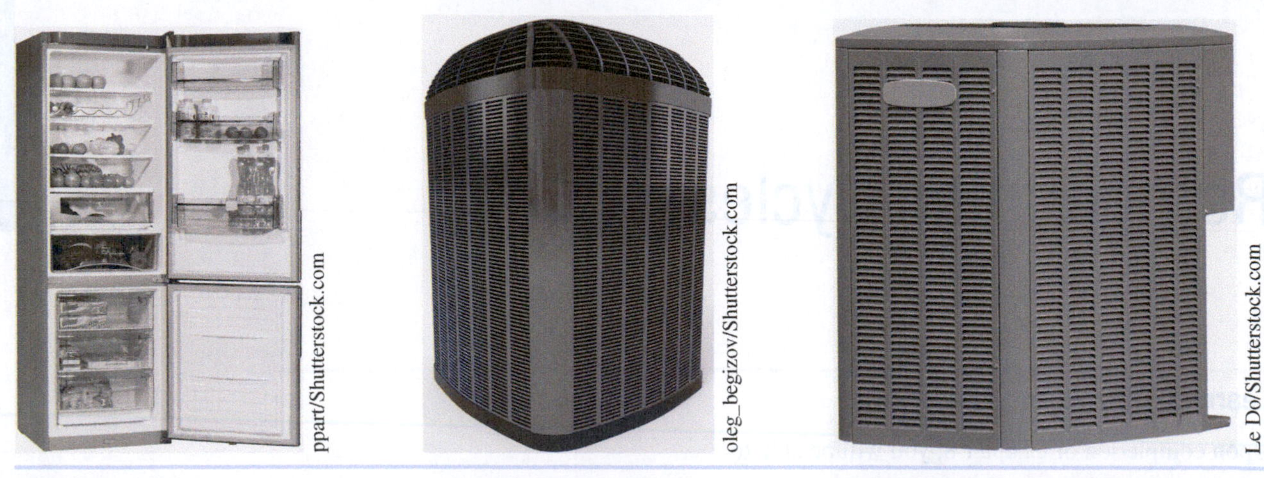

ppart/Shutterstock.com
oleg_begizov/Shutterstock.com
Le Do/Shutterstock.com

FIGURE 8.2 Refrigeration cycles are used in household refrigerators, air conditioning units, and heat pump systems.

the temperatures used in the calculations. Recall that Eqs. (5.6) and (5.7) are the expressions for the maximum possible coefficient of a refrigerator and a heat pump, respectively:

$$\beta_{\max} = \frac{T_C}{T_H - T_C} \tag{5.6}$$

and

$$\gamma_{\max} = \frac{T_H}{T_H - T_C} \tag{5.7}$$

Also, keep in mind that T_C represents the temperature of the cold thermal reservoir from which heat is removed, and T_H is the temperature of the hot thermal reservoir to which heat is rejected. As we saw in Chapter 7 in our discussion of power cycles, a working fluid flows through the cycle, gains energy through heat transfer from the hot reservoir, and is cooled by rejecting heat to a cooler cold reservoir. The working fluid temperature and the reservoir temperature must be such that heat transfer can occur in the proper direction—as a result, the value of T_C might not correspond to the coldest working fluid temperature, and T_H might not correspond to the hottest working fluid temperature in the cycle.

For example, consider the scenario shown in Figure 8.3. The working fluid in the cycle enters a heat exchanger at 40°C and exits at 30°C as it rejects heat to the hot reservoir. In order for heat transfer to occur out of the working fluid, the temperature of the reservoir must always be less than the temperature of the working fluid; therefore, the reservoir temperature can be no higher than 30°C. The appropriate value to use for T_H is 30°C (303 K), not the 40°C that corresponds to the highest temperature in the cycle. Similarly, consider the heat addition process for the working fluid shown in Figure 8.4. Here, the working fluid is being warmed because it is colder than the cold reservoir. The fluid enters the heat exchanger at 0°C and exits at 10°C. The working fluid must always be colder than the thermal reservoir to receive heat from the reservoir, so the coldest the reservoir can be is 10°C.

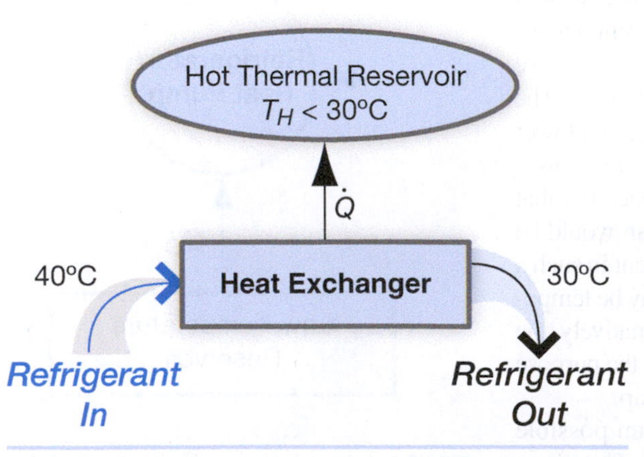

FIGURE 8.3 Heat can be transferred out of one medium to a thermal reservoir that is at a lower temperature.

Therefore, T_C should be 10°C (283 K) rather than 0°C, which corresponds to the coldest temperature in the cycle.

There are both vapor refrigeration cycles (which involve a phase change between the liquid and vapor phases) and gas refrigeration cycles (which involve only the gaseous phase throughout the whole cycle). Although water could be used as the working fluid for vapor refrigeration cycles, it is not normally employed because the pressure–temperature phase-change characteristics of water are not particularly good for most refrigeration operations—generally, liquid water exists at very low pressures where refrigeration processes are concerned, and this can lead to inefficiencies and poor operation. In addition, if we want to lower the temperature of a space below 0°C, the water will freeze and it becomes very difficult to move solid substances through a series of devices.

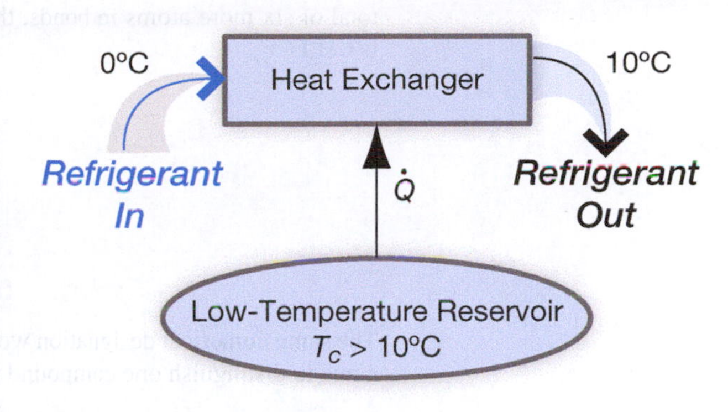

FIGURE 8.4 A fluid can receive heat from a thermal reservoir provided the fluid is at a lower temperature than the reservoir.

As such, other fluids, such as ammonia, propane, and chemical compounds known as refrigerants, are used in vapor refrigeration cycles. These substances have boiling/condensation pressure–temperature characteristics that are more appropriate for achieving the heat transfers needed at desired temperatures without excessively high or low pressures. Refrigerant compounds are numbered in a standard way: the first digit on the right represents the number of fluorine atoms, the second digit from the right represents one plus the number of hydrogen atoms, the third digit from the right represents the number of carbon atoms minus 1 (and is ignored if there is only one carbon atom), and the fourth number from the right represents the number of double or triple carbon–carbon bonds in the compound (and is ignored if all the bonds are single bonds). If there are any open spots in the compound that are not filled by fluorine or hydrogen, they are considered chlorine atoms by default (unless specified as some other atom).

▶ EXAMPLE 8.1

Determine the chemical formula for, and draw a diagram of, the compound for the refrigerants R-12 and R-134a.

Solution:

R-12: Using the numbering system described in the text, the right digit represents the number of fluorine atoms, so there are two F atoms. The numeral 1 represents the number of hydrogen atoms plus 1, so there are 0 hydrogen atoms. There is no third digit from the right, so that digit is by default 0, and 0 is equal to the number of carbon atoms minus 1; that is, there is one carbon atom. A single carbon atom can have four bonds, so there are two remaining spaces, which are filled by chlorine. Therefore, the chemical formula for R-12 is CF_2Cl_2.

```
     F
     |
Cl − C − Cl
     |
     F
```

R-12

R-134a: The "a" refers to the possibility of one or more chemical compounds taking on the same numerical designation. The 4 indicates that there are four fluorine atoms, and the 3 is equal to 1 plus the number of hydrogen atoms; that is, there are two H atoms. The numeral 1 is equal to the number of carbon atoms minus 1, so there are two C atoms. Finally, there is no fourth digit, so all of the carbon bonds are single bonds. Because two carbons atoms, singly bonded to each other, can hold a total of six more atoms in bonds, there is no space for more chlorine atoms. The chemical formula is CH_2FCF_3.

```
    H   F
    |   |
F - C - C - F
    |   |
    H   F
```

R-134a

The same numerical designation would work for CHF_2CHF_2, which necessitates the use of "a" in the name to distinguish one compound from the other.

QUESTION FOR THOUGHT/DISCUSSION

While some categories of refrigerants have outstanding properties for both refrigeration systems and short-term safety, they have been found to be detrimental to the planet when used in large quantities for a long period of time. In what ways can engineers balance the positive and negative aspects of a material or product and thus arrive at a solution that benefits both the need for the product and the needs of the environment?

8.2 THE VAPOR-COMPRESSION REFRIGERATION CYCLE

Using a phase change for a substance is an effective method of accomplishing a substantial amount of heat transfer without having to be concerned with large temperature changes of a working fluid being used to remove or deliver heat. We have seen this technique used in the Rankine power cycle. The technique is also used in the vapor-compression refrigeration cycle, which is also known as the reversed Rankine cycle. This is the most common thermodynamic cycle used for refrigeration purposes, because it is simple, effective, and relatively inexpensive.

The vapor-compression refrigeration cycle consists of four processes. For an ideal cycle, these four processes are as follows:

Process 1-2: Isentropic compression
Process 2-3: Constant-pressure heat removal
Process 3-4: Throttling process
Process 4-1: Constant-pressure heat addition

These processes are accomplished in a compressor, condenser, throttling valve, and an evaporator, respectively, as shown in **Figure 8.5**. A *T-s* diagram for the cycle is shown in **Figure 8.6**. The working fluid undergoes a phase change, both in the condenser, where heat is removed from the working fluid and delivered to the warm space, and in the evaporator, where the working fluid becomes a vapor by receiving a heat input from the cold space.

Unless otherwise stated, we assume that the working fluid leaves the evaporator as a saturated vapor ($x_1 = 1$) and that it leaves the condenser as a saturated liquid ($x_3 = 0$). It is assumed that $P_1 = P_4$ and $P_2 = P_3$, because these processes are both constant-pressure by definition. Furthermore, we will assume that there is no power used in the evaporator, condenser, or throttling valve ($\dot{W}_e = \dot{W}_{cond} = \dot{W}_{tv} = 0$), no heat transfer in the compressor or throttling valve ($\dot{Q}_{comp} = \dot{Q}_{tv} = 0$), and no changes in kinetic or potential energy for all devices. Using the first law for steady-state, steady-flow, single-inlet, single-outlet systems, Eq. (4.15),

$$\dot{Q} - \dot{W} = \dot{m}\left(h_{out} - h_{in} + \frac{V_{out}^2 - V_{in}^2}{2} + g(z_{out} - z_{in})\right) \quad (4.15)$$

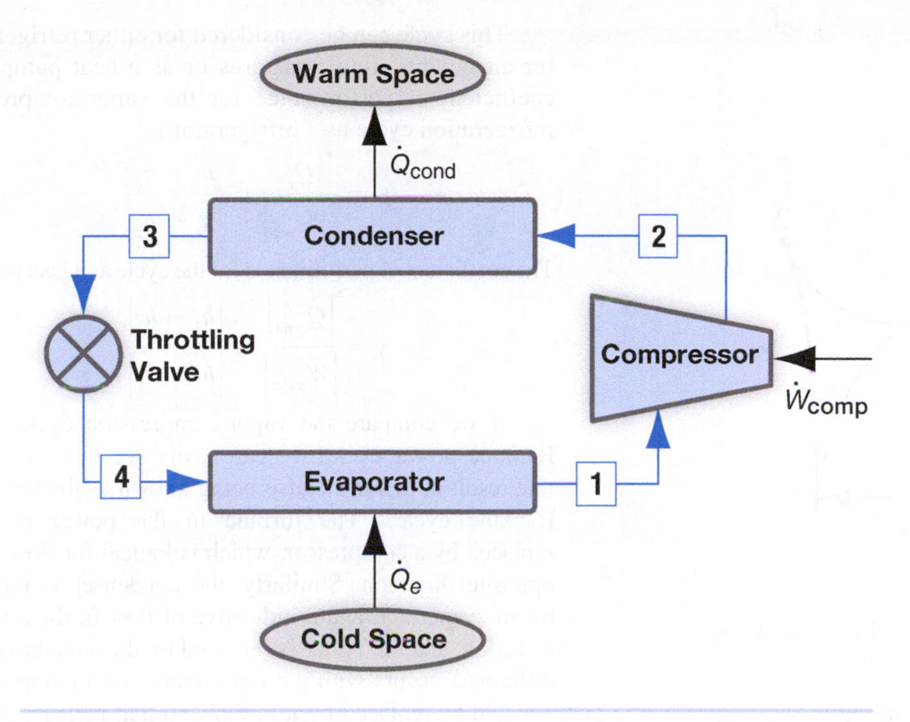

FIGURE 8.5 A component diagram of a vapor-compression refrigeration cycle.

an expression describing the behavior of each device can be derived.

Evaporator: $$\dot{Q}_e = \dot{m}(h_1 - h_4) \tag{8.1}$$

(Note: This value is sometimes referred to as the rate of refrigeration or refrigeration capacity, and the value is sometimes converted to kW of refrigeration.)

Compressor: $$\dot{W}_{comp} = \dot{m}(h_1 - h_2) \tag{8.2}$$

Condenser: $$\dot{Q}_{cond} = \dot{m}(h_3 - h_2) \tag{8.3}$$

Throttling valve: $$h_3 = h_4 \tag{8.4}$$

For the ideal vapor-compression refrigeration cycle, the compressor is considered isentropic: $s_1 = s_2$. However, actual compressors are not isentropic, and so by incorporating an isentropic efficiency for the compressor into the device, a more realistic assessment of a cycle can be made. The isentropic efficiency of the compressor is

$$\eta_{comp} = \frac{h_1 - h_{2s}}{h_1 - h_2} \tag{8.5}$$

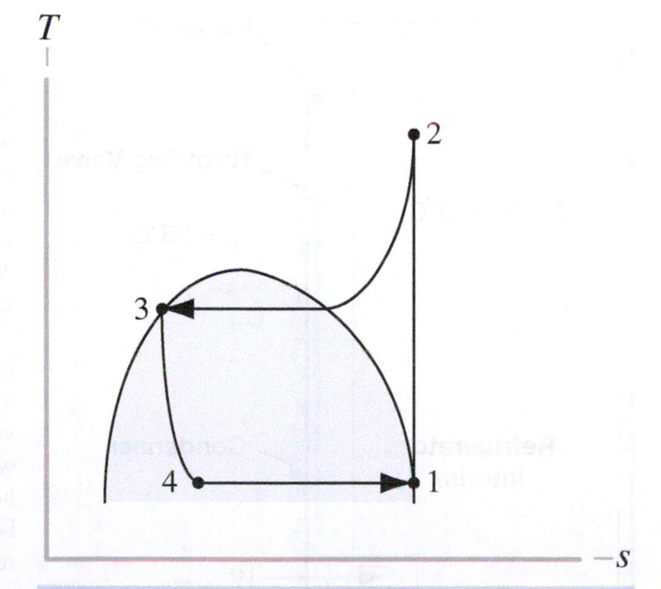

FIGURE 8.6 A *T-s* diagram of an ideal vapor-compression refrigeration cycle.

and **Figure 8.7** shows a *T-s* diagram for a nonideal vapor-compression refrigeration cycle. Note that in Eq. (8.5), $s_{2s} = s_1$ and $P_{2s} = P_2$.

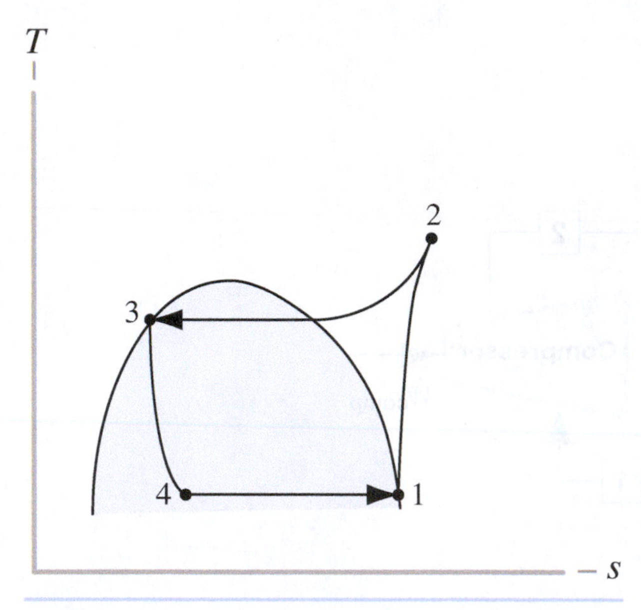

FIGURE 8.7 A *T-s* diagram of a nonideal vapor-compression refrigeration cycle.

This cycle can be considered for either refrigeration (or air conditioning) purposes or as a heat pump. The coefficient of performance for the vapor-compression refrigeration cycle as a refrigerator is

$$\beta = \frac{\dot{Q}_e}{|\dot{W}_{comp}|} = \frac{h_1 - h_4}{|h_1 - h_2|} \tag{8.6}$$

The coefficient of performance for the cycle as a heat pump is

$$\gamma = \frac{|\dot{Q}_{cond}|}{|\dot{W}_{comp}|} = \frac{|h_3 - h_2|}{|h_1 - h_2|} \tag{8.7}$$

If we compare the vapor-compression cycle to the Rankine power cycle, we can easily see the similarities that result in this cycle also being known as the "reversed Rankine cycle." The turbine in the power cycle is replaced by a compressor, which is logical for flow in the opposite direction. Similarly, the condenser is replaced by an evaporator, again indicative of flow in the reversed direction, and the boiler is replaced by the condenser. The difference occurs with the replacement of a pump with a throttling valve. Logically, the pump would be replaced by a turbine, so that power could be recovered in the process of lowering the pressure of the working fluid. In some cases, this is done and the refrigeration cycle has a turbine rather than a throttling valve. If we encounter this situation, we analyze the turbine just as we analyze other turbines, and the net power ($\dot{W}_{net} = \dot{W}_{comp} + \dot{W}_t$) is considered in the expressions for the coefficients of performance, rather than just the compressor power. However, use of a turbine is not common because (a) a turbine is much more expensive initially than a throttling valve and is more expensive to maintain, and (b) the turbine does not recover a large amount of power because it involves the expansion of a liquid into a saturated mixture, which tends to lead toward an inefficient device. Therefore, the use of a turbine typically only makes sense in very large systems.

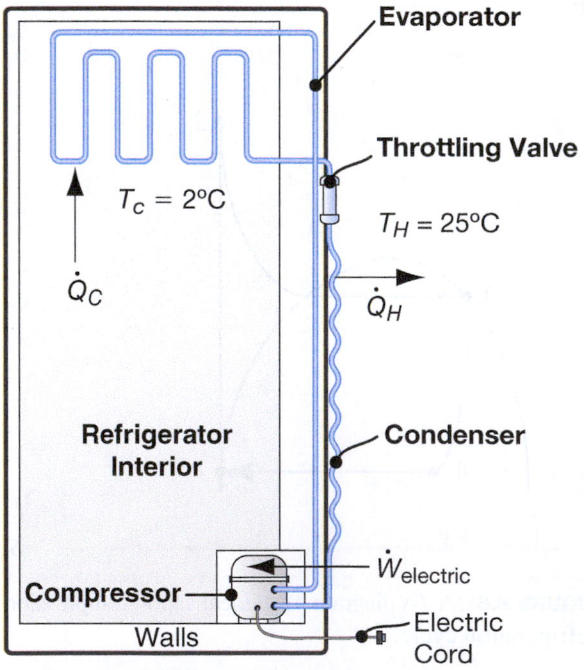

FIGURE 8.8 A cut-away schematic diagram of a household refrigerator. The refrigerant in contact with the interior must be at a lower temperature than the interior, and the refrigerant in contact with the exterior must be at a higher temperature than the exterior.

It can be difficult to understand how a vapor-compression refrigeration cycle achieves its purpose of removing heat from a cold space and/or delivering heat to a warm space. To help visualize this, consider the diagram shown in Figure 8.8, where the cycle is shown as it may be applied in a common household refrigerator. The interior of the refrigerator is to be maintained at 2°C, and the outside room air is 25°C. The refrigerant to be used is R-134a. The evaporator coils are embedded into the walls of the refrigerator, and the condenser coils are placed on the outside of the refrigerator—they may be covered for protection or exposed directly to the room. The compressor is on the outside of the refrigerator, typically near the bottom of the refrigerator, and the expansion valve is located in the tubing between the end of the condenser coils and the beginning of the evaporator coils.

To achieve cooling, the refrigerant must be below 2°C. This can be achieved with an evaporator pressure for the

R-134a of 200 kPa. The saturation temperature, which corresponds to the temperature of the R-134a in the evaporator, is −10.09°C; because this is below 2°C, heat can be transferred from the interior of the refrigerator to the refrigerant, causing the R-134a to boil. To transfer the heat to the surroundings, the R-134a temperature must remain above 25°C as it passes through the entire condenser. This can be achieved with a condenser pressure of 800 kPa, for which the saturation temperature (and therefore the temperature at which condensation occurs in the condenser) is 31.33°C. So, because it has a compressor that increases the R-134a pressure from 200 kPa to 800 kPa, this refrigerator can transfer heat from the cold space inside the refrigerator to the surrounding room.

STUDENT EXERCISE

Develop a computer model for the vapor-compression refrigeration cycle, allowing for both its use as a refrigerator and as a heat pump, and also allowing for both nonideal and ideal operation. Be sure that the model can accept different types of refrigerants as its working fluid. If provided with appropriate states of the working fluid, the model should be able to calculate any remaining unknown property states, the heat transfer rate to the evaporator, the heat transfer rate from the condenser, the power required by the compressor, and the coefficient of performance of the system as a refrigerator and as a heat pump.

▶ EXAMPLE 8.2

A vapor-compression refrigeration cycle uses R-134a as its refrigerant. The R-134a enters the compressor as a saturated vapor at 200 kPa, and exits the condenser as a saturated liquid at 800 kPa. The rate of refrigeration of the cycle is to be 10.55 kW of refrigeration. Determine the coefficient of performance of the device as a refrigerator and the mass flow rate of R-134a needed for (a) an isentropic compressor and (b) an isentropic compressor efficiency of 0.75.

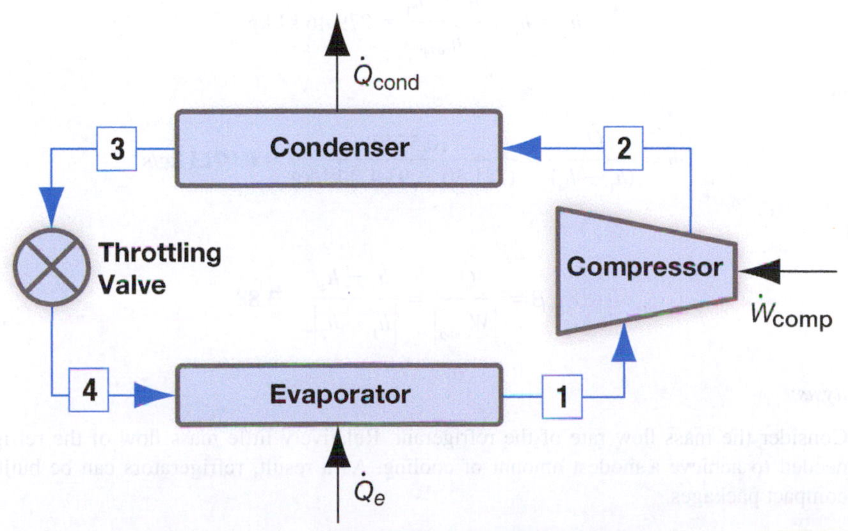

Given: $x_1 = 1.0$, $P_1 = 200$ kPa, $x_3 = 0.0$, $P_3 = 800$ kPa, $\dot{Q}_e = 10.55$ kW

Find: β, $\dot{m}$

Solution: The cycle to be analyzed is the vapor-compression refrigeration cycle. In part (a) the cycle is ideal, while in part (b) the compressor is nonisentropic. The working fluid is the refrigerant R-134a, for which we will use a computer program for property data.

Assume: $P_1 = P_4$, $P_2 = P_3$, $\dot{W}_e = \dot{W}_{cond} = \dot{W}_{tv} = \dot{Q}_{tv} = \dot{Q}_{comp} = 0$, $\Delta KE = \Delta PE = 0$ for each device. Steady-state, steady-flow processes for each device.

(a) The compressor is isentropic. The enthalpies at each state are found via computer programs:

$$h_1 = h_g\,(@P_1) = 241.30 \text{ kJ/kg}$$
$$s_1 = s_g\,(@P_1) = 0.9253 \text{ kJ/kg} \cdot \text{K}$$
$$s_2 = s_1 = 0.9253 \text{ kJ/kg} \cdot \text{K}$$
$$h_2 = 269.92 \text{ kJ/kg}$$
$$h_3 = h_f\,(@P_3) = 93.42 \text{ kJ/kg}$$
$$h_4 = h_3 = 93.42 \text{ kJ/kg}$$

From Eq. (8.1):

$$\dot{m} = \frac{\dot{Q}_e}{(h_1 - h_4)} = \frac{10.55 \text{ kW}}{(241.30 - 93.42) \text{ kJ/kg}} = \mathbf{0.0713 \text{ kg/s}}$$

The coefficient of performance is found from Eq. (8.6):

$$\beta = \frac{\dot{Q}_e}{\left|\dot{W}_{\text{comp}}\right|} = \frac{h_1 - h_4}{\left|h_1 - h_2\right|} = \mathbf{5.17}$$

(b) The nonisentropic compressor has $h_{\text{comp}} = 0.75$.

From part (a), the enthalpies at states 1, 3, and 4 remain unchanged and the state 2 enthalpy is considered the isentropic state 2 enthalpy:

$$h_1 = 241.30 \text{ kJ/kg} \qquad h_{2s} = 269.92 \text{ kJ/kg}$$
$$h_3 = 93.42 \text{ kJ/kg} \qquad h_4 = 93.42 \text{ kJ/kg}$$

Using the isentropic efficiency of the compressor (Eq. (8.5)) yields the actual outlet state 2 enthalpy:

$$h_2 = h_1 + \frac{h_{2s} - h_1}{\eta_{\text{comp}}} = 279.46 \text{ kJ/kg}$$

Then

$$\dot{m} = \frac{\dot{Q}_e}{(h_1 - h_4)} = \frac{10.55 \text{ kW}}{(241.30 - 93.42) \text{ kJ/kg}} = \mathbf{0.0713 \text{ kg/s}}$$

and

$$\beta = \frac{\dot{Q}_e}{\left|\dot{W}_{\text{comp}}\right|} = \frac{h_1 - h_4}{\left|h_1 - h_2\right|} = \mathbf{3.88}$$

Analysis:

(1) Consider the mass flow rate of the refrigerant. Relatively little mass flow of the refrigerant is needed to achieve a modest amount of cooling. As a result, refrigerators can be built in very compact packages.

(2) The isentropic efficiency of the compressor does not impact the amount of heat that is removed from the cold space, but will impact the amount of heat rejected to a warm environment.

(3) An inefficient compressor can rapidly reduce the efficiency (seen through the coefficient of performance) of the entire system.

▶ EXAMPLE 8.3

A vapor-compression refrigeration cycle is to be used as a heat pump. The refrigerant used is ammonia, and its mass flow rate is 2 kg/s. The ammonia enters the compressor as a saturated vapor at 345 kPa, and exits the condenser as a saturated liquid at 1.725 MPa. The isentropic efficiency of the compressor is 0.78. Determine the rate of heat delivered to the warm space, and the coefficient of performance of the heat pump.

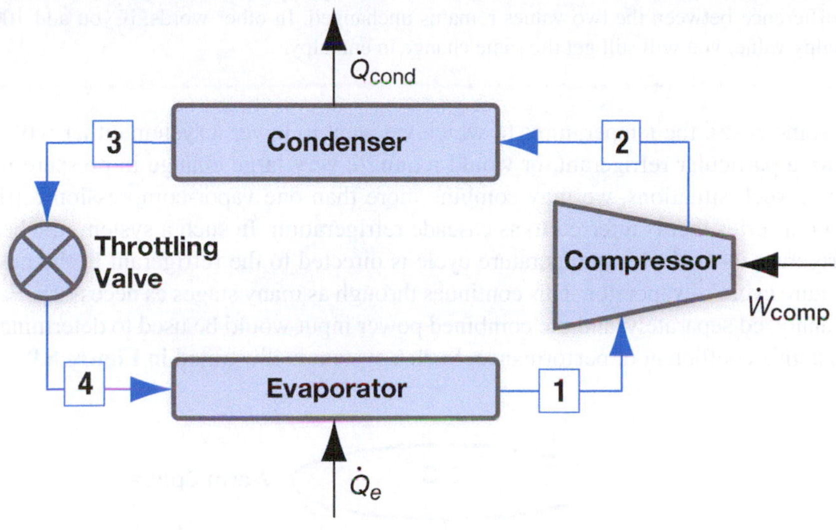

Given: $x_1 = 1.0$, $P_1 = 345$ kPa, $x_3 = 0.0$, $P_3 = 1.725$ MPa, $\dot{m} = 2.0$ kg/s, $\eta_{comp} = 0.78$

Find: $\dot{Q}_{cond}$ and γ

Solution: The cycle to be analyzed is the vapor-compression refrigeration cycle that is being used as a heat pump. The cycle is nonideal, as the compressor is nonisentropic. The working fluid is ammonia, for which we will use a computer program for property data.

Assume: $P_1 = P_4$ and $P_2 = P_3$, $\dot{W}_e = \dot{W}_{cond} = \dot{W}_{tv} = \dot{Q}_{tv} = \dot{Q}_{comp} = 0$, $\Delta KE = \Delta PE = 0$ for each device. Steady-state, steady-flow processes for each device.

It is first necessary to find the enthalpy values at each state, using the properties found with a computer program:

$$h_1 = h_g(@P_1) = 1437 \text{ kJ/kg}$$

$$s_1 = s_g(@P_1) = 5.4106 \text{ kJ/kg} \cdot \text{K}$$

$$s_{2s} = s_1: h_{2s} = 1675 \text{ kJ/kg}$$

For the nonisentropic compressor,

$$h_2 = h_1 + \frac{h_{2s} - h_1}{\eta_{comp}} = 1742 \text{ kJ/kg}$$

$$h_3 = h_f(@P_3) = 390.2 \text{ kJ/kg}$$

$$h_4 = h_3 = 390.2 \text{ kJ/kg}$$

The heat removed from the refrigerant in the condenser is found from Eq. (8.3):

$$\dot{Q}_{cond} = \dot{m}(h_3 - h_2) = \mathbf{-2.70\ MW}$$

and so 2.70 MJ/s of heat is delivered to the warm space.

The coefficient of performance of the heat pump is found from Eq. (8.7):

$$\gamma = \frac{|\dot{Q}_{cond}|}{|\dot{W}_{comp}|} = \frac{|h_3 - h_2|}{|h_1 - h_2|} = \mathbf{4.42}$$

Analysis: The coefficient of performance of the device as a heat pump is always larger than the coefficient of performance as a refrigerator: $\gamma = \beta + 1$. Also, note that the actual value of the enthalpy of a refrigerant can depend on what computer program is used. In particular, different programs may use different reference states to which a value of "0" is assigned the enthalpy. This is more likely to occur for refrigerants than water/steam. If you obtain different enthalpy values than those shown here, remember that the difference in enthalpy is what is important in the calculations. If a different arbitrary 0-point is used, such that both enthalpy values are changed by the same amount, the difference between the two values remains unchanged. In other words, if you add 100 to each enthalpy value, you will still get the same change in enthalpy.

In some cases, the temperatures to which we want to lower a system either will not work well with a particular refrigerant, or would require a very large change in pressure in a compressor. In such situations, we may combine more than one vapor-compression refrigeration cycle into a series that is referred to as cascade refrigeration. In such a system, the heat out of the refrigerant in the lowest temperature cycle is directed to the refrigerant in the next lowest temperature cycle's evaporator. This continues through as many stages as necessary. Each cycle can be analyzed separately, and the combined power input would be used to determine the cascade system's coefficient of performance. Such a system is illustrated in **Figure 8.9**.

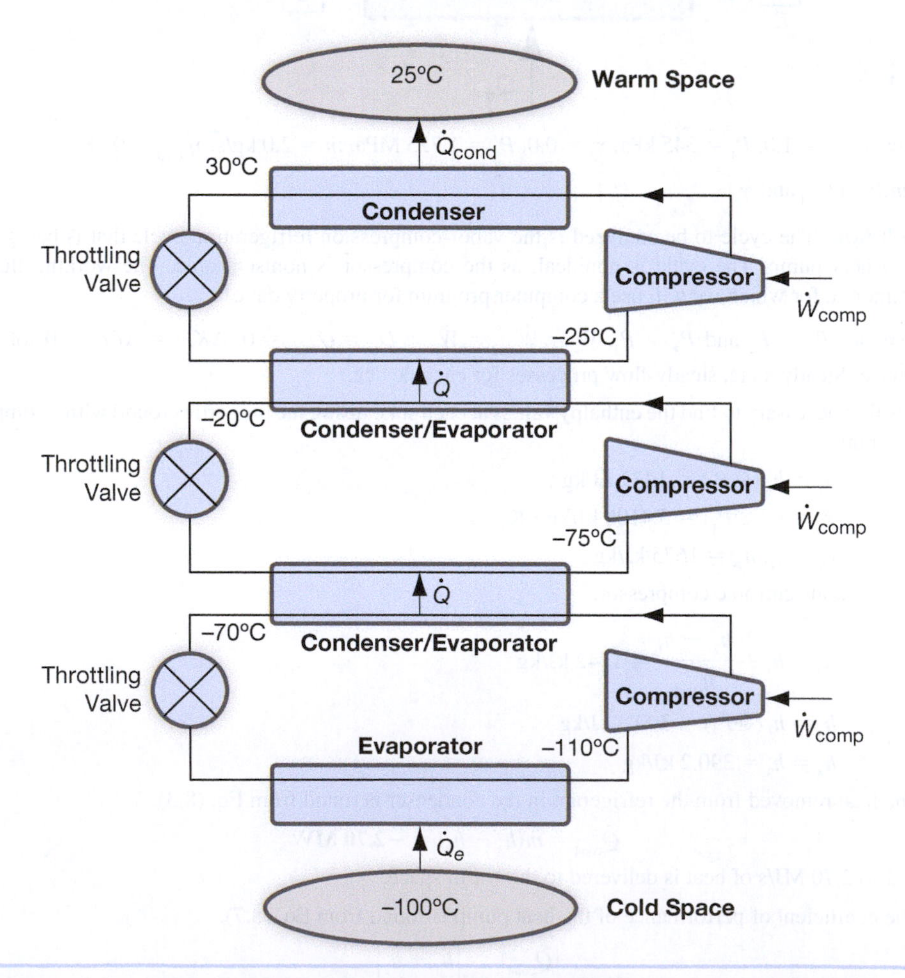

FIGURE 8.9 An example of how three vapor-compression refrigeration cycles can be combined into a cascade refrigeration system.

QUESTION FOR THOUGHT/DISCUSSION

Heat pumps are a generally cost-effective method for heating or cooling buildings. What can be done by the government and utilities companies to encourage wider adoption of such systems?

8.3 ABSORPTION REFRIGERATION

As noted in Chapter 7, compressors use significantly more power to pressurize a gas than pumps use to pressurize a liquid. As a result, for large refrigeration applications, the power consumption needed to move the refrigerant can become very large and require either multiple compressors or very large compressors. The cost of operating such a system also becomes large. Yet the compressors are necessary if a liquid is boiled in the process of absorbing heat from a cold space. It would be of interest from an operational cost standpoint to reduce the power consumption through using a pump. This is a fundamental idea behind the concept of an absorption refrigeration cycle.

In an absorption refrigeration cycle, the refrigerant is sent through a system as in a vapor-compression refrigeration cycle, but instead of being directed from the evaporator to a compressor, the refrigerant vapor is absorbed into a carrier liquid. This creates a liquid solution that can be sent through a pump and pressurized as a liquid. After being pressurized, the refrigerant is separated from the carrier liquid through the addition of heat in a generator. The refrigerant can then proceed to the condenser, throttling valve, and evaporator of a normal vapor-compression refrigeration cycle. A simple form of this absorption refrigeration system is shown in **Figure 8.10**.

The absorption of the refrigerant vapor by the carrier liquid occurs in the absorber. Cooling water may be used in the absorber, because more refrigerant can be dissolved into the carrier liquid at lower temperatures. The release of the refrigerant occurs in the generator, with heat coming from some external source to aid in the release of the refrigerant. Additional components can be included in the basic system to improve the performance of the system. These include a heat exchanger, which transfers heat from the carrier fluid exiting the generator

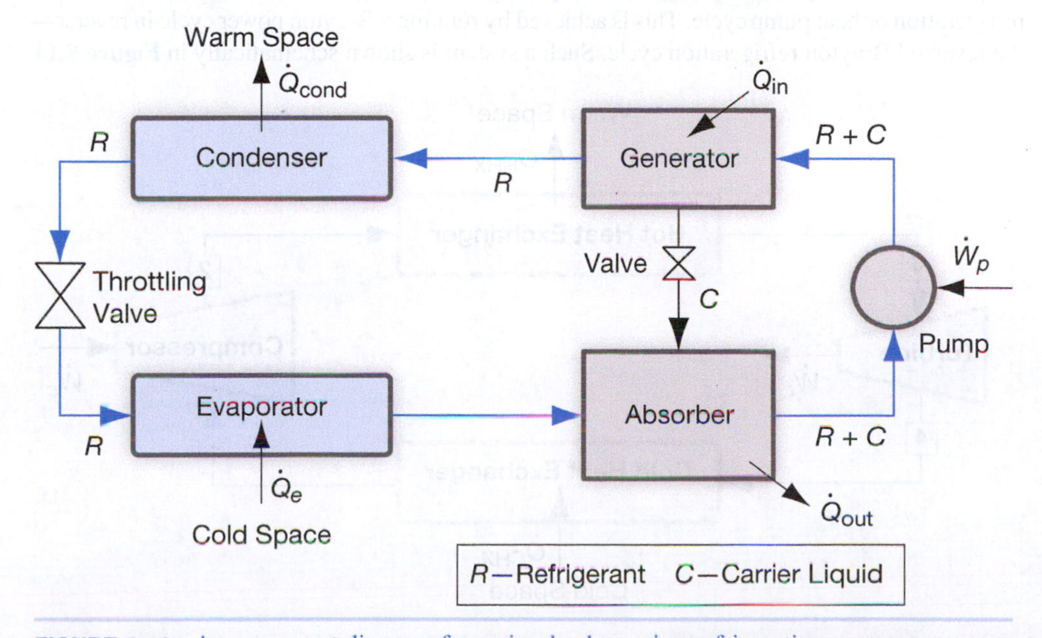

FIGURE 8.10 A component diagram for a simple absorption refrigeration system.

to the solution entering the generator, thereby reducing the amount of heat needed to be added in the generator; and a rectifier, which can be used to remove any remaining traces of carrier liquid in the refrigerant vapor before it enters the condenser (the presence of a carrier liquid may result in ice forming in the throttling valve or evaporator).

One combination of fluids used in this type of system is ammonia as the refrigerant and water as the carrier liquid. Although absorption refrigeration systems are used in some applications, particularly large systems, the performance of the systems tends to be lower than that of vapor-compression refrigeration systems. This is a result of the greater energy demands of the system. Although the power input is reduced by using a pump rather than a compressor, the mass flow rate through the pump will be much higher than the mass flow rate through the compressor, offsetting some of this benefit. In addition, the heat that must be added to the fluid in the generator increases the required energy input for the system, thereby lowering the coefficient of performance. However, if waste heat is available that can be used for this purpose, then the overall cost of implementing the system may not be as large—essentially the heat is "free" energy at that point. Although the cycle's coefficient of performance may not be as high as the coefficient of performance of a vapor-compression refrigeration system, the cost of the energy to operate the system may be low enough to warrant its use.

QUESTION FOR THOUGHT/DISCUSSION

For what characteristics of an air conditioning system may an absorption refrigeration system best be suited?

8.4 REVERSED BRAYTON REFRIGERATION CYCLE

Gases that do not undergo a phase change can be used for refrigeration purposes. We have already seen how the temperature of a gas increases as its pressure rises through a compressor; this enables heat to be removed from a cold space with a cold gas, but rejected to a warm space from the hotter gas exiting the compressor. In addition, we have seen that turbines will lower the temperature of a gas as the pressure decreases. Appropriate choices of pressures and temperatures of the gases entering and exiting a turbine or compressor can lead to an acceptable refrigeration or heat pump cycle. This is achieved by running a Brayton power cycle in reverse—the reversed Brayton refrigeration cycle. Such a system is shown schematically in **Figure 8.11**.

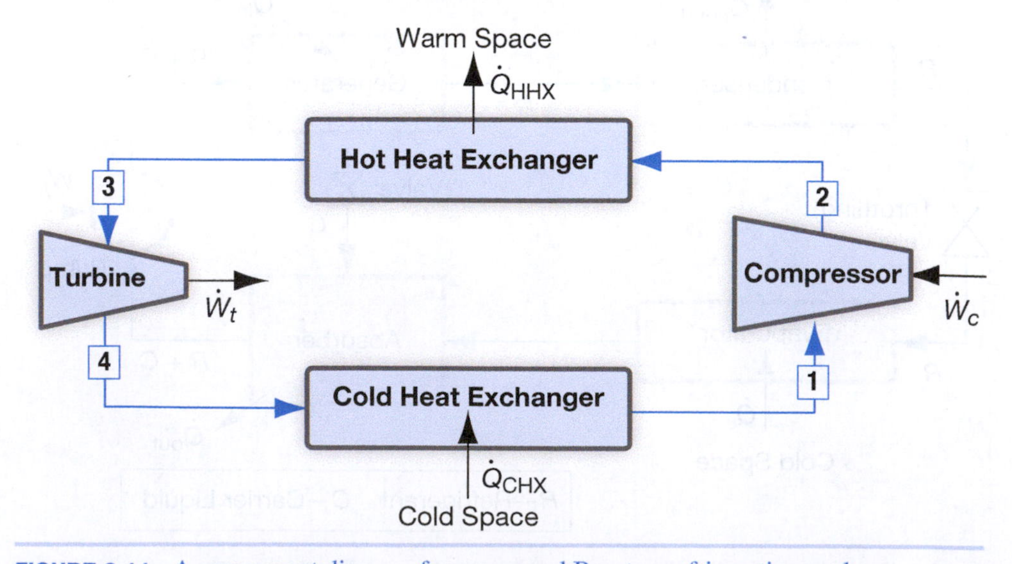

FIGURE 8.11 A component diagram for a reversed Brayton refrigeration cycle.

The ideal reversed Brayton refrigeration cycle consists of four processes, described below and shown in the *T*-*s* diagram in **Figure 8.12**.

Process 1-2: Isentropic compression
Process 2-3: Constant-pressure heat rejection
Process 3-4: Isentropic expansion
Process 4-1: Constant-pressure heat addition

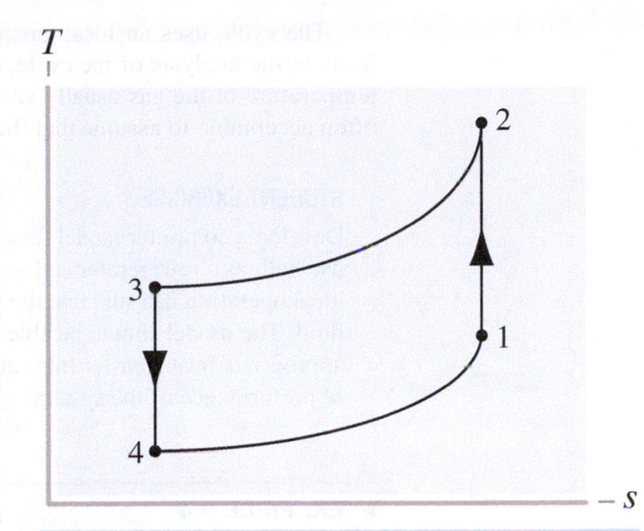

FIGURE 8.12 A *T*-*s* diagram for an ideal reversed Brayton cycle.

To analyze each process, we assume the following: (1) There is no work used in the high-temperature and low-temperature heat exchangers ($\dot{W}_{HHX} = \dot{W}_{CHX} = 0$). (2) There is no heat transfer in the compressor or turbine ($\dot{Q}_c = \dot{Q}_t = 0$). (3) There is no change in kinetic or potential energy through each device.

Applying the first law for open systems at steady state and steady flow with a single inlet and single outlet (Eq. (4.15)) to each device yields the following:

Compressor: $$\dot{W}_c = \dot{m}(h_1 - h_2) \tag{8.8}$$

High-temperature heat exchanger: $$\dot{Q}_{HHX} = \dot{m}(h_3 - h_2) \tag{8.9}$$

Turbine: $$\dot{W}_t = \dot{m}(h_3 - h_4) \tag{8.10}$$

Low-temperature heat exchanger: $$\dot{Q}_{CHX} = \dot{m}(h_1 - h_4) \tag{8.11}$$

The net power consumed for the cycle is $\dot{W}_{net} = \dot{W}_c + \dot{W}_t$. The coefficients of performance for a refrigerator and a heat pump are

$$\beta = \frac{\dot{Q}_{CHX}}{|\dot{W}_{net}|} = \frac{h_1 - h_4}{|h_1 - h_2 + h_3 - h_4|} \tag{8.12}$$

and

$$\gamma = \frac{|\dot{Q}_{HHX}|}{|\dot{W}_{net}|} = \frac{|h_3 - h_2|}{|h_1 - h_2 + h_3 - h_4|} \tag{8.13}$$

respectively. To assist in relating states, remember that $P_1 = P_4$ and $P_2 = P_3$ by the definitions of the processes in the cycle.

In practice, we know that the compressor and turbine will not be isentropic, and the analysis of the nonideal compressor and turbine incorporates the appropriate isentropic efficiencies:

$$\eta_c = \frac{h_1 - h_{2s}}{h_1 - h_2} \quad \text{and} \quad \eta_t = \frac{h_3 - h_4}{h_3 - h_{4s}}$$

Figure 8.13 presents a typical *T*-*s* diagram for this nonideal reversed Brayton refrigeration cycle.

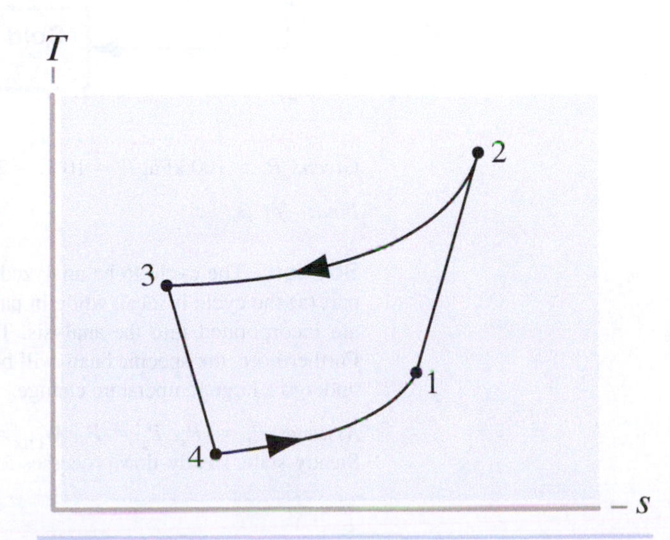

FIGURE 8.13 A *T*-*s* diagram for a nonideal reversed Brayton cycle.

The cycle uses an ideal gas, so we can either consider the variability of the specific heats in the analysis of the cycle, or assume the specific heats to be constant. Because the temperature of the gas usually varies only by a few hundred degrees during the cycle, it is often acceptable to assume that the specific heats are constant.

STUDENT EXERCISE

Develop a computer model for the reversed Brayton refrigeration cycle, allowing for its use both as a refrigerator and as a heat pump, and also allowing for both ideal and non-ideal operation. Be sure that the model can accept different types of gases as its working fluid. The model should be able to determine unknown thermodynamic property states, appropriate heat transfer rates and compressor and turbine powers, and the coefficients of performance of the system.

▶ EXAMPLE 8.4

Nitrogen gas is used as the working fluid in a reversed Brayton refrigeration cycle. The cycle operates as a closed loop, with the nitrogen entering the compressor at 100 kPa and 10°C, and entering the turbine at 600 kPa and 50°C. The cycle is to be used as a refrigerator and is to remove 20 kW of heat from a cold region. Assume the N_2 behaves as an ideal gas with constant specific heats. Determine the mass flow rate of the N_2 through the cycle and the coefficient of performance of the cycle if (a) the compressor and turbine are isentropic, and (b) the isentropic efficiencies of both the turbine and compressor are 0.78.

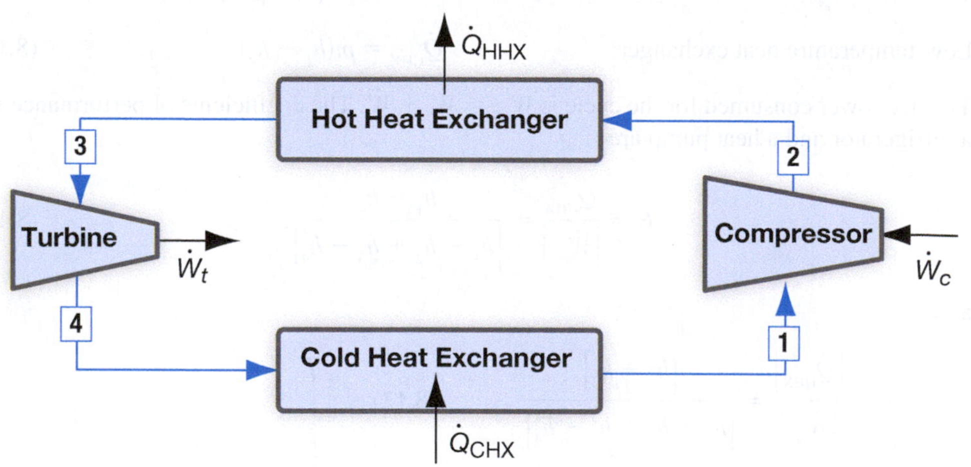

Given: $P_1 = 100$ kPa, $T_1 = 10°C = 283$ K, $P_3 = 600$ kPa, $T_3 = 50°C = 323$ K, $\dot{Q}_{CHX} = 20$ kW

Find: $\dot{m}, \beta$

Solution: The cycle to be analyzed is a reversed Brayton refrigeration cycle in a refrigerator. In part (a) the cycle is ideal, while in part (b) a nonisentropic compressor and a non-isentropic turbine are incorporated into the analysis. The working fluid is N_2, which we can treat as an ideal gas. Furthermore, the specific heats will be treated as constant because we do not expect that the N_2 will undergo a large temperature change.

Assume: $P_1 = P_4$, $P_2 = P_3$, $\dot{W}_{CHX} = \dot{W}_{HHX} = \dot{Q}_t = \dot{Q}_c = 0$, $\Delta KE = \Delta PE = 0$ for each device. Steady-state, steady-flow processes for each device.

For the N_2, use $c_p = 1.039$ kJ/kg · K, $k = 1.405$ (values at 300 K).

First, find the temperature at states 2 and 4.

(a) For an isentropic compressor: $T_2 = T_1(P_2/P_1)^{(k-1)/k} = 474$ K

For an isentropic turbine: $T_4 = T_3(P_4/P_3)^{(k-1)/k} = 192.7$ K

The mass flow rate of the refrigerant can be found using the heat absorbed through the low-temperature heat exchanger, as described in Eq. (8.11):

$$\dot{m} = \frac{\dot{Q}_{CHX}}{(h_1 - h_4)} = \frac{\dot{Q}_{CHX}}{c_p(T_1 - T_4)} = \mathbf{0.213\ kg/s}$$

The coefficient of performance is found from Eq. (8.12):

$$\beta = \frac{\dot{Q}_{CHX}}{|\dot{W}_{net}|} = \frac{h_1 - h_4}{|h_1 - h_2 + h_3 - h_4|} = \frac{c_p(T_1 - T_4)}{|c_p(T_1 - T_2 + T_3 - T_4)|} = \mathbf{1.49}$$

(b) Incorporating $\eta_c = \eta_t = 0.78$ involves taking the outlet states found in part (a) as isentropic outlet states:

$$T_{2s} = 474 \text{ K}; T_{4s} = 192.7 \text{ K}$$

Using the isentropic efficiencies yields the actual outlet temperatures:

$$T_2 = T_1 + (T_{2s} - T_1)/\eta_c = 527.9 \text{ K}$$

$$T_4 = T_3 - (T_3 - T_{4s})\eta_t = 221 \text{ K}$$

The mass flow rate of the N_2 and the coefficient of performance are then

$$\dot{m} = \frac{\dot{Q}_{CHX}}{(h_1 - h_4)} = \frac{\dot{Q}_{CHX}}{c_p(T_1 - T_4)} = \mathbf{0.310\ kg/s}$$

$$\beta = \frac{\dot{Q}_{CHX}}{|\dot{W}_{net}|} = \frac{h_1 - h_4}{|h_1 - h_2 + h_3 - h_4|} = \frac{c_p(T_1 - T_4)}{|c_p(T_1 - T_2 + T_3 - T_4)|} = \mathbf{0.434}$$

Analysis: The coefficient of performance of a reversed Brayton cycle is typically not as high as that of a vapor-compression refrigeration cycle. Also note that the presence of the turbine causes inefficiencies in the system to impact the mass flow rate needed to achieve a certain amount of cooling, because the turbine outlet conditions impact the rate of refrigeration.

An open-loop reversed Brayton "cycle" can also be used, wherein fresh air is continually ingested into the compressor, sent to the high-temperature heat exchanger, and then sent through the turbine to recover some power. This is shown in Figure 8.14. Because the working fluid is being continually replaced, this is not an exact thermodynamic cycle, but it can be adequately modeled as one. This heat pump can be an effective means of heating a room space, because the high-temperature air produced by the compressor can deliver a reasonable amount of heat to a space. We could also design an open system to have fresh air brought into the turbine, but such a system requires the compressor to drive a turbine to produce a subatmospheric pressure gas—a process that is not particularly efficient. Although the cold air produced in such a way could remove considerable amounts of heat from a cold space, there are generally more effective refrigeration means available for use.

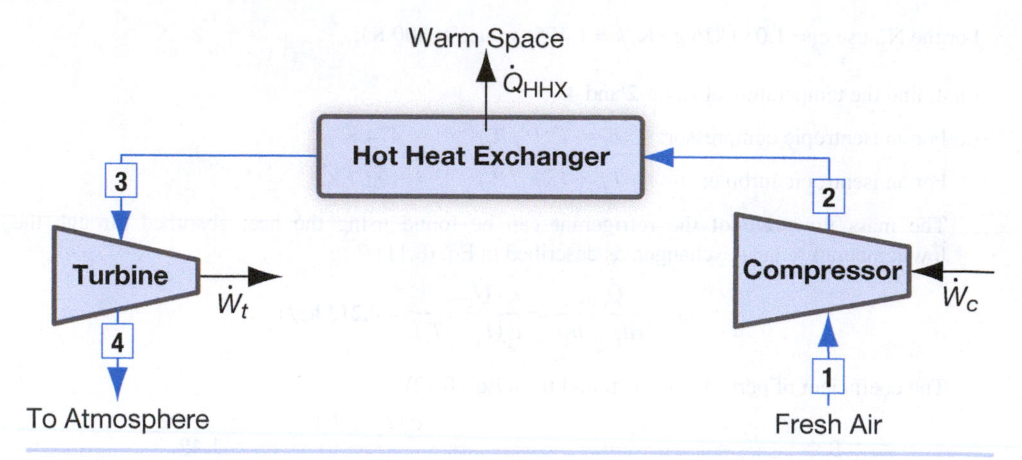

FIGURE 8.14 A component diagram of a reversed Brayton heat pump system.

QUESTION FOR THOUGHT/DISCUSSION

Under what conditions or for which applications does using a reversed Brayton cycle make sense instead of a vapor-compression refrigeration cycle?

Summary

In this chapter, we have explored the fundamentals of common refrigeration cycles. These cycles are used both for cooling a space and for heating a space, with the heating process cycle being termed a heat pump cycle. Other technologies can be used for achieving cooling, such as thermoelectric cooling, evaporative cooling, and radiation cooling. Although all of these technologies are used in some applications, the cycles discussed in this chapter are the most commonly used mechanical refrigeration processes today.

PROBLEMS

8.1 A refrigerator is to be used to keep food at a temperature of 2°C while placed in a room where the ambient temperature is 21°C. (a) Design a basic scheme for the working fluid of a refrigerator that will allow for this process to occur. (b) What is the Carnot coefficient of performance for this situation?

8.2 A freezer, located in a room that has a temperature of 24°C, is designed to keep food at –12°C. (a) Design a basic scheme for the working fluid of a refrigerator that will allow this process to occur. (b) What is the Carnot coefficient of performance for this situation?

8.3 An air conditioning unit in a building is to cool air inside the building to 20°C while the outdoor air temperature is 32°C. (a) Design a basic scheme for the working fluid of an air conditioner that will allow this process to occur. (b) What is the maximum possible coefficient of performance for the air conditioner?

8.4 A refrigerator is designed to liquefy N_2 gas by cooling the gas to –196°C. This refrigerator is operating with the ambient temperature at 20°C. What is the maximum possible coefficient of performance for this refrigerator?

8.5 A heat pump is designed to transfer heat into a building at a temperature of 25°C and receive heat from underground at a temperature of 8°C. What is the maximum possible coefficient of performance of this heat pump?

8.6 Determine the chemical formula and draw a possible chemical structure diagram of the following refrigerant compounds: (a) R-11, (b) R-22, (c) R-115, (d) R-216.

8.7 Determine the chemical formula and draw a possible chemical structure diagram of the following refrigerant compounds: (a) R-12, (b) R-50, (c) R-143, (d) R-1120.

8.8 Determine the chemical formula and draw a possible chemical structure diagram of the following refrigerant compounds: (a) R-13, (b) R-124, (c) R-23, (d) R-114.

8.9 A vapor-compression refrigeration cycle uses R-22 as a refrigerant. The cycle removes 15 kW of heat from the cold space. The R-22 enters the compressor as a saturated vapor at 250 kPa and is compressed to 1000 kPa. The R-22 exits the condenser as a saturated liquid. Determine the mass flow rate of the R-22, the compressor power, and the coefficient of performance of the cycle as a refrigerator if (a) the compressor is isentropic, and (b) the compressor has an isentropic efficiency of 0.81.

8.10 A vapor-compression refrigeration cycle uses R-22 as a refrigerant. The mass flow rate of the R-22 is 0.20 kg/s. The R-22 enters the compressor as a saturated vapor at 300 kPa and exits the condenser as a saturated liquid at 1200 kPa. Determine the rate at which heat is removed from the cold space through the evaporator and the coefficient of performance of the cycle as a refrigerator if (a) the compressor is isentropic, and (b) the isentropic efficiency of the compressor is 0.75.

8.11 A vapor-compression refrigeration cycle uses R-134a as a refrigerant. The mass flow rate of the R-134a is 0.15 kg/s. The R-134a exits the evaporator as a saturated vapor at 140 kPa absolute and enters the throttling valve as a saturated liquid at 700 kPa absolute. Determine the rate at which heat is removed from the cold space through the evaporator and the coefficient of performance of the cycle as a refrigerator if (a) the compressor is isentropic, and (b) the isentropic efficiency of the compressor is 0.78. (c) Determine the maximum possible coefficient of performance for the cycle.

8.12 A vapor-compression refrigeration cycle uses R-134a as a refrigerant. The cycle is designed to act as a heat pump, and it delivers 25 kW of heat to a warm space. The R-134a enters the compressor as a saturated vapor at 350 kPa and exits the condenser as a saturated liquid at 1 MPa. Determine the mass flow rate of the R-134a and the coefficient of performance of the cycle as a heat pump if (a) the compressor is isentropic, and (b) the isentropic efficiency of the compressor is 0.74. (c) Determine the maximum possible coefficient of performance for the cycle.

8.13 Consider a vapor-compression refrigeration cycle using R-134a as the refrigerant. It is proposed to use a turbine in place of the throttling valve. The R-134a enters the compressor as a saturated vapor at 200 kPa and exits the condenser as a saturated liquid at 900 kPa. The cycle is to remove 15 kW of heat from the cold space through the evaporator. The isentropic efficiency of the compressor is 0.75, and the proposed turbine has an isentropic efficiency of 0.60. Determine the mass flow rate of R-134a needed and the coefficient of performance of the cycle as a refrigerator for (a) the cycle employing a throttling valve, and (b) the cycle employing the proposed turbine in place of the throttling valve. (c) Determine the maximum possible coefficient of performance for the cycle.

8.14 A vapor-compression refrigeration cycle uses R-134a in a heat pump format. The mass flow rate of the R-134a is 300 g/s, and the R-134a enters the compressor as a saturated vapor at –15°C and exits the condenser as a saturated liquid at 30°C. The isentropic efficiency of the compressor is 0.80. Determine the rate at which heat is delivered to the warm space through the condenser and the coefficient of performance of the cycle as a heat pump for (a) the cycle employing a throttling valve, and (b) the cycle employing a turbine with an isentropic efficiency of 0.70 in place of the throttling valve.

8.15 A vapor-compression refrigeration cycle is to use R-134a as a refrigerant. The R-134a is to enter the compressor as a saturated vapor at 200 kPa with a mass flow rate of 0.75 kg/s. The R-134a enters the throttling valve as a saturated liquid. The isentropic efficiency of the compressor is 0.75. Plot the rate of refrigeration, the coefficient of performance of the cycle as a refrigerator, and the maximum possible coefficient of performance of the cycle as a refrigerator for condenser pressures ranging from 400 kPa to 1200 kPa.

8.16 Ammonia is used as the refrigerant in a vapor-compression refrigerant cycle operating as a freezer. The ammonia enters the compressor as a saturated vapor at –25°C and exits the condenser as a saturated liquid. The unit must remove heat at a rate of 15 kW from the freezer compartment. The isentropic efficiency of the compressor is 0.78. Plot the mass flow rate of the ammonia, the compressor exit temperature, and the coefficient of performance of the freezer for compressor exit pressures ranging from 800 kPa to 1600 kPa.

8.17 A vapor-compression refrigeration cycle uses R-22 as a refrigerant. The R-22 enters the compressor as a saturated vapor at 350 kPa and exits the condenser as a saturated liquid. The cycle is designed to be a heat pump, and it delivers 25 kW of heat to a warm space. The isentropic efficiency of the compressor is 0.78. Plot the mass flow rate of the R-22, the coefficient of performance of the cycle as a heat pump, and the maximum possible coefficient of performance of the cycle as a heat pump for condenser pressures ranging between 700 kPa and 2000 kPa.

8.18 A vapor-compression refrigeration cycle uses R-134a as a refrigerant. The R-134a enters the compressor as a saturated vapor and exits the condenser as a saturated liquid at 1.1 MPa absolute. The mass flow rate of the R-134a is 0.10 kg/s. The isentropic efficiency of the compressor is 0.80. Plot the heat removed from the cold space through the condenser, the coefficient of performance of the cycle as a refrigerator, and the maximum possible coefficient of performance of the cycle as a refrigerator for evaporator pressures ranging between 100 kPa absolute and 600 kPa absolute.

8.19 A vapor-compression refrigeration cycle uses R-134a as the refrigerant. The R-134a enters the compressor as a saturated vapor and exits the condenser as a saturated liquid at 1000 kPa. The mass flow rate of the R-134a is 1.25 kg/s. The isentropic efficiency of the compressor is 0.80. Plot the heat delivered to the warm space, the coefficient of performance of the cycle as a heat pump, and the maximum possible coefficient of performance of the cycle as a heat pump for evaporator pressures ranging between 100 kPa and 500 kPa.

8.20 A vapor-compression refrigeration cycle has its refrigerant entering the compressor as a saturated vapor at –10°C and has saturated liquid entering the throttling valve at 40°C. The cycle is to remove 50 kW of heat from the cold space through the evaporator. Determine the mass flow rate of the refrigerant and the coefficient of performance of the cycle as a refrigerator if the refrigerant is (a) R-22, (b) R-134a, and (c) ammonia.

8.21 25 kW of heat is to be removed from a cold space through heat transfer via the evaporator in a vapor-compression refrigeration cycle. The cycle uses ammonia as its refrigerant. The ammonia enters the compressor as a saturated liquid at 200 kPa and exits the condenser as a saturated liquid at 1800 kPa. Plot the mass flow rate of the refrigerant and the coefficient of performance of the cycle as a refrigerator for isentropic compressor efficiencies varying between 0.50 and 1.0.

8.22 A heat pump, operating as a vapor-compression refrigeration cycle, is to deliver 75 kW of heat to a building. The cycle uses R-134a, and the R-134a enters the compressor as a saturated vapor at 140 kPa absolute and exits the condenser as a saturated liquid at 1000 kPa absolute. Plot the mass flow rate of the R-134a and the coefficient of performance of the cycle as a heat pump for isentropic compressor efficiencies ranging between 0.50 and 1.0.

8.23 A dirty vapor-compression refrigeration cycle is experiencing problems with the condenser, such that the refrigerant is not completely condensing inside the condenser. The mass flow rate of R-134a through the cycle is 0.50 kg/s. The evaporator pressure is 200 kPa and the condenser pressure is 1100 kPa. The isentropic efficiency of the compressor is 0.76. Plot the rate of refrigeration and the coefficient of performance of the cycle as a refrigerator for condenser exit qualities ranging between 0.0 and 0.5.

8.24 A cascade refrigeration system uses two vapor-compression refrigeration cycles to achieve cooling to a lower temperature. The low-temperature cycle uses ammonia to remove heat from a cold space. The mass flow rate of the ammonia is 0.50 kg/s, and the ammonia enters the compressor as a saturated vapor at 50 kPa and exits the condenser as a saturated liquid at 500 kPa. Heat is rejected to the high-temperature cycle through the low-temperature cycle's condenser (and into the high-temperature cycle's evaporator). The high-temperature cycle uses R-134a, with saturated vapor entering its compressor at 240 kPa and exiting the condenser as a saturated liquid at 1000 kPa. The isentropic efficiency of each compressor is 0.78. Determine the mass flow rate of the R-134a and the overall cycle's coefficient of performance as a refrigerator.

8.25 A cascade refrigeration system uses two vapor-compression refrigeration cycles to cool a space. The low-temperature cycle uses propane as the refrigerant and removes 25 kW of heat from the cold space. The propane enters the compressor at 20 kPa absolute as a saturated vapor and exits the condenser as a saturated liquid at 200 kPa absolute. Heat is transferred to the evaporator of the high-temperature cycle through the low-temperature cycle's condenser. R-22 flows through the high-temperature cycle, exiting the evaporator as a saturated vapor at 175 kPa absolute and exiting the condenser as a saturated liquid at 350 kPa absolute. The isentropic efficiency of each compressor is 0.75. Determine the mass flow rates of the R-22 and of the propane and the overall cycle's coefficient of performance as a refrigerator.

8.26 A cascade refrigeration system using two vapor-compression refrigeration cycles is to make dry ice by cooling CO_2. Propane will be used in the low-temperature loop, with saturated propane vapor at –80°C exiting the evaporator. The propane is compressed to 200 kPa with a compressor whose isentropic efficiency is 0.77. The propane is then condensed to a saturated liquid in its condenser, which serves as the high-temperature loop and uses R-134a. The R-134a exits its evaporator at –30°C as a saturated vapor and is compressed to 700 kPa by a compressor whose isentropic efficiency is 0.80. Suppose 10 kW of heat is to be removed from the CO_2. Determine the mass flow rate of each refrigerant and the overall cycle's coefficient of performance as a refrigerator.

8.27 A cascade refrigeration system uses two vapor-compression refrigeration cycles. The low-temperature cycle uses R-22 to remove 20 kW of heat from a cold space, with the R-22 entering the compressor as a saturated vapor at 100 kPa and exiting the condenser as a saturated liquid. R-134a flows through the high-temperature cycle, leaving the evaporator as a saturated vapor and exiting the condenser as a saturated liquid at 1000 kPa. The temperature of the condensing R-22 in its condenser must be 5°C higher than the boiling R-134a temperature in its evaporator in order to achieve sufficient heat transfer rates. The isentropic efficiency of each compressor is 0.80. Plot the overall coefficient of performance of the system, the mass flow rate of the R-22, and the mass flow rate of the R-134a for R-22 condenser pressures ranging between 200 kPa and 1400 kPa.

8.28 A reversed Brayton refrigeration cycle uses air as its working fluid in a closed loop. The air enters the compressor at 100 kPa and –10°C and enters the turbine at 1000 kPa and 30°C. The mass flow rate of the air through the cycle is 1.5 kg/s. Determine the rate at which heat is removed from the cold space and the coefficient of performance of the cycle if (a) the turbine and compressor are isentropic, and (b) the turbine and compressor both have isentropic efficiencies of 0.80. Assume that the specific heats of the air are constant.

8.29 Repeat Problem 8.28, but take into account the variability of the specific heats of the air.

8.30 Repeat Problem 8.28, but use the following working fluids: (a) nitrogen gas (N_2), (b) carbon dioxide (CO_2), (c) argon (Ar), and (d) methane (CH_4).

8.31 A reversed Brayton cycle refrigerator operates in an open loop and uses air as the working fluid. The air enters the turbine at 20°C and 101 kPa and expands to 20 kPa. The air exits the cold heat exchanger at 0°C and is compressed to 101 kPa. The mass flow rate of the air is 0.25 kg/s. Determine the heat removed from the cold space and the coefficient of performance of the cycle as a refrigerator if (a) the turbine and compressor are isentropic, and (b) the isentropic efficiency of both the turbine and compressor is 0.80. Consider the air to behave as an ideal gas with constant specific heats.

8.32 An open-loop reversed Brayton cycle using air acts as a heat pump. The air enters the compressor at 100 kPa and 18°C and enters the turbine at 1 MPa and 30°C. The cycle is to deliver 50 kW of heat to a warm space. Determine the coefficient of performance of the cycle as a heat pump and the mass flow rate of the air if (a) the turbine and compressor are isentropic, and (b) the isentropic efficiency of the compressor is 0.75 and the isentropic efficiency of the turbine is 0.80. Treat the air as an ideal gas with constant specific heats.

8.33 An open-loop reversed Brayton cycle using air serves as a heat pump. The volumetric flow rate of air entering the compressor is 1 m^3/s. The air enters the compressor at 10°C and 100 kPa and enters the turbine at 1200 kPa and 40°C. The isentropic efficiency of the compressor is 0.78, and the isentropic efficiency of the turbine is 0.80. Determine the rate of heat delivered to the warm space and the coefficient of performance of the cycle as a heat pump if (a) the specific heats are considered to be constant, and (b) the specific heats of the air are considered to be variable.

8.34 An open-loop reversed Brayton cycle uses air in a heat pump mode. The air enters the compressor at 5°C and 101 kPa at 0.25 m^3/s and then enters the turbine at 1000 kPa and 30°C. Determine the rate of heat delivery to the warm space and the coefficient of performance of the cycle as a heat pump, assuming constant specific heats for the air, if (a) the compressor isentropic efficiency is 0.70 and the turbine isentropic efficiency if 0.80, and (b) the compressor isentropic efficiency is 0.80 and the turbine isentropic efficiency is 0.70.

8.35 A reversed Brayton cycle, operating in a closed loop, employs nitrogen gas as the working fluid. The N_2 enters the compressor at 0°C and 100 kPa, and enters the turbine at 40°C and 1200 kPa. The isentropic efficiency of the turbine is 0.80. The mass flow rate of the nitrogen is 1.25 kg/s. Plot the rate at which heat is removed from the cold space, the rate at which heat is rejected to the warm space, and the coefficient of performance of the cycle as a refrigerator for compressor isentropic efficiencies ranging between 0.50 and 1.0. Assume that the nitrogen behaves as an ideal gas with constant specific heats.

8.36 A reversed Brayton cycle, operating in a closed loop, employs argon gas as the working fluid. The Ar enters the compressor at –4°C and 100 kPa absolute and enters the turbine at 30°C and 1.25 MPa. The isentropic efficiency of the compressor is 0.77. The mass flow rate of the argon is 350 g/s. Plot the rate at which heat is removed from the cold space, the rate at which heat is rejected to the warm space, and the coefficient of performance of the cycle as a refrigerator for turbine isentropic efficiencies ranging between 0.50 and 1.0.

8.37 A closed-loop reversed Brayton cycle uses carbon dioxide gas as the working fluid. The CO_2 enters the compressor at 5°C and 100 kPa, where it is compressed to 1000 kPa. The CO_2 later enters the turbine at 35°C, and the turbine isentropic efficiency is 0.79. The CO_2 mass flow rate is 0.35 kg/s. Plot the rate at which heat is removed from the cold space, the rate of heat rejection to the warm space, and the coefficient of performance of the unit as a refrigerator for compressor isentropic efficiencies ranging between 0.50 and 1.0.

8.38 A reversed Brayton cycle, operating in a closed loop, uses nitrogen gas as the working fluid. The N_2 enters the compressor at 0°C and 120 kPa. The N_2 enters the turbine at 40°C and the compressor outlet pressure. The mass flow rate of the N_2 is 1.5 kg/s. The isentropic efficiencies of the compressor and turbine are both 0.80. Plot the rate at which heat is removed from the cold space through the cold heat exchanger and the coefficient of performance of the cycle as a refrigerator for compressor outlet pressures ranging between 400 kPa and 1500 kPa. Assume the nitrogen behaves as an ideal gas with constant specific heats.

8.39 A reversed Brayton cycle, operating in a closed loop, uses carbon dioxide gas as the working fluid. The CO_2 enters the compressor at 0°C. The CO_2 enters the turbine at 40°C and 1200 kPa. The mass flow rate of the CO_2 is 2.0 kg/s. The isentropic efficiencies of the compressor and turbine are both 0.80. Plot the rate at which heat is rejected to the warm space through the hot heat exchanger and the coefficient of performance of the cycle as a heat pump for compressor inlet pressures ranging between 50 kPa and 500 kPa. Assume the CO_2 behaves as an ideal gas with constant specific heats.

8.40 A reversed Brayton cycle heat pump, operating in a closed loop, is to deliver 100 kW of heat to a warm space, using air as the working fluid. The pressure entering the compressor is 150 kPa and the pressure entering the turbine is 1400 kPa. The air enters the turbine at 30°C after delivering the heat to the warm space. The compressor isentropic efficiency is 0.75, and the turbine isentropic efficiency is 0.82. The temperature of the cold thermal reservoir varies, such that the temperature of the air entering the compressor can vary between –25°C and 5°C. Plot the mass flow rate of the air and the coefficient of performance of the cycle as a heat pump for compressor inlet temperatures in this range. Assume the air behaves as an ideal gas with constant specific heats.

8.41 Repeat Problem 8.40, but incorporate the variability of the specific heats of air.

8.42 A reversed Brayton cycle, operating in a closed loop and using argon as the working fluid, is to remove 40 kW of heat from a cold source. The Ar enters the compressor at 0°C and 100 kPa and exits the compressor at 1200 kPa. The compressor isentropic efficiency is 0.77, and the turbine isentropic efficiency is 0.84. The temperature of the hot thermal reservoir to which heat is rejected varies, which causes the Ar to enter the turbine at temperatures ranging between 15°C and 50°C. Plot the mass flow rate of the Ar and the coefficient of performance of the cycle as a refrigerator for turbine inlet temperatures in this range.

8.43 Repeat Problem 8.42, but use nitrogen gas rather than argon as the working fluid.

8.44 An open-loop reversed Brayton cycle is used as a heat pump in the winter. Outside air enters the compressor at 100 kPa and exits at 700 kPa. The air enters the turbine at 700 kPa and 30°C. The volumetric flow rate of air entering the compressor is 1.1 m^3/s. The isentropic efficiencies of both the turbine and the compressor are 0.75. Plot the rate of heat rejection to the warm space and the coefficient of performance of the cycle as a heat pump for temperatures of air entering the compressor varying between –30°C and 10°C. Assume the air behaves as an ideal gas with constant specific heats.

8.45 Repeat Problem 8.44, but incorporate the variability of the specific heats of air.

8.46 You wish to investigate using an open-loop Brayton cycle, operating on air, to provide refrigeration at a very cold temperature. Your system will induct air into the turbine inlet, with the compressor exhausting air to the atmosphere. The air is to enter the turbine at 100 kPa, and will enter the compressor at 20 kPa and –50°C. The air will exit the compressor at 100 kPa. The volumetric flow rate of air entering the turbine is 1.5 m^3/s. The isentropic efficiency of the compressor is 0.70 and the isentropic efficiency of the turbine is 0.75. Considering the specific heats of the air to be constant, plot the rate of heat removal from the cold space through the cold heat exchanger and the coefficient of performance of the cycle as a refrigerator for temperatures of air entering the turbine ranging between 0°C and 30°C.

DESIGN/OPEN-ENDED PROBLEMS

8.47 Design a vapor-compression refrigeration cycle to be used as a refrigerator. You may choose any refrigerant available. The isentropic efficiency of the compressor available to you is 0.75. The cold space is to be maintained at 3°C, and the warm space available for heat rejection is at 24°C. The refrigerator must be able to remove 30 kW of heat from the cold space. Assume that saturated vapor enters the compressor and saturated liquid exits the condenser. Your design should include your choice of refrigerant, the required mass flow rate of the refrigerant, and the condenser and evaporator pressures. Evaluate your design through the coefficient of performance.

8.48 You wish to design a system to rapidly deep-freeze tissue samples using a vapor-compression refrigeration system. The freezer temperature is to be maintained at –40°C, while the room temperature in which the freezer is located is at 20°C. The system must remove 10 kW of heat from the cold space when set for maximum cooling. You may use any refrigerant, although the maximum pressure the system can hold is 1500 kPa. The isentropic efficiency of the compressor should be considered 0.80. Assume that you have a saturated liquid exiting the condenser and a saturated vapor exiting the compressor. Your design should specify the refrigerant, the mass flow rate of the refrigerant at maximum cooling, and the condenser and evaporator pressures. Provide the coefficient of performance of your system.

8.49 Design a vapor-compression refrigeration cycle to be used as a heat pump. You may choose any refrigerant available. The isentropic efficiency of the compressor is 0.80. The cold space is maintained at 10°C, and the warm space to which heat is delivered is at 22°C. The heat pump must supply 100 kW of heat to the warm space. Saturated vapor enters the compressor and saturated liquid exits the condenser. Your design should specify the refrigerant chosen, the required mass flow rate of the refrigerant, and the condenser and evaporator pressures. Evaluate your design through the coefficient of performance.

8.50 Design a two-stage or three-stage vapor-compression cascade refrigeration system to be used to create very low temperatures for liquefying gases in a laboratory setting. The cold space in the refrigerator must be maintained at no higher than –170°C and the environment temperature is 25°C. The system must remove 3 kW of heat from the cold space. Your design should specify the refrigerants used, the temperatures and pressures entering and exiting each component in the system, the power requirement, and the coefficient of performance of the refrigeration system.

8.51 Design a reversed Brayton cycle closed-loop heat pump system. The isentropic efficiency of the turbine is 0.80 and the isentropic efficiency of the compressor is 0.78. The gas is to leave the high-temperature heat exchanger at a temperature of 30°C after delivering 150 kW of heat to the warm space. The gas can exit the cold heat exchanger at a temperature no higher than 0°C. The mass flow rate of the gas can be no larger than 2 kg/s. Choose the gas to be used, and the pressures and temperatures entering and exiting the turbine and compressor. Determine the necessary mass flow rate and the coefficient of performance of the cycle as a heat pump. Justify your choices of design parameters.

CHAPTER 9

Ideal Gas Mixtures

Learning Objectives

Upon completion of Chapter 9, you will be able to

9.1 Apply various methods of describing the composition of gas mixtures;

9.2 Calculate thermodynamic properties for ideal gas mixtures; and

9.3 Explain some of the differences between real gas mixtures and ideal gas mixtures.

9.1 INTRODUCTION

To this point, we have been concerned with working fluids that consist of a single, unchanging component, such as water, the various refrigerants, or single-component gases such as N_2 and CO_2. The exception to this has been our consideration of "air," or, more appropriately, "dry air," as a working fluid. Air is not a single-component gas, but rather is a mixture of gases. But because air is such a common mixture of gases, it is natural that the properties of air have been determined and that air is often treated as a single-component gas. However, there are many practical systems that use a mixture of gases (other than air) as the working fluid, as shown in **Figure 9.1**. For example, a gas turbine engine experiences the gaseous products of a combustion process flowing through the turbine section. A home furnace uses combustion products as a heat source from which heat is transferred to air. Some material processing systems may require an environment consisting of a mixture of gases devoid of oxygen. Some medical treatments will require patients to receive a gas mixture with an oxygen concentration higher than that of ordinary air.

Although air may be the most commonly experienced mixture of gases, it is clear that many other applications require mixtures of gases other than air, and that these gas mixtures may have a wide variety of compositions. For instance, the products of combustion processes will vary with the type of fuel used and the amount of air initially present in the reactants. There are an infinite number of ways in which gases can be combined into a mixture, so it is not practical to expect to have pre-determined thermodynamic properties available for all possible gas mixtures. We need calculation methods we can use in determining necessary properties of gas mixtures. In this chapter we develop such calculation methods. By the end of this chapter, you should be able to determine the thermodynamic properties of any mixture of ideal gases. Your study of this chapter will also help you understand and solve basic thermodynamic problems involving ideal gas mixtures.

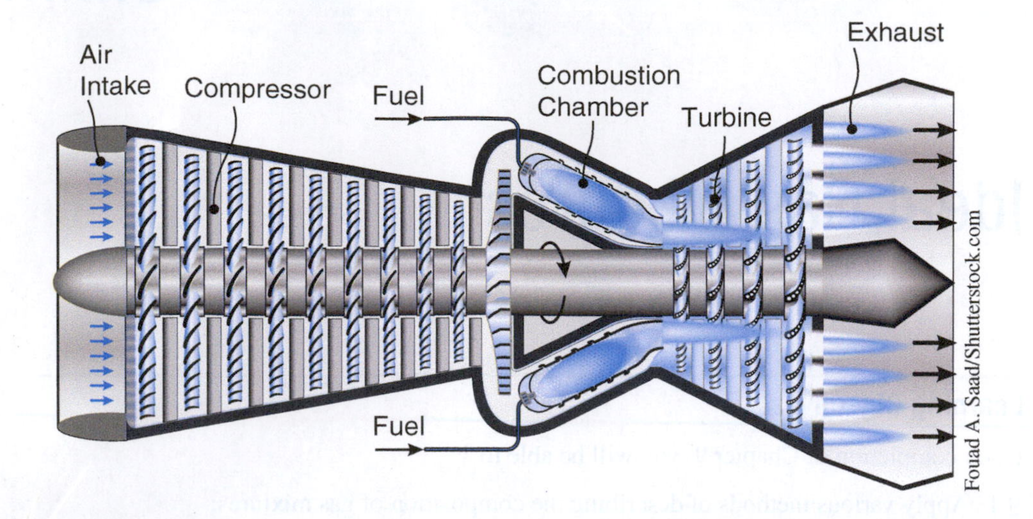

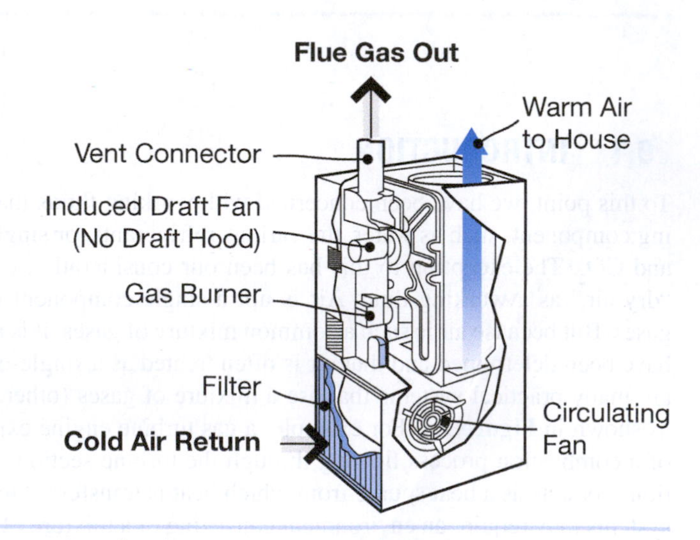

FIGURE 9.1 Mixtures of gases can occur in many forms and applications, such as in combustion products in gas turbines or furnaces, and as special mixtures to facilitate materials processing.

> **QUESTION FOR THOUGHT/DISCUSSION**
>
> In addition to air and the products of combustion, where else might gas mixtures be used?

9.2 DEFINING THE COMPOSITION OF A GAS MIXTURE

To determine the properties of a gas mixture, we must first specify the composition of the mixture. Let us consider the mixture of gases to consist of j gas components, with each component having a mass m_i. For example, Figure 9.2 illustrates a system with $j = 3$ components. Each component also has a particular molecular mass, M_i, such that the number of moles of each component, n_i, can be found from $n_i = m_i/M_i$. The total mass of the gas mixture, m, and the

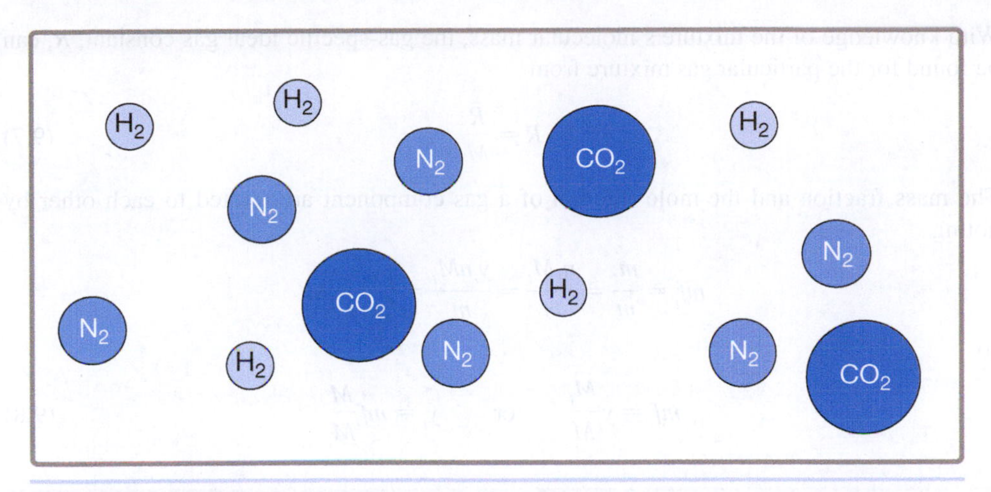

FIGURE 9.2 A gas mixture contains many different species in one system.

total number of moles present in the mixture, n, are found by adding these quantities for the component gases:

$$m = \sum_{i=1}^{j} m_i \tag{9.1}$$

and

$$n = \sum_{i=1}^{j} n_i \tag{9.2}$$

We can determine the fraction of a component gas in a total gas mixture by considering the component's mass or number of moles. The mass fraction of component i, mf_i, is given as

$$mf_i = \frac{m_i}{m} \tag{9.3}$$

and the mole fraction of component i, y_i, is found from

$$y_i = \frac{n_i}{n} \tag{9.4}$$

Both the mass fractions and the mole fractions of the gas components of a mixture sum to 1:

$$\sum_{i=1}^{j} mf_i = \sum_{i=1}^{j} y_i = 1$$

The molecular mass of the mixture can be determined by dividing the mass of the mixture by the number of moles in the mixture:

$$M = \frac{m}{n} \tag{9.5}$$

It can be shown through substitution that the molecular mass of the mixture can also be determined from knowledge of the mixture's composition, as specified by either mass fractions or mole fractions:

$$M = \sum_{i=1}^{j} y_i M_i = \left[\sum_{i=1}^{j} \frac{mf_i}{M_i} \right]^{-1} \tag{9.6}$$

With knowledge of the mixture's molecular mass, the gas-specific ideal gas constant, R, can be found for the particular gas mixture from

$$R = \frac{\overline{R}}{M} \tag{9.7}$$

The mass fraction and the mole fraction of a gas component are related to each other by noting

$$mf_i = \frac{m_i}{m} = \frac{n_i M_i}{m} = \frac{y_i n M_i}{m} = \left(\frac{n}{m}\right) y_i M_i$$

so

$$mf_i = y_i \frac{M_i}{M} \quad \text{or} \quad y_i = mf_i \frac{M}{M_i} \tag{9.8}$$

▶ EXAMPLE 9.1

Consider a gas mixture to consist of 3 kg H_2, 5 kg CO_2, 4 kg O_2, and 8 kg N_2. Determine the mass and mole fractions, the molecular mass, and the gas-specific ideal gas constant for this mixture.

Given: $m_{H_2} = 3$ kg $m_{CO_2} = 5$ kg

$m_{O_2} = 4$ kg $m_{N_2} = 8$ kg

Find: mf_i (mass fraction of each gas), y_i (mole fraction of each gas), M, R

Solution: The four substances in the problem are gases. These, and their resulting mixture, will be considered ideal gases for determination of the gas-specific ideal gas constant, although the rest of the calculations are independent of whether or not the gases are ideal.

The total mass is found by adding the individual component masses using Eq. (9.1):

$$m = m_{H_2} + m_{CO_2} + m_{O_2} + m_{N_2} = 20 \text{ kg}$$

The mass fraction for each component is then found using Eq. (9.3):

$$\mathbf{mf_{H_2}} = \frac{m_{H_2}}{m} = \mathbf{0.15}$$

$$\mathbf{mf_{CO_2} = 0.25}$$

$$\mathbf{mf_{O_2} = 0.20}$$

$$\mathbf{mf_{N_2} = 0.40}$$

There are different paths to finding mole fractions. Two methods will be used here so that it is clear that the results from the different approaches are the same.

Method 1: Determine the number of moles of each component, and then find the mole fractions using these values.

$$n_i = m_i/M_i$$

The molecular masses of each gas can be taken as $M_{H_2} = 2$ kg/kmole, $M_{CO_2} = 44$ kg/kmole, $M_{O_2} = 32$ kg/kmole, and $M_{N_2} = 28$ kg/kmole. So

$$n_{H_2} = 3 \text{ kg}/2 \text{ kg/kmole} = 1.5 \text{ kmole}$$

$$n_{CO_2} = 0.1136 \text{ kmole}$$

$$n_{O_2} = 0.125 \text{ kmole}$$

$$n_{N_2} = 0.2857 \text{ kmole}$$

The total number of moles is then found from Eq. (9.2): $n = n_{H_2} + n_{CO_2} + n_{O_2} + n_{N_2} = 2.024$ kmoles.

The mole fractions are then found from Eq. (9.4):

$$y_{H_2} = \frac{n_{H_2}}{n} = \mathbf{0.741}$$

$$y_{CO_2} = \mathbf{0.0561}$$

$$y_{O_2} = \mathbf{0.0617}$$

$$y_{N_2} = \mathbf{0.141}$$

Method 2: Find the molecular mass of the mixture from the mass fractions, and then use Eq. (9.8) to find the mole fractions.

From Eq. (9.6),

$$M = \left[\sum_{i=1}^{4} \frac{mf_i}{M_i}\right]^{-1} = \left[\frac{mf_{H_2}}{M_{H_2}} + \frac{mf_{CO_2}}{M_{CO_2}} + \frac{mf_{O_2}}{M_{O_2}} + \frac{mf_{N_2}}{M_{N_2}}\right]^{-1}$$

$$= \left[\frac{0.15}{2 \text{ kg/kmole}} + \frac{0.25}{44 \text{ kg/kmole}} + \frac{0.20}{32 \text{ kg/kmole}} + \frac{0.40}{28 \text{ kg/kmole}}\right]^{-1}$$

$$\mathbf{M = 9.88\ kg/kmole}$$

Then

$$y_i = mf_i \frac{M}{M_i}$$

so

$$y_{H_2} = mf_{H_2}(M/M_{H_2}) = (0.15)(9.88 \text{ kg/kmole}/2 \text{ kg/kmole}) = \mathbf{0.741}$$

$$y_{CO_2} = \mathbf{0.0561}$$

$$y_{O_2} = \mathbf{0.0617}$$

$$y_{N_2} = \mathbf{0.141}$$

These are the same values as found with Method 1.

Finally, the gas-specific ideal gas constant is found from Eq. (9.7):

$$R = \frac{\overline{R}}{M} = \frac{8.314 \text{ kJ/kmole} \cdot \text{K}}{9.88 \text{ kg/kmole}} = \mathbf{0.842\ kJ/kg \cdot K}$$

Analysis: As you can see, the mole fractions are the same using both methods. Therefore, using either method is acceptable. In general, as the number of components in the mixture increases, Method 2 will be the shorter approach to use.

9.2.1 The Composition of Dry Air

We have encountered numerous problems in which dry air, usually referred to simply as "air," has been the working fluid. As we will discuss further in Chapter 10, the air surrounding us is a combination of numerous gaseous molecules and water vapor. The water vapor content changes regularly, but the composition of the remaining molecules remains nearly constant for different locations and different times. The term *dry air* refers to this mixture of molecules in the air with the water vapor removed. The composition of the molecules making up the lesser components of the dry air does vary slightly with time and space, but the combination of molecules making up over 99% of the dry air is, for all practical purposes, unchanging. The best-known example of the changing nature of the minor components of dry air is the slow increase in CO_2 concentrations that have been observed since the 1950s. This increase, as measured at the Mauna Loa Observatory in Hawaii, is illustrated in **Figure 9.3**. As of May 2020, the mole fraction of CO_2 at Mauna Loa was 417 ppm. For comparison, in 1988 the CO_2

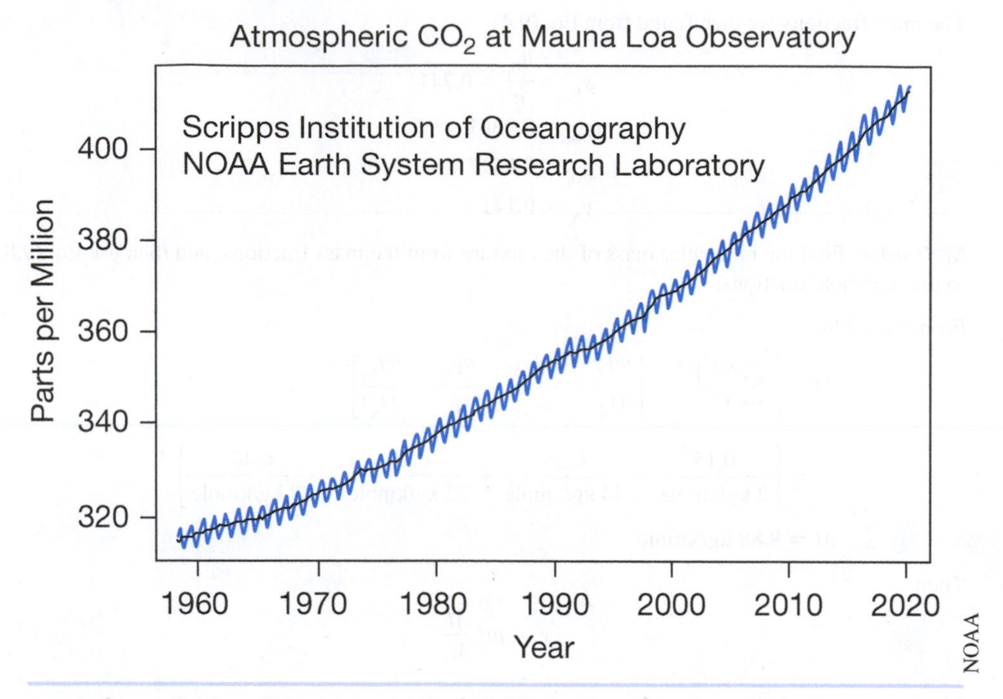

FIGURE 9.3 A plot of the concentration of CO_2 in the atmosphere at the Mauna Loa Observatory in Hawaii.

concentration was approximately 350 ppm. Although this increase may have a dramatic effect on the planet's climate, the increase in CO_2 concentration is still slow enough that it will not impact the engineering thermodynamics properties of dry air for our purposes because dry air is still overwhelmingly composed of nitrogen and oxygen.

The composition of dry air can be taken to be, on a mole fraction basis,

$$y_{N_2} = 0.7807$$
$$y_{O_2} = 0.2095$$
$$y_{Ar} = 0.0093$$
$$y_{CO_2} = 0.0004$$
$$y_{other} = 0.0001$$

The other gases include neon, helium, sulfur dioxide, and methane—essentially all other stable gases. As can be seen with this composition, 99% of dry air consists of just N_2 and O_2, and when argon is included, approximately 99.95% of dry air is accounted for (on a molar basis).

The molecular mass of dry air can be found using Eq. (9.6), and by considering more precise values of the molecular masses than used in Example 9.1. Here, we will use M_{N_2} = 28.02 kg/kmole, M_{O_2} = 32.00 kg/kmole, M_{Ar} = 39.94 kg/kmole, and M_{CO_2} = 44.01 kg/kmole. The contribution of CO_2 is very small, so we will ignore the other gases present in dry air. Using Eq. (9.6), the molecular mass of dry air is

$$M = \sum_{i=1}^{j} y_i M_i = (0.7807)(28.02) + (0.2095)(32.00) + (0.0093)(39.94) + (0.0004)(44.01)$$
$$= 28.97 \text{ kg/kmole}$$

Although this number may not look familiar, calculating the gas-specific ideal gas constant for dry air yields a more recognizable value:

$$R = 8.314 \text{ kJ/kmole} \cdot \text{K}/28.97 \text{ kg/kmole} = 0.287 \text{ kJ/kg} \cdot \text{K}$$

From this, two things should be apparent. First, dry air is in fact a mixture of gases, and the properties that have been used for dry air directly stem from this fact. Second, if we have been solving problems using dry air, methods of solving problems using other gas mixtures should be very similar. It may be necessary to derive the properties describing a mixture of gases other than dry air (those properties are readily available due to the common nature of such substances), but the fundamental methods used in solving thermodynamic applications involving gas mixtures should not change. To support this, we will develop techniques that allow us to find the thermodynamic properties of any ideal gas mixture, and then demonstrate their use through relatively common application problems.

9.3 IDEAL GAS MIXTURES

Although the mass and mole fraction designations can be used to describe the composition of any (real or ideal) gas mixture, other properties may vary between real gas mixtures and ideal gas mixtures. As we have noted, in a real gas, interactions between molecules can change the property values that are correct. For instance, consideration of the physical size of the molecule and the attractive forces between molecules leads to different values of specific volume being predicted for a gas using an equation of state, such as the van der Waals equation, and the ideal gas law. This will also hold true for mixtures of real gases (i.e., real gas mixtures), although additional complications arise between dissimilar molecules. As a result, evaluating properties of real gas mixtures can become quite complicated. But, unless we are at very high pressures or low temperatures, the influence of the effects of real gas mixtures on an engineering design is relatively small. Most engineering applications that use gas mixtures can be adequately modeled, at least in the early stages of design and analysis, as ideal gas mixtures (i.e., mixtures of ideal gases). If typical safety factors are used in a design, ideal gas mixture analysis is adequate for the final designs of most practical systems.

An ideal gas mixture takes on the features of a specific ideal gas—this is to be expected because we have been treating dry air as an ideal gas. The most important feature of an ideal gas mixture is that there is no interaction between the molecules. As a result, an ideal gas mixture will follow all of the equations of state previously developed for ideal gases, such as the following:

$$\text{Ideal gas law: } P\forall = mRT \quad (\text{also: } Pv = RT;\ P\forall = n\overline{R}T;\ P\overline{v} = \overline{R}T)$$

Property changes:

$$\Delta u = \int_{T_1}^{T_2} c_v dT;\ \Delta h = \int_{T_1}^{T_2} c_p dT;\ \Delta s = \int_{T_1}^{T_2} \frac{c_v}{T} dT + R \ln \frac{v_2}{v_1} = \int_{T_1}^{T_2} \frac{c_p}{T} dT - R \ln \frac{P_2}{P_1}$$

These last equations simplify with an assumption of constant specific heats. If we use a constant specific heat analysis, it will be necessary to find the values of the specific heats for a mixture.

To move toward that goal, let us first consider two principles involving the partial pressures and the partial volumes of an ideal gas mixture.

9.3.1 Dalton's Law of Partial Pressure

The partial pressure of a component of an ideal gas mixture, P_i, is the pressure that a component would have if it was alone in the volume occupied by the mixture at the mixture temperature. Dalton's law of partial pressure, named after scientist John Dalton of the 18th and 19th centuries, can be stated as follows:

The pressure of an ideal gas mixture is equal to the sum of the partial pressures of the component gases of the mixture.

In equation form, this can be written as

$$P = \sum_{i=1}^{j} P_i \tag{9.9}$$

Considering an ideal gas mixture, the partial pressure of each component is found from the ideal gas law as

$$P_i = \frac{n_i \bar{R} T}{V\!\!\!\!-} \tag{9.10}$$

where n_i is the number of moles of the component, and T and $V\!\!\!\!-$ are the mixture temperature and volume, respectively. Also from the ideal gas law, the total pressure of the mixture is

$$P = \frac{n \bar{R} T}{V\!\!\!\!-} \tag{9.11}$$

where n is the total number of moles of gases in the mixture. Taking the ratio of Eq. (9.10) to Eq. (9.11) yields an expression for the *partial pressure ratio* of component i:

$$\frac{P_i}{P} = \frac{\dfrac{n_i \bar{R} T}{V\!\!\!\!-}}{\dfrac{n \bar{R} T}{V\!\!\!\!-}} = \frac{n_i}{n} = y_i \tag{9.12}$$

For an ideal gas mixture, the partial pressure ratio is equal to the mole fraction for a gas component. In turn, this indicates that if we know the total pressure of the mixture and the mole fraction composition of the mixture, we can quickly calculate the partial pressure of an individual gas. Such information is important; for example, it can help us determine when a component of a gas mixture may begin to condense out of the mixture.

Figure 9.4 illustrates how the partial pressure of a mixture can be interpreted. Suppose there is a mixture of dry air at a total pressure of 100 kPa, as shown in Figure 9.4a. A filtering system is developed that removes all but one type of gas molecule without affecting the system volume or temperature. The system is first set up to remove all gas components other than N_2. Considering the molar composition of dry air given in Section 9.2.1, the mole fraction of N_2 is $y_{N_2} = 0.7807$. Multiplying this by the total mixture pressure yields a partial pressure of 78.07 kPa, as shown on the barometer in Figure 9.4b. Next, the filtering system is applied to the original mixture, but it is set to remove everything except O_2. The mole fraction of O_2 in dry air is 0.2095, which yields a partial pressure of O_2 of 20.95 kPa for this situation. The O_2 is the only gas left in the space, and the barometer in Figure 9.4c shows a pressure of 20.95 kPa. Similarly, when set to leave only Ar, the barometer reads 0.93 kPa, as shown in Figure 9.4d; this value corresponds to the Ar partial pressure. Finally, when the filtering system is set to remove everything except CO_2, nearly all of the gas is removed from the container, yielding a barometric pressure of 0.04 kPa or 40 Pa—which is the partial pressure of CO_2—as shown in Figure 9.4e.

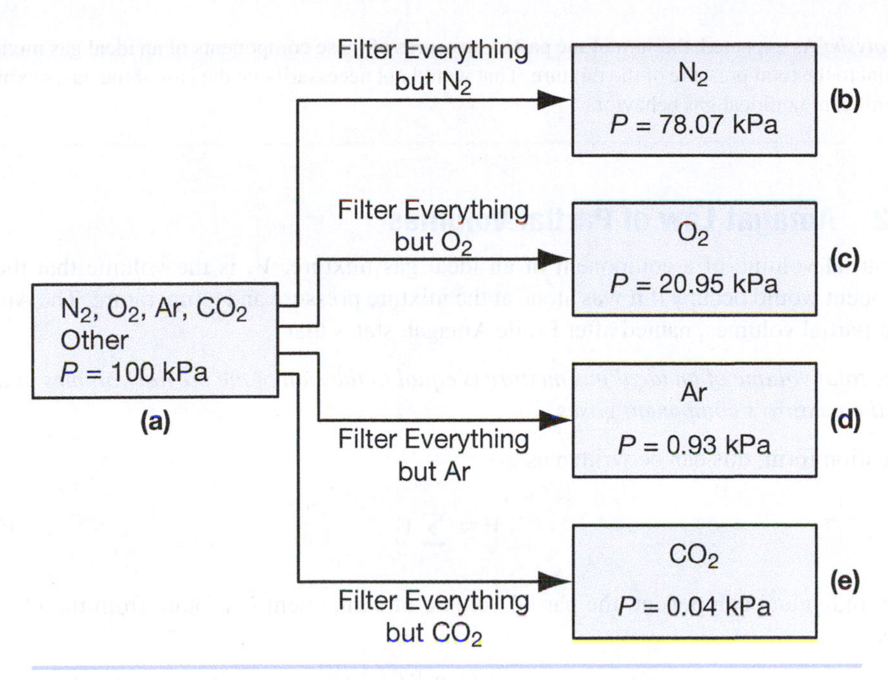

FIGURE 9.4 The pressures that we would expect to get if we took dry air at 100 kPa, considered it to be an ideal gas, and removed each component but one from the original volume (at the original mixture temperature). For example, if all the gases but O_2 were removed, the expected pressure in the container would be 20.95 kPa.

Considering these results, we can see that for an ideal gas mixture the total pressure is additive of the partial pressures of the gas as stated by Dalton. Keep in mind that gas mixtures that are considered real will not have the same partial pressure values as those derived for an ideal gas mixture; the deviation will be greater for mixtures that display behavior far from ideal.

▶ EXAMPLE 9.2

A mixture of gases consisting of 1.5 kg of CH_4 and 4.0 kg of N_2 has a total pressure of 290 kPa. Determine the partial pressure of each component, considering the mixture to be ideal.

Given: $m_{CH_4} = 1.5\text{ kg}$, $m_{N_2} = 4.0\text{ kg}$, $P = 290\text{ kPa}$

Find: P_{CH_4}, P_{N_2}

Solution: The working fluid in this problem is an ideal gas mixture consisting of two components. The number of moles of each component is found from $n_i = m_i/M_i$. Taking $M_{CH_4} = 16$ kg/kmole and $M_{N_2} = 28$ kg/kmole yields

$n_{CH_4} = 0.09375\text{ kmole}$ $\quad$ $n_{N_2} = 0.1429\text{ kmole}$

The total number of moles in the mixture is $n = n_{CH_4} + n_{N_2} = 0.2366\text{ kmole}$.

The mole fraction of each component, $y_i = n_i/n$, is

$y_{CH_4} = 0.3962$ $\quad$ $y_{N_2} = 0.6038$

The partial pressure of each component is found from $P_i = y_iP$. Therefore,

$\mathbf{P_{CH_4} = 115\text{ kPa}}$ $\quad$ $\mathbf{P_{N_2} = 175\text{ kPa}}$

Analysis: As expected, the sum of the partial pressures of these components of an ideal gas mixture is equal to the total pressure of the mixture. That would not necessarily be the case if the gases exhibited significant nonideal gas behavior.

9.3.2 Amagat Law of Partial Volumes

The partial volume of a component of an ideal gas mixture, V_i, is the volume that the gas component would occupy if it was alone at the mixture pressure and temperature. The Amagat law of partial volumes, named after Emile Amagat, states that

The total volume of an ideal gas mixture is equal to the sum of the partial volumes of all of the mixture's component gases.

In equation form, this can be written as

$$V = \sum_{i=1}^{j} V_i \tag{9.13}$$

For an individual component, the partial volume of component i is found from the ideal gas law as

$$V_i = \frac{n_i \overline{R} T}{P} \tag{9.14}$$

where P and T are the mixture pressure and temperature, respectively. The total volume of the mixture is also found with the ideal gas law:

$$V = \frac{n \overline{R} T}{P} \tag{9.15}$$

A *volume fraction* for component i can be found from the ratio of Eqs. (9.14) and (9.15):

$$\frac{V_i}{V} = \frac{\dfrac{n_i \overline{R} T}{P}}{\dfrac{n \overline{R} T}{P}} = \frac{n_i}{n} = y_i \tag{9.16}$$

For an ideal gas mixture, the volume fraction of component i is also equal to the mole fraction, just as the partial pressure ratio was.

To help visually illustrate the partial volume concept, consider the ideal gas mixture of dry air shown in **Figure 9.5**. In Figure 9.5a, there are four component gases filling an entire volume given as 1 m^3. Now, suppose that a sorting machine is developed that can separate each component molecule and place it in a bin. The bins can expand to fill just the entire size necessary to hold the molecules, at the mixture temperature and pressure. Figure 9.5b shows these four bins for N_2, with a volume of 0.7807 m^3 (equal to the product of the mixture volume and the mole fraction of N_2 in dry air); for O_2, with a volume of 0.2095 m^3; for Ar, with a volume of 0.0093 m^3; and for CO_2, with a volume of 0.0004 m^3. In Figure 9.5c, the bins are placed next to each other in the original container, and it can be seen that the vast majority of the volume or the original mixture contains N_2 or O_2 molecules.

Just as for the partial pressure, the total volume of an ideal gas mixture is the summation of the partial volumes of the components as stated by Amagat. Again, keep in mind that this applies only for ideal gas mixtures, and that real gas mixtures will deviate somewhat, although usually not dramatically, from this behavior.

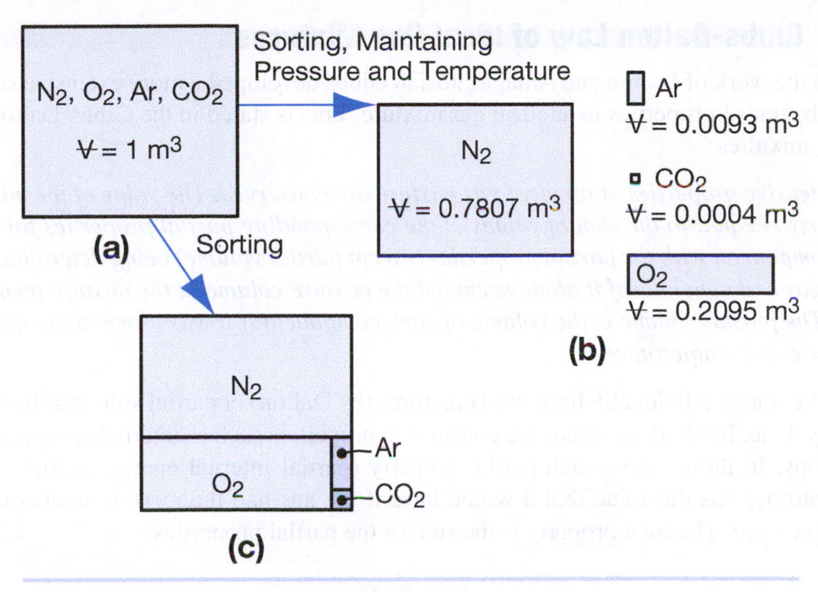

FIGURE 9.5 (a) An original mixture of the gases that compose dry air in a volume of 1.0 m^3; the mixture is considered to be an ideal gas. (b) The volumes of the containers that would be needed to hold each of the specific gases at the original gas temperature and pressure. (c) The distribution of gases in the original volume if the gases were sorted into spaces containing specific gas components.

▶ EXAMPLE 9.3

The mixture described in Example 9.2 is at a temperature of 50°C. Determine the partial volume of CH_4 and N_2 for the system.

Given: $m_{CH_4} = 1.5$ kg, $m_{N_2} = 4.0$ kg, $T = 50°C = 323$ K, $P = 290$ kPa

Find: $\forall_{CH_4}$, $\forall_{N_2}$

Solution: The working fluid is an ideal gas mixture consisting of two components.

From Example 9.2, the number of moles in the mixture and the mole fractions of each component are

$$n = 0.2366 \text{ kmole} \qquad y_{CH_4} = 0.3962 \qquad y_{N_2} = 0.6038$$

The total volume of the mixture can be found from the ideal gas law:

$$\forall = \frac{n\bar{R}T}{P} = \frac{(0.2366 \text{ kmole})(8.314 \text{ kJ/kmole} \cdot \text{K})(323 \text{ K})}{290 \text{ kPa}} = 2.191 \text{ m}^3$$

The partial volumes are then found from

$$\forall_i = y_i \forall$$

So

$$\mathbf{\forall_{CH_4} = 0.868 \text{ m}^3} \qquad \mathbf{\forall_{N_2} = 1.32 \text{ m}^3}$$

Analysis: Notice that, as expected, the sum of the two partial volumes in this ideal gas mixture is equal to the total volume of the mixture.

9.3.3 Gibbs-Dalton Law of Ideal Gas Mixtures

Based on the work of Dalton and Amagat, Josiah Gibbs developed a more extensive statement of the behavior of properties in an ideal gas mixture. This is stated in the Gibbs-Dalton law of ideal gas mixtures:

> *All extensive properties of an ideal gas mixture are conserved. The value of the mixture property is equal to the sum of values of the corresponding partial properties for each gas component, with the partial properties (except partial volume) being determined for each gas component as if it alone occupied the mixture volume at the mixture temperature. The partial volume is the volume of each component if it was alone at the mixture pressure and temperature.*

This can be stated individually for partial pressures (by Dalton) or partial volumes (by Amagat) as already done. But it also extends the concepts to properties such as internal energy, enthalpy, and entropy. In these cases, each partial property (partial internal energy, partial enthalpy, partial entropy) has the value that it would have if the gas had the mixture temperature and mixture pressure. The total property is the sum of the partial properties:

$$U = \sum_{i=1}^{j} U_i \tag{9.17}$$

$$H = \sum_{i=1}^{j} H_i \tag{9.18}$$

$$S = \sum_{i=1}^{j} S_i \tag{9.19}$$

Although it is a useful concept on its own, the Gibbs-Dalton law becomes more powerful when these expressions are expanded and the intensive properties of the mixture are determined. To illustrate this, consider enthalpy. The following development could be done for any extensive property, and enthalpy is used as an example. The total partial enthalpy can be written as a product of the mass of the gas and the mass-specific enthalpy, h (or as a product of the number of moles of the gas and the molar-specific enthalpy, $\overline{h}$, where $\overline{h} = H/n$). Expanding Eq. (9.18) yields

$$H = \sum_{i=1}^{j} H_i = \sum_{i=1}^{j} m_i h_i = \sum_{i=1}^{j} n_i \overline{h}_i \tag{9.20}$$

Dividing Eq. (9.20) by the mass of the mixture yields the mass-specific enthalpy of the mixture, and dividing Eq. (9.20) by the number of moles of the mixture yields the molar-specific enthalpy of the mixture:

$$h = \frac{H}{m} = \sum_{i=1}^{j} \frac{m_i}{m} h_i = \sum_{i=1}^{j} mf_i h_i \tag{9.21a}$$

and

$$\overline{h} = \frac{H}{n} = \sum_{i=1}^{j} \frac{n_i}{n} \overline{h}_i = \sum_{i=1}^{j} y_i \overline{h}_i \tag{9.21b}$$

There is nothing particular to specific enthalpy for this derivation, so we can conclude that the concept applies to other specific properties. Therefore, the mass-intensive properties of the mixture are the sum of the mass-fraction-weighted specific properties, and the molar-intensive properties of the mixture are the sum of the mole-fraction-weighted molar-specific properties. So

$$v = \frac{V}{m} = \sum_{i=1}^{j} mf_i v_i; \qquad u = \frac{U}{m} = \sum_{i=1}^{j} mf_i u_i; \qquad s = \frac{S}{m} = \sum_{i=1}^{j} mf_i s_i \tag{9.22a}$$

and

$$\bar{v} = \frac{V}{n} = \sum_{i=1}^{j} y_i \bar{v}_i; \qquad \bar{u} = \frac{U}{n} = \sum_{i=1}^{j} y_i \bar{u}_i; \qquad \bar{s} = \frac{S}{n} = \sum_{i=1}^{j} y_i \bar{s}_i \tag{9.22b}$$

Although we may not usually think of them as such, the specific heats are also specific intensive properties. Therefore, we can conclude that

$$c_p = \sum_{i=1}^{j} mf_i c_{p,i} \qquad c_v = \sum_{i=1}^{j} mf_i c_{v,i} \tag{9.23a}$$

and

$$\bar{c}_p = \sum_{i=1}^{j} y_i \bar{c}_{p,i} \qquad \bar{c}_v = \sum_{i=1}^{j} y_i \bar{c}_{v,i} \tag{9.23b}$$

Application of Eqs. (9.23) in solving thermodynamic problems will be easier if the specific heats of the different gases are assumed to be constant; however, the equations also will work when variable specific heats are considered. In those cases, however, it is often easier to apply Eqs. (9.21) and (9.22) as needed.

▶ EXAMPLE 9.4

For the mixture composition of dry air described in Section 9.2.1, determine the value of the constant-pressure specific heat, c_p, at 300 K for dry air.

Given: Dry air: $y_{N_2} = 0.7807$; $y_{O_2} = 0.2095$; $y_{Ar} = 0.0093$; $y_{CO_2} = 0.0004$

$T = 300$ K

Find: c_p

Solution: The working fluid is an ideal gas mixture with the composition of dry air. As we are more accustomed to using mass-based specific heats, we will solve this problem by first converting the mole fractions listed above to mass fractions.

Previously, we found that $M = 28.97$ kg/kmole.

Using Eq. (9.8), $mf_i = y_i \frac{M_i}{M}$, we can find the mass fractions as follows:

$mf_{N_2} = 0.7552$; $mf_{O_2} = 0.2314$; $y_{Ar} = 0.0128$; $y_{CO_2} = 0.0006$

At 300 K, the gases have the following values of c_p:

$c_{p,N_2} = 1.039$ kJ/kg · K $\qquad c_{p,O_2} = 0.918$ kJ/kg · K $\qquad c_{p,Ar} = 0.520$ kJ/kg · K

$c_{p,CO_2} = 0.846$ kJ/kg · K

From Eq. (9.23a), $c_p = \sum_{i=1}^{j} mf_i c_{p,i} = 1.004$ kJ/kg · K.

Analysis: This calculated value is very close to the experimental value of 1.005 kJ/kg · K that is often used. An alternative solution procedure is to find $\bar{c}_p$ and then use the molecular mass of the mixture to convert this to c_p.

With this ability to determine the specific heats of an ideal gas mixture, it is possible to solve thermodynamic problems involving gases for any possible ideal gas mixture. We illustrate this in the next section.

QUESTION FOR THOUGHT/DISCUSSION

From an engineering standpoint, under what conditions is it prudent *not* to treat a mixture of gases as an ideal gas mixture?

9.4 SOLUTIONS OF THERMODYNAMIC PROBLEMS INCORPORATING IDEAL GAS MIXTURES

We have solved thermodynamic problems involving the First Law of Thermodynamics and the entropy balance using air, and the basic procedures for solving such problems will not change when we consider ideal gas mixtures. However, because predetermined properties do not exist for every ideal gas mixture potentially used in an application, the specific heats for a particular mixture must be found in order to complete the solution. Although it is possible to use relationships such as Eqs. (9.21) and (9.22) to incorporate variable specific heats into a solution procedure, for now we will focus on using an approximation of constant specific heats. When we study combustion in Chapter 11, we will incorporate solutions using variable specific heats. This will be done because the great temperature changes experienced in combustion processes lead to unacceptably large errors if a constant specific heat solution is used.

The solution procedure we will use is very similar to the procedures we used previously. The primary change is that the mixture composition needs to be determined, and the values of the constant specific heats of the ideal gas mixture must be found before the problem can be solved. This procedure is illustrated in the following examples.

STUDENT EXERCISE

If your computer models of the various devices and systems that you have developed are not able to work with any ideal gas, modify the models to accept, as input, values for the specific heats for any ideal gas mixture.

► EXAMPLE 9.5

An ideal gas mixture consisting of 2.0 kg of N_2 and 4.0 kg of CO_2 is to be heated in a piston–cylinder assembly at constant pressure. The mixture is initially at 300 K and is heated to a temperature of 500 K. The gas pressure is 750 kPa. Assume the mixture has constant specific heats, determined at 300 K. Determine the heat input to the mixture during the process.

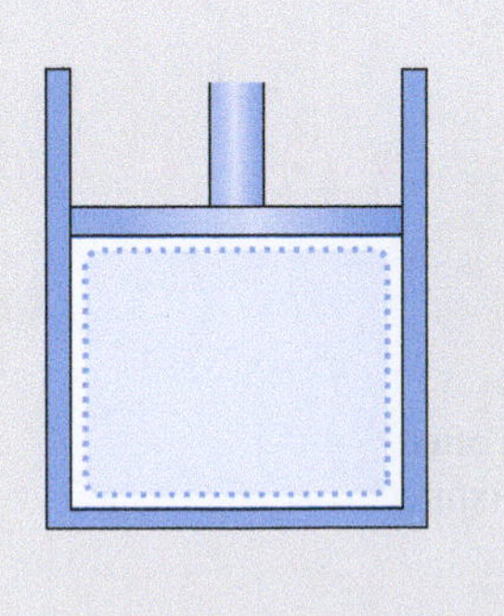

Given: $m_{N_2} = 2.0\text{ kg}$, $m_{CO_2} = 4.0\text{ kg}$, $P = 750\text{ kPa (constant)}$, $T_1 = 300\text{ K}$, $T_2 = 500\text{ K}$

Find: Q

Solution: The system is a closed system, with the working fluid being heated at constant pressure; thus we would expect the volume to expand. The working fluid is a two-component ideal gas mixture.

Assume: $\Delta KE = \Delta PE = 0$. The gases behave as ideal gases with constant specific heats.

To solve this problem, we will use the first law for closed systems. Equation (4.32) is

$$Q - W = m\left(u_2 - u_1 + \frac{V_2^2 - V_1^2}{2} + g(z_2 - z_1)\right)$$

Assuming no changes in kinetic or potential energy, and assuming constant specific heats,

$$Q - W = m(u_2 - u_1) = mc_v(T_2 - T_1)$$

There will be moving boundary work in this problem, with a constant pressure:

$$W = \int_1^2 P\,d\forall = P(\forall_2 - \forall_1)$$

From these equations, it is clear that we will need to determine the volumes at the initial and final states, and we will need the mass of the mixture and the c_v value of the mixture. First, we find the total mass of the mixture, and we find the mass fractions:

$$m = m_{N_2} + m_{CO_2} = 6.0 \text{ kg}$$

$$mf_{N_2} = \frac{m_{N_2}}{m} = 0.333 \qquad mf_{CO_2} = \frac{m_{CO_2}}{m} = 0.667$$

With $M_{N_2} = 28$ kg/kmole and $M_{CO_2} = 44$ kg/kmole, the molecular mass of the mixture is

$$M = \left[\sum_{i=1}^{j} \frac{mf_i}{M_i}\right]^{-1} = 36.97 \text{ kg/kmole}$$

and the gas-specific ideal gas constant is

$$R = \frac{\overline{R}}{M} = 0.225 \text{ kJ/kg} \cdot \text{K}$$

At 300 K, $c_{v,N_2} = 0.743$ kJ/kg · K, and $c_{v,CO_2} = 0.657$ kJ/kg · K, so for the mixture using Eq. (9.23a),

$$c_v = \sum_{i=1}^{j} mf_i c_{v,i} = 0.6856 \text{ kJ/kg} \cdot \text{K}$$

The initial and final volumes are found from the ideal gas law:

$$\forall_1 = \frac{mRT_1}{P} = 0.540 \text{ m}^3 \qquad \forall_2 = \frac{mRT_2}{P} = 0.900 \text{ m}^3$$

The moving boundary work is $W = P(\forall_2 - \forall_1) = 270$ kJ.
The heat transfer is $Q = mc_v(T_2 - T_1) + W =$ **1090 kJ.**

Analysis: Only experience with this type of problem will allow you to determine if these numbers are reasonable. From this problem, we can observe that the values are logical. The work is positive, indicating that the system did work on the surroundings as it expanded. The heat transfer is positive, indicating that there was energy added to the system, as we would expect if the system was producing work and the working fluid was also increasing in temperature.

▶ EXAMPLE 9.6

A mixture of two parts H_2 and one part O_2 (by molar composition) is to be accelerated through an insulated nozzle before entering an engine. The mixture enters the nozzle at 40°C and exits at 30°C. If the mixture enters at a velocity of 2 m/s, what is the exit velocity? Assume that there are no work interactions or changes in potential energy, and that the specific heats of the mixture can be considered constant.

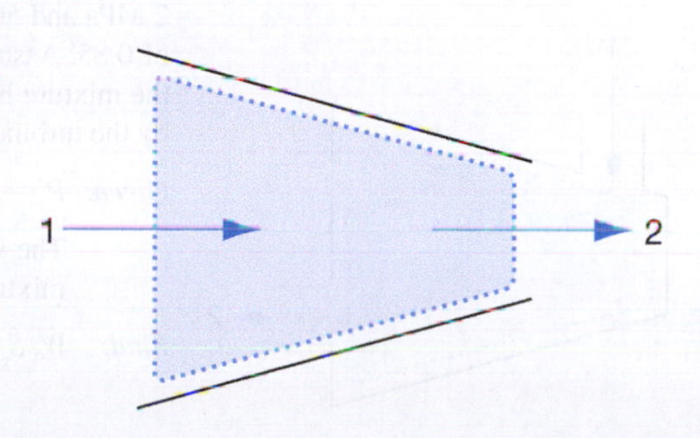

Given: $T_1 = 40°\text{C}$, $T_2 = 30°\text{C}$, $\dot{Q} = 0$ (insulated), $V_1 = 2$ m/s

With a total of three "parts" of the mixture, the mole fractions given are $y_{H_2} = 0.667$ and $y_{O_2} = 0.3333$.

Find: V_2

Solution: The system to be analyzed is a nozzle, which can be viewed as a steady-state, steady-flow, single-inlet, single-outlet open system. The working fluid is an ideal gas mixture, which will be assumed to have constant specific heats.

Assume: $\dot{W} = 0$, $\Delta PE = 0$, ideal gas mixture with constant specific heats.

To find the exit velocity, we will use the first law for a single-inlet, single-outlet, steady-state, steady-flow system (Eq. (4.15a)):

$$\dot{Q} - \dot{W} = \dot{m}\left(h_2 - h_1 + \frac{V_2^2 - V_1^2}{2} + g(z_2 - z_1)\right)$$

With no heat transfer, work, or changes in potential energy, and with an ideal gas with constant specific heats, Eq. (4.15a) reduces to

$$h_2 - h_1 + \frac{V_2^2 - V_1^2}{2} = c_p(T_2 - T_1) + \frac{V_2^2 - V_1^2}{2} = 0$$

The constant-pressure specific heat, c_p, is needed for the mixture. To find c_p, we first find the molecular mass of the mixture, knowing that $M_{H_2} = 2$ kg/kmole and $M_{O_2} = 32$ kg/kmole:

$$M = \sum_{i=1}^{j} y_i M_i = 12 \text{ kg/kmole}$$

Then we find the mass fractions using $mf_i = y_i \frac{M_i}{M}$, so

$$mf_{H_2} = 0.111, \text{ and } mf_{O_2} = 0.889$$

At 40°C, $c_{p,H_2} = 14.34$ kJ/kg · K and $c_{p,O_2} = 0.921$ kJ/kg · K, so from Eq. (9.23a),

$$c_p = \sum_{i=1}^{j} mf_i c_{p,i} = 2.41 \text{ kJ/kg} \cdot \text{K}$$

Solving for V_2,

$$V_2 = \left[2g_c c_p(T_1 - T_2) + V_1^2\right]^{1/2} = \left[2\left(2410\frac{\text{J}}{\text{kg} \cdot \text{K}}\right)(40°\text{C} - 30°\text{C}) + \left(2\frac{\text{m}}{\text{s}}\right)^2\right]^{1/2}$$

$$\mathbf{V_2 = 220 \text{ m/s}}$$

Analysis: As we are multiplying c_p by the temperature difference, the same result will be obtained using °C or K, as the difference in temperature remains 10°C or 10 K. The exit velocity is much higher than the entrance velocity, as we would expect for a nozzle.

► EXAMPLE 9.7

An ideal gas mixture has a volumetric composition of 25% N_2, 50% O_2, and 25% CO_2. The mixture flows through an insulated turbine at a rate of 12 kg/s. The mixture enters the turbine at 2 MPa and 500°C, and exits the turbine at 300 kPa. The turbine has an isentropic efficiency of 0.85. Assuming that the changes in kinetic and potential energy are negligible, and that the mixture has constant specific heats (evaluated at 600 K), determine the power produced by the turbine and the entropy generation rate for the process.

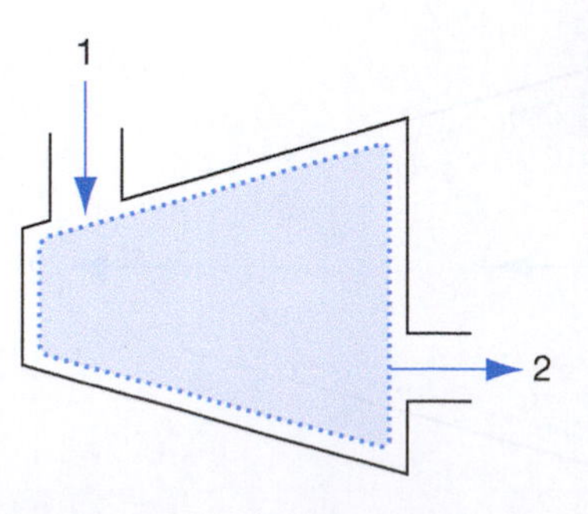

Given: $P_1 = 2$ MPa, $T_1 = 500°\text{C} = 773$ K, $P_2 = 300$ kPa, $\dot{m} = 12$ kg/s, $\eta_{s,t} = 0.85$.

The volumetric composition is equivalent to the molar composition for an ideal gas mixture, so $y_{N_2} = 0.25$, $y_{O_2} = 0.50$, $y_{CO_2} = 0.25$.

Find: $\dot{W}$, $\dot{S}_{gen}$

Solution: The turbine can be viewed as a single-inlet, single-outlet, steady-state, steady-flow system. The working fluid is an ideal gas mixture. The properties for the mixture will be found with a constant specific heat.

Assume: $\Delta KE = \Delta PE = 0$, $\dot{Q} = 0$ (insulated), constant specific heats evaluated at 600 K.

Considering the first law for a single-inlet, single-outlet, steady-state, steady-flow system (Eq. (4.15)),

$$\dot{Q} - \dot{W} = \dot{m}\left(h_2 - h_1 + \frac{V_2^2 - V_1^2}{2} + g(z_2 - z_1)\right)$$

With no changes in kinetic or potential energy, and no heat transfer, and with an ideal gas with constant specific heats, the first law reduces to

$$\dot{W} = \dot{m}(h_1 - h_2) = \dot{m}c_p(T_1 - T_2)$$

Similarly, the entropy rate balance (Eq. (6.32)) reduces to

$$\dot{S}_{gen} = \dot{m}(s_2 - s_1) = \dot{m}\left[c_p \ln\frac{T_2}{T_1} - R\ln\frac{P_2}{P_1}\right]$$

From these equations, it is apparent that values of c_p and R for the mixture are needed, and that the exit temperature must be found. First, to obtain the mixture properties, we will convert the mole fractions to mass fractions.

$$M_{N_2} = 28 \text{ kg/kmole} \qquad M_{O_2} = 32 \text{ kg/kmole} \qquad M_{CO_2} = 44 \text{ kg/kmole}$$

$$M = \sum_{i=1}^{j} y_i M_i = (0.25)(28) + (0.50)(32) + (0.25)(44) = 34 \text{ kg/kmole}$$

We then find the mass fractions using $mf_i = y_i\frac{M_i}{M}$, so

$$mf_{N_2} = 0.2059 \qquad mf_{O_2} = 0.4706 \qquad mf_{CO_2} = 0.3235$$

With $c_{p,N_2} = 1.075$ kJ/kg · K, $c_{p,O_2} = 1.003$ kJ/kg · K, and $c_{p,CO_2} = 1.075$ kJ/kg · K (at 600 K) Equation (9.23a) yields

$$c_p = \sum_{i=1}^{j} mf_i c_{p,i} = 1.041 \text{ kJ/kg} \cdot \text{K}$$

In addition,

$$R = \frac{\bar{R}}{M} = 0.2445 \text{ kJ/kg} \cdot \text{K}$$

Knowing that $c_v = c_p - R$, we can find that $c_v = 0.796$ kJ/kg · K, and $k = c_p/c_v = 1.307$.

With k, we can find the isentropic outlet temperature from the turbine:

$$T_{2s} = T_1\left(\frac{P_2}{P_1}\right)^{\frac{k-1}{k}} = 495 \text{ K}$$

The isentropic efficiency of the turbine leads to the calculation of T_2, the actual outlet temperature:

$$\eta_{s,t} = \frac{h_1 - h_2}{h_1 - h_{2s}} = \frac{c_p(T_1 - T_2)}{c_p(T_1 - T_{2s})}$$

Solving yields $T_2 = 537$ K.

(Note: Although the choice of specific heats at 600 K is not quite the average temperature, it is a reasonably close to average value and can be used with good accuracy in the calculations.)

Completing the final calculations,

$$\dot{W} = \dot{m}c_p(T_1 - T_2) = \mathbf{2950\ kW}$$

$$\dot{S}_{gen} = \dot{m}\left[c_p \ln\frac{T_2}{T_1} - R\ln\frac{P_2}{P_1}\right] = \mathbf{1.02\ kW/K}$$

Analysis: This solution procedure was very similar to the type of solution used in Chapter 6, but it includes calculation of the mixture properties. In addition, the calculated results appear to be consistent with what we may expect from a turbine with a rather small mass flow rate.

It is important to note that not all problems that appear to involve mixtures are easier to solve using mixture properties. For instance, in some cases, the problem involves the creation of the mixture, and there is no clear entrance condition for the mixture. In such a problem it is often easier to treat the components of the mixture as separate unmixed gases and to add the results for each gas flowing through the system separately. This is possible because the ideal gas mixture experiences no interactions between the components, and so the final results can be the additive result of several similar problems involving each gas separately, such as shown in **Figure 9.6**. This is illustrated in Example 9.8.

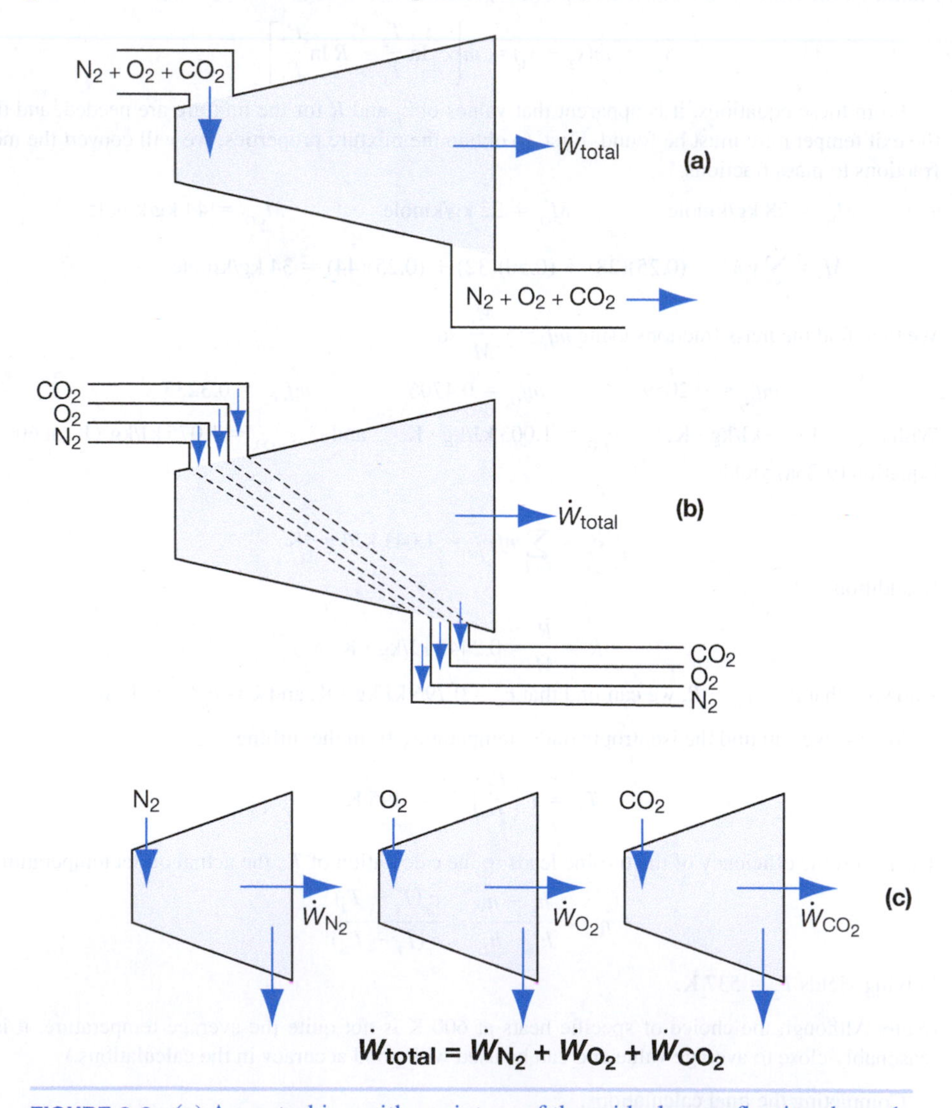

FIGURE 9.6 (a) A gas turbine with a mixture of three ideal gases flowing through it. (b) Because ideal gas mixtures experience no effects of being in a mixture, the turbine can be thought of as having three separate streams of gases flowing through it. (c) Alternatively, the problem can be thought of as three separate turbine problems with one gas flowing through each, with the total work being the sum of the work produced by the three gases.

▶ EXAMPLE 9.8

A mixing chamber receives 1.5 kg/s of CO_2 at a temperature of 20°C, and 2.0 kg/s of CH_4 at 0°C. In the mixing chamber, the two streams are combined and heated, and the combined stream exits at 30°C. Assuming that the gases behave as ideal gases with constant specific heats, and that the process is at steady state, determine the rate of heat transfer in the mixing chamber.

Given: Inlet 1 (CO_2): $T_1 = 20°C$, $\dot{m}_1 = 1.5$ kg/s, $c_{p,CO_2} = 0.846$ kJ/kg · K
Inlet 2 (CH_4), $T_2 = 0°C$, $\dot{m}_2 = 2.0$ kg/s, $c_{p,CH_4} = 2.254$ kJ/kg · K
Outlet 3 (CO_2 and CH_4): $T_3 = 30°C$

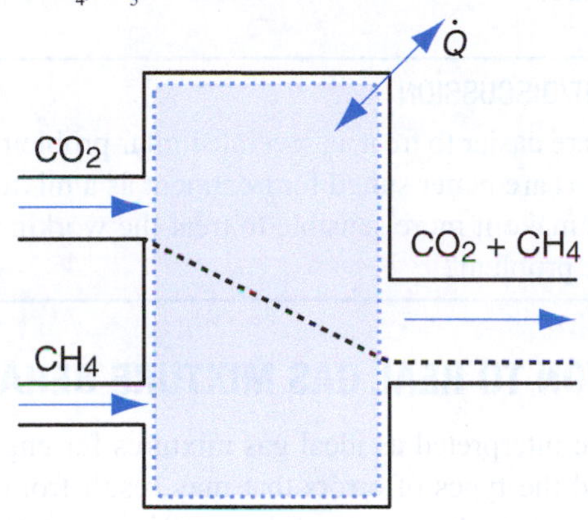

Find: $\dot{Q}$

Solution: The system is an open system with two inlets and one outlet. It operates at steady-state, steady-flow conditions. The working fluids involved are ideal gases.

Assume: No changes in kinetic energy or potential energy of the gases. $\dot{W} = 0$. The specific heats of the gases are constant.

Treating this as a mixture from beginning to end is made difficult by the two gases entering at different temperatures. We do not know what temperature to use for an "inlet" temperature of the combined streams. We could use the first law to determine the temperature of the mixture before it begins to be heated in the chamber. But if that calculation is to be included, it is probably easier overall to treat the problem as two separate problems (one involving heating CO_2 from 20°C to 30°C, and one involving heating CH_4 from 0°C to 30°C) and add the results. The assumption of an ideal gas mixture indicates that there is no interaction between the molecules, and the gases behave as if there is nothing else present. Therefore, each gas is unaffected by the presence of the other gas.

Each problem becomes a single-inlet, single-outlet, steady-state, steady-flow system. Assuming no power and no changes in kinetic or potential energy, the first law (Eq. (4.15))

$$\dot{Q} - \dot{W} = \dot{m}\left(h_{out} - h_{in} + \frac{V_{out}^2 - V_{in}^2}{2} + g(z_{out} - z_{in})\right)$$

becomes for an ideal gas with constant specific heats

$$\dot{Q} = \dot{m}(h_{out} - h_{in}) = \dot{m}c_p(T_{out} - T_{in})$$

For the CO_2: $\dot{Q}_{CO_2} = \dot{m}_1 c_{p,CO_2}(T_3 - T_1) = 12.69$ kW

For the CH_4: $\dot{Q}_{CH_4} = \dot{m}_2 c_{p,CH_4}(T_3 - T_2) = 135.24$ kW

The total heat transfer rate is then

$$\dot{Q} = \dot{Q}_{CO_2} + \dot{Q}_{CH_4} = \mathbf{148\ kW}$$

Analysis: It is logical that more heat would be needed to increase the temperature of the CH_4 than the CO_2 because the flow rate for the methane is larger, as is the necessary temperature increase. You may want to try to figure out this problem by treating it as a mixture throughout. To do so, you would need to perform an additional energy balance. The energy balance would consider just the mixing of the two inlet streams. This analysis of the mixing would assume that no heat transfer or work is being done.

In general, it will be computationally simpler to treat a mixture with common initial and final conditions as a mixture utilizing mixture properties. However, ideal gas mixtures allow a process to be treated as a series of separate gases undergoing the process and the total result is the sum of the parts; this is the approach we should use when the problem is too difficult to solve as a single gas mixture.

QUESTION FOR THOUGHT/DISCUSSION

While some problems are easier to treat as several similar problems, each involving one ideal gas, other problems are better suited for treatment as a mixture. What characteristics of a system would make it more sensible to treat the working fluid as an ideal gas mixture throughout the problem?

9.5 INTRODUCTION TO REAL GAS MIXTURE BEHAVIOR

Many gas mixtures can be interpreted as ideal gas mixtures for engineering purposes, but it is important to understand the types of errors that may result from such an approximation. Therefore, in this section we introduce some of the variations in real gas mixture behavior.

In analyzing the pressure-volume-temperature (*P-v-T*) behavior of a real gas mixture, we can use complex equations of state, such as the van der Waals equation, and use correction factors derived for the specific gas mixture. In such a process, the constants are derived for the mixture as

$$a = \left(\sum_{i=1}^{j} y_i a_i^{1/2}\right)^2 \tag{9.24a}$$

and

$$b = \sum_{i=1}^{j} y_i b_i \tag{9.24b}$$

where a_i and b_i are the van der Waals constants for each specific gas in the mixture.

A simpler approach is to use the compressibility factor for each gas, and then use a mole-fraction-weighted summation of the individual gas components' compressibility factor:

$$Z = \sum_{i=1}^{j} y_i Z_i \tag{9.25}$$

where Z_i is the compressibility factor of component i of the mixture. It is usually best to find these compressibility factors using the mixture temperature and mixture pressure; that is, using Amagat's law. This approach will find $T_{R,i} = T/T_{cr,i}$ and $P_{R,i} = P/P_{cr,i}$, where T and P are the mixture temperature and pressure, respectively. Alternatively, we can find the compressibility factor using the mixture temperature and volume, or we can use Kay's rule, which states that a pseudo-critical pressure and pseudo-critical temperature of the mixture can be found as a mole-fraction-weighted sum of the critical pressures and critical temperatures of the individual components; we then use those values to find the mixture's compressibility factor. Once the compressibility factor is found, the gas mixture's behavior follows:

$$P\mathcal{V} = Zn\overline{R}T \tag{9.26}$$

The differences that can be experienced are illustrated in Example 9.9.

▶ EXAMPLE 9.9

A real gas mixture, consisting of 35% N_2 and 65% C_2H_4 (on a molar basis), has a pressure of 18 MPa and a temperature of 300 K. Using the compressibility factor of the mixture, determine the mixture's molar-specific volume, and compare this value to that predicted using the ideal gas law.

Given: $y_{N_2} = 0.35$, $y_{C_2H_4} = 0.65$, $P = 18{,}000$ kPa, $T = 300$ K

Find: $\bar{v}$ (using both the compressibility factor and the ideal gas law)

Solution: First, determine the compressibility factor of each gas using the mixture temperature and the mixture pressure (following Amagat's model).

For N_2, $T_{c,N_2} = 126.2$ K, $P_{c,N_2} = 3390$ kPa:

$$T_{R,N_2} = T/T_{c,N_2} = 2.38$$

$$P_{R,N_2} = P/P_{c,N_2} = 5.31$$

From the compressibility chart, $Z_{N_2} = 1.03$.

For C_3H_8, $T_{c,C_2H_4} = 282.4$ K, $P_{c,C_2H_4} = 5120$ kPa:

$$T_{R,C_2H_4} = T/T_{c,C_2H_4} = 1.06$$

$$P_{R,C_2H_4} = P/P_{c,C_2H_4} = 3.52$$

From the compressibility chart, $Z_{C_2H_4} = 0.52$.

Using Eq. (9.25),

$$Z = \sum_{i=1}^{2} y_i Z_i = y_{N_2} Z_{N_2} + y_{C_2H_4} Z_{C_2H_4} = 0.699$$

Then

$$\bar{v} = \frac{Z\bar{R}T}{P} = \frac{(0.699)(8.314 \text{ kJ/kmole} \cdot \text{K})(300 \text{ K})}{18{,}000 \text{ kPa}} = \mathbf{0.0968\ m^3/kmole}$$

If the ideal gas law is used, the result is

$$\bar{v} = \frac{\bar{R}T}{P} = \frac{(8.314 \text{ kJ/kmole} \cdot \text{K})(300 \text{ K})}{18{,}000 \text{ kPa}} = \mathbf{0.139\ m^3/kmole}$$

Analysis: The ideal gas law overpredicts the molar-specific volume in this case by ~43%. While the value found using the compressibility factor may not be exact, it is much closer than the value found using the ideal gas law. The compressibility chart is a good tool that can be used to gauge how close the gas mixture's behavior is to ideal. For example, if the gas mixture pressure had been 100 kPa, the compressibility factor approach would determine that the compressibility factor was very close to 1, and the gas mixture would be considered ideal.

Keep in mind that, just as with a single-component gas, extreme conditions must be in place to move the real gas behavior far from that predicted by the ideal gas law. In this case, the pressure is 18 MPa. This results in a much higher density than is typically seen in engineering applications. You can still treat many, if not most, gas mixture problems as ideal gas mixtures for engineering purposes. However, you should also recognize that very high pressures or very low gas temperatures will result in a gas whose density is large enough to push the gas behavior away from ideal conditions.

Relationships between real gas properties in mixtures become more complex as properties such as the specific internal energy, u, and the specific enthalpy, h, are also functions of pressure for real gases. Therefore, if the real gas behavior impacts the pressure, the real gas behavior also impacts these property values. As Example 9.9 shows, it is not expected that the deviation from ideal gas behavior will be great for many engineering problems until rather extreme conditions are met. As a result, methods for finding real gas mixture properties such as enthalpy and entropy are left for a more advanced course on thermodynamics.

Summary

In this chapter, we have explored concepts surrounding ideal gas mixtures. We have encountered the different methods of describing a gas mixture's composition, and learned how to find thermodynamic properties for any ideal gas mixture. We have seen that the methods of solving thermodynamic problems do not significantly change when using an ideal gas mixture as the working fluid: the primary change is that thermodynamic properties will need to be determined on a case-by-case basis. Finally, we have learned some of the differences between the behavior of ideal gas mixtures and real gas mixtures. With this knowledge, we are now ready to consider two situations where gas mixtures are common—psychrometrics and combustion.

KEY EQUATIONS

Mass Fraction:
$$mf_i = \frac{m_i}{m} \tag{9.3}$$

Mole Fraction:
$$y_i = \frac{n_i}{n} \tag{9.4}$$

Molecular Mass of the Mixture:
$$M = \sum_{i=1}^{j} y_i M_i = \left[\sum_{i=1}^{j} \frac{mf_i}{M_i}\right]^{-1} \tag{9.6}$$

Specific Heats of the Mixture:
$$c_p = \sum_{i=1}^{j} mf_i c_{p,i} \qquad c_v = \sum_{i=1}^{j} mf_i c_{v,i} \tag{9.23a}$$
$$\bar{c}_p = \sum_{i=1}^{j} y_i \bar{c}_{p,i} \qquad \bar{c}_v = \sum_{i=1}^{j} y_i \bar{c}_{v,i} \tag{9.23b}$$

PROBLEMS

9.1 A gas mixture consists of 2 kg H_2, 7 kg CO_2, and 5 kg N_2. Determine the mass and mole fractions of each component, the molecular mass of the mixture, and the gas-specific ideal gas constant for this mixture.

9.2 A gas mixture consists of 5 g O_2, 3 g CO, 1 g CH_4, and 2 g CO_2. Determine the mass and mole fractions of each component, the molecular mass of the mixture, and the gas-specific ideal gas constant for this mixture.

9.3 A gas mixture consists of 2 kmoles N_2, 3 kmoles C_3H_8, and 1 kmole H_2. Determine the mass and mole fractions of each component, the molecular mass of the mixture, and the gas-specific ideal gas constant for this mixture.

9.4 A gas mixture consists of 5 kmoles O_2, 2 kmoles CH_4, and 6 kmoles CO_2. Determine the mass and mole fractions of each component, the molecular mass of the mixture, and the gas-specific ideal gas constant for this mixture.

9.5 A manufacturing process must be performed in an environment consisting of the following ideal gas mixture: 0.25 m^3 He, 0.35 m^3 N_2, and 0.10 m^3 Ar. The pressure of the mixture is 101 kPa, and the temperature of the mixture is 298 K. Determine the mass and mole fractions of each component, the molecular mass of the mixture, and the gas-specific ideal gas constant for this mixture.

9.6 A gas mixture consists of 2 kg N_2 and a variable amount of H_2. Plot the mass and mole fractions of each component of the mixture, the molecular mass of the mixture, and the gas-specific ideal gas constant for this mixture for masses of H_2 ranging between 0.1 kg and 5 kg.

9.7 A gas mixture consists of 5 kmoles O_2 and a variable amount of CH_4. Plot the mass and mole fractions of each component of the mixture, the molecular mass of the mixture, and the gas-specific ideal gas constant for this mixture for molar amounts of CH_4 ranging between 1 kmole and 20 kmoles.

9.8 A gas mixture consists of 1.5 kg of He, 2.0 kg of Ar, and a variable amount of Ne. Plot the mass and mole fractions of each component of the mixture and the molecular mass of the mixture for masses of Ne ranging between 0 kg and 10 kg.

9.9 A mass-based analysis of a gas mixture finds that the mixture is 15% He, 25% O_2, and 60% of an unknown gas. It is determined that the molecular mass of the mixture is 12.08 kg/kmole. Determine the molecular mass of the unknown gas and propose a reasonable possibility for the identity of the unknown gas.

9.10 A gravimetric (mass-based) analysis of a gas mixture yields that the mixture has 25% N_2, 60% CO_2, and 15% of an unknown gas. It is determined that the gas-specific ideal gas constant of the mixture is 0.227 kJ/kg · K. Determine the molecular mass of the unknown gas, and propose a reasonable possibility for the identity of the unknown gas.

9.11 A molar analysis of a gas mixture yields that the mixture is 30% O_2, 45% C_3H_8, and 25% of an unknown gas. It is determined that the gas-specific ideal gas constant of the mixture is 0.249 kJ/kg · K. Determine the molecular mass of the unknown gas, and propose a reasonable possibility for the identity of the unknown gas.

9.12 A chemical analysis of a gas mixture determines that it is composed of only He and Ar. The gas-specific ideal gas constant for the mixture is determined to be 0.285 kJ/kg · K. Determine the mole fractions and the mass fractions of He and Ar present in the mixture.

9.13 You are handed a mixture of CH_4 and air, and you must determine if the mixture is flammable. You learn that the mixture is flammable if the ratio of the mass of air to the mass of fuel is between 10.5 and 37.4. You perform an experiment on the mixture, and you determine that the gas-specific ideal gas constant for the mixture is 0.295 kJ/kg · K. Is the mixture flammable?

9.14 Repeat Problem 9.13, but you are given two other mixtures of CH_4 and air, and you determine that these mixtures have gas-specific ideal gas constants of 0.354 kJ/kg · K and 0.495 kJ/kg · K, respectively. Are either or both of these mixtures flammable?

9.15 Consider a room in which there is a leak of carbon monoxide (CO). While the impacts on an individual human will vary, suppose that someone exposed to 800 ppm (on a volumetric basis, 0.08%) will die within 3 hours. Initially the room is empty of CO and contains 21 kg of air. The CO leak adds 0.1 kg of CO to the room each hour. Ignore any air or CO exiting the room. How long will the leak need to go undetected for the molar CO concentration to reach 800 ppm?

9.16 Rework Problem 9.15, but consider that air (with no CO) exits the room at the same rate at which the CO is added: 0.1 kg air/h.

9.17 An ideal gas mixture has, on a molar basis, a composition of 25% N_2, 5% H_2, 30% Ar, and 40% CO_2. The total pressure of the mixture is 250 kPa. Determine the partial pressure of each component.

9.18 An ideal gas mixture has, on a mass basis, a composition of 50% N_2, 40% O_2, and 10% CH_4. The total pressure of the mixture is 450 kPa absolute. Determine the partial pressure of each component of the mixture.

9.19 An ideal gas mixture consists of 40% Ar, 40% O_2, and 20% Ne, on a molar basis. The total volume occupied by the mixture is 0.75 m^3. Determine the partial volume of each component of the mixture.

9.20 An ideal gas mixture consists of 70% N_2, 25% H_2, and 5% CO_2 on a mass basis. The total volume occupied by the mixture is 1.25 m^3. Determine the partial volume of each component of the mixture.

9.21 A container with a volume of 0.2 m^3 holds an ideal gas mixture consisting of 20% Ne, 40% CO_2, and 40% N_2 on a mass basis. Determine the partial volume of each component of the mixture.

9.22 An ideal gas mixture is held in a rigid container. Initially, the partial pressures of the components of the mixture are 8.5 kPa N_2, 6.2 kPa CO_2, and 1.5 kPa Ar. The mixture temperature is initially 15°C. The mixture is heated until the temperature is 80°C. Determine the partial pressures of the N_2, CO_2, and Ar after heating.

9.23 A container is divided into three sections by partitions. Section A of the container holds 0.25 m^3 of N_2, Section B contains 0.35 m^3 of He, and Section C contains 0.10 m^3 of CO_2. The pressure and temperature of each gas are 125 kPa and 300 K, respectively. The partitions are removed, and the three gases are allowed to mix. Assuming the gases all behave as ideal gases, determine the molecular mass of the final mixture and the constant-pressure specific heat for the final mixture.

9.24 An initially evacuated container is to be filled with three ideal gases. The volume of the container is 0.25 m^3. After each gas is added, a pressure measurement is made of the system. First, O_2 is added until the absolute pressure inside the container is 125 kPa. Next, Ar is added until the pressure inside the container is 350 kPa. Finally, CH_4 is added until the pressure inside the container is 380 kPa. The final temperature of the mixture is 150°C, and assume that each gas is added at 150°C. For the final mixture, determine the partial pressure of each component and the total mass of the mixture in the container.

9.25 An ideal gas mixture consists of 1.5 kg of N_2, 2.5 kg of O_2, and 0.15 kg of CO_2. Determine the molecular mass of the mixture, and both the constant-pressure and constant-volume specific heats for the mixture at 300 K.

9.26 A gas mixture consisting of 2.5 kg of CO_2, 0.5 kg of N_2, and 0.75 kg of H_2 is to be sent through a compressor. Assuming the mixture is ideal, determine the molecular mass of the mixture, and both the constant-pressure and constant-volume specific heats for the mixture at 300 K.

9.27 0.15 kmole of O_2 is mixed with 0.05 kmole of CH_4 and 0.20 kmole of CO_2. Determine the gas-specific ideal gas constant and the specific heat ratio (k) of the mixture.

9.28 An ideal gas mixture consists of 0.25 kmole H_2, 0.50 kmole CH_4, and 1.5 kmole Ne. Determine the molecular mass of the mixture and the mass-based constant-pressure specific heat of the mixture at 300 K.

9.29 By mass, an ideal gas mixture consists of 25% N_2, 60% CO_2, and 15% O_2. Determine the mass-based constant-pressure specific heat and constant-pressure specific heat at 300 K.

9.30 By volume, an ideal gas mixture consists of 15% N_2, 25% CO_2, 40% Ar, and 20% CH_4. Determine both the constant-pressure and constant-volume specific heats for the mixture, in both mass-based and mole-based units at 300 K.

9.31 An ideal gas mixture consists of N_2 and CO_2. The mixture temperature is 300 K. Plot the mass-based constant-pressure specific heat, constant-volume specific heat, specific heat ratio, and gas-specific ideal gas constant as the mass fraction of N_2 varies between 0 and 1.0.

9.32 An ideal gas mixture consists of CH_4 and O_2. The mixture temperature is 300 K. Plot the molar constant-pressure specific heat, molar constant-volume specific heat, and specific heat ratio as the mole fraction of CH_4 varies between 0 and 1.0.

9.33 An ideal gas mixture initially consists of 1.0 kg of N_2 and 1.0 kg of H_2. O_2 is then added to the mixture. The mixture temperature is maintained at 300 K. Plot the mass fraction of N_2 and the mass fraction of O_2 in the mixture, the constant-pressure specific heat of the mixture, the gas-specific ideal gas constant for the mixture, and the specific heat ratio as the mass of O_2 added to the mixture varies from 0 kg to 10 kg.

9.34 An ideal gas mixture consists of 0.25 kg of CO_2 and 0.10 kg of H_2. N_2 gas is then added to this mixture while the mixture temperature is maintained at 298 K. Plot the constant-volume specific heat and the specific heat ratio of the mixture as a function of the N_2 added to the mixture for N_2 amounts ranging from 0 kg to 1 kg.

9.35 Plot the constant-pressure specific heat for an ideal gas mixture of N_2 and Ar at 300 K, as a function of the N_2 mass fraction as it varies between 0.0 and 1.0. Determine the mixture composition for measured c_p values of (a) 0.98 kJ/kg · K, (b) 0.73 kJ/kg · K, and (c) 0.59 kJ/kg · K.

9.36 You ask a laboratory technician to prepare an ideal gas mixture consisting of N_2, O_2, and H_2. In order to be used in a process, the mass fraction of O_2 cannot exceed 30%. The technician gives you the mixture and informs you that he added O_2 and H_2 to a container that already held 0.5 kg of N_2. However, he did not keep track of how much O_2 and H_2 he added. You determine that the mass of the final mixture is 1.75 kg, and through experimentation you determine that the constant-volume specific heat of the mixture is 2.25 kJ/kg · K. What are the mass fractions of each gas in the final mixture, and is the mixture safe to use in the process?

9.37 For the situation in Problem 9.36, you ask the technician to make a new N_2, O_2, and H_2 mixture. He again is somewhat careless in his technique, and he brings you a new 1.75-kg mixture (having started with 0.5 kg of N_2). You measure the constant-volume specific heat of this mixture to be 5.75 kJ/kg · K. Is this mixture safe for use in the process?

9.38 Referring to Problems 9.36 and 9.37, you again try to have the technician make the appropriate mixture. This time, he starts with 0.50 kg of O_2 and adds N_2 and H_2, again producing a 1.75-kg mixture without keeping careful records of what amounts were added. You measure the constant-volume specific heat of the mixture at 300 K to be 2.90 kJ/kg · K. Is this mixture safe for use in the process?

9.39 You are given an ideal gas mixture consisting of N_2, O_2, and CO_2 in an unknown combination. You measure the constant-pressure specific heat of the mixture to be 0.942 kJ/kg · K, and the constant-volume specific heat of the mixture to be 0.699 kJ/kg · K at 25°C. Determine the composition of the mixture in terms of mass fractions.

9.40 An ideal gas mixture consists of three unknown gases. A label on the container states "This contains a flammable mixture diluted by a noble gas." You are able to perform a mass analysis by separating each component, and you determine that gas A has a mass fraction of 27%, gas B has a mass fraction of 61%, and gas C has a mass fraction of 12%. You are also able to determine that the mixture has $c_p = 0.971$ kJ/kg · K and $c_v = 0.694$ kJ/kg · K at 300 K. Propose likely gases for the three unknown components, and determine the partial pressure of each component if the total mixture pressure is 150 kPa.

9.41 An ideal gas mixture consisting of 1.5 kg N_2 and 0.60 kg Ar is to be compressed in a piston–cylinder assembly, following the relationship $Pv^{1.3}$ = constant. Initially, the mixture pressure is 100 kPa and the mixture temperature is 25°C. After compression, the mixture pressure and temperature are 2000 kPa and 300°C, respectively. Determine the work and heat transfer for the process.

9.42 An ideal gas mixture consisting of 1.1 kg CH_4, 2.0 kg N_2, and 1.5 kg CO_2 is placed in a rigid container whose volume is 0.5 m³. The initial temperature of the mixture is 10°C. 250 kJ of heat is added to the mixture. Determine the final temperature and pressure of the mixture.

9.43 A balloon contains 0.5 mole of O_2 and 0.25 mole of CO_2 at a pressure of 140 kPa absolute and a temperature of 15°C. Heat is added to the ideal gas mixture, at constant pressure, until the temperature is 40°C. Determine the amount of heat added to the mixture and the change in volume for the process.

9.44 A balloon contains 0.25 kg of He, 0.10 kg of N_2, and 0.05 kg of Ar at a pressure of 200 kPa and a temperature of 50°C. Heat is removed from the ideal gas mixture, at constant pressure, until the temperature is 30°C. Determine the amount of heat removed from the mixture and the change in volume for the process.

9.45 A rigid container with a volume of 50 L holds an ideal gas mixture consisting of 20% N_2, 5% H_2, and 75% CO_2 (by mass) initially at 10°C and 350 kPa absolute. The container is placed into a large vat of ice water (at 0°C), and heat is transferred from the gas mixture until its temperature is 0°C. Determine the amount of heat transfer for the process and the amount of entropy generation for the process.

9.46 2 kg of an ideal gas mixture consisting of 50% N_2 and 50% O_2 (by volume) is placed inside an insulated piston–cylinder device. Initially the mixture has a volume of 0.45 m³ and a temperature of 300°C. The mixture expands until the volume is 0.65 m³. The surface temperature is 25°C. Determine the work done and entropy generation during the process and the final pressure and temperature of the mixture (a) if the expansion is isentropic, (b) if the expansion follows the relationship $Pv^{1.8}$ = constant, and (c) if the expansion is at constant pressure.

9.47 A rigid container with a volume of 0.25 m³ contains an ideal gas mixture consisting of He and CO_2. Initially the mixture temperature and pressure are 300 K and 150 kPa, and after heating the mixture temperature is 500 K. Plot the heat transfer required for the process and the final mixture pressure for mixture compositions ranging from pure He to pure CO_2 as functions of (a) mass fraction of He, and (b) mole fraction of He.

9.48 A piston–cylinder device contains 0.75 kg of an ideal gas mixture consisting of Ne and O_2 gases. The pressure of the mixture is initially 700 kPa absolute, and the initial temperature is 25°C. Heat is added to the mixture, causing it to double in volume at constant pressure. Plot the heat transfer for the process, the work done in the process, and the final temperature of the mixture as the composition of the mixture varies from pure Ne to pure O_2 as a function of (a) O_2 mass fraction, and (b) O_2 mole fraction.

9.49 A piston–cylinder device contains 0.25 kmole of an ideal gas mixture which consists of CH_4 and N_2 gases. The initial pressure and temperature of the mixture are 1200 kPa and 600 K. The mixture is cooled at constant pressure, which causes it to have a final volume that is 60% of its initial volume. Plot the heat transfer for the process, the work done in the process, and the final temperature of the mixture as the composition of the mixture varies from pure CH_4 to pure N_2 as a function of (a) N_2 mass fraction, and (b) N_2 mole fraction.

9.50 An ideal gas mixture in an adiabatic piston–cylinder assembly is to undergo an isentropic expansion from a volume of 0.30 m^3 to a volume of 0.50 m^3. The mixture is to consist of H_2 and Ar, and initially the pressure is 1000 kPa and the temperature is 200°C. Plot the work done for the process, the final pressure, and the final temperature after the process as a function of the mass fraction of H_2 as the mass fraction of H_2 varies between 0 and 1.

9.51 An adiabatic piston–cylinder assembly contains a mixture of He and N_2. The mixture initially has a volume of 40 L at 500 kPa absolute and a temperature of 25°C, and then undergoes an isentropic compression to a pressure of 1.75 MPa. Plot the work done for the process, the change in volume during the process, and the final temperature after the process as a function of the mass fraction of N_2 as the mass fraction of N_2 varies between 0 and 1.

9.52 An ideal gas mixture consisting of 25% N_2, 20% O_2, and 55% CO_2 (by mass) enters an insulated turbine at a mass flow rate of 5.5 kg/s, a pressure of 2.0 MPa, and a temperature of 700°C. The gases exit the turbine at a pressure of 100 kPa. Assuming constant specific heats for the mixture, determine the exit temperature and the power produced if (a) the process is isentropic, and (b) the isentropic efficiency of the turbine is 0.80.

9.53 An ideal gas mixture consisting of 30% CO_2, 50% N_2, and 20% Ne (by mass) enters an insulated compressor at a steady volumetric flow rate of 0.25 m^3/s, a pressure of 100 kPa, and a temperature of 20°C. The gases exit the compressor at 800 kPa. Assuming constant specific heats for the mixture, determine the exit temperature and the power consumed if (a) the process is isentropic, and (b) the isentropic efficiency of the compressor is 0.75.

9.54 An ideal gas mixture consisting of 25% CH_4, 40% N_2, and 35% Ar (by mass) enters an insulated turbine at a pressure of 1.1 MPa, a temperature of 500°C, and a volumetric flow rate of 275 L/s. The mixture exits the turbine at 150 kPa absolute. Assuming constant specific heats for the mixture, determine the power produced and the exit temperature if (a) the process is isentropic, and (b) the isentropic efficiency of the turbine is 0.77.

9.55 A turbine with an isentropic efficiency of 0.82 receives an ideal gas mixture consisting of 50% CO_2, 25% N_2, and 25% O_2 (by volume). The mixture enters with a volumetric flow rate of 7.5 m^3/s, a pressure of 1.5 MPa, and a temperature of 1000°C. The mixture exits at a pressure of 125 kPa. Assuming that the mixture has constant specific heats, determine the exit temperature, the power produced, and the entropy generation rate if (a) the turbine is insulated, and (b) there is a heat loss of 500 kW from the turbine, whose surface temperature is maintained at 320 K.

9.56 An ideal gas mixture consisting of H_2 and Ar enters an insulated turbine at 900 kPa and 800°C. The mixture exits the turbine at 200 kPa. The isentropic efficiency of the turbine is 0.78. The power produced by the turbine is 1.20 MW. Plot the exit temperature and the necessary mass flow rate of the mixture for (a) H_2 mass fractions varying between 0 and 1, and (b) H_2 mole fractions varying between 0 and 1. Assume that the ideal gas mixture has constant specific heats.

9.57 A CH_4–N_2 ideal gas mixture enters an insulated turbine at 1.25 MPa absolute and 700°C and exits at 125 kPa absolute. The isentropic efficiency of the turbine is 0.85. The turbine receives 650 L/s of gas through its intake. Plot the exit temperature and the power produced for (a) CH_4 mass fractions varying between 0 and 1, and (b) CH_4 mole fractions varying between 0 and 1, assuming that the ideal gas mixture has constant specific heats.

9.58 An insulated gas compressor receives an ideal gas mixture consisting of 15% H_2, 40% N_2, and 45% CO_2, by mass. The gas mixture enters at a temperature of 17°C and a pressure of 100 kPa, and exits at a pressure of 1200 kPa. The mass flow rate of the mixture is 2.5 kg/s. Assuming that the gas mixture has constant specific heats, determine the exit temperature and power consumed if (a) the compressor is isentropic, and (b) the compressor has an isentropic efficiency of 0.78.

9.59 An ideal gas mixture consisting of 25% N_2, 50% O_2, and 25% CO, by volume, enters an insulated gas compressor at a temperature of 15°C and a pressure of 100 kPa absolute. The mass flow rate of the mixture is 2 kg/s. The mixture exits the compressor at 900 kPa absolute. Determine the exit temperature and the power consumed if (a) the compressor is isentropic, and (b) the compressor has an isentropic efficiency of 0.80.

9.60 An uninsulated compressor used 1500 kW of power to compress 4.0 kg/s of an ideal gas mixture from 100 kPa to 1000 kPa, with the mixture entering at 300 K. The mixture consists of 30% CH_4, 60% CO, and 10% Ar, by volume. The surface of the compressor is maintained at 310 K. The isentropic efficiency of the compressor is 82%. Determine the exit temperature, the rate of heat transfer, and the entropy generation rate for the compressor.

9.61 An ideal gas mixture of CH_4 and O_2 enters an insulated compressor at 140 L/s, 20°C, and and 100 kPa absolute. The mixture exits the compressor at 850 kPa absolute. The isentropic efficiency of the compressor is 0.82. Assuming the mixture has constant specific heats, plot the exit temperature and power consumed for (a) O_2 mass fractions ranging from 0 to 1, and (b) O_2 mole fractions ranging from 0 to 1.

9.62 An H_2–Ar ideal gas mixture enters an insulated compressor at 25°C, 100 kPa, and 1.5 m^3/s. The mixture exits the compressor, whose isentropic efficiency is 0.75, at 1100 kPa. Assuming the mixture has constant specific heats, plot the exit temperature and power consumed for (a) H_2 mass fractions ranging from 0 to 1, and (b) H_2 mole fractions ranging from 0 to 1.

9.63 A mixture of 15% H_2 and 85% O_2, by mass, enters a nozzle at 20°C, 100 kPa, and a velocity of 5 m/s. The cross-sectional area of the nozzle inlet is 0.10 m^2. The mixture exits the nozzle at 250 m/s, a pressure of 92 kPa, and a temperature of 18°C. The surface of the nozzle is maintained at 20°C. Assuming that the mixture has constant specific heats, determine the rate of heat transfer and the rate of entropy generation for the process.

9.64 An insulated nozzle is used to accelerate an ideal gas mixture from 3 m/s to 100 m/s. The mass flow rate of the mixture is 0.15 kg/s. The mixture consists of 30% Ar and 70% CH_4, by mass. The entrance temperature and pressure are 15°C and 300 kPa, respectively. The mixture exits at 200 kPa. Assuming the mixture has constant specific heats, determine the exit temperature of the mixture and the rate of entropy generation for the process.

9.65 An insulated nozzle receives 1.5 kg/s of an ideal gas mixture of 40% CO_2, 40% O_2, and 20% N_2, by volume. The mixture enters at 2 m/s and 20°C and exits at 175 m/s. Determine the exit temperature of the mixture, assuming the mixture has constant specific heats.

9.66 An insulated diffuser receives 35 L/s of an ideal gas mixture of 60% N_2, 30% H_2, and 10% O_2, by volume, at 130 m/s and 7°C. The mixture exits the diffuser at 10°C. Determine the exit velocity of the mixture, assuming that the mixture has constant specific heats.

9.67 An insulated heat exchanger is used to condense water by transferring heat to an ideal gas mixture. The water enters at 150 kPa with a quality of 0.85, and exits as a saturated liquid at 150 kPa. The mass flow rate of the water is 15 kg/s. The ideal gas mixture enters at 20°C and exits at 110°C. Determine the mass flow rate of the ideal gas mixture for mixture compositions of (a) 30% N_2 and 70% CO_2 (by volume), (b) 30% N_2 and 70% O_2 (by volume), and (c) 30% N_2 and 70% H_2 (by volume).

9.68 An insulated heat exchanger is used to cool oil (whose specific heat is 1.8 kJ/kg · K) from 140°C to 40°C, at constant pressure, by transferring heat to an ideal gas mixture. The mass flow rate of the oil is 2 kg/s. The ideal gas mixture enters at 15°C and exits at 115°C, and flows at a constant pressure of 200 kPa absolute. Determine the mass flow rate of the mixture, the inlet volumetric flow rate of the mixture, and the rate of entropy generation for the process for mixture compositions of (a) 50% N_2 and 50% H_2 (by volume), (b) 50% N_2 and 50% CO (by volume), and (c) 50% N_2 and 50% CO_2 (by volume).

9.69 An insulated heat exchanger is used to cool water from 95°C to 30°C, at a pressure of 100 kPa, by transferring heat to an ideal gas mixture that consists of N_2 and CO_2. The mass flow rate of the water is 10 kg/s. The ideal gas mixture enters at 10°C and exits at 70°C, flowing at a constant pressure of 250 kPa. Plot the mass flow rate of the mixture and the inlet volumetric flow rate of the mixture for (a) N_2 mass fractions ranging from 0 to 1, and (b) N_2 mole fractions ranging from 0 to 1.

9.70 An adiabatic mixing chamber receives 2.5 kg/s of N_2 at 25°C and 110 kPa through one inlet and 4.0 kg/s of CO_2 at 70°C and 110 kPa through a second inlet. The combined N_2–CO_2 stream exits at 105 kPa. Assuming that the specific heats of the gases are constant, determine the exit temperature and rate of entropy generation for the process.

9.71 An adiabatic mixing chamber is used to combine three ideal gases into one outlet stream. 0.25 kg/s of N_2 enters at 10°C and 150 kPa through one inlet, 0.15 kg/s of CO enters at 30°C and 150 kPa through a second inlet, and 0.30 kg/s of CO_2 enters at 50°C and 150 kPa through a third inlet. The combined stream exits at 140 kPa. Assuming that the specific heats of the gases are constant, determine the exit temperature and rate of entropy generation for the process.

9.72 An uninsulated mixing chamber, whose surface temperature is 20°C, receives 40 L/s of O_2 at 40°C and 175 kPa through one inlet, 15 L/s of C_3H_8 at 25°C and 175 kPa through a second inlet, and 35 L/s of Ar at 42°C and 175 kPa through a third inlet. The combined stream exits at 30°C and 175 kPa. Assuming constant specific heats, determine the rate of heat transfer and the rate of entropy generation for the process.

9.73 An uninsulated separation machine receives 1.5 m^3/s of a mixture of He and O_2 at 40°C and 100 kPa. The mixture is initially 30% He and 70% O_2 by volume. The machine separates the two gases into separate streams and uses 10 kW of power in the process. Determine the heat transfer rate for the machine if (a) the He and O_2 both exit at 30°C, (b) the He exits at 20°C and the O_2 exits at 40°C, and (c) the He exits at 40°C and the O_2 exits at 20°C. Solve part (a) by treating the whole process as an ideal gas mixture and by treating the gases as two separate streams throughout.

9.74 An ideal gas mixture consisting of 30% N_2, 50% O_2, and 20% CO_2 (by volume) enters an insulated turbine at 1350 K and 1.50 MPa. The mixture exits at 850 K. The mass flow rate of the mixture is 7.5 kg/s. Determine the power produced (a) assuming constant specific heats for the gases, evaluated at 300 K, (b) assuming constant specific heats for the gases, evaluated at 1100 K, and (c) considering the specific heats to be variable.

9.75 A gas compressor receives an ideal gas mixture of 60% N_2, 30% O_2, and 10% CO (by volume) at a rate of 100 L/s. The mixture enters at 20°C and 100 kPa and exits at 120°C, experiencing a heat loss of 70 kJ/kg during the process. Determine the power consumed by the compressor (a) assuming constant specific heats for the gases, and (b) considering the specific heats to be variable.

9.76 An insulated gas compressor inducts an ideal gas mixture consisting of 40% N_2 and 60% CO_2 (by mass) at a rate of 1.5 m^3/s. The mixture enters at 298 K and 101 kPa, and exits at 550 K. Determine the power consumed by the compressor (a) assuming constant specific heats for the gases, and (b) considering the specific heats to be variable.

9.77 An insulated turbine is to produce 50 MW of power using an ideal gas mixture of N_2 and CO_2. The mixture enters at 1000 K and 1200 kPa and exits at 700 K. Considering the specific heats to be variable, plot the mixture mass flow rate and inlet volumetric flow rate required as (a) the mass fraction of N_2 varies between 0 and 1, and (b) the mole fraction of N_2 varies between 0 and 1.

9.78 An adiabatic gas compressor receives 2.0 m³/s of an ideal gas mixture at 300 K and 100 kPa. The mixture consists of O_2 and CO_2, and it exits the compressor at 610 K. Considering the specific heats to be variable, plot the power consumed by the compressor as (a) the mass fraction of O_2 varies from 0 to 1, and (b) the mole fraction of O_2 varies between 0 and 1.

9.79 An ideal gas mixture consisting of 25% N_2, 30% O_2, and 45% CO_2 (by mass) flows through an insulated turbine, entering at 1100 K and exiting at 750 K. The mass flow rate of the mixture is 15.0 kg/s. Assume the gases have constant specific heats. Determine the power produced by the turbine by (a) treating the mixture as a mixture through the whole solution, and (b) treating each gas separately through the turbine and adding the results.

9.80 An ideal gas mixture consisting of 10% H_2, 70% Ar, and 20% O_2 (by volume) enters an adiabatic turbine at 666 K and exits at 390 K. The mass flow rate of the mixture is 11 kg/s. Assume the gases have constant specific heats. Determine the power produced by the turbine by (a) treating the mixture as a mixture through the whole solution, and (b) treating each gas separately through the turbine and adding the results.

9.81 An insulated compressor is used to increase the pressure of an ideal gas mixture. The mixture enters at 300 K and 100 kPa, and exits at 700 K and 800 kPa. The mixture consists of, by volume, 50% H_2, 10% N_2, 15% O_2, and 25% CO_2. The mass flow rate of the mixture is 12.0 kg/s. Assume the gases have constant specific heats. Determine the power consumed by the compressor and the entropy generation rate for the process by (a) treating the mixture as a mixture through the whole solution, and (b) treating each gas separately through the turbine and adding the results.

9.82 An insulated compressor is used to pressurize an ideal gas mixture, which consists of 35% N_2 and 65% O_2, by mass. The mixture enters at 0.75 m³/s, 300 K, and 100 kPa, and exits at 750 K and 900 kPa. Assume the gases have constant specific heats. Determine the power consumed by the compressor and the entropy generation rate for the process by (a) treating the mixture as a mixture through the whole solution, and (b) treating each gas separately through the turbine and adding the results.

9.83 A gas mixture is comprised of 35% O_2, 10% CH_4, and 55% CO, by volume. The mixture temperature is 250 K and the mixture pressure is 325 kPa. Determine the molar-specific volume of the mixture by (a) using the ideal gas model, (b) using the van der Waals equation, and (c) using the compressibility factor following Amagat's model.

9.84 A gas mixture is comprised of 25% N_2, 50% CH_4, and 25% CO_2, by volume. The mixture temperature is 320 K and the mixture pressure is 175 kPa. Determine the molar-specific volume of the mixture by (a) using the ideal gas model, (b) using the van der Waals equation, and (c) using the compressibility factor following Amagat's model.

9.85 A gas mixture consists of 40% H_2 and 60% N_2, by volume. Determine the molar-specific volume of the mixture using both the ideal gas model and the compressibility factor for the following conditions: (a) T = 200 K, P = 15 MPa; (b) T = 200 K, P = 150 kPa; (c) T = 1000 K, P = 15 MPa; and (d) T = 1000 K, P = 150 kPa.

9.86 A gas mixture consists of 30% CH_4, 10% CO, and 60% CO_2, by volume. Determine the molar-specific volume of the mixture using both the ideal gas model and the compressibility factor for the following conditions: (a) $T = 275$ K, $P = 10$ MPa; (b) $T = 1100$ K, $P = 10$ MPa; (c) $T = 275$ K, $P = 50$ kPa; and (d) $T = 1100$ K, $P = 50$ kPa.

9.87 A gas mixture consists of 50% N_2 and 50% CO_2, by mass. The mixture is in a container whose volume is 0.5 m^3. The mixture temperature is 300 K. Using both the ideal gas model and the van der Waals equation, plot the pressure of the mixture as the mass varies between 0.25 kg and 50 kg.

9.88 A gas mixture consists of 20% N_2, 30% O_2, and 50% CH_4, by volume. The mixture is in a container of volume 75 liters and is at a temperature of 300 K. Using both the ideal gas model and the van der Waals equation, plot the mass of the mixture in the container as the pressure varies between 25 kPa absolute and 7 MPa.

DESIGN/OPEN-ENDED PROBLEMS

9.89 You wish to design a system to produce compressed gas; the gas will be used in a manufacturing process. The process requires that the compressed gas has no oxygen in it, either in the form of oxygen molecules or in the form of oxygen atoms in other compounds. In addition, the process requires that the concentration of CH_4 be at least 20% but no more than 40% (by volume). The compressor system must produce 0.25 kg/s of gas for the process at a temperature of 50°C and a pressure of 700 kPa. The isentropic efficiency of the compressor is 0.75. Assume that the gases used in the compressor enter at 25°C and 100 kPa. Choose a mixture of gases that meets these requirements, and design a compressor and cooling system to meet the process needs. Analyze the current cost of operating such a system.

9.90 An ideal gas mixture consists of 60% N_2, 20% CO_2, 10% CO, and 10% O_2, by volume. A mixture flow rate of 0.5 m^3/s must be cooled from 150°C and 250 kPa to 50°C and 250 kPa by passing it through a heat exchanger. The coolant may not exceed 40°C in temperature, and it must enter the heat exchanger at a temperature no cooler than 10°C. Choose a coolant, and specify the flow rate and inlet and outlet conditions of the coolant passing through the heat exchanger.

9.91 Design a flow control system for the reactants of a combustion process. In the system, the amounts of N_2 and O_2 are controlled separately and combined with the fuel in the mixing chamber. The system must be able to deliver a reactant mixture at rates between 3 L/s and 30 L/s, at a pressure of 350 kPa absolute. The volumetric ratio between the N_2 and O_2 must be able to be varied between 2 and 3.76. The fuel to be used is CH_4, and the volumetric ratio between the CH_4 and the O_2 must be able to vary between 0.4 and 0.7. The gases originally come from separate tanks at a pressure of 1.750 MPa. Specify the flow equipment to be used, describe how it will be controlled, and identify any regulators needed and the piping requirements.

9.92 A particular factory has available copious supplies of N_2, CO_2, and Ar gas. A certain process in the factory requires water to be cooled from 75°C to 30°C. The plant needs approximately 10 kg/s of water cooled for the process. The plant manager decides to use the available gases in a heat exchanger to cool the water. The gases are all available for use at 0°C and 125 kPa. Design a gas mixture and identify the flow rates and temperatures required in the heat exchanger.

9.93 Design a closed-loop reversed Brayton cycle heat pump that will deliver 20 kW of heat to a warm space. The heat must be delivered to a large space where the temperature is to be no cooler than 20°C, and the heat source in the cold-temperature heat exchanger is the outside air at 0°C. You are to use an ideal gas mixture consisting of N_2, Ne, and CO_2. The isentropic efficiency of both the turbine and the compressor is 0.80. Include in your design the composition of the mixture, the temperatures and pressures throughout the cycle, and the necessary mass flow rate of the mixture.

9.94 Discuss some of the design considerations to be used when choosing the composition of a gas mixture for a process. Choose two or three potential uses for a gas mixture of unspecified composition, and discuss what properties of the gases should be included in the design process.

CHAPTER 10

Psychrometrics: The Study of "Atmospheric Air"

Learning Objectives

Upon completion of Chapter 10, you will be able to

10.1 Describe the terminology surrounding mixtures of dry air and water vapor;

10.2 Explain the importance of such mixtures in engineering;

10.3 Recognize some of the methods used to determine the humidity of moist air; and

10.4 Apply the principles of thermodynamics to the analysis of a variety of psychrometric applications.

10.1 INTRODUCTION

Our comfort level is directly impacted by a fluid that surrounds us constantly—the air that is present in the atmosphere. A large fraction of the energy used in many homes and businesses is consumed in processes designed to alter the condition of the air, typically for the purpose of making people comfortable. For example, approximately 51% of the energy used in residential units in the United States in 2015 was used for heating or cooling the air in buildings. The air that surrounds us is a mixture of "dry air" (which is a mixture of N_2, O_2, Ar, CO_2, etc., and is what we have been referring to in previous chapters as "air") and water vapor. The components of dry air are essentially constant with time and most locations, although the percentages of each component can vary slightly with altitude and time. However, the amount of water vapor present in air can change dramatically in short time periods. For example, as shown in **Figure 10.1**, if a cold front moves through an area on a hot, humid day, the water vapor content in the air at a given location can easily be halved in a matter of minutes as the cooler, drier air moves in. The study of this combination of dry air and water vapor (i.e., the study of "atmospheric air") is known as *psychrometrics*.

In a psychrometric analysis, we will typically take the dry air and water vapor to be an ideal gas mixture. You should recall that water is not treated as an ideal gas. Yet here we are saying that water vapor will be treated as a component in an ideal gas mixture and therefore is to be treated as an ideal gas. Is this acceptable? The answer is yes, because in this situation the water vapor is at a very low concentration. The partial pressure of water vapor in air at atmospheric pressure is usually lower than 4 kPa, and often much lower than that. Water

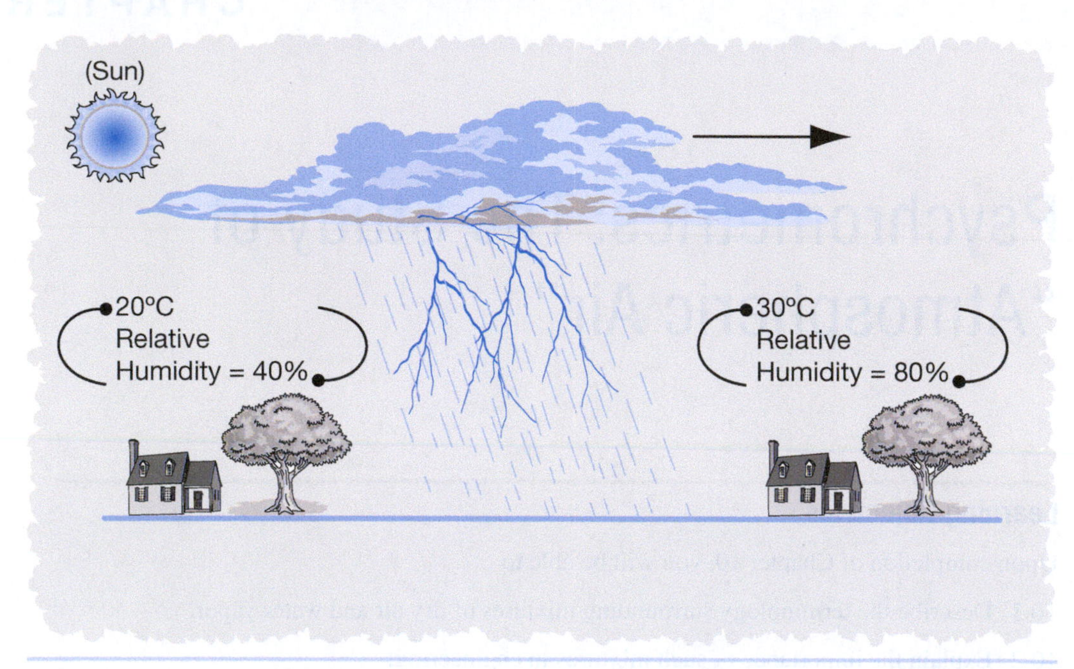

FIGURE 10.1 The water vapor content of air (humidity) can change quickly as a front passes through an area. In this case, a cold front changes hot, humid air to warm, drier air.

vapor, on a volumetric basis, only occupies a few percent of the air's volume. At these low partial pressures, the behavior of the water vapor is close to that of an ideal gas. Furthermore, any deviations from ideal gas behavior result in little impact on the overall mixture properties because the water vapor is such a minor component of air. Any errors introduced in an analysis by assuming that the mixture of dry air and water vapor is an ideal gas mixture are easily handled by designing systems that alter the composition of the mixture; the errors are too small to be of consequence on an engineering basis. Therefore, we will simplify the complexity of the mixture analysis and treat the water vapor as an ideal gas.

Even though water vapor is a minor component of "atmospheric air" (water vapor usually makes up a slightly larger percentage of the air than argon gas), it plays a crucial role in determining our comfort level. Evaporation of perspiration from our skin, as shown in **Figure 10.2**, is an important cooling mechanism for our bodies. In this process, liquid water evaporates from the skin into the air, and this evaporation results in a cooling of the skin. When the partial pressure of the water vapor in the air is high (relatively speaking), the evaporation of water from skin is slower and the rate of evaporative cooling is reduced. On hot, humid days, not only is the rate of evaporation reduced, but the body also needs cooling at a faster rate due to the higher external temperatures. We often feel sweaty, sticky, and uncomfortable at such times because our perspiration is not evaporating quickly. On the other hand, if the air is too "dry," evaporation can occur very quickly, which leads to rapid overcooling of our bodies and causes us to feel too cold. In Section 10.4, we briefly discuss the conditions in which our bodies feel most comfortable.

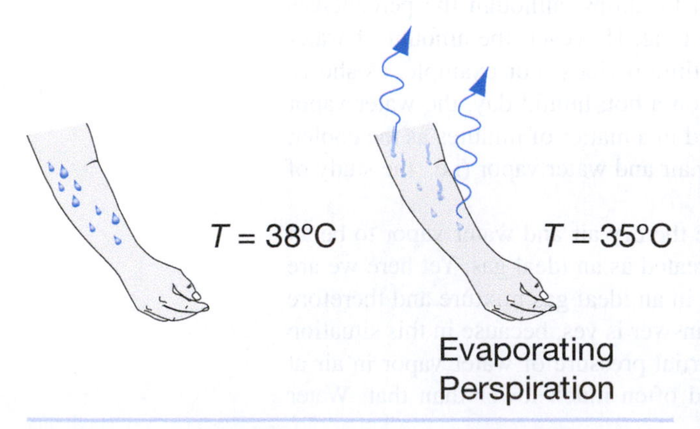

FIGURE 10.2 Evaporative cooling takes place as perspiration evaporates off of skin.

Because the condition of air is so important to human comfort, a branch of engineering known as heating, ventilating, and air conditioning, or HVAC, has developed. HVAC specializes in changing the "condition" of the air via the cooling and dehumidifying (removing water vapor from the air) processes commonly thought of as "air conditioning," but it can also involve heating, humidifying (adding water vapor to the air), and purifying/cleaning the air. Later in this chapter, we will apply psychrometric analysis to some of these processes. We will present calculations for determining heat removal or heat addition rates, as well as determining the mass flow rates of water necessary as water is added or removed from the air. Other common psychrometric processes include mixing streams of air and using cooling towers to transport energy from a stream of one fluid, such as water, to the atmosphere. These processes will also be discussed in this chapter. However, before we consider these applications, we first present various concepts and terms relating to psychrometrics.

10.2 BASIC CONCEPTS AND TERMINOLOGY OF PSYCHROMETRICS

Because we are treating atmospheric air as an ideal gas mixture of dry air and water vapor, the total pressure of the mixture, P, is

$$P = P_{da} + P_v \tag{10.1}$$

where P_{da} is the partial pressure of the dry air component, and P_v is the partial pressure of the water vapor in the mixture. The partial pressure of the water vapor, P_v, must be less than the saturation pressure of water at the air temperature. Recall from Chapter 3, and as shown in **Figure 10.3**, that when the pressure of water (or any substance) is above that of the saturation pressure at the temperature of the water, the water is in liquid form and is considered a compressed liquid. In order for the water to be in vapor form, and in the air, it must have a pressure at or below the saturation pressure.

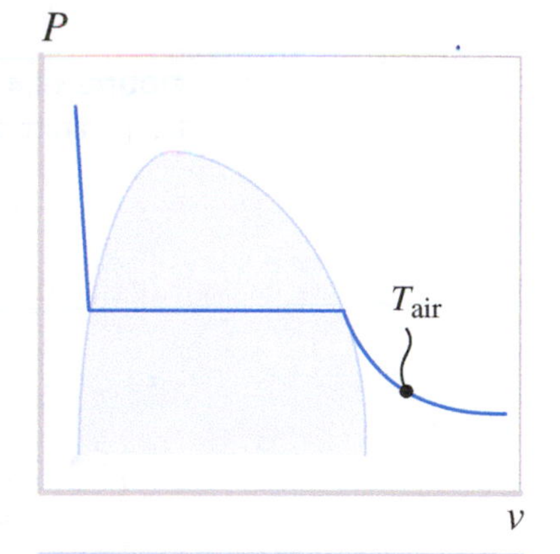

FIGURE 10.3 A line of constant temperature on a *P*-*v* diagram of water around the liquid–vapor phase-change region.

One method of characterizing the amount of water vapor present in the air is by specifying the air's *relative humidity*, ϕ. The relative humidity is defined as the ratio between the partial pressure of the water vapor in the air and the saturation pressure of water at the temperature of the air, $P_{sat}(T)$:

$$\phi = \frac{P_v}{P_{sat}(T)} \tag{10.2}$$

Air that has a relative humidity of 1 (i.e., 100%) is considered to be "saturated air" because the partial pressure of the water vapor is as high as it can be for that temperature ($P_v = P_{sat}(T)$), and the air is holding as much water vapor as possible. Although relative humidity is a popular method of characterizing water vapor content, it has scientific limitations. Knowing the relative humidity without knowing the temperature does not allow us to determine the actual amount of water vapor present in the air. The saturation pressure of water increases as the temperature increases, as shown in **Figure 10.4**, so warmer air has the ability to hold more water vapor. As a result, the same amount of water vapor in the air will have a different relative humidity depending on the temperature. For example, as illustrated in **Figure 10.5**, in cold climates in winter the outdoor air may be saturated (i.e., $\phi_{outdoor} = 100\%$), but the relative humidity of the same mixture of dry air and water vapor will be lower indoors after it has been heated to a comfortable temperature—the indoor relative humidity might be quite low; in fact, around 20 or 30% is common.

Relative humidity is useful for providing information regarding comfort levels and the likelihood of precipitation. Drier air promotes evaporation of sweat more readily than air that is close to saturated, so relative humidity can be used to help gauge our body's comfort level. Likewise, precipitation forms in the atmosphere when the amount of water at a location exceeds the saturation pressure of the water in the air. Although precipitation (rain or snow) can reach the ground if the relative humidity at ground level is less than 100%, it becomes increasingly

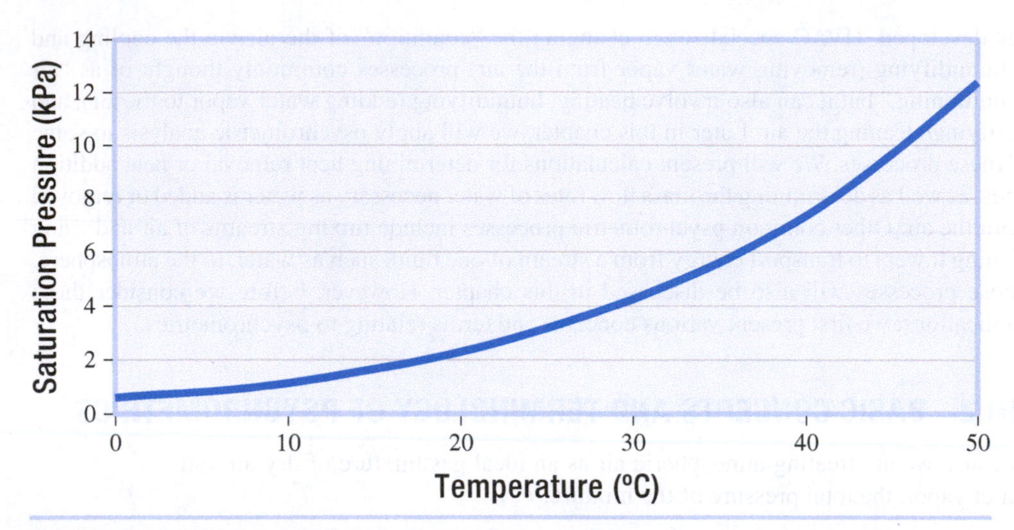

FIGURE 10.4 The saturation pressure of water as a function of temperature. A higher temperature air–water vapor mixture can hold a larger amount of water vapor.

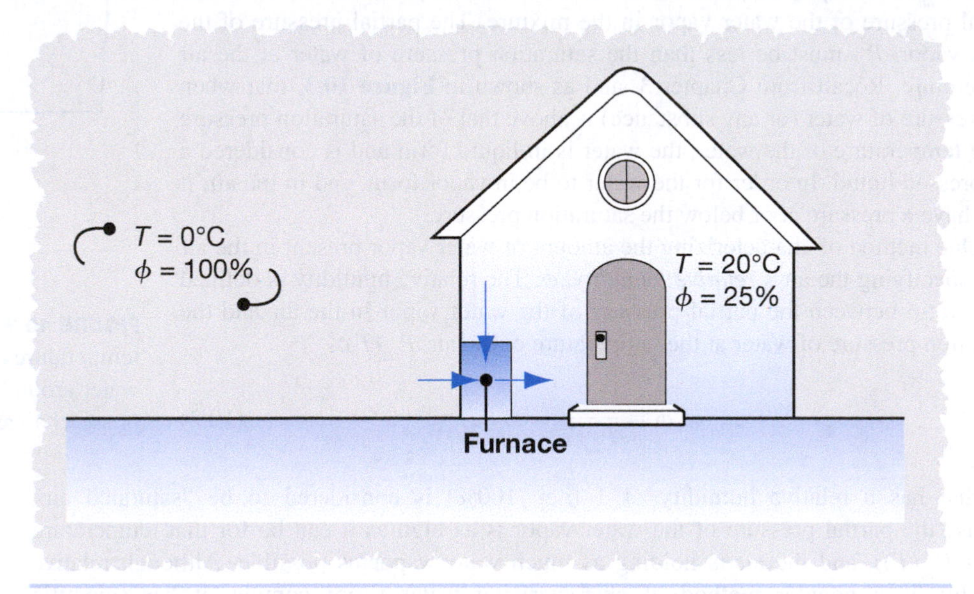

FIGURE 10.5 When saturated cold air is heated, the resulting mixture may contain an uncomfortably low amount of water vapor. This situation is shown here with a furnace heating cold, saturated air and producing warm air with a low relative humidity.

difficult for rain or snow to fall as the humidity drops farther below 100%. So precipitation is more likely to occur if the relative humidity is above 95% at ground level. Furthermore, phenomena such as fog also usually only occur with a relative humidity at 100%. However, these factors do not aid in calculating the properties of air as needed in engineering design. Therefore, more scientific methods of characterizing the mixture composition are desirable.

QUESTION FOR THOUGHT/DISCUSSION

Consider the evaporation and condensation of water in the atmosphere. What happens to the water if rain or snow forms in the atmosphere and begins to fall but no precipitation reaches the ground?

One method for characterizing the mixture composition of air independent of temperature is known as the *absolute humidity*, *specific humidity*, or *humidity ratio*, ω. This value, which we will refer to as the humidity ratio, is the ratio of the mass of water vapor, m_v, to the mass of dry air, m_{da}, present in the mixture:

$$\omega = \frac{m_v}{m_{da}} \tag{10.3}$$

The mass of water vapor in air compared to the amount of dry air is usually on the order of a few percent, and so values of the humidity ratio are often around 0.01 kg water vapor/kg dry air. As we will see later in this chapter, significant changes in water vapor content, from a comfort perspective, will show little change in the value of the humidity ratio, and so, in general, the general public has not adopted everyday use of the humidity ratio.

The humidity ratio can be directly related to the relative humidity. Keeping in mind the ideal gas law,

$$m_v = \frac{P_v \forall}{\frac{R}{M_v}T} \quad \text{and} \quad m_{da} = \frac{P_{da} \forall}{\frac{R}{M_{da}}T} = \frac{(P - P_v)\forall}{\frac{R}{M_{da}}T} \tag{10.4a, b}$$

we can derive through substitution in Eq. (10.3) that

$$\omega = \frac{M_v P_v}{M_{da}(P - P_v)} \tag{10.5}$$

Note: $M_v = 18.02$ kg/kmole, and $M_{da} = 28.97$ kg/kmole, so $M_v/M_{da} = 0.622$. Substituting this value in Eq. (10.5) and recalling Eq. (10.2) yields

$$\omega = 0.622\frac{\phi P_{sat}(T)}{P - \phi P_{sat}(T)} \tag{10.6}$$

A second method of quantifying the water vapor content in air independent of temperature is the *dew point temperature*, T_{dp}. The dew point temperature is defined as the temperature at which water will begin to condense out of air as the temperature of the air is lowered. This can be represented symbolically as

$$T_{dp} = T_{sat}(P_v) \tag{10.7}$$

and is shown graphically in **Figure 10.6**. When air, typically warm air with a high relative humidity, is cooled, the air will reach a point where the partial pressure of the water vapor is equal to the saturation pressure of water at the air temperature. If the air is to be cooled any further, water must condense out of the air, which will lower the partial pressure of the water vapor in the air. If this happens outdoors on warm days/nights, dew will form on grass, automobiles, and so on, and the dew is the liquid water that was formed as a portion of the water vapor condensed from the air as it was cooled. If this process happens on a cold day/night, frost (solid water) will be deposited directly on surfaces.

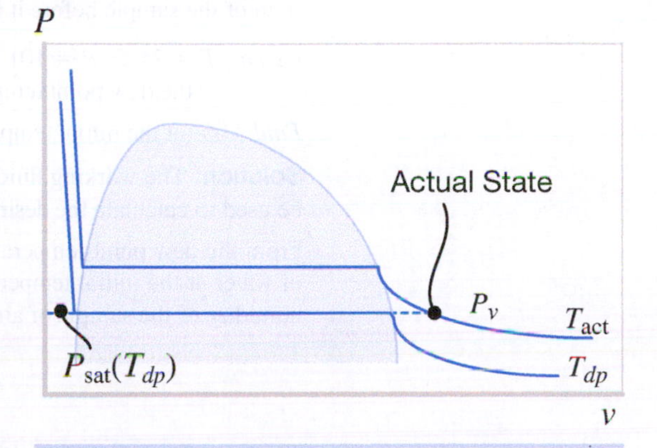

FIGURE 10.6 An illustration of the dew point temperature. The dew point temperature is the saturation temperature corresponding to the partial pressure of the water vapor in the given actual state.

As a side note, if a dramatic change in the water vapor content of air is not expected to occur overnight (i.e., if no warm or cold fronts are expected to pass through an area), a good estimate of how cold a night may become is the previous evening's dew point temperature. If the temperature tries to drop below the dew point temperature, condensation, and the accompanying release of heat, will occur—and this minimizes the likelihood of the temperature dropping much below the previous evening's dew point temperature.

The dew point temperature and relative humidity share a commonality: If air is at a relative humidity of 100%, then the air is saturated and the actual temperature is equal to the dew point temperature.

▶ EXAMPLE 10.1

On a warm summer afternoon, the air temperature is 28°C and the relative humidity is 55%. The air pressure is 101.325 kPa. Determine the partial pressure of the water vapor, the humidity ratio, and the dew point temperature of the air.

Given: $T = 28°\text{C}$, $\phi = 55\%$, $P = 101.325$ kPa

Find: P_v, ω, T_{dp}

Solution: The working fluid in this problem is atmospheric air—an ideal gas mixture of dry air and water vapor. As such, psychrometric relationships will be used to calculate the desired quantities.

At 28°C, $P_{sat}(T) = 3.782$ kPa. Therefore, from Eq. (10.2),

$$P_v = \phi P_{sat}(T) = (0.55)(3.782\text{ kPa}) = \mathbf{2.08\ kPa}$$

From Eq. (10.5),

$$\omega = 0.622\frac{P_v}{P - P_v} = 0.622\frac{2.08\text{ kPa}}{101.325 - 2.08\text{ kPa}} = \mathbf{0.0130\ kg\ H_2O/kg\ dry\ air}$$

Using property calculation software, the temperature for which the saturation pressure is 2.08 kPa is

$$T_{dp} = T_{sat}(P_v) = \mathbf{18.1°C}$$

▶ EXAMPLE 10.2

A sample of air is initially measured to have a temperature of 24°C and a pressure of 101 kPa. It is then cooled, at constant pressure, and water begins to condense out of the air at a temperature of 18°C. Determine the relative humidity of the sample at its initial temperature, and the humidity ratio of the sample before it is cooled.

Given: $T = 24°\text{C}$, $P = 101$ kPa, $T_{dp} = 18°\text{C}$, as the temperature at which water begins to condense is the dew point temperature

Find: ϕ (at the initial temperature), ω (before water condenses out)

Solution: The working fluid in this problem is atmospheric air, so psychrometric relationships will be used to calculate the desired quantities.

From the dew point temperature, it can be determined that $P_v = 2.035$ kPa. The saturation pressure of water at the initial temperature, $P_{sat}(T) = 3.065$ kPa. The ratio of these two values is the relative humidity of the sample of air, initially:

$$\phi = \frac{P_v}{P_{sat}(T)} = \frac{2.035\text{ kPa}}{3.065\text{ kPa}} = 0.664 = \mathbf{66.4\%}$$

The humidity ratio is found from Eq. (10.5):

$$\omega = 0.622\frac{P_v}{P - P_v} = 0.622\frac{2.035\text{ kPa}}{101\text{ kPa} - 2.035\text{ kPa}} = \mathbf{0.0128\ kg\ H_2O\ vapor/kg\ dry\ air}$$

Analysis: First, the dew point temperature is a given quantity. The phrase in the problem statement, "and water begins to condense out of the air at a temperature of," indicates the dew point temperature as water condenses out of air when it is cooled to its dew point. Second, note that while the relative humidity is not dependent on the total system pressure, the humidity ratio will be affected by the total pressure of a mixture.

When performing a psychrometric analysis, one quantity that is often constant is the amount of dry air present in a system. The amount of water vapor may change during a dehumidifying or humidifying process, but the amount of dry air is typically constant in such processes. As such, we will define psychrometric properties, which are specific properties based on the mass of dry air. The psychrometric specific volume, for example, is defined as

$$v_a = \mathcal{V}/m_{da} \tag{10.8}$$

The psychrometric specific enthalpy is derived from the total enthalpy of the mixture. The total enthalpy is the sum of the enthalpy of the dry air and the enthalpy of the water vapor:

$$H = H_{da} + H_v$$

$$H = m_{da}h_{da} + m_v h_v$$

Dividing by the mass of dry air, an expression for the psychrometric enthalpy, h_a, is developed:

$$h_a = \frac{H}{m_{da}} = h_{da} + \frac{m_v}{m_{da}} h_v$$

Recalling the definition of the humidity ratio (Eq. (10.3)),

$$h_a = h_{da} + \omega h_v \tag{10.9}$$

The specific enthalpy of the dry air is determined at the temperature of the mixture. The specific enthalpy of the water vapor is a function of both the water vapor partial pressure and the air temperature; however, the functional dependence of the specific enthalpy on pressure is small at the low partial pressures of the water vapor in air, and so we will take $h_v = h_g(T)$; that is, the specific enthalpy of the water vapor is equal to that of saturated water vapor at the temperature of the mixture.

If air is in contact with liquid water, water will evaporate until the air is saturated. It may require a large amount of time for the air to become saturated, but the time can be shortened by increasing the contact surface area between the liquid water and the air, and by stirring or agitating the air such that the water vapor can be quickly transported through the system rather than having to slowly diffuse from the surface through the entire system. This characteristic of liquid water–air systems can be exploited by various techniques to determine the humidity level of the air.

10.3 METHODS OF DETERMINING HUMIDITY

The level of the humidity of air is measured constantly, yet it does not lend itself to a simple technique. Common, inexpensive devices used to measure humidity usually rely on changes in the properties of a substance as it is exposed to a moist air environment. For example, if a substance that can absorb water, such as salt, is placed on a substance that does not absorb water, the system's shape can be deformed as the humidity level changes; this change in shape can be recorded by a dial indicator. Other systems measure humidity by using the changes in a substance's resistance or capacitance as it absorbs or releases water vapor from or into the air. These types of systems tend to be reasonably accurate for most purposes; better accuracy is often possible with more expensive devices. However, these systems can also be somewhat

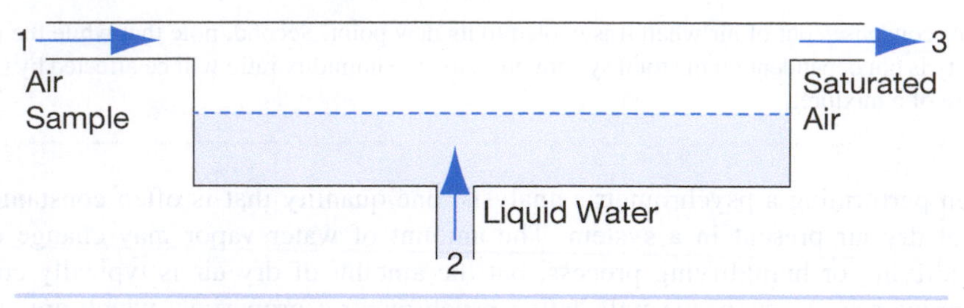

FIGURE 10.7 A schematic diagram of an adiabatic saturator.

slow to respond to changes in humidity levels. Furthermore, they need to be calibrated to devices that determine humidity levels with a high level of accuracy.

One device that can obtain a more accurate measure of humidity is the *adiabatic saturator*. **Figure 10.7** shows a schematic diagram of an adiabatic saturator. This type of device is insulated to minimize any possible heat transfer losses, and it uses a First Law of Thermodynamics analysis coupled with the conservation of mass to determine the level of humidity in a sample of air. As shown in Figure 10.7, the sample of air with an unknown humidity enters the device and passes over the liquid water. The level of the liquid water is held constant, and the amount of liquid water needed to be added to maintain a constant level is measured. Provided that the residence time for which the air is in contact with the water is sufficiently long, the stream of air exiting will be saturated (i.e., the relative humidity of the exiting air is 100%). By measuring the temperatures of the air and water entering, and the air leaving, we can determine the relative humidity of the inlet air.

Analysis of this device requires some assumptions. The first assumption is that the exit air relative humidity is 100%: $\phi_3 = 100\%$. Second, it is assumed that there is no heat transfer from the system, that any work interactions are negligible, and that there are no effects from kinetic or potential energy in the system. The system is assumed to be at steady state and steady flow. Because the thermal mass of the water is large in comparison to the inlet air, it is assumed that the water inlet temperature is the same as the water in the system, and that this value is constant. Applying these assumptions, the first law for an open system with multiple inlets and outlets at steady state, Eq. (4.13), reduces to

$$\dot{m}_{da,1}h_{da,1} + \dot{m}_{v,1}h_{v,1} + \dot{m}_{w,2}h_{w,2} = \dot{m}_{da,3}h_{da,3} + \dot{m}_{v,3}h_{v,3} \tag{10.10}$$

where the dry air and water vapor components of the moist air stream are considered separately. The subscript w for state 2 refers to properties of liquid water. From the conservation of mass for dry air, $\dot{m}_{da,1} = \dot{m}_{da,3} = \dot{m}_{da}$. Dividing Eq. (10.10) by this mass flow rate of dry air yields

$$h_{da,1} + \omega_1 h_{v,1} + \frac{\dot{m}_{w,2}}{\dot{m}_{da}}h_{w,2} = h_{da,3} + \omega_3 h_{v,3} \tag{10.11}$$

At this point, we assume that the enthalpy of the liquid water is equal to that of a saturated liquid at the water temperature ($h_{w,2} = h_{f,2}(T_2)$), and the enthalpy of the water vapor is equal to that of a saturated vapor at the moist air temperature ($h_v = h_g(T)$). Furthermore, the temperature change of the dry air is not expected to be large, so it is acceptable to treat the dry air as having constant specific heats: $h_{da,3} - h_{da,1} = c_{p,da}(T_3 - T_1)$.

The conservation of mass of water can be used to simplify Eq. (10.11) further. The conservation of mass of water yields

$$\dot{m}_{v,1} + \dot{m}_{w,2} = \dot{m}_{v,3} \tag{10.12}$$

which allows for the mass flow rate of liquid water to be determined as

$$\dot{m}_{w,2} = \dot{m}_{v,3} - \dot{m}_{v,1}$$

Dividing this by the mass flow rate of dry air yields

$$\frac{\dot{m}_{w,2}}{\dot{m}_{da}} = \omega_3 - \omega_1 \tag{10.13}$$

Equation (10.13) as well as the assumptions discussed can then be substituted into Eq. (10.11). Rearranging the resulting equation allows for an expression to be developed for ω_1, the inlet air stream humidity ratio:

$$\omega_1 = \frac{c_{p,da}(T_3 - T_1) + \omega_3(h_{g,3} - h_{f,2})}{h_{g,1} - h_{f,2}} \tag{10.14}$$

If not all of the temperatures are measured, it is often assumed that the exit air temperature and the water temperature are the same as a result of the long residence time needed to saturate the air—sufficiently long to bring the air and water into thermal equilibrium. If $T_2 = T_3$, then $h_{g,3} - h_{f,2} = h_{fg,2}$—the difference in the specific enthalpy of water between a saturated liquid and a saturated vapor at the water temperature (i.e., the enthalpy of vaporization). If this result is applied, then the inlet air stream humidity ratio can be found from

$$\omega_1 = \frac{c_{p,da}(T_3 - T_1) + \omega_3 h_{fg,2}}{h_{g,1} - h_{f,2}} \tag{10.15}$$

It should be noted that the enthalpy values for water are known for given temperatures, and the specific heat of dry air is known. The exit air stream humidity ratio can be easily calculated based on the assumption of a relative humidity of 100% from Eq. (10.6). So, with a properly functioning adiabatic saturator, the humidity ratio of the air can be determined through temperature measurements. It must be remembered, though, that this method of determining humidity ratio depends on the air becoming saturated, which requires time. This time must be gained by either a very slowly moving air stream or a very long contact length between the air and liquid water—sometimes both. This limits the practicality of the adiabatic saturator for regular use.

▸ EXAMPLE 10.3

An adiabatic saturator is used to determine the humidity ratio of a moist air sample. The air enters the adiabatic saturator at 22°C and exits at 15°C. The water is also added to the saturator at 15°C. The total air pressure is constant at 101.3 kPa. Assuming that the air exiting the device is saturated, determine the humidity ratio and the relative humidity of the inlet air sample.

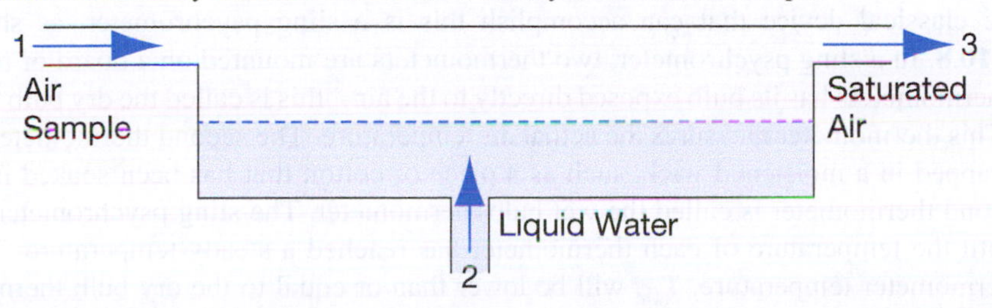

Given: $T_1 = 22°C$, $T_2 = T_3 = 15°C$, $\phi_3 = 100\%$

Find: ω_1 and ϕ_1

Solution: The fluid under consideration is moist (atmospheric) air, and the system under consideration is an adiabatic saturator. Therefore, the adiabatic saturator relationships will be used in the analysis.

Assume: Steady-state, steady-flow, exit air is saturated, $Q = 0$, $h_{w,2} = h_{f,2}(T_2)$, $h_v = h_g(T)$, dry air has constant specific heats.

The humidity ratio can be found from Eq. (10.15):

$$\omega_1 = \frac{c_{p,da}(T_3 - T_1) + \omega_3 h_{fg,2}}{h_{g,1} - h_{f,2}}$$

For dry air, take $c_{p,da} = 1.005$ kJ/kg · K. From a computer program for determining psychrometric properties, it can be found that $h_{g,1} = 2541.7$ kJ/kg, $h_{f,2} = 62.99$ kJ/kg, and $h_{fg,2} = 2465.9$ kJ/kg.

From Eq. (10.6),

$$\omega_3 = 0.622\frac{\phi_3 P_{sat,3}(T)}{P - \phi_3 P_{sat,3}(T)}$$

where, from the program, $P_{sat,3}(T) = 1.705$ kPa.

Solving for the humidity ratio at state 3, $\omega_3 = 0.01065$ kg H_2O/kg dry air.

Then, solving for the humidity ratio at state 1,

$$\omega_1 = \frac{1.005 \text{ kJ/kg dry air}(15°\text{C} - 22°\text{C}) + (0.01065 \text{ kg H}_2\text{O/kg dry air})(2465.9 \text{ kJ/kg H}_2\text{O})}{(2541.7 - 62.99) \text{ kJ/kg H}_2\text{O}}$$

$$\boldsymbol{\omega_1 = 0.00776 \text{ kg H}_2\text{O/kg dry air}}$$

Equation (10.6) can then be used to find the relative humidity of the inlet air:

$$\omega_1 = 0.622\frac{\phi_1 P_{sat,1}(T)}{P - \phi_1 P_{sat,1}(T)} \quad \text{with } P_{sat,1}(T) = 2.645 \text{ kPa}$$

Solving for the relative humidity yields

$$\boldsymbol{\phi_1 = 0.472 = 47.2\%}.$$

Analysis: Both of these values are typical of what would be expected from air with properties approximately equal to the given inlet conditions.

It is desirable to reduce the residence time required for a system to develop saturated air. One classical device that can accomplish this is a sling psychrometer, as shown in **Figure 10.8**. In a sling psychrometer, two thermometers are mounted on a board or rod. One of the thermometers has its bulb exposed directly to the air—this is called the dry bulb thermometer. This thermometer measures the actual air temperature. The second thermometer has its bulb wrapped in a moistened wick, such as a piece of cotton that has been soaked in water. The second thermometer is called the wet bulb thermometer. The sling psychrometer is then spun until the temperature of each thermometer has reached a steady temperature. The wet bulb thermometer temperature, T_{wb}, will be lower than or equal to the dry bulb thermometer temperature, T_{db}, because there will be evaporative cooling of the wet bulb from evaporation of water from the wick into the air. If the relative humidity is 100%, then $T_{wb} = T_{db}$ because no evaporation from the wick takes place. But if the relative humidity is less than 100%, then

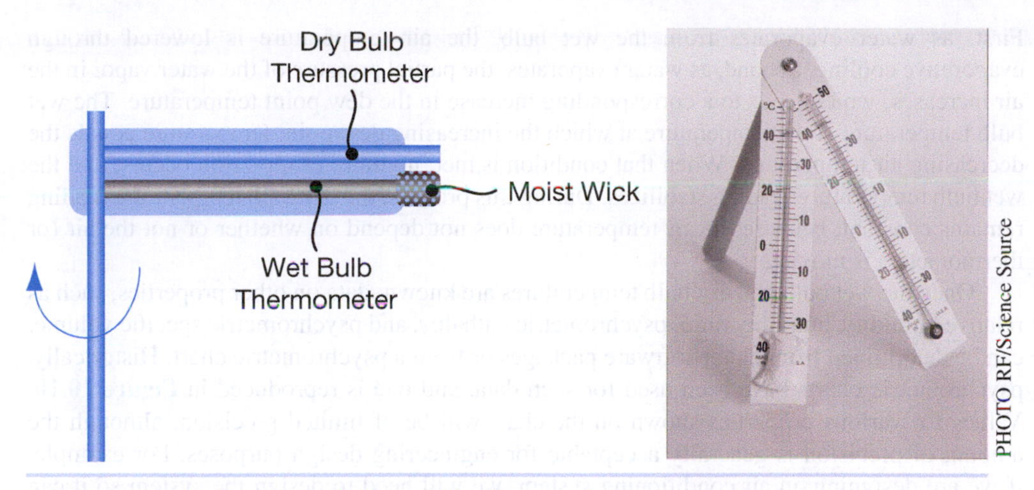

FIGURE 10.8 A diagram and picture of a sling psychrometer.

$T_{wb} < T_{db}$. It should be noted that the psychrometer could be used in a stationary position, but by swinging the psychrometer the evaporation of the water can occur more quickly and produce a measurement of the humidity in less time.

Some people mistake the wet bulb temperature for the dew point temperature, but the wet bulb temperature will be higher than the dew point temperature as long as the relative humidity is less than 100%. The wet bulb temperature represents the temperature to which air can be cooled through the evaporation of water, whereas the dew point temperature represents the temperature at which water would condense from the air if it were cooled. Although the dew point temperature is found by simply cooling the air, there are two opposing factors at work in the process of establishing the wet bulb temperature, as illustrated in **Figure 10.9**.

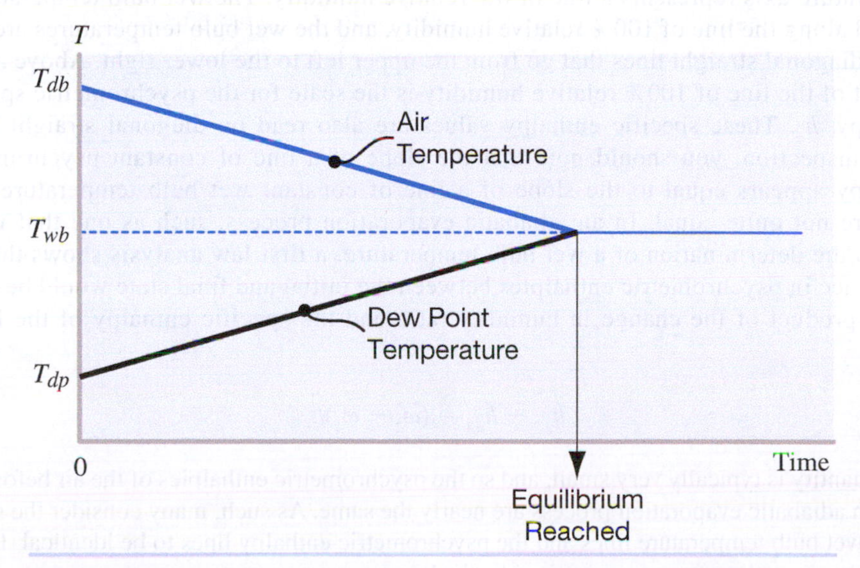

FIGURE 10.9 As water evaporates into an air–water vapor mixture, the mixture is cooled, resulting in a decrease in the dry bulb temperature. At the same time, the partial pressure of the water vapor increases, causing the dew point temperature of the mixture to increase. When the dry bulb and dew point temperatures become equal, the evaporation stops and the mixture temperature is the wet bulb temperature.

First, as water evaporates from the wet bulb, the air temperature is lowered through evaporative cooling. Second, as water evaporates, the partial pressure of the water vapor in the air increases, which leads to a corresponding increase in the dew point temperature. The wet bulb temperature is the temperature at which the increasing dew point temperature equals the decreasing air temperature. When that condition is met, no more evaporation occurs, and the wet bulb temperature reading stabilizes. During this process, the dry bulb temperature reading remains constant, because the air temperature does not depend on whether or not the air (or thermometer) is moving.

Once the wet bulb and dry bulb temperatures are known, data on other properties, such as relative humidity, humidity ratio, psychrometric enthalpy, and psychrometric specific volume, can be determined from either software packages or from a psychrometric chart. Historically, psychrometric charts have been used for such data, and one is reproduced in **Figure 10.10**. Values for various properties shown on the chart will be of limited precision, although the amount of precision is generally acceptable for engineering design purposes. For example, if we are designing an air conditioning system, we will need to design the system so it can work under a range of input and output conditions. As such, great precision is not needed for design calculations because the system will need to be controlled through sensor input for precise performance. The design calculations must be accurate enough to assure that the system will meet the user's needs, but it is not likely that the system will be controlled through such calculations.

Although computers have made the use of such charts unnecessary for determining psychrometric data, the charts are still useful for illustrating the processes that occur in psychrometric systems. Psychrometric charts can be easier to use depending on which data are available for a particular state. As such, let us explore the format of the psychrometric chart in Figure 10.10. Along the bottom of the chart is plotted the dry bulb temperature, which is the actual air temperature. Moving up from this axis is a series of curved lines that represent lines of constant relative humidity. The dry bulb temperature axis represents a line of 0% relative humidity. The wet bulb temperature is plotted along the line of 100% relative humidity, and the wet bulb temperatures are read along diagonal straight lines that go from the upper left to the lower right. Above and to the left of the line of 100% relative humidity is the scale for the psychrometric specific enthalpy, h_a. These specific enthalpy values are also read on diagonal straight lines. Upon inspection, you should note that the slope of a line of constant psychrometric enthalpy appears equal to the slope of a line of constant wet bulb temperature—but they are not quite equal. In an adiabatic evaporation process, such as one that would lead to the determination of a wet bulb temperature, a first law analysis shows that the difference in psychrometric enthalpies between the initial and final state would be equal to the product of the change in humidity ratio and the specific enthalpy of the liquid water:

$$h_{a,2} - h_{a,1} = (\omega_2 - \omega_1)h_f$$

This quantity is typically very small, and so the psychrometric enthalpies of the air before and after an adiabatic evaporation process are nearly the same. As such, many consider the slopes of the wet bulb temperature lines and the psychrometric enthalpy lines to be identical. But be aware that in reality they are not quite identical.

Along the right side of the psychrometric chart is a scale for the humidity ratio. The humidity ratio is read horizontally on the chart. If we take the humidity ratio and find its intersection point with the 100% relative humidity line, the resulting temperature that is read is the dew point temperature for the set of conditions that share that particular humidity ratio. Finally, the chart contains diagonal lines that slope more steeply than the wet bulb temperature lines—these are lines of constant psychrometric specific volume.

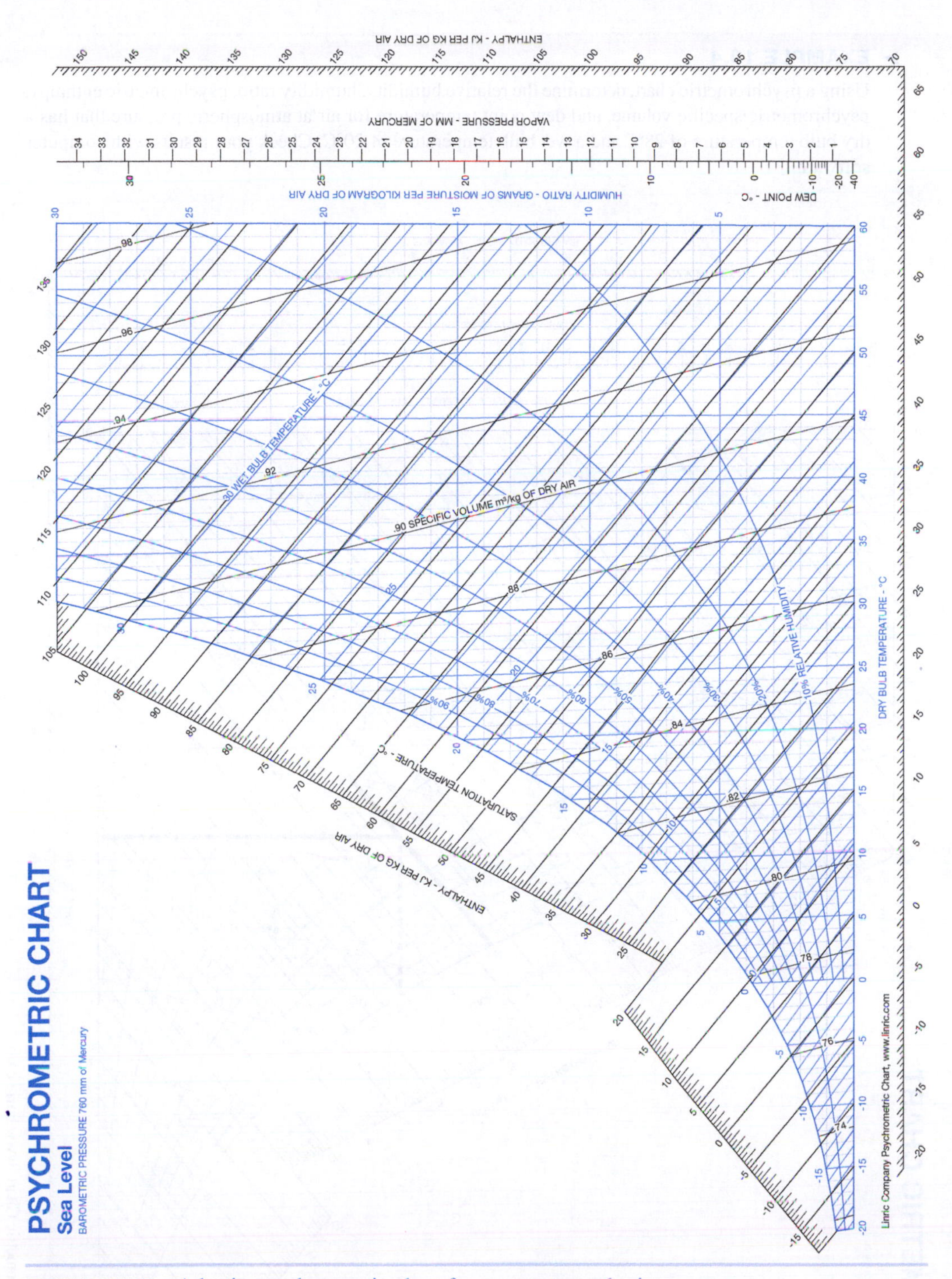

FIGURE 10.10 A basic psychrometric chart for use at atmospheric pressure.

Source: Linric Company Psychrometric Chart, www.linric.com

It can be helpful to identify a computer program or tablet app to assist in finding psychrometric property data. These are easy to find through an Internet search for "psychrometric calculator." If you are using a particular equation solver, you may also wish to add code for finding the psychrometric properties to that solver.

▶ EXAMPLE 10.4

Using a psychrometric chart, determine the relative humidity, humidity ratio, psychrometric enthalpy, psychrometric specific volume, and dew point temperature for air at atmospheric pressure that has a dry bulb temperature of 28°C and a wet bulb temperature of 20°C. Check your results with computer software.

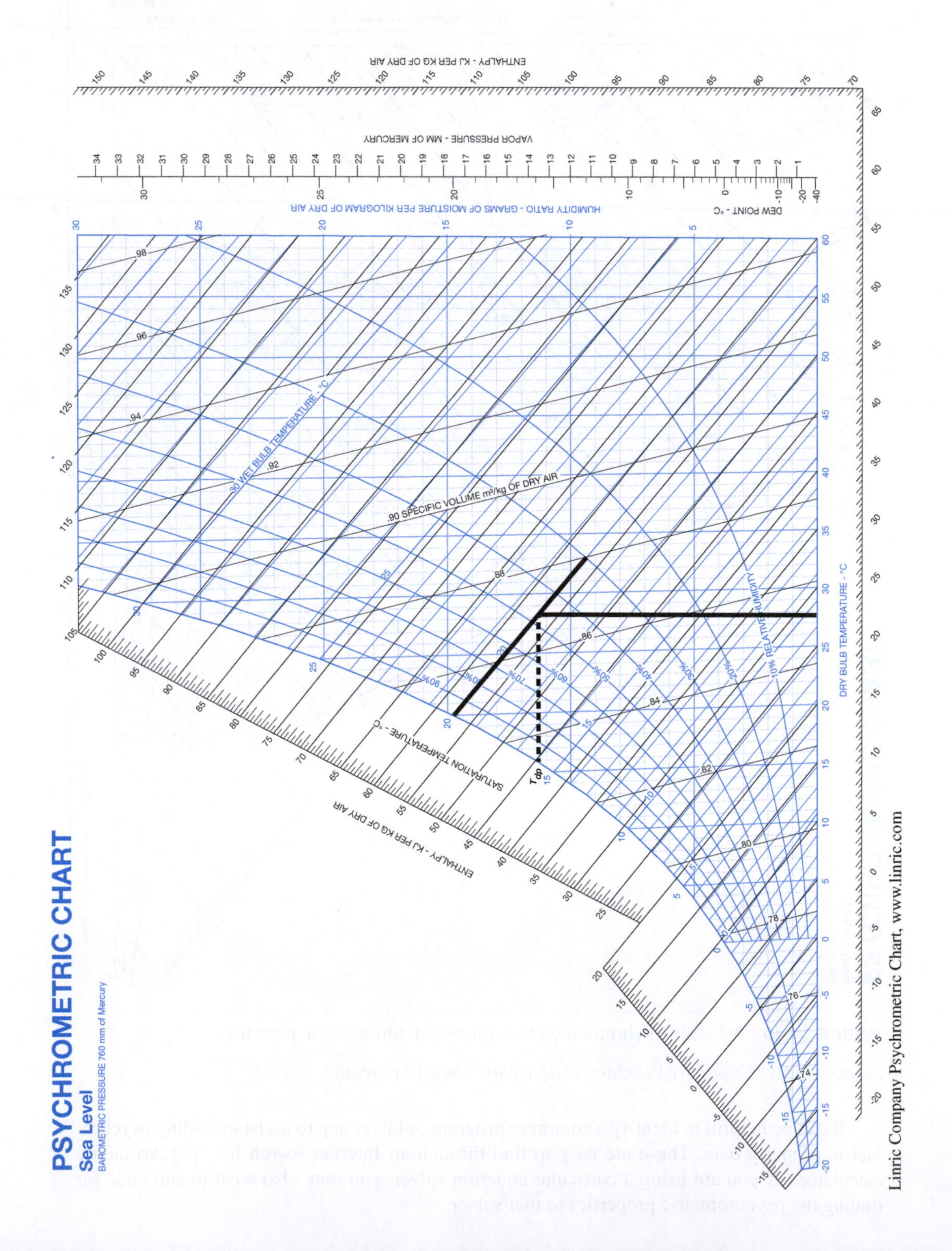

Linric Company Psychrometric Chart, www.linric.com

Given: $T_{db} = 28°C$, $T_{wb} = 20°C$

Find: ϕ, ω, T_{dp}, h_a, v_a

Solution: Here we will use two methods of finding psychrometric properties for this moist air.

Using the psychrometric chart, the following points can be read:

$\phi = 48\%$ $\omega = 0.0115$ **kg** H_2O**/kg dry air** $T_{dp} = 16°C$

$h_a = 57.5$ **kJ/kg dry air** $v_a = 0.869$ **m**^{3}**/kg dry air**

A software application yields

$\phi = 48.6\%$ $\omega = 0.0117$ **kg** H_2O**/kg dry air** $T_{dp} = 16.2°C$

$h_a = 58.07$ **kJ/kg dry air** $v_a = 0.87$ **m**^{3}**/kg dry air**

Analysis: While reading such values off a psychrometric chart requires practice, good approximate values can be found from the psychrometric chart. With the widespread availability of technology (i.e., computers and smartphones) that can give psychrometric properties nearly instantly, you might question the need for psychrometric charts. Even though psychrometric charts are not necessary in many cases for actual values of properties, they can still be useful in helping to visualize the process that will occur in a psychrometric application, and in some cases (such as mixing of two streams) they might provide a quicker answer than if a problem is solved by hand.

Note that some of the values on the psychrometric chart are dependent on the total pressure. Therefore, different psychrometric charts exist for different total pressures.

It can be helpful to identify a computer program or tablet app to assist in finding psychrometric property data. These are easy to find through an Internet search for "psychrometric calculator." If you are using a particular equation solver, you may also wish to add to that solver the code for finding psychrometric properties.

QUESTION FOR THOUGHT/DISCUSSION

To what degree of precision must an engineer determine the relative humidity of air in a room when designing an air conditioning system?

10.4 COMFORT CONDITIONS

Altering the condition of moist air is the engineering purpose most closely associated with psychrometrics, so it is important to understand the temperature and humidity ranges that most people find comfortable. Overall air quality is also important in determining the level of comfort people feel in an environment, but here we will assume that the air is of a good quality. To get good air quality, it may be necessary to filter the air or to mix cleaner air with lower quality air—such modifications to improve air quality are not the focus of psychrometrics.

Many factors influence the ranges of temperature and relative humidity that most people find comfortable. In addition, individual preferences may fall outside the ranges most people consider comfortable, and we may need to adjust the desired end conditions to accommodate such preferences. ASHRAE (the American Society of Heating, Refrigeration, and Air-Conditioning Engineers) is a professional organization that has developed standards for the conditions that are judged comfortable by most people. One significant factor influencing the acceptable conditions is the season. In colder climates, people tend to wear heavier clothing in winter and lighter clothing in summer. Therefore, conditions that are considered comfortable for most people are generally cooler in winter than summer. As seen in **Figure 10.11**, for

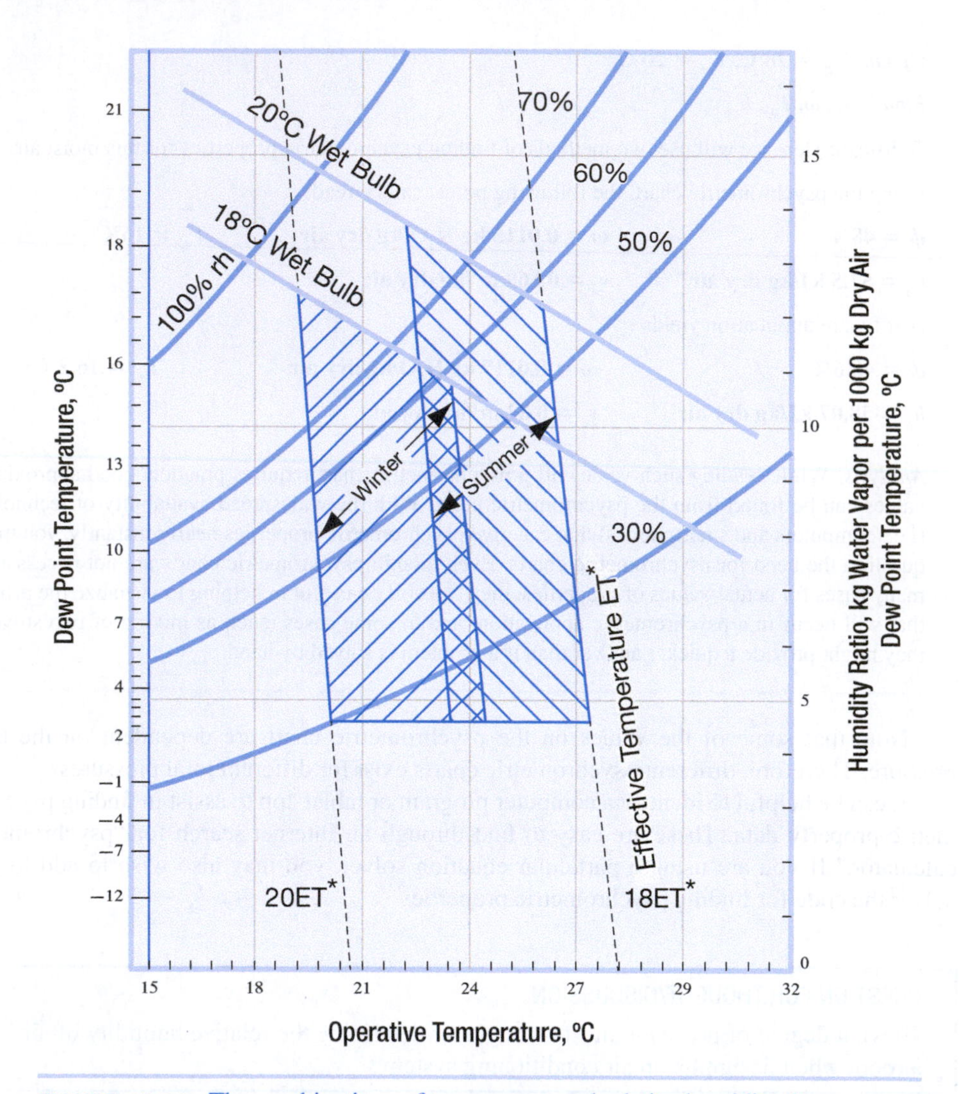

FIGURE 10.11 The combinations of temperature and relative humidity that most people consider to be comfortable, both for winter and summer conditions. The effective temperature takes into account air temperature, humidity, and air movement.

primarily sedentary activity, most people judge the conditions of optimal comfort while wearing summer clothing in the range of 22–26°C at a relative humidity of 50%. Recalling that the humidity level impacts the rate of cooling from a body, the temperature range shifts lower (by about 1°C) for a higher relative humidity of ~60%, and shifts higher by a similar amount for a lower relative humidity of ~30%. Based purely on comfort considerations, a reasonable target point for the temperature and relative humidity of air in summer is 24°C and 45%, respectively.

In winter, people are more comfortable at lower temperatures, with a temperature range of 20–23°C at a relative humidity of 50%, again with a shift toward slightly lower temperatures at higher values of relative humidity, and a shift toward slightly higher temperatures at lower values of relative humidity. Again, from a purely comfort level standpoint, a reasonable target point for the temperature and relative humidity in winter is about 21°C with a relative humidity of 45%. Most humans are generally not comfortable at very high relative humidity levels, above 75%, nor are most humans comfortable at low relative humidity levels, below about 25%.

Other factors must be considered when determining the comfort level of people. Recall that comfort level is often determined by the rate of cooling of the body. If there is a breeze or draft in a system, higher air temperatures may be needed to counteract the additional convective heat losses from the body. Similarly, if a large potential for thermal radiation losses exists in a space (such as through windows in winter), higher temperatures may be needed to counteract the additional radiation heat transfer. If a space is to be used for intense exercise, such as in a gymnasium, we would expect that the occupants will be generating extra heat and would prefer a somewhat cooler temperature for the space in order to be comfortable. A more extreme example is an indoor ice rink, where the users of the rink are both exercising and generally wearing heavier clothes. In such a space, the air temperatures may be kept at what is considered to be quite low for indoor spaces. Clearly, we must consider the physical nature of the space as well as the intended use of the space in order to determine an appropriate target temperature and relative humidity for the space.

This discussion has concentrated on the conditions that most people find comfortable. Unfortunately, conditioning the air to these conditions requires energy, and as the cost of energy rises, there is increased emphasis on reducing the amount of energy used in conditioning a space. As mentioned earlier, heating of air in winter and cooling of air in summer can be the primary use of energy in a home or commercial business and can also be a significant cost in some industrial facilities. To reduce energy use and cost, it often is necessary to accept warmer temperatures in summer and cooler temperatures in winter. Such considerations should be considered in light of the comfort levels of people, in general. Although the occupants of a home might save energy by heating their home to only 15°C in the winter, the level of discomfort may not be worth the money saved. However, lowering the temperature in winter and raising the temperature in summer of spaces when they are not occupied, or at night, is an effective way to reduce energy consumption. These savings result from a reduction in the amount of heat conduction through walls to the exterior because the temperature difference between the two is reduced, which leads to lower energy requirements for heating and cooling to maintain the interior temperature. Because the space is unoccupied, or because blankets are in use for sleeping at night in winter, there is less need to worry about the comfort of people at these times.

Now that we have considered the comfort level of people, the next two sections focus on the psychrometric processes most commonly used for conditioning air.

QUESTION FOR THOUGHT/DISCUSSION

What ranges of temperature and relative humidity do you find comfortable, and how do your personal preferences compare with the ASHRAE analysis of what most people find comfortable?

10.5 COOLING AND DEHUMIDIFYING OF MOIST AIR

The process most commonly thought of as an air conditioning process is the cooling and dehumidifying of moist air. This process is typically used to make hot, humid air more comfortable by cooling the air and removing water vapor from the air. At times, air can just be cooled, but it is unlikely that the resulting relative humidity will be at a desirable level, and so some dehumidification is usually conducted. A device that is commonly used for the cooling and dehumidification process is shown schematically in **Figure 10.12**. The cooling and dehumidification processes are accomplished by transferring heat from the air to a fluid passing through a heat exchanger. Usually, a refrigerant flows through a cooling coil, and the air passes over the outside of the tubes of the cooling coil. This cooling coil, depending on the system used, can be the "evaporator" of a vapor-compression refrigeration cycle, as studied in Chapter 8. The refrigerant receives as its heat input in the evaporator the heat that is being removed from the air.

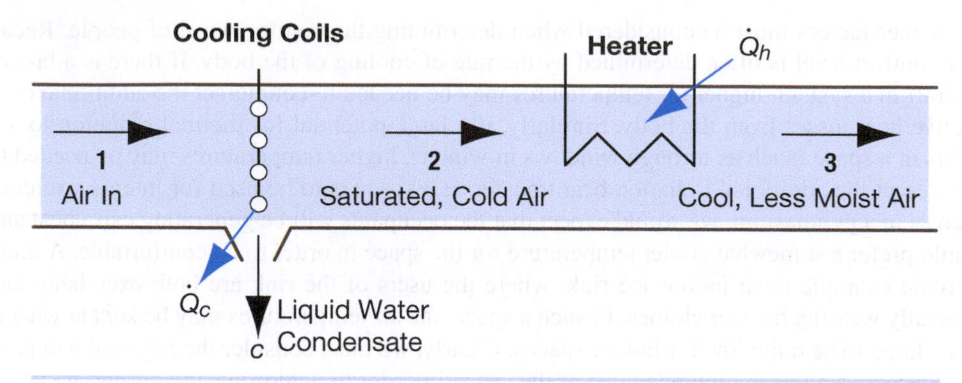

FIGURE 10.12 A schematic diagram of a cooling and dehumidifying process. The moist air is first cooled and dehumidified by passing over cooling coils, and the cold saturated air is subsequently heated to increase its condition to a more comfortable combination of temperature and relative humidity. Condensed water ("*c*") is drained from the system by the cooling coils.

To understand how this process achieves cooling and dehumidification, it is useful to show the process on a psychrometric chart. An example is presented in **Figure 10.13a**. In this case, hot, humid air is brought into contact with the cooling coil, and the temperature of the air decreases at a constant humidity ratio. The air cools until it becomes saturated, with a relative humidity at 100%. At this point, the air may be at or below the desired final temperature, but it is unlikely that the water vapor content will be acceptable because a relative humidity of 100% is generally not desired from a comfort perspective. Further cooling of the air, however, will require water to condense out of the air, because air at lower temperatures cannot hold as much water vapor as air at higher temperatures. So, continued contact with the cooling coils results in the air cooling at a constant relative humidity of 100%, and with liquid water being removed from the air stream as the condensate. This process continues until the desired final humidity ratio is achieved.

The system depicted in Figure 10.12 places a heater after the cooling coils. The heating provided is necessary because the air stream exiting the cooling coils is usually a cool, saturated air stream that would be uncomfortable for most people. As a result, this air stream is

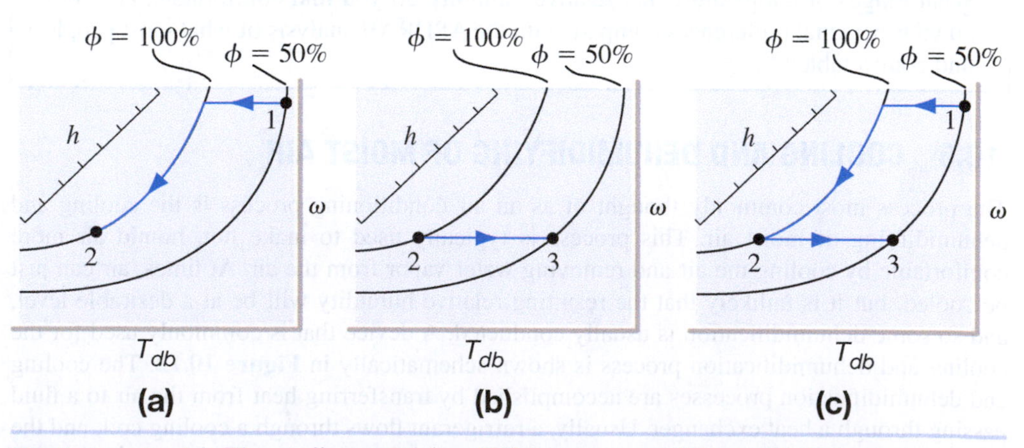

FIGURE 10.13 (a) The cooling and dehumidifying process of Figure 10.12 drawn on a psychrometric chart. (b) The heating process shown. (c) The complete process of Figure 10.12 shown on a psychrometric chart.

heated at a constant humidity ratio until the desired final temperature and relative humidity is achieved. This heating process is shown on a psychrometric chart in Figure 10.13b, and the combined two-step process is shown in Figure 10.13c. The desired final temperature and relative humidity fix the final humidity ratio, and in turn this fixes how cool the air becomes in the cooling coil.

Analyses of the cooling and dehumidification process and of the heating process are typically performed separately. Let us first consider the cooling and dehumidification process. We will assume that the process is a steady-state, steady-flow process in a device with a single inlet and two outlets. We will also assume that there is no work, and no effects from kinetic or potential energy. With these assumptions, Eq. (4.13) can be reduced to

$$\dot{Q}_c = \dot{m}_{da,2}h_{da,2} + \dot{m}_{v,2}h_{v,2} + \dot{m}_{w,c}h_{w,c} - (\dot{m}_{da,1}h_{da,1} + \dot{m}_{v,1}h_{v,1}) \tag{10.16}$$

where $\dot{Q}_c$ is the rate of heat transfer in the cooling coil section and $\dot{m}_{w,c}$ is the mass flow rate of the condensed water leaving the system. The conservation of mass applied to the water can be used to show that the mass flow rate of the condensed water is equal to

$$\dot{m}_{w,c} = \dot{m}_{v,1} - \dot{m}_{v,2} \tag{10.17}$$

In Eq. (10.16) the dry air and water vapor components of the air stream are kept separate. Psychrometric enthalpies for the moist air can also be used, considering that the mass flow rate of the dry air is constant:

$$\dot{Q}_c = \dot{m}_{da}(h_{a,2} - h_{a,1}) + \dot{m}_{w,c}h_{w,c} \tag{10.18}$$

One point of concern with using Eqs. (10.16) and (10.18) involves the specific enthalpy of the condensed water. Inspection of a psychrometric chart of the process shows that water is condensing at a range of temperatures, and as such there is a range of possible values for the enthalpy of the water. Fortunately, the contribution to the heat transfer rate from this term is generally small, because the mass flow rate of the condensed water is usually very small (on the order of 1%) compared to the mass flow rate of the dry air. Therefore, extreme accuracy on the choice of the enthalpy of the condensed water is not needed for engineering purposes. We will normally assume that the liquid water exiting the system has cooled to the temperature of the air exiting the cooling section (T_2); then the choice for the enthalpy of the condensed water leaving the system will be that of a saturated liquid at T_2: $h_{w,c} = h_{f,2}$. If we have a temperature measurement of the condensed water leaving the system, we should use that temperature instead of T_2.

Equation (10.18) can also be written on a per unit mass flow of dry air basis by dividing the equation by the mass flow rate of the dry air:

$$\frac{\dot{Q}_c}{\dot{m}_{da}} = (h_{a,2} - h_{a,1}) + (\omega_1 - \omega_2)h_{w,c} \tag{10.19}$$

where Eq. (10.17) has been used for the mass flow rate of the condensed water.

Analysis of the heating section of the system is simpler because it can be modeled as a single-inlet, single-outlet, steady-state, steady-flow open system. We assume that there is no work interaction, and that the changes in kinetic and potential energy are negligible. Equation (4.15) then reduces to

$$\dot{Q}_h = \dot{m}_{da,3}h_{da,3} + \dot{m}_{v,3}h_{v,3} - (\dot{m}_{da,2}h_{da,2} + \dot{m}_{v,2}h_{v,2}) \tag{10.20}$$

where $\dot{Q}_h$ is the heat transfer rate through the heating section. Again, the mass flow rate of the dry air is constant, and Eq. (10.20) can be written in terms of psychrometric enthalpies:

$$\dot{Q}_h = \dot{m}_{da}(h_{a,3} - h_{a,2}) \tag{10.21}$$

On a per mass flow rate of dry air basis, Eq. (10.21) becomes

$$\frac{\dot{Q}_h}{\dot{m}_{da}} = h_{a,3} - h_{a,2} \tag{10.22}$$

When analyzing the cooling and dehumidification process, remember the following assumptions:

$$\omega_3 = \omega_2 \quad \text{and typically} \quad \phi_2 = 100\%$$

STUDENT EXERCISE

Develop a computer model for the platform of your choice that will facilitate the calculations for cooling and dehumidification processes. These calculations may be needed to determine the state of air entering or exiting the system, the rates of heat transfer in the cooling and heating sections, or the amount of water removed from the system.

▶ EXAMPLE 10.5

A moist air stream with a temperature of 29°C and relative humidity of 70% is to be conditioned to 21°C with a relative humidity of 40%. The mass flow rate of dry air to be conditioned is 1.1 kg/s. Determine the rate at which water is condensed from the air by the cooling coil.

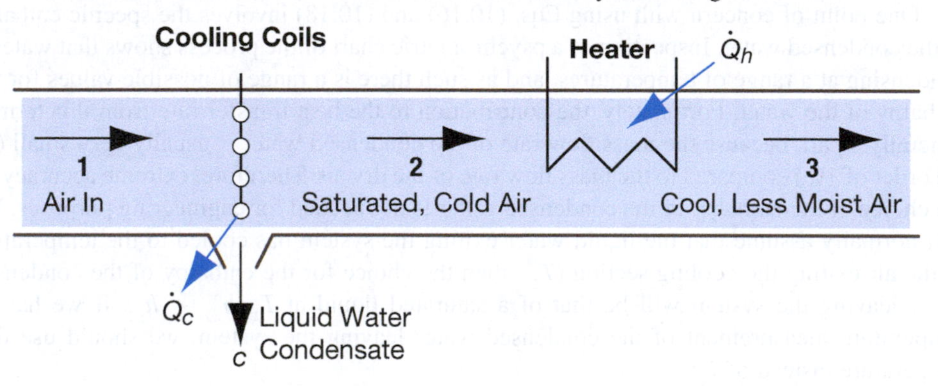

Given: $T_1 = 29°C$, $\phi_1 = 70\%$, $T_3 = 21°C$, $\phi_3 = 40\%$, $\dot{m}_{da} = 1.1$ kg/s

Find: $\dot{m}_{w,c}$

Solution: Using Eqs. (10.3) and (10.17),

$$\dot{m}_{w,c} = \dot{m}_{v,1} - \dot{m}_{v,2} = \dot{m}_{da}(\omega_1 - \omega_2)$$

Recall that $\omega_2 = \omega_3$. Using computer psychrometric property software, the humidity ratios at the beginning and end states can be found:

$$\omega_1 = 0.0185 \text{ kg H}_2\text{O/kg dry air} \qquad \omega_3 = \omega_2 = 0.00632 \text{ kg H}_2\text{O/kg dry air}$$

Substitution yields $\dot{m}_{w,c} = \dot{m}_{da}(\omega_1 - \omega_2) =$ **0.0134 kg H_2O/s.**

Analysis: The mass flow rate of the condensed water is 1.2% of the mass flow rate of the dry air. Such a small amount will have only a small impact on the heat removal rate from the air through the cooling coils.

▶ EXAMPLE 10.6

On a summer day, 2.0 m^3/s of moist air with a temperature of 30°C and a relative humidity of 65% enter an air conditioning system such as shown in Figure 10.12. The air exits the heating section at 20°C with a relative humidity of 40%. Determine (a) the rate at which water vapor is condensed from the air, (b) the rate of heat transfer in the cooling section, and (c) the rate of heat addition in the heating section.

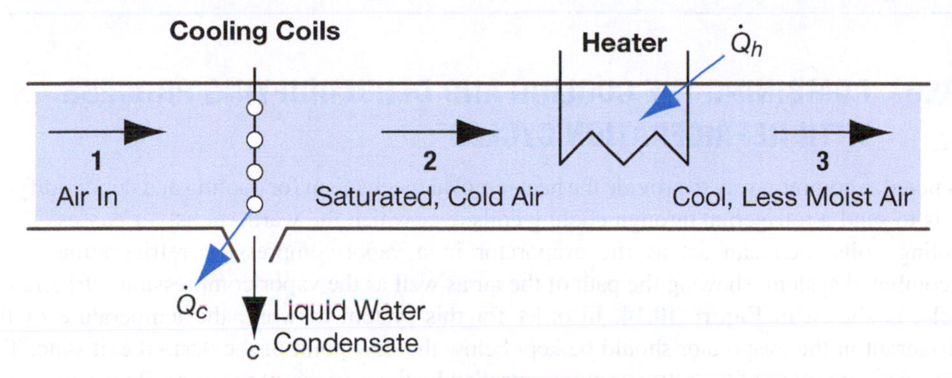

Given: $T_1 = 30°C$, $\phi_1 = 65\%$, $T_3 = 20°C$, $\phi_3 = 40\%$, $\dot{V}_1 = 2.0\ m^3/s$

Find: (a) $\dot{m}_{w,c}$, (b) $\dot{Q}_c$, (c) $\dot{Q}_h$

Solution: The working fluids are moist air and liquid water, and the system for analysis is the typical cooling and dehumidying system described.

Assume: Processes are steady-state, steady-flow, in an open system. Assume no impacts from kinetic or potential energy, and that no work is used. $\omega_3 = \omega_2$, and upon inspection of the process on a psychrometric chart, $\phi_2 = 100\%$. $T_c = T_2 = T_{dp,3}$.

Using computer psychrometric property software, the psychrometric enthalpies and humidity ratios at states 1 and 3 can be determined:

$$h_{a,1} = 75.5 \text{ kJ/kg dry air} \qquad \omega_1 = 0.0177 \text{ kg } H_2O\text{/kg dry air}$$

$$h_{a,3} = 35.1 \text{ kJ/kg dry air} \qquad \omega_3 = 0.0059 \text{ kg } H_2O\text{/kg dry air}$$

Furthermore, the psychrometric specific volume at state 1 is $v_{a,1} = 0.882\ m^3$/kg dry air. The dew point temperature for state 3 is 6°C. Therefore, $T_2 = 6°C$ and $\phi_2 = 100\%$. With this information,

$$h_{a,2} = 20.9 \text{ kJ/kg dry air} \qquad \omega_2 = 0.0059 \text{ kg } H_2O\text{/kg dry air}$$

In addition, $h_c = h_f(6°C) = 25.2$ kJ/kg H_2O.

The mass flow rate of dry air can be found from $\dot{m}_{da} = \dfrac{\dot{V}_1}{v_{a,1}} = 2.268$ kg/s.

(a) $\dot{m}_{w,c} = \dot{m}_{v,1} - \dot{m}_{v,2} = \dot{m}_{da}(\omega_1 - \omega_2) = \mathbf{0.0268\ kg\ H_2O/s}$

(b) Equation (10.18) can be used to determine the heat transfer in the cooling coil:

$$\dot{Q}_c = \dot{m}_{da}(h_{a,2} - h_{a,1}) + \dot{m}_{w,c}h_{w,c} = \mathbf{-123\ kW}$$

(c) The rate of heat transfer in the heater can be found from Eq. (10.21):

$$\dot{Q}_h = \dot{m}_{da}(h_{a,3} - h_{a,2}) = \mathbf{32.2\ kW}$$

Analysis: As expected, more heat is removed from the air through the cooling coil than is added by the heating element. The contribution of the condensing of liquid water to the heat removal in the cooling coil is approximately 0.5%. It can be argued that this is small enough to be ignored, but the contribution from condensing water can become more significant when more water is being removed from the air. The higher the temperature of the water leaving the cooling coils, the more significant this factor becomes.

QUESTION FOR THOUGHT/DISCUSSION

Cooling and dehumidifying air requires a large amount of electricity, and wide-scale use of such systems in buildings in the United States is a relatively recent phenomenon, occurring over the past 60 years. How has the widespread availability of these systems changed the United States (e.g., culturally, where people live, etc.)?

10.6 COMBINING THE COOLING AND DEHUMIDIFYING PROCESS WITH REFRIGERATION CYCLES

As noted, a common way to provide the heat removal mechanism for cooling and dehumidifying air is to send a refrigerant through cooling coils over which the warm, moist air passes. These cooling coils then can act as the evaporator in a vapor-compression refrigeration cycle. A combined system, showing the path of the air as well as the vapor-compression refrigeration cycle, is shown in **Figure 10.14**. In order for this system to work, the temperature of the refrigerant in the evaporator should be kept below the dew point of the desired exit state. The desired temperature of the refrigerant is controlled by the refrigerant pressure. Because it is not easy to adjust this parameter in an operating system, we must know the lowest desired dew point temperature to be achieved by the cooling and dehumidifying system before it is designed and constructed. This temperature will fix the maximum pressure of the refrigerant in the evaporator.

Analysis of this combined system relates the mass flow rates of the air and the refrigerant through the heat transfer rate in the evaporator. The direction of the heat transfer in the cooling coils/evaporator is out of the air and into the refrigerant—so the magnitude of the heat transfer rate is the same, but the sign changes in accordance with the sign convention for heat transfer:

$$\dot{Q}_e = -\dot{Q}_c$$

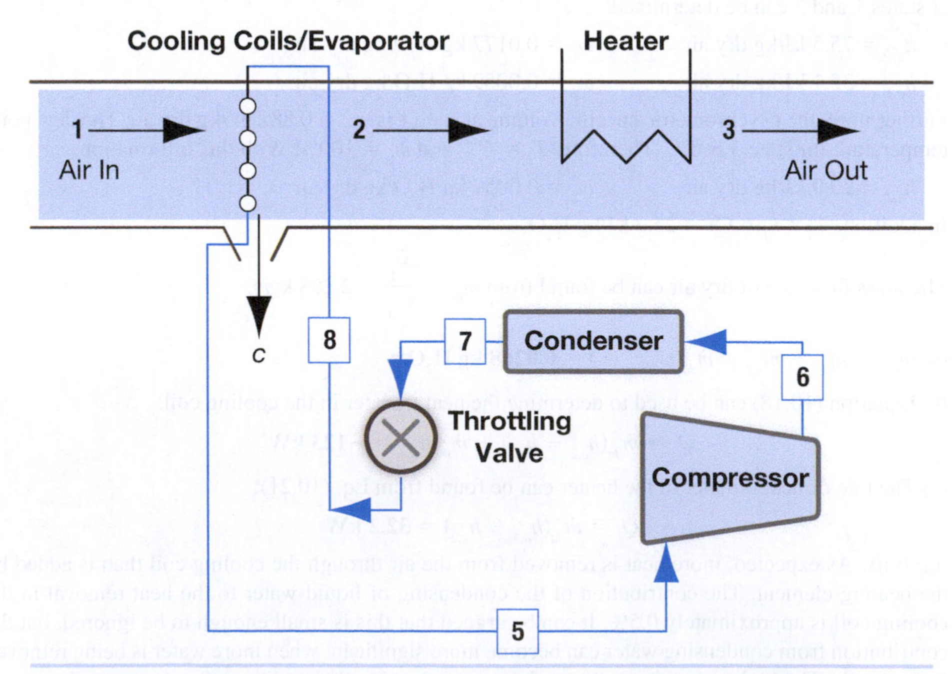

FIGURE 10.14 The cooling coils of the process shown in Figure 10.12 often contain refrigerant and serve as the evaporator in a vapor-compression refrigeration cycle. The two systems are shown combined here.

where $\dot{Q}_e$ is the heat transfer rate in the evaporator for the refrigerant and $\dot{Q}_c$ is the heat transfer rate in the cooling coils for the air. Recalling the vapor-compression refrigeration cycle analysis in Chapter 8, we see that

$$\dot{m}_R(h_5 - h_8) = -[\dot{m}_{da}(h_{a,2} - h_{a,1}) + \dot{m}_{w,c}h_{w,c}] \tag{10.23}$$

where $\dot{m}_R$ is the mass flow rate of the refrigerant and where the vapor-compression cycle state numbers have been renumbered to reflect the diagram in Figure 10.14.

STUDENT EXERCISE

Develop a computer model that will combine the analysis of a cooling and dehumidifying process with that of a vapor-compression refrigeration cycle. The module should allow you to determine heat loads from the cooling and dehumidifying process for use in your previously developed vapor-compression refrigeration cycle module; the module should also allow you to determine the amount of air that can be treated by a set vapor-compression refrigeration cycle.

▶ EXAMPLE 10.7

The cooling and dehumidification process of Example 10.6 is to be accomplished with the cooling coils acting as the evaporator of a vapor-compression refrigeration cycle. R-134a acts as the refrigerant in the cycle; it enters the evaporator at 200 kPa and exits the evaporator as a saturated vapor. The R-134a is compressed in a compressor with an isentropic efficiency of 0.80 to a pressure of 1000 kPa, and it exits the condenser as a saturated liquid. Determine the power required by the compressor to operate this system.

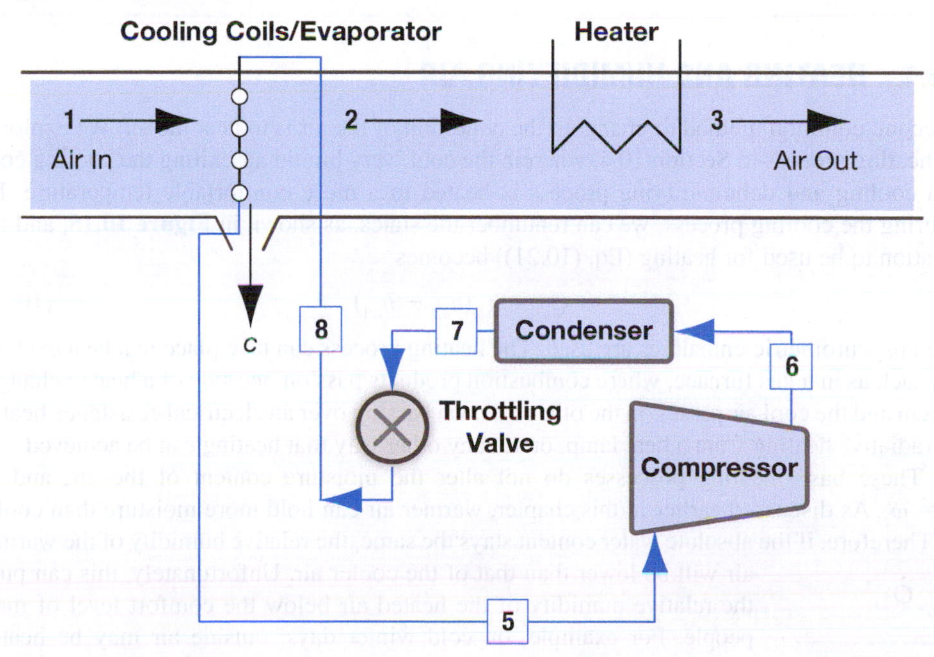

Given: From Example 10.6, the heat transfer rate from the air to the refrigerant in the cooling coils: $\dot{Q}_c = -123$ kW.

$P_5 = P_8 = 200$ kPa, $P_6 = P_7 = 1000$ kPa, $\eta_c = 0.80$, $x_5 = 1$, $x_7 = 0$

Find: $\dot{W}_c$ (the compressor power)

Solution: Having already solved the psychrometric part of the problem, the portion under consideration is the vapor-compression refrigeration cycle. The working fluid is R-134a, and its properties can be found from computer programs that provide refrigerant property data.

Assume: $P_5 = P_8$, $P_6 = P_7$, $\dot{W}_e = \dot{W}_{cond} = \dot{W}_{tv} = \dot{Q}_{tv} = \dot{Q}_{comp} = 0$, $\Delta KE = \Delta PE = 0$ for each device. Steady-state, steady-flow processes for each device.

From a refrigerant property program, the following enthalpy (and entropy) states for the refrigerant can be found (following the vapor-compression refrigeration cycle analysis outlined in Chapter 8):

$$h_5 = 241.30 \text{ kJ/kg}$$
$$s_5 = 0.9253 \text{ kJ/kg} \cdot \text{K} = s_{6s}$$

With s_{6s} and $P_{6s} = P_6$: $h_{6s} = 274.63$ kJ/kg

$$h_7 = 105.29 \text{ kJ/kg} = h_8$$

Using the isentropic efficiency of the compressor, $\eta_c = \frac{h_5 - h_{6s}}{h_5 - h_6}$, the actual outlet enthalpy from the compressor can be found: $h_6 = 282.96$ kJ/kg.

The heat transfer rate into the refrigerant is $\dot{Q}_e = -\dot{Q}_c = 123$ kW.

For the cycle, $\dot{Q}_e = \dot{m}_R(h_5 - h_8)$, so $\dot{m}_R = 0.904$ kg/s.

The compressor power is found as

$$\dot{W}_c = \dot{m}_R(h_5 - h_6) = \mathbf{-37.7\ kW}$$

So the compressor requires 37.7 kW of power to allow the cycle to perform the desired cooling and dehumidifying of the air stream.

Analysis: You can see how the cooling and dehumidifying of air in a summer air conditioning process can easily become a major expense for the owner of a large building. The amount of power to run the compressor in a large system can be extensive and expensive. Even for smaller buildings, such as a home, the cost of energy to cool and dehumidify air can become problematic.

10.7 HEATING AND HUMIDIFYING AIR

A second common method of changing the condition of the air is to heat the air. We explored the heating process in Section 10.4, wherein the cool, very humid air exiting the cooling coils of a cooling and dehumidifying process is heated to a more comfortable temperature. By ignoring the cooling process, we can renumber the states, as shown in **Figure 10.15**, and the equation to be used for heating (Eq. (10.21)) becomes

$$\dot{Q}_h = \dot{m}_{da}(h_{a,2} - h_{a,1}) \tag{10.24}$$

when psychrometric enthalpies are used. The heating process can take place in a heat exchanger, such as in a gas furnace, where combustion products pass on one side of a heat exchanger system and the cool air passes on the other side, via passing over an electrical-resistance heater, via radiative heating from a heat lamp, or via any other way that heating can be achieved.

These basic heating processes do not alter the moisture content of the air, and so $\omega_1 = \omega_2$. As discussed earlier in this chapter, warmer air can hold more moisture than cooler air. Therefore, if the absolute water content stays the same, the relative humidity of the warmer air will be lower than that of the cooler air. Unfortunately, this can push the relative humidity of the heated air below the comfort level of most people. For example, on cold winter days, outside air may be heated for use in a building. Although the outside air may have a high relative humidity, the relative humidity of the heated air with the same moisture content is uncomfortably low. Therefore, a humidification process is often used to increase the relative humidity of the air.

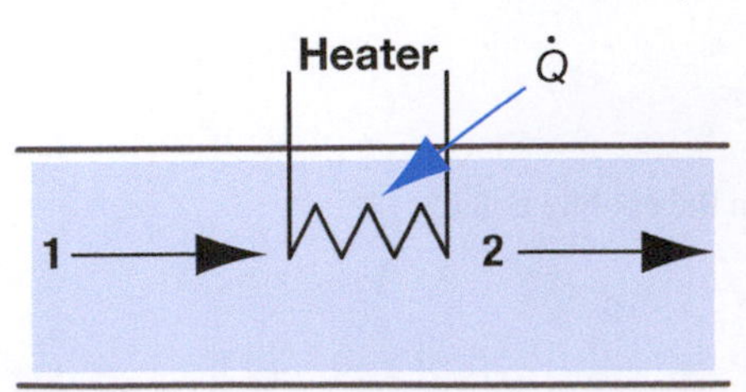

FIGURE 10.15 A schematic diagram of air passing over a heating element.

Two basic methods can be used to humidify air in a cost-effective manner: (1) passing the air over liquid water, such as passing the air through a porous pad that has been saturated with liquid water; and (2) adding

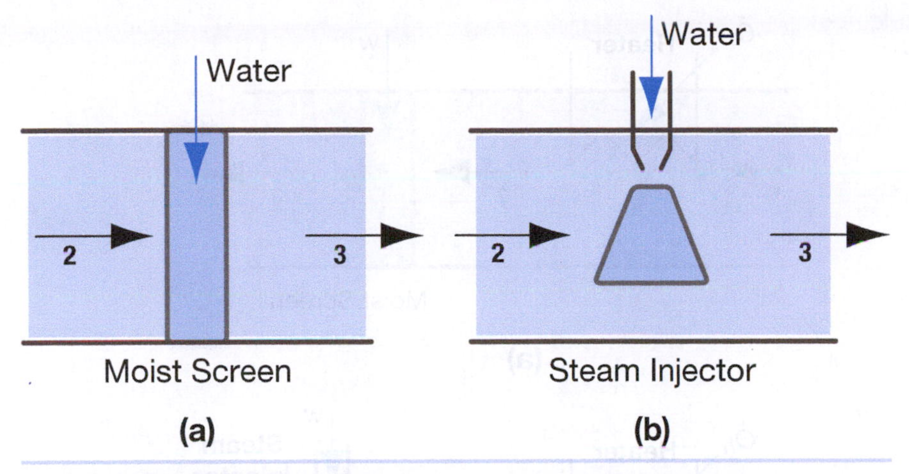

FIGURE 10.16 Methods of adding moisture to air include (a) passing the air over a moist screen holding liquid water, and (b) injecting steam into the air.

steam to the air. These two processes are illustrated schematically in **Figure 10.16**. These two processes have opposite effects on the air with respect to temperature. Liquid water must evaporate in the first method, so evaporative cooling of the air takes place, which can lower the temperature of the air. Therefore, if this method is used, we may wish to heat the air above its final desired temperature in order to accommodate the cooling. For the second process of adding water vapor, the steam will generally be hotter than the air temperature. Therefore, steam added to the air tends to increase the temperature of the air, and so we may choose to heat the air to a lower temperature before adding steam to the air so that the final desired temperature is still reached. In both cases, keep in mind that the amount of water added is small compared to the amount of air present, and temperature changes of the air will be small as a result. However, based on the small range of temperatures that are comfortable for most people, even this small change can push the temperature to an uncomfortable condition.

The humidification process can be modeled as a steady-state, steady-flow process with no heat transfer, no work, and no impacts from kinetic or potential energy effects. The mass flow rate of the dry air will stay constant during the process. As a result, for two inlets and one outlet, the first law for a humidification process can be written as

$$\dot{m}_{da}h_{da,2} + \dot{m}_{v,2}h_{v,2} + \dot{m}_w h_w = \dot{m}_{da}h_{da,3} + \dot{m}_{v,3}h_{v,3} \tag{10.25}$$

The conservation of the mass of water yields

$$\dot{m}_w = \dot{m}_{v,3} - \dot{m}_{v,2} \tag{10.26}$$

Combining Eqs. (10.25) and (10.26), and dividing by the mass flow rate of dry air, yields the first law as

$$h_{da,2} + \omega_2 h_{v,2} + (\omega_3 - \omega_2)h_w = h_{da,3} + \omega_3 h_{v,3} \tag{10.27}$$

or, in terms of psychrometric enthalpies,

$$h_{a,2} + (\omega_3 - \omega_2)h_w = h_{a,3} \tag{10.28}$$

In Eqs. (10.25), (10.27), and (10.28), the enthalpy of the water can be either equal to that of liquid water ($h_w = h_f(T)$) if liquid water is added to the system, or to water vapor ($h_w = h_g(T)$) if steam is added for humidification purposes.

The heating and humidifying processes can be combined, as shown in **Figure 10.17**. In such an analysis, Eqs. (10.25) through (10.28) are each used sequentially to analyze the system. **Figure 10.18** shows two generic psychrometric charts to illustrate the heating and humidifying processes (a) when liquid water is added, and (b) when water vapor is added.

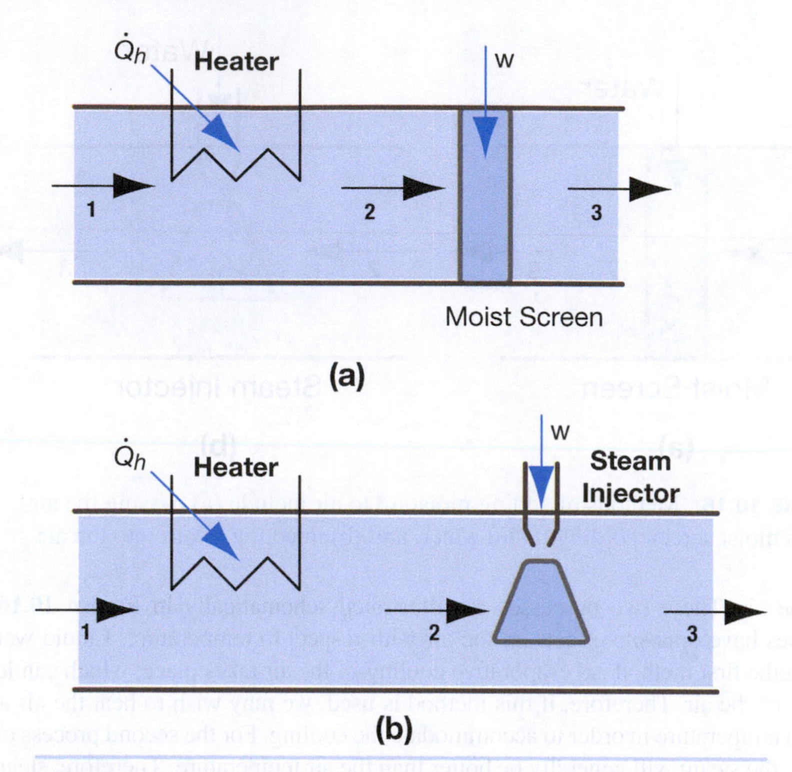

FIGURE 10.17 (a) A system for heating and humidifying air via passing the air over a moist screen. (b) A system for heating and humidifying air via steam injection.

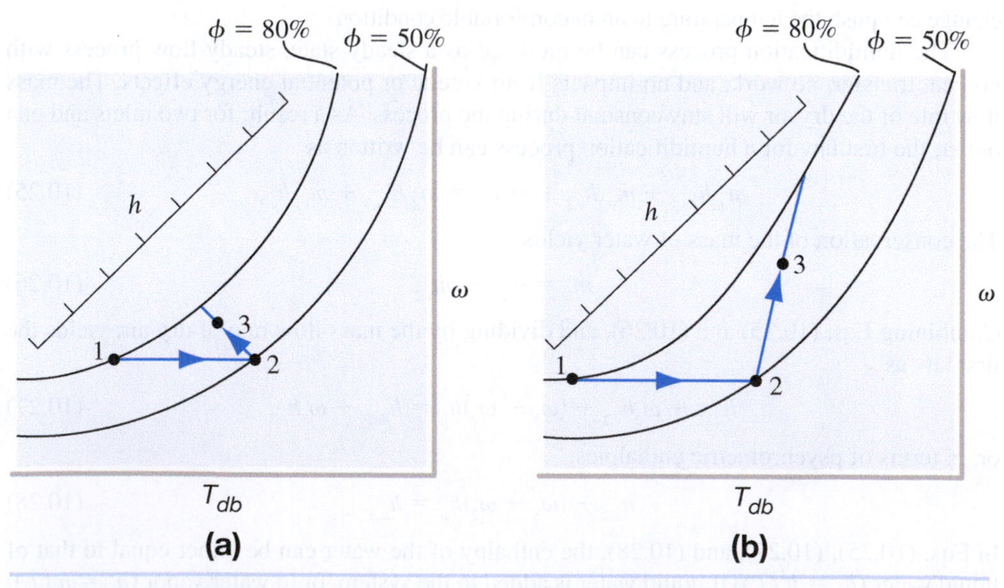

FIGURE 10.18 Example processes for heating and humidifying as shown on psychrometric charts. (a) Water is added by evaporation from a moist screen. (b) Water is added by steam injection.

STUDENT EXERCISE

Develop a computer model that will allow the analysis of heating and humidification systems. The module should allow for the determination of moist air properties, heating loads, or the amount of water needed to be added, and it should be able to work with both liquid water and water vapor humidification systems.

▶ EXAMPLE 10.8

Cold moist air is to be heated and humidified. The air enters at 5°C with a relative humidity of 80%, and the air after humidification should be 22°C with a relative humidity of 40%. Humidification is achieved by passing the air over a moist screen saturated with liquid water at 18°C. The mass flow rate of dry air through the system is 5 kg/s. Determine the amount of heat added during the heating process, and the mass flow rate of liquid water added during humidification. The pressure is constant at 101.325 kPa.

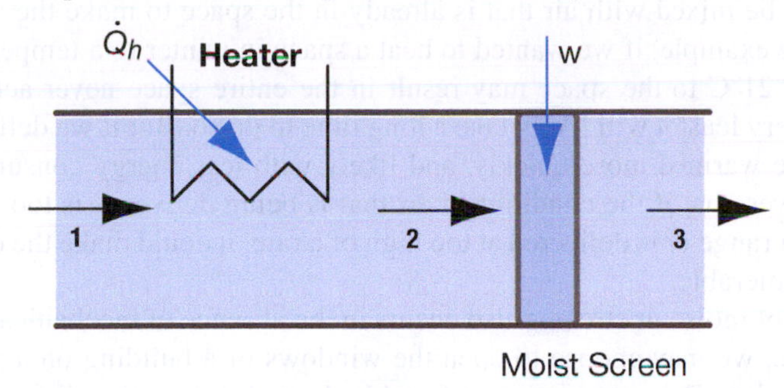

Given: $T_1 = 5°C, \phi_1 = 80\%, T_3 = 22°C, \phi_3 = 40\%, T_w = 18°C, \dot{m}_{da} = 5$ kg dry air/s, $P = 101.325$ kPa

Find: $\dot{Q}_h, \dot{m}_w$

Solution: The working fluids are moist air and liquid water, and the system for analysis is the heating and humidying system using a moist screen.

Assume: Processes are steady-state, steady-flow, in open systems. Assume no impacts from kinetic or potential energy, and that no work is used. $\omega_1 = \omega_2$.

First, we must determine the enthalpy of the air after heating. We know that $\omega_1 = \omega_2$ during the heating process. This information, combined with Eq. (10.28), will allow us to determine the psychrometric enthalpy at state 2. From a computer psychrometric property program, the following psychrometric properties can be found:

$h_{a,1} = 16.0$ kJ/kg dry air $\qquad \omega_1 = 0.00439$ kg H_2O/kg dry air

$h_{a,3} = 39.1$ kJ/kg dry air $\qquad \omega_3 = 0.00668$ kg H_2O/kg dry air

Also, for the liquid water, $h_w = h_f(T_w) = 75.58$ kJ/kg H_2O.

Knowing $\omega_2 = \omega_1 = 0.00439$ kg H_2O/kg dry air, Eq. (10.28) can be used to find $h_{a,2}$.

$$h_{a,2} = h_{a,3} - (\omega_3 - \omega_2)h_w = 38.9 \text{ kJ/kg dry air}$$

(Note: The temperature and relative humidity that corresponds to this combination of ω_2 and $h_{a,2}$ is 28°C and 18% relative humidity. This is not in the comfort zone for most people.)

Now we can use Eq. (10.24) to find the heating load of the process:

$$\dot{Q}_h = \dot{m}_{da}(h_{a,2} - h_{a,1}) = \mathbf{115\ kW}$$

We can use Eq. (10.26) to find the amount of water added through the humidification process:

$$\dot{m}_w = \dot{m}_{v,3} - \dot{m}_{v,2} = \dot{m}_{da}(\omega_3 - \omega_2) = 0.0115 \text{ kg } H_2O/\text{s}$$

Analysis: Although this is a small amount of liquid, it is nontrivial. It corresponds to approximately 11.5 cm^3/s of a volumetric flow—in just over one day, a cubic meter of water would need to be evaporated into the air to maintain the desired humidity level.

QUESTION FOR THOUGHT/DISCUSSION

Why might someone want to humidify a building in winter rather than only heat the air?

10.8 MIXING OF MOIST AIR STREAMS

It is not common for all of the air in a facility to be conditioned instantaneously. It is more common during air conditioning processes that either only a portion of air from a space is removed from a space to be conditioned, or fresh outside air is brought into the facility and conditioned. This air is then mixed with the air already in the space. Although this process may be happening continuously, the air that is present in the space is the result of the mixing of conditioned and unconditioned air. Therefore, when developing conditioned air, we may wish to produce air that is outside the comfort level of people in the space, because the air we produce will be mixed with air that is already in the space to make the overall mixture comfortable. For example, if we wanted to heat a space in winter to a temperature of 21°C, delivering air at 21°C to the space may result in the entire space never actually reaching 21°C, or at the very least it will likely take a long time to do so. But if we deliver air at 30°C, the space will be warmed more quickly, and likely with less energy consumption overall. Be aware, however, that if the conditioned air that is being delivered is too far out of most people's comfort range or is delivered at too high of a rate, it could make the conditions near the inlet vent intolerable.

The mixing of moist air streams also occurs in the absence of mechanical air conditioning. For example, we may choose to open the windows of a building on a pleasant day in order to make the interior space more comfortable. In such a case, the air from the outside is mixing with the inside air. Or we may constantly bring outside air into a facility to maintain air quality, such as in **Figure 10.19**. In this case, the air is often mixed with some inside air before being conditioned, or the outside air may be mixed directly with the inside air already present.

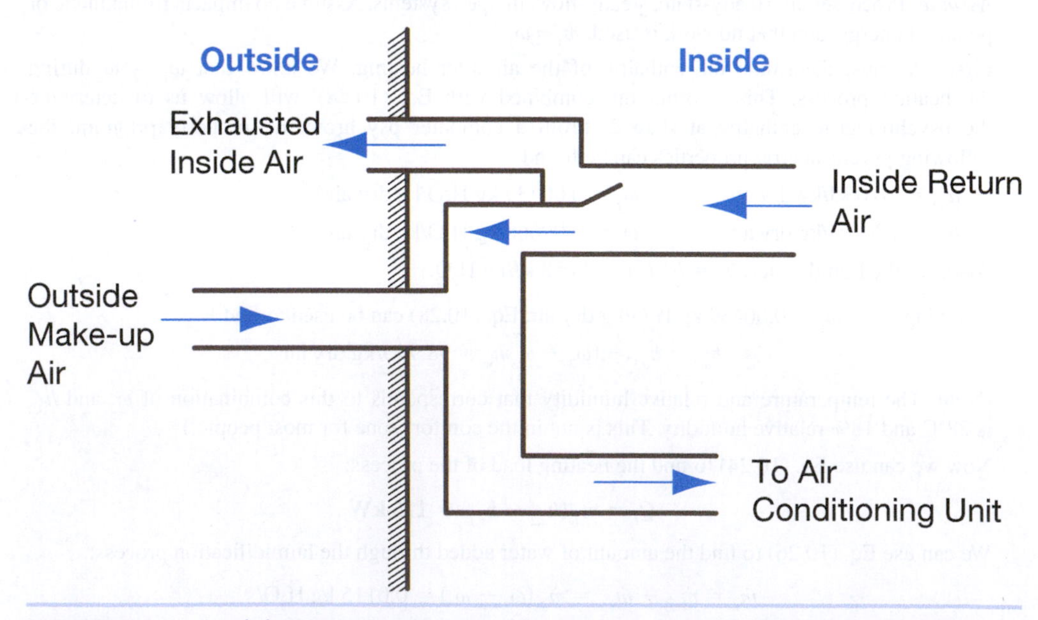

FIGURE 10.19 A mixing process where outside make-up air is combined with a portion of return air from inside a building. The mixture is then sent to an air conditioning unit. The rest of the return air is exhausted to the outdoors.

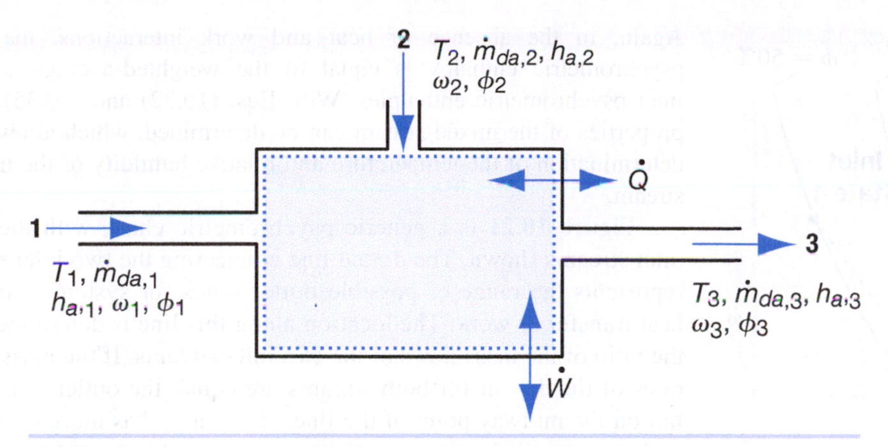

FIGURE 10.20 A mixing process with two streams of moist air being combined to form a third stream of moist air. Heat transfer and work interactions can occur in the system as well.

We use the conservation of mass of dry air, the conservation of mass of water, and the First Law of Thermodynamics to analyze the mixing of moist air stream processes. Figure 10.20 shows a mixing process in which two streams are combined to create one exit stream. In this analysis, we assume that there are no effects of kinetic or potential energy, and that the system is at steady state. The conservation of mass of dry air yields

$$\dot{m}_{da,3} = \dot{m}_{da,1} + \dot{m}_{da,2} \tag{10.29}$$

The water will only be in vapor form, and so the conservation of mass of water is

$$\dot{m}_{v,3} = \dot{m}_{v,1} + \dot{m}_{v,2} \tag{10.30}$$

Equation (10.30) can also be written in terms of humidity ratios:

$$\dot{m}_{da,3}\omega_3 = \dot{m}_{da,1}\omega_1 + \dot{m}_{da,2}\omega_2 \tag{10.31}$$

and dividing by the mass flow rate of dry air exiting yields

$$\omega_3 = \frac{\dot{m}_{da,1}}{\dot{m}_{da,3}}\omega_1 + \frac{\dot{m}_{da,2}}{\dot{m}_{da,3}}\omega_2 \tag{10.32}$$

From Eq. (10.32), we can see that the humidity ratio of the mixed stream is equal to the weighted-average humidity ratios of the inlet streams, where the weighting is from the mass flow rates of dry air through each inlet.

The first law, with the assumptions made, is

$$\dot{Q} - \dot{W} + \dot{m}_{da,1}h_{a,1} + \dot{m}_{da,2}h_{a,2} = \dot{m}_{da,3}h_{a,3} \tag{10.33}$$

where psychrometric enthalpies are used. Often, there is little heat transfer, and the work interactions can be ignored. If the system is assumed to be adiabatic with no significant work interactions, Eq. (10.33) reduces to

$$\dot{m}_{da,1}h_{a,1} + \dot{m}_{da,2}h_{a,2} = \dot{m}_{da,3}h_{a,3} \tag{10.34}$$

Dividing by the mass flow rate of the dry air in the outlet stream yields

$$h_{a,3} = \frac{\dot{m}_{da,1}}{\dot{m}_{da,3}}h_{a,1} + \frac{\dot{m}_{da,2}}{\dot{m}_{da,3}}h_{a,2} \tag{10.35}$$

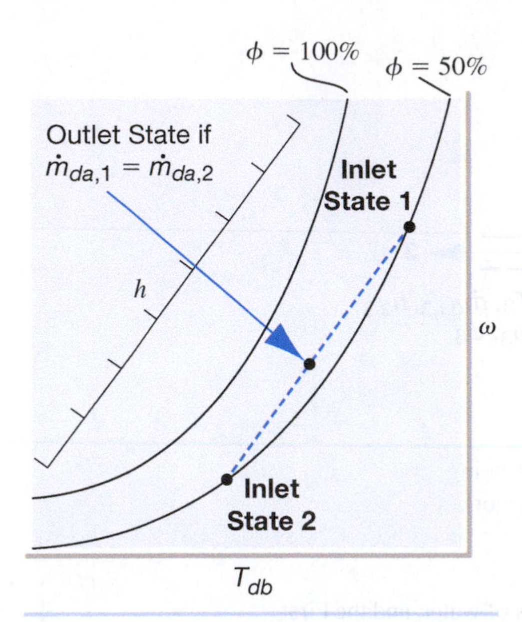

FIGURE 10.21 If two states are mixed in a constant-pressure, steady-state process without heat transfer or work, the resulting outlet stream will fall along the line connecting the two states on a psychrometric chart. The position on the line is determined by the relative mass flow rates of dry air of the two inlet streams. If the mass flow rates of the two streams are the same, the outlet state will fall on the center point of the line.

Again, in the absence of heat and work interactions, the exit psychrometric enthalpy is equal to the weighted-average of the inlet psychrometric enthalpies. With Eqs. (10.32) and (10.35), two properties of the mixed stream can be determined, which allows for determination of the temperature and relative humidity of the mixed stream.

Figure 10.21 is a generic psychrometric chart, with the two inlet streams shown. The dotted line connecting the two inlet states represents the range of possible outlet states for systems with no heat transfer or work. The location along this line is determined by the ratio of the mass flows of the two inlet streams. If the mass flow rates of the dry air for both streams are equal, the outlet state will fall on the midway point of the line. If stream 1 has more flow, the outlet state will be proportionally closer to state 1, and if stream 2 has more flow, the outlet state will be proportionally closer to state 2. By controlling the relative flow rates of the two streams, we can control the outlet state.

STUDENT EXERCISE

Develop a computer model that will allow for analysis of a mixing process for moist air streams. The module should allow determination of the outlet state if the inlet state is known, or the condition of one inlet state if the condition of the outlet state and the other inlet state is known. Furthermore, the module should allow for the flow rates of the stream to be determined.

▶ EXAMPLE 10.9

Two moist air streams are to be mixed together before being sent into a manufacturing facility. Stream 1, which is pulled from inside the building, has a temperature of 32°C, a relative humidity of 60%, and a mass flow rate of dry air of 5 kg dry air/s. Stream 2 is outside air and has a temperature of 4°C and a relative humidity of 70%, with a mass flow rate of dry air of 15 kg dry air/s. Determine the temperature and relative humidity of the mixed stream. The overall pressure is maintained at 101.325 kPa.

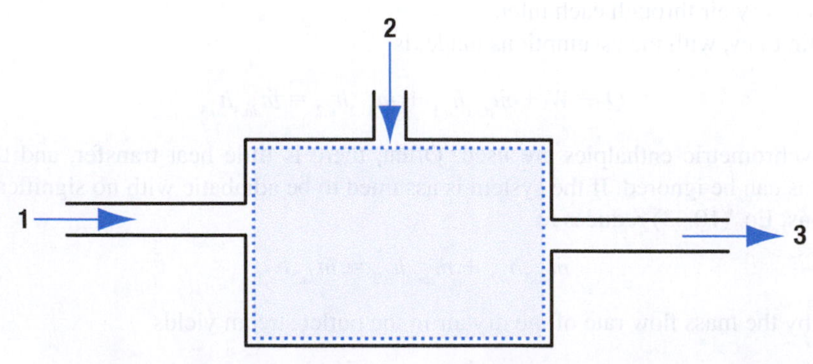

Given: $T_1 = 32°C$, $\phi_1 = 60\%$, $\dot{m}_{da,1} = 5$ kg dry air/s, $T_2 = 4°C$, $\phi_2 = 70\%$, $\dot{m}_{da,2} = 15$ kg dry air/s

Find: T_3, ϕ_3

Solution: The working fluids are moist air, and the system for analysis is a simple mixing process.

Assume: The mixing process is steady-state, steady-flow, in an open system. Assume no impacts from kinetic or potential energy, and $\dot{Q} = \dot{W} = 0$.

First, the mass flow rate of the dry air in the mixed stream is found from Eq. (10.29):

$$\dot{m}_{da,3} = \dot{m}_{da,1} + \dot{m}_{da,2} = 20 \text{ kg dry air/s}$$

The psychrometric enthalpies and humidity ratios of the inlet states are found from computer programs:

$h_{a,1} = 97.0$ kJ/kg dry air $\qquad \omega_1 = 0.0183$ kg H_2O/kg dry air

$h_{a,2} = 31.4$ kJ/kg dry air $\qquad \omega_2 = 0.00362$ kg H_2O/kg dry air

Equation (10.32) is used to find the mixed stream humidity ratio:

$$\omega_3 = \frac{\dot{m}_{da,1}}{\dot{m}_{da,3}}\omega_1 + \frac{\dot{m}_{da,2}}{\dot{m}_{da,3}}\omega_2 = 0.00729 \text{ kg } H_2O\text{/kg dry air}$$

Equation (10.35) is used to find the mixed stream psychrometric enthalpy:

$$h_{a,3} = \frac{\dot{m}_{da,1}}{\dot{m}_{da,3}}h_{a,1} + \frac{\dot{m}_{da,2}}{\dot{m}_{da,3}}h_{a,2} = 47.8 \text{ kJ/kg dry air}$$

Using ω_3 and $h_{a,3}$, the mixed stream temperature and relative humidity can be found as

$$\mathbf{T_3 = 11.6°C, \qquad \phi_3 = 86\%}$$

Analysis: Because actual water vapor contents are being mixed, the mixed stream may have a relative humidity higher than the relative humidity of either inlet stream. The temperature of the mixed stream will be between the temperatures of the inlet streams.

10.9 COOLING TOWER APPLICATIONS

Cooling towers, such as those in **Figure 10.22**, are systems used to transfer large amounts of heat to the air from a cooling fluid—typically water. Although cooling towers can be designed in a "dry" mode, where the air passes over pipes containing the cooling fluid, the most efficient cooling towers work by having direct contact between the air and the water. One obvious

FIGURE 10.22 Cooling towers take many forms, including (a) a natural-draft cooling tower and (b) a mechanical-draft cooling tower.

application for cooling towers is in a power plant. As we saw in Chapter 7, vast amounts of heat from Rankine cycle power plants must be removed from the steam in the condenser. The condenser usually uses water as the heat removal medium, and then this cooling water must transfer the heat to the environment. In some cases, this is done by directly emptying the cooling water into a natural body of water, but many power plants use a cooling tower to transfer heat from the cooling water to the air. Cooling towers can use fans to force the flow of air through the system (such systems are considered "mechanical-draft") or can use the changing buoyancy of the air through the system to develop a flow of air (such systems are "natural-draft"). Mechanical-draft cooling towers can usually be smaller than natural-draft cooling towers, because they do not need large height changes to achieve adequate density differences between the inlet and outlet to promote air flow. However, mechanical-draft cooling towers also use more power and require more maintenance, and therefore are often better suited to smaller applications. In addition, there are many other types of design choices to be made when designing cooling towers, including whether the flows should be cross-flow or counter-flow; the type and arrangement of the "fill," which is the contact surface for the water as it enters the air; and the type of water distribution system to be used. Here, we are concerned with the psychrometric analysis of cooling towers, rather than a comprehensive design of the cooling towers. Such design requires more details in heat transfer than you have learned so far.

Wet cooling towers are more efficient than dry cooling towers because they transfer heat between the water and the air through two mechanisms: (a) convective heat transfer between the two fluids, and (b) evaporative cooling of liquid water as it evaporates into the air. Dry cooling towers lack this second mechanism, which can easily provide more than half of the overall cooling of the water in a wet cooling tower. The disadvantage of the wet cooling tower is that cooling water is continuously lost to the air in the form of water vapor, and this cooling water must be replaced; doing so requires a body of water available nearby.

Figure 10.23 presents a schematic diagram of a wet cooling tower to be used in the psychrometric analysis of cooling towers. The hot cooling water enters at state A, and the cooled cooling water exits at state B. The ambient air enters at 1, and the heated, humidified air exits at state 2. With wet cooling towers, due to the residence time of the air in contact with the water, the outlet air is often saturated. In our analysis, we will consider the system to be at steady state, and we will assume that there are no influences from kinetic or potential energy changes. We will neglect any external heat transfer because we do not expect such a transfer to be large compared to the internal transfer of heat between fluids. The mass flow rate of dry air entering at state 1 is equal to that exiting at state 2, and so the mass flow rate of dry air is simply written as $\dot{m}_{da}$. The conservation of mass of water for this multiple-inlet, multiple-outlet system is

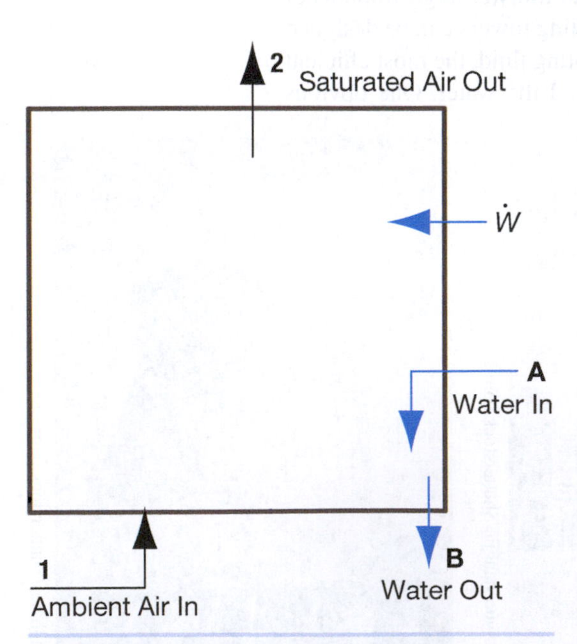

FIGURE 10.23 A schematic diagram of a cooling tower, showing the water inlet and outlet and the air inlet and outlet.

$$\dot{m}_{v,1} + \dot{m}_{w,A} = \dot{m}_{v,2} + \dot{m}_{w,B} \tag{10.36}$$

We will introduce the term Ω, which is similar to the humidity ratio for water vapor but is the mass of liquid water divided by the mass of dry air: $\Omega = \frac{\dot{m}_w}{\dot{m}_{da}}$. With this term, if Eq. (10.36) is divided by the mass flow rate of dry air, the resulting expression is

$$\omega_1 + \Omega_A = \omega_2 + \Omega_B \tag{10.37}$$

Neglecting external heat transfer, but still allowing for possible work interactions, we can write the First Law of Thermodynamics as

$$-\dot{W} + \dot{m}_{da}h_{a,1} + \dot{m}_{w,A}h_{w,A} = \dot{m}_{da}h_{a,2} + \dot{m}_{w,B}h_{w,B} \tag{10.38}$$

where psychrometric enthalpies are used. The enthalpy of the liquid water can be approximated as $h_w = h_f(T)$, as previously done in this chapter. With this substitution, and dividing by the mass flow rate of dry air, we can rewrite Eq. (10.38) as

$$-\frac{\dot{W}}{\dot{m}_{da}} + h_{a,1} + \Omega_A h_{f,A} = h_{a,2} + \Omega_B h_{f,B} \tag{10.39}$$

In some cooling tower analyses, it may be advantageous to split the psychrometric enthalpy into its component terms. If this is done, the change in specific enthalpy of the dry air can be found using constant specific heats ($h_{da,2} - h_{da,1} = c_{p,da}(T_2 - T_1)$), and the specific enthalpy of the water vapor can be taken as that of saturated water vapor at the air temperature ($h_v = h_g(T)$). With these modifications, Eqs. (10.37) and (10.39) can be combined to yield

$$-\frac{\dot{W}}{\dot{m}_{da}} + \omega_1 h_{g,1} + \Omega_A h_{f,A} = c_{p,da}(T_2 - T_1) + \omega_2 h_{g,2} + [\Omega_A - (\omega_2 - \omega_1)]h_{f,B} \tag{10.40}$$

which allows us to analyze the cooling tower without knowing how much cooling water is being returned from the cooling tower.

STUDENT EXERCISE

Develop a computer model that can be used to analyze cooling towers, based on a variety of inputs, and that can provide a variety of outputs, including temperatures, humidities, and flow rates.

▶ EXAMPLE 10.10

A natural-draft cooling tower at a power plant receives cooling water from the condenser at 30°C and cools it to 15°C. The rate at which water is received by the cooling tower is 800,000 kg/min. The ambient air conditions are 10°C and the relative humidity is 40%. The air exits the cooling tower as saturated air at 25°C. Consider the air pressure to be constant at 101.325 kPa. Find (a) the amount of make-up water needed to compensate for evaporation, and (b) the volumetric flow rate of air needed to enter the system.

Given: $T_A = 30°C$, $T_B = 15°C$, $T_1 = 10°C$, $\phi_1 = 40\%$, $T_2 = 25°C$, $\phi_2 = 100\%$,
$\dot{m}_{w,A} = 800{,}000$ kg H_2O/min

Find: (a) $\dot{m}_{\text{make-up}}$, (b) $\dot{V}_1$

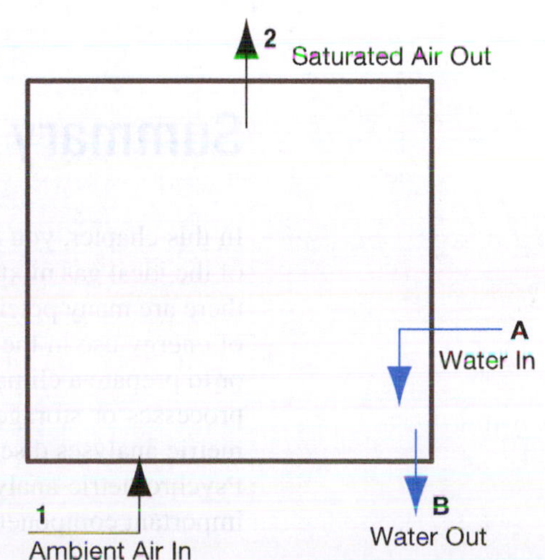

Solution: The working fluids are moist air and liquid water. The system under consideration is a standard natural-draft cooling tower.

Assume: $\dot{Q} = \dot{W} = 0$, no impacts of kinetic or potential energy of the streams. Steady-state, steady-flow open system.

From computer programs, the following humidity ratios for the inlet and outlet air, as well as the enthalpies of the liquid water and water vapor, can be found:

$\omega_1 = 0.00304$ kg H_2O/kg dry air $\omega_2 = 0.0201$ kg H_2O/kg dry air
$h_{g,1} = 2519$ kJ/kg H_2O $h_{g,2} = 2547$ kJ/kg H_2O
$h_{f,A} = 125.71$ kJ/kg H_2O $h_{f,B} = 62.97$ kJ/kg H_2O
$c_{p,da} = 1.005$ kJ/kg dry air · K

With these values, we can use Eq. (10.40),

$$-\frac{\dot{W}}{\dot{m}_{da}} + \omega_1 h_{g,1} + \Omega_A h_{f,A} = c_{p,da}(T_2 - T_1) + \omega_2 h_{g,2} + [\Omega_A - (\omega_2 - \omega_1)]h_{f,B}$$

to solve for $\Omega_A = 0.904$ kg H_2O/kg dry air.

Then, $\dot{m}_{da} = \frac{\dot{m}_{w,A}}{\Omega_A} = 884{,}956$ kg dry air/min.

(a) The amount of make-up water needed is equal to that lost to evaporation. The amount lost to evaporation is the difference in the mass flow rates of the water vapor entering and exiting:

$$\dot{m}_{\text{make-up}} = \dot{m}_{\text{evap}} = \dot{m}_{v,2} - \dot{m}_{v,1} = \dot{m}_{da}(\omega_2 - \omega_1) = \mathbf{15{,}100\ kg\ H_2O/min}$$

(b) From the computer program, the psychrometric specific volume of the ambient air is

$$v_{a,1} = 0.805\ \text{m}^3/\text{kg dry air}$$

Then the volumetric flow rate of air needed to enter the system is

$$\dot{V}_1 = \dot{m}_{da} v_{a,1} = \mathbf{713{,}000\ m^3/min = 11{,}900\ m^3/s}$$

Analysis: This gives you an idea of the sheer size of power plants. The cooling water can be considered to be extracting 836 MW of heat from the condensing steam. If the power plant has a thermal efficiency of 40%, the amount of power produced by the facility is approximately 557 MW. While this is large, many steam power plants produce over 1000 MW of electricity. Yet even this plant requires a huge amount of cooling air to operate. To accommodate the air and the water, the cooling tower must be very large.

QUESTION FOR THOUGHT/DISCUSSION

What are some of the local environmental impacts of cooling towers, and what, if any, global impacts can be attributed to cooling towers?

Summary

In this chapter, you have learned the fundamentals behind psychrometrics, which is the study of the ideal gas mixture of dry air and water vapor. This mixture is the air that surrounds us, so there are many potential applications that require psychrometric analysis. One prominent area of energy use in the world is the alteration of the condition of air to improve human comfort or to prepare a climate-controlled environment for applications such as certain manufacturing processes or storage systems. Engineers working on such applications rely on the psychrometric analyses discussed in this chapter to assist in the design of appropriate HVAC systems. Psychrometric analyses can also reach into such fields as cooling tower design, which is an important component of many power plant designs today.

KEY EQUATIONS

Relative Humidity:

$$\phi = \frac{P_v}{P_{sat}(T)} \quad (10.2)$$

Humidity Ratio:

$$\omega = 0.622\frac{\phi P_{sat}(T)}{P - \phi P_{sat}(T)} \quad (10.6)$$

Dew Point Temperature:

$$T_{dp} = T_{sat}(P_v) \quad (10.7)$$

Cooling/Dehumidifying Processes:

$$\dot{m}_{w,c} = \dot{m}_{v,1} - \dot{m}_{v,2} \quad (10.17)$$

$$\dot{Q}_c = \dot{m}_{da}(h_{a,2} - h_{a,1}) + \dot{m}_{w,c}h_{w,c} \quad (10.18)$$

$$\dot{Q}_h = \dot{m}_{da}(h_{a,3} - h_{a,2}) \quad (10.21)$$

Heating/Humidifying Processes:

$$\dot{Q}_h = \dot{m}_{da}(h_{a,2} - h_{a,1}) \quad (10.24)$$

$$\dot{m}_w = \dot{m}_{v,3} - \dot{m}_{v,2} \quad (10.26)$$

Mixing of Two Moist Air Streams:

$$\omega_3 = \frac{\dot{m}_{da,1}}{\dot{m}_{da,3}}\omega_1 + \frac{\dot{m}_{da,2}}{\dot{m}_{da,3}}\omega_2 \quad (10.32)$$

$$h_{a,3} = \frac{\dot{m}_{da,1}}{\dot{m}_{da,3}}h_{a,1} + \frac{\dot{m}_{da,2}}{\dot{m}_{da,3}}h_{a,2} \quad (10.35)$$

Cooling Towers:

$$-\frac{\dot{W}}{\dot{m}_{da}} + \omega_1 h_{g,1} + \Omega_A h_{f,A} = c_{p,da}(T_2 - T_1) + \omega_2 h_{g,2} + [\Omega_A - (\omega_2 - \omega_1)]h_{f,B} \quad (10.40)$$

PROBLEMS

10.1 Determine the relative humidity of moist air mixtures having the following properties: (a) $T = 20°C$, $P_v = 0.725$ kPa; (b) $T = 30°C$, $y_v = 0.025$, $P = 101$ kPa; (c) $T = 10°C$, $y_v = 0.005$, $P = 150$ kPa; (d) $T = 40°C$, $y_v = 0.025$, $P = 250$ kPa.

10.2 Determine the relative humidity of moist air mixtures having the following properties: (a) $T = 25°C$, $P_v = 2.5$ kPa absolute; (b) $T = 5°C$, $P_v = 0.5$ kPa absolute; (c) $T = 30°C$, $y_v = 0.017$, $P = 200$ kPa absolute; (d) $T = 65°C$, $y_v = 0.021$, $P = 700$ kPa absolute.

10.3 Determine the partial pressure of the water vapor in moist air at the following conditions: (a) $T = 25°C$, $\phi = 75\%$, $P = 101$ kPa; (b) $T = 35°C$, $\phi = 25\%$, $P = 101$ kPa; (c) $T = 45°C$, $\phi = 50\%$, $P = 300$ kPa; (d) $T = 12°C$, $\phi = 100\%$, $P = 101$ kPa; (e) $T = 30°C$, $\phi = 40\%$, $P = 101$ kPa; (f) $T = 30°C$, $\phi = 40\%$, $P = 550$ kPa.

10.4 Determine the humidity ratio and dew point temperature for moist air under the following conditions: (a) $T = 15°C$, $\phi = 85\%$, $P = 100$ kPa; (b) $T = 35°C$, $\phi = 85\%$, $P = 100$ kPa; (c) $T = 35°C$, $\phi = 30\%$, $P = 100$ kPa; (d) $T = 15°C$, $\phi = 85\%$, $P = 500$ kPa.

10.5 Determine the humidity ratio and dew point temperature for moist air under the following conditions: (a) $T = 15°C$, $\phi = 80\%$, $P = 100$ kPa; (b) $T = 35°C$, $\phi = 80\%$, $P = 100$ kPa; (c) $T = 35°C$, $\phi = 30\%$, $P = 100$ kPa; (d) $T = 15°C$, $\phi = 80\%$, $P = 1$ MPa.

10.6 Determine the relative humidity and the dew point temperature for moist air under the following conditions (all at 101.325 kPa): (a) $T = 20°C$, $\omega = 0.0075$ kg H_2O/kg dry air; (b) $T = 30°C$, $\omega = 0.0115$ kg H_2O/kg dry air; (c) $T = 15°C$, $\omega = 0.004$ kg H_2O/kg dry air; (d) $T = 35°C$, $\omega = 0.015$ kg H_2O/kg dry air.

10.7 Plot the relative humidity and the dew point temperature for moist air at 35°C and 101.325 kPa as the humidity ratio varies between 0 and 0.035 kg H_2O/kg dry air.

10.8 Plot the humidity ratio and dew point temperature for moist air at 30°C and 101.325 kPa as the relative humidity varies between 0 and 100%.

10.9 Plot the relative humidity for moist air at 25°C and 101.325 kPa as the humidity ratio varies between 0 and 0.020 kg H_2O/kg dry air.

10.10 Plot the air temperature for moist air with a dew point of 10°C and a pressure of 101.325 kPa as the relative humidity varies between 5 and 100%.

10.11 Plot the air temperature for moist air with a humidity ratio of 0.007 kg H_2O/kg dry air and a pressure of 101.325 kPa as the relative humidity varies from 5 to 100%.

10.12 Choose a combination of temperature and relative humidity that you consider to be comfortable in the summer, and a combination of these properties that you consider to be comfortable in the winter. Determine the dew point temperature for these two conditions.

10.13 Moist air enters a pipe at a volumetric flow rate of 0.5 m^3/s. The air has a temperature of 25°C, a relative humidity of 30%, and a pressure of 101.325 kPa. Determine the mass flow rate of dry air, the psychrometric specific volume, and the psychrometric specific enthalpy of the mixture.

10.14 For the moist air in Problem 10.13, plot the mass flow rate of dry air, the psychrometric specific volume, and the psychrometric specific enthalpy for relative humidities ranging between 20 and 100%, while the air temperature is maintained at 25°C.

10.15 Moist air flows through an HVAC duct at a volumetric flow rate of 0.25 m^3/s. The air has a temperature of 15°C, a relative humidity of 70%, and a pressure of 100.9 kPa. Determine the mass flow rate of dry air, the psychrometric specific volume, and the psychrometric specific enthalpy of the mixture.

10.16 For the moist air in Problem 10.15, plot the mass flow rate of dry air, the psychrometric specific volume, and the psychrometric specific enthalpy for relative humidities ranging between 10 and 100% while the air temperature is maintained at 15°C.

10.17 Moist air is extracted from a factory at a rate of 560 L/s. The air has a temperature of 35°C and a relative humidity of 70%. Determine the mass flow rate of dry air, the psychrometric specific volume, and the psychrometric specific enthalpy of the mixture.

10.18 Using a psychrometric chart, determine the relative humidity, humidity ratio, psychrometric specific enthalpy, and dew point temperature for atmospheric-pressure moist air at the following conditions: (a) $T_{db} = 25°C$, $T_{wb} = 20°C$; (b) $T_{db} = 25°C$, $T_{wb} = 10°C$; (c) $T_{db} = 25°C$, $T_{wb} = 25°C$; (d) $T_{db} = 15°C$, $T_{wb} = 5°C$. Check your answers with a software program.

10.19 Using a psychrometric chart, determine the relative humidity, humidity ratio, psychrometric specific enthalpy, and dew point temperature for atmospheric-pressure moist air at the following conditions: (a) T_{db} = 25°C, T_{wb} = 15°C; (b) T_{db} = 25°C, T_{wb} = 10°C; (c) T_{db} = 25°C, T_{wb} = 25°C; (d) T_{db} = 12°C, T_{wb} = 2°C. Check your answers with a software program.

10.20 Using a psychrometric chart, determine the wet bulb temperature, humidity ratio, psychrometric specific enthalpy, and dew point temperature for atmospheric-pressure moist air at the following conditions: (a) T_{db} = 25°C, ϕ = 20%; (b) T_{db} = 22°C, ϕ = 40%; (c) T_{db} = 27°C, ϕ = 70%; (d) T_{db} = 15°C, ϕ = 45%. Check your answers with a software program.

10.21 Using a psychrometric chart, determine the wet bulb temperature, humidity ratio, psychrometric specific enthalpy, and relative humidity for atmospheric-pressure moist air at the following conditions: (a) T_{db} = 25°C, T_{dp} = 20°C; (b) T_{db} = 21°C, T_{dp} = 10°C; (c) T_{db} = 24°C, T_{dp} = 15°C; (d) T_{db} = 15°C, T_{dp} = 14°C. Check your answers with a software program.

10.22 Using a psychrometric chart, determine the wet bulb temperature, humidity ratio, psychrometric specific enthalpy, and dew point temperature for atmospheric-pressure moist air at the following conditions: (a) T_{db} = 11°C, ϕ = 57%; (b) T_{db} = 22°C, ϕ = 82%; (c) T_{db} = 24°C, ϕ = 73%; (d) T_{db} = 6°C, ϕ = 57%. Check your answers with a software program.

10.23 An adiabatic saturator is used to determine the humidity of inlet air. The water temperature and exit air temperatures are both 25°C, and the exit air is saturated. The air inlet temperature is measured as 30°C. Determine the humidity ratio and relative humidity of the inlet air stream, assuming the air pressure is 101.325 kPa.

10.24 An adiabatic saturator receives air at 20°C, and saturated air exits the device at 10°C. The water temperature is also 10°C. Determine the inlet air humidity ratio and the relative humidity if the air pressure is 101.325 kPa.

10.25 An adiabatic saturator receives air at 25°C and has a water temperature of 10°C. The saturated air exits at 10°C. Determine the inlet air humidity ratio and the relative humidity if the air pressure is 0.3 MPa.

10.26 Hot, humid air at 30°C and 60% relative humidity enters a cooling and dehumidifying device at a mass flow rate of dry air of 2.5 kg/s. After passing over the cooling coils, the saturated air is heated to 22°C and a relative humidity of 30%. The air pressure is 101.325 kPa. Determine the heat transfer rates through the cooling coils and the heating element, and determine the rate of condensed water leaving the system.

10.27 Moist air at 30°C and a relative humidity of 40% enters a cooling and dehumidifying system. The air exits the system at 20°C and 40% relative humidity. The air pressure is 101.325 kPa, and the mass flow rate of dry air is 2 kg/s. Determine (a) the rate at which water is condensed from the air, (b) the heat transfer rate through the cooling coils, and (c) the heat transfer rate through the heating element.

10.28 Moist air at atmospheric pressure and a volumetric flow rate of 3.5 m^3/s enters a cooling and dehumidifying system at 25°C and a relative humidity of 90%. The air exits the system at 20°C with a relative humidity of 25%. Determine (a) the rate at which water is condensed from the air, (b) the heat transfer rate through the cooling coils, and (c) the heat transfer rate through the heating element.

10.29 An air conditioning system is to cool and dehumidify moist air. The air enters the system at 32°C with a relative humidity of 50% at a rate of 0.55 m^3/s. The air is to exit the system at 18°C with a relative humidity of 40%. Determine (a) the rate at which water is condensed from the air, (b) the heat transfer rate through the cooling coils, and (c) the heat transfer rate through the heating element.

10.30 Moist air at atmospheric pressure enters a cooling and dehumidifying system at 32°C and a relative humidity of 50%. The volumetric flow rate of the air entering the system is 1.5 m^3/s. The air is to exit the system at a temperature of 18°C. Plot the mass flow rate of condensed water, the heat transfer rate through the cooling coil, and the heat transfer rate through the heating element for exit state relative humidities ranging between 35% and 100%.

10.31 Repeat Problem 10.30, but with an inlet air state of 27°C and a relative humidity of 60%.

10.32 Moist air at atmospheric pressure enters a cooling and dehumidifying system at 32°C and a relative humidity of 50%. The volumetric flow rate of the air entering the system is 1.5 m^3/s. The air is to exit the system at a relative humidity of 40%. Plot the mass flow rate of condensed water, the heat transfer rate through the cooling coil, and the heat transfer rate through the heating element for exit state temperatures ranging between 5°C and 30°C.

10.33 Moist air at atmospheric pressure enters a cooling and dehumidifying system at 30°C and a relative humidity of 70%. The volumetric flow rate of the air entering the system is 140 L/s. The air is to exit the system at a temperature of 20°C. Plot the mass flow rate of condensed water, the heat transfer rate through the cooling coil, and the heat transfer rate through the heating element for exit state relative humidities ranging between 35% and 100%.

10.34 Moist air at atmospheric pressure enters a cooling and dehumidifying system at 35°C and a volumetric flow rate of 2.0 m^3/s. The air exiting the system is to have a temperature of 20°C and a relative humidity of 50%. Plot the mass flow rate of condensed water, the heat transfer rate through the cooling coil, and the heat transfer rate through the heating element for values of the inlet air relative humidities between 30% and 80%.

10.35 Moist air at atmospheric pressure enters a cooling and dehumidifying system at a relative humidity of 80% and a volumetric flow rate of 2.0 m^3/s. The air exiting the system is to have a temperature of 20°C and a relative humidity of 50%. Plot the mass flow rate of condensed water, the heat transfer rate through the cooling coil, and the heat transfer rate through the heating element for values of the inlet temperature ranging between 15°C and 35°C.

For Problems 10.36 through 10.42, assume that the refrigerant exits the evaporator as a saturated vapor and that the refrigerant exits the condenser as a saturated liquid.

10.36 A vapor-compression refrigeration cycle is to provide the heat removal mechanism for the cooling coils described in Problem 10.26. The refrigerant used in the system is R-134a. The evaporator pressure is 210 kPa and the condenser pressure is 1200 kPa. Determine the power required by the compressor of the refrigeration cycle if (a) the compressor is isentropic, and (b) the isentropic efficiency of the compressor is 80%.

10.37 Repeat Problem 10.36 if the refrigerant is ammonia instead of R-134a.

10.38 A vapor-compression refrigeration cycle is to provide the heat removal mechanism for the cooling coils described in Problem 10.27. The refrigerant used in the system is R-134a. The evaporator pressure is 140 kPa absolute and the condenser pressure is 0.98 MPa. Determine the power required by the compressor of the refrigeration cycle if (a) the compressor is isentropic, and (b) the isentropic efficiency of the compressor is 80%.

10.39 A vapor-compression refrigeration cycle is to provide the heat removal mechanism for the cooling coils described in Problem 10.28. The refrigerant used in the system is R-134a. The evaporator pressure is 150 kPa and the condenser pressure is 1000 kPa. Plot the power required by the compressor of the refrigeration cycle as a function of the isentropic efficiency of the compressor, with a range of isentropic efficiencies between 0.40 and 1.0.

10.40 A vapor-compression refrigeration cycle is to provide the heat removal mechanism for the cooling coils described in Problem 10.26. The refrigerant used in the system is R-134a. The condenser pressure is 1000 kPa. The isentropic efficiency of the compressor is 0.80. Plot the power required by the compressor of the refrigeration cycle for evaporator pressures ranging between 100 kPa and 300 kPa.

10.41 A vapor-compression refrigeration cycle is to provide the heat removal mechanism for the cooling coils described in Problem 10.27. The refrigerant used in the system is R-134a. The evaporator pressure is 275 kPa absolute. The isentropic efficiency of the compressor is 0.80. Plot the power required by the compressor of the refrigeration cycle for condenser pressures ranging between 830 kPa absolute and 2 MPa.

10.42 Repeat Problem 10.41, but use ammonia as the refrigerant instead of R-134a.

10.43 0.45 m^3/s of moist air enters a furnace at a temperature of 15°C, a relative humidity of 55%, and a pressure of 101.325 kPa. The air is heated to 22°C. Determine the exit state relative humidity and the rate of heat transfer to the air.

10.44 A moist air stream enters a heating unit at 5°C, 101.325 kPa, and a relative humidity of 70%. The volumetric flow rate of this air stream is 2.5 m^3/s. Determine the exit state relative humidity and the rate of heat transfer in order for the air to exit at a temperature of 25°C.

10.45 Moist air, with a temperature of −4°C, a pressure of 101.325 kPa, and a relative humidity of 60%, enters a heater at a volumetric flow rate of 140 L/s. The air is heated to 25°C. What is the relative humidity of the exiting air and the rate of heat addition for the process?

10.46 Moist air, at 15°C, 101.325 kPa, and a relative humidity of 90%, enters a heater at a volumetric flow rate of 0.5 m^3/s. Plot the exit state relative humidity and the rate of heat transfer for the process for exit temperatures ranging between 18°C and 35°C.

10.47 Saturated air at 12°C and 101.325 kPa is to be heated, with a volumetric flow rate of 0.75 m^3/s entering the heater. Plot the exit state temperature and the rate of heat transfer for the process for exit state relative humidities varying between 10% and 90%.

10.48 A cool stream of air enters a heater and humidification system at a temperature of 5°C, a pressure of 101.325 kPa, a relative humidity of 60%, and a volumetric flow rate of 0.25 m^3/s. The air is to exit the system at 22°C and a relative humidity of 45%. The humidification process is to use evaporation of liquid water whose temperature is 20°C. Determine the temperature and relative humidity of the air between the heater and humidifier, the rate of heat transfer to the air through the heating element, and the rate of water addition to the air through the humidifier section.

10.49 Repeat Problem 10.48, but with the humidification process being performed through the addition of steam at 100°C.

10.50 Moist air enters a heater and humidification system at a temperature of 2°C, a pressure of 101.325 kPa, a relative humidity of 70%, and a volumetric flow rate of 110 L/s. The air is to exit the system at 24°C and a relative humidity of 40%. The humidification process is to use evaporation of liquid water whose temperature is 15°C. Determine the temperature and relative humidity of the air between the heater and humidifier, the rate of heat transfer to the air through the heating element, and the rate at which water is added to the air through the humidifier section.

10.51 Repeat Problem 10.50, but with the humidification process being performed through the addition of steam at 100°C.

10.52 Air enters a heating/humidifying system that uses the evaporation of liquid water to humidify the air. The entrance conditions are 5°C, a relative humidity of 80%, a pressure of 101.325 kPa, and a volumetric flow rate of 0.25 m^3/s. The final exit state is to be 23°C with a relative humidity of 50%. The temperature of the liquid water that is evaporated is 20°C. Determine the rate of heat addition in the heating element and the rate at which water is added in the humidifier.

10.53 Moist air at 10°C, 101.325 kPa, and a relative humidity of 40% enters a heating and humidifying system at a volumetric flow rate of 2.0 m^3/s. The air is to be humidified using liquid water, and the final exit state of the air is to be 25°C and 40% relative humidity. Plot the rate of heat addition, the temperature of the air between the heating and humidification systems, and the amount of water needed to be added as the temperature of the liquid water varies between 10°C and 50°C.

10.54 Moist air at 8°C, 101.325 kPa, and a relative humidity of 50% enters a heating and humidifying system at a volumetric flow rate of 1.0 m^3/s. The air is to be humidified using liquid water, and the final exit state of the air is to be 25°C and 50% relative humidity. Plot the rate of heat addition, the relative humidity of the air between the heating and humidification systems, and the amount of water needed to be added as the temperature of the liquid water varies between 10°C and 50°C.

10.55 Moist air at 4°C and 101.325 kPa enters a heating and humidifying system at a volumetric flow rate of 140 L/s. The air is to be humidified using liquid water at 15°C. The final state of the air is 24°C and a relative humidity of 50%. Plot the rate of heat addition, the air temperature of the intermediate state between the heating and humidification systems, and the amount of water needed to be added as the inlet relative humidity varies between 10% and 100%.

10.56 Moist air with a relative humidity of 70% and a pressure of 101.325 kPa enters a heating and humidifying system at a volumetric flow rate of 1.5 m^3/s. The air is to be humidified using steam at 100°C. The final state of the air is 25°C and a relative humidity of 50%. Plot the rate of heat addition, the air temperature of the intermediate state between the heating and humidification systems, and the amount of water needed to be added for inlet temperatures ranging from 0°C to 15°C.

10.57 Air is to be humidified using liquid water at 10°C. Before humidification, the air is at 40°C, 101.325 kPa, and a relative humidity of 10%. The mass flow rate of the dry air in the system is 1 kg/s. Plot the exit temperature and exit relative humidity of the air for mass flow rates of evaporating water between 0.001 kg/s and 0.005 kg/s.

10.58 Air is to be humidified using steam at 100°C. Prior to humidification, the air has a temperature of 16°C, a pressure of 101.325 kPa, and a relative humidity of 20%. The mass flow rate of the dry air in the system is 0.5 kg/s. Plot the exit temperature and the exit relative humidity of the air for mass flow rates of steam between 0.5 g/s and 4 g/s.

10.59 Two streams of moist air are to be mixed. The first stream enters the mixing chamber with a dry air mass flow rate of 5 kg/s, a temperature of 30°C, and a relative humidity of 40%. The second stream enters the chamber with a dry air mass flow rate of 3 kg/s, a temperature of 10°C, and a relative humidity of 80%. The pressure of the system is maintained at 101.325 kPa. What are the temperature and relative humidity of the mixed stream?

10.60 A stream of moist air with a temperature of 15°C and a relative humidity of 70% enters a mixing chamber at a volumetric flow rate of 0.50 m^3/s. A second stream of moist air enters the mixing chamber at a flow rate of 0.20 m^3/s, a temperature of 33°C, and a relative humidity of 25%. The pressure of the system is 101.325 kPa. What are the temperature, relative humidity, and dew point temperature of the mixed stream?

10.61 Air from an office building is to be recirculated and mixed with outside air. The air from inside the office building enters the mixing chamber at a volumetric flow rate of 5 m^3/s, a temperature of 24°C, and a relative humidity of 40%. The outside air has a volumetric flow rate of 2 m^3/s, a temperature of 12°C, and a relative humidity of 80%. The pressures of all streams are 101.325 kPa. Determine the temperature and relative humidity of the mixed stream.

10.62 To keep air fresh in a factory, recirculated air is mixed with external air. The air from the factory has a volumetric flow rate of 3 m^3/s, a temperature of 32°C, and a relative humidity of 60%. The outside air has a volumetric flow rate of 1 m^3/s, a temperature of 20°C, and a relative humidity of 60%. (a) Determine the temperature and relative humidity of the mixed stream. (b) It is desired to return the air to the factory at 15°C and a relative humidity of 40%. To do this, the mixed stream is sent through a cooling and dehumidifying system. Determine the heat transfer rates for the cooling coil section and the heating element section for this system.

10.63 In winter, fresh air is needed in an office building to maintain air quality. Engineers are considering mixing fresh air with recirculated air from the building before heating the air in order to reduce heating loads. The outside air is −1°C with a relative humidity of 70%, and 280 L/s of this air is needed. The inside air enters the mixing chamber at 25°C and a relative humidity of 30% at a volumetric flow rate of 225 L/s. The mixed stream is then to be heated and humidified and returned to the building at 23°C and a relative humidity of 50%. The whole system is maintained at 101.325 kPa. Determine the heating load for (a) the outside air alone, and (b) the mixed stream of outside and interior air. The humidification is to occur with cool water, so do not be concerned with the energy requirements of humidification. Is it more or less energy intensive to heat the mixed stream as opposed to the fresh air alone?

10.64 Three streams, all at 101.325 kPa pressure, are to be mixed together. Stream 1 has a dry air mass flow rate of 2 kg/s, a temperature of 20°C, and a relative humidity of 30%. Stream 2 has a dry air mass flow rate of 5 kg/s, a temperature of 25°C, and a relative humidity of 70%. Stream 3 has a dry air mass flow rate of 3 kg/s, a temperature of 5°C, and a relative humidity of 40%. Determine the temperature and relative humidity of the mixed stream.

10.65 Four streams of atmospheric-pressure moist air are mixed together. The volumetric flow rates, temperatures, and relative humidities of the four streams are listed in the accompanying table. Determine the temperature and relative humidity of the mixed stream.

Stream	Temperature (°C)	Relative Humidity	Volumetric Flow Rate (L/s)
A	5	40	140
B	10	80	200
C	25	50	85
D	30	30	110

10.66 Two moist air streams are to be mixed together to produce a stream with a temperature of 20°C and a relative humidity of 40%. One inlet stream is at 15°C and a relative humidity of 50%. The second stream has a temperature of 24°C but a variable relative humidity. Determine the ratio of the mass flow rates of dry air for the two inlet streams and the relative humidity of the second stream. The pressure of each stream is 101.325 kPa.

10.67 Two moist air streams are mixed together to produce a stream with a temperature of 25°C and a relative humidity of 60%. One inlet stream is known to have a temperature of

30°C and a relative humidity of 50%. The other inlet stream is unknown. It is also known that the mass flow rate of dry air of the known inlet stream is twice that of the unknown inlet stream. All pressures are 101.325 kPa. Determine the temperature, humidity ratio, and relative humidity of the unknown inlet stream.

10.68 Two atmospheric-pressure inlet streams are mixed together. Stream 1 has a temperature of 10°C, a relative humidity of 80%, and a volumetric flow rate of 2 m^3/s. Stream 2 has a temperature of 30°C and a relative humidity of 60%. Plot the temperature and relative humidity of the exit streams as the volumetric flow rate of the second stream varies between 1 m^3/s and 10 m^3/s.

10.69 Two atmospheric-pressure inlet streams are mixed together. Stream 1 has a temperature of 2°C, a relative humidity of 70%, and a volumetric flow rate of 55 L/s. Stream 2 has a temperature of 30°C and a relative humidity of 65%. Plot the temperature and relative humidity of the exit streams as the volumetric flow rate of the second stream varies between 12.5 L/s and 250 L/s.

10.70 Outside air is to be mixed with interior air and then sent through a cooling and dehumidifying system whose cooling coil acts as the evaporator of a vapor-compression refrigeration cycle. The outside air has a temperature of 28°C and a relative humidity of 60%. The interior air to be mixed with the outside air has a temperature of 24°C and a relative humidity of 35%. The mixed stream is to be returned to the space at a temperature and relative humidity of 20°C and 50%, at a volumetric flow rate of 5 m^3/s. All pressures of the moist air streams are 101.325 kPa. The vapor-compression refrigeration cycle uses R-134a, and the R-134a exits the evaporator as a saturated vapor at 200 kPa and exits the condenser as a saturated liquid at 1000 kPa. The isentropic efficiency of the compressor is 0.78. Plot the power consumed by the compressor as a function of the fraction of the interior air used in the mixing (from 0% to 100% interior air).

10.71 Repeat Problem 10.70, but consider the outside air to have a temperature of 33°C and a relative humidity of 30%.

10.72 Outside air is to be mixed with interior air and then sent to a heating and humidifying system. The humidifier uses evaporation of liquid water to achieve humidification. The outside air has a temperature of 2°C and a relative humidity of 60%. The interior air has a temperature of 20°C and a relative humidity of 40%. The heating and humidifying system is to deliver 1.5 m^3/s of air at 24°C and a relative humidity of 60% to the space. The pressures of all the moist air streams are 101.325 kPa. The heating system uses heat produced in the combustion of a fuel. The combustion process provides 28,500 kJ/kg of fuel. The cost of the fuel is $1.60/kg of fuel. Plot the cost of the fuel as a function of outside air used in the mixed stream, as the percentage of the outside air varies between 0% and 100%.

10.73 A wet cooling tower is used to transfer heat from cooling water to the air. The air enters the cooling tower at 101.325 kPa, 15°C, and a relative humidity of 40%. The air exits as saturated air at 101.325 kPa and 32°C. The cooling water enters the cooling tower at 35°C and exits the cooling tower at 20°C. The mass flow rate of the cooling water is 500 kg/s. Determine the rate at which water must be replaced in the cooling water returning to the process, and the volumetric flow rate of the incoming moist air to the cooling tower.

10.74 A wet cooling tower is used to cool water that cycles through a condenser of an HVAC plant for a factory. The air enters the cooling tower at 20°C and a relative humidity of 60%. The air exits the cooling tower saturated at 35°C. The cooling water enters the cooling tower at 38°C and returns to the condenser after exiting the cooling tower at 21°C. The mass flow rate of the cooling water is 150 kg/s. Determine the rate at which water must be replaced in the cooling water that returns to the condenser, and the volumetric flow rate of the air that enters the cooling tower.

10.75 A small power plant uses a wet cooling tower to transfer heat from cooling water to the air. The air enters the cooling tower at 101.325 kPa, 4°C, and a relative humidity of 70%. The air exits as saturated air at 101.325 kPa and 25°C. The cooling water enters the cooling tower at 30°C and exits the cooling tower at 7°C. The mass flow rate of the cooling water is 180 kg/s. Determine the rate at which water must be replaced in the cooling water returning to the process, and the volumetric flow rate of the incoming moist air to the cooling tower.

10.76 A wet cooling tower receives air at 20°C and a relative humidity of 50%. The air exits as saturated air at 30°C. The air pressure is 101.325 kPa. Cooling water enters the tower at 35°C and exits at 22°C. Plot the volumetric flow rate of air entering the cooling tower as the cooling water flow rate entering the tower varies between 100 kg/s and 10,000 kg/s.

10.77 A wet cooling tower is used to cool water that enters at a rate of 1000 kg/s and 35°C. The air enters the cooling tower at 10°C and a relative humidity of 40%. The cooling water is to exit the cooling tower at 15°C. Saturated air is to exit the cooling tower. Plot the volumetric flow rate of air entering the cooling tower and the rate of evaporation of the cooling water as the air exit temperature varies between 15°C and 30°C.

10.78 A wet cooling tower receives water at a rate of 900 kg/s and a temperature of 30°C. The water leaves the cooling tower at 15°C. The air exits the cooling tower at a temperature of 29°C and a relative humidity of 100%. The moist air entering the cooling tower is at 13°C. Plot the volumetric flow rate of air entering the cooling tower and the rate of evaporation of the cooling water as the relative humidity of the moist air entering the tower varies between 10% and 100%.

10.79 A wet cooling tower is located in an area which is frequently foggy in the early morning. Water enters the cooling tower at a rate of 500 kg/s and a temperature of 30°C, and must exit the cooling tower at 15°C. Saturated air enters the cooling tower, and saturated air at a temperature of 27°C exits the tower. Plot the volumetric flow rate of air entering the cooling tower and the rate of evaporation of the cooling water as the temperature of the saturated air entering the tower varies between 2°C and 15°C.

10.80 A wet cooling tower receives air with a relative humidity of 75% at a rate of 12,000 m^3/s. The air exits at a relative humidity of 100% and a temperature of 35°C. The cooling water enters at a temperature of 40°C and exits at a temperature equal to 2°C above the ambient temperature. Plot the mass flow rate of cooling water entering the tower and the rate of evaporation of the cooling water as the temperature of the moist air entering the tower varies between 2°C and 20°C.

10.81 A large cooling tower serving a Rankine-cycle power plant is to be placed in the desert. The ambient air can be assumed to have a relative humidity of 10%. The cooling tower receives water at a temperature of 45°C, and returns the water to the condenser of the power plant at the ambient air temperature. Saturated air exits the cooling tower at 40°C. The cooling tower receives 10,000 m^3/s of ambient air. Power plant operation is affected by the ambient temperature throughout the day. Plot the rate at which make-up water must be added to the water returning to the condenser from the cooling tower, and the percentage of water entering the cooling tower that is lost to evaporation, as the ambient air temperature varies between 5°C and 40°C. Explain the results of the analysis. Discuss some ways in which the ambient air temperature will impact the performance of the power plant.

DESIGN/OPEN-ENDED PROBLEMS

10.82 It is often necessary to bring a supply of fresh air into a building to maintain air quality. Suppose that a factory needs to be cooled in the winter, and that it is desired to achieve this cooling by mixing in fresh air from the outside. To maintain air quality, at least 10% of the air inside the building must be replaced every hour with air from outside the building. The processes in the factory cause the air to be heated to 35°C before it is removed from the building. It is desired to have the air cooled to 25°C and to maintain the humidity in the building at 40%. Specify the basic requirements of a system that will allow for this process to occur under different outdoor air conditions. Consider that the outdoor air may vary in temperature (in winter) between –15°C and 10°C, and that the relative humidity of the outside air is typically 60%.

10.83 A college student lives in an apartment without a humidifier. As a result, the air in her apartment is usually 22°C with a relative humidity of 20% in the winter. Design a system that will allow the student to humidify the air to a relative humidity of 50% without significantly lowering the temperature and without purchasing a humidifier. The volume of the student's apartment is 60 m^3.

10.84 In a cooking class, students are learning how to cook pasta and are boiling large amounts of water. This increases the relative humidity in the classroom to an uncomfortable level. In particular, the room, which has a volume of 50 m^3, has a temperature of 25°C and a relative humidity of 90% during the class. Without buying a dehumidifier, the instructor would like to reduce the relative humidity in the classroom to 50% in one hour, so that the next class can experience a more pleasant working environment. Design a system that could accomplish this dehumidification.

10.85 A cooling tower is to be built at an office building in a desert environment. When the tower is in operation, it is expected that the outdoor air conditions will be 35°C with a 10% relative humidity. Some water is available nearby, and the water temperature is 30°C. Design the thermodynamic characteristics of a cooling tower that will remove 50 MW of heat from the cooling system of the building to the environment.

10.86 A Rankine cycle power plant produces 400 MW of power at a thermal efficiency of 40%. The plant uses external cooling water in its condenser to condense the exiting steam into a condensed liquid. The external cooling water then transfers its heat to the environment through a wet cooling tower. The outside air will enter the cooling tower at 5°C with a relative humidity of 60%. The cooling water exits the condenser at 35°C. Design a cooling tower that will accomplish this task. Your design should include the cooling water exit temperature, the air exit temperature and relative humidity, the volumetric flow rate of the air, and the cooling water make-up rate.

10.87 Consider the cooling coil of Example 10.5. This cooling coil is to be the evaporator of a vapor-compression refrigeration cycle. Design a vapor-compression refrigeration cycle that will accommodate this evaporator, and discuss the practicality of the designed cycle.

CHAPTER 11

Combustion Analysis

Learning Objectives

Upon completion of Chapter 11, you will be able to

11.1 Explain the chemical fundamentals of combustion processes;

11.2 Construct the global chemical reactions of combustion processes;

11.3 Calculate the heat released during combustion;

11.4 Recognize the irreversible nature of combustion;

11.5 Discuss the fundamentals of fuel cells; and

11.6 Employ fundamental chemical equilibrium analysis.

11.1 INTRODUCTION

Combustion is the highly irreversible, heat-releasing chemical reaction between a fuel and an oxidizer. The chemical reaction changes the composition of the fuel and oxidizer (the "reactants"), creating new chemical species (the "products") in the process. Humans have used combustion for thousands of years to produce light and release heat in tasks such as cooking food. In modern times, combustion processes produce heat that is used to power equipment for a range of purposes, including creating electricity and moving vehicles. Most conventional power plants produce electricity by combustion: the heat of combustion burns a fuel and releases the chemical energy bound in the fuel compound; this energy is used to boil water or produce high-temperature gases that can operate a turbine. In turn, this turbine powers a rotating shaft inside an electrical generator. Similarly, most forms of mechanized transportation (automobiles, trucks, trains, airplanes) are powered by an internal-combustion engine in which a combustion process produces high-temperature, high-pressure gases to power a prime mover. Considering all of the applications which use combustion, including those shown in **Figure 11.1**, combustion clearly is extremely important in today's world.

Combustion is a simple, compact process by which the chemical energy present in fuels can be released to perform some meaningful task. Renewable energy sources such as wind and solar power are not harnessed on a large scale because, despite their potential to produce a great deal of energy, they are diffuse sources: that is, harvesting energy from these sources requires large amounts of land and equipment. In contrast, the chemical energy in fuels is dense. Electricity-generating power plants that use combustion to release the chemical energy in fuels as their

Paket/Shutterstock.com

Pixel Embargo/Shutterstock.com

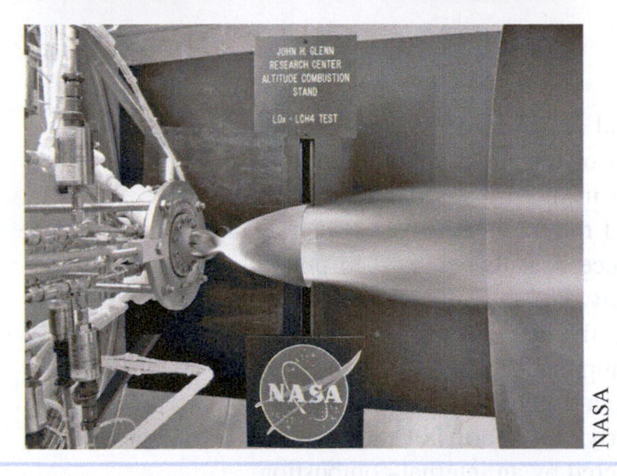

NASA

FIGURE 11.1 Various examples of combustion applications: a gas stove flame, a candle flame, and a rocket engine.

energy source are much smaller than solar- or wind-powered facilities producing the same amount of electricity. Fuels can be transported onboard vehicles to serve as the energy source to power those vehicles. Consider the difference in the amount of energy available in 1 m^3 of air at standard temperature and pressure, moving at a very high velocity of 30 m/s, and the chemical energy available for release from the combustion of 1 m^3 of methane gas at standard temperature and pressure. The kinetic energy of the moving air is approximately 0.5 kJ. The amount of energy that can be released by burning methane is approximately 32,500 kJ.

The harnessing of energy produced by combustion has led to wide-scale industrialization and tremendous improvements in quality of life, but these improvements are not without cost to our environment. Air pollution from combustion processes has led to increases in smog, has damaged the atmosphere's protective ozone layer, and has led to the phenomenon of acid rain. The CO_2 produced in the combustion of carbon-based fuels has slowly increased the CO_2 concentration in the atmosphere. Rising CO_2 levels in the air have been scientifically linked to global warming, and the potentially damaging climate change associated with such warming. Furthermore, humanity's insatiable quest for energy has resulted in the vast use of nonrenewable fossil fuels. Once fossil fuels are used, they are gone forever. A limited supply of resources and a seemingly unlimited demand for those resources have increased fuel prices in recent years (and, as we know, high fuel prices are reflected in high costs of not only the gasoline we purchase at the pump, but also the food we put on our tables). As fossil fuel supplies diminish, the price of extraction increases and the supply available decreases despite increased demand—both of these factors suggest further increases in energy prices in the future. The earth may yet hold enough fossil fuels to meet humanity's demands for hundreds of years to come, but the cost of recovering those fuels will continue to increase, and the soaring prices for obtaining and purchasing such fuels will likely make their use too costly long before their supply is depleted.

Engineers must consider these issues as they work on developing new methods of transforming energy into useful forms. Some people advocate cessation of fossil fuel combustion and consumption, but the reality is that the combustion of fossil fuels will be an important component of the world's energy usage for decades to come. Furthermore, the combustion of renewable sources may continue far into the future. In this chapter, we will learn the basics of combustion analysis so that, as an engineer, you will be able evaluate and design combustion processes. We develop methods for determining the basic composition of combustion products based on the reactants used, methods for determining the combustion reactants based on the products of combustion, and methods for determining the approximate heat released during a combustion process. We will simplify our analysis to aid in comprehension and to fit within the parameters of this text, and we will discuss the implications of these simplifications later in the chapter.

QUESTION FOR THOUGHT/DISCUSSION

Think of some places or processes where combustion is used. What is the purpose of combustion in those processes? What are some alternatives to combustion that could be used to accomplish that purpose?

11.2 THE COMPONENTS OF THE COMBUSTION PROCESS

As mentioned, combustion involves the chemical reaction of a fuel and an oxidizer into combustion products:

$$\text{Fuel} + \text{Oxidizer} \rightarrow \text{Products}$$

Fuels can be solid, liquid, or gaseous in nature, as can oxidizers, although oxidizers are predominantly in gaseous form. The products are primarily in gaseous form, although some products can liquefy if the products are cooled sufficiently; and some ash or soot residue may be present in solid form. In this chapter, we will primarily consider the products to be ideal gas mixtures in the gaseous state. In addition, we will generally be concerned with the oxidizer being either a single ideal gas or an ideal gas mixture. Depending on the fuel and combustor design, the fuel can either be combined into an ideal gas mixture with the oxidizer for the entire reactant stream, or treated as a separate reactant stream entering the combustion process. The use of these ideal gas mixture approximations will involve some of the elements discussed in Chapter 9.

11.2.1 Oxidizers

As the name of this reactant suggests, molecular oxygen (O_2) is the oxidizer present in most combustion processes. Although O_2 gas can be used by itself as an oxidizer in a combustion process, O_2 is most commonly delivered to a combustion process as a component of air. In such a case, we may consider air to be the oxidizer, although the components of air other than O_2 contribute very little to the combustion process. For the most part, the other main component, N_2, does little in the combustion process other than help create the pollutant NO_x, lower the process temperature, and reduce the amount of heat that is available for use in a process. For our purposes, we will generally ignore the water vapor present in the atmosphere, and instead model the air used in a combustion process as dry air. If we consider the water vapor present in the air, we can assume here that the water vapor passes unaltered into the products.

As discussed in Chapter 9, the composition of dry air, on a molar basis, is ~78.07% N_2, 20.95% O_2, and ~1% other—primarily Ar. The argon, CO_2, and so on in these "other" gases will either be unreactive or be in such low concentrations that they will have no significant impact on the analysis, so we choose to group these together with the N_2 gas and assume that the N_2 is unreactive and passes straight from the reactants to the products. This will allow us to model the dry air, on a molar basis, as being approximately 79% N_2 and 21% O_2. We could write the combustion reaction as ($0.21\ O_2 + 0.79\ N_2$) as representative of the air. However, we will normally be concerned with the number of moles of O_2 present, so it is easier to divide this expression by 0.21 and represent the air as ($O_2 + 3.76\ N_2$).

11.2.2 Fuels

Although many substances burn, most of the fuels we will consider are in the most commonly used class of fuels: hydrocarbons. Hydrocarbons are compounds that consist of mixtures of some number of carbon atoms and some number of hydrogen atoms. Analysis of pure carbon and pure H_2 fuels can follow the same analysis techniques used with hydrocarbons, so we

TABLE 11.1 Some Classifications of Hydrocarbon Families. Often, hydrocarbons of the same families have similar combustion characteristics.

Classification	Chemical Formula/Arrangement	Examples
Alkanes	C_nH_{2n+2}/Straight or branched chain of C-atoms	Methane (CH_4), Propane (C_3H_8), Butane (C_4H_{10}), Octane (C_8H_{18})
Alkenes	C_nH_{2n}/Straight or branched chain of C-atoms	Ethene (Ethylene) (C_2H_4), Propene (C_3H_6)
Alkynes	C_nH_{2n-2}/Straight or branched chain of C-atoms	Acetylene (C_2H_2)
Cycloparaffins	C_nH_{2n}/Ring of C-atoms	Cyclopropane (C_3H_6), Cyclobutane (C_4H_8)
Aromatics	C_nH_{2n-6}/Based on ring of 6 C-atoms with alternating single and double bonds	Benzene (C_6H_6), Toluene (C_7H_8)

n is a positive integer.

can group those into our consideration of fuels as well. We will also consider alcohol fuels, which are hydrocarbon fuels with an oxygen atom included in the compound. However, some of the generic expressions we will develop later in this chapter will not be applicable to alcohol fuels.

Hydrocarbon fuels fall into various categories based on the relative number of hydrogen and carbon atoms present in the compound as well as the orientation of the atoms. Some of the common classifications and examples of the types are found in **Table 11.1**. The energy released by the combustion of these fuels is present in the chemical bonds in the compounds. Breaking these bonds will release the energy, although some energy is used to create the bonds in the new compounds of the products.

QUESTION FOR THOUGHT/DISCUSSION

Petroleum and natural gas are common sources of both combustion fuels and feedstocks for industries such as plastics. Do you think that humans are best utilizing petroleum and natural gas by burning them?

11.2.3 Products

When a hydrocarbon fuel burns, the carbon reacts with the oxidizer to create first carbon monoxide (CO) and then carbon dioxide (CO_2)—note that although CO is a product of hydrocarbon combustion, it is also a fuel in its own right because it can further oxidize to CO_2. The hydrogen in the fuel reacts to form water, H_2O. As will be discussed later, there are many intermediate chemical reactions present in these reactions, but for most engineering purposes it is only necessary to know the beginning and ending compounds in the processes.

There are some terms we will use when describing combustion processes, and these are defined as follows. *Complete combustion* refers to the condition of having the fuel completely oxidized; that is, all of the carbon in the fuel becomes carbon dioxide (CO_2), and all of the hydrogen in the fuel becomes water (H_2O). If there are any other atoms in the compound that can be oxidized to release heat, they too are completely oxidized (such as sulfur becoming sulfur dioxide (SO_2)). *Perfect combustion* refers to the condition of complete combustion occurring with no additional oxygen present. Perfect combustion can also be referred to as *stoichiometric combustion*, and it requires the reactant mixture to be a specific composition (termed a *stoichiometric mixture*) for a given fuel. A stoichiometric mixture does not need to result in perfect combustion, but perfect combustion must begin with a stoichiometric mixture. The products of perfect combustion of a hydrocarbon fuel in air are considered to be only CO_2, H_2O, and N_2.

If there is more oxygen (or air) present in the reactants than what is required to produce complete combustion, the combustion is termed *lean* combustion (or fuel-lean combustion). If there is complete combustion with a lean reactant mixture between a hydrocarbon and air, the products of the combustion process will be CO_2, H_2O, O_2, and N_2. The same amount of CO_2 and H_2O will be created as in the complete combustion of a stoichiometric mixture, but the amount of N_2 will increase due to the larger amount of air, and the O_2 in the products will represent the excess oxygen left over after all of the oxidizer needed by the carbon and hydrogen was consumed.

If there is less oxygen (or air) present in the reactants than what is required to produce complete combustion, the combustion is termed *rich* combustion (or fuel-rich combustion). Because there is insufficient oxygen present to produce complete combustion, the products of rich combustion are a much more complicated blend. Depending on how little oxygen is present, we may expect to find CO_2, CO, H_2O, H_2, N_2, and unburned fuel. The unburned fuel may either be in the form of the original fuel compound, or in the form of other hydrocarbon compounds that were formed in the breakdown of the original fuel. For example, if we are burning a very rich mixture of propane and air, we may find that the products contain not only propane, but also methane, ethane, and ethylene, among other hydrocarbons.

11.3 A BRIEF DESCRIPTION OF THE COMBUSTION PROCESS

At this point, it may be helpful to present a brief qualitative description of what occurs as a hydrocarbon fuel burns with oxygen. We begin with a mixture of fuel and oxygen (or air). This mixture can be either already thoroughly mixed together (which results in what is called a premixed flame) or in separate volumes (which results in a non-premixed flame) that must diffuse together at the interface between the fuel and air. The combustion that takes place in a typical gasoline engine is an example of the former, whereas a candle can be thought of as an example of the latter. At this point, there needs to be an ignition source, which is simply an application of heat so that the combustion reactions can begin. This source can take a variety of forms, including a spark (such as from a spark plug or a match), a hot surface, and hot gases (such as compressed air in a compression-ignition engine). The purpose of this heat source is to locally increase the temperature of the fuel and oxygen so that chemical reactions can begin to proceed at a more rapid pace. Although fuel and oxygen molecules will react at low temperatures, the reactions proceed too slowly to provide the heat necessary to warm nearby molecules and speed the combustion process.

As the small volume of fuel and oxygen are heated, chemical reactions begin and pyrolysis of the fuel occurs. This process involves breaking the fuel molecules into smaller compounds, some of which are stable and some of which are unstable. The unstable atoms and molecules are called radicals. At the same time, the oxygen molecules will break into unstable oxygen atoms. The unstable radicals can much more easily be combined than stable molecules, and thus additional chemical reactions occur. Many of these reactions release heat, which increases the local temperature and increases the rates of the chemical reactions. The higher temperatures also warm the surrounding fuel and oxygen so that the combustion process can begin in those regions. Fuel and oxygen are rapidly reacting chemically, thus producing high-temperature products. The chemical reactions continue until the stable products are formed, until the temperatures are lowered too much through heat transfer to the surroundings, or until insufficient oxygen or fuel is available to proceed. These processes are illustrated in **Figure 11.2**.

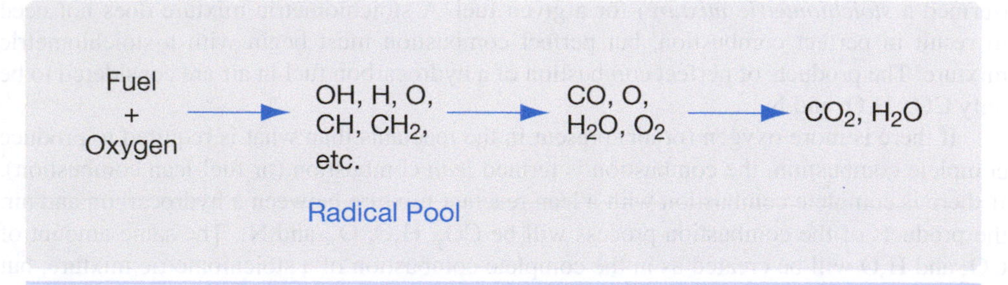

FIGURE 11.2 As fuel and oxygen are burned, the larger molecules first break into unstable radicals, which then start to combine into combustion products, with the formation of CO_2 tending to be slower than the formation of H_2O.

As the combustion process proceeds, the stable molecules CO and H_2O are formed first among the products. Additional oxygen then reacts with the CO to create CO_2. Knowing this will help you understand the nature of the products of rich combustion. If there is only a small amount of excess fuel, then there will nearly be enough O_2 present to complete the combustion of CO to CO_2, and so the amount of CO will be relatively small. As more excess fuel is present, we tend to see more CO and less CO_2 formed, although the hydrogen tends to form H_2O. As the amount of excess fuel increases, we continue to see more CO and less CO_2, but also an increasing amount of unburned hydrocarbon compounds, as well as some H_2 that did not encounter sufficient O_2 to form H_2O.

Not every mixture of fuel and oxygen (or air) will combust. Every fuel has a specific range of the amounts of oxygen that must be present in order to sustain combustion—this is the flammability range of the fuel, with the ends of the range being called the lean flammability limit and the rich flammability limit. This intuitively should make sense. If we have a room full of air and we release a test tube full of methane into the room, thoroughly mix the gases, and then use a spark to ignite the mixture, it will not burn. The fuel is too diffuse to allow for a sustained chemical reaction. If a few methane molecules did combust, the heat produced would be too small to ignite the distant neighboring fuel molecules. Similarly, if we have a tank of fuel and introduce a tiny amount of air into the tank, the fuel will not ignite. In this case, there is too little air to provide the oxygen necessary to sustain a chemical reaction.

11.4 BALANCING COMBUSTION REACTIONS

To determine the global chemical reaction governing a combustion process, we must consider the appropriate expected products described above, and the number of atoms of each type must be equal in the products and in the reactants. This process, called atom balancing, allows us to determine the coefficients in front of each compound in the chemical equation. To illustrate the process, we begin by determining the chemical equation describing the perfect combustion of methane in oxygen—that is, the equation that describes the burning of a stoichiometric mixture of methane and oxygen completely. In such a case, we know that the only products are CO_2 and H_2O (as in this case we are concerned with oxygen alone as the oxidizer, so there is no N_2 in the reactants or products). We will consider 1 mole of CH_4 as the basis for the process, and we will see that it is often easiest to make the assumption that the combustion reaction is using 1 mole (or 1 kmole) of fuel. In this case, the chemical reaction is written

$$CH_4 + aO_2 \rightarrow bCO_2 + cH_2O$$

where a, b, and c are unknown coefficients that must be determined through the atom-balancing procedure. There are three atom balances that can be written: one for carbon atoms, one for oxygen atoms, and one for hydrogen atoms. The balances are written as the number of atoms of the specific type in the reactants being equal to the number of atoms of that type in the products; we must consider all of the species that contain a specific atom, and that the number of atoms of a certain type in that species is equal to the product of the number of moles of the species and the number of atoms of that type in the compound:

C-balance: $(1)(1) = b(1)$ (1 carbon atom each in CH_4 and CO_2)

H-balance: $(1)(4) = c(2)$ (4 H in CH_4 and 2 H in H_2O)

O-balance: $(a)(2) = b(2) + c(1)$ (2 O in O_2 and CO_2 and 1 O in H_2O)

Solving these equations yields $b = 1$, $c = 2$, and $a = 2$. Therefore, the chemical equation for the perfect combustion of 1 mole of CH_4 with O_2 is

$$CH_4 + 2O_2 \rightarrow CO_2 + 2H_2O$$

This process is illustrated in Figure 11.3.

Now, if the oxidizer is carried in as air, this expression can be extended by replacing the O_2 with the $(O_2 + 3.76N_2)$ approximation for dry air developed in Section 11.2.1:

$$CH_4 + 2(O_2 + 3.76N_2) \rightarrow CO_2 + 2H_2O + dN_2$$

The coefficients in front of the O_2, CO_2, and H_2O do not change, because the addition of N_2 has no effect on these numbers. We need a nitrogen balance to solve for d to complete the chemical equation:

N-balance: $2(3.76)(2) = d(2)$

Solving yields $d = 7.52$, and the chemical equation for the perfect combustion of 1 mole of CH_4 in air is

$$CH_4 + 2(O_2 + 3.76N_2) \rightarrow CO_2 + 2H_2O + 7.52N_2$$

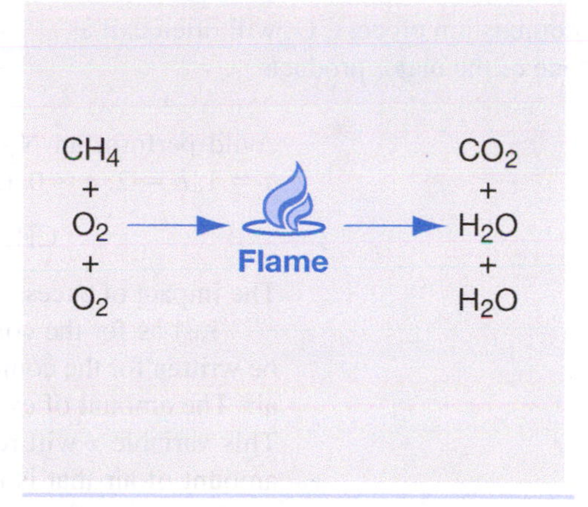

FIGURE 11.3 As methane and oxygen are burned in a flame, CO_2 and H_2O are formed.

Please remember that this equation does not describe the combustion of an individual molecule of methane. Of course, it is not possible to have 7.52 N_2 molecules. Rather, we are considering the combustion of 1 mole, or 1 kmole, of the fuel. One mole contains an Avogadro's number of molecules of that substance: 6.0225×10^{23} molecules. Therefore, this equation states that we are involving 7.52 moles, or over 4.5×10^{24} molecules, of N_2. Although there still must be an integer number of molecules involved, we cannot control the reaction to that level of precision when dealing with moles of substances. This is true of all the global chemical reactions we will be concerned with in this chapter.

Investigation of this chemical reaction balancing procedure should suggest that a generic equation for the perfect combustion of any hydrocarbon compound can be created. Suppose that the compound has the formula C_yH_z, where y and z represent the number of carbon and hydrogen atoms in the fuel, respectively. For CH_4, $y = 1$ and $z = 4$. These numbers need not be integers, because the fuel could be a mixture of compounds, with y and z being average values. If we perform atom balances on the C, H, O, and N in the chemical reaction describing the perfect combustion of 1 mole of C_yH_z, we derive the following equation:

$$C_yH_z + \left(y + \frac{z}{4}\right)(O_2 + 3.76N_2) \rightarrow yCO_2 + \frac{z}{2}H_2O + 3.76\left(y + \frac{z}{4}\right)N_2 \qquad (11.1)$$

It is helpful to keep this equation handy so that you do not need to perform new atom balances on every possible hydrocarbon compound, and the factor $(y + z/4)$ describing the coefficient in front of the air for a stoichiometric mixture will be particularly useful to remember.

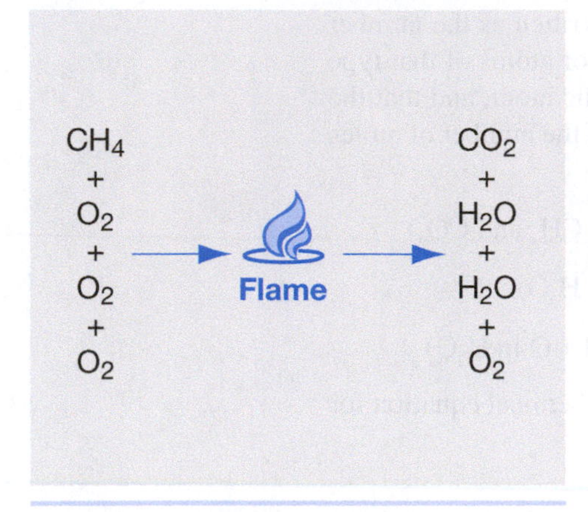

FIGURE 11.4 If excess O_2 is present in a combustion process, O_2 will often exit as one of the major products.

The chemical balancing of the complete combustion of lean mixtures is much like that of stoichiometric mixtures, except now O_2 also appears in the products. In addition, it will be common that the amount of excess air will be known, or that it can be determined from information derived from the known products. Let us consider the combustion of 1 mole of methane, but with 20% more air present than in a stoichiometric mixture. The 20% additional air will change the coefficient in front of the air from 2 in a stoichiometric mixture to 2.4. Therefore, the chemical equation can be initially written as

$$CH_4 + 2.4(O_2 + 3.76N_2) \rightarrow aCO_2 + bH_2O + cO_2 + dN_2$$

The atom balances are written as follows:

C-balance: $(1)(1) = a(1)$

H-balance: $(1)(4) = b(2)$

O-balance: $(2.4)(2) = a(2) + b(1) + c(2)$

N-balance: $(2.4)(3.76)(2) = d(2)$

(You may notice that if nitrogen only appears in the form of N_2, we could perform an N_2-balance rather than an N-balance.) Solving the atom balances yields $a = 1$, $b = 2$, $c = 0.4$, and $d = 9.024$, resulting in the following chemical equation:

$$CH_4 + 2.4(O_2 + 3.76N_2) \rightarrow CO_2 + 2H_2O + 0.4O_2 + 9.024N_2$$

The impact of excess O_2 on a combustion process is illustrated in **Figure 11.4**.

Just as for the complete combustion of a stoichiometric mixture, a general equation can be written for the complete combustion of a generic hydrocarbon, C_yH_z, in a lean mixture with air. The amount of excess air needs to be specified, and will be done so through a variable x. This variable x will represent the % theoretical air/100%, where the % theoretical air is the amount of air that is present in a stoichiometric mixture of fuel and air. So, if there is 20% excess air, this corresponds to 120% theoretical air, which would give $x = 1.2$. For this lean combustion, the complete combustion can be described as

$$C_yH_z + (x)\left(y + \frac{z}{4}\right)(O_2 + 3.76N_2) \rightarrow yCO_2 + \frac{z}{2}H_2O + (x - 1)\left(y + \frac{z}{4}\right)O_2 + 3.76(x)\left(y + \frac{z}{4}\right)N_2 \quad (11.2)$$

If $x = 1$, representing a stoichiometric mixture, this equation reduces to Eq. (11.1).

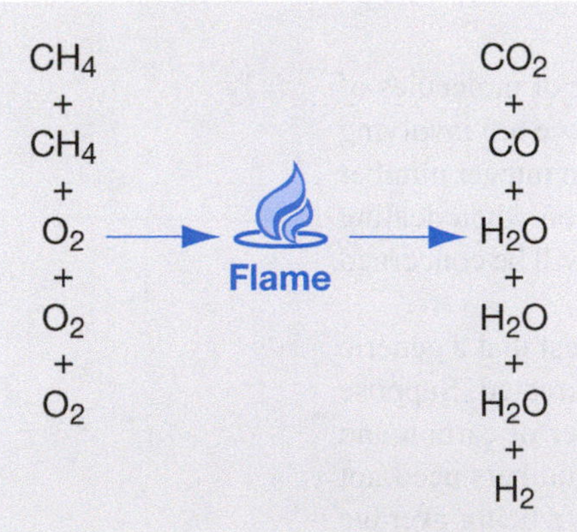

FIGURE 11.5 If the reactants contain an insufficient amount of O_2 to lead to complete combustion, CO and H_2 (and possibly unburned fuel) may become significant products.

Rich combustion mixtures present a greater problem for the development of chemical reactions describing such reactions. Not including unburned hydrocarbons, a set of products of CO_2, CO, H_2O, H_2, and N_2 will require the determination of five coefficients if the reactant mixture is known. Yet, there are only four atom balances available. Therefore, such a problem is not solvable. Later in this chapter, we will present a method for approximating the products. Until then, we will have to consider that for rich combustion processes, the products of combustion will have to be measured. The impact of deficit O_2 experienced in rich combustion is illustrated in **Figure 11.5**.

▶ EXAMPLE 11.1

Write the chemical reaction for the complete combustion of octane, C_8H_{18}, with (a) a stoichiometric mixture of fuel and air, and (b) 25% excess air. First use the atom-balancing approach, and then check these results with the general equations for perfect combustion and for complete combustion of a lean mixture.

Given: Complete combustion of C_8H_{18} with air

Find: The chemical reaction describing the complete combustion of octane with (a) 100% theoretical air and (b) 25% excess air

Solution: This problem will be solved using atom balancing. We solve a set of equations that equate the number of each type of atom in both the reactants and the products.

(a) For a stoichiometric mixture reaching perfect combustion with air and octane, there is 100% theoretical air and the only products are CO_2, H_2O, and N_2:

$$C_8H_{18} + a(O_2 + 3.76N_2) \rightarrow bCO_2 + cH_2O + dN_2$$

For each type of atom, the number of that type of atom must be the same in the reactants and products. The number of atoms of a particular type in either the reactants or the products is equal to the number of moles of a particular species (molecule or atom) containing that atom times the number of atoms of that type in that species. For example, for carbon, the reactants have only one species containing carbon—C_8H_{18}—and one species in the products containing carbon—CO_2. In the preceding reaction, there is one mole of C_8H_{18}, and b moles of CO_2. The C_8H_{18} molecule contains 8 carbon atoms, and the CO_2 contains 1 carbon atom. Equating the number of carbon atoms in the reactants and products yields the carbon atom balance:

C-balance: $(1)(8) = b(1)$

The remaining atom balances are as follows:

H-balance: $(1)(18) = c(2)$

O-balance: $(a)(2) = b(2) + c(1)$

N-balance: $(3.76a)(2) = d(2)$

Solving these equations yields $a = 12.5$, $b = 8$, $c = 9$, and $d = 47$, giving the chemical equation

$$\mathbf{C_8H_{18} + 12.5(O_2 + 3.76N_2) \rightarrow 8CO_2 + 9H_2O + 47N_2}$$

We can then use the expression for the perfect combustion of a generic hydrocarbon, Eq. (11.1), with $y = 8$ and $z = 18$, to confirm this result:

$$C_yH_z + \left(y + \frac{z}{4}\right)(O_2 + 3.76N_2) \rightarrow yCO_2 + \frac{z}{2}H_2O + 3.76\left(y + \frac{z}{4}\right)N_2$$

Substitution gives

$$C_8H_{18} + 12.5(O_2 + 3.76N_2) \rightarrow 8CO_2 + 9H_2O + 47N_2$$

(b) For lean combustion with 25% excess air (or 125% theoretical air), the coefficient in front of the air becomes

$(12.5)(1.25) = 15.63$

as the multiplier of the coefficient for 100% theoretical air is equal to the $\frac{\%\text{ theoretical air}}{100\%}$. The excess air also leads to the presence of O_2 in the products (in addition to the three products already expected from complete combustion). The chemical equation, with unknown coefficients in the products, is

$$C_8H_{18} + 15.63(O_2 + 3.76N_2) \rightarrow aCO_2 + bH_2O + cO_2 + dN_2$$

The atom-balancing yields the following four equations that can be solved for the four unknowns:

C-balance: $(1)(8) = a(1)$

H-balance: $(1)(18) = b(2)$

O-balance: $(15.63)(2) = a(2) + b(1) + c(2)$

N-balance: $((15.63)(3.76))(2) = d(2)$

Solving these equations yields $a = 8$, $b = 9$, $c = 3.13$, and $d = 58.77$, giving the chemical reaction

$$\mathbf{C_8H_{18} + 15.63(O_2 + 3.76N_2) \rightarrow 8CO_2 + 9H_2O + 3.13O_2 + 58.77N_2}$$

This can be confirmed with Eq. (11.2), with $y = 8$, $z = 18$, and $x = 1.25$ (representing 125% theoretical air or 25% excess air):

$$C_yH_z + (x)\left(y + \frac{z}{4}\right)(O_2 + 3.76N_2) \rightarrow yCO_2 + \frac{z}{2}H_2O + (x - 1)\left(y + \frac{z}{4}\right)O_2 + 3.76(x)\left(y + \frac{z}{4}\right)N_2$$

Substitution gives

$$C_8H_{18} + 15.63(O_2 + 3.76N_2) \rightarrow 8CO_2 + 9H_2O + 3.13O_2 + 58.77N_2$$

Analysis: While the general expressions work well, it is important to understand the process that led to the equations, and it also must be remembered that they only work for hydrocarbon fuels.

▶ EXAMPLE 11.2

Write the chemical reaction for the perfect combustion of methanol, CH_3OH, in air.

Given: Combustion of CH_3OH with air.

Find: The chemical reaction describing the perfect combustion of methanol in air (100% theoretical air is needed for perfect combustion).

Solution: Methanol is an alcohol rather than a pure hydrocarbon, so Eq. (11.1) cannot be used, and instead, the atom-balancing approach must be employed. Writing the equation with only the products of CO_2, H_2O, and N_2 (as representative of perfect combustion):

$$CH_3OH + a(O_2 + 3.76N_2) \rightarrow bCO_2 + cH_2O + dN_2$$

The four atom balances are

C-balance: $(1)(1) = b(1)$

H-balance: $(1)(3 + 1) = c(2)$

O-balance: $(1)(1) + (a)(2) = b(2) + c(1)$

N-balance: $(3.76a)(2) = d(2)$

Solving these equations yields $a = 1.5$, $b = 1$, $c = 2$, and $d = 5.64$. Therefore, the chemical reaction is

$$\mathbf{CH_3OH + 1.5(O_2 + 3.76N_2) \rightarrow CO_2 + 2H_2O + 5.64N_2}$$

Analysis: Notice that less air is needed for the perfect combustion of an alcohol in comparison to its hydrocarbon base, in this case methane (CH_4). This is the concept that is used in getting "cleaner" emissions from a spark-ignition engine by using an oxygenated fuel (such as gasoline with 10% ethanol). Such engines often operate fuel-rich and produce considerable CO and unburned hydrocarbon emissions as a result. By adding oxygen to the fuel, the overall reaction is closer to a stoichiometric mixture, and so less CO and unburned hydrocarbons are produced. In automobiles, the CO and unburned hydrocarbons are often subsequently treated in a catalytic converter. The catalytic converter can reduce their concentrations by more than 95%. If we produce less CO and unburned hydrocarbons leaving the combustion process, the converter is able to operate on a smaller number of molecules, and so the overall pollutant emissions are even lower.

QUESTION FOR THOUGHT/DISCUSSION

Why would using Eq. (11.2) not work for finding the global chemical reaction for the combustion of an alcohol with air?

11.5 METHODS OF CHARACTERIZING THE REACTANT MIXTURE

Although we could always refer to a reactant mixture by the number of moles of each component, such a system would be cumbersome and would often make it difficult for others to know what to expect from the mixture. For example, if someone states that 1 mole of a fuel with which you are unfamiliar reacts with 3 moles of O_2, you will not readily know whether the process is lean or rich, and therefore you will not know what to expect of the products or whether or not the process is likely to be stable. As a result, several methods of characterizing a reactant mixture composition have been developed. The first two of these have already been introduced.

11.5.1 Percent Theoretical Air

As discussed previously, "theoretical air" is the amount of air that is necessary to burn a fuel to completion with no excess oxygen available; that is, the amount of air in a stoichiometric mixture. The term *100% theoretical air* corresponds to a stoichiometric mixture for a particular fuel. Values greater than 100% indicate that there is excess air, and the reactant mixture is lean, whereas values less than 100% indicate that there is a deficit of air (or excess of fuel) and that the reactant mixture is rich. So, if someone states that a fuel is burning with 120% excess air, it is known that the combustion process is lean and that the products are likely to be primarily CO_2, H_2O, O_2, and N_2.

11.5.2 Percent Excess Air or Percent Deficit Air

This method, also mentioned previously, is very similar to the percent theoretical air approach, but instead gives a value that is in comparison to the percent theoretical air. The percent excess air is used for lean combustion and is related to the percent theoretical air through

$$\text{Percent Excess Air} = \text{Percent Theoretical Air} - 100\%$$

So, if a reactant mixture is 115% theoretical air, it may also be referred to as 15% excess air.

The percent deficit air designation is used for rich combustion environments. It is related to the percent theoretical air through

$$\text{Percent Deficit Air} = 100\% - \text{Percent Theoretical Air}$$

Therefore, if a reactant mixture is 90% theoretical air, it could also be characterized as 10% deficit air.

11.5.3 Air–Fuel Ratio and Fuel–Air Ratio

The air–fuel ratio provides the amount of air divided by the amount of fuel present in the reactant mixture. For any given combustion situation, two air–fuel ratios exist: one based on the number of moles of each substance and one based on the mass of each substance in the reactants. Although either form can be used for any combustion situation, we will tend to find the molar value used more for gaseous fuels, whereas the mass value is used more for liquid or solid fuels. This is because we are much more likely to know the volume, which is closely related to moles, of a gaseous fuel than the volume of a liquid or solid fuel, and we are more likely to know the mass of a liquid or solid than of a gas. The mass-based air–fuel ratio is

$$AF = \frac{m_{\text{air}}}{m_{\text{fuel}}} \tag{11.3}$$

and the mole-based air–fuel ratio is

$$\overline{AF} = \frac{n_{\text{air}}}{n_{\text{fuel}}} \tag{11.4}$$

When computing these values, remember that the air consists of both the O_2 and the N_2. Although it is easy to think of just taking the coefficient in front of the air as the number of moles of air, remember that that coefficient is multiplied by (1 + 3.76) inside the term representing air. Therefore, for the method used in writing the chemical equations to this point, the number of moles of air is the coefficient in front of the air multiplied by 4.76.

The fuel–air ratio can also be on a mass or molar basis. It simply is the inverse of the appropriate air–fuel ratio:

$$FA = (AF)^{-1} \quad \text{and} \quad \overline{FA} = (\overline{AF})^{-1} \tag{11.5}$$

The air–fuel ratio tends to give numbers greater than 1, but less than 30, whereas the fuel–air ratio tends to give small decimal values. As a result, differences in the air–fuel ratios of reactant mixtures are more pronounced than the corresponding difference in the fuel–air ratios.

A disadvantage of either the air–fuel ratio or the fuel–air ratio is that the number, by itself, does not reveal any details of the combustion process. We must know the air–fuel (or fuel–air) ratio corresponding to the case of 100% theoretical air for the fuel in question in order to interpret what to expect from the reactant mixture. In a field or industry that uses a common fuel, the characterization of the reactant mixture with the air–fuel or fuel–air ratio is more likely to be encountered because the engineers in that field will be knowledgeable of what the numbers imply. For example, gasoline has an $AF = 14.6$ for a stoichiometric mixture. Therefore, if we indicate that the AF in a gasoline-fueled spark-ignition engine is 13, we would know that the mixture has less air than a stoichiometric mixture and the combustion is rich. Similarly, if the AF is given as 15.8, we would know that the combustion is lean because there is excess air. But without knowing the air–fuel ratio of the stoichiometric mixture, we would not know what to expect from the combustion process.

11.5.4 Equivalence Ratio

In an effort to aid in the interpretation of the air–fuel ratio, the concept of the equivalence ratio, ϕ, was developed. The equivalence ratio is defined as the ratio of the fuel–air ratio of the actual mixture to that of a stoichiometric mixture for that fuel. The definition works for both mass-based and mole-based quantities, because the ratio of the molecular weights of the fuel and air used to convert between mass and moles will cancel in the equivalence ratio calculation. Representing the actual reactant mixture with a subscript a and the stoichiometric mixture with a subscript st, the equivalence ratio can be found from

$$\phi = \frac{FA_a}{FA_{st}} = \frac{\overline{FA}_a}{\overline{FA}_{st}} = \frac{AF_{st}}{AF_a} = \frac{\overline{AF}_{st}}{\overline{AF}_a} \tag{11.6}$$

The inverse relationship between the fuel–air and air–fuel ratios leads to the stoichiometric quantity for the air–fuel ratio appearing in the numerator. Upon inspection, you may find that the molar quantities of these ratios may be easily determined from the chemical reaction. Furthermore, if we write the actual and stoichiometric chemical reactions based on 1 mole of fuel, we can find the equivalence ratio from the ratio of the coefficients in front of the air quantity—for a hydrocarbon fuel, this quantity can be quickly found from the $(y + z/4)$ term from Eq. (11.1) for the stoichiometric case.

If $\phi = 1$, the mixture is stoichiometric; if $\phi < 1$, the mixture is lean; and if $\phi > 1$, the mixture is rich. Although the results will not be identical, we can expect similar products for the same equivalence ratio for different fuels. For example, both propane and octane combustion with air at the same equivalence ratio will give a similar set of products.

▶ EXAMPLE 11.3

Propane, C_3H_8, burns to completion with 110% theoretical air. Write the chemical reaction for this process and characterize the reactant mixture in terms of percent excess (or deficit) air, mass-based and mole-based air–fuel ratio, mass-based and mole-based fuel–air ratio, and equivalence ratio.

Given: C_3H_8 burning to completion with 110% theoretical air

Find: The global chemical reaction for the process, % excess or deficit air in the reactants, AF, $\overline{AF}$, FA, $\overline{FA}$, ϕ (equivalence ratio)

Solution: With 110% theoretical air, the combustion is lean. For lean combustion, 110% theoretical air = **10% excess air.**

Using Eq. (11.2) with $x = 1.1$ (representing 10% excess air), $y = 3$, and $z = 8$ yields

$$\mathbf{C_3H_8 + 5.5(O_2 + 3.76N_2) \rightarrow 3CO_2 + 4H_2O + 0.5O_2 + 20.68N_2}$$

We will take the molecular masses of the reactants to be $M_{C_3H_8} = 44$, $M_{O_2} = 32$, and $M_{N_2} = 28$. Considering 1 kmole of fuel being burned, the mass of the fuel is $m_{fuel} = n_{fuel}M_{fuel} = n_{C_3H_8}M_{C_3H_8} = 44$ kg.

For the air, $m_{air} = m_{O_2} + m_{N_2} = n_{O_2}M_{O_2} + n_{N_2}M_{N_2} = (5.5)(32) + (5.5)(3.76)(28) = 755$ kg.

Therefore,

$$AF = \frac{m_{air}}{m_{fuel}} = \mathbf{17.2}$$

The number of moles of the fuel is 1 kmole, and the number of moles of the air is $n_{air} = n_{O_2} + n_{N_2} = 5.5 + 5.5(3.76) = 26.18$ kmole.

So

$$\overline{AF} = \frac{n_{air}}{n_{fuel}} = \mathbf{26.2}$$

The respective fuel–air ratios are just the inverses of these numbers:

$$FA = (AF)^{-1} = \mathbf{0.0581} \quad \text{and} \quad \overline{FA} = (\overline{AF})^{-1} = \mathbf{0.0382}$$

The equivalence ratio requires knowledge of the stoichiometric reactant mixture. Using Eq. (11.1), the stoichiometric reactant mixture is $C_3H_8 + 5\,(O_2 + 3.76N_2)$.

The molar air fuel ratio for this is

$$\overline{AF}_{st} = 5(1 + 3.76)/1 = 23.8$$

Knowing that the actual molar air–fuel ratio is 26.2, the equivalence ratio can be found as

$$\phi = \frac{\overline{AF}_{st}}{\overline{AF}_{a}} = \frac{23.8}{26.2} = \mathbf{0.909}$$

As mentioned, because both the actual and stoichiometric reactants were based on the same number of moles of fuel (1 kmole in this case), the value could have been found from a ratio of the coefficients in front of the air: $5/5.5 = 0.909$.

11.6 DETERMINING REACTANTS FROM KNOWN PRODUCTS

You might think that someone operating a device that includes a combustion process will always know the composition of the reactants. However, that is not the case; often the controls on the air into the process are somewhat imprecise, and the composition of a fuel may not be completely known. Keep in mind that common fuels such as gasoline, diesel fuel, and natural gas consist of many compounds and are variable in composition. Therefore, engineers will sometimes try to deduce the reactant composition by measuring the combustion products. After measuring the products, atom balances can be performed to determine the reactants. If the fuel composition is known, such an analysis can provide the air–fuel ratio on both a mass and molar basis, whereas if the fuel composition is unknown, the analysis can provide a ratio of the hydrogen to carbon for the fuel (with the assumption that the fuel is a hydrocarbon) and a mass-based air–fuel ratio.

Major combustion products can be measured relatively easily, and routinely such measurements are made with a device such as a five-gas analyzer, as shown in Figure 11.6. A five-gas analyzer measures the concentrations of CO_2, CO, unburned hydrocarbons, NO_x, and O_2. The CO_2 and CO are commonly measured with nondispersive infrared detectors, the unburned hydrocarbons with a flame ionization detector, the NO_x with a chemiluminescent analyzer, and the O_2 with a paramagnetic analyzer. If this simple detection system is used, the remainder of the dry products (the products minus the water) are assumed to be N_2. More detailed analysis of the products can be obtained with such systems as a Fourier transform infrared (FTIR) system, which can provide more details on the hydrocarbon species, or a gas chromatograph, which can provide a detailed analysis of nearly all components in the combustion products. These two techniques are more expensive and time consuming than the five-gas analyzer technique and so are used more for research projects than routine measurements.

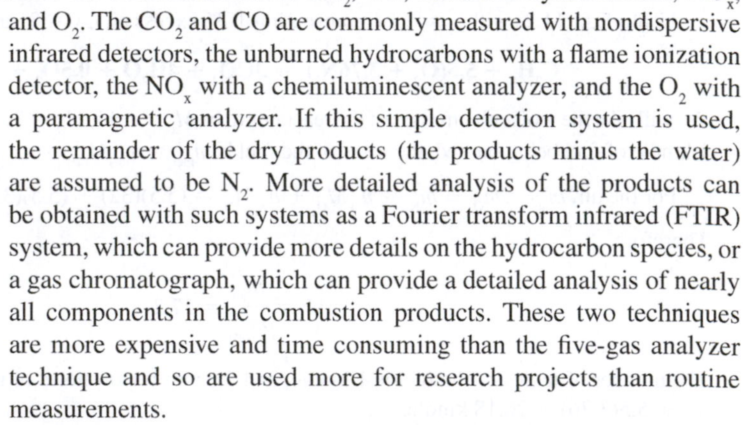

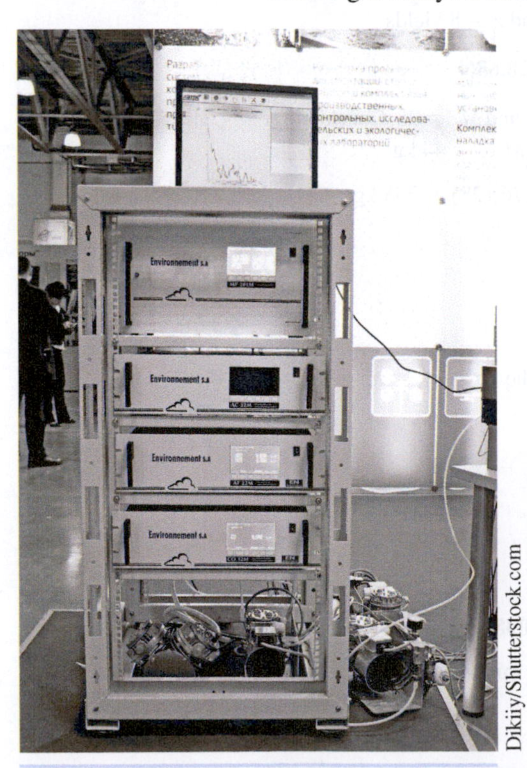

Dikiiy/Shutterstock.com

FIGURE 11.6 A simple five-gas analyzer control and readout panel.

Before performing an analysis of the products, the products are often dried so as to remove the water vapor. Water vapor, if it condenses into water, can damage the measurement equipment. When the products are dried, the analysis is considered to be a "dry products analysis," and the dry products analysis will provide the mole fractions of the nonwater components of the products. A useful procedure for setting up the chemical reaction is to consider the analysis based on 100 moles (or kmoles) of dry products. If given as molar percentages, the coefficients in front of the dry products are simply the value listed as the percentage. There is also an unknown amount of water, and the reactants are unknown. The following two examples illustrate the procedure for determining the reactants based on whether the fuel composition is known or unknown.

▸ **EXAMPLE 11.4**

A dry products analysis is performed on a combustion process where propane (C_3H_8) is burned with air. The dry products analysis gave the following volumetric percentages for the composition: 9.85% CO_2, 4.93% CO, 0.62% O_2, 1.23% H_2, and 83.37% N_2. Determine the chemical reaction describing this combustion process, the molar air–fuel ratio of the reactants, and the equivalence ratio of the reactants.

Given: Fuel: C_3H_8. Oxidizer: Air. Dry products analysis mole fractions:

$$y_{CO_2} = 0.0985,\ y_{CO} = 0.0493,\ y_{O_2} = 0.0062,\ y_{H_2} = 0.0123,\ y_{N_2} = 0.8337$$

Find: Global chemical reaction describing the combustion, $\overline{AF}$, ϕ

Solution: We are dealing with an ideal gas mixture in this dry products analysis. The given volumetric percentages are equal to the mole fractions for an ideal gas mixture.

Assume: The water originally in the products has been removed prior to the dry products analysis.

We will first write the chemical reaction based on 100 moles of dry products, which requires there to be an unknown amount of the known fuel present. Multiplying the mole fractions by 100 gives the following global reaction:

$$aC_3H_8 + b(O_2 + 3.76N_2) \rightarrow 9.85CO_2 + 4.93CO + 1.23H_2 + cH_2O + 0.62O_2 + 83.37N_2$$

Note that there is water in the products, and at an unknown quantity. There are four atom balances available, and three unknowns. To find the unknowns, the C-balance, H-balance, and N-balance will be used. (The O-balance could be used later as a check.)

C-balance: $3a = 9.85(1) + 4.93(1)$

H-balance: $8a = 1.23(2) + 2c$

N-balance: $3.76b(2) = 83.37(2)$

Solving yields $a = 4.927$, $b = 22.17$, and $c = 18.48$. Therefore, the chemical equation for this process is

$$4.927C_3H_8 + 22.17(O_2 + 3.76N_2) \rightarrow 9.85CO_2 + 4.93CO + 1.23H_2 + 18.48H_2O + 0.62O_2 + 83.37N_2$$

or, dividing the coefficients by 4.927 to put the reaction in terms of 1 mole of fuel,

$$\mathbf{C_3H_8 + 4.5(O_2 + 3.76N_2) \rightarrow 2CO_2 + CO + 0.25H_2 + 3.75H_2O + 0.125O_2 + 16.92N_2}$$

The molar air–fuel ratio is found as

$$\overline{AF} = \frac{n_{air}}{n_{fuel}} = \frac{n_{O_2} + n_{N_2}}{n_{C_3H_8}} = \frac{4.5 + 4.5(3.76)}{1} = \mathbf{21.4}$$

For 1 mole of fuel, the reactants for a stoichiometric mixture can be found with Eq. (11.1), using $y = 3$ and $z = 8$:

$$C_3H_8 + (3 + 8/4)(O_2 + 3.76\,N_2):\quad C_3H_8 + 5(O_2 + 3.76\,N_2)$$

The stoichiometric mixture molar air–fuel ratio is then $\overline{AF}_{st} = 5(1 + 3.76)/1 = 23.8$.

The equivalence ratio is then calculated as $\phi = \dfrac{\overline{AF}_{st}}{\overline{AF}_a} = \dfrac{23.8}{21.4} = \mathbf{1.11}$

Analysis: The combustion process is slightly rich, as suggested by the presence of a considerable amount of CO (although much less than the amount of CO_2) and relatively little H_2 or O_2 in the products.

As you can see in the previous example, this method can be used for analyzing both complete and incomplete combustion. The products in Example 11.4 contained CO and H_2 that both required oxidation, and some O_2 that, given enough time and mixing, could have oxidized some of these products—although not all of the CO and H_2. It should be noted that most real combustion processes are incomplete. Complete combustion requires thorough mixing and adequate time for the process to be at high temperatures to complete the reactions. So it is common that lean flames will have some CO in their products, although the amount of CO is likely to be small in comparison to the amount of CO_2. In fact, the amount of CO will become more prominent as the reactant mixture becomes very lean. In such cases, although there is more than enough O_2 to fully oxidize all the carbon, the combustion temperatures are lower and the gases cool too quickly for all of the oxidation to occur. With experience, however, you will be able to interpret incomplete combustion products and determine whether the process is rich or lean. For example, if there is a substantial amount of O_2 in the products, and relatively little CO and unburned hydrocarbons, the mixture is likely lean, whereas if the opposite is true, the reactant mixture is likely rich. This analysis can also provide insight into the overall design of the combustor and combustion process. For example, if the products indicate that the reactant mixture should have an equivalence ratio of 0.9, but there is more CO than CO_2 present, there was likely either inadequate mixing or insufficient time to complete the combustion process.

▶ EXAMPLE 11.5

An unknown hydrocarbon burns in air, resulting in the following dry products analysis (percentages given are on a molar basis): 12.68% CO_2, 0.67% CO, 1.33% O_2, and 85.32% N_2. Determine a hydrocarbon model for the fuel, the mass-based air–fuel ratio of the reactants, and the equivalence ratio for the process.

Given: Oxidizer: Air. Dry products analysis mole fractions: $y_{CO_2} = 0.1268$, $y_{CO} = 0.0067$, $y_{O_2} = 0.0133$, $y_{N_2} = 0.8532$

Find: Hydrocarbon model of the fuel (C_yH_z), AF, ϕ

Solution: We are dealing with an ideal gas mixture in this dry products analysis.

Assume: The water originally in the products has been removed prior to the dry products analysis.

The unknown fuel will be modeled as C_yH_z, and we will consider that 1 mole of this unknown fuel is burned. The chemical equation is then written based upon 100 moles of dry products:

$$C_yH_z + a(O_2 + 3.76N_2) \rightarrow 12.68CO_2 + 0.67CO + bH_2O + 1.33O_2 + 85.32N_2$$

There are four unknowns (a, b, y, z) and four atom-balancing equations that can be used:

C-balance: $(1)y = 12.68(1) + (0.67)(1)$

H-balance: $(1)z = 2b$

O-balance: $2a = (12.68)(2) + (0.67)(1) + b + 1.33(2)$

N-balance: $3.76a(2) = 85.32(2)$

Solving these equations yields $a = 22.69$, $b = 16.69$, $y = 13.35$, and $z = 33.38$.

With these values, the hydrocarbon model of the fuel is $\mathbf{C_{13.35}H_{33.38}}$, which clearly does not represent a situation with 1 mole of a single hydrocarbon compound. This does indicate that the ratio of H-atoms to C-atoms in the fuel is $33.38/13.35 = 2.5$. This may represent a single fuel (in this case, it does represent butane C_4H_{10}) or a mixture of fuels that give that overall H/C ratio. Because we don't know which is the case, we cannot determine how many moles of the fuel were used in the chemical reaction, and so we are unable to determine a molar air–fuel ratio; that is, n_{fuel} is unknown. However, we can still determine the mass-based air–fuel ratio.

The mass of the air is

$$m_{air} = m_{O_2} + m_{N_2} = n_{O_2}M_{O_2} + m_{N_2}M_{N_2} = (22.69 \text{ moles})(32 \text{ g/mole}) + ((22.69)(3.76) \text{ moles})(28 \text{ g/mole})$$
$$= 3114.9 \text{ g}$$

For the mass of the fuel, we need the molecular weight of the compound:

$$M_{fuel} = (\text{number of carbon atoms})(M_C) + (\text{number of hyrdogen atoms})(M_H)$$
$$= (13.35)(12) + (33.38)(1) = 193.6 \text{ g/mole}$$

With $n_{fuel} = 1$ mole, then $m_{fuel} = 193.6$ g.

The mass-based air–fuel ratio is then found as $AF = \frac{m_{air}}{m_{fuel}} = \mathbf{16.1.}$

From Eq. (11.1), for 1 mole of fuel, the coefficient in front of the air for a stoichiometric mixture is $y + z/4 = 21.7$.

The mass of the air for such a reactant mixture is 2979.0 g, and the stoichiometric mass-based air–fuel ratio is

$$AF_{st} = \frac{m_{air}}{m_{fuel}} = 15.39$$

The equivalence ratio is then calculated to be $\phi = \frac{AF_{st}}{AF_a} = \mathbf{0.96.}$

Analysis: From the product composition, it is not surprising that the mixture is only slightly lean, because the amount of CO in comparison to CO_2 is small, and the amount of O_2 is also small.

You may also calculate the equivalence ratio just using the coefficients in front of the air for the stoichiometric and actual cases, because the number of moles of fuel in each case is 1:

$$\phi = 15.39/16.1 = 0.96$$

QUESTION FOR THOUGHT/DISCUSSION

When might it be easier to determine information about the reactants of a combustion process by measuring the products rather than carefully measuring the reactants?

11.7 ENTHALPY OF A COMPOUND AND THE ENTHALPY OF FORMATION

When performing energy analyses, we have often used internal energy and enthalpy values that are in actuality relative to a specific reference point. For example, the enthalpy values of water that we have used are relative to an arbitrary 0 point for the specific internal energy as saturated liquid at 0.01°C. Yet, even solid water has internal energy, and so clearly the internal energy and enthalpy are not actually 0 at such a point. However, we have only been concerned with differences in these properties between states. The difference found using relative values is the same as the difference found using an absolute internal energy or enthalpy—the absolute value of the property at the reference state will be subtracted off in the calculation. This approach is perfectly acceptable when the working fluid substances are not changing chemical composition. But in combustion processes and other chemical reactions, the chemical composition of the compounds is changing, and the use of relative values for internal energy and enthalpy will result in errors. A consideration of how much energy is required or released as a compound is created or destroyed is needed.

To do this, we can create a reference state to which enthalpy values of different compounds can be related that is not completely absolute. This reference state will be the natural state of the chemical elements at standard temperature and pressure (STP) (25°C, 101.325 kPa). Setting this as a reference state, we can derive a quantity that represents the amount of energy stored or released during the process of creating a compound from the natural states of its elements. By creating a reference state, we incorporate into our analysis the amount of energy that is required or released during the creation or destruction of a compound. This procedure presumes that chemical elements in their natural states at standard temperature and pressure require no energy to exist at that state and if left alone have no ability to do work at that state.

What is the natural state of a chemical element? It is not necessarily the atom as considered on the periodic table. For example, it is highly unusual to see the element oxygen in atomic form, particularly at standard temperature and pressure. Oxygen can exist as ozone, O_3, but again, that isn't its most common state. The most common form—that is, the natural form—of the element oxygen is diatomic oxygen, O_2. Similarly, the natural state of hydrogen is H_2, and the natural state of nitrogen is N_2. The other element of concern with combustion processes here is carbon, and the natural state of carbon is taken as graphite (as opposed to other states of carbon, such as diamond), C(s).

The energy required or released when compounds are created or destroyed is derived from the creation of the compounds from their elements in their natural state. This energy is the energy present in chemical bonds, and the energy associated with being released as bonds are broken and consumed as bonds are created. The sum of the energy associated with these processes is called the enthalpy of formation and is given the symbol $\overline{h}_f^o$. In this symbol, the bar indicates that it is given as a molar-specific enthalpy value (there is also a mass-specific value, but this will not be used here), the subscript f refers to "formation," and the superscript o refers to standard pressure (101.325 kPa). We will use this value as determined at the reference temperature, T_{ref}, which is the standard temperature (25°C), but the value could be determined at other temperatures as well. We will be focusing on ideal gases, so the pressure dependence will not be important for the enthalpy. But if we were considering real gases, we would need to account for the effects on the enthalpy of changing the pressure from the reference pressure.

The enthalpy of formation can be determined experimentally or theoretically. Most engineers will never need to determine the enthalpy of formation for a compound in such a way, and instead they will use tabulated values for the compounds of interest; as such, we will not elaborate here on the methods used to determine the values. Inspection of a table will show that some compounds have positive enthalpies of formation, and some have negative enthalpies of formation. A positive value means that heat was required to create the compound from its elements, whereas a negative value indicates that heat is released in the process. For example, the enthalpy of formation of CO_2 is −393,520 kJ/kmole. This means that as graphite and oxygen were combined (such as in a combustion process), CO_2 was created and −393,520 kJ/kmole of heat was released as the CO_2 was cooled to 25°C. Conversely, the enthalpy of formation of atomic hydrogen, H, is +218,000 kJ/kmole. H is created by breaking the bond in the H_2 molecule, and this requires an input of energy. Therefore, the enthalpy of formation of H is positive.

The enthalpy of formation of the elements in their natural state at STP is considered to be 0:

$$\overline{h}_{fH_2}^o = \overline{h}_{fO_2}^o = \overline{h}_{fN_2}^o = \overline{h}_{fC(s)}^o = 0$$

Values for the enthalpy of formation of a variety of compounds can be found in Table A.10, or via information available from various sources online.

An important characteristic of the enthalpy of formation is that the enthalpy for a substance at a given temperature and pressure is equal to the sum of the enthalpy of formation and the difference in enthalpy between the reference state at which the enthalpy of formation was determined and the actual state:

$$\bar{h}_i(T, P) = \bar{h}^o_{f,i}(T_{ref}) + \left(\bar{h}_{i,T,P} - \bar{h}_{i,T_{ref},P^o}\right) \tag{11.7}$$

where T_{ref} is the reference temperature and P^o is the reference pressure of 101.325 kPa. If you are not concerned with a compound changing forms, then the change in enthalpy between two states reverts to being a difference in the relative enthalpy between two states because the enthalpy of formation and the enthalpy at the reference state cancel in the calculation.

In our combustion analyses, we will be considering ideal gases, and so the pressure dependency of the enthalpy will not exist. We will also be considering the reference temperature to be the standard temperature, T^o (25°C or 298 K). With these considerations, the molar-specific enthalpy values become

$$\bar{h}_i(T) = \bar{h}^o_{f,i}(T^o) + (\bar{h}_{i,T} - \bar{h}_{i,T^o}) \tag{11.8}$$

The first term on the right-hand side of Eq. (11.8) is the enthalpy of formation of the compound, and the term in parentheses is often referred to as the sensible enthalpy. Values for the specific enthalpy as a function of temperature for various compounds often found in combustion calculations are available in Table A.4 in the appendix. Values can also be found for these compounds from various Internet sources; however, be aware that different sources may use different 0 points, so when taking the difference between the two terms in the sensible enthalpy, be sure that the values are from the same source.

11.8 FURTHER DESCRIPTION OF THE COMBUSTION PROCESS

In Section 11.2, we briefly described the chemistry of the combustion process, but we barely considered the heat produced. We know that most combustion processes are used to produce heat. As the carbon and hydrogen in the fuel are oxidized, heat is released. Some of this is used to heat surrounding fuel and oxygen to spread the combustion process to those regions. But more heat is released than is needed to sustain the combustion process. Overall, the combustion products are raised to a high temperature. At the same time, heat is lost from the gases through conduction, convection, and radiation. Depending on the design of the combustor, one of these mechanisms may be dominant, and the amount of heat lost to the surroundings may be more or less severe. For example, if combustion takes place on a water-cooled burner, the heat lost from the combustion gases to the burner will result in cooler gas temperatures than would be expected if the combustion was taking place on a burner that was not water-cooled. Competing processes are occurring. Energy released from the combustion reactions is added to the gases to raise their temperature, and at the same time heat is lost to the surroundings from the high-temperature gases, as shown in **Figure 11.7**. The relative rates of these processes will determine how hot the product gases can become.

In a combustion process, the combustion products will be very hot immediately following the primary chemical reactions. Subsequently, the gases cool. As the gases cool from their peak temperature to the lower temperature at which they leave the combustor, heat is released to the surroundings. This heat may be used, for example, to boil water in the steam generator of a Rankine cycle power plant. The cooler the gases are allowed to become before leaving the system, the more heat is released for use in a process. Therefore, for a given fuel and air mixture entering at a certain temperature, there will be greater heat released if the products leave the system at 500 K as opposed to 700 K. The amount of heat released is a combination of the heat that is lost during the combustion processes even as the gases become hotter and the heat that is lost after the combustion products have reached their maximum temperature as they cool to their exit temperature. For most purposes, it is assumed that the most heat that can be released is the amount that is removed from the products to cool the products to the

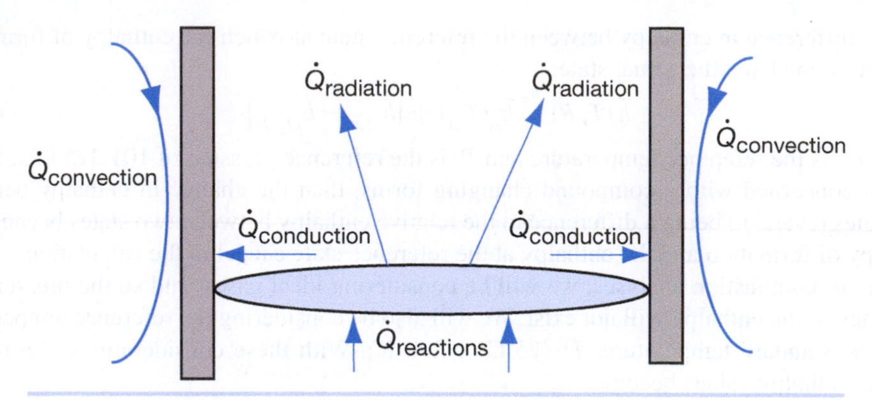

FIGURE 11.7 Heat can be added to and removed from combustion gases through conduction and radiation. In addition, convective heat transfer can remove heat from the entire system.

standard temperature of 25°C. If the combustion surroundings are colder, say 0°C, adjustments could be made in the calculations to account for this, but the change in the heat release would not normally be significant enough to be concerned with such a small change.

11.9 HEAT OF REACTION

The heat of reaction is the amount of heat that is released during a chemical reaction process, from the time the reactants enter the combustor to the time the products leave at some later time. When applied to a combustion process, this is also referred to as the *heat of combustion*, and the terms *enthalpy of reaction* or *enthalpy of combustion* may also be used. The determination of this quantity is done simply as an application of the First Law of Thermodynamics, with the reactants being considered as the initial state and the products as the final state. For our purposes, most combustion processes are analyzed as an open system using enthalpy values, because there is often a stream of fuel and oxidizer entering the system and a stream of products exiting the system.

11.9.1 The Open Combustion System

We will assume that our open systems with combustion processes are occurring at steady state with steady flows. Consider the open system containing a combustion process shown in Figure 11.8. In this case, the inlet state consists of the reactants (subscript R) at the reactant temperature and pressure, and the exit state consists of the products (subscript P) at the product temperature and pressure. The size of the system can be adjusted to be as small as the combustion process itself. It can also be adjusted to contain a mixing area for the fuel and oxidizer, to include the entire combustor, or perhaps to contain other pieces of equipment. As we will see, the changes to the enthalpy will easily overwhelm changes in kinetic and potential energy in nearly all combustion processes, and so we will ignore the effects of changes in kinetic and potential energy in our

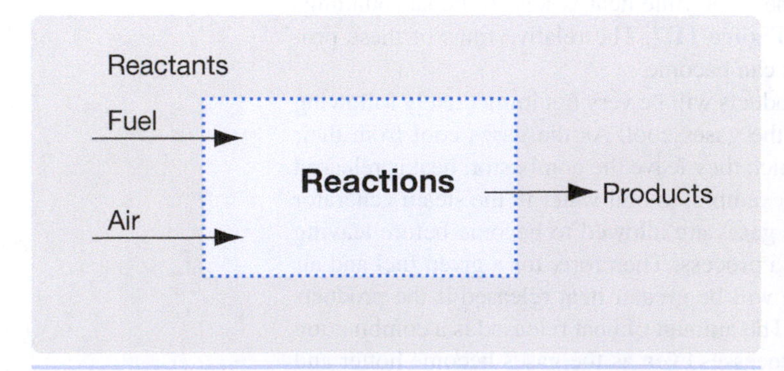

FIGURE 11.8 In a flowing combustion process, the system experiences fuel and air entering as reactants and the products of combustion exiting the system.

analysis. For this system, with negligible changes in kinetic and potential energy, the First Law of Thermodynamics for open systems at steady state can be written as

$$\dot{Q} - \dot{W} = \dot{H}_P - \dot{H}_R \tag{11.9}$$

where $\dot{H}$ is the rate of entropy either entering the system with the reactants or leaving the system with the products. This value is determined as the sum of the enthalpies flowing in or out of the system of the individual components:

$$\dot{H}_R = \sum_R \dot{n}_i \bar{h}_i = \sum_R \dot{n}_i \left(\bar{h}^o_{f,i} + \bar{h}_{i,T_R} - \bar{h}_{i,T^o}\right) \tag{11.10}$$

and

$$\dot{H}_P = \sum_P \dot{n}_i \bar{h}_i = \sum_P \dot{n}_i \left(\bar{h}^o_{f,i} + \bar{h}_{i,T_P} - \bar{h}_{i,T^o}\right) \tag{11.11}$$

where T_R is the reactant temperature entering the system and T_P is the product temperature leaving the system. In most cases, the power, $\dot{W}$, is not included in the calculation because the analysis is concentrating on either the flame or on the combustor itself, and there is usually negligible power consumed in the combustor in comparison to the heat release. However, if the system is enlarged to include a turbine, for example, and if we were to model both the combustor and the turbine of a gas turbine combustor at the same time, we may need to retain the power term in Eq. (11.9).

For the much more common situation of assuming the power to be 0, the first law can be solved for the heat of reaction, $\dot{Q}$, after substituting Eqs. (11.10) and (11.11) into Eq. (11.9):

$$\dot{Q} = \sum_P \dot{n}_i \left(\bar{h}^o_{f,i} + \bar{h}_{i,T_P} - \bar{h}_{i,T^o}\right) - \sum_R \dot{n}_i \left(\bar{h}^o_{f,i} + \bar{h}_{i,T_R} - \bar{h}_{i,T^o}\right) \tag{11.12}$$

It is often convenient to solve for the heat transfer rate per kmole of fuel. Such a quantity can then be used to quickly determine how much fuel is needed in a particular combustion process to yield a desired amount of a rate of heat transfer. This quantity is found by dividing Eq. (11.12) by the molar flow rate of the fuel:

$$\frac{\dot{Q}}{\dot{n}_{\text{fuel}}} = \sum_P \frac{\dot{n}_i}{\dot{n}_{\text{fuel}}} \left(\bar{h}^o_{f,i} + \bar{h}_{i,T_P} - \bar{h}_{i,T^o}\right) - \sum_R \frac{\dot{n}_i}{\dot{n}_{\text{fuel}}} \left(\bar{h}^o_{f,i} + \bar{h}_{i,T_R} - \bar{h}_{i,T^o}\right) \tag{11.13}$$

Equation (11.13) has an added benefit in terms of calculation efficiency. If the chemical reaction under consideration is written in terms of 1 kmole (or mole) of fuel, then the coefficients in front of each enthalpy term in Eq. (11.13) are simply the coefficients in front of that particular reactant or product in the chemical equation because the quantity of kilomoles of fuel in such case is 1 kmole/s.

One factor that will need to be determined when calculating the heat of combustion is the phase of the water in the products. For most combustion processes, the water will leave in vapor form, but if the products are cooled to a low enough temperature (below the dew point temperature of water based on the partial pressure of the water vapor in the products) in a particular combustor under analysis, water can exit in both liquid and vapor phases. Unless very high total pressures are being used, it is generally safe to assume that the water is in vapor form as long as the product temperature is above 350 K. If the products are cooled to sufficiently low temperatures so that some water condenses, then we will still normally treat the water as either all liquid or all vapor in the products rather than being concerned with the exact amount of water that has become liquid.

The following examples illustrate the methods used in calculating the heat of combustion using Eq. (11.13).

▶ EXAMPLE 11.6

A stoichiometric mixture of ethane and air is burned to completion. The reactants enter at 298 K, and the products exit at 298 K. Determine the heat released per kmole of ethane if (a) the water is considered all vapor in the products, and (b) the water is considered all liquid in the products.

Given: Fuel is ethane, C_2H_6. $\phi = 1$ (stoichiometric), $T_R = T_P = 298$ K

Find: $\dfrac{\dot{Q}}{\dot{n}_{\text{fuel}}}$ for the cases where the water in the products is (a) all vapor, and (b) all liquid.

Solution: The working fluid is an ideal gas, except for the liquid water in the products of part (b).

Assume: $\dot{W} = \Delta KE = \Delta PE = 0$. Steady-state, steady-flow open system.

The first step is to write the chemical reaction describing the complete combustion of a stoichiometric ethane–air mixture. From Eq. (11.1), with $y = 2$ and $z = 6$ for ethane,

$$C_2H_6 + 3.5(O_2 + 3.76N_2) \rightarrow 2CO_2 + 3H_2O + 13.16N_2$$

Equation (11.13) will be used to determine the heat released per kmole of ethane. In Eq. (11.13), because both $T_R = T^o = 298$ K and $T_P = T^o = 298$ K, the sensible enthalpy component $(\bar{h}_{i,T} - \bar{h}_{i,T^o})$ for each reactant and product is 0. Therefore, Eq. (11.13) reduces to

$$\frac{\dot{Q}}{\dot{n}_{\text{fuel}}} = \sum_P \frac{\dot{n}_i}{\dot{n}_{\text{fuel}}}\left(\bar{h}^o_{f,i}\right) - \sum_R \frac{\dot{n}_i}{\dot{n}_{\text{fuel}}}\left(\bar{h}^o_{f,i}\right)$$

This expands to

$$\frac{\dot{Q}}{\dot{n}_{C_2H_6}} = \left[\frac{\dot{n}_{CO_2}}{\dot{n}_{C_2H_6}}\bar{h}^o_{f,CO_2} + \frac{\dot{n}_{H_2O}}{\dot{n}_{C_2H_6}}\bar{h}^o_{f,H_2O} + \frac{\dot{n}_{N_2}}{\dot{n}_{C_2H_6}}\bar{h}^o_{f,N_2}\right] - \left[\frac{\dot{n}_{C_2H_6}}{\dot{n}_{C_2H_6}}\bar{h}^o_{f,C_2H_6} + \frac{\dot{n}_{O_2}}{\dot{n}_{C_2H_6}}\bar{h}^o_{f,O_2} + \frac{\dot{n}_{N_2}}{\dot{n}_{C_2H_6}}\bar{h}^o_{f,N_2}\right]$$

The enthalpies of formation are (from Table A.10)

$$\bar{h}^o_{f,CO_2} = -393{,}520 \text{ kJ/kmole } CO_2$$

$$\bar{h}^o_{f,C_2H_6} = -84{,}680 \text{ kJ/kmole } C_2H_6$$

$$\bar{h}^o_{f,O_2} = \bar{h}^o_{f,N_2} = 0$$

and $\bar{h}^o_{f,H_2O}$ depends on whether the water is in vapor or liquid form.

(a) For the case where all the water in the products is considered as a vapor, $\bar{h}^o_{f,H_2O(g)} = -241{,}820$ kJ/kmole H_2O.

Substituting,

$$\frac{\dot{Q}}{\dot{n}_{C_2H_6}} = \left[\frac{2 \text{ kmole } CO_2}{\text{kmole } C_2H_6}(-393{,}520 \text{ kJ/kmole } CO_2) + \frac{3 \text{ kmole } H_2O}{\text{kmole } C_2H_6}(-241{,}820 \text{ kJ/kmole } H_2O)\right.$$

$$\left. + \frac{13.16 \text{ kmole } N_2}{\text{kmole } C_2H_6}(0)\right] - \left[(1)(-84{,}680 \text{ kJ/kmole } C_2H_6)\right.$$

$$\left. + \frac{3.5 \text{ kmole } O_2}{\text{kmole } C_2H_6}(0) + \frac{13.16 \text{ kmole } N_2}{\text{kmole } C_2H_6}(0)\right]$$

$$\frac{\dot{Q}}{\dot{n}_{C_2H_6}} = \mathbf{-1{,}428{,}000 \text{ kJ/kmole } C_2H_6}$$

Notice how the units cancel in the equation so that each term ends up as an energy term divided by the kmole of fuel. This will happen in all such calculations, so we will not include the units of each term in each example. But keep in mind that the units cancel as shown here.

(b) For the case where all the water in the products is considered as a vapor, $\bar{h}^o_{f,H_2O(l)} = -285{,}830$ kJ/kmole H_2O.

(The difference between the enthalpy of formation of the liquid and the vapor is the enthalpy of condensation of water at 298 K: $-h_{fg} = 2442.3$ kJ/kg, or, on a molar basis: $-\overline{h}_{fg} = -44{,}010$ kJ/kmole.) Substituting this value for the enthalpy of formation of water in the preceding equation yields

$$\frac{\dot{Q}}{\dot{n}_{C_2H_6}} = -1{,}560{,}000 \text{ kJ/kmole } C_2H_6$$

Analysis: The heat transfer rate is negative because the heat is out of the reaction and our sign convention is such that the numerical answer is negative for heat out of a system. This heat can then be transferred to another system as a heat input. For example, in a steam generator in a Rankine cycle steam power plant, analysis of the combustion process will produce a negative heat transfer because it represents heat out of the combustion process. This heat is then used as a heat input to the water, which transforms from a compressed liquid into a vapor.

In addition, more heat is released as the water condenses into a liquid; however, as discussed previously, most combustion processes experience the water exiting in vapor form due to the higher product temperatures.

Example 11.6 introduces a calculation that is used to compare the heating potential of different fuels. By having the reactants enter at standard temperature, it is considered that there is not energy being added with the fuel to the combustion process. By having the products cooled to the standard temperature, it is considered that the maximum amount of heat that can possibly be extracted from the combustion products under standard conditions is being removed. A stoichiometric mixture allows for the minimum amount of air necessary for complete combustion, and so there is no excess air in the products absorbing energy that could have been transferred to other systems. By considering complete combustion, the most amount of heat to be released by the oxidation of the fuel is being determined. This set of conditions allows all fuels to be compared on an equal basis. The quantity that is determined from this analysis is the heating value of the fuel and represents the most heat that could be generated during the combustion of a fuel at standard conditions. The heating value is normally listed as a positive quantity and is often expressed on a mass basis $\left(\text{HV} = -\frac{\dot{Q}}{\dot{m}_{fuel}}\right)$.

The product temperature is 298 K, and two heating values are being considered. The lower heating value (LHV) is calculated assuming that all of the water is in vapor form, and the higher heating value (HHV) is calculated assuming that all of the water is in liquid form. For ethane, LHV = 47,480 kJ/kg and HHV = 51,870 kJ/kg. If the values from Example 11.6 are converted to heating values, the calculated values are LHV = 47,483 kJ/kg and HHV = 51,874 kJ/kg, which clearly are within the precision of the values used in the calculations. Table 11.2 lists the higher and lower heating values for various fuels.

The heating values allow for quick estimates of how much fuel may be needed to provide a desired amount of heat, and also allow for estimates of the costs that may be incurred using different fuels. For example, suppose we wanted to consider operating an engine on either pure octane or pure ethanol, both in liquid form. The LHV of octane is 1.66 times larger than the LHV of ethanol. If octane costs less than 66% more per kg than ethanol, it would be more economical to operate the engine on octane. But if octane costs more than 66% more per kg, it would be more economical to operate the engine on ethanol. Obviously, other factors (such as product temperature) come into making such a decision, but this general statement provides an estimate for beginning the analysis.

Example 11.6 is useful to illustrate how the heating value of a fuel is calculated, but it is not common for the product temperature to be lowered to 25°C before exiting the combustor. The following example illustrates the more common scenario of the reactants entering at standard temperature but the products exiting at an elevated temperature.

TABLE 11.2 Higher and Lower Heating Values (kJ/kg) for Various Fuels

Substance	Formula	Higher Heating Value (kJ/kg)	Lower Heating Value (kJ/kg)
Carbon (s)	C	32,800	32,800
Hydrogen (g)	H_2	141,800	120,000
Carbon Monoxide (g)	CO	10,100	10,100
Methane (g)	CH_4	55,530	50,050
Methanol (l)	CH_3OH	22,660	19,920
Acetylene (g)	C_2H_2	49,970	48,280
Ethane (g)	C_2H_6	51,900	47,520
Ethanol (l)	C_2H_5OH	29,670	26,810
Propane (l)	C_3H_8	50,330	46,340
Butane (l)	C_4H_{10}	49,150	45,370
Toluene (l)	C_7H_8	42,400	40,500
Octane (l)	C_8H_{18}	47,890	44,430
Gasoline (l)[a]		47,300	44,000
Light Diesel (l)[a]		46,100	43,200
Natural Gas (g)[a]		50,000	45,000

[a]Heating values for this fuel depend on the exact composition of the fuel, but the values given are values for typical compositions.

QUESTION FOR THOUGHT/DISCUSSION

Hydrogen has such a large heating value in comparison to other fuels listed in Table 11.2. What are some limitations that prevent H_2 from being used as the dominant fuel in combustion processes?

▶ EXAMPLE 11.7

Methane with 15% excess air enters a combustor at 25°C and burns to completion. The products exit the system at 720 K. Determine (a) the heat released per kmole of CH_4, and (b) the rate of heat release for 0.25 m³/s of methane if the methane is at standard temperature and pressure.

Given: Fuel: CH_4. Oxidizer: Air. $T_R = 25°C = 298$ K, $T_P = 720$ K, 15% excess air (115% theoretical air), $\dot{V}_{CH_4} = 0.25$ m³/s, $P = 101.325$ kPa

Find: (a) $\frac{\dot{Q}}{\dot{n}_{CH_4}}$, (b) $\dot{Q}$

Solution: The working fluid is an ideal gas throughout—the temperature of the products assures that the water in the products will be in vapor form.

Assume: $\dot{W} = \Delta KE = \Delta PE = 0$. Steady-state, steady-flow, open system.

The chemical reaction can be determined from Eq. (11.2) with $y = 1$, $z = 4$, and $x = 1.15$:

$$CH_4 + 2.3(O_2 + 3.76N_2) \rightarrow CO_2 + 2H_2O + 0.3O_2 + 8.65N_2$$

Because $T_R = T^o = 298$ K, the sensible enthalpy of the reactants is 0, as in Example 11.6. But the sensible enthalpy terms are still present for the products. Expansion of Eq. (11.13) yields

$$\frac{\dot{Q}}{\dot{n}_{CH_4}} = \left(1\,\frac{\text{kmole CO}_2}{\text{kmole CH}_4}\right)\left(\bar{h}_f^o + \bar{h}_{720K} - \bar{h}_{298K}\right)_{CO_2} + \left(2\,\frac{\text{kmole H}_2\text{O}}{\text{kmole CH}_4}\right)\left(\bar{h}_f^o + \bar{h}_{720K} - \bar{h}_{298K}\right)_{H_2O}$$

$$\times \left(0.3\,\frac{\text{kmole O}_2}{\text{kmole CH}_4}\right)\left(\bar{h}_f^o + \bar{h}_{720K} - \bar{h}_{298K}\right)_{O_2} + \left(8.65\,\frac{\text{kmole N}_2}{\text{kmole CH}_4}\right)\left(\bar{h}_f^o + \bar{h}_{720K} - \bar{h}_{298K}\right)_{N_2}$$

$$-\left[\left(1\,\frac{\text{kmole CH}_4}{\text{kmole CH}_4}\right)\bar{h}^o_{f,CH_4} + \left(2.3\,\frac{\text{kmole O}_2}{\text{kmole CH}_4}\right)\bar{h}^o_{f,O_2} + \left(8.65\,\frac{\text{kmole N}_2}{\text{kmole CH}_4}\right)\bar{h}^o_{f,N_2}\right]$$

The water in the products will be in vapor form, because the temperature is too high for liquid water to exist.

The values to be substituted can be found in the following summary table (from Tables A.4 and A.10):

Compound	$\bar{h}_f^o$ (kJ/kmole)	$\bar{h}_{720K}$ (kJ/kmole)	$\bar{h}_{298K}$ (kJ/kmole)
CO_2	−393,520	28,121	9364
H_2O	−241,820	24,840	9904
O_2	0	21,845	8682
N_2	0	21,220	8669
CH_4	−74,850	—	—

Because there is no sensible enthalpy consideration for CH_4 in this example, there is no need to be concerned with the values related to the sensible enthalpy of CH_4 in the table.

Upon substitution,

$$\frac{\dot{Q}}{\dot{n}_{CH_4}} = \mathbf{-641{,}200\ kJ/kmole\ CH_4}$$

(b) Standard temperature and pressure is $T = 298$ K and $P = 101.325$ kPa. The ideal gas law can be used to determine the number of kmoles of CH_4 in 0.25 m^3:

$$n = \frac{PV}{\bar{R}T} = \frac{(101.325\text{ kPa})(0.25\text{ m}^3)}{(8.314\text{ kJ/kmole}\cdot\text{K})(298\text{ K})} = 0.0102\text{ kmole}$$

As the volumetric flow rate is 0.25 m^3/s, this number of moles of CH_4 will be burned each second. Therefore,

$$\dot{n}_{CH_4} = 0.0102\text{ kmole/s}$$

The heat release rate is then

$$\dot{Q} = \dot{n}_{CH_4}\frac{\dot{Q}}{\dot{n}_{CH_4}} = \mathbf{-6560\ kW}$$

Analysis: Depending on how cool the products are allowed to be, there can be large amounts of heat produced in the combustion of relatively small amounts of hydrocarbon fuels. If the product temperature is increased, the amount of heat released decreases, and if the product temperature is decreased, more heat is released. Note that gaseous CH_4 has a fairly low density for a gas, and therefore there is a relatively low mass of CH_4 in the given volume.

Only strictly when a reactant temperature is equal to the reference temperature does the sensible enthalpy of a reactant equal 0. But if the reactants enter at slightly higher or lower temperatures, but still at what would be considered the ambient temperature, relatively little error would be produced by assuming the sensible enthalpy of the reactants to be 0 and using the approach in Example 11.7. There are enough assumptions being made in the calculation that small changes in the calculation of the heat of reaction would fall within the uncertainty expected in the practical application of the combustion process. But there are times when the reactants are intentionally preheated and a sensible enthalpy component of the reactants is required. For example, we may preheat the reactants by sending them through a heat exchanger to recover some heat from combustion products exiting a furnace—these gases will have already accomplished their purpose and are being sent to the exhaust at an elevated temperature. Capturing some of this heat in the reactants reduces the amount of fuel that must be added to reach a certain temperature in the combustion process. Another example is the steam generator in a power plant. The gases exit the economizer at an elevated temperature, and some of that energy can be recovered, rather than being dumped into the atmosphere, by sending the gases through an air preheater before going to the smokestack.

In some cases, only the air is preheated and the fuel is still brought into the process at ambient conditions. In this case, there is no sensible enthalpy contribution for the fuel, but there is for the air. In other cases, both the air and fuel are heated, at which point the sensible enthalpy of both must be considered. In some cases, data are not readily available for the molar-specific enthalpy of the fuel. In such cases, it is acceptable to approximate the sensible enthalpy of the fuel as if the fuel has constant specific heats. Although it is preferable not to mix constant and variable specific heat approaches, in this case the amount of fuel is generally small in comparison to the air, and the temperature difference between the fuel and the reference temperature often does not introduce substantial errors in the calculation:

$$\left(\bar{h}_{T_R} - \bar{h}_{T^o}\right)_{\text{fuel}} \approx \bar{c}_{p,\text{fuel}}\left(T_R - T^o\right) \tag{11.14}$$

▶ EXAMPLE 11.8

Propane, C_3H_8, is to burn to completion with 12% excess air. The fuel and air enter the combustor at 450 K, and the products exit at 1000 K. Determine the rate of heat released per kmole of propane.

Given: Fuel: C_3H_8. Oxidizer: Air. $T_R = 450$ K, $T_P = 1000$ K, 12% excess air (112% theoretical air)

Find: $\dfrac{\dot{Q}}{\dot{n}_{C_3H_8}}$

Solution: The working fluid is an ideal gas throughout—the temperature of the products assures that the water in the products will be in vapor form.

Assume: $\dot{W} = \Delta KE = \Delta PE = 0$. Steady-state, steady-flow open system. The propane is a gas.

Using Eq. (11.2) with $y = 3$, $z = 8$, and $x = 1.12$, the chemical reaction for the combustion process is found as

$$C_3H_8 + 5.6(O_2 + 3.76N_2) \rightarrow 3CO_2 + 4H_2O + 0.6O_2 + 21.06N_2$$

Keeping in mind that $T^o = 298$ K, Eq. (11.13) is used to determine the heat release per kmole of propane. Equation (11.14) is used for the sensible enthalpy of the propane:

$$\frac{\dot{Q}}{\dot{n}_{C_3H_8}} = \left(3\frac{\text{kmole CO}_2}{\text{kmole C}_3\text{H}_8}\right)(\bar{h}_f^o + \bar{h}_{1000K} - \bar{h}_{298K})_{CO_2} + \left(4\frac{\text{kmole H}_2\text{O}}{\text{kmole C}_3\text{H}_8}\right)(\bar{h}_f^o + \bar{h}_{1000K} - \bar{h}_{298K})_{H_2O}$$
$$+ \left(0.6\frac{\text{kmole O}_2}{\text{kmole C}_3\text{H}_8}\right)(\bar{h}_f^o + \bar{h}_{1000K} - \bar{h}_{298K})_{O_2} + \left(21.06\frac{\text{kmole N}_2}{\text{kmole C}_3\text{H}_8}\right)(\bar{h}_f^o + \bar{h}_{1000K} - \bar{h}_{298K})_{N_2}$$
$$- \left[\left(1\frac{\text{kmole C}_3\text{H}_8}{\text{kmole C}_3\text{H}_8}\right)(\bar{h}_{f,C_3H_8}^o + \bar{c}_{p,C_3H_8}(T_R - T^o)) + \left(5.6\frac{\text{kmole O}_2}{\text{kmole C}_3\text{H}_8}\right)(\bar{h}_f^o + \bar{h}_{450K} - \bar{h}_{298K})_{O_2}\right.$$
$$\left. + \left(21.06\frac{\text{kmole N}_2}{\text{kmole C}_3\text{H}_8}\right)(\bar{h}_f^o + \bar{h}_{450K} - \bar{h}_{298K})_{N_2}\right]$$

The water in the products will be in vapor form due to the high temperature.

The properties to be inserted can be found in the following table (from Tables A.4 and A.10):

Compound	$\bar{h}_f^o$ (kJ/kmole)	$\bar{h}_{1000K}$ (kJ/kmole)	$\bar{h}_{450K}$ (kJ/kmole)	$\bar{h}_{298K}$ (kJ/kmole)
CO_2	−393,520	42,769	—	9364
H_2O	−241,820	35,882	—	9904
O_2	0	31,389	13,228	8682
N_2	0	30,129	13,105	8669
C_3H_8	−103,850	—	—	—

For the specific heat of the propane,

$$\bar{c}_{p,C_3H_8} = 73.49 \text{ kJ/kmole} \cdot \text{K}$$

Calculating yields

$$\frac{\dot{Q}}{\dot{n}_{C_3H_8}} = \mathbf{-1{,}504{,}000\ kJ/kmole\ C_3H_8}$$

Analysis: Preheating the reactants leads to a greater heat release from the combustion product (for the same product temperature). If there had been no preheating of the products (i.e., $T_R = 298$ K), the rate of heat release would be $\frac{\dot{Q}}{\dot{n}_{C_3H_8}} = -1{,}374{,}000$ kJ/kmole C_3H_8. Therefore, less fuel is needed to provide the same amount of heat of reaction when preheating is used.

STUDENT EXERCISE

Create a computer model that will calculate the heat of reaction per kmole of fuel when given the reactant and product temperature, as well as the chemical reaction. Modify the model so that it can be used to determine either the heat released for a known amount of fuel or the amount of fuel needed to provide a specified heat release. Test the module through comparison to Examples 11.6 through 11.8.

11.9.2 Heat of Reaction: Closed Systems at Constant Pressure

Some combustion systems can be best modeled as closed systems. In such an analysis, we are looking for the total amount of heat released in a combustion process, rather than for a rate of heat release. One closed system analysis of interest is a system with constant pressure. As developed in Chapter 4, the first law for a closed system can be written as

$$Q - W = m\left(u_2 - u_1 + \frac{V_2^2 - V_1^2}{2} + g(z_2 - z_1)\right) \tag{4.32}$$

For a combustion process, we will consider the initial state to be the reactants, R, and the final state to be the products, P. Just as with the open system, the changes in kinetic and potential energy can generally be considered negligible. It is also more useful to write the internal energy on a molar basis. Rewriting Eq. (4.32) with these considerations yields

$$Q - W = U_P - U_R = \sum_P n_i \bar{u}_i - \sum_R n_i \bar{u}_i \tag{11.15}$$

For a constant-pressure process, the moving boundary work becomes

$$W = \int P\,d\forall = P(\forall_P - \forall_R) = P\forall_P - P\forall_R = \sum_P P n_i \bar{v}_i - \sum_R P n_i \bar{v}_i \tag{11.16}$$

Substituting Eq. (11.16) into Eq. (11.15) results in

$$Q = \sum_P n_i \bar{u}_i - \sum_R n_i \bar{u}_i + \sum_P P n_i \bar{v}_i - \sum_R P n_i \bar{v}_i \tag{11.17}$$

Recall that $\bar{h} = \bar{u} + P\bar{v}$. With this consideration, the heat transfer for a closed system undergoing combustion at constant pressure can be found from

$$Q = \sum_P n_i \bar{h}_i - \sum_R n_i \bar{h}_i \tag{11.18}$$

where $\bar{h}_i(T) = \bar{h}^o_{f,i}(T^o) + (\bar{h}_{i,T} - \bar{h}_{i,T^o})$ from Eq. (11.8). If written on a per kmole of fuel basis, Eq. (11.18) becomes

$$\frac{Q}{n_{\text{fuel}}} = \sum_P \frac{n_i}{n_{\text{fuel}}}\left(\bar{h}^o_{f,i} + \bar{h}_{i,T_P} - \bar{h}_{i,T^o}\right) - \sum_R \frac{n_i}{n_{\text{fuel}}}\left(\bar{h}^o_{f,i} + \bar{h}_{i,T_R} - \bar{h}_{i,T^o}\right) \tag{11.19}$$

which is nearly identical to Eq. (11.13). Eq. (11.13) involves rates of heat transfer and flow, whereas Eq. (11.19) involves totals of these quantities.

11.9.3 Heat of Reaction for Closed Systems at Constant Volume

When the closed system under consideration is a closed volume, such as in a bomb calorimeter (a device used to measure the heat of reaction from the combustion of a fuel), a similar analysis can be performed as that in Section 11.9.2. However, because the volume is constant, there is no moving boundary work: $W = 0$. For reference, a schematic diagram of a constant-volume combustion chamber can be found in **Figure 11.9**. Equation (11.15) reduces to

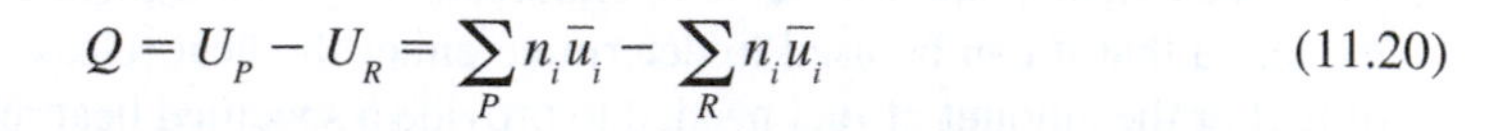

$$Q = U_P - U_R = \sum_P n_i \bar{u}_i - \sum_R n_i \bar{u}_i \tag{11.20}$$

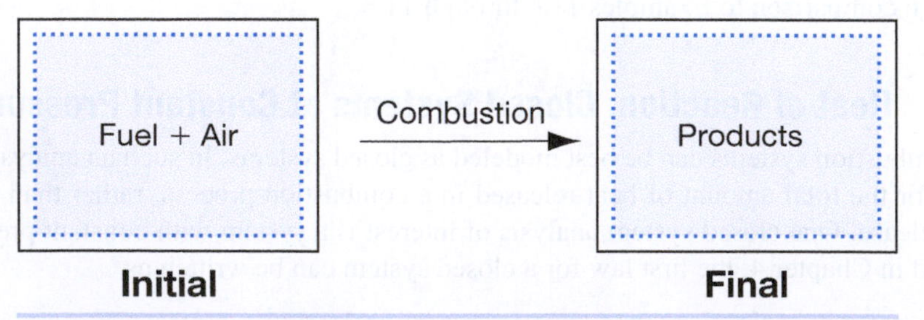

FIGURE 11.9 In a closed system combustion process, the fuel and air will react in the container to form combustion products.

Computation of the heat of reaction from this equation is complicated by data for the internal energy of formation often being less readily available than for the enthalpy of formation. Sensible internal energy values may be found easily, though. This problem can also be solved by recalling that $U = H - P\forall$. For an ideal gas mixture, $P\forall = n\bar{R}T$, and so $U = H - n\bar{R}T$. With this consideration, the heat of reaction for a combustion process in a closed system at constant volume can be found from

$$Q = \left[\sum_P n_i(\bar{h}^o_{f,i} + \bar{h}_{i,T_P} - \bar{h}_{i,T^o}) - n_P\bar{R}T_P\right] - \left[\sum_R n_i(\bar{h}^o_{f,i} + \bar{h}_{i,T_R} - \bar{h}_{i,T^o}) - n_R\bar{R}T_R\right] \quad (11.21)$$

On a per kmole of fuel basis, Eq. (11.21) becomes

$$\frac{Q}{n_{\text{fuel}}} = \left[\sum_P \frac{n_i}{n_{\text{fuel}}}(\bar{h}^o_{f,i} + \bar{h}_{i,T_P} - \bar{h}_{i,T^o}) - \frac{n_P}{n_{\text{fuel}}}\bar{R}T_P\right] - \left[\sum_R \frac{n_i}{n_{\text{fuel}}}(\bar{h}^o_{f,i} + \bar{h}_{i,T_R} - \bar{h}_{i,T^o}) - \frac{n_R}{n_{\text{fuel}}}\bar{R}T_R\right] \quad (11.22)$$

▶ EXAMPLE 11.9

0.25 kg of liquid methanol is placed inside a rigid vessel with 110% theoretical air at atmospheric pressure. The reactants are at 298 K before ignition. After combustion, the products are cooled to 700 K. Determine the amount of heat transfer that occurs for this process.

Given: Fuel: CH_3OH. Oxidizer: Air. $m_{CH_3OH} = 0.25$ kg, 110% theoretical air, $T_R = 298$ K, $T_P = 700$ K, closed system at constant volume (rigid vessel)

Find: Q

Solution: The air and products are ideal gases. The methanol fuel is liquid, although that will not impact the solution procedure.

Assume: $W = \Delta KE = \Delta PE = 0$

First, determine the number of kmole of ethanol: $n_{CH_3OH} = m_{CH_3OH}/M_{CH_3OH} = 0.00780$ kmole, because $M_{CH_3OH} = 32.04$ kg/kmole.

Next, determine the chemical equation for the combustion of 1 kmole of methanol with 100% theoretical air. Because methanol contains oxygen, it is not a true hydrocarbon, and this equation will be found by performing atom balancing:

$$CH_3OH + a(O_2 + 3.76N_2) \rightarrow bCO_2 + cH_2O + dN_2$$

C-balance: $(1)(1) = b(1)$

H-balance: $(1)(3 + 1) = c(2)$

O-balance: $(1)(1) + 2a = 2b + c$

N-balance: $3.76a(2) = 2d$

Solving yields $a = 1.5$, $b = 1$, $c = 2$, and $d = 5.64$.

For 110% theoretical air, the chemical reaction can then be found:

$$CH_3OH + (1.1)(1.5)(O_2 + 3.76N_2) \rightarrow CO_2 + 2H_2O + 0.15O_2 + 6.204N_2$$

Per kmole of fuel, the number of moles of reactants and products can be determined from this reaction:

$$n_R = 1 + (1.1)(1.5)(1 + 3.76) = 8.854 \text{ kmole}$$

$$n_P = 1 + 2 + 0.15 + 6.204 = 9.354 \text{ kmole}$$

Equation (11.22) is now used to determine the heat of reaction per kmole of methanol:

$$\frac{Q}{n_{CH_3OH}} = \left[\sum_P \frac{n_i}{n_{CH_3OH}}(\bar{h}^o_{f,i} + \bar{h}_{i,T_P} - \bar{h}_{i,T^o}) - \frac{n_P}{n_{fuel}}\bar{R}T_P\right] - \left[\sum_R \frac{n_i}{n_{CH_3OH}}(\bar{h}^o_{f,i} + \bar{h}_{i,T_R} - \bar{h}_{i,T^o}) - \frac{n_R}{n_{fuel}}\bar{R}T_R\right]$$

$$= \left[1(\bar{h}^o_f + \bar{h}_{T_P} - \bar{h}_{T^o})_{CO_2} + 2(\bar{h}^o_f + \bar{h}_{T_P} - \bar{h}_{T^o})_{H_2O} + 0.15(\bar{h}^o_f + \bar{h}_{T_P} - \bar{h}_{T^o})_{O_2} + 6.204(\bar{h}^o_f + \bar{h}_{T_P} - \bar{h}_{T^o})_{N_2} - n_P\bar{R}T_P\right] - \left[(\bar{h}^o_f)_{CH_3OH} - n_R\bar{R}T_R\right]$$

Note that the sensible enthalpy terms are 0 for the reactants because $T_R = T^o$.

The following data is to be substituted into the equation (from Tables A.4 and A.10):

Compound	$\bar{h}^o_f$ (kJ/kmole)	$\bar{h}_{700K}$ (kJ/kmole)	$\bar{h}_{298K}$ (kJ/kmole)
CO_2	−393,520	27,125	9364
H_2O	−241,820	24,088	9904
O_2	0	21,184	8682
N_2	0	20,604	8669
$CH_3OH(l)$	−238,810	—	—

Solving yields $\dfrac{Q}{n_{CH_3OH}} = -548{,}800$ kJ/kmole.

The total heat transfer for the 0.25 kg of methanol is then

$$Q = n_{CH_3OH}\frac{Q}{n_{CH_3OH}} = \mathbf{-4280\ kJ}$$

QUESTION FOR THOUGHT/DISCUSSION

How and why does increasing the amount of air in the reactants of a lean combustion process decrease the amount of heat released in the process? How and why does lowering the product temperature increase the amount of heat released in the process?

11.9.4 Heat of Reaction and Power Cycles

In Chapter 7, we analyzed many power cycles. A commonality among these power cycles is that they have a heat input. In Chapter 7, we generally assumed that this heat came from somewhere, but we were not concerned specifically with determining the source. In most of the power cycles, with the notable exception of the nuclear energy used in some Rankine cycle power plants, the heat source is a fuel burning in a combustion process, as illustrated in **Figure 11.10**. In some cases, the heat source is in a separate stream from the working fluid of the cycle. This is clearly the case with the Rankine cycle, where the steam is generated by heat transfer through pipes into the working fluid from the hot gases produced in a combustion process. In other practical devices, the combustion process is in the actual cycle, such as the combustion process in a cylinder in a spark-ignition or compression-ignition engine, or in the combustor in a Brayton cycle gas turbine system. Because we are often interested in knowing how much fuel is required in a combustion process to provide the necessary heat input in a power cycle, we may combine the heat of reaction calculation with the cycle analysis.

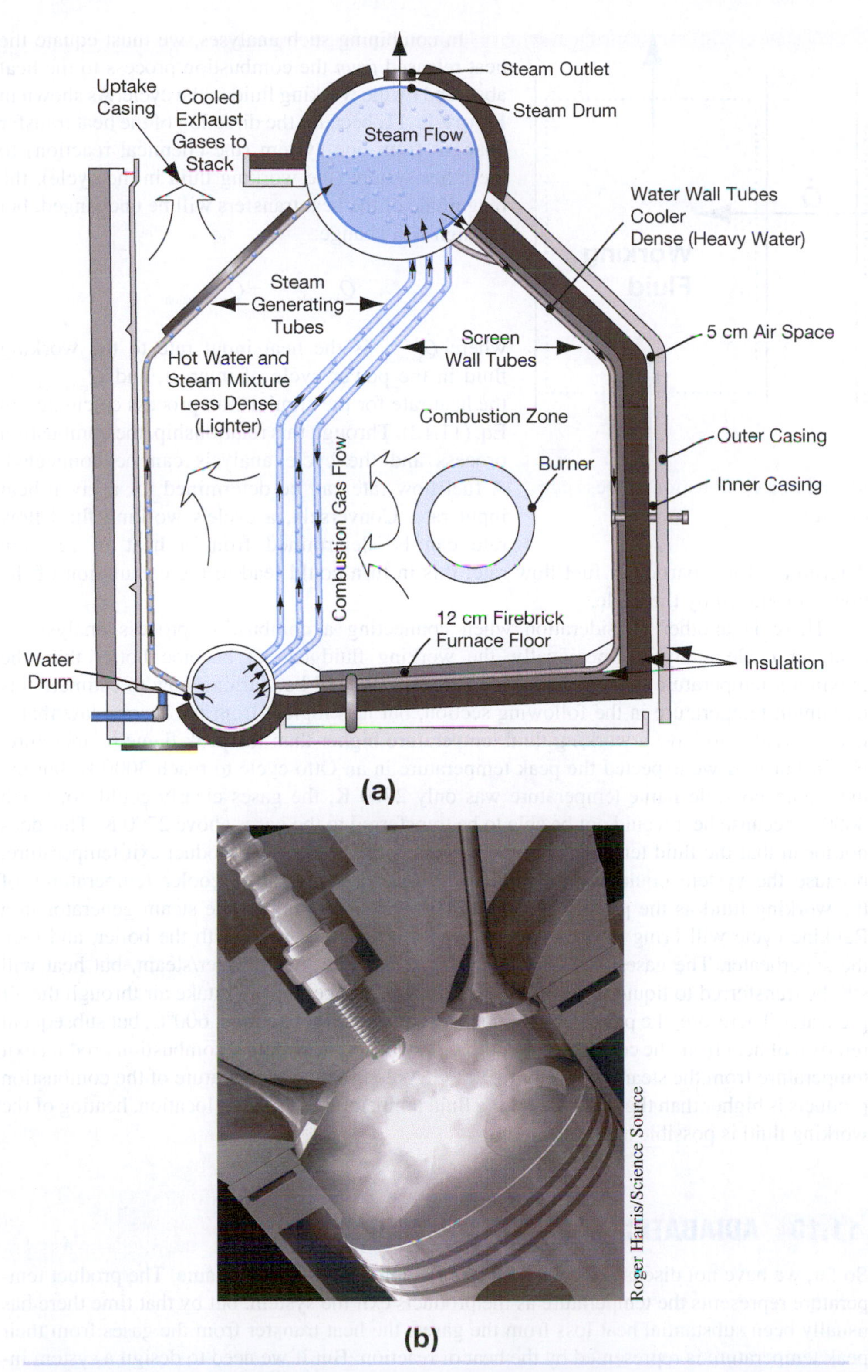

FIGURE 11.10 (a) In a combustion-powered Rankine cycle power plant, the combustion occurs in the steam generator, surrounded by water tubes to remove heat. (b) The combustion in a spark-ignition engine occurs inside the cylinders, near the valves.

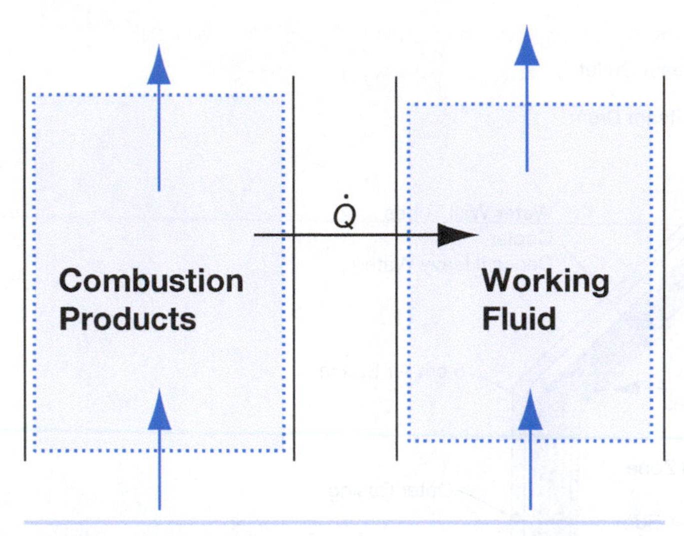

FIGURE 11.11 Heat produced in a combustion process can be transferred to a working fluid.

In combining such analyses, we must equate the heat released *from* the combustion process to the heat absorbed by the working fluid *in* the cycle. As shown in **Figure 11.11**, because the direction of the heat transfer changes from one system (the chemical reaction) to the other system (the working fluid in the cycle), the magnitude of the heat transfers will be unchanged, but the sign will change:

$$\dot{Q}_{\text{in,cycle}} = -\dot{Q}_{\text{combustion}}$$

where $\dot{Q}_{\text{in,cycle}}$ is the heat input rate to the working fluid in the power cycle of interest, and $\dot{Q}_{\text{combustion}}$ is the heat rate for the combustion process calculated in Eq. (11.12). Through this relationship, the combustion process and the cycle analysis can be connected. A fuel flow rate can be determined for a given heat input rate. Conversely, a cycle's working fluid flow rate can be determined from a heat of reaction determined for a particular fuel flow rate; this in turn could lead to the calculation of the power produced by the cycle.

There is another consideration when connecting a combustion process analysis to a power cycle analysis. Specifically, the working fluid cannot become hotter than the maximum temperature of the combustion process. We will be learning how to estimate this maximum temperature in the following section, but it is logical from the second law that it is not possible to have a working fluid temperature higher than the peak flame temperature. For instance, if we expected the peak temperature in an Otto cycle to reach 3000 K, but the maximum possible flame temperature was only 2700 K, the gases clearly could not reach 3000 K because heat would not be able to be transferred to the gases above 2700 K. This does not mean that the fluid temperature must be less than the eventual product exit temperature, because the system could be designed to utilize heat transfer to cooler temperatures of the working fluid as the product temperature drops. For example, the steam generator in a Rankine cycle will bring the combustion products first in contact with the boiler, and then the superheater. The gases will cool as they transfer heat to the water/steam, but heat will still be transferred to liquid water in the economizer, and perhaps to intake air through the air preheater. Therefore, the peak steam temperature may be, for example, 600°C, but subsequent removal of heat from the combustion products may lower the eventual combustion product exit temperature from the steam generator to 300°C. As long as the temperature of the combustion products is higher than the cycle's working fluid temperature at a given location, heating of the working fluid is possible at that location.

11.10 ADIABATIC FLAME TEMPERATURE

So far, we have not discussed in detail the actual temperature of the flame. The product temperature represents the temperature as the products exit the system, but by that time there has usually been substantial heat loss from the gases: the heat transfer from the gases from their peak temperature is represented by the heat of reaction. But if we need to design a system involving a combustion process and it is important to know how hot the gases get in that system, we must know the maximum temperature of the flame. For example, if we are going to use the hot gases from a combustion process to transfer heat to another fluid flowing through a thin tube passing through the gases, we will want to know if the tube may melt or at least become damaged from the high temperatures.

Unfortunately, in actuality it is difficult to obtain an accurate prediction of a flame temperature or even an accurate measurement. A common degree of uncertainty in many temperature measurement techniques for flame temperatures is ±5%. Suppose we measure a temperature of 2000 K. With this uncertainty, we can be confident that the temperature of the flame is somewhere between 1900 K and 2100 K. Clearly, that is a large difference, and one that goes beyond just having a precise understanding of the temperature for material concerns. At such temperatures, the amount formed of the pollutant NO increases with temperature; it approximately doubles (under some flame conditions) for each 100 K increase in temperature. Therefore, four times as much NO would be formed at 2100 K as at 1900 K.

Theoretical calculations of the maximum temperature are complicated by the need to thoroughly understand and model the heat transfer environment. In many combustors, this involves a complicated geometry, with the environment made more difficult by the presence of turbulent flames. However, it is relatively simple to calculate the maximum possible temperature that can be generated from the combustion of a particular set of reactants possessing a specific reactant temperature. This maximum temperature is known as the *adiabatic flame temperature* (AFT). In practice, this temperature will not be achieved due to the fact that there will be some heat losses from the combustion gases, the combustion is likely not to be fully complete in real systems, and processes that occur at high temperatures, such as dissociation and ionization of the products, lower the gas temperature. **Figure 11.12** shows the adiabatic flame temperature as a function of equivalence ratio for several fuels in air, with the reactants entering at 298 K.

The adiabatic flame temperature assumes that there is no heat transfer from the combustion process. It is found by setting the heat transfer in Eq. (11.13) to 0 and solving for the product temperature, T_P.

$$0 = \sum_P \frac{\dot{n}_i}{\dot{n}_{\text{fuel}}}\left(\bar{h}^o_{f,i} + \bar{h}_{i,T_P} - \bar{h}_{i,T^o}\right) - \sum_R \frac{\dot{n}_i}{\dot{n}_{\text{fuel}}}\left(\bar{h}^o_{f,i} + \bar{h}_{i,T_R} - \bar{h}_{i,T^o}\right) \tag{11.23}$$

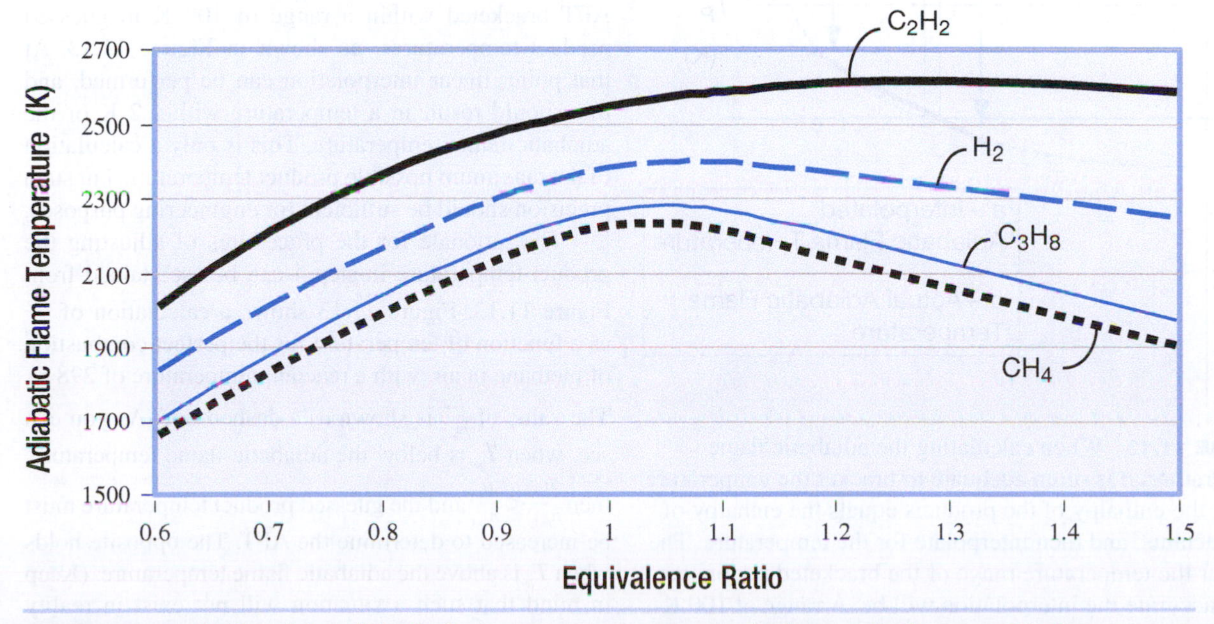

FIGURE 11.12 A plot of the adiabatic flame temperature versus equivalence ratio for several fuels.

Clearly, there is a problem with this simple statement of the solution procedure, because T_P only appears as a subscript and it is not possible to solve directly for a subscript. The solution procedure is indirect and is iterative in approach.

From Eq. (11.23), we can see that the solution procedure seeks the product temperature that will set the enthalpy of the products equal to that of the reactants:

$$\frac{\dot{H}_P}{\dot{n}_{\text{fuel}}} = \sum_P \frac{\dot{n}_i}{\dot{n}_{\text{fuel}}}\left(\overline{h}^o_{f,i} + \overline{h}_{i,T_P} - \overline{h}_{i,T^o}\right) = \sum_R \frac{\dot{n}_i}{\dot{n}_{\text{fuel}}}\left(\overline{h}^o_{f,i} + \overline{h}_{i,T_R} - \overline{h}_{i,T^o}\right) = \frac{\dot{H}_R}{\dot{n}_{\text{fuel}}} \tag{11.24}$$

To do this, we can use the following procedure when we have determined the chemical reaction:

1. Calculate $\frac{\dot{H}_R}{\dot{n}_{\text{fuel}}} = \sum_R \frac{\dot{n}_i}{\dot{n}_{\text{fuel}}}\left(\overline{h}^o_{f,i} + \overline{h}_{i,T_R} - \overline{h}_{i,T^o}\right)$ for the known reactant mixture and reactant temperature(s).
2. Guess a value of T_P.
3. Calculate $\frac{\dot{H}_P}{\dot{n}_{\text{fuel}}} = \sum_P \frac{\dot{n}_i}{\dot{n}_{\text{fuel}}}\left(\overline{h}^o_{f,i} + \overline{h}_{i,T_P} - \overline{h}_{i,T^o}\right)$ for the value of T_P.
4. Compare $\frac{\dot{H}_P}{\dot{n}_{\text{fuel}}}$ to $\frac{\dot{H}_R}{\dot{n}_{\text{fuel}}}$. Adjust T_P as necessary.
   (a) If $\frac{\dot{H}_P}{\dot{n}_{\text{fuel}}} = \frac{\dot{H}_R}{\dot{n}_{\text{fuel}}}$, then $T_P = AFT$.
   (b) If $\frac{\dot{H}_P}{\dot{n}_{\text{fuel}}} < \frac{\dot{H}_R}{\dot{n}_{\text{fuel}}}$, then increase T_P.
   (c) If $\frac{\dot{H}_P}{\dot{n}_{\text{fuel}}} > \frac{\dot{H}_R}{\dot{n}_{\text{fuel}}}$, then decrease T_P.
5. Return to step 3 and repeat until the AFT is found.

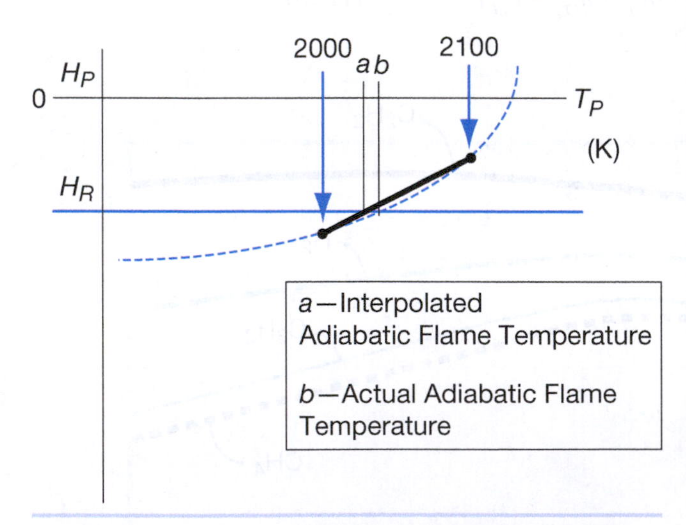

FIGURE 11.13 When calculating the adiabatic flame temperature, it is often adequate to bracket the temperature where the enthalpy of the products equals the enthalpy of the reactants, and then interpolate for the temperature. The smaller the temperature range of the bracketed region, the more accurate the interpolation will be. A range of 100 K is typically adequate.

It is unlikely that a guessed value will result in something that exactly gives the enthalpy of the products equal to the enthalpy of the reactants. So it is normally acceptable to continue the iteration until you have the AFT bracketed within a range of 100 K in guessed product temperatures, as shown in **Figure 11.13**. At that point, linear interpolation can be performed, and this should result in a temperature within 2 K of the adiabatic flame temperature. This is only a calculation of the maximum possible product temperature, but such precision should be sufficient for engineering purposes.

The rationale for the procedures of adjusting the product temperature in step 4 can be ascertained from Figure 11.13. Figure 11.13 shows a calculation of $\frac{\dot{H}_P}{\dot{n}_{\text{fuel}}}$ as a function of temperature for the perfect combustion of methane in air, with a reactant temperature of 298 K. The value of $\frac{\dot{H}_R}{\dot{n}_{\text{fuel}}}$ is shown as a dashed line. As you can see, when T_P is below the adiabatic flame temperature, then $\frac{\dot{H}_P}{\dot{n}_{\text{fuel}}} < \frac{\dot{H}_R}{\dot{n}_{\text{fuel}}}$ and the guessed product temperature must be increased to determine the AFT. The opposite holds when T_P is above the adiabatic flame temperature. (Keep in mind that such a situation will not exist in reality unless additional heat is added to the process.)

▶ EXAMPLE 11.10

Determine the adiabatic flame temperature for the complete combustion of gaseous propane and air in a stoichiometric mixture, with a reactant temperature of 298 K.

Given: Fuel: C_3H_8(g). Oxidizer: Air. $T_R = 298$ K, $\phi = 1$

Find: AFT

Solution: This is a standard adiabatic flame temperature calculation, and the working fluid is an ideal gas in the products.

Assume: $\dot{Q} = \dot{W} = \Delta KE = \Delta PE = 0$

First, the chemical reaction can be determined from Eq. (11.1) because the equivalence ratio is 1 for a stoichiometric mixture. In Eq. (11.1), $y = 3$ and $z = 8$ for propane:

$$C_3H_8 + 5(O_2 + 3.76N_2) \rightarrow 3CO_2 + 4H_2O + 18.8N_2$$

Because $T_R = T^o$, the reactants have no sensible enthalpy. The enthalpy of the reactants is then found from

$$\frac{\dot{H}_R}{\dot{n}_{C_3H_8}} = \sum_R \frac{\dot{n}_i}{\dot{n}_{C_3H_8}}\left(\bar{h}^o_{f,i} + \bar{h}_{i,T_R} - \bar{h}_{i,T^o}\right) = \sum_R \frac{\dot{n}_i}{\dot{n}_{C_3H_8}}\left(\bar{h}^o_{f,i}\right) = \bar{h}^o_{f,C_3H_8} = -103{,}850 \text{ kJ/kmole } C_3H_8$$

The enthalpy of the products will be found from

$$\frac{\dot{H}_P}{\dot{n}_{C_3H_8}} = \sum_P \frac{\dot{n}_i}{\dot{n}_{C_3H_8}}\left(\bar{h}^o_{f,i} + \bar{h}_{i,T_P} - \bar{h}_{i,T^o}\right) = 3\left(\bar{h}^o_f + \bar{h}_{T_P} - \bar{h}_{T^o}\right)_{CO_2} + 4\left(\bar{h}^o_f + \bar{h}_{T_P} - \bar{h}_{T^o}\right)_{H_2O} + 18.8\left(\bar{h}^o_f + \bar{h}_{T_P} - \bar{h}_{T^o}\right)_{N_2}$$

This calculation is performed much like the calculation in Examples 11.7 and 11.8.

The following values are constant for all guessed temperatures:

Compound	$\bar{h}^o_f$ (kJ/kmole)	$\bar{h}_{298K}$ (kJ/kmole)
CO_2	−393,520	9364
H_2O	−241,820	9904
N_2	0	8669

First, guess $T_P = 2200$ K. For this product temperature, the following values hold:

Compound	$\bar{h}_{2200K}$ (kJ/kmole)
CO_2	112,939
H_2O	92,940
N_2	72,040

Then

$$\frac{\dot{H}_P}{\dot{n}_{C_3H_8}} = -313{,}596 \text{ kJ/kmole } C_3H_8$$

Because $\frac{\dot{H}_P}{\dot{n}_{C_3H_8}} < \frac{\dot{H}_R}{\dot{n}_{C_3H_8}}$, T_P will be increased to 2300 K.

Compound	$\bar{h}_{2300K}$ (kJ/kmole)
CO_2	119,035
H_2O	98,199
N_2	75,676

Then

$$\frac{\dot{H}_P}{\dot{n}_{C_3H_8}} = -205{,}915 \text{ kJ/kmole } C_3H_8$$

Again, this is too low, and so T_p will be increased to 2400 K:

Compound	$\bar{h}_{2400K}$ (kJ/kmole)
CO_2	125,152
H_2O	103,508
N_2	79,320

Then

$$\frac{\dot{H}_P}{\dot{n}_{C_3H_8}} = -97{,}821 \text{ kJ/kmole } C_3H_8$$

This value is now above the enthalpy of the reactants. To find the AFT, we can use linear interpolation to find the value of T_P when $\frac{\dot{H}_P}{\dot{n}_{\text{fuel}}} = \frac{\dot{H}_R}{\dot{n}_{\text{fuel}}}$. This occurs at **AFT = 2394 K.**

Analysis: Clearly, very high temperatures can be attained by the burning of a hydrocarbon fuel with no heat transfer. Also, this value may differ from tabulated values that are available. Such values are often found when considering dissociation of the products, but this analysis does not include that phenomenon.

QUESTION FOR THOUGHT/DISCUSSION

If the adiabatic flame temperature is not reached in practice, why might an engineer be interested in knowing the value of the AFT for a given set of reactants?

11.11 ENTROPY BALANCE FOR COMBUSTION PROCESSES

As mentioned at the start of this chapter, a combustion process is highly irreversible. Although a fuel and oxidizer can be burned to produce products and heat, if we take the same amount of heat and the same products, the mixture will not return to the original fuel and oxidizer, alone, and at the original temperature. Some of the products may be induced to react to form a fuel and an oxidizer, but most of the products will not revert to the form of fuel and oxidizer. Thus, the process is highly irreversible.

When discussing irreversibility, the concept of entropy generation often arises. As discussed in Chapter 6, entropy is generated during irreversible processes, and the farther from reversible the process is, the more entropy is generated. Although we do not need to determine if a combustion process can occur, it may be helpful to compare the amounts of entropy generation between different chemical reactions in proposed combustion systems to determine which one is less irreversible.

For this analysis, we will focus on open systems, such as in **Figure 11.14**. Modifications for closed systems can be done as in Section 11.8 and Chapter 6. We also focus on steady-state systems with isothermal boundaries, with the reactants entering and the products exiting. For our purposes, Eq. (6.32) can be modified to

$$\dot{S}_{\text{gen}} = \sum_P \dot{n}_i \bar{s}_i - \sum_R \dot{n}_i \bar{s}_i - \sum_{j=1}^{n} \frac{\dot{Q}_j}{T_{b,j}} \tag{11.25}$$

or, on a unit molar flow of fuel basis,

$$\frac{\dot{S}_{\text{gen}}}{\dot{n}_{\text{fuel}}} = \sum_P \frac{\dot{n}_i}{\dot{n}_{\text{fuel}}} \bar{s}_i - \sum_R \frac{\dot{n}_i}{\dot{n}_{\text{fuel}}} \bar{s}_i - \sum_{j=1}^{n} \frac{\dot{Q}_j/\dot{n}_{\text{fuel}}}{T_{b,j}} \tag{11.26}$$

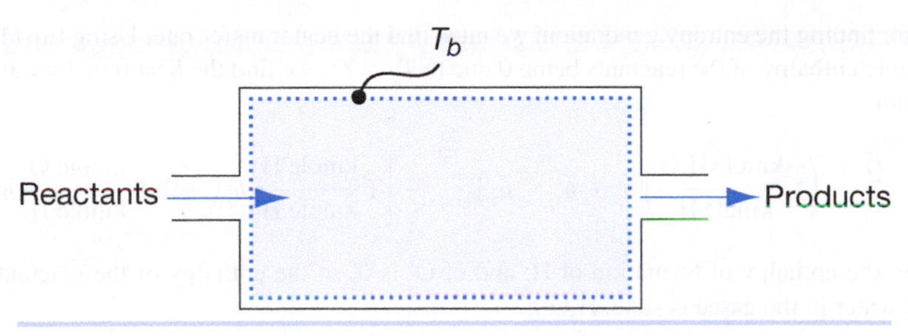

FIGURE 11.14 A basic open system, with an isothermal boundary at temperature T_b.

Unlike the enthalpy, there is no entropy of formation because entropy values are already referenced to a common zero point as described by the Third Law of Thermodynamics (see Chapter 6). However, most data for the entropy are determined for standard pressure (101.325 kPa) and need to be adjusted if the pressure of that component is not 101.325 kPa (such as if the total pressure of a mixture of gases is 101.325 kPa, but each component has a smaller partial pressure). To convert the specific entropy at standard pressure to that at the actual partial pressure of the component, the following expression can be used for components of an ideal gas mixture:

$$\bar{s}_i(T, P) = \bar{s}_i^o(T) - \bar{R}\ln\frac{P_i}{P^o} = \bar{s}_i^o(T) - \bar{R}\ln\frac{y_i P}{P^o} \tag{11.27}$$

The impact of partial pressure does have implications based on how reactants are brought into a system. If the fuel and oxidizer are brought in separately, then the fuel will have a pressure equal to its total pressure and will not be part of the overall reactant mixture with an associated mole fraction. If the fuel and oxidizer are mixed before entering the system, then the fuel will indeed have some partial pressure less than the total reactant pressure. Similarly, if liquid water is in the products, it would not have a mole fraction associated with being part of an ideal gas mixture and would instead have the total pressure of the water as its pressure. (We would not use water vapor data for the entropy of the water if the water is in liquid form.)

▶ EXAMPLE 11.11

Hydrogen is mixed with a stoichiometric amount of O_2, and the mixture enters a combustor at 298 K. The mixture burns to completion, and the products exit at a temperature of 500 K. The process occurs at 101.325 kPa pressure. If the surface temperature of the combustor is maintained at 450 K, determine the rate of entropy generation per kmole of H_2.

Given: Fuel: H_2. Oxidizer: O_2. $T_R = 298$ K, $T_P = 500$ K, $T_b = 450$ K, $P = 101.325$ kPa

Find: $\dfrac{\dot{S}_{gen}}{\dot{n}_{H_2}}$

Solution: The working fluids are ideal gases, and the system is a steady-state, steady-flow open system.

Assume: $\dot{W} = \Delta KE = \Delta PE = 0$

Keeping in mind that the only product of H_2–O_2 perfect combustion is H_2O, using atom balances we can determine that the chemical reaction is

$$H_2 + 0.5O_2 \rightarrow H_2O$$

At 500 K, the water will be in vapor form at atmospheric pressure.

Before finding the entropy generation, we must find the heat transfer rate. Using Eq. (11.13), with the sensible enthalpy of the reactants being 0 due to $T_R = T^o$, we find the heat transfer rate per kmole of H_2 from

$$\frac{\dot{Q}}{\dot{n}_{H_2}} = \left(1\frac{\text{kmole } H_2O}{\text{kmole } H_2}\right)(\bar{h}_f^o + \bar{h}_{T_P} - \bar{h}_{T^o})_{H_2O(g)} - \left[1\frac{\text{kmole } H_2}{\text{kmole } H_2}\bar{h}^o_{f,H_2} + 0.5\frac{\text{kmole } O_2}{\text{kmole } H_2}\bar{h}^o_{f,O_2}\right]$$

However, the enthalpy of formation of H_2 and of O_2 is 0, so the enthalpy of the reactants at 298 K is 0. For water in the gaseous state, $H_2O(g)$,

Compound	$\bar{h}_f^o$ (kJ/kmole)	$\bar{h}_{500K}$ (kJ/kmole)	$\bar{h}_{298K}$ (kJ/kmole)
H_2O	−241,820	16,828	9904

So

$$\frac{\dot{Q}}{\dot{n}_{H_2}} = -234{,}896 \text{ kJ/kmole } H_2$$

For the entropy generation rate per kmole of H_2, we can use Eq. (11.26):

$$\frac{\dot{S}_{gen}}{\dot{n}_{H_2}} = \frac{\dot{n}_{H_2O}}{\dot{n}_{H_2}}\bar{s}_{H_2O} - \left[\frac{\dot{n}_{H_2}}{\dot{n}_{H_2}}\bar{s}_{H_2} + \frac{\dot{n}_{O_2}}{\dot{n}_{H_2}}\bar{s}_{O_2}\right] - \frac{\dot{Q}/\dot{n}_{H_2}}{T_b}$$

Values for the temperature-dependent portion of the entropy can be found in Table A.10.

Because water is the only product and its pressure is 101.325 kPa, the molar-specific entropy is determined at 101.325 kPa and product temperature 500 K:

$$\bar{s}_{H_2O} = \bar{s}^o_{H_2O} = 206.413 \text{ kJ/kmole} \cdot \text{K}$$

For the products, the mole fraction of each component is $y_{H_2} = 0.667$, $y_{O_2} = 0.333$.

With Eq. (11.27), the molar-specific entropies of the reactants can be found ($P = P^o = 101.325$ kPa, and $T_R = 298$ K):

$$\bar{s}_{H_2} = \bar{s}^o_{H_2} - \bar{R}\ln\frac{y_{H_2}P}{P^o} = 130.574 - (8.314)\ln\frac{(0.667)(101.325 \text{ kPa})}{101.325 \text{ kPa}} = 133.941 \text{ kJ/kmole} \cdot \text{K}$$

$$\bar{s}_{O_2} = \bar{s}^o_{O_2} - \bar{R}\ln\frac{y_{O_2}P}{P^o} = 205.033 - (8.314)\ln\frac{(0.333)(101.325 \text{ kPa})}{101.325 \text{ kPa}} = 214.175 \text{ kJ/kmole} \cdot \text{K}$$

Substituting,

$$\frac{\dot{S}_{gen}}{\dot{n}_{H_2}} = (1)\bar{s}_{H_2O} - \left[(1)\bar{s}_{H_2} + (0.5)\bar{s}_{O_2}\right] - \frac{\dot{Q}/\dot{n}_{H_2}}{T_b} = \mathbf{435 \text{ kJ/kmole } H_2 \cdot K}$$

Analysis: Although we are not used to entropy generation rates per molar units, we can see that this would be a large entropy generation rate if converted for common flow rates. Therefore, the combustion process is highly irreversible—often more so than the mechanical processes examined in Chapter 6.

QUESTION FOR THOUGHT/DISCUSSION

Someone proposes using H_2–O_2 combustion in a spark-ignition engine, capturing the water that is produced and then reforming the water into H_2 and O_2 using the power of the engine—thus producing an engine that operates forever. Why won't this process work in practice?

The highly irreversible nature of combustion processes has led people to seek alternative means of releasing the chemical energy contained in a fuel. One such method is a fuel cell, which will be discussed shortly. However, before the fuel cell can be studied, we must introduce a new thermodynamic property: the Gibbs function.

11.12 THE GIBBS FUNCTION

A thermodynamic property that we have not yet encountered but that is often used in chemical studies is the Gibbs function, G. The Gibbs function is equal to the enthalpy minus the product of the temperature and entropy of a system:

$$G = H - TS \tag{11.28}$$

On a mass-specific or molar-specific basis, the specific Gibbs functions are

$$g = \frac{G}{m} = h - Ts \quad \text{and} \quad \bar{g} = \frac{G}{n} = \bar{h} - T\bar{s} \tag{11.29}$$

The Gibbs function is a combination of three thermodynamic properties, so it too should be recognized as a thermodynamic property, much like how enthalpy is recognized as a thermodynamic property.

11.13 FUEL CELLS

A fuel cell is a device that is designed to release the chemical energy bound in a fuel through a controlled reaction between a fuel and an oxidizer. The reaction is performed in an electrochemical device and generates an electrical current that can be used to power an external circuit. As with a combustion process, the fuel cell does produce products, but by not bringing the fuel and oxidizer directly together the system generates substantially less heat and has a much lower entropy generation rate. A schematic diagram of a simple fuel cell is shown in Figure 11.15. This particular type is a proton exchange membrane fuel cell, but other types of fuel cell systems exist. Hydrogen is the most commonly used fuel for fuel cells because it offers, in general, the best performance. Carbon compounds have a tendency

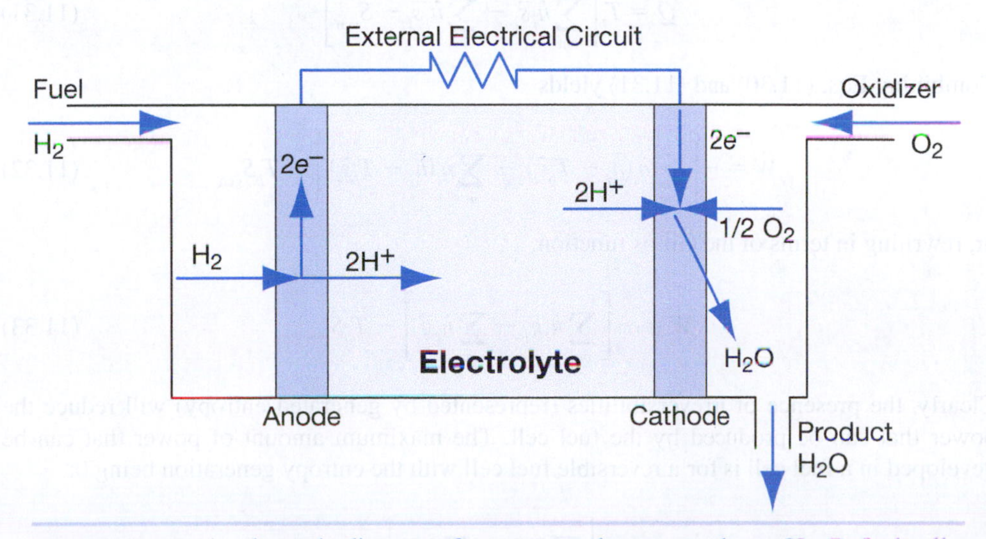

FIGURE 11.15 A schematic diagram of a proton exchange membrane H_2–O_2 fuel cell with the appropriate chemical reactions shown.

to reduce the performance of the fuel cell. Hydrogen does not exist in plentiful quantities on earth; it must be generated from other compounds. Hydrogen generation is accomplished either near the fuel cell in a re-forming process or as a stand-alone system.

As shown in Figure 11.15, the fuel (in this case H_2) reacts at the anode to produce electrons and hydrogen ions. The electrons flow through a circuit, which is connected to an external load, toward the cathode. The ions flow through an electrolyte toward the cathode. The oxidizer (in this case, O_2) is introduced near the cathode, and the oxidizer, hydrogen ions, and electrons recombine on the cathode to produce the product of the reaction: water. The overall chemical reaction is the same as would be experienced in a combustion process, but the process in a fuel cell is slower and more controlled. The fuel cell shares similarities to batteries, although a fuel cell uses a steady stream of reactants rather than the fixed amount of reactants initially placed inside the battery.

The purpose of the fuel cell is generally not to produce heat, but rather to produce the electron stream that can be used for powering electronic devices. The current can be small, and as a result fuel cells must often be grouped together to provide adequate power. This grouping increases the costs of the systems and renders fuel cells most appropriate for specialized applications (such as on spacecraft) rather than for widespread use (such as in automobiles). However, fuel cells do result in lower pollutant emissions than combustion processes do, and because fuel cells are not incorporated into heat engines (as is often the case with combustion processes), a much higher percentage of the chemical energy present in the fuel can be ultimately delivered to a process. Research is ongoing to improve fuel cells and make them more attractive options for transportation and power generation applications.

Analysis of a fuel cell centers on determining the electomotive potential, or voltage, of the fuel cell. If we consider a steady-state, steady-flow, open system with no changes in kinetic or potential energy, we can use the first law of thermodynamics to develop an expression for the power:

$$\dot{W} = -\left[\sum_{\text{out}} \dot{n}_i \bar{h}_i - \sum_{\text{in}} \dot{n}_i \bar{h}_i\right] + \dot{Q} \tag{11.30}$$

For a system with a single isothermal boundary, the heat transfer rate can be written from the entropy balance as

$$\dot{Q} = T_b\left[\sum_{\text{out}} \dot{n}_i \bar{s}_i - \sum_{\text{in}} \dot{n}_i \bar{s}_i - \dot{S}_{\text{gen}}\right] \tag{11.31}$$

Combining Eqs. (11.30) and (11.31) yields

$$\dot{W} = -\left[\sum_{\text{out}} \dot{n}_i(\bar{h}_i - T_b\bar{s}_i) - \sum_{\text{in}} \dot{n}_i(\bar{h}_i - T_b\bar{s}_i)\right] - T_b\dot{S}_{\text{gen}} \tag{11.32}$$

or, rewriting in terms of the Gibbs function,

$$\dot{W} = -\left[\sum_{\text{out}} \dot{n}_i \bar{g}_i - \sum_{\text{in}} \dot{n}_i \bar{g}_i\right] - T_b\dot{S}_{\text{gen}} \tag{11.33}$$

Clearly, the presence of irreversibilities (represented by generated entropy) will reduce the power that can be produced by the fuel cell. The maximum amount of power that can be developed in a fuel cell is for a reversible fuel cell with the entropy generation being 0:

$$\dot{W}_{\text{max}} = -\left[\sum_{\text{out}} \dot{n}_i \bar{g}_i - \sum_{\text{in}} \dot{n}_i \bar{g}_i\right] = -\Delta\dot{G} \tag{11.34}$$

Although tabulated values of the Gibbs function exist, it may be necessary to calculate values for the Gibbs function from the enthalpy and the entropy of a substance at a particular temperature and pressure. In such a case, remember to include the enthalpy of formation of the substance, as well as the partial pressure of the substance in the calculation.

Electrical work is also equal to the current multiplied by the electrical potential (voltage) of the system. The current is equal to the number of electrons flowing through the circuit multiplied by the charge of an electron. So, the electrical work can also be written as

$$\dot{W} = \Phi I = \Phi \dot{n}_e N_A e \tag{11.35}$$

where Φ is the electrical potential, I is the current, $\dot{n}_e$ is the rate of kmole of electrons flowing through the circuit, N_A is Avogadro's number (6.022×10^{26} electrons/kmole), and e is the charge of an electron ($e = 1.602 \times 10^{-23}$ kJ/eV). Combining Eqs. (11.34) and (11.35), an expression for the maximum voltage of a fuel cell can be written:

$$\Phi_{max} = \frac{-\Delta G}{96{,}485\, n_e} = \frac{-\Delta \dot{G}}{96{,}485\, \dot{n}_e} \tag{11.36}$$

where G is in kJ.

▸ EXAMPLE 11.12

An H_2–O_2 fuel cell operates at 400 K and 101.325 kPa. The product of the cell is water vapor. Determine the maximum voltage for the fuel cell.

Given: Fuel: H_2. Oxidizer: O_2. $T = 400$ K, $P = 101.325$ kPa. Product: H_2O(g)

Find: Φ_{max}

Solution: This is a basic H_2–O_2 fuel cell. The only product is water vapor. While the system is an open system, we will ignore the flow rates and only deal with total quantities.

Assume: The working fluids are all ideal gases.

For the overall reaction, $H_2 + ½O_2 \rightarrow H_2O(g)$. However, it is useful to consider only the anode in determining the number of kmoles of electrons, based on 1 kmole of fuel:

$$H_2 \rightarrow 2H^+ + 2e^-, \text{ and so } n_e = 2 \text{ kmoles}$$

For each compound,

$$\bar{g} = \bar{h} - T\bar{s} = \bar{h}_f^o + \bar{h}_T - \bar{h}_{T^o} - T\left(\bar{s}_T^o - \bar{R}\ln\frac{yP}{P^o}\right)$$

but because each substance has a mole fraction of 1 where it enters and exits, and because $P = P^o$, there is no need for a pressure correction of the entropy component.

Therefore, at 400 K, the following values are found:

$$\bar{g}_{H_2} = (0 + 11{,}426 - 8468)\frac{\text{kJ}}{\text{kmole } H_2} - (400\text{ K})\left(139.106\frac{\text{kJ}}{\text{kmole } H_2 \cdot \text{K}}\right) = -52{,}684.4 \text{ kJ/kmole } H_2$$

$$\bar{g}_{O_2} = (0 + 11{,}711 - 8682)\frac{\text{kJ}}{\text{kmole } O_2} - (400\text{ K})\left(213.765\frac{\text{kJ}}{\text{kmole } O_2 \cdot \text{K}}\right) = -82{,}477.0 \text{ kJ/kmole } O_2$$

$$\bar{g}_{H_2O} = (-241{,}820 + 13{,}356 - 9904)\frac{\text{kJ}}{\text{kmole } H_2O} - (400\text{ K})\left(198.673\frac{\text{kJ}}{\text{kmole } H_2O \cdot \text{K}}\right)$$

$$= -317{,}837.2 \text{ kJ/kmole } H_2O$$

The negative of the change in the Gibbs function is then found from

$$-\Delta G = -\left[\sum_{out} n_i \bar{g}_i - \sum_{in} n_i \bar{g}_i\right] = -\left[n_{H_2O}\bar{g}_{H_2O} - \left(n_{H_2}\bar{g}_{H_2} + n_{O_2}\bar{g}_{O_2}\right)\right]$$

$$= -\left[(1)(-317{,}837.2) - \left((1)(-52{,}684.4) + \left(\frac{1}{2}\right)(-82{,}477.0)\right)\right] = 223{,}914.3 \text{ kJ}$$

Then

$$\Phi_{max} = \frac{-\Delta G}{96{,}485 n_e} = \mathbf{1.16\ V}$$

Analysis: Clearly, although they are a promising power source, fuel cells must be combined to generate enough power for widespread practical application.

11.14 INTRODUCTION TO CHEMICAL EQUILIBRIUM

In previous sections of this chapter, we have alluded to various events that may impact the composition of a product mixture. In particular, we have mentioned the process of dissociation, but we have also hinted at the complex chemical processes that may lead to incomplete combustion. Dissociation is the dissolution of a compound into smaller components, such as an oxygen molecule (O_2) splitting into two oxygen atoms (2 O) at high temperatures. Although some dissociation can occur at any temperature, the process becomes more pronounced at higher temperatures, when compounds have higher energy and are therefore more prone to breaking chemical bonds. Such processes occur to all of the compounds at high temperatures, and recombination of the products can also subsequently occur. The global reactions we have discussed thus far do reasonably well at describing the product composition, but the discrepancy between the actual product composition and the simple theoretical composition becomes greater at higher temperatures. In turn, this discrepancy can impact the amount of heat released from a combustion process.

The combustion products will undergo many fundamental chemical reactions until a state of chemical equilibrium is reached. Like thermal equilibrium or mechanical equilibrium, chemical equilibrium is the state where the composition of the chemical mixture will not spontaneously change. The concentrations of the components, at the particular temperature and pressure under consideration, are steady when a system is at chemical equilibrium. However, changing the pressure or temperature, or adding or removing a component, may force the system to seek a new chemical equilibrium state. Part of a system being at thermodynamic equilibrium includes the condition of the system being at chemical equilibrium.

We will not go into the details here, but it is important to understand that a system is considered to be at chemical equilibrium when the system's Gibbs function is minimized. Computer programs are readily available that can calculate the composition of a mixture of gases at chemical equilibrium at a particular temperature and pressure. These programs rely on finding the composition of the mixture that gives a minimum Gibbs function but includes the appropriate number of atoms of each type in the composition.

Chemical equilibrium can also be studied on an individual reaction basis. It would be tedious to solve a series of these equations simultaneously to determine the equilibrium composition of a set of combustion products, but applying this technique can lead to greater understanding of which factors affect chemical equilibrium and whether such calculations are necessary. For example, if the combustion products are at a relatively low temperature, performing a chemical equilibrium calculation will have little impact on the product composition, whereas such a calculation may be necessary at very high temperatures.

Consider the following chemical equilibrium reaction:

$$\nu_A A + \nu_B B \leftrightarrow \nu_C C + \nu_D D$$

where *A*, *B*, *C*, and *D* are different chemical species and the coefficients represent the relative number of each component in a chemical equilibrium mixture. The dual-direction arrow indicates that the reaction proceeds in both directions: *A* and *B* react to create *C* and *D*, whereas *C* and *D* react to create *A* and *B*. Thus, although all species are both reactants and products, we will refer to the species on the left-hand side of this equation as the reactants, and the species on the right-hand side as the products.

At equilibrium, the rate at which a species is destroyed is equal to the rate at which a species is created. For example, in a room full of air, some O_2 molecules may be dissociating into O atoms at any given time, but at the same time some O atoms may be recombining to form O_2. The total concentration of O_2 and O is constant (and is predominantly O_2), but any given molecule may be in the process of dissociating at any time. At a temperature of 3000 K, the rates of O_2 destruction are increased, but when chemical equilibrium is reached, the rates of O_2 creation are also increased, again leading to steady concentrations. However, the amount of O needed to create enough O_2 to maintain a steady concentration must be higher; this requirement leads to a different equilibrium mixture than present at room temperature.

For a reaction at chemical equilibrium, an equilibrium constant can be written. In terms of the partial pressures of the components, this can be written as

$$K_P = \frac{\left(\frac{P_C}{P^o}\right)^{\nu_C}\left(\frac{P_D}{P^o}\right)^{\nu_D}}{\left(\frac{P_A}{P^o}\right)^{\nu_A}\left(\frac{P_B}{P^o}\right)^{\nu_B}} \tag{11.37}$$

For an ideal gas mixture, $P_i = y_i P$, where *P* is the total mixture pressure. With this substitution, the expression for the equilibrium constant can be rewritten as

$$K_P = \frac{y_C^{\nu_C} y_D^{\nu_D}}{y_A^{\nu_A} y_B^{\nu_B}}\left(\frac{P}{P^o}\right)^{\nu_C + \nu_D - (\nu_A + \nu_B)} \tag{11.38}$$

or more generally as

$$K_P = \frac{\prod_{P'} y_i^{\nu_i}}{\prod_{R'} y_i^{\nu_i}}\left(\frac{P}{P^o}\right)^{\sum_{P'} \nu_i - \sum_{R'} \nu_i} \tag{11.39}$$

where P' refers to the compounds on the right-hand side of the equilibrium reaction and R' refers to the compounds on the left-hand side. The capital pi (Π) is a mathematical symbol similar in concept to the capital sigma for a summation but represents taking a product of the terms.

The equilibrium constant is found from the change in the Gibbs function that would occur if the equilibrium reaction proceeded completely from the reactants to the products at standard pressure:

$$\Delta G^o = \sum_{P'} \nu_i \bar{g}_i - \sum_{R'} \nu_i \bar{g}_i \tag{11.40}$$

and then the equilibrium constant at the temperature of interest is

$$K_p(T) = \exp \frac{-\Delta G^o}{RT} \tag{11.41}$$

Values for the equilibrium constant, or the logarithm of the equilibrium constant, are readily available for many common equilibrium reactions. The values for some equilibrium reactions can be found in Table A.11. If values are unavailable, the equilibrium constant can be calculated with enthalpy and entropy data.

A common goal of a chemical equilibrium analysis is to determine the mole fractions of each component in the mixture at chemical equilibrium, or possibly the total number of moles of each component. Such values can be found using Eq. (11.39). But you may quickly notice that there are as many unknowns in Eq. (11.39) as species of interest, yet there is only one equation. As such, the problem would be unsolvable. The key to solving this problem is to relate all of the mole fractions to a single variable—a progress of reaction variable, ξ. To do this, the following steps are necessary:

(1) Write expressions for the initial number of moles present for each species: $n_{i,i}$.

(2) Determine a change in the number of moles of each species (Δn_i) by assigning the progress of reaction variable to one species and writing the expressions for the other based on the relationship among the species in the equilibrium reaction. Some species will be created (a positive change) and some will be destroyed (a negative change), with the amounts of the change equal to the ratios of the coefficients in front of the species in the equilibrium reactions.

(3) With the initial number of moles and the change in the number of moles, write expressions for the final number of moles of each species at chemical equilibrium: $n_{f,i} = n_{i,i} + \Delta n_i$.

(4) Add the number of moles of each species at chemical equilibrium, $n_T = \Sigma n_{f,i}$, and determine expressions for the mole fraction of each species at equilibrium, y_i. Each of these expressions should be in terms of the unknown ξ.

(5) Insert these expressions into Eq. (11.39), and solve for ξ.

(6) Using ξ, solve for either the mole fractions, y_i, or the number of moles of each species, $n_{f,i}$, depending on the desired end result.

This solution procedure is most easily understood through its use in an application. The following example provides such an application.

▶ EXAMPLE 11.13

1 kmole of CO_2 is rapidly heated to a temperature near 2000 K. Some of the CO_2 dissociates into O_2 and CO, and an equilibrium mixture is reached following the reaction

$$CO_2 \leftrightarrow ½\,O_2 + CO$$

Determine the number of moles of each compound when chemical equilibrium is achieved at a temperature of 2000 K and a pressure of 101.325 kPa.

Given: $n_{i,CO_2} = 1$ kmole, $T = 2000$ K, $P = 101.325$ kPa

Find: n_{f,CO_2}, n_{f,O_2}, $n_{f,CO}$ at chemical equilibrium

Solution: This is a chemical equilibrium calculation problem. The three substances involved are ideal gases.

The coefficient of the reactant is $\nu_{CO_2} = 1$, while the coefficients for the products are $\nu_{O_2} = 1/2$ and $\nu_{CO} = 1$.

To begin this analysis, we write the number of moles of each substance initially:

$$n_{CO_{2,i}} = 1 \text{ kmole}$$
$$n_{O_{2,i}} = n_{CO,i} = 0$$

Next, we assign a progress of reaction variable. Because CO_2 is destroyed, let us consider that the change in CO_2 is equal to a negative ξ: $\Delta n_{CO_2} = -\xi$.

According to the equilibrium reaction, for every mole of CO_2 destroyed, ½ mole of O_2 and 1 mole of CO are created. Therefore, the changes in these compounds as chemical equilibrium is reached are

$$\Delta n_{O_2} = -\frac{1}{2}\xi \quad \text{and} \quad \Delta n_{CO} = -\xi$$

The final number of moles of each compound is found from $n_{f,i} = n_{i,i} + \Delta n_i$:

$$n_{f,CO_2} = 1 - \xi$$
$$n_{f,O_2} = 0 + ½\xi = ½\xi$$
$$n_{f,CO} = 0 + \xi = \xi$$

Adding these expressions yields the total number of moles present at equilibrium: $n_T = 1 + ½\xi$.

Expressions for the mole fractions are then found from $y_i = n_{f,i}/n_T$:

$$y_{CO_2} = \frac{1 - \xi}{1 + \frac{1}{2}\xi} \qquad y_{O_2} = \frac{\frac{1}{2}\xi}{1 + \frac{1}{2}\xi} \qquad y_{CO} = \frac{\xi}{1 + \frac{1}{2}\xi}$$

With these, Eq. (11.39) can be written:

$$K_P = \frac{\prod_{P'} y_i^{\nu_i}}{\prod_{R'} y_i^{\nu_i}}\left(\frac{P}{P^o}\right)^{\sum_{P'}\nu_i - \sum_{R'}\nu_i} = \frac{y_{O_2}^{\nu_{O_2}} y_{CO}^{\nu_{CO}}}{y_{CO_2}^{\nu_{CO_2}}}\left(\frac{P}{P^o}\right)^{\nu_{O_2} + \nu_{CO} - \nu_{CO_2}}$$

with $P = P^o = 101.325$ kPa.

$$K_P = \frac{\left(\frac{\xi/2}{1 + \frac{\xi}{2}}\right)^{1/2}\left(\frac{\xi}{1 + \frac{\xi}{2}}\right)^{1}}{\left(\frac{1 - \xi}{1 + \frac{\xi}{2}}\right)^{1}}\left(\frac{101.325 \text{ kPa}}{101.325 \text{ kPa}}\right)^{(\frac{1}{2} + 1 - 1)}$$

For this reaction at 2000 K, $\ln K_P = -6.641$, so $K_P = 0.001306$.

Clearly, this equation is complex and may need to be solved numerically. In this case, the equation is solved for $\xi = 0.0674$.

The final number of kmoles of each compound may be found for the mixture at chemical equilibrium:

$$n_{CO_2} = 1 - \xi = \mathbf{0.933 \text{ kmole}}$$
$$n_{O_2} = ½\xi = \mathbf{0.0337 \text{ kmole}}$$
$$n_{CO} = \xi = \mathbf{0.0674 \text{ kmole}}$$

Analysis: As expected, at this elevated level there is some dissociation of the CO_2 into CO and O_2. You may notice that the dissociation changed the total number of moles. When a chemical equilibrium reaction occurs that results in a change in the number of moles, there often is a pressure dependency in the equilibrium constant expression. In this case, there was no pressure effect on the equilibrium constant because the pressure was that of the reference pressure.

As the previous example indicates, if we were to use the demonstrated solution process for many chemical equilibrium processes simultaneously, the solution procedure would quickly become computationally difficult if not overwhelming. As a result, the computer programs designed to seek a minimum Gibbs function for the mixture at chemical equilibrium were developed and are used.

In the problems at the end of this chapter, we explore the ways in which temperature, pressure, and the presence of additional species impact chemical equilibrium. In general, more species tend to exist in greater quantities as the temperature increases, pressure impacts some equilibrium processes (when the number of moles is changing in the equilibrium reaction) but not others (when the numbers of moles in the reactants and products of an equilibrium reaction are the same), and additional species will impact the total number of moles present, which can alter the composition of the mixture even if that species is not involved in the equilibrium reaction.

QUESTION FOR THOUGHT/DISCUSSION

How much dissociation of the products of combustion do you think is necessary before an engineer should be concerned with the impact of the revised product mixture on a heat of reaction calculation?

11.15 THE WATER–GAS SHIFT REACTION AND RICH COMBUSTION

As discussed earlier in this chapter, we cannot use atom balances alone to predict the products from a rich combustion process of a hydrocarbon in air. With the expected presence of CO, CO_2, H_2O, H_2, and N_2 in the products, the four atom-balancing equations would be expected to find five unknown coefficients. Nonetheless, it is useful to predict the approximate composition of a rich mixture. Obviously, this becomes extremely complex when a variety of unburned hydrocarbons are included, but if we exclude the possibility of unburned hydrocarbons being present in a significant quantity, we can use a chemical equilibrium reaction to assist in predicting the products from rich combustion. The reaction of interest is known as the water–gas shift reaction, and it describes chemical equilibrium among CO, CO_2, H_2O, and H_2:

$$CO + H_2O \leftrightarrow CO_2 + H_2$$

Suppose we are analyzing a rich combustion process of a generic hydrocarbon, C_yH_z, burning in air at a known equivalence ratio, ϕ. A generic expression for this process can be written as

$$C_yH_z + \frac{y + \frac{z}{4}}{\phi}(O_2 + 3.76N_2) \rightarrow aCO_2 + bCO + cH_2O + dH_2 + eN_2$$

Four atom balances can be developed, and the water–gas shift reaction equation for chemical equilibrium can be employed as a fifth equation. If we write the mole fractions for the products and insert those into Eq. (11.39), we derive the following expression:

$$K_P = \frac{a \cdot d}{b \cdot c} \tag{11.42}$$

Solving the atom balances and Eq. (11.42) yields the following expressions:

$$a = \frac{2f(K_p - 1) + y + \frac{z}{2}}{2(K_p - 1)} - \frac{1}{2(K_p - 1)}\left[\left(2f(K_p - 1) + y + \frac{z}{2}\right)^2 - 4K_p(K_p - 1)(2fy - y^2)\right]^{1/2} \tag{11.43a}$$

$$b = y - a \tag{11.43b}$$

$$c = 2f - a - y \tag{11.43c}$$

$$d = -2f + a + y + z/2 \tag{11.43d}$$

$$e = 3.76f \tag{11.43e}$$

where $f = \dfrac{y + \frac{z}{4}}{\phi}$.

Values for K_p for the water–gas shift reaction at various temperatures can be found in Table 11.3. You will note that the relationship between T and K_p is very nonlinear at lower temperatures. However, because the calculations being performed are estimates of the products, it is acceptable to use linear interpolation in Table 11.3 for values of the equilibrium constant if the desired temperature falls between tabulated values. Alternatively, you may consider that the product temperature in practice often has considerable uncertainty, and so you may end up just using a value at the closest temperature.

TABLE 11.3 Values of K_p for the water–gas shift reaction: $CO + H_2O \leftrightarrow CO_2 + H_2$

T (K)	K_p
298	104,200
400	1540
500	137.7
600	28.3
800	4.22
1000	1.442
1200	0.733
1400	0.465
1600	0.336
1800	0.265
2000	0.220
2200	0.192
2400	0.172
2600	0.158
2800	0.147
3000	0.139
3500	0.125

Clearly, this is not as simple of a calculation as that of the complete combustion of a lean or stoichiometric mixture. Furthermore, we should not expect it to be quite as accurate, and it generally is better if $\phi > 1.15$. Also, we must specify a temperature for the products so the correct value of K_P can be used.

▶ EXAMPLE 11.14

Using the water–gas shift reaction approach, determine the expected products of the combustion of 1 kmole of octane in air, at an equivalence ratio of 1.2, at 2000 K.

Given: Fuel: C_8H_{18}. Oxidizer: Air. $\phi = 1.2$, $T = 2000$ K

Find: Expected products of this combustion

Solution: For this rich combustion problem, we will use the water–gas shift reaction approach.

We use Eqs. (11.43). First, for octane, $y = 8$ and $z = 18$. In this case, $\phi = 1.2$, and so $f = 10.42$.

At 2000 K, $K_P = 0.2200$.

Solving Eqs. (11.43) yields

$a = 4.926$

$b = 3.074$

$c = 7.914$

$d = 1.086$

$e = 39.18$

This yields the predicted chemical reaction of

$$\mathbf{C_8H_{18} + 10.42(O_2 + 3.76N_2) \rightarrow 4.93CO_2 + 3.07CO + 7.91H_2O + 1.09H_2 + 39.18N_2}$$

Analysis: This reaction would change at a different product temperature. In addition, this reaction illustrates how the combustion process progresses. Initially, H_2O and CO are predominantly formed, and then the CO subsequently oxidizes to CO_2. If we compare the amount of H_2O to H_2 and the amount of CO_2 to CO, we see that there is much more H_2O than H_2, whereas the amounts of CO_2 and CO are more evenly distributed. This is consistent with the concept of the CO_2 forming later, when sufficient O_2 is present.

Summary

In this chapter, we have explored a number of concepts pertaining to combustion processes, and we have considered these processes primarily from an engineering viewpoint. As such, we have concentrated on global chemical reactions which describe the approximate set of products expected from a combustion process, and we have made assumptions about the completeness of the combustion processes. For the most part, such analysis methods will give approximations for expected products and heat releases that are adequate for engineering purposes. However, practical combustion systems may not be well modeled in these ways at all times. For example, in a spark-ignition engine, there is often insufficient time for the fuel and oxidizer to complete combustion before the temperature drops to a point where CO oxidation becomes slow; as a result, engine cylinders release large amounts of CO emissions. In other cases, the item of interest for modeling will be lost. For example, if we want to know how much NO_x emissions to expect from a process, the global chemical reaction (in which N_2 does not react) will not provide the necessary information. Furthermore, although the adiabatic flame temperature calculation approach used in this chapter is useful, it will generally overpredict the adiabatic flame temperature that would be calculated if we combined a chemical equilibrium of the products analysis with calculation of the adiabatic flame temperature.

More complex modeling techniques are used to handle such situations. For example, we may need to use a chemical kinetics modeling approach to fully understand the amounts of pollutants that should be expected, because NO and unburned hydrocarbons are best modeled using detailed chemical kinetics. Chemical kinetics modeling recognizes that the combustion process does not consist of a single global reaction. Rather, depending on the fuel and oxidizer used, there are tens or hundreds of elementary chemical reactions that describe the reactions among intermediate species, of which there may be 20, 30, 40, or more. By tracking the creation and destruction of all these species, we can more accurately determine the concentrations of minor species, such as NO and unburned hydrocarbons, in the products.

If we are modeling the behavior of a turbulent combustion process in an actual combustor, we may choose to use a set of reduced chemical reactions to model the chemistry combined with a computational fluid dynamics (CFD) package to model the fluid flow. Attempting to model a full chemical kinetics model with a difficult turbulent flow environment is too computationally intensive.

In the end, the models are likely to miss some details of an actual combustion system. In such a case, models can be used to assist in developing a combustion system, and they can provide reasonable estimates of what to expect from a combustor. But the design may need to be studied experimentally and may need to be adjusted through trial-and-error to reach the best performing device. For example, perhaps you are interested in using a furnace in a manufacturing process, and you calculate that you need a mass flow rate of methane of 2.5 kg/s to meet your heat release needs. In practice, you find that you need a little more methane flow, and you adjust that flow to 2.6 kg/s. The analysis techniques learned here, and more advanced techniques, provide the approximate level of fuel flow needed, but in practice, you have to make some adjustments to meet your goals.

KEY EQUATIONS

Perfect Combustion—Generic Hydrocarbon:

$$C_yH_z + \left(y + \frac{z}{4}\right)(O_2 + 3.76N_2) \rightarrow yCO_2 + \frac{z}{2}H_2O + 3.76\left(y + \frac{z}{4}\right)N_2 \tag{11.1}$$

Complete Combustion—Lean Mixture, Generic Hydrocarbon:

$$C_yH_z + (x)\left(y + \frac{z}{4}\right)(O_2 + 3.76N_2) \rightarrow yCO_2 + \frac{z}{2}H_2O + (x-1)\left(y + \frac{z}{4}\right)O_2 + 3.76(x)\left(y + \frac{z}{4}\right)N_2 \tag{11.2}$$

x is the $\frac{\%\ \text{theoretical air}}{100\%}$.

Equivalence Ratio:

$$\phi = \frac{FA_a}{FA_{st}} = \frac{\overline{FA}_a}{\overline{FA}_{st}} = \frac{AF_{st}}{AF_a} = \frac{\overline{AF}_{st}}{\overline{AF}_a} \quad (11.6)$$

Heat of Reaction—Open System, Steady-State, Steady-Flow:

$$\frac{\dot{Q}}{\dot{n}_{\text{fuel}}} = \sum_P \frac{\dot{n}_i}{\dot{n}_{\text{fuel}}}\left(\overline{h}^o_{f,i} + \overline{h}_{i,T_P} - \overline{h}_{i,T^o}\right) - \sum_R \frac{\dot{n}_i}{\dot{n}_{\text{fuel}}}\left(\overline{h}^o_{f,i} + \overline{h}_{i,T_R} - \overline{h}_{i,T^o}\right) \quad (11.13)$$

Heat of Reaction—Closed Systems at Constant Pressure:

$$\frac{Q}{n_{\text{fuel}}} = \sum_P \frac{n_i}{n_{\text{fuel}}}\left(\overline{h}^o_{f,i} + \overline{h}_{i,T_P} - \overline{h}_{i,T^o}\right) - \sum_R \frac{n_i}{n_{\text{fuel}}}\left(\overline{h}^o_{f,i} + \overline{h}_{i,T_R} - \overline{h}_{i,T^o}\right) \quad (11.19)$$

Heat of Reaction—Closed Systems at Constant Volume:

$$\frac{Q}{n_{\text{fuel}}} = \left[\sum_P \frac{n_i}{n_{\text{fuel}}}\left(\overline{h}^o_{f,i} + \overline{h}_{i,T_P} - \overline{h}_{i,T^o}\right) - \frac{n_P}{n_{\text{fuel}}}\overline{R}T_P\right] - \left[\sum_R \frac{n_i}{n_{\text{fuel}}}\left(\overline{h}^o_{f,i} + \overline{h}_{i,T_R} - \overline{h}_{i,T^o}\right) - \frac{n_R}{n_{\text{fuel}}}\overline{R}T_R\right] \quad (11.22)$$

Entropy Generation Rate—Open Systems, Steady-State, Steady-Flow, Isothermal Boundaries:

$$\frac{\dot{S}_{\text{gen}}}{\dot{n}_{\text{fuel}}} = \sum_P \frac{\dot{n}_i}{\dot{n}_{\text{fuel}}}\overline{s}_i - \sum_R \frac{\dot{n}_i}{\dot{n}_{\text{fuel}}}\overline{s}_i - \sum_{j=1}^{n} \frac{\dot{Q}_j/\dot{n}_{\text{fuel}}}{T_{b,j}} \quad (11.26)$$

Fuel Cells—Maximum Electrical Potential:

$$\Phi_{\max} = \frac{-\Delta G}{96{,}485 n_e} = \frac{-\Delta \dot{G}}{96{,}485 \dot{n}_e} \quad (11.36)$$

PROBLEMS

11.1 Write the chemical reaction describing the perfect combustion (complete combustion of a stoichiometric mixture) for (a) C_3H_8 with O_2 as the oxidizer, (b) C_6H_{14} with air as the oxidizer, and (c) C_2H_5OH with air as the oxidizer.

11.2 Write the chemical reaction describing the perfect combustion (complete combustion of a stoichiometric mixture) for (a) C_2H_2 with air as the oxidizer, (b) $C_{10}H_{22}$ with air as the oxidizer, and (c) CH_3OH with O_2 as the oxidizer.

11.3 Write the chemical reaction describing the complete combustion of the following combinations of fuel and air: (a) CH_4 with 125% theoretical air, (b) C_8H_{18} with 5% excess air, (c) C_2H_4 with 30% excess air.

11.4 Write the chemical reaction describing the complete combustion of the following combinations of fuel and air: (a) H_2 with 15% excess air, (b) C_3H_8 with 200% theoretical air, (c) CO with 15% excess air, (d) C_4H_{10} with 120% theoretical air.

11.5 Ethane, C_2H_6, burns to completion with 20% excess air. Determine (a) the chemical reaction describing this process, (b) the mass-based air–fuel ratio, (c) the molar air–fuel ratio, and (d) the equivalence ratio of the process.

11.6 Butane (C_4H_{10}) burns to completion with 10% excess air. Determine (a) the chemical reaction describing this process, (b) the mass-based air–fuel ratio, (c) the molar air–fuel ratio, and (d) the equivalence ratio of the process.

11.7 Acetylene (C_2H_2) and air are mixed together with a molar air–fuel ratio of 14.3, and the mixture is then ignited and burned to completion. Determine (a) the chemical reaction describing the process, (b) the % theoretical air, (c) the % excess or deficit air, (d) the mass-based air–fuel ratio, and (e) the equivalence ratio of the process.

11.8 A mixture of propane and air with a molar air–fuel ratio of 25 burns to completion. Determine (a) the chemical reaction describing the process, (b) the % theoretical air, (c) the % excess or deficit air, (d) the mass-based air–fuel ratio, and (e) the equivalence ratio of the process.

11.9 A mixture of octane and air with a mass-based air–fuel ratio of 16 burns to completion. Determine (a) the chemical reaction describing the process, (b) the % theoretical air, (c) the % excess or deficit air, (d) the molar air–fuel ratio, and (e) the equivalence ratio of the process.

11.10 A mixture of methane and air with an equivalence ratio of 0.82 burns to completion. Determine (a) the chemical reaction describing the process, (b) the % theoretical air and the % excess air, (c) the mass-based air–fuel ratio, and (d) the molar air–fuel ratio.

11.11 A mixture of butane and air has an equivalence ratio of 1.2. Determine (a) the % theoretical and the % deficit air, (b) the mass-based air–fuel ratio, and (c) the molar air–fuel ratio.

11.12 A mixture of heptane and air has an equivalence ratio of 1.25. Determine (a) the % theoretical and the % deficit air, (b) the mass-based air–fuel ratio, and (c) the molar air–fuel ratio.

11.13 A mixture of ethanol and air has an equivalence ratio of 1.15. Determine (a) the % theoretical and the % deficit air, (b) the mass-based air–fuel ratio, and (c) the molar air–fuel ratio.

11.14 Suppose a fuel consists of 50% CH_4 and 50% C_2H_6 by volume. This fuel is mixed with air, and the resulting equivalence ratio of the mixture is 1.12. Determine (a) the % theoretical and the % deficit air, (b) the mass-based air–fuel ratio, and (c) the molar air–fuel ratio.

11.15 A mixture of acetylene and air has a molar fuel–air ratio of 0.080. Determine (a) the chemical reaction describing the complete combustion of this mixture, (b) the mass-based fuel–air ratio of the mixture, (c) the equivalence ratio of the mixture, and (d) the % theoretical air.

11.16 Plot the molar air–fuel ratio for a methane–air mixture as the equivalence ratio varies between 0.50 and 1.50.

11.17 Plot the mass-based air–fuel ratio of an octane–air mixture as the equivalence ratio varies between 0.50 and 1.50.

11.18 Plot the mass-based air–fuel ratio and the molar air–fuel ratio for a stoichiometric mixture with air as the oxidizer for the alkane series of fuels from methane (CH_4) through dodecane ($C_{12}H_{26}$).

11.19 Plot the mass-based air–fuel ratio and the molar air–fuel ratio for a stoichiometric mixture with air as the oxidizer for the alkene series of fuels from ethene (C_2H_4) through dodecene ($C_{12}H_{24}$).

11.20 Plot the mass-based air–fuel ratio and the molar air–fuel ratio for a stoichiometric mixture with air as the oxidizer for the alkanes from methane (CH_4) through hexane (C_6H_{14}) and their corresponding alcohols methanol (CH_3OH) through hexanol ($C_6H_{13}OH$).

11.21 Write the chemical reaction describing the complete combustion of a stoichiometric mixture of air with a fuel that consists of 25% methane and 75% propane on a molar basis.

11.22 A fuel consists of 90% octane and 10% ethanol on a mass basis. This fuel is placed with air such that the mass-based air–fuel ratio is 15.5. Write the chemical reaction describing the complete combustion of this mixture. Determine the % theoretical air and the equivalence ratio of the process.

11.23 A fuel consists of, on a molar basis, 30% ethene and 70% propane. This fuel is put into a mixture with air such that the equivalence ratio of the mixture is 0.88. Write the chemical reaction describing the complete combustion of this mixture, and determine the mass-based and molar fuel–air ratios of the mixture.

11.24 A dry products analysis of the combustion of ethane (C_2H_6) with air yields the following molar percentages: 11.4% CO_2, 2.85% O_2, and 85.75% N_2. Write the chemical reaction describing this combustion process. Determine the molar air–fuel ratio and the equivalence ratio of the reactant mixture.

11.25 A dry products analysis is performed on the combustion of butane (C_4H_{10}) in dry air. On a volumetric basis, the dry products are 8.31% CO_2, 2.77% CO, 5.54% O_2, and 83.37% N_2. Write the chemical reaction describing this combustion process. Determine the mass-based air–fuel ratio and the equivalence ratio of the reactant mixture.

11.26 Ethane (C_2H_6) is burned in dry air. A dry products analysis of the process yields (on a volumetric basis) 11.9% CO_2, 1.32% CO, 4.63% H_2, 1.32% O_2, and 80.83% N_2. Write the chemical reaction describing this combustion process. Determine the mass-based air–fuel ratio and the equivalence ratio of the reactant mixture.

11.27 Propane (C_3H_8) burns with air, and a volumetric analysis of the dry products is 8.10% CO_2, 7.10% CO, 3.55% H_2, 1.27% O_2, and 79.98% N_2. Write the chemical reaction describing this combustion process, and determine the molar air–fuel ratio, the equivalence ratio, and the % theoretical air for the process.

11.28 Pentane (C_5H_{12}) burns with air. A dry products analysis determines that the composition of these products on a molar basis is 12.3% CO_2, 0.5% CO, 2.3% O_2, and 84.9% N_2. Write the chemical reaction describing this combustion process, and determine the mass-based air–fuel ratio, the equivalence ratio, and the % theoretical air for the process.

11.29 A mixture of methane (CH_4) and air combusts, yielding a dry products analysis (on a molar basis) of 7.46% CO_2, 4.97% CO, 6.21% H_2, 1.87% O_2, and 79.49% N_2. Write the chemical reaction describing this combustion process, and determine the molar air–fuel ratio, the equivalence ratio, and the % theoretical air for the process. If the total pressure of the products is 101.325 kPa, determine the dew point temperature of the water in the products.

11.30 Octane (C_8H_{18}) is to be used in an engine, and burns with air. A five-gas analyzer is used to determine mole fractions of the dry products, and it is determined that the dry products consist of 9.70% CO_2, 1.39% CO, 5.54% O_2, and 83.37% N_2. Write the chemical reaction describing this combustion process, and determine the mass-based air–fuel ratio, the equivalence ratio, and the % theoretical air for the process. If the total pressure of the products is 101.325 kPa, determine the dew point temperature of the water in the products.

11.31 Butane (C_4H_{10}) is burned with air, yielding a dry products analysis result (on a molar basis) of 4.0% CO_2, 10.0% CO, 2.0% CH_4, 4.0% H_2, 5.0% O_2, and 75.0% N_2. Write the chemical reaction describing this combustion process, and determine the molar air–fuel ratio, the equivalence ratio, and the % theoretical air for the process. If the total pressure of the products is 101.325 kPa, determine the dew point temperature of the water in the products.

11.32 Ethanol (C_2H_5OH) burns with air, and an analysis of the dry products is performed. It is determined that on a molar basis, the dry products consist of 11.3% CO_2, 1.26% CO, 0.63% H_2, 4.10% O_2, and 82.71% N_2. Write the chemical reaction describing this combustion process, and determine the molar air–fuel ratio, the equivalence ratio, and the % theoretical air for the process.

11.33 An unknown hydrocarbon fuel burns with air. A volumetric analysis of the dry products yields 14.05% CO_2, 1.41% O_2, and 84.54% N_2. Determine a model for the hydrocarbon (C_yH_z), the mass-based air–fuel ratio, the equivalence ratio, and the % theoretical air for the process. If the total pressure of the products is 101.325 kPa, determine the dew point temperature of the water in the products.

11.34 An unknown hydrocarbon fuel burns with air. A volumetric analysis of the dry products yields 13.11% CO_2, 1.61% CO, 4.01% H_2, 0.80% O_2, and 80.47% N_2. Determine a model for the hydrocarbon (C_yH_z), the mass-based air–fuel ratio, the equivalence ratio, and the % theoretical air for the process. If the total pressure of the products is 101.325 kPa, determine the dew point temperature of the water in the products.

11.35 An unknown hydrocarbon fuel burns with air. A volumetric analysis of the dry products yields 12.76% CO_2, 0.93% CO, 0.70% H_2, 0.46% CH_4, 2.32% O_2, and 82.83% N_2. Determine a model for the hydrocarbon (C_yH_z), the mass-based air–fuel ratio, the equivalence ratio, and the % theoretical air for the process. If the total pressure of the products is 101.325 kPa, determine the dew point temperature of the water in the products.

11.36 A dry volumetric products analysis from the combustion of an unknown hydrocarbon fuel and air yields the following: 9.86% CO_2, 4.93% CO, 1.23% H_2, 0.62% O_2, and 83.4% N_2. Determine a model for the hydrocarbon (C_yH_z), the mass-based air–fuel ratio, the equivalence ratio, and the % theoretical air for the process. If the total pressure of the products is 101.325 kPa, determine the dew point temperature of the water in the products.

11.37 Ethane burns to completion with 100% theoretical air in a steady process. The reactants enter at 298 K and the products exit at 298 K. Determine the heat of reaction (per kmole of fuel) for this process (a) assuming the water in the products is all vapor, and (b) assuming the water in the products is all liquid. Compare these results to published values for the lower heating value and the higher heating value for ethane.

11.38 Butane burns to completion with 100% theoretical air in a steady process. The reactants enter at 298 K and the products exit at 298 K. Determine the heat of reaction (per kmole of fuel) for this process (a) assuming the water in the products is all vapor, and (b) assuming the water in the products is all liquid. Compare these results to published values for the lower heating value and the higher heating value for butane.

11.39 Acetylene burns to completion with 100% theoretical air. The reactants enter at 25°C and the products exit at 25°C. Determine the heat of reaction (per mole of fuel) for this process (a) assuming the water in the products is all vapor, and (b) assuming the water in the products is all liquid. Compare these results to published values for the lower heating value and the higher heating value for acetylene.

11.40 Liquid hexane burns to completion with 100% theoretical air in a steady process. The reactants enter at 298 K and the products exit at 298 K. Determine the heat of reaction (per kmole of fuel) for this process (a) assuming the water in the products is all vapor, and (b) assuming the water in the products is all liquid. Compare these results to published values for the lower heating value and the higher heating value for hexane.

11.41 Octane burns to completion with air at an equivalence ratio of 0.90 in a steady process. The mass flow rate of the octane is 0.15 kg/s. The reactants enter at 298 K and the products exit at 298 K. Determine the heat of reaction for this process (a) assuming the water in the products is all vapor, and (b) assuming the water in the products is all liquid.

11.42 A fuel mixture consisting of 60% heptane and 40% ethanol, on a molar basis, burns to completion with air at an equivalence ratio of 0.85 in a steady process. The reactants enter at 298 K and the products exit at 298 K. The mass flow rate of the fuel is 0.10 kg/s. Determine the heat of reaction for this process (a) assuming the water in the products is all vapor, and (b) assuming the water in the products is all liquid.

11.43 Methane enters a combustion chamber with air at an equivalence ratio of 0.85 and burns to completion in a steady process. The reactants enter at 298 K and exit at 666 K. Determine the heat of reaction (per mole of methane) for the process.

11.44 Propane enters a combustion chamber with 20% excess air and burns to completion in a steady process. The reactants enter at 298 K and exit at 1000 K. Determine the heat of reaction (per kmole of propane) for the process.

11.45 A reactant mixture of octane and air with an equivalence ratio of 0.92 burns to completion in a steady process. The reactants enter at 298 K and exit at 800 K. Determine the heat of reaction (per kmole of octane) for the process.

11.46 Ethane enters a furnace at a volumetric flow rate of 0.25 m^3/s, a temperature of 298 K, and a pressure of 101 kPa. In the furnace, the ethane mixes with 10% excess air and burns to completion. The products exit at 700 K. Determine the rate of heat release from this combustion process.

11.47 Methane enters a furnace at a volumetric flow rate of 1 m^3/s, a temperature of 298 K, and a pressure of 99 kPa absolute. In the furnace, the methane mixes with 10% excess air and burns to completion. The products exit at 666 K. Determine the rate of heat release from this combustion process.

11.48 Consider the combustion of propane and air in a steady process. The reactant mixture has an equivalence ratio of 0.86, and the reactants enter at 298 K. The system pressure is 101.325 kPa. Plot the heat of reaction, per kmole of fuel, for product temperatures varying between 500 K and 1500 K.

11.49 Consider the combustion of octane and air in a steady process. The reactant mixture has an equivalence ratio of 0.96, and the reactants enter at 298 K. The system pressure is 101.325 kPa. Plot the heat of reaction, per kmole of fuel, for product temperatures varying between 500 K and 1500 K.

11.50 Butane and air burn to completion in a steady process. The reactants enter at 298 K and exit at 700 K. Plot the heat of reaction, per kmole of fuel, for percentages of excess air varying between 0% and 100%.

11.51 Ethylene and air burn to completion in a steady process. The reactants enter at 298 K and exit at 800 K. Plot the heat of reaction, per kmole of fuel, for equivalence ratios of the reactant mixture varying between 0.60 and 1.0.

11.52 A mixture of methane with 10% excess air is heated to 450 K before entering the combustor in a furnace in a steady process. The mixture burns to completion, and the products exit the furnace at 800 K. Determine the heat of reaction per kmole of fuel.

11.53 Propane and air, at an equivalence ratio of 0.86, enter a combustor at a temperature of 120°C and burn to completion. The mass flow rate of the propane is 0.35 kg/s. The products exit at 650°C. Determine the rate of heat release for the combustion process.

11.54 Ethylene (C_2H_4) and air enter a combustor at a temperature of 400 K and at an equivalence ratio of 0.95. The mass flow rate of the ethylene is 0.05 kg/s. The mixture burns to completion, and the products exit the combustor at 700 K. Determine the rate of heat release for the combustion process.

11.55 Octane (C_8H_{18}) burns with 130% theoretical air, with the octane entering at a mass flow rate of 1.2 kg/s and at a temperature of 298 K. The air is preheated before entering the combustor, and enters at 500 K. The mixture burns to completion, and the products exit at 650 K. Determine the rate of heat release for the combustion process.

11.56 Butane (C_4H_{10}) enters a combustor at a rate of 0.50 kg/s at a temperature of 25°C. The butane is mixed with 15% excess air, which enters the combustor at 247°C. The mixture burns until all of the hydrogen is converted to water vapor, but only 90% of the carbon is converted to CO_2, while the remaining carbon forms CO. The products exit the combustor at 577°C. Determine the rate of heat release for the combustion process.

11.57 A mixture of ethane (C_2H_6) and 20% excess air enters a combustion chamber with an ethane molar flow rate of 0.25 kmole/s. The mixture burns to completion and exits at 800 K. Plot the heat released during the combustion process as a function of reactant temperature, as the reactant temperature varies between 298 K and 600 K.

11.58 Propane (C_3H_8) burns to completion with 10% excess air. The flow rate of the fuel is 0.70 kg/s. The products exit at 666 K. Plot the heat released during the combustion process as a function of reactant temperature, as the reactant temperatures vary between 298 K and 500 K.

11.59 Methane (CH_4) burns to completion with 15% excess air. The methane flow rate is 0.25 kg/s. The products exit at 1200 K. Plot the heat released as a function of reactant temperature for (a) a situation where the methane enters at 298 K but the air reactant temperature varies between 298 and 600 K, and (b) a situation where both the methane and air enter at temperatures that vary between 298 and 600 K.

11.60 A Rankine cycle power plant has a thermal efficiency of 40% and produces 750 MW of power. The heat is to be supplied to the water in the steam generator through the combustion of methane (CH_4) with 10% excess air. The methane and air both enter the steam generator at 298 K, burn to completion, and the products exit the steam generator at 650 K. The combustion process takes place at 101 kPa. Determine the required mass flow rate of the methane, the corresponding volumetric flow rate of the fuel, and the volumetric flow rate of the air entering the steam generator.

11.61 Repeat Problem 11.60, but determine the flow rates required for (a) propane (C_3H_8), (b) octane (C_8H_{18}), and (c) coal (modeled as solid carbon) if they are used as the fuels.

11.62 It is decided to capture some of the energy left in the products of the process described in Problem 11.60; the energy will be used to preheat the air being directed toward the combustion process. Through the addition of an air preheater in the steam generator, the air temperature (but not the methane temperature) is raised to 450 K, and the product temperature is decreased to 475 K. Determine the required mass flow rate of the methane, the corresponding volumetric flow rate of the fuel, and the volumetric flow rate of the air entering the steam generator with this modification in place.

11.63 Model the combustion process in a spark-ignition engine as an open system process. An Otto cycle with a compression ratio of 9 produces 75 kW of power. Model the Otto cycle as using an ideal gas with constant specific heats. The heat is provided by the combustion of octane with air at an equivalence ratio of 1. The air and fuel begin the combustion process at a temperature of 700 K, and the products exit the combustion process at 1400 K. Determine the mass flow rate of octane needed to produce the necessary heat in this combustion process.

11.64 A Diesel cycle with a compression ratio of 20 and a cut-off ratio of 2.5 produces 200 kW of power. Model the Diesel cycle as using an ideal gas with constant specific heats and with the heat provided in a constant-pressure closed system process. The air enters the compression stroke at 300 K, and the combustion process begins at the end of the compression stroke. The fuel is decane and enters the process at 500 K. (Note that this is not a perfect model for the complicated process in an actual engine.) Consider the overall equivalence ratio of the process to be 0.6. The products exit the combustion process at 1300 K. Determine the mass flow rate of decane needed to produce the necessary heat for this combustion process.

11.65 Rework Problem 11.63, but instead of using pure octane as the fuel, use a mixture of 20% ethanol and 80% octane, by volume on a gaseous basis, and find the mass flow rate of the ethanol and the octane.

11.66 A Brayton cycle has air entering the compressor at 100 kPa and 25°C. The air exits the compressor at a pressure of 1100 kPa. The "air" (in reality, combustion products, but model them as air for the Brayton cycle) enters the turbine at a temperature of 1400 K. The power produced by the cycle is 50 MW. Consider the air to have variable specific heats. The heat is provided through the combustion of octane (C_8H_{18}) with air, with the air being supplied from a portion of the air exiting the compressor. The reactant mixture is at a stoichiometric level. The fuel enters at 298 K and burns to completion. The product temperature is 1400 K. Determine the mass flow rate of octane needed for the process and the fraction of air exiting the compressor needed for the combustion process for (a) an isentropic turbine and compressor, and (b) an isentropic turbine efficiency of 0.75 and an isentropic compressor efficiency of 0.75.

11.67 Consider the Brayton cycle described in Problem 7.62. The heat is to be supplied by the combustion of octane (C_8H_{18}) with air, with 10% excess air being used in the combustion and the combustion proceeding to completion. The air enters the process at the exit temperature from the compressor, the fuel enters at 25°C, and the products exit at 1200°C. For both parts (a) and (b) of Problem 7.62, determine the mass flow rate of octane needed for the combustion process.

11.68 Propane and air are to burn at constant pressure in a piston–cylinder device. The mixture initially is at 298 K and contains 0.25 m^3 of the propane–air mixture (equivalence ratio of 1) at a pressure of 150 kPa. The mixture burns to completion in a constant-pressure process, and the products are cooled to 1500 K. Determine the amount of heat released during the process.

11.69 Ethane (C_2H_6) is placed in a rigid container with air at 298 K at an equivalence ratio of 1. There is 0.10 kmole of ethane initially. The mixture is burned to completion, and the products are cooled to 700 K. Determine the amount of heat released during the process.

11.70 0.08 kmole of propane (C_3H_8) is placed in a rigid container with 15% excess air at 298 K. The mixture is burned to completion, and the products are cooled to 1000 K. Determine the amount of heat released during the process.

11.71 Butane (C_4H_{10}) is placed in a rigid container with 10% excess air at 298 K. Initially, there is 0.25 mole of butane. The mixture is burned to completion, and the products are cooled to 777 K. Determine the amount of heat released during the process.

11.72 Methane is to be burned with 10% excess air, with the reactants initially being at 298 K. Determine the adiabatic flame temperature for the process.

11.73 Determine the adiabatic flame temperature for the combustion of ethanol (C_2H_5OH) with 15% excess air and the reactants originally at 298 K.

11.74 Hexane (C_6H_{14}) is to be burned with 120% theoretical air, with the reactants initially being at 298 K. Determine the adiabatic flame temperature for the process.

11.75 Determine the adiabatic flame temperature for the combustion of ethane (C_2H_6) with the following oxidizers: (a) 100% theoretical air, (b) 100% excess air. In both cases, the reactants are at 298 K.

11.76 Plot the adiabatic flame temperature for the combustion of propane (C_3H_8) with air, with the reactants initially at 298 K, as a function of the equivalence ratio, with the equivalence ratio varying between 0.5 and 1.0.

11.77 Plot the adiabatic flame temperature for the combustion of octane (C_8H_{18}) with air, with the reactants initially at 298 K, as a function of the equivalence ratio, with the equivalence ratio varying between 0.5 and 1.0.

11.78 Plot the adiabatic flame temperature for the combustion of propane (C_3H_8) with air, with the reactant mixture being stoichiometric as a function of reactant temperature, with the reactant temperature varying between 298 K and 700 K.

11.79 A mixture of methane (CH_4) and air, with an equivalence ratio of 0.85, burns to completion in a furnace. The reactants enter at 298 K, and the products exit at 1000 K. The heat is transferred to a process across a surface whose temperature is 800 K. The methane enters the furnace at a rate of 0.10 m^3/s and at a pressure of 101 kPa. Determine the rate of entropy generation for the combustion process.

11.80 Octane (C_8H_{18}) burns to completion with air in a piston–cylinder device. The mass of octane to be burned is 25 grams, and the process occurs with 10% excess air. The process is assumed to be at a constant pressure of 200 kPa, with the reactants initially at 600 K. After the combustion process is complete, the products are cooled to 1200 K. The walls of the cylinder are maintained at 1000 K. Determine the amount of entropy generation during the process.

11.81 A mixture of propane and air is to be burned in a gas grill. The mass flow rate of the propane is 0.02 kg/s, and the mixture has an equivalence ratio of 0.90. Assume the process to be at constant pressure (101 kPa), and assume that the reactants are initially at 298 K. The products leave the burner at 1100 K. The walls of the burner are maintained at 600 K. Determine the rate of entropy generation during the process.

11.82 Liquid ethanol (C_2H_5OH) burns to completion with 15% excess air. The reactants enter at 298 K. The surface temperature of the combustor is maintained at 800 K. The pressure is constant at 101.325 kPa, and the fuel and air enter the combustor separately. Plot the entropy generation rate per kmole of ethanol for product temperatures ranging between 800 K and 1500 K.

11.83 Liquid heptane (C_7H_{16}) burns to completion with 20% excess air. The reactants enter at 298 K, and the products exit at 1000 K. The pressure is 101.325 kPa, and the fuel and air enter the combustor separately. Plot the entropy generation per kmole of heptane for combustor surface temperatures ranging from 300 K to 1000 K.

11.84 Hexane (C_6H_{14}) burns to completion with air. The reactants enter at 298 K, the products exit at 900 K, and the combustor surface temperature is maintained at 600 K. The pressure of the process is 101.325 kPa. Plot the entropy generation per kmole of hexane for reactant equivalence ratios ranging from 0.50 to 1.0.

11.85 A fuel cell uses H_2 and O_2 to produce electricity. The fuel cell's product is liquid water. The cell operates at 298 K and 101.325 kPa. Determine the maximum voltage possible for the fuel cell.

11.86 A fuel cell uses H_2 and air to produce electricity. The products of the fuel cell are liquid water and nitrogen. The cell operates at 298 K. Determine the maximum voltage for the fuel cell if (a) the fuel cell pressure is 101.325 kPa, and (b) the fuel cell pressure is 1 MPa.

11.87 An H_2–air fuel cell produces electricity. The products of the fuel cell are water vapor and nitrogen. The cell operates at 400 K and 101.325 kPa. Determine the maximum voltage for the fuel cell.

11.88 An H_2–O_2 fuel cell produces electricity and water vapor. The cell operates at 900 K and 500 kPa. Determine the maximum voltage for the fuel cell.

11.89 Suppose that the temperature of an H_2–O_2 fuel cell is varied at a pressure of 101.325 kPa. Plot the maximum voltage of the fuel cell for operation at atmospheric pressure, but with the cell temperature varying between 400 K and 1000 K.

11.90 Suppose that the pressure of an H_2–O_2 fuel cell is varied. Plot the maximum voltage of the fuel cell for operation at 430 K, but with cell pressures varying between 20 kPa and 500 kPa.

11.91 0.5 kmole of O_2 is heated to 1500 K and 101.325 kPa, at which point dissociation occurs until chemical equilibrium is reached at that temperature. Determine the number of kmoles of O_2 and O that exist once chemical equilibrium is achieved.

11.92 0.25 kmole of N_2 is heated to 2200 K and 101.325 kPa, at which point dissociation occurs until chemical equilibrium is reached. Determine the number of kmoles of N_2 and N that exist once chemical equilibrium is achieved.

11.93 Consider the dissociation of three different diatomic gases at high temperature. For each of the following three gases, consider that 1 kmole of the gas is initially heated to 2500 K at a pressure of 101.325 kPa. Determine the resulting mixture of gases (the diatomic and monatomic forms) once chemical equilibrium is reached at 2500 K and 101.325 kPa for (a) H_2, (b) O_2, and (c) N_2.

11.94 1 kmole of CO_2 is rapidly heated to 2200 K at a pressure of 200 kPa. Chemical reactions occur until an equilibrium mixture of CO_2, CO, and O_2 is reached. Determine the molar composition of the resulting mixture at chemical equilibrium.

11.95 A mixture initially consisting of 1 kmole of O_2 and 1 kmole of N_2 is at 1000 K and 101.325 kPa. The mixture reaches chemical equilibrium with NO. (a) Determine the molar composition of the O_2, N_2, and NO mixture that exists at chemical equilibrium. (b) Repeat the process for a mixture temperature of 3000 K.

11.96 A mixture initially consists of 1 kmole of CO_2 and 2 kmoles of Ar. The mixture is brought into an equilibrium mixture of CO_2, CO, O_2, and Ar at a temperature of 2200 K and a pressure of 101.325 kPa. Determine the molar composition of this equilibrium mixture.

11.97 2 kmoles of O_2 are mixed with 3 kmoles of He and heated to 2500 K. Some of the O_2 dissociates, and a mixture of O_2, O, and He exists at chemical equilibrium. Determine the composition of this mixture if (a) the mixture pressure is 101.325 kPa, and (b) the mixture pressure is 1 MPa.

11.98 Consider a system consisting of 1 kmole of O_2 initially. This mixture is heated and reaches chemical equilibrium at each test temperature. Plot the mole fraction of O_2 and the mole fraction of O for a mixture of O_2 and O at chemical equilibrium for a mixture pressure of 101.325 kPa and a temperature varying between 298 K and 3000 K. Repeat this calculation for pressures of 10 kPa and 2 MPa.

11.99 Consider the chemical equilibrium reaction among O_2, N_2, and NO. Initially, the mixture consists of 1 kmole of O_2 and 3.76 kmoles of N_2. Plot the mole fractions of all three components of the mixture at chemical equilibrium as the temperature of the mixture varies between 298 K and 3000 K. Would the mole fractions change if the mixture pressure changes?

11.100 1 kmole of CO_2 is heated and reaches a chemical equilibrium mixture with CO and O_2 at 2000 K. Plot the mole fractions of the three components as a function of pressure, with the pressure ranging between 10 kPa and 2.5 MPa.

11.101 A mixture initially consists of 1 kmole of CO_2, 0.75 kmole of CO, and 0.25 kmole of O_2. This mixture is heated to 2300 K. Determine the composition of the mixture when chemical equilibrium is reached if the mixture pressure is 101.325 kPa.

11.102 1 kmole of NO is heated to 1000 K. A mixture of O_2, N_2, and NO is formed at a pressure of 101.325 kPa and a temperature of 1000 K. Determine the molar composition of this mixture once chemical equilibrium is reached.

11.103 A set of combustion products, consisting of 0.5 kmole CO_2, 0.25 kmole CO, 0.75 kmole H_2O, and 0.10 kmole H_2, is brought to chemical equilibrium at 1800 K and 101.325 kPa, following the water–gas shift reaction. Determine the composition of this equilibrium mixture.

11.104 1 kmole of H_2O is mixed with some amount of Ar gas. The mixture is heated, and a state of chemical equilibrium among H_2O, H_2, O_2, and Ar is reached at a temperature of 2000 K and a pressure of 101.325 kPa. Plot the number of moles of each component for initial amounts of Ar ranging between 0.05 kmole and 5 kmoles.

11.105 1 mole of CO_2 is heated and chemical equilibrium is reached among CO_2, CO, and O_2 at a pressure of 250 kPa. Plot the mole fractions of each component of the mixture at chemical equilibrium for mixture temperatures ranging between 555 K and 3000 K.

11.106 Consider the combustion of propane (C_3H_8) with air at an equivalence ratio of 1.3. Using the combined atom balance and water–gas shift reaction approach, determine the chemical reaction describing this process for a product temperature of 1500 K.

11.107 Butane burns with air at an equivalence ratio of 1.20. Using the combined atom balance and water–gas shift reaction approach, determine the chemical reaction describing this process for a product temperature of 1200 K.

11.108 Ethane (C_2H_6) burns with air at an equivalence ratio of 1.25. Using the combined atom balance and water–gas shift reaction approach, determine the chemical reaction describing this process for a product temperature of 1200 K.

11.109 Octane (C_8H_{18}) burns with air at a mass-based air–fuel ratio of 13. Using the combined atom balance and water–gas shift reaction approach, determine the chemical reaction describing this process for a product temperature of (a) 500 K, (b) 1000 K, and (c) 1500 K.

11.110 Butane (C_4H_{10}) burns with 80% theoretical air. Using the combined atom balance and water–gas shift reaction approach, determine the chemical reaction describing this process for a product temperature of (a) 500 K, (b) 1000 K, and (c) 1500 K.

11.111 Methane (CH_4) burns with 25% deficit air. The reactants enter at 298 K, and the products exit at 1000 K. Determine the chemical reaction describing this combustion using the combined atom balance and water–gas shift reaction approach, and then determine the heat of reaction per kmole of fuel.

11.112 Ethane (C_2H_6) burns with air at an equivalence ratio of 1.50. The reactants enter at 298 K and the products exit at 1000 K. Determine the chemical reaction describing this combustion using the combined atom balance and water–gas shift reaction approach, and then determine the heat of reaction per kmole of fuel.

11.113 Propane (C_3H_8) burns with air at an equivalence ratio of 1.35. The reactants enter at 298 K. To determine the products, use the combined atom balance and water–gas shift reaction approach. Consider the volumetric flow rate of the propane entering the combustor to be 0.05 m^3/s and the inlet pressure to be 101 kPa. Plot the rate of heat release for this combustion for product temperatures varying between 500 K and 1600 K. (Note that the composition of the products will change as the temperature changes.)

11.114 Propane (C_3H_8) burns with air in a rich combustion process. The reactants enter at 298 K, and the products exit at 800 K. The propane enters the combustor at atmospheric pressure and a volumetric flow rate of 0.25 m^3/s. To determine the products, use the combined atom balance and water–gas shift reaction approach. Plot the rate of heat release for this process as a function of reactant mixture equivalence ratio, with the equivalence ratio varying between 1.1 and 1.5.

11.115 Methane burns with air at atmospheric pressure at an equivalence ratio of 1.25. The reactants enter at 400 K, and the volumetric flow rate of the methane is 42 L/s. To determine the products, use the combined atom balance and water–gas shift reaction approach. Plot the rate of heat release for this combustion for product temperatures varying between 500 K and 1250 K. (Note that the composition of the products will change as the temperature changes.)

11.116 Repeat Problem 11.60, but consider the combustion to occur with 20% deficit air. Solve for the requested flow rates.

11.117 Repeat Problem 11.63, but consider the combustion to occur with an equivalence ratio of 1.2. Solve for the requested quantities.

11.118 Repeat Problem 11.67, but consider the combustion to occur at a mass-based air–fuel ratio of 12.5. Solve for the requested quantities.

DESIGN/OPEN-ENDED PROBLEMS

11.119 A boiler is required to produce 1 kg/s of saturated steam at atmospheric pressure, with liquid water entering the boiler at 20°C. There are two potential fuels readily available for use: natural gas (modeled as methane) and propane. Assume that the fuel and air would enter the combustor of the boiler at 25°C and atmospheric pressure. Design a combustion process (choose the equivalence ratio and the product exit temperature) that will provide sufficient heat for the boiler. Assume that 95% of the heat released in the combustion process is transferred to the water. Using current prices for natural gas and propane in your area, determine the daily fuel cost for your combustion process (assuming that the process runs 24 hours a day).

11.120 Design a combustion process for a home furnace that can provide a steady supply of heated air to the home. The air is heated through a heat exchanger, with the other working fluid being the combustion products. The heated air is to be supplied at 35°C and 101.325 kPa, at a rate of 0.25 m^3/s. The cold-air return brings air from the rest of the home to the furnace at a temperature of 18°C. For safety purposes, your reactant mixture should be overall lean, but with an equivalence ratio of at least 0.80. Your design should include the fuel used, the air–fuel ratio, the product temperatures both entering and exiting the heat exchanger, and the flow rates of the air and fuel.

11.121 Design a combustion process (specifying the fuel, equivalence ratio, and product exit temperature) to provide the heat required by the modified Rankine cycle described in Example 7.4. Based on current fuel prices, estimate the daily cost of the fuel for the cycle. Assume the fuel and air enter at 25°C.

11.122 Design a combustion process (specifying the fuel, equivalence ratio, and product exit temperature) to provide the heat required by the modified Rankine cycle described in Example 7.5. Based on current fuel prices, estimate the daily cost of the fuel for the cycle. Assume the fuel and air enter at 25°C.

11.123 Repeat Problem 11.122, but now add an air-preheat section to the steam generator in the Rankine cycle, such that the air to be used in the combustion process is heated to 400 K by the gases exiting the combustion process. This will require increasing the required heat load in the steam generator to include the heating of the air.

11.124 Design a combustion process (specifying the fuel, equivalence ratio, and product exit temperature) to provide the heat required by the Brayton cycle described in Example 7.10. Based on current fuel prices, estimate the daily cost of the fuel for the cycle. Assume the fuel and air enter at 25°C.

11.125 Consider Example 10.8. The heating of the air is to be accomplished in a furnace, where there is a 95% efficiency of transferring heat from the combustion process to the air. Design a combustion process (specifying the fuel, equivalence ratio, and product exit temperature) to provide the heat required. Assume that the fuel and the air enter at 5°C.

11.126 Consider a tank-less water heater. The water heater is to supply 1 liter/s of water at 45°C, using water that is supplied at 10°C. The fuel to be used in the heater is natural gas, which you can approximate at 93% CH_4 and 7% C_2H_6 by volume. The combustion products must exit the system at a temperature no lower than 60°C. The air and fuel both enter the system at 10°C but can be preheated with exhaust gases prior to the combustion process. Consider the efficiency of the combustion products–water heat exchanger to be 95%. Design a combustion process that will provide the necessary heat to the water.

TABLE A.1 Properties of Some Ideal Gases

Substance	Chemical Formula	Molecular Mass, M (kg/kmol)	Gas Constant, R (kJ/kg · K)	c_p (kJ/kg · K)	c_v (kJ/kg · K)	k	Critical Temperature, T_c (K)	Critical Pressure, P_c (MPa)
Air		28.97	0.2870	1.005	0.718	1.400	132.5	3.77
Ammonia	NH_3	17.03	0.4882	2.16	1.67	1.29	405.5	11.28
Argon	Ar	39.948	0.2081	0.5203	0.3122	1.667	151	4.86
Butane	C_4H_{10}	58.124	0.1433	1.7164	1.5734	1.091	425.2	3.80
Carbon Dioxide	CO_2	44.01	0.1889	0.846	0.657	1.289	304.2	7.39
Carbon Monoxide	CO	28.011	0.2968	1.040	0.744	1.400	133	3.50
Ethane	C_2H_6	30.070	0.2765	1.7662	1.4897	1.186	305.5	4.48
Helium	He	4.003	2.0769	5.1926	3.1156	1.667	5.3	0.23
Hydrogen	H_2	2.016	4.124	14.307	10.183	1.405	33.3	1.30
Methane	CH_4	16.043	0.5182	2.2537	1.7354	1.299	191.1	4.64
Neon	Ne	20.183	0.4119	1.0299	0.6179	1.667	44.5	2.73
Nitrogen	N_2	28.013	0.2968	1.039	0.743	1.400	126.2	3.39
Octane	C_8H_{18}	114.22	0.0728	1.7113	1.6385	1.044	569	2.49
Oxygen	O_2	31.999	0.2598	0.918	0.658	1.395	154.8	5.08
Propane	C_3H_8	44.097	0.1885	1.6794	1.4909	1.126	370	4.26
Sulfur Dioxide	SO_2	64.063	0.1298	0.64	0.51	1.29	430.7	7.88
Water vapor	H_2O	18.015	0.4615	1.8723	1.4108	1.327	647.1	22.06

The values for c_p and c_v are taken at 300 K.

TABLE A.2 Values of the Specific Heats at Different Temperatures for Common Ideal Gases (kJ/kg · K)

Temp. (K)	Air			Nitrogen, N_2			Oxygen, O_2		
	c_p	c_v	k	c_p	c_v	k	c_p	c_v	k
250	1.003	0.716	1.401	1.039	0.742	1.400	0.913	0.653	1.398
300	1.005	0.718	1.400	1.039	0.743	1.400	0.918	0.658	1.395
350	1.008	0.721	1.398	1.041	0.744	1.399	0.928	0.668	1.389
400	1.013	0.726	1.395	1.044	0.747	1.397	0.941	0.681	1.382
450	1.020	0.733	1.391	1.049	0.752	1.395	0.956	0.696	1.373
500	1.029	0.742	1.387	1.056	0.759	1.391	0.972	0.712	1.365
550	1.040	0.753	1.381	1.065	0.768	1.387	0.988	0.728	1.358
600	1.051	0.764	1.376	1.075	0.778	1.382	1.003	0.743	1.350
650	1.063	0.776	1.370	1.086	0.789	1.376	1.017	0.758	1.343
700	1.075	0.788	1.364	1.098	0.801	1.371	1.031	0.771	1.337
750	1.087	0.800	1.359	1.110	0.813	1.365	1.043	0.783	1.332
800	1.099	0.812	1.354	1.121	0.825	1.360	1.054	0.794	1.327
900	1.121	0.834	1.344	1.145	0.849	1.349	1.074	0.814	1.319
1000	1.142	0.855	1.336	1.167	0.870	1.341	1.090	0.830	1.313
	Carbon Dioxide, CO_2			Carbon Monoxide, CO			Hydrogen, H_2		
	c_p	c_v	k	c_p	c_v	k	c_p	c_v	k
250	0.791	0.602	1.314	1.039	0.743	1.400	14.051	9.927	1.416
300	0.846	0.657	1.288	1.040	0.744	1.399	14.307	10.183	1.405
350	0.895	0.706	1.268	1.043	0.746	1.398	14.427	10.302	1.400
400	0.939	0.750	1.252	1.047	0.751	1.395	14.476	10.352	1.398
450	0.978	0.790	1.239	1.054	0.757	1.392	14.501	10.377	1.398
500	1.014	0.825	1.229	1.063	0.767	1.387	14.513	10.389	1.397
550	1.046	0.857	1.220	1.075	0.778	1.382	14.530	10.405	1.396
600	1.075	0.886	1.213	1.087	0.790	1.376	14.546	10.422	1.396
650	1.102	0.913	1.207	1.100	0.803	1.370	14.571	10.447	1.395
700	1.126	0.937	1.202	1.113	0.816	1.364	14.604	10.480	1.394
750	1.148	0.959	1.197	1.126	0.829	1.358	14.645	10.521	1.392
800	1.169	0.980	1.193	1.139	0.842	1.353	14.695	10.570	1.390
900	1.204	1.015	1.186	1.163	0.866	1.343	14.822	10.698	1.385
1000	1.234	1.045	1.181	1.185	0.888	1.335	14.983	10.859	1.380

TABLE A.3 Ideal Gas Properties of Air

T (K)	h (kJ/kg)	P_r	u (kJ/kg)	v_r	s^o (kJ/kg · K)	T (K)	h (kJ/kg)	P_r	u (kJ/kg)	v_r	s^o (kJ/kg · K)
200	199.97	0.3363	142.56	1707	1.29559	780	800.03	43.35	576.12	51.64	2.69013
220	219.97	0.4690	156.82	1346	1.39105	820	843.98	52.49	608.59	44.84	2.74504
240	240.02	0.6355	171.13	1084	1.47824	860	888.27	63.09	641.40	39.12	2.79783
260	260.09	0.8405	185.45	887.8	1.55848	900	932.93	75.29	674.58	34.31	2.84856
280	280.13	1.0889	199.75	738.0	1.63279	940	977.92	89.28	708.08	30.22	2.89748
290	290.16	1.2311	206.91	676.1	1.66802	980	1023.25	105.2	741.98	26.73	2.94468
300	300.19	1.3860	214.07	621.2	1.70203	1020	1068.89	123.4	776.10	23.72	2.99034
310	310.24	1.5546	221.25	572.3	1.73498	1060	1114.86	143.9	810.62	21.14	3.03449
320	320.29	1.7375	228.43	528.6	1.76690	1100	1161.07	167.1	845.33	18.896	3.07732
340	340.42	2.149	242.82	454.1	1.82790	1140	1207.57	193.1	880.35	16.946	3.11883
360	360.58	2.626	257.24	393.4	1.88543	1180	1254.34	222.2	915.57	15.241	3.15916
380	380.77	3.176	271.69	343.4	1.94001	1220	1301.31	254.7	951.09	13.747	3.19834
400	400.98	3.806	286.16	301.6	1.99194	1260	1348.55	290.8	986.90	12.435	3.23638
420	421.26	4.522	300.69	266.6	2.04142	1300	1395.97	330.9	1022.82	11.275	3.27345
440	441.61	5.332	315.30	236.8	2.08870	1340	1443.60	375.3	1058.94	10.247	3.30959
460	462.02	6.245	329.97	211.4	2.13407	1380	1491.44	424.2	1095.26	9.337	3.34474
480	482.49	7.268	344.70	189.5	2.17760	1420	1539.44	478.0	1131.77	8.526	3.37901
500	503.02	8.411	359.49	170.6	2.21952	1460	1587.63	537.1	1168.49	7.801	3.41247
520	523.63	9.684	374.36	154.1	2.25997	1500	1635.97	601.9	1205.41	7.152	3.44516
540	544.35	11.10	389.34	139.7	2.29906	1540	1684.51	672.8	1242.43	6.569	3.47712
560	565.17	12.66	404.42	127.0	2.33685	1580	1733.17	750.0	1279.65	6.046	3.50829
580	586.04	14.38	419.55	115.7	2.37348	1620	1782.00	834.1	1316.96	5.574	3.53879
600	607.02	16.28	434.78	105.8	2.40902	1660	1830.96	925.6	1354.48	5.147	3.56867
620	628.07	18.36	450.09	96.92	2.44356	1700	1880.1	1025	1392.7	4.761	3.5979
640	649.22	20.65	465.05	88.99	2.47716	1800	2003.3	1310	1487.2	3.944	3.6684
660	670.47	23.13	481.01	81.89	2.50985	1900	2127.4	1655	1582.6	3.295	3.73541
680	691.82	25.85	496.62	75.50	2.54175	2000	2252.1	2068	1678.7	2.776	3.7994
700	713.27	28.80	512.33	69.76	2.57277	2100	2377.4	2559	1775.3	2.356	3.8605
720	734.82	32.02	528.14	64.53	2.60319	2200	2503.2	3138	1872.4	2.012	3.9191
740	756.44	35.50	544.02	59.82	2.63280						

Based on data from J. H. Keenan and J. Kaye, Gas Tables, Wiley, New York, 1945.

TABLE A.4 Ideal Gas Properties of Nitrogen, N_2

$\bar{h}_f^o = 0$ kJ/kmol

T (K)	$\bar{h}$ (kJ/kmol)	$\bar{u}$ (kJ/kmol)	$\bar{s}^o$ (kJ/kmol · K)	T (K)	$\bar{h}$ (kJ/kmol)	$\bar{u}$ (kJ/kmol)	$\bar{s}^o$ (kJ/kmol · K)
0	0	0	0	1000	30 129	21 815	228.057
220	6391	4562	182.639	1020	30 784	22 304	228.706
240	6975	4979	185.180	1040	31 442	22 795	229.344
260	7558	5396	187.514	1060	32 101	23 288	229.973
280	8141	5813	189.673	1080	32 762	23 782	230.591
298	8669	6190	191.502	1100	33 426	24 280	231.199
300	8723	6229	191.682	1120	34 092	24 780	231.799
320	9306	6645	193.562	1140	34 760	25 282	232.391
340	9888	7061	195.328	1160	35 430	25 786	232.973
360	10 471	7478	196.995	1180	36 104	26 291	233.549
380	11 055	7895	198.572	1200	36 777	26 799	234.115
400	11 640	8314	200.071	1240	38 129	27 819	235.223
420	12 225	8733	201.499	1260	38 807	28 331	235.766
440	12 811	9153	202.863	1280	39 488	28 845	236.302
460	13 399	9574	204.170	1300	40 170	29 361	236.831
480	13 988	9997	205.424	1320	40 853	29 878	237.353
500	14 581	10 423	206.630	1340	41 539	30 398	237.867
520	15 172	10 848	207.792	1360	42 227	30 919	238.376
540	15 766	11 277	208.914	1380	42 915	31 441	238.878
560	16 363	11 707	209.999	1400	43 605	31 964	239.375
580	16 962	12 139	211.049	1440	44 988	33 014	240.350
600	17 563	12 574	212.066	1480	46 377	34 071	241.301
620	18 166	13 011	213.055	1520	47 771	35 133	242.228
640	18 772	13 450	214.018	1560	49 168	36 197	243.137
660	19 380	13 892	214.954	1600	50 571	37 268	244.028
680	19 991	14 337	215.866	1700	54 099	39 965	246.166
700	20 604	14 784	216.756	1800	57 651	42 685	248.195
720	21 220	15 234	217.624	1900	61 220	45 423	250.128
740	21 839	15 686	218.472	2000	64 810	48 181	251.969
760	22 460	16 141	219.301	2100	68 417	50 957	253.726
780	23 085	16 599	220.113	2200	72 040	53 749	255.412
800	23 714	17 061	220.907	2300	75 676	56 553	257.02
820	24 342	17 524	221.684	2400	79 320	59 366	258.580
840	24 974	17 990	222.447	2500	82 981	62 195	260.073
860	25 610	18 459	223.194	2600	86 650	65 033	261.512
880	26 248	18 931	223.927	2700	90 328	67 880	262.902
900	26 890	19 407	224.647	2800	94 014	70 734	264.241
920	27 532	19 883	225.353	2900	97 705	73 593	265.538
940	28 178	20 362	226.047	3000	101 407	76 464	266.793
960	28 826	20 844	226.728	3100	105 115	79 341	268.007
980	29 476	21 328	227.398	3200	108 830	82 224	269.186

Based on data from JANAF Thermochemical Tables, NSRDS-NBS-37, 1971

TABLE A.4 Ideal Gas Properties of Oxygen, O_2

$\bar{h}_f^o = 0$ kJ/kmol

T (K)	$\bar{h}$ (kJ/kmol)	$\bar{u}$ (kJ/kmol)	$\bar{s}^o$ (kJ/kmol·K)	T (K)	$\bar{h}$ (kJ/kmol)	$\bar{u}$ (kJ/kmol)	$\bar{s}^o$ (kJ/kmol·K)
0	0	0	0	1020	32 088	23 607	244.164
220	6404	4575	196.171	1040	32 789	24 142	244.844
240	6984	4989	198.696	1060	33 490	24 677	245.513
260	7566	5405	201.027	1080	34 194	25 214	246.171
280	8150	5822	203.191	1100	34 899	25 753	246.818
298	8682	6203	205.033	1120	35 606	26 294	247.454
300	8736	6242	205.213	1140	36 314	26 836	248.081
320	9325	6664	207.112	1160	37 023	27 379	248.698
340	9916	7090	208.904	1180	37 734	27 923	249.307
360	10 511	7518	210.604	1200	38 447	28 469	249.906
380	11 109	7949	212.222	1220	39 162	29 018	250.497
400	11 711	8384	213.765	1240	39 877	29 568	251.079
420	12 314	8822	215.241	1260	40 594	30 118	251.653
440	12 923	9264	216.656	1280	41 312	30 670	252.219
460	13 535	9710	218.016	1300	42 033	31 224	252.776
480	14 151	10 160	219.326	1320	42 753	31 778	253.325
500	14 770	10 614	220.589	1340	43 475	32 334	253.868
520	15 395	11 071	221.812	1360	44 198	32 891	254.404
540	16 022	11 533	222.997	1380	44 923	33 449	254.932
560	16 654	11 998	224.146	1400	45 648	34 008	255.454
580	17 290	12 467	225.262	1440	47 102	35 129	256.475
600	17 929	12 940	226.346	1480	48 561	36 256	257.474
620	18 572	13 417	227.400	1520	50 024	37 387	258.450
640	19 219	13 898	228.429	1540	50 756	37 952	258.928
660	19 870	14 383	229.430	1560	51 490	38 520	259.402
680	20 524	14 871	230.405	1600	52 961	39 658	260.333
700	21 184	15 364	231.358	1700	56 652	42 517	262.571
720	21 845	15 859	232.291	1800	60 371	45 405	264.701
740	22 510	16 357	233.201	1900	64 116	48 319	266.722
760	23 178	16 859	234.091	2000	67 881	51 253	268.655
780	23 850	17 364	234.960	2100	71 668	54 208	270.504
800	24 523	17 872	235.810	2200	75 484	57 192	272.278
820	25 199	18 382	236.644	2300	79 316	60 193	273.981
840	25 877	18 893	237.462	2400	83 174	63 219	275.625
860	26 559	19 408	238.264	2500	87 057	66 271	277.207
880	27 242	19 925	239.051	2600	90 956	69 339	278.738
900	27 928	20 445	239.823	2700	94 881	72 433	280.219
920	28 616	20 967	240.580	2800	98 826	75 546	281.654
940	29 306	21 491	241.323	2900	102 793	78 682	283.048
960	29 999	22 017	242.052	3000	106 780	81 837	284.399
980	30 692	22 544	242.768	3100	110 784	85 009	285.713
1000	31 389	23 075	243.471	3200	114 809	88 203	286.989

Based on data from JANAF Thermochemical Tables, NSRDS-NBS-37, 1971

TABLE A.4 Ideal Gas Properties of Carbon Dioxide, CO_2

$\bar{h}_f^o = -393\ 520$ kJ/kmol

T (K)	$\bar{h}$ (kJ/kmol)	$\bar{u}$ (kJ/kmol)	$\bar{s}^o$ (kJ/kmol · K)	T (K)	$\bar{h}$ (kJ/kmol)	$\bar{u}$ (kJ/kmol)	$\bar{s}^o$ (kJ/kmol · K)
0	0	0	0	1020	43 859	35 378	270.293
220	6601	4772	202.966	1040	44 953	36 306	271.354
240	7280	5285	205.920	1060	46 051	37 238	272.400
260	7979	5817	208.717	1080	47 153	38 174	273.430
280	8697	6369	211.376	1100	48 258	39 112	274.445
298	9364	6885	213.685	1120	49 369	40 057	275.444
300	9431	6939	213.915	1140	50 484	41 006	276.430
320	10 186	7526	216.351	1160	51 602	41 957	277.403
340	10 959	8131	218.694	1180	52 724	42 913	278.361
360	11 748	8752	220.948	1200	53 848	43 871	279.307
380	12 552	9392	223.122	1220	54 977	44 834	280.238
400	13 372	10 046	225.225	1240	56 108	45 799	281.158
420	14 206	10 714	227.258	1260	57 244	46 768	282.066
440	15 054	11 393	229.230	1280	58 381	47 739	282.962
460	15 916	12 091	231.144	1300	59 522	48 713	283.847
480	16 791	12 800	233.004	1320	60 666	49 691	284.722
500	17 678	13 521	234.814	1340	61 813	50 672	285.586
520	18 576	14 253	236.575	1360	62 963	51 656	286.439
540	19 485	14 996	238.292	1380	64 116	52 643	287.283
560	20 407	15 751	239.962	1400	65 271	53 631	288.106
580	21 337	16 515	241.602	1440	67 586	55 614	289.743
600	22 280	17 291	243.199	1480	69 911	57 606	291.333
620	23 231	18 076	244.758	1520	72 246	59 609	292.888
640	24 190	18 869	246.282	1560	74 590	61 620	294.411
660	25 160	19 672	247.773	1600	76 944	63 741	295.901
680	26 138	20 484	249.233	1700	82 856	68 721	299.482
700	27 125	21 305	250.663	1800	88 806	73 840	302.884
720	28 121	22 134	252.065	1900	94 793	78 996	306.122
740	29 124	22 972	253.439	2000	100 804	84 185	309.210
760	30 135	23 817	254.787	2100	106 864	89 404	312.160
780	31 154	24 669	256.110	2200	112 939	94 648	314.988
800	32 179	25 527	257.408	2300	119 035	99 912	317.695
820	33 212	26 394	258.682	2400	125 152	105 197	320.302
840	34 251	27 267	259.934	2500	131 290	110 504	322.308
860	35 296	28 125	261.164	2600	137 449	115 832	325.222
880	36 347	29 031	262.371	2700	143 620	121 172	327.549
900	37 405	29 922	263.559	2800	149 808	126 528	329.800
920	38 467	30 818	264.728	2900	156 009	131 898	331.975
940	39 535	31 719	265.877	3000	162 226	137 283	334.084
960	40 607	32 625	267.007	3100	168 456	142 681	336.126
980	41 685	33 537	268.119	3200	174 695	148 089	338.109
1000	42 769	34 455	269.215				

Based on data from JANAF Thermochemical Tables, NSRDS-NBS-37, 1971

TABLE A.4 Ideal Gas Properties of Carbon Monoxide, CO

$\bar{h}_f^o = -110\ 530$ kJ/kmol

T (K)	$\bar{h}$ (kJ/kmol)	$\bar{u}$ (kJ/kmol)	$\bar{s}^o$ (kJ/kg · K)	T (K)	$\bar{h}$ (kJ/kmol)	$\bar{u}$ (kJ/kmol)	$\bar{s}^o$ (kJ/kg · K)
0	0	0	0	1040	31 688	23 041	235.728
220	6391	4562	188.683	1060	32 357	23 544	236.364
240	6975	4979	191.221	1080	33 029	24 049	236.992
260	7558	5396	193.554	1100	33 702	24 557	237.609
280	8140	5812	195.713	1120	34 377	25 065	238.217
300	8723	6229	197.723	1140	35 054	25 575	238.817
320	9306	6645	199.603	1160	35 733	26 088	239.407
340	9889	7062	201.371	1180	36 406	26 602	239.989
360	10 473	7480	203.040	1200	37 095	27 118	240.663
380	11 058	7899	204.622	1220	37 780	27 637	241.128
400	11 644	8319	206.125	1240	38 466	28 426	241.686
420	12 232	8740	207.549	1260	39 154	28 678	242.236
440	12 821	9163	208.929	1280	39 844	29 201	242.780
460	13 412	9587	210.243	1300	40 534	29 725	243.316
480	14 005	10 014	211.504	1320	41 226	30 251	243.844
500	14 600	10 443	212.719	1340	41 919	30 778	244.366
520	15 197	10 874	213.890	1360	42 613	31 306	244.880
540	15 797	11 307	215.020	1380	43 309	31 836	245.388
560	16 399	11 743	216.115	1400	44 007	32 367	245.889
580	17 003	12 181	217.175	1440	45 408	33 434	246.876
600	17 611	12 622	218.204	1480	46 813	34 508	247.839
620	18 221	13 066	219.205	1520	48 222	35 584	248.778
640	18 833	13 512	220.179	1560	49 635	36 665	249.695
660	19 449	13 962	221.127	1600	51 053	37 750	250.592
680	20 068	14 414	222.052	1700	54 609	40 474	252.751
700	20 690	14 870	222.953	1800	58 191	43 225	254.797
720	21 315	15 328	223.833	1900	61 794	45 997	256.743
740	21 943	15 789	224.692	2000	65 408	48 780	258.600
760	22 573	16 255	225.533	2100	69 044	51 584	260.370
780	23 208	16 723	226.357	2200	72 688	54 396	262.065
800	23 844	17 193	227.162	2300	76 345	57 222	263.692
820	24 483	17 665	227.952	2400	80 015	60 060	265.253
840	25 124	18 140	228.724	2500	83 692	62 906	266.755
860	25 768	18 617	229.482	2600	87 383	65 766	268.202
880	26 415	19 099	230.227	2700	91 077	68 628	269.596
900	27 066	19 583	230.957	2800	94 784	71 504	270.943
920	27 719	20 070	231.674	2900	98 495	74 383	272.249
940	28 375	20 559	232.379	3000	102 210	77 267	273.508
960	29 033	21 051	233.072	3100	105 939	80 164	274.730
980	29 693	21 545	233.752	3150	107 802	81 612	275.326
1000	30 355	22 041	234.421	3200	109 667	83 061	275.914
1020	31 020	22 540	235.079				

Based on data from JANAF Thermochemical Tables, NSRDS-NBS-37, 1971

TABLE A.4 Ideal Gas Properties for Hydrogen, H_2

$\bar{h}_f^o = 0$ kJ/kmol

T (K)	$\bar{h}$ (kJ/kmol)	$\bar{u}$ (kJ/kmol)	$\bar{s}^o$ (kJ/kmol · K)	T (K)	$\bar{h}$ (kJ/kmol)	$\bar{u}$ (kJ/kmol)	$\bar{s}^o$ (kJ/kmol · K)
0	0	0	0	1440	42 808	30 835	177.410
260	7370	5209	126.636	1480	44 091	31 786	178.291
270	7657	5412	127.719	1520	45 384	32 746	179.153
280	7945	5617	128.765	1560	46 683	33 713	179.995
290	8233	5822	129.775	1600	47 990	34 687	180.820
298	8468	5989	130.574	1640	49 303	35 668	181.632
300	8522	6027	130.754	1680	50 622	36 654	182.428
320	9100	6440	132.621	1720	51 947	37 646	183.208
340	9680	6853	134.378	1760	53 279	38 645	183.973
360	10 262	7268	136.039	1800	54 618	39 652	184.724
380	10 843	7684	137.612	1840	55 962	40 663	185.463
400	11 426	8100	139.106	1880	57 311	41 680	186.190
420	12 010	8518	140.529	1920	58 668	42 705	186.904
440	12 594	8936	141.888	1960	60 031	43 735	187.607
460	13 179	9355	143.187	2000	61 400	44 771	188.297
480	13 764	9773	144.432	2050	63 119	46 074	189.148
500	14 350	10 193	145.628	2100	64 847	47 386	189.979
520	14 935	10 611	146.775	2150	66 584	48 708	190.796
560	16 107	11 451	148.945	2200	68 328	50 037	191.598
600	17 280	12 291	150.968	2250	70 080	51 373	192.385
640	18 453	13 133	152.863	2300	71 839	52 716	193.159
680	19 630	13 976	154.645	2350	73 608	54 069	193.921
720	20 807	14 821	156.328	2400	75 383	55 429	194.669
760	21 988	15 669	157.923	2450	77 168	56 798	195.403
800	23 171	16 520	159.440	2500	78 960	58 175	196.125
840	24 359	17 375	160.891	2550	80 755	59 554	196.837
880	25 551	18 235	162.277	2600	82 558	60 941	197.539
920	26 747	19 098	163.607	2650	84 368	62 335	198.229
960	27 948	19 966	164.884	2700	86 186	63 737	198.907
1000	29 154	20 839	166.114	2750	88 008	65 144	199.575
1040	30 364	21 717	167.300	2800	89 838	66 558	200.234
1080	31 580	22 601	168.449	2850	91 671	67 976	200.885
1120	32 802	23 490	169.560	2900	93 512	69 401	201.527
1160	34 028	24 384	170.636	2950	95 358	70 831	202.157
1200	35 262	25 284	171.682	3000	97 211	72 268	202.778
1240	36 502	26 192	172.698	3050	99 065	73 707	203.391
1280	37 749	27 106	173.687	3100	100 926	75 152	203.995
1320	39 002	28 027	174.652	3150	102 793	76 604	204.592
1360	40 263	28 955	175.593	3200	104 667	78 061	205.181
1400	41 530	29 889	176.510	3250	106 545	79 523	205.765

Based on data from JANAF Thermochemical Tables, NSRDS-NBS-37, 1971

TABLE A.4 Ideal Gas Properties for Water Vapor, H_2O

$\overline{h}_f^o = -241\ 810$ kJ/kmol

T (K)	$\overline{h}$ (kJ/kmol)	$\overline{u}$ (kJ/kmol)	$\overline{s}^o$ (kJ/kmol · K)	T (K)	$\overline{h}$ (kJ/kmol)	$\overline{u}$ (kJ/kmol)	$\overline{s}^o$ (kJ/kmol · K)
0	0	0	0	1020	36 709	28 228	233.415
220	7295	5466	178.576	1040	37 542	28 895	234.223
240	7961	5965	181.471	1060	38 380	29 567	235.020
260	8627	6466	184.139	1080	39 223	30 243	235.806
280	9296	6968	186.616	1100	40 071	30 925	236.584
298	9904	7425	188.720	1120	40 923	31 611	237.352
300	9966	7472	188.928	1140	41 780	32 301	238.110
320	10 639	7978	191.098	1160	42 642	32 997	238.859
340	11 314	8487	193.144	1180	43 509	33 698	239.600
360	11 992	8998	195.081	1200	44 380	34 403	240.333
380	12 672	9513	196.920	1220	45 256	35 112	241.057
400	13 356	10 030	198.673	1240	46 137	35 827	241.773
420	14 043	10 551	200.350	1260	47 022	36 546	242.482
440	14 734	11 075	201.955	1280	47 912	37 270	243.183
460	15 428	11 603	203.497	1300	48 807	38 000	243.877
480	16 126	12 135	204.982	1320	49 707	38 732	244.564
500	16 828	12 671	206.413	1340	50 612	39 470	245.243
520	17 534	13 211	207.799	1360	51 521	40 213	245.915
540	18 245	13 755	209.139	1400	53 351	41 711	247.241
560	18 959	14 303	210.440	1440	55 198	43 226	248.543
580	19 678	14 856	211.702	1480	57 062	44 756	249.820
600	20 402	15 413	212.920	1520	58 942	46 304	251.074
620	21 130	15 975	214.122	1560	60 838	47 868	252.305
640	21 862	16 541	215.285	1600	62 748	49 445	253.513
660	22 600	17 112	216.419	1700	67 589	53 455	256.450
680	23 342	17 688	217.527	1800	72 513	57 547	259.262
700	24 088	18 268	218.610	1900	77 517	61 720	261.969
720	24 840	18 854	219.668	2000	82 593	65 965	264.571
740	25 597	19 444	220.707	2100	87 735	70 275	267.081
760	26 358	20 039	221.720	2200	92 940	74 649	269.500
780	27 125	20 639	222.717	2300	98 199	79 076	271.839
800	27 896	21 245	223.693	2400	103 508	83 553	274.098
820	28 672	21 855	224.651	2500	108 868	88 082	276.286
840	29 454	22 470	225.592	2600	114 273	92 656	278.407
860	30 240	23 090	226.517	2700	119 717	97 269	280.462
880	31 032	23 715	227.426	2800	125 198	101 917	282.453
900	31 828	24 345	228.321	2900	130 717	106 605	284.390
920	32 629	24 980	229.202	3000	136 264	111 321	286.273
940	33 436	25 621	230.070	3100	141 846	116 072	288.102
960	34 247	26 265	230.924	3150	144 648	118 458	288.9
980	35 061	26 913	231.767	3200	147 457	120 851	289.884
1000	35 882	27 568	232.597	3250	150 250	123 250	290.7

Based on data from JANAF Thermochemical Tables, NSRDS-NBS-37, 1971

TABLE A.5 Thermodynamic Properties of Select Solids and Liquids

Substance	Specific Heat, c (kJ/kg · K)	Density, ρ (kg/m³)	Thermal Conductivity, κ (W/m · K)
Solids, 300 K			
Aluminum	0.903	2700	237
Brick, common	0.835	1920	0.72
Copper	0.385	8930	401
Cork	1.800	120	0.039
Glass, plate	0.750	2500	1.4
Granite	0.775	2630	2.79
Iron	0.447	7870	80.2
Lead	0.129	11 300	35.3
Silver	0.235	10 500	429
Steel (AISI 302)	0.480	8060	15.1
Tin	0.227	7310	66.6
Liquids (Saturated), 300 K			
Ammonia	4.818	599.8	0.465
Mercury	0.139	13 529	8.540
Engine Oil	1.909	884.1	0.145
Water	4.180	996.5	0.613

Note: the data are taken from various sources and are representative of property data. The data vary with environmental conditions, and material composition in the case of substances such as brick.

Based on data from Keenan, Keyes, Hill, and Moore, Steam Tables, Wiley, New York, 1969; G. J. Van Wylen and R. E. Sonntag, Fundamentals of Classical Thermodynamics, Wiley, New York, 1973.

TABLE A.6 Properties of Saturated Water (Liquid-Vapor)—Temperature

		v (m³/kg)		u (kJ/kg)		h (kJ/kg)			s (kJ/kg · K)		
T (°C)	P (MPa)	v_f	v_g	u_f	u_g	h_f	h_{fg}	h_g	s_f	s_{fg}	s_g
0.01	0.000611	0.001000	206.1	0.0	2375.3	0.0	2501.3	2501.3	0.0000	9.1571	9.1571
2	0.0007056	0.001000	179.9	8.4	2378.1	8.4	2496.6	2505.0	0.0305	9.0738	9.1043
5	0.0008721	0.001000	147.1	21.0	2382.2	21.0	2489.5	2510.5	0.0761	8.9505	9.0266
10	0.001228	0.001000	106.4	42.0	2389.2	42.0	2477.7	2519.7	0.1510	8.7506	8.9016
15	0.001705	0.001001	77.93	63.0	2396.0	63.0	2465.9	2528.9	0.2244	8.5578	8.7822
20	0.002338	0.001002	57.79	83.9	2402.9	83.9	2454.2	2538.1	0.2965	8.3715	8.6680
25	0.003169	0.001003	43.36	104.9	2409.8	104.9	2442.3	2547.2	0.3672	8.1916	8.5588
30	0.004246	0.001004	32.90	125.8	2416.6	125.8	2430.4	2556.2	0.4367	8.0174	8.4541
35	0.005628	0.001006	25.22	146.7	2423.4	146.7	2418.6	2565.3	0.5051	7.8488	8.3539
40	0.007383	0.001008	19.52	167.5	2430.1	167.5	2406.8	2574.3	0.5723	7.6855	8.2578
45	0.009593	0.001010	15.26	188.4	2436.8	188.4	2394.8	2583.2	0.6385	7.5271	8.1656
50	0.01235	0.001012	12.03	209.3	2443.5	209.3	2382.8	2592.1	0.7036	7.3735	8.0771
55	0.01576	0.001015	9.569	230.2	2450.1	230.2	2370.7	2600.9	0.7678	7.2243	7.9921
60	0.01994	0.001017	7.671	251.1	2456.6	251.1	2358.5	2609.6	0.8310	7.0794	7.9104
65	0.02503	0.001020	6.197	272.0	2463.1	272.06	2346.2	2618.3	0.8935	6.9375	7.8310
70	0.03119	0.001023	5.042	292.9	2469.5	293.0	2333.8	2626.8	0.9549	6.8012	7.7561
75	0.03858	0.001026	4.131	313.9	2475.9	313.9	2321.4	2635.3	1.0155	6.6678	7.6833
80	0.04739	0.001029	3.407	334.8	2482.2	334.9	2308.8	2643.7	1.0754	6.5376	7.6130
85	0.05783	0.001032	2.828	355.8	2488.4	355.9	2296.0	2651.9	1.1344	6.4109	7.5453
90	0.07013	0.001036	2.361	376.8	2494.5	376.9	2283.2	2660.1	1.1927	6.2872	7.4799
95	0.08455	0.001040	1.982	397.9	2500.6	397.9	2270.2	2668.1	1.2503	6.1664	7.4167
100	0.1013	0.001044	1.673	418.9	2506.5	419.0	2257.0	2676.0	1.3071	6.0486	7.3557
110	0.1433	0.001052	1.210	461.1	2518.1	461.3	2230.2	2691.5	1.4188	5.8207	7.2395
120	0.1985	0.001060	0.8919	503.5	2529.2	503.7	2202.6	2706.3	1.5280	5.6024	7.1304
130	0.2701	0.001070	0.6685	546.0	2539.9	546.3	2174.2	2720.5	1.6348	5.3929	7.0277
140	0.3613	0.001080	0.5089	588.7	2550.0	589.1	2144.8	2733.9	1.7395	5.1912	6.9307
150	0.4758	0.001090	0.3928	631.7	2559.5	632.2	2114.2	2746.4	1.8422	4.9965	6.8387
160	0.6178	0.001102	0.3071	674.9	2568.4	675.5	2082.6	2758.1	1.9431	4.8079	6.7510
170	0.7916	0.001114	0.2428	718.3	2576.5	719.2	2049.5	2768.7	2.0423	4.6249	6.6672
180	1.002	0.001127	0.1941	762.1	2583.7	763.2	2015.0	2778.2	2.1400	4.4466	6.5866
190	1.254	0.001141	0.1565	806.2	2590.0	807.5	1978.8	2786.4	2.2363	4.2724	6.5087
200	1.554	0.001156	0.1274	850.6	2595.3	852.4	1940.8	2793.2	2.3313	4.1018	6.4331
210	1.906	0.001173	0.1044	895.5	2599.4	897.7	1900.8	2798.5	2.4253	3.9340	6.3593
220	2.318	0.001190	0.08620	940.9	2602.4	943.6	1858.5	2802.1	2.5183	3.7686	6.2869
230	2.795	0.001209	0.07159	986.7	2603.9	990.1	1813.9	2804.0	2.6105	3.6050	6.2155

(*Continued*)

TABLE A.6 *(Continued)*

T (°C)	P (MPa)	v (m³/kg)		u (kJ/kg)		h (kJ/kg)			s (kJ/kg · K)		
		v_f	v_g	u_f	u_g	h_f	h_{fg}	h_g	s_f	s_{fg}	s_g
240	3.344	0.001229	0.05977	1033.2	2604.0	1037.3	1766.5	2803.8	2.7021	3.4425	6.1446
250	3.973	0.001251	0.05013	1080.4	2602.4	1085.3	1716.2	2801.5	2.7933	3.2805	6.0738
260	4.688	0.001276	0.04221	1128.4	2599.0	1134.4	1662.5	2796.9	2.8844	3.1184	6.0028
270	5.498	0.001302	0.03565	1177.3	2593.7	1184.5	1605.2	2789.7	2.9757	2.9553	5.9310
280	6.411	0.001332	0.03017	1227.4	2586.1	1236.0	1543.6	2779.6	3.0674	2.7905	5.8579
290	7.436	0.001366	0.02557	1278.9	2576.0	1289.0	1477.2	2766.2	3.1600	2.6230	5.7830
300	8.580	0.001404	0.02168	1332.0	2563.0	1344.0	1405.0	2749.0	3.2540	2.4513	5.7053
310	9.856	0.001447	0.01835	1387.0	2546.4	1401.3	1326.0	2727.3	3.3500	2.2739	5.6239
320	11.27	0.001499	0.01549	1444.6	2525.5	1461.4	1238.7	2700.1	3.4487	2.0883	5.5370
330	12.84	0.001561	0.01300	1505.2	2499.0	1525.3	1140.6	2665.9	3.5514	1.8911	5.4425
340	14.59	0.001638	0.01080	1570.3	2464.6	1594.2	1027.9	2622.1	3.6601	1.6765	5.3366
350	16.51	0.001740	0.008815	1641.8	2418.5	1670.6	893.4	2564.0	3.7784	1.4338	5.2122
360	18.65	0.001892	0.006947	1725.2	2351.6	1760.5	720.7	2481.2	3.9154	1.1382	5.0536
370	21.03	0.002213	0.004931	1844.0	2229.0	1890.5	442.2	2332.7	4.1114	0.6876	4.7990
374.14	22.088	0.003155	0.003155	2029.6	2029.6	2099.3	0.0	2099.3	4.4305	0.0000	4.4305

TABLE A.7 Properties of Saturated Water (Liquid-Vapor)—Pressure

		v (m³/kg)		u (kJ/kg)		h (kJ/kg)			s (kJ/kg · K)		
P (MPa)	T (°C)	v_f	v_g	u_f	u_g	h_f	h_{fg}	h_g	s_f	s_{fg}	s_g
0.0006	0.01	0.001000	206.1	0.0	2375.3	0.0	2501.3	2501.3	0.0000	9.1571	9.1571
0.0008	3.8	0.001000	159.7	15.8	2380.5	15.8	2492.5	2508.3	0.0575	9.0007	9.0582
0.001	7.0	0.001000	129.2	29.3	2385.0	29.3	2484.9	2514.2	0.1059	8.8706	8.9765
0.0012	9.7	0.001000	108.7	40.6	2388.7	40.6	2478.5	2519.1	0.1460	8.7639	8.9099
0.0014	12.0	0.001001	93.92	50.3	2391.9	50.3	2473.1	2523.4	0.1802	8.6736	8.8538
0.0016	14.0	0.001001	82.76	58.9	2394.7	58.9	2468.2	2527.1	0.2101	8.5952	8.8053
0.002	17.5	0.001001	67.00	73.5	2399.5	73.5	2460.0	2533.5	0.2606	8.4639	8.7245
0.003	24.1	0.001003	45.67	101.0	2408.5	101.0	2444.5	2545.5	0.3544	8.2240	8.5784
0.004	29.0	0.001004	34.80	121.4	2415.2	121.4	2433.0	2554.4	0.4225	8.0529	8.4754
0.006	36.2	0.001006	23.74	151.5	2424.9	151.5	2415.9	2567.4	0.5208	7.8104	8.3312
0.008	41.5	0.001008	18.10	173.9	2432.1	173.9	2403.1	2577.0	0.5924	7.6371	8.2295
0.01	45.8	0.001010	14.67	191.8	2437.9	191.8	2392.8	2584.6	0.6491	7.5019	8.1510
0.012	49.4	0.001012	12.36	206.9	2442.7	206.9	2384.1	2591.0	0.6961	7.3910	8.0871
0.014	52.6	0.001013	10.69	220.0	2446.9	220.0	2376.6	2596.6	0.7365	7.2968	8.0333
0.016	55.3	0.001015	9.433	231.5	2450.5	231.5	2369.9	2601.4	0.7719	7.2149	7.9868
0.018	57.8	0.001016	8.445	241.9	2453.8	241.9	2363.9	2605.8	0.8034	7.1425	7.9459
0.02	60.1	0.001017	7.649	251.4	2456.7	251.4	2358.3	2609.7	0.8319	7.0774	7.9093
0.03	69.1	0.001022	5.229	289.2	2468.4	289.2	2336.1	2625.3	0.9439	6.8256	7.7695
0.04	75.9	0.001026	3.993	317.5	2477.0	317.6	2319.1	2636.7	1.0260	6.6449	7.6709
0.06	85.9	0.001033	2.732	359.8	2489.6	359.8	2293.7	2653.5	1.1455	6.3873	7.5328
0.08	93.5	0.001039	2.087	391.6	2498.8	391.6	2274.1	2665.7	1.2331	6.2023	7.4354
0.1	99.6	0.001043	1.694	417.3	2506.1	417.4	2258.1	2675.5	1.3029	6.0573	7.3602
0.12	104.8	0.001047	1.428	439.2	2512.1	439.3	2244.2	2683.5	1.3611	5.9378	7.2980
0.14	109.3	0.001051	1.237	458.2	2517.3	458.4	2232.0	2690.4	1.4112	5.8360	7.2472
0.16	113.3	0.001054	1.091	475.2	2521.8	475.3	2221.2	2696.5	1.4553	5.7472	7.2025
0.18	116.9	0.001058	0.9775	490.5	2525.9	490.7	2211.1	2701.8	1.4948	5.6683	7.1631
0.2	120.2	0.001061	0.8857	504.5	2529.5	504.7	2201.9	2706.6	1.5305	5.5975	7.1280
0.3	133.5	0.001073	0.6058	561.1	2543.6	561.5	2163.8	2725.3	1.6722	5.3205	6.9927
0.4	143.6	0.001084	0.4625	604.3	2553.6	604.7	2133.8	2738.5	1.7770	5.1197	6.8967
0.6	158.9	0.001101	0.3157	669.9	2567.4	670.6	2086.2	2756.8	1.9316	4.8293	6.7609
0.8	170.4	0.001115	0.2404	720.2	2576.8	721.1	2048.0	2769.1	2.0466	4.6170	6.6636
1	179.9	0.001127	0.1944	761.7	2583.6	762.8	2015.3	2778.1	2.1391	4.4482	6.5873
1.2	188.0	0.001139	0.1633	797.3	2588.8	798.6	1986.2	2784.8	2.2170	4.3072	6.5242
1.4	195.1	0.001149	0.1408	828.7	2592.8	830.3	1959.7	2790.0	2.2847	4.1854	6.4701
1.6	201.4	0.001159	0.1238	856.9	2596.0	858.8	1935.2	2794.0	2.3446	4.0780	6.4226

(*Continued*)

TABLE A.7 *(Continued)*

		v (m³/kg)		u (kJ/kg)		h (kJ/kg)			s (kJ/kg · K)		
P (MPa)	T (°C)	v_f	v_g	u_f	u_g	h_f	h_{fg}	h_g	s_f	s_{fg}	s_g
2	212.4	0.001177	0.09963	906.4	2600.3	908.8	1890.7	2799.5	2.4478	3.8939	6.3417
4	250.4	0.001252	0.04978	1082.3	2602.3	1087.3	1714.1	2801.4	2.7970	3.2739	6.0709
6	275.6	0.001319	0.03244	1205.4	2589.7	1213.3	1571.0	2784.3	3.0273	2.8627	5.8900
8	295.1	0.001384	0.02352	1305.6	2569.8	1316.6	1441.4	2758.0	3.2075	2.5365	5.7440
10	311.1	0.001452	0.01803	1393.0	2544.4	1407.6	1317.1	2724.7	3.3603	2.2546	5.6149
12	324.8	0.001527	0.01426	1472.9	2513.7	1491.3	1193.6	2684.9	3.4970	1.9963	5.4933
14	336.8	0.001611	0.01149	1548.6	2476.8	1571.1	1066.5	2637.6	3.6240	1.7486	5.3726
16	347.4	0.001711	0.00931	1622.7	2431.8	1650.0	930.7	2580.7	3.7468	1.4996	5.2464
18	357.1	0.001840	0.00749	1698.9	2374.4	1732.0	777.2	2509.2	3.8722	1.2332	5.1054
20	365.8	0.002036	0.00583	1785.6	2293.2	1826.3	583.7	2410.0	4.0146	0.9135	4.9281
22.09	374.14	0.003155	0.003155	2029.6	2029.6	2099.3	0.0	2099.3	4.4305	0.0	4.4305

TABLE A.8 Properties of Superheated Water Vapor

T (°C)	v (m³/kg)	u (kJ/kg)	h (kJ/kg)	s (kJ/kg · K)	v (m³/kg)	u (kJ/kg)	h (kJ/kg	s (kJ/kg · K)	v (m³/kg)	u (kJ/kg)	h (kJ/kg)	s (kJ/kg · K)
	P = 0.010 MPa (45.81°C)				P = 0.050 MPa (81.33°C)				P = 0.10 MPa (99.63°C)			
Sat.	14.674	2437.9	2584.7	8.1502	3.240	2483.9	2645.9	7.5939	1.6940	2506.1	2675.5	7.3594
50	14.869	2443.9	2592.6	8.1749								
100	17.196	2515.5	2687.5	8.4479	3.418	2511.6	2682.5	7.6947	1.6958	2506.7	2676.2	7.3614
150	19.512	2587.9	2783.0	8.6882	3.889	2585.6	2780.1	7.9401	1.9364	2582.8	2776.4	7.6134
200	21.825	2661.3	2879.5	8.9038	4.356	2659.9	2877.7	8.1580	2.172	2658.1	2875.3	7.8343
250	24.136	2736.0	2977.3	9.1002	4.820	2735.0	2976.0	8.3556	2.406	2733.7	2974.3	8.0333
300	26.445	2812.1	3076.5	9.2813	5.284	2811.3	3075.5	8.5373	2.639	2810.4	3074.3	8.2158
400	31.063	2968.9	3279.6	9.6077	6.209	2968.5	3278.9	8.8642	3.103	2967.9	3278.2	8.5435
500	35.679	3132.3	3489.1	9.8978	7.134	3132.0	3488.7	9.1546	3.565	3131.6	3483.1	8.8342
600	40.295	3302.5	3705.4	10.1608	8.057	3302.2	3705.1	9.4178	4.028	3301.9	3704.7	9.0976
700	44.911	3479.6	3928.7	10.4028	8.981	3479.4	3928.5	9.6599	4.490	3479.2	3928.2	9.3398
800	49.526	3663.8	4159.0	10.6281	9.904	3663.6	4158.9	9.8852	4.952	3663.5	4158.6	9.5652
900	54.141	3855.0	4396.4	10.8396	10.828	3854.9	4396.3	10.0967	5.414	3854.8	4396.1	9.7767
1000	58.757	4053.0	4640.6	11.0393	11.751	4052.9	4640.5	10.2964	5.875	4052.8	4640.3	9.9764
1100	63.372	4257.5	4891.2	11.2287	12.674	4257.4	4891.1	10.4859	6.337	4257.3	4891.0	10.1659
1200	67.987	4467.9	5147.8	11.4091	13.597	4467.8	5147.7	10.6662	6.799	4467.7	5147.6	10.3463
1300	72.602	4683.7	5409.7	11.5811	14.521	4683.6	5409.6	10.8382	7.260	4683.5	5409.5	10.5183
	P = 0.20 MPa (120.23°C)				P = 0.30 MPa (133.55°C)				P = 0.40 MPa (143.63°C)			
Sat.	0.8857	2529.5	2706.7	7.1272	0.6058	2543.6	2725.3	6.9919	0.4625	2553.6	2738.6	6.8959
150	0.9596	2576.9	2768.8	7.2795	0.6339	2570.8	2761.0	7.0778	0.4708	2564.5	2752.8	6.9299
200	1.0803	2654.4	2870.5	7.5066	0.7163	2650.7	2865.6	7.3115	0.5342	2646.8	2860.5	7.1706
250	1.1988	2731.2	2971.0	7.7086	0.7964	2728.7	2967.6	7.5166	0.5951	2726.1	2964.2	7.3789
300	1.3162	2808.6	3071.8	7.8926	0.8753	2806.7	3069.3	7.7022	0.6548	2804.8	3066.8	7.5662
400	1.5493	2966.7	3276.6	8.2218	1.0315	2965.6	3275.6	8.0330	0.7726	2964.4	3273.4	7.8985
500	1.7814	3130.8	3487.1	8.5133	1.1867	3130.0	3486.0	8.3251	0.8893	3129.2	3484.9	8.1913
600	2.013	3301.4	3704.0	8.7770	1.3414	3300.8	3703.2	8.5892	1.0055	3300.2	3702.4	8.4558
700	2.244	3478.8	3927.6	9.0194	1.4957	3478.4	3927.1	8.8319	1.1215	3477.9	3926.5	8.6987
800	2.475	3663.1	4158.2	9.2449	1.6499	3662.9	4157.8	9.0576	1.2372	3662.4	4157.3	8.9244
900	2.706	3854.5	4395.8	9.4566	1.8041	3854.2	4395.4	9.2692	1.3529	3853.9	4395.1	9.1362
1000	2.937	4052.5	4640.0	9.6563	1.9581	4052.3	4639.7	9.4690	1.4685	4052.0	4639.4	9.3360
1100	3.168	4257.0	4890.7	9.8458	2.1121	4256.5	4890.4	9.6585	1.5840	4256.5	4890.2	9.5256
1200	3.399	4467.5	5147.3	10.0262	2.2661	4467.2	5147.1	9.8389	1.6996	4467.0	5146.8	9.7060
1300	3.630	4683.2	5409.3	10.1982	2.4201	4683.0	5409.0	10.0110	1.8151	4682.8	5408.8	9.8780

(Continued)

Based on data from Keenan, Keyes, Hill, and Moore, Steam Tables, Wiley, New York, 1969; G. J. Van Wylen and R. E. Sonntag, Fundamentals of Classical Thermodynamics, Wiley, New York, 1973.

TABLE A.8 *(Continued)*

T (°C)	v (m³/kg)	u (kJ/kg)	h (kJ/kg)	s (kJ/kg · K)	v (m³/kg)	u (kJ/kg)	h (kJ/kg	s (kJ/kg · K)	v (m³/kg)	u (kJ/kg)	h (kJ/kg)	s (kJ/kg · K)
	P = 0.50 MPa (151.86°C)				P = 0.60 MPa (158.85°C)				P = 0.80 MPa (170.43°C)			
Sat.	0.3749	2561.2	2748.7	6.8213	0.3157	2567.4	2756.8	6.7600	0.2404	2576.8	2769.1	6.6628
200	0.4249	2642.9	2855.4	7.0592	0.3520	2638.9	2850.1	6.9665	0.2608	2630.6	2839.3	6.8158
250	0.4744	2723.5	2960.7	7.2709	0.3938	2720.9	2957.2	7.1816	0.2931	2715.5	2950.0	7.0384
300	0.5226	2802.9	3064.2	7.4599	0.4344	2801.0	3061.6	7.3724	0.3241	2797.2	3056.5	7.2328
350	0.5701	2882.6	3167.7	7.6329	0.4742	2881.2	3165.7	7.5464	0.3544	2878.2	3161.7	7.4089
400	0.6173	2963.2	3271.9	7.7938	0.5137	2962.1	3270.3	7.7079	0.3843	2959.7	3267.1	7.5716
500	0.7109	3128.4	3483.9	8.0873	0.5920	3127.6	3482.8	8.0021	0.4433	3126.0	3480.6	7.8673
600	0.8041	3299.6	3701.7	8.3522	0.6697	3299.1	3700.9	8.2674	0.5018	3297.9	3699.4	8.1333
700	0.8969	3477.5	3925.9	8.5952	0.7472	3477.0	3925.3	8.5107	0.5601	3476.2	3924.2	8.3770
800	0.9896	3662.1	4156.9	8.8211	0.8245	3661.8	4156.5	8.7367	0.6181	3661.1	4155.6	8.6033
900	1.0822	3853.6	4394.7	9.0329	0.9017	3853.4	4394.4	8.9486	0.6761	3852.8	4393.7	8.8153
1000	1.1747	4051.8	4639.1	9.2328	0.9788	4051.5	4638.8	9.1485	0.7340	4051.0	4638.2	9.0153
1100	1.2672	4256.3	4889.9	9.4224	1.0559	4256.1	4889.6	9.3381	0.7919	4255.6	4889.1	9.2050
1200	1.3596	4466.8	5146.6	9.6029	1.1330	4466.5	5146.3	9.5185	0.8497	4466.1	5145.9	9.3855
1300	1.4521	4682.5	5408.6	9.7749	1.2101	4682.3	5408.3	9.6906	0.9076	4681.8	5407.9	9.5575
	P = 1.00 MPa (179.91°C)				P = 1.20 MPa (187.99°C)				P = 1.40 MPa (195.07°C)			
Sat.	0.19444	2583.6	2778.1	6.5865	0.16333	2588.8	2784.8	6.5233	0.14084	2592.8	2790.0	6.4693
200	0.2060	2621.9	2827.9	6.6940	0.16930	2612.8	2815.9	6.5898	0.14302	2603.1	2803.3	6.4975
250	0.2327	2709.9	2942.6	6.9247	0.19234	2704.2	2935.0	6.8294	0.16350	2698.3	2927.2	6.7467
300	0.2579	2793.2	3051.2	7.1229	0.2138	2789.2	3045.8	7.0317	0.18228	2785.2	3040.4	6.9534
350	0.2825	2875.2	3157.7	7.3011	0.2345	2872.2	3153.6	7.2121	0.2003	2869.2	3149.5	7.1360
400	0.3066	2957.3	3263.9	7.4651	0.2548	2954.9	3260.7	7.3774	0.2178	2952.5	3257.5	7.3026
500	0.3541	3124.4	3478.5	7.7622	0.2946	3122.8	3476.3	7.6759	0.2521	3321.1	3474.1	7.6027
600	0.4011	3296.8	3697.9	8.0290	0.3339	3295.6	3696.3	7.9435	0.2860	3294.4	3694.8	7.8710
700	0.4478	3475.3	3923.1	8.2731	0.3729	3474.4	3922.0	8.1881	0.3195	3473.6	3920.8	8.1160
800	0.4943	3660.4	4154.7	8.4996	0.4118	3659.7	4153.8	8.4148	0.3528	3659.0	4153.0	8.3358
900	0.5407	3852.2	4392.9	8.7118	0.4505	3851.6	4392.2	8.6272	0.3861	3851.1	4391.5	8.5556
1000	0.5871	4050.5	4637.6	8.9119	0.4892	4050.0	4637.0	8.8274	0.4192	4049.5	4636.4	8.7559
1100	0.6335	4255.1	4888.6	9.1017	0.5278	4254.6	4888.0	9.0172	0.4524	4254.1	4887.5	8.9457
1200	0.6798	4465.6	5145.4	9.2822	0.5665	4465.1	5144.9	9.1977	0.4855	4464.7	5144.4	9.1262
1300	0.7261	4681.3	5407.4	9.4543	0.6051	4680.9	5407.0	9.3698	0.5186	4680.4	5406.5	9.2984

TABLE A.8 *(Continued)*

T (°C)	v (m³/kg)	u (kJ/kg)	h (kJ/kg)	s (kJ/kg · K)	v (m³/kg)	u (kJ/kg)	h (kJ/kg)	s (kJ/kg · K)	v (m³/kg)	u (kJ/kg)	h (kJ/kg)	s (kJ/kg · K)
	P = 1.60 MPa (201.41°C)				P = 1.80 MPa (207.15°C)				P = 2.0 MPa (212.42°C)			
Sat.	0.12380	2596.0	2794.0	6.4218	0.11042	2598.4	2797.1	6.3794	0.09963	2600.3	2799.5	6.3409
225	0.13287	2644.7	2857.3	6.5518	0.11673	2636.6	2846.7	6.4808	0.10377	2628.3	2835.8	6.4147
250	0.14184	2692.3	2919.2	6.6732	0.12497	2686.0	2911.0	6.6066	0.11144	2679.6	2902.5	6.5453
300	0.15862	2781.1	3034.8	6.8844	0.14021	2776.9	3029.2	6.8226	0.12547	2772.6	3023.5	6.7664
350	0.17456	2866.1	3145.4	7.0694	0.15457	2863.0	3141.2	7.0100	0.13857	2859.8	3137.0	6.9563
400	0.19005	2950.1	3254.2	7.2374	0.16847	2947.7	3250.9	7.1794	0.15120	2945.2	3247.6	7.1271
500	0.2203	3119.5	3472.0	7.5390	0.19550	3117.9	3469.8	7.4825	0.17568	3116.2	3467.6	7.4317
600	0.2500	3293.3	3693.2	7.8080	0.2220	3292.1	3691.7	7.7523	0.19960	3290.9	3690.1	7.7024
700	0.2794	3472.7	3919.7	8.0535	0.2482	3471.8	3918.5	7.9983	0.2232	3470.9	3917.4	7.9487
800	0.3086	3658.3	4152.1	8.2808	0.2742	3657.6	4151.2	8.2258	0.2467	3657.0	4150.3	8.1765
900	0.3377	3850.5	4390.8	8.4935	0.3001	3849.9	4390.1	8.4386	0.2700	3849.3	4389.4	8.3895
1000	0.5668	4049.0	4635.8	8.6938	0.3260	4048.5	4635.2	8.6391	0.2933	4048.0	4634.6	8.5901
1100	0.3958	4253.7	4887.0	8.8837	0.3518	4253.2	4886.4	8.8290	0.3166	4252.7	4885.9	8.7800
1200	0.4248	4464.2	5143.9	9.0643	0.3776	4463.7	5143.4	9.0096	0.3398	4463.3	5142.9	8.9607
1300	0.4538	4679.9	5406.0	9.2364	0.4034	4679.5	5405.6	9.1818	0.3631	4679.0	5405.1	9.1329
	P = 2.50 MPa (233.99°C)				P = 3.00 MPa (233.90°C)				P = 3.50 MPa (242.60°C)			
Sat.	0.07998	2603.1	2803.1	6.2575	0.06668	2604.1	2804.2	6.1869	0.05707	2603.7	2803.4	6.1253
225	0.08027	2605.6	2806.3	6.2639								
250	0.08700	2662.6	2880.1	6.4085	0.07058	2644.0	2855.8	6.2872	0.05872	2623.7	2829.2	6.1749
300	0.09890	2761.6	3008.8	6.6438	0.08114	2750.1	2993.5	6.5390	0.06842	2738.0	2977.5	6.4461
350	0.10976	2851.9	3126.3	6.8403	0.09053	2843.7	3115.3	6.7428	0.07678	2835.3	3104.0	6.6579
400	0.12010	2939.1	3239.3	7.0148	0.09936	2932.8	3230.9	6.9212	0.08453	2926.4	3222.3	6.8405
450	0.13014	3025.5	3350.8	7.1746	0.10787	3020.4	3344.0	7.0834	0.09196	3015.3	3337.2	7.0052
500	0.13998	3112.1	3462.1	7.3234	0.11619	3108.0	3456.5	7.2338	0.09918	3103.0	3450.9	7.1572
600	0.15930	3288.0	3686.3	7.5960	0.13243	3285.0	3682.3	7.5085	0.11324	3282.1	3678.4	7.4339
700	0.17832	3468.7	3914.5	7.8435	0.14838	3466.5	3911.7	7.7571	0.12699	3464.3	3908.8	7.6837
800	0.19716	3655.3	4148.2	8.0720	0.16414	3653.5	4145.9	7.9862	0.14056	3651.8	4143.7	7.9134
900	0.21590	3847.9	4387.6	8.2853	0.17980	3846.5	4385.9	8.1999	0.15402	3845.0	4384.1	8.1276
1000	0.2346	4046.7	4633.1	8.4861	0.19541	4045.4	4631.6	8.4009	0.16743	4044.1	4630.1	8.3288
1100	0.2532	4251.5	4884.6	8.6762	0.21098	4250.3	4883.3	8.5912	0.18080	4249.2	4881.9	8.5192
1200	0.2718	4462.1	5141.7	8.8569	0.22652	4460.9	5140.5	8.7720	0.19415	4459.8	5139.3	8.7000
1300	0.2905	4677.8	5404.0	9.0291	0.24206	4676.6	5402.8	8.9442	0.20749	4675.5	5401.7	8.8723
	P = 4.0 MPa (250.40°C)				P = 4.5 MPa (257.49°C)				P = 5.0 MPa (263.99°C)			
Sat.	0.04978	2602.3	2801.4	6.0701	0.04406	2600.1	2798.3	6.0198	0.03944	2597.1	2794.3	5.9734
275	0.05457	2667.9	2886.2	6.2285	0.04730	2650.3	2863.2	6.1401	0.04141	2631.3	2838.3	6.0544
300	0.05884	2725.3	2960.7	6.3615	0.05135	2712.0	2943.1	6.2828	0.04532	2698.0	2924.5	6.2084
350	0.06645	2826.7	3092.5	6.5821	0.05840	2817.8	3080.6	6.5131	0.05194	2808.7	3068.4	6.4493
400	0.07341	2919.9	3213.6	6.7690	0.06475	2913.3	3204.7	6.7047	0.05781	2906.6	3195.7	6.6459

(Continued)

TABLE A.8 *(Continued)*

T (°C)	v (m³/kg)	u (kJ/kg)	h (kJ/kg)	s (kJ/kg · K)	v (m³/kg)	u (kJ/kg)	h (kJ/kg)	s (kJ/kg · K)	v (m³/kg)	u (kJ/kg)	h (kJ/kg)	s (kJ/kg · K)
	P = 4.0 MPa (250.40°C)				P = 4.5 MPa (257.49°C)				P = 5.0 MPa (263.99°C)			
450	0.08002	3010.2	3330.3	6.9363	0.07074	3005.0	3323.3	6.8746	0.06330	2999.7	3316.2	6.8186
500	0.08643	3099.5	3445.3	7.0901	0.07651	3095.3	3439.6	7.0301	0.06857	3091.0	3433.8	6.9759
600	0.09885	3279.1	3674.4	7.3688	0.08765	3276.0	3670.5	7.3110	0.07869	3273.0	3666.5	7.2589
700	0.11095	3462.1	3905.9	7.6198	0.09847	3459.9	3903.0	7.5631	0.08849	3457.6	3900.1	7.5122
800	0.12287	3650.0	4141.5	7.8502	0.10911	3648.3	4139.3	7.7942	0.09811	3646.6	4137.1	7.7440
900	0.13469	3843.6	4382.3	8.0647	0.11965	3842.2	4380.6	8.0091	0.10762	3840.7	4378.8	7.9593
1000	0.14645	4042.9	4628.7	8.2662	0.13013	4041.6	4627.2	8.2108	0.11707	4040.4	4625.7	8.1612
1100	0.15817	4248.0	4880.6	8.4567	0.14056	4246.8	4879.3	8.4015	0.12648	4245.6	4878.0	8.3520
1200	0.16987	4458.6	5138.1	8.6376	0.15098	4457.5	5136.9	8.5825	0.13587	4456.3	5135.7	8.5331
1300	0.18156	4674.3	5400.5	8.8100	0.16139	4673.1	5399.4	8.7549	0.14526	4672.0	5398.2	8.7055
	P = 6.0 MPa (275.64°C)				P = 7.0 MPa (285.88°C)				P = 8.0 MPa (295.06°C)			
Sat.	0.03244	2589.7	2784.3	5.8892	0.02737	2580.5	2772.1	5.8133	0.02352	2569.8	2758.0	5.7432
300	0.03616	2667.2	2884.2	6.0674	0.02947	2632.2	2838.4	3.9305	0.02426	2590.9	2785.0	5.7906
350	0.04223	2789.6	3043.0	6.3335	0.03524	2769.4	3016.0	6.2283	0.02995	2747.7	2987.3	6.1301
400	0.04739	2892.9	3177.2	6.5408	0.03993	2878.6	3158.1	6.4478	0.03432	2863.8	3138.3	6.3634
450	0.05214	2988.9	3301.8	6.7193	0.04416	2978.0	3287.1	6.6327	0.03817	2966.7	3272.0	6.5551
500	0.05665	3082.2	3422.2	6.8803	0.04814	3073.4	3410.3	6.7975	0.04175	3064.3	3398.3	6.7240
550	0.06101	3174.6	3540.6	7.0288	0.05195	3167.2	3530.9	6.9486	0.04516	3159.8	3521.0	6.8778
600	0.06525	3266.9	3658.4	7.1677	0.05565	3260.7	3650.3	7.0894	0.04845	3254.4	3642.0	7.0206
700	0.07352	3453.1	3894.2	7.4234	0.06283	3448.5	3888.3	7.3476	0.05481	3443.9	3882.4	7.2812
800	0.08160	3643.1	4132.7	7.6566	0.06981	3639.5	4128.2	7.5822	0.06097	3636.0	4123.8	7.5173
900	0.08958	3837.8	4375.3	7.8727	0.07669	3835.0	4371.8	7.7991	0.06702	3832.1	4368.3	7.7351
1000	0.09749	4037.8	4622.7	8.0751	0.08350	4035.3	4619.8	8.0020	0.07301	4032.8	4616.9	7.9384
1100	0.10536	4243.3	4875.4	8.2661	0.09027	4240.9	4872.8	8.1933	0.07896	4238.6	4870.3	8.1300
1200	0.11321	4454.0	5133.3	8.4474	0.09703	4451.7	5130.9	8.3747	0.08489	4449.5	5128.5	8.3115
1300	0.12106	4669.6	5396.0	8.6199	0.10377	4667.3	5393.7	8.5473	0.09080	4665.0	5391.5	8.4842
	P = 9.0 MPa (303.40°C)				P = 10.0 MPa (311.06°C)				P = 12.5 MPa (327.89°C)			
Sat.	0.02048	2557.8	2742.1	5.6772	0.018026	2544.4	2724.7	5.6141	0.013495	2505.1	2673.8	5.4624
325	0.02327	2646.6	2856.0	5.8712	0.019861	2610.4	2809.1	5.7568				
350	0.02580	2724.4	2956.6	6.0361	0.02242	2699.4	2923.4	5.9443	0.016126	2624.6	2826.2	5.7118
400	0.02993	2848.4	3117.8	6.2854	0.02641	2832.4	3096.5	6.2120	0.02000	2789.3	3039.3	6.0417
450	0.03350	2955.2	3256.6	6.4844	0.02975	2943.4	3240.9	6.4190	0.02299	2912.5	3199.8	6.2719
500	0.03677	3055.2	3336.1	6.6576	0.03279	3045.8	3373.7	6.5966	0.02560	3021.7	3341.8	6.4618
550	0.03987	3152.2	3511.0	6.8142	0.03564	3144.6	3500.9	6.7561	0.02801	3125.0	3475.2	6.6290
600	0.04285	3248.1	3633.7	6.9589	0.03837	3241.7	3625.3	6.9029	0.03029	3225.4	3604.0	6.7810
650	0.04574	3343.6	3755.3	7.0943	0.04101	3338.2	3748.2	7.0398	0.03248	3324.4	3730.4	6.9218
700	0.04857	3439.3	3876.5	7.2221	0.04358	3434.7	3870.5	7.1687	0.03460	3422.9	3855.3	7.0536
800	0.05409	3632.5	4119.3	7.4596	0.04859	3628.9	4114.8	7.4077	0.03869	3620.0	4103.6	7.2965
900	0.05950	3829.2	4364.3	7.6783	0.05349	3826.3	4361.2	7.6272	0.04267	3819.1	4352.5	7.5182

TABLE A-8 *(Continued)*

T (°C)	v (m³/kg)	u (kJ/kg)	h (kJ/kg)	s (kJ/kg · K)	v (m³/kg)	u (kJ/kg)	h (kJ/kg)	s (kJ/kg · K)	v (m³/kg)	u (kJ/kg)	h (kJ/kg)	s (kJ/kg · K)
	P = 9.0 MPa (303.40°C)				P = 10.0 MPa (311.06°C)				P = 12.5 MPa (327.89°C)			
1000	0.06485	4030.3	4614.0	7.8821	0.05832	4027.8	4611.0	7.8315	0.04658	4021.6	4603.8	7.7237
1100	0.07016	4236.3	4867.7	8.0740	0.06312	4234.0	4865.1	8.0237	0.05045	4228.2	4858.8	7.9165
1200	0.07544	4447.2	5126.2	8.2556	0.06789	4444.9	5123.8	8.2055	0.05430	4439.3	5118.0	8.0987
1300	0.08072	4662.7	5389.2	8.4284	0.07265	4460.5	5387.0	8.3783	0.05813	4654.8	5381.4	8.2717
	P = 15.0 MPa (342.24°C)				P = 17.5 MPa (354.75°C)				P = 20.0 MPa (365.81°C)			
Sat.	0.010337	2455.5	2610.5	5.3098	0.007920	2390.2	2528.8	5.1419	0.005834	2293.0	2409.7	4.9269
350	0.011470	2520.4	2692.4	5.4421								
400	0.015649	2740.7	2975.5	5.8811	0.012447	2685.0	2902.9	5.7213	0.009942	2619.3	2818.1	5.5540
450	0.018445	2879.5	3156.2	6.1404	0.015174	2844.2	3109.7	6.0184	0.012695	2806.2	3060.1	5.9017
500	0.02080	2996.6	3308.6	6.3443	0.017358	2970.3	3274.1	6.2383	0.014768	2942.9	3238.2	6.1401
550	0.02293	3104.7	3448.6	6.5199	0.019288	3083.9	3421.4	6.4230	0.016555	3062.4	3393.5	6.3348
600	0.02491	3208.6	3582.3	6.6776	0.02106	3191.5	3560.2	6.5866	0.018178	3174.0	3537.6	6.5048
650	0.02680	3310.3	3712.3	6.8224	0.02274	3296.0	3693.9	6.7357	0.019693	3281.4	3675.3	6.6582
700	0.02861	3410.9	3840.1	6.9572	0.02434	3398.7	3824.6	6.8736	0.02113	3386.4	3809.0	6.7993
800	0.03210	3610.9	4092.4	7.2040	0.02738	3601.8	4081.1	7.1244	0.02385	3592.7	4069.7	7.0544
900	0.03546	3811.9	4343.8	7.4279	0.03031	3804.7	4335.1	7.3507	0.02645	3797.5	4326.4	7.2830
1000	0.03875	4015.4	4596.6	7.6348	0.03316	4009.3	4589.5	7.5589	0.02897	4003.1	4582.5	7.4925
1100	0.04200	4222.6	4852.6	7.8283	0.03597	4216.9	4846.4	7.7531	0.03145	4211.3	4840.2	7.6874
1200	0.04523	4433.8	5112.3	8.0108	0.03876	4428.3	5106.6	7.9360	0.03391	4422.8	5101.0	7.8707
1300	0.04845	4649.1	5376.0	8.1840	0.04154	4643.5	5370.5	8.1093	0.03636	4638.0	5365.1	8.0442
	P = 25.0 MPa				P = 30.0 MPa				P = 40.0 MPa			
375	0.0019731	1798.7	1848.0	4.0320	0.001789	1737.8	1791.5	3.9305	0.0016407	1677.1	1742.8	3.8290
400	0.006004	2430.1	2580.2	5.1418	0.002790	2067.4	2151.1	4.4728	0.0019077	1854.6	1930.9	4.1135
425	0.007881	2609.2	2806.3	5.4723	0.005303	2455.1	2614.2	5.1504	0.002532	2096.9	2198.1	4.5029
450	0.009162	2720.7	2949.7	5.6744	0.006735	2619.3	2821.4	5.4424	0.003693	2365.1	2512.8	4.9459
500	0.011123	2884.3	3162.4	5.9592	0.008678	2820.7	3081.1	5.7905	0.005622	2678.4	2903.3	5.4700
550	0.012724	3017.5	3335.6	6.1765	0.010168	2970.3	3275.4	6.0342	0.006984	2869.7	3149.1	5.7785
600	0.014137	3137.9	3491.4	6.3602	0.011446	3100.5	3443.9	6.2331	0.008094	3022.6	3346.4	6.0114
650	0.015433	3251.6	3637.4	6.5229	0.012596	3221.0	3598.9	6.4058	0.009063	3158.0	3520.6	6.2054
700	0.016646	3361.3	3777.5	6.6707	0.013661	3335.8	3745.6	6.5606	0.009941	3283.6	3681.2	6.3750
800	0.018912	3574.3	4047.1	6.9345	0.015623	3555.5	4024.2	6.8332	0.011523	3517.8	3978.7	6.6662
900	0.021045	3783.0	4309.1	7.1680	0.017448	3768.5	4291.9	7.0718	0.012962	3739.4	4257.9	6.9150
1000	0.02310	3990.9	4568.5	7.3802	0.019196	3978.8	4554.7	7.2867	0.014324	3954.6	4527.6	7.1356
1100	0.02512	4200.2	4828.2	7.5765	0.020903	4189.2	4816.3	7.4845	0.015642	4167.4	4793.1	7.3364
1200	0.02711	4412.0	5089.9	7.7605	0.022589	4401.3	5079.0	7.6692	0.016940	4380.1	5057.7	7.5224
1300	0.02910	4626.9	5354.4	7.9342	0.024266	4616.0	5344.0	7.8432	0.018229	4594.3	5323.5	7.6969

TABLE A.9 Properties of Compressed Liquid Water

T (°C)	v (m³/kg)	u (kJ/kg)	h (kJ/kg)	s (kJ/kg · K)	v (m³/kg)	u (kJ/kg)	h (kJ/kg)	s (kJ/kg · K)	v (m³/kg)	u (kJ/kg)	h (kJ/kg)	s (kJ/kg · K)
	P = 5 MPa (264.0°C)				P = 10 MPa (311.1°C)				P = 15 MPa (342.4°C)			
0	0.000998	0.04	5.04	0.0001	0.000995	0.09	10.04	0.0002	0.000993	0.15	15.05	0.0004
20	0.001000	83.65	88.65	0.296	0.000997	83.36	93.33	0.2945	0.000995	83.06	97.99	0.2934
40	0.001006	167.0	172.0	0.570	0.001003	166.4	176.4	0.5686	0.001001	165.8	180.78	0.5666
60	0.001015	250.2	255.3	0.828	0.001013	249.4	259.5	0.8258	0.001010	248.5	263.67	0.8232
80	0.001027	333.7	338.8	1.072	0.001024	332.6	342.8	1.0688	0.001022	331.5	346.81	1.0656
100	0.001041	417.5	422.7	1.303	0.001038	416.1	426.5	1.2992	0.001036	414.7	430.28	1.2955
120	0.001058	501.8	507.1	1.523	0.001055	500.1	510.6	1.5189	0.001052	498.4	514.19	1.5145
140	0.001077	586.8	592.2	1.734	0.001074	584.7	595.4	1.7292	0.001071	582.7	598.72	1.7242
160	0.001099	672.6	678.1	1.938	0.001095	670.1	681.1	1.9317	0.001092	667.7	684.09	1.9260
180	0.001124	759.6	765.2	2.134	0.001120	756.6	767.8	2.1275	0.001116	753.8	770.50	2.1210
200	0.001153	848.1	853.9	2.326	0.001148	844.5	856.0	2.3178	0.001143	841.0	858.2	2.3104
	P = 20 MPa (365.8°C)				P = 30 MPa				P = 50 MPa			
0	0.000990	0.19	20.01	0.0004	0.000986	0.25	29.82	0.0001	0.000977	0.20	49.03	0.0014
20	0.000993	82.77	102.6	0.2923	0.000989	82.17	111.8	0.2899	0.000980	81.00	130.02	0.2848
40	0.000999	165.2	185.2	0.5646	0.000995	164.0	193.9	0.5607	0.000987	161.9	211.21	0.5527
60	0.001008	247.7	267.8	0.8206	0.001004	246.1	276.2	0.8154	0.000996	243.0	292.79	0.8052
80	0.001020	330.4	350.8	1.0624	0.001016	328.3	358.8	1.0561	0.001007	324.3	374.70	1.0440
100	0.001034	413.4	434.1	1.2917	0.001029	410.8	441.7	1.2844	0.001020	405.9	456.89	1.2703
120	0.001050	496.8	517.8	1.5102	0.001044	493.6	524.9	1.5018	0.001035	487.6	539.39	1.4857
140	0.001068	580.7	602.0	1.7193	0.001062	576.9	608.8	1.7098	0.001052	569.8	622.35	1.6915
160	0.001088	665.4	687.1	1.9204	0.001082	660.8	693.3	1.9096	0.001070	652.4	705.92	1.8891
180	0.001112	751.0	773.2	2.1147	0.001105	745.6	778.7	2.1024	0.001091	735.7	790.25	2.0794
200	0.001139	837.7	860.5	2.3031	0.001130	831.4	865.3	2.2893	0.001115	819.7	875.5	2.2634

TABLE A.10 Enthalpy of Formation, Gibbs Function of Formation, Entropy, Molecular Mass, and Specific Heat of Common Substances at 25°C and 1 atm

Substance	Formula	Molecular Mass, M (kg/kmol)	$\bar{h}_f^o$ (kJ/kmol)	$\bar{g}_f^o$ (kJ/kmol)	$\bar{s}^o$ (kJ/kmol · K)	c_p (kJ/kg · K)
Carbon	C (s)	12.011	0	0	5.74	0.708
Hydrogen	H_2 (g)	2.016	0	0	130.68	14.4
Nitrogen	N_2 (g)	28.01	0	0	191.61	1.039
Oxygen	O_2 (g)	32.00	0	0	205.04	0.918
Carbon Monoxide	CO (g)	28.013	−110,530	−137,150	197.65	1.05
Carbon Dioxide	CO_2 (g)	44.01	−393,520	−394,360	213.80	0.846
Water Vapor	H_2O (g)	18.02	−241,820	−228,590	188.83	1.87
Water (liquid)	H_2O (l)	18.02	−285,830	−237,180	69.92	4.18
Methane	CH_4 (g)	16.043	−74,850	−50,790	186.16	2.20
Ethane	C_2H_6 (g)	30.070	−84,680	−32,890	229.49	1.75
Ethylene	C_2H_4 (g)	28.05	52,280	68,120	219.83	1.55
Acetylene	C_2H_2 (g)	26.038	226,730	209,170	200.85	1.69
Propane	C_3H_8 (g)	44.097	−103,850	−23,490	269.91	1.67
n-Butane	C_4H_{10} (g)	58.123	−126,150	−15,710	310.12	1.73
n-Octane	C_8H_{18} (l)	114.231	−249,950	6610	360.79	2.23
Methanol	CH_3OH (l)	32.042	−238,660	−166,360	126.80	2.53
Ethanol	C_2H_5OH (l)	46.069	−277,690	−174,890	160.70	2.44

TABLE A.11 Values of the Natural Logarithm of the Equilibrium Constant, ln K_P, for Various Chemical Equilibrium Reactions

The equilibrium constant K_P for the reaction $\nu_A A + \nu_B B \Leftrightarrow \nu_C C + \nu_D D$ is defined as $K_p = \dfrac{P_C^{\nu_C} P_D^{\nu_D}}{P_A^{\nu_A} P_B^{\nu_B}}$

Temp., K	$H_2 \Leftrightarrow 2H$	$O_2 \Leftrightarrow 2O$	$N_2 \Leftrightarrow 2N$	$H_2O \Leftrightarrow H_2 + \frac{1}{2}O_2$	$H_2O \Leftrightarrow \frac{1}{2}H_2 + HO$	$CO_2 \Leftrightarrow CO + \frac{1}{2}O_2$	$\frac{1}{2}N_2 + \frac{1}{2}O_2 \Leftrightarrow NO$
298	−164.005	−186.975	−367.480	−92.208	−106.208	−103.762	−35.052
500	−92.827	−105.630	−213.373	−52.691	−60.281	−57.616	−20.295
1000	−39.803	−45.150	−99.127	−23.163	−26.034	−23.529	−9.388
1200	−30.874	−35.605	−80.611	−18.182	−20.283	−17.871	−7.569
1400	−24.463	−27.742	−66.329	−14.609	−16.099	−13.842	−6.270
1600	−19.637	−22.285	−56.055	−11.921	−13.066	−10.830	−5.294
1800	−15.866	−18.030	−48.051	−9.826	−10.657	−8.497	−4.536
2000	−12.840	−14.622	−41.645	−8.145	−8.728	−6.635	−3.931
2200	−10.353	−11.827	−36.391	−6.768	−7.148	−5.120	−3.433
2400	−8.276	−9.497	−32.011	−5.619	−5.832	−3.860	−3.019
2600	−6.517	−7.521	−28.304	−4.648	−4.719	−2.801	−2.671
2800	−5.002	−5.826	−25.117	−3.812	−3.763	−1.894	−2.372
3000	−3.685	−4.357	−22.359	−3.086	−2.937	−1.111	−2.114
3200	−2.534	−3.072	−19.937	−2.451	−2.212	−0.429	−1.888
3400	−1.516	−1.935	−17.800	−1.891	−1.576	0.169	−1.690
3600	−0.609	−0.926	−15.898	−1.392	−1.088	0.701	−1.513
3800	0.202	−0.019	−14.199	−0.945	−0.501	1.176	−1.356
4000	0.934	0.796	−12.660	−0.542	−0.044	1.599	−1.216
4500	2.486	2.513	−9.414	0.312	0.920	2.490	−0.921
5000	3.725	3.895	−6.807	0.996	1.689	3.197	−0.686
5500	4.743	5.023	−4.665	1.560	2.318	3.771	−0.497
6000	5.590	5.963	−2.865	2.032	2.843	4.245	−0.341

Based on data from Gordon J. Van Wylen and Richard E. Sonntag, Fundamentals of Classical Thermodynamics, English/SI Version. 3rd ed. (New York: John Wiley & Sons, 1986), p 723, Table A.14. Based on thermodynamic data given in JANAF, Thermochemical Tables (Midland, MI: Thermal Research Laboratory, The Dow Chemical Company, 1971).

Index

A

Absolute pressure, 23
Absolute temperature scales, 17–19
Absorption refrigeration cycle, 337–338
Adiabatic flame temperature (AFT), 458–462
  hydrocarbon fuel burning, 411
Adiabatic process, 47, 218–219
Adiabatic saturator, 390–392
Air. *See also* Psychrometrics
  atmospheric contents of, 383–384
  combustion reactant mixtures and, 437–442
  comfort level and, 383–385, 398–399
  composition of dry, 355–357
  cooling and dehumidifying moist air, 399–404
  dry, 355–357, 383–384, 385–387, 389, 390–391
  evaporation and conditions of, 384–385
  heating and humidifying air, 406–410
  humidity of, 385–386, 389–397
  humidity ratio (ω), 387
  mixing moist air streams, 410–413
  mixture composition, 387–389
  moist, 389–391, 399–401, 410–413
  percent deficit, 437
  percent excess, 437
  percent theoretical, 437
  saturated, 385–386
  specific properties of, 389
  study of atmospheric, 383–417
  water vapor content, 355, 383–388
Air compressors, 140–141
Air conditioners, 153, 327
Air standard cycles (ASC), 287–288
  assumptions for, 287
  Brayton cycle, 288–297
  Diesel cycle, 288, 302–305
  dual cycle, 307–309
  Otto cycle, 288, 297–302
Air-conditioning systems, 153
  cooling and dehumidifying process, 399–404
  heat pumps, 153, 327
  heating and humidifying processes, 406–410
  mixing moist air streams, 410–413
  psychrometric charts for, 399–400, 408–409
  refrigeration cycles for, 404–406
  refrigerators, 153
  use of devices and cycles, 327
Air–fuel and fuel–air ratios, 437–438
Amagat law of partial volumes, 360–361
American Society of Heating, Refrigeration, and Air-Conditioning Engineers (ASHRAE), 397
Atkinson cycle, 310
Atmospheric air, contents of, 383–384. *See also* Psychrometrics
Atmospheric pressure, 22
Avogadro's number, 19

B

Balancing combustion reactions, 432–434
Blackbody radiation, 45
Brayton cycle, 288–297. *See also* Reversed Brayton refrigeration cycle
  assumptions for, 290
  constant-specific heat approach, 290
  intercooling with, 293–295
  processes of, 288–289
  Rankine cycle compared to, 292–293
  regeneration heat transfer for, 296–297
  reheat stage for, 295–296
  thermal efficiency (η) of, 290–293
  use of, 288

C

Carnot efficiencies, 182–185
Carnot power cycle, 253–255
  gas, 253–254
  processes of, 253–254
  reversibility of, 253–254
  thermal efficiency (η) of, 254–255
  vapor, 253–254
Cascade refrigeration system, 336
Celsius (°C) scale, 16–18
Chemical reactions
  balancing for combustion, 432–434
  combustion process, 431–432, 445
  equilibrium of combustion reactions, 468–472
  heat of, 446–458
  reactant mixtures, 437–440
  reactants from known products, 440–443
  water–gas shift, 472–473
Chemical work, 55
Clausius inequality, 195–199
  entropy and, 195–198
  heat engines, 195–198
  irreversible cycles, 196–199
  reversible cycles, 195–198
  thermal reservoirs, 195–196
Closed systems
  combustion, heat of reactions in, 453–456
  constant pressure, 453–454
  constant volume, 454–456
  enthalpy of reactions, 452–456
  entropy balance in, 214–217
  first law of thermodynamics in, 144–150
  Otto cycle, 148–149
  piston-cylinder devices, 144–147
  Rankine cycle, 148–149
  system boundary, 6
  thermodynamic cycle, 148–149
  use of, 7–8
Coefficient of performance, 153–154
  heat pumps (γ), 155, 332, 339
  refrigerators (β), 153, 331, 339
  reversed Brayton refrigeration cycle, 339
  thermal efficiency (η) and, 153–154
  vapor-compression refrigeration cycle, 330–332
Combustion, 427–476
  adiabatic flame temperature (AFT), 458–461
  applications of, 427–428
  balancing reactions, 432–434
  chemical equilibrium and, 468–472
  closed systems at constant pressure, 453–454
  closed systems at constant volume, 454–456
  complete, 430–431
  components of process, 429–431
  defined, 427
  enthalpy of compounds and formation, 443–445
  entropy balance for, 462–464
  fuel cells, 465–468
  fuels for, 429–430
  Gibbs function for, 465
  heat of reactions, 446–458

Combustion—*Cont.*
heat transfer (Q) of process, 447, 454, 456, 458–462
heating values (HV), 449
lean, 431
open systems, 446–453
oxidizers for, 429
perfect, 430–431
power cycles and, 456–460
process of, 431–432, 445–446
products of, 430–431, 440–443
reactant mixtures, 437–439
reactants from known products, 440–443
rich, 431, 472–473
spark-ignition engines, 288, 298–302
stoichiometric, 430
water–gas shift reaction, 472–473
Combustion engines, 3, 52, 150–151, 456–458
Complete combustion, 430
Compressed liquids, 74–76, 93–94, 97
Compressibility factor (Z), 87–88, 370
Compression ratio (r), 300–301
Compressors
isentropic efficiency of, 224–228
single-inlet, single-outlet open systems, 119–122
steady-state, steady-flow process of, 121–122
thermal efficiency (η) and, 153
Computational fluid dynamics (CFD) software, 475
Condensers, 256, 258
Conduction, 42–44
Fourier's law for, 43
heat transfer (Q) by, 42–44
thermal conductivity (κ) of, 43
Conservation of energy, 107. *See also* First law of thermodynamics
Conservation of mass, 108–112, 390
Constant specific heats
Brayton cycle approach, 290
constant-pressure specific heat, 78–79
constant-volume specific heat, 79
entropy changes and, 202–204
equations of state, 83–85, 91–92
isentropic efficiency and, 219–220
specific heat ratio, 79
Constant-pressure closed combustion systems, 453–454
Constant-pressure processes, 74–75
Constant-temperature processes, 74–76
Constant-volume closed combustion systems, 454–455
Convection, heat transfer (Q) by, 44–45
Conversion factor (g), 20
Cooling and dehumidifying moist air, 399–403
air-conditioning process, 399–402
cooling process analysis, 400–402
heat transfer (Q) for, 401–402
heating process analysis, 401
psychrometric charts for, 400
refrigeration cycles combined with, 404–405
vapor-compression refrigeration cycle for, 404–405
Cooling towers, 413–415
analysis of, 414–415
efficiency of, 414
types of, 414
use of, 413–415
Counter-flow heat exchangers, 213–214
Critical point, 71–75
Crystalline solids, entropy of, 196–197

D

Dalton's law of partial pressure, 358–359
Dehumidifiers, 132
Dehumidifying. *See* Cooling and dehumidifying moist air
Density
comparison for phase states, 72, 77
phase diagrams and, 73, 77
Dependent properties, 14
Dew point temperature, 387–388, 393
Diesel cycle, 288, 303–306
four-stroke, compression-ignition engine, 303–304
processes of, 304–305
thermal efficiency (η) of, 305
thermodynamic analysis of, 304–306
use of, 288
Diffusers, 115–116
Displacement, 48
Dry air, 355–357
atmospheric constant of, 383, 388–389
composition of, 355–357, 387
specific quantities based on, 389
Dry bulb temperature, 393–395
Dry cooling tower, 414
Dry products analysis, 440
Dual cycle, 307–308
processes of, 307–308
thermal efficiency (η) of, 307
thermodynamic analysis of, 307–308
use of, 307

E

Efficiency
Carnot, 182–183
coefficients of performance, 153
cooling towers, 414
entropy and, 219–225, 232
first law of thermodynamics, 150–153
irreversibility and, 232–233
isentropic, 218–227
second law of thermodynamics, 181–183, 232–233
thermal (η), 150, 151, 153, 155–156
Electrical work, 53
Emissivity (ε), 45
Energy
balance, 107
change in a system, 107
conservation of, 108
conversion, 15
defined, 35
entropy and quality of, 197
first law of thermodynamics for, 107
heat transfer, 41–48
importance and demand of, 1–2
internal (U), 38–39
kinetic (KE), 37–38
magnitude of, 39
mass transfer, 57–59
nature of, 35–61
potential (PE), 37
specific (e), 39
thermodynamic system and process analysis, 59–60
total (E), 36, 39
transport of, 40–59
types of, 36–39
units of measurement, 36
use of, 35–36
work transfer, 48–56
Energy transfer, 107
closed systems, 144–150
first law of thermodynamics for, 107, 112–113, 144–147
open systems, 112–115
piston–cylinder devices, 144–147
rate basis of, 112–114
Engineering accuracy, 24
Engines. *See also* Heat engines
combustion, 3, 52, 151, 454–456
compression-ignition, 303–304
four-stroke, 298, 303–304
spark-ignition, 148–149, 456–457
Enthalpy (H), 57
air mixture composition and, 389
closed systems at constant pressure, 453–454
closed systems at constant volume, 454–455
combustion compounds and formation, 443–445
combustion reactions, 445
equations of state, 82–86, 92

Gibbs-Dalton law and, 362–363
ideal gas changes, 82–86
ideal gas mixtures, 357–361
incompressible substances, 91–92
mass transfer, 57–58
open system, 122, 138, 142–143, 150
psychrometric, 389
specific (h), 57–58, 91, 362, 389
Enthalpy of formation, 443–445
Entropy, 195–235
Clausius inequality and, 195–198
closed systems, 214–216
compressors, 224–225
consistency of analysis, 228–230
crystalline solids, 197
energy quality and, 197
evaluating changes of in a system, 201–205
exergy, 233
fossil fuels and, 218
generation, 198–199, 201, 206
Gibbs equations for, 202–203
heat engines, 195–198
heat exchangers, 213–214
ideal gases, 202–204, 219–221
incompressible substances, 205
iron, 185
irreversible cycles, 197–198, 230–233
isentropic efficiencies, 218–228
open systems, 206–212, 235
pumps, 224–228
reversible cycles, 198
second law of thermodynamics for, 213–215, 232–233
single-inlet, single outlet open systems, 230–231
specific heats and, 202–204, 219–220
thermal reservoirs, 195–196
thermodynamic cycles, 196
third law of thermodynamics for, 197
transport, 187
turbines, 221–224
Entropy balance, 206–216
closed systems, 214–215
combustion processes, 462–464
entropy generation for, 217–218
equations, 235
equilibrium and, 202, 207
heat exchangers, 213–215
open systems, 206–212
second law of thermodynamics for, 213–215
Entropy generation
calculation of, 198–201
Clausius inequality and, 195–196
entropy balance and, 214–216
irreversibility and, 230–233
process possibility and, 199
Environmental impact of energy, 1–3
Equations of state, 14, 69–98
compressed liquids data for, 97
compressibility factor (Z), 87–88
constant specific heats and, 83–85, 91–92
constant-pressure specific heat, 78–79
constant-volume specific heat, 79
enthalpy changes and, 82–86, 92
ideal gas law, 14, 80–82
ideal gases, 80–91
incompressible substances, 91–92
phase diagrams for, 69–77
principle of corresponding states, 87
refrigerant properties, 92–94
saturated property tables for, 95
specific heat ratio, 79
state postulate, 78
steam tables for, 93
superheated vapor tables for, 95–96
thermodynamic properties and, 14, 69–103
van der Waals equation, 89–90, 98
vapor dome and, 92–93
water properties, 92–94
Equilibrium, 12–13
chemical reactions, 468–472
combustion processes, 468–472
entropy balance and, 206–208
thermodynamic, 12–13
Equilibrium constant, 469–470
Equivalence ratio (ϕ), 438–439
Evaporation and air conditions, 384
Exergy, 233
Expanding ice cubes, 173–175
Extensive properties, 13–14
Externally reversible process, 180–181

F

Fahrenheit (°F) scale, 17
Feedwater heaters, 276–285
closed, 280–287
open, 276–280
Rankine cycle, 273–283
First law efficiency. *See* Thermal efficiency
First law of thermodynamics, 107–155
closed systems, 144–150
conservation of energy, 107
conservation of mass, 108–112
energy balance and, 107
energy transfer, 107, 112–113
mass transfer and, 108–110
open systems, 109–110, 112–144
rate basis of, 109, 112–114
steady-state, steady-flow processes, 109–110, 113–140
thermal efficiency (η) and, 150–153
transient (unsteady) systems, 109 140–144
Five-gas analyzer, 440
Flow work, 57–58
Fluids. *See* Working fluids
Force (F), 20–22
mass (m) and, 20–21
Newton's second law of motion and, 20
pressure (P) and, 21–23
units of measurement, 19–20
weight (W) as, 20–21
work resulting from displacement, 48
Fourier transform infrared (FTIR) system, 440
Fourier's law, 43
Four-stroke, compression-ignition engine, 303–304
Four-stroke, spark-ignition engine, 298
Freezing, phase diagrams and, 70–72
Fuel cells, 465–467
Fuels for combustion, 429–430

G

Gage pressure, 22–23
Gas mixtures, 351–371
Amagat law of partial volumes, 360–361
behavior of, 370–371
composition of, 352–354
compressibility factor (Z) of, 370
Dalton's law of partial pressure, 358–359
dry air composition, 355–357
enthalpy (H) and, 362–363
Gibbs-Dalton law of ideal gas mixtures, 362–363
ideal, 357–371
ideal gas constant (R) for, 354, 357
mass fraction of, 353–354
mole fraction of, 353–356
molecular mass of, 353, 356
partial pressure ratio for, 358
solutions of thermodynamic problems, 364–370
volume fraction of, 360
working fluids as, 351–352
Gas phase, 26
characteristics of, 26
phase diagrams for, 70–74
three-dimensional (3-D) diagrams for, 70–71
vapor as, 26, 72, 74
Gas power cycle, 252–253, 287
Gas refrigeration cycle, 329
Gases, irreversibility of expansion, 180–181

Gibbs equations, 202
Gibbs function, 465
Gibbs-Dalton law of ideal gas mixtures, 362–364

H

Heat engines
Carnot efficiency of, 182–183
Clausius inequality for, 195–198
entropy of, 195–198
first law of thermodynamics for, 150–151
irreversible cycles, 197–198
Kelvin-Planck statement of, 176–177
reversible cycles, 195–196
second law of thermodynamics for, 176–178, 181–182
thermal efficiency (η) of, 150–151
Heat exchangers
counter-flow, 213–214
entropy balance in, 214
first law of thermodynamics for, 132–137
parallel-flow, 213–214
second law of thermodynamics for, 173–175, 213–214
Heat of combustion reactions, 446–458
closed systems at constant pressure, 453–454
closed systems at constant volume, 454–455
open systems, 446–452
power cycles and, 456–458
Heat pumps
Carnot efficiency of, 182–183
coefficient of performance (γ), 153
first law of thermodynamics applied to, 152–153
second law of thermodynamics applied to, 182–183
thermal efficiency (η) of, 152–153
Heat transfer (Q), 41–48
adiabatic process and, 47
Brayton cycle regeneration, 296–297
combustion processes, 445–446, 464
conduction, 42–43
convection, 44
cooling and dehumidifying moist air, 399–402
energy transport by, 41–48
heating and humidifying air, 406–408
psychrometrics and, 399–406
radiation, 45–46
Rankine cycle with regeneration, 275–276, 277–278, 281–283, 285
sign convention for, 47–48
total, 47–48
Heating and humidifying air, 406–408
combined processes, 407–408
heat transfer (Q) for, 406, 408
humidifying methods, 406–408
psychrometric charts for, 408
Heating values (HV), 449
Heating, ventilating, and air conditioning (HVAC), 384–385
Humidifying methods, 406–408
Humidity, 385–386, 389–399. *See also* Moist air
adiabatic saturator for, 390–391
air component analysis, 389–392
assumptions of, 390
comfort conditions and, 397–400
conservation of mass for, 390–391
determination of, 389–397
dry bulb temperature, 392–395
measurement of, 389–397
precipitation formation and, 386
psychrometric chart for, 394–398
psychrometrics of, 385, 387–398
relative (ϕ), 385–387, 390
saturated air, 385–386
sling psychrometer for, 392–393
temperature measurement of, 393–398
wet bulb temperature, 393–395
Humidity ratio (ω), 387, 391

I

Ideal gas constant (R, $\bar{R}$), 14, 80, 354, 357
Ideal gas law, 14, 80–81
Ideal gases, 80–91
compressibility factor (Z), 87–88
constant specific heats of, 83–85, 202–203, 219–220
enthalpy changes for, 82–86
entropy changes for, 202–204
equations of state, 80–91
internal energy changes for, 82–86
isentropic efficiency of, 219–220
linear interpolation for, 86
mixtures, 357–370
principle of corresponding states, 87
pseudo-reduced specific volume, 88
software for property values, 85–86
van der Waals equation for, 89–90
variable specific heats of, 85–86, 203–205, 219
Incompressible substances
entropy changes for, 205
equations of state, 91–92
Independent properties, 14
Intensive properties, 13–14
Intercooling, Brayton cycle, 293–294
Internal energy (U), 38–39, 82–87
changes in for ideal gases, 82–87
constant specific heats and, 83–85
magnitude of, 39–40
molecular levels of, 38–39
variable specific heats and, 85–86
Internally reversible process, 180
Irreversible processes
Clausius inequality, 197–198
combustion, 462–465
entropy balance for, 462–464
entropy generation and, 230–231
entropy of, 197–198
exergy (availability), 233
second law of thermodynamics for, 179–182, 221–222
single-inlet, single-outlet open systems, 230–231
Isentropic efficiencies, 218–227
compressors, 224–228
defined, 218
entropy and, 219–228
ideal gases, 219–221
isobaric process, 35
isochoric process, 24
isolated system, 7–8
isothermal process, 24
pumps, 224–228
specific heats and, 219–220
turbine outlet state, 221–222
turbines, 221–224

K

Kelvin (K) scale, 17–19
Kelvin-Planck statement, 176–177
Key equations
absolute pressure, 27
changes in enthalpy and internal energy, 98
closed systems, 156
coefficient of performance, refrigerator, 156
coefficient of performance, heat pump, 156
complete combustion, 475
compressibility factor, 98
conservation of mass, 155
convective heat transfer rate, 61
cooling/dehumidifying process, 417
cooling towers, 417
dew point temperature, 417
electrical power, 61
entropy change, 234–235
entropy generation rate, 476
entropy rate balance, 235
equivalence ration, 476
first law of thermodynamics, 155

force, 27
fuel cells maximum electrical potential, 476
Gibbs, 234
heat conduction rate, 61
heating/humidifying processes, 417
heat of reaction, 476
humidity ratio, 417
ideal gas law, 98
isentropic compression, 235
isentropic efficiency, 235
isentropic processes, 235
kinetic energy, 61
mass flow rate, 155
mass fraction, 372
maximum coefficient of performance, heat pump, 187
maximum coefficient of performance, refrigerator, 187
maximum thermal efficiency, heat engine, 187
mixing of two moist air streams, 417
mole fraction, 372
molecular mass of mixture, 372
moving boundary work, 61
net radiative heat transfer rate, 61
perfect combustion 475
potential energy, 61
rate of energy transfer via mass flow, 61
relative humidity, 417
specific heats of mixture, 372
steady-state, steady-flow, multiple-inlet, multiple-outlet open systems, 156
steady-state, steady-flow, single-inlet, single-outlet open systems, 156
temperature scale conversion, 27
thermal efficiency–heat engine, 156
total energy, 61
van der Waals equation of state, 98
Kinetic energy (KE), 36–38
magnitude of, 39
motion and, 37–38

L

Lean combustion, 431
Light-emitting diodes (LED), 3
Linear interpolation, 86
Liquid phase, 26
characteristics of, 26
compressed (subcooled), 74–76, 93–94, 97
density comparison for, 72, 77
phase diagrams for, 70–77
property comparison for phase determination, 93–94
saturated, 74–77, 93–95
saturated property tables, 95
three-dimensional (3-D) diagrams for, 70–71
transition regions, 72–74

M

Magnetic work, 55
Magnitude of energy, 39–40
Manometer, 22–23
Mass (*m*), 19–20
Avogadro's number, 19–20
conservation of, 108–112, 390
force (*F*) and, 20–21
gas mixtures, 353–354
molecular (*M*), 20, 353–354
moles (*n*), 19–20
units of measurement, 19–20
weight (*W*) and, 20–21
Mass fraction, 353–354
Mass transfer, 57–59
conservation of mass, 108–110
energy transport by, 57–59
enthalpy (*H*) of, 57–58
first law of thermodynamics and, 108–112
flow work, 57–58
open systems, 109–110
rate basis of, 109–110
specific enthalpy (*h*) of, 57–58
steady-state, steady-flow processes, 109
transient (unsteady) systems, 109–110
Mean effective pressure (mep), 301
Mechanical-draft cooling tower, 413–414
Miller cycle, 310
Mixing chambers, 132, 137–140
Moist air
analysis of components, 389–391
comfort conditions and, 398–400
cooling and dehumidifying, 399–402
mixing moist air streams, 410–413
Mole fraction, 353–356
dry air, 355–356
gas mixtures, 353–354
Molecular mass (*M*), 19–20
determination of, 19–20
dry air, 356
gas mixtures, 352–353
Moles (*n*), 19–20
Moving boundary work, 48–50
Multiple inlet or outlet systems, 131–140
dehumidifiers, 132
first law of thermodynamics in, 131–140
heat exchangers, 132–137
mixing chambers, 132, 137–140

N

National Institute of Standards and Technology (NIST) software, 86
Natural state of chemical elements, 444
Natural-draft cooling tower, 413–414
Newton's second law of motion, 20
Nozzles, 115–119

O

Open systems
combustion, 446–453, 462–464
energy transfer, 112–113
enthalpy of reactions, 446–453
entropy balance in, 206–213, 462–464
first law of thermodynamics in, 112–144
heat of combustion reactions in, 446–453
mass transfer of, 109–110
multiple inlet or outlet, 131–140
single-inlet, single-outlet, 113–130
steady-state, steady-flow processes, 109, 114–140
system boundary, 6–7
transient (unsteady) processes, 109–110, 140–144
use of, 9–10
Open-loop reversed Brayton cycle, 341–342
Otto cycle
compression ratio (*r*) of, 300–301
first law of thermodynamic applied to, 148
four-stroke, spark-ignition engine, 298
mean effective pressure (mep), 301
piston-cylinder device with, 149–150
processes of, 298–299
spark-ignition engines, 148–149, 456–457
thermal efficiency (η) of, 300
thermodynamic analysis of, 298–303
use of, 300–301
Oxidizers for combustion, 429

P

P-T diagram, 71–72
P-v diagram, 71–74
P-v-T diagram, 70–71
Parallel-flow heat exchangers, 213–214
Partial pressure ratio, 358, 385
Path function, 48
Peacesoftware, 86
Percent deficit air, 437
Percent excess air, 437

Percent theoretical air, 434, 437
Perfect combustion, 430–431
Perpetual motion machines (PMM), 185–186
Phase changes, 74–75
Phase diagrams, 69–77
  compressed (subcooled) liquids, 74–76
  critical point, 71–73, 75–76
  density comparison from, 72, 77
  freezing and, 70–71
  gas and vapor distinction, 72, 74
  P-T, 71–72
  P-v, 71–74
  P-v-T, 70–71
  phase changes and, 75–77
  quality of a substance, 76–77
  saturated substances, 74–77
  saturation pressure, 75
  saturation temperature, 75
  superheated vapor, 74–76
  T-v, 72
  three-dimensional (3-D) diagrams for, 70–71
  transition regions, 72–74
  triple line, 72
  triple point, 71–72
  use of, 69–70
  vapor dome, 74
Phase equilibrium, 27
Phases of matter, 25–26
Piston-cylinder devices
  energy transfer from, 144–147
  first law of thermodynamics applied to, 144–147
  Kelvin-Planck statement for, 177
  moving boundary work of, 48–50
  Otto cycle for, 148–149
  Rankine cycle for, 148–149
  second law of thermodynamics for, 176–117
Plasma phase, 25
Point function, 48
Polytropic process, 50
Potential energy (PE), 36, 37, 38
  gravitational fields and, 37
  magnitude of, 39–40
Power, 48
  electrical, 53
Power cycles, 251–310
  air standard cycle (ASC), 287–288
  Brayton cycle, 288–297
  Carnot cycle, 253–255
  combustion, heat of reactions and, 456–458
  Diesel cycle, 288, 303–307
  dual cycle, 307–308
  fossil fuels, 253
  gas, 252–253, 287–288
  Otto cycle, 148, 288, 297–303
  Rankine cycle, 148, 255–287
  thermodynamic cycles, 251–253
  use of, 251–253
  vapor, 252–253, 255–256
  working fluids in, 251–253, 456–458
Precipitation, formation of, 385–386
Pressure (*P*), 21–24
  absolute, 23
  atmospheric, 22–23
  constant-pressure processes, 24, 74–75
  Dalton's law of partial pressure, 358–359
  force (*F*) and, 21–23
  ideal gas mixtures and, 357–364
  mechanical equilibrium and, 22
  partial pressure ratio, 358, 360
  psychrometrics and, 385–389
  saturation, 74, 75, 77, 95, 385–387
  three-dimensional (3-D) diagrams for, 70–71
  units of measurement, 22
Principle of corresponding states, 87–88
Problem analysis and solution procedure, 59. *See also* Thermodynamic problem analysis
Properties, thermodynamic, 13–15, 16–24
  dependent, 14
  equations of state and, 14, 69–106
  extensive, 13–14
  force, 19–23
  independent, 14
  intensive, 13–14
  mass (*m*), 13–14, 17, 19–21
  molecular mass (*M*), 20
  phase diagrams for, 69–77
  pressure (*P*), 21–24
  specific, 14, 17, 21
  temperature (*T*), 16–19
  volume (*V*), 13–14, 21
  water and, 92–97
  weight, 20–21
Property comparison approach for phase determination, 93–94
Property determination for water and refrigerants, 92–97
  vapor dome and, 92–93
Pseudo-reduced specific volume, 88
Psychrometric chart, 394–398
Psychrometrics, 383–426
  comfort level and, 383–386, 397–399
  cooling and dehumidifying moist air, 399–404
  cooling tower applications, 413–416
  dew point temperature, 387–388, 393–394
  dry air, 383–384, 387, 389
  dry bulb temperature, 392–394
  evaporation and, 384
  heat transfer (*Q*), 390, 399–407
  heating and humidifying air, 406–410
  humidity, 385–388, 389–397
  humidity ratio (ω), 387, 391, 394
  mixing moist air streams, 410–413
  moist air, 389–391, 397–404, 410–413
  partial pressure ratio, 358, 360
  precipitation, formation of, 385–386
  psychrometric calculator, 395
  refrigeration cycles and, 404–405
  relative humidity (ϕ), 385–388, 390–391
  saturation pressure, 385–387
  specific properties and, 389
  study of, 383–385
  water vapor, 383–390
  wet bulb temperature, 393–394
Pumps
  isentropic efficiency of, 224–228
  single-inlet, single-outlet open systems, 113–131
  steady-state, steady-flow process of, 113–131
Pure substance, 25

Q

Quality of a mixture, 76–77

R

Radiation, 45–47
  blackbody, 45
  emissivity (ε) of a surface and, 45
  heat transfer (*Q*) by, 45–47
  Stefan-Boltzmann constant (σ) and, 45
Rankine (R) scale, 18–19
Rankine cycle, 148, 255–287
  assumptions for, 257–258
  Brayton cycle compared to, 288, 293
  Carnot cycle compared to, 262, 265
  closed feedwater heater for, 280–287
  closed system, as example of a, 148
  condensers in a, 256–258
  first law of thermodynamics applied to, 148
  ideal, 255–262
  non-ideal, 263–266
  open feedwater heater for, 276–280
  piston-cylinder devices with, 149
  processes in ideal, basic, 256, 263
  regeneration heat transfer for, 275–287
  reheat stage for, 270–275
  steam generators, 256–257
  superheat for, 266–270, 276

thermal efficiency (η) of, 260–265, 266, 270, 273, 275
use of, 255
working fluids in, 255–256, 273
Reactants, 440–443
air–fuel and fuel–air ratios, 437–438
combustion, 440–442
dry products analysis for, 440–443
equivalence ratio (ϕ), 438–440
known products as, 440–443
mixture characterization, 440–442
percent deficit air, 437
percent excess air, 437
percent theoretical air, 437
Refrigerants
ammonia as, 338
compound numbering, 329
compressed liquids property data, 97
property determination of, 92–97
saturated property tables, 95
steam tables, 93
superheated vapor tables, 95–97
vapor dome and, 92–93
Refrigeration cycles, 327–342
absorption, 337–338
cooling and dehumidifying process combined with, 399–404
gas, 329
reversed Brayton, 338–342
use of, 327–328
vapor, 329
vapor-compression, 330–337
working fluid in, 328–329
Refrigerators
Carnot efficiency of, 182–185
Clausius statement of the second law of thermodynamics and, 177–178
coefficient of performance (β), 153
defined, 327
first law of thermodynamics applied to, 152–153
second law of thermodynamics and, 178, 182–183
thermal efficiency (η) of, 150–153
Regeneration heat transfer
Brayton cycle, 296–297
Rankine cycle, 275–286
Reheat
Brayton cycle, 295–296
Rankine cycle, 270–274
Relative humidity (ϕ), 385–388, 390–391
Relative temperature scales, 17
Reversed Brayton refrigeration cycle, 338–342
coefficients of performance for, 339
open-loop cycle, 341
processes of, 339
thermal efficiency (η) of, 339
thermodynamic analysis of, 339
Reversible processes
Clausius inequality of, 195–198
entropy of, 195–198
externally, 180–181
internally, 180–181
second law of thermodynamics for, 179–181
thermal efficiency (η) of, 181–185
Rich combustion, 431, 472–473
Rotating shaft power, 53
Rotating shaft work, 51–53

S

Saturated air, 385–386
Saturated property tables, 95
Saturated substances, 74–77, 93–94
Saturation pressure, 74–75, 385–387
Saturation temperature, 74, 76
Scientific laws that define thermodynamics, 5
Second law efficiency, 232–233
Second law of thermodynamics, 173–187
Carnot efficiencies, 182–185
Clausius statement of, 177–178
entropy balance and, 214–216
expanding ice cubes, 173–175
heat engines, 175–178
heat exchangers, 173–175, 213–214
irreversibility and efficiency, 232–233
irreversible processes, 179–181
Kelvin-Planck statement of, 176–177
nature of, 173–175
perpetual motion machines and, 185–186
piston-cylinder devices, 176–178
refrigerators and, 178
reversible processes, 179–181
thermodynamic temperature scale from, 181–182
uses of, 175–176
Shaft work machines, 51–53, 119–128
Shaft work machines open systems, 121–122
Single-inlet, single-outlet open systems
compressors, 121–122
diffusers, 115–117
entropy generation, 230–231
first law of thermodynamics in, 113–130
irreversibility of, 230–231
nozzles, 115–119
pumps, 121–122
shaft work machines, 119–128
throttling devices, 129–130
turbines, 121–124
Sling psychrometer, 392–393
Software
analysis of thermodynamic systems using, 60
computational fluid dynamics (CFD), 475
ideal gas property values from, 85–86
National Institute of Standards and Technology (NIST), 86
Peace, 86
Wolfram Alpha, 86
Solid phase, characteristics of, 25
Spark-ignition engines, 148–149, 456–457
combustion of, 456–457
first law of thermodynamics and, 148–149
heat of reaction in, 456–457
piston–cylinder device in, 148–149
Specific heats, *see* Constant specific heats; Variable specific heats
Specific properties
air mixture composition and, 386, 387
constant specific heats, 82–83, 92, 202–203, 219–220
constant-pressure specific heat, 78–79
constant-volume specific heat, 79
energy (e), 39
enthalpy, 57–58, 78–79, 86, 92, 362, 389
entropy (s) changes and, 202–205
equations of state, 79–80, 82–86, 91–92
heat ratio, 79
humidity (ratio), 387, 389
ideal gas mixtures, 357–363
ideal gases, 82–86, 202–204, 219–220
incompressible substances, 91–92
isentropic efficiency and, 219–220
pseudo-reduced specific volume, 88
psychrometrics and, 389
thermodynamics and, 14
variable specific heat, 85–86, 203–204, 220
volume (v), 14, 21, 389
Spring work, 54
Standard temperature and pressure (STP), 444
State postulate, 78
Steady-state, steady-flow processes, 109, 113–140
compressors, 119–122
dehumidifiers, 132
diffusers, 115–117
first law of thermodynamics in, 113–140
heat exchangers, 132–138
mass transfer in, 108
mixing chambers, 132, 137–140

Steady-state, steady-flow processes—*Cont.*
  multiple inlet or outlet systems, 131–134
  nozzles, 115–119
  pumps, 121–122
  shaft work machines, 119–128
  single-inlet, single-outlet open systems, 113–131
  throttling devices, 129–130
  turbines, 121–124
Steam generators, 256–257
Steam tables, 93
Stefan-Boltzmann constant (σ), 45
Stoichiometric combustion, 430
Superheated vapor
  compression of, 76
  equations of state and, 95–97
  phase diagrams, 74–76
  properties of water vapor, 95–97
  Rankine cycle with, 266–268
  thermal efficiency (η) of processes, 264–267
  use of table values, 95–96
Surface tension, work from, 55
Surroundings, 6
System boundary, 6

T

T-v diagram, 74–75
Tank-filling processes, 141–144
Temperature (T), 16–19
  absolute scales, 17–19
  adiabatic flame (AFT), 458–461
  comfort conditions and, 397–400
  constant-temperature processes, 74–76
  dew point, 387–388
  dry bulb, 392–394
  humidity measurement by, 390–394
  phase changes and, 26
  psychrometric chart, 394–398
  relative scales, 17
  saturation, 74, 76
  second law of thermodynamics and, 182
  thermodynamic property of, 16–19
  thermodynamic scale of, 181–182
  three-dimensional (3-D) diagrams for, 70–71
  units of measurement, 16–18
  wet bulb, 393–394
  Zeroth law application, 25
Thermal conductivity (κ), 43
Thermal efficiency (η)
  Brayton cycle, 290–293
  Carnot efficiency, 182–185
  Carnot power cycle, 253–255
  coefficients of performance, 153–154
  Diesel cycle, 305–306
  dual cycle, 308
  first law of thermodynamics and, 150–153
  heat engines, 150–151, 182–183
  heat pumps, 150, 152–153, 182–183
  maximum (reversible), 182–183
  Otto cycle, 300
  Rankine cycle, 260–266, 270, 273
  Rankine cycle with superheat, 266, 270
  refrigerators, 150, 152–153, 182–183
  reversed Brayton refrigeration cycle, 338–341
  second law of thermodynamics for, 182–185
  vapor-compression refrigeration cycle, 331
Thermal reservoirs, 195–196
Thermodynamic cycle, 12
  closed systems, 148–150
  defined, 12
  entropy of, 198–201
  first law of thermodynamics applied to, 148–150
  power cycles, 251–253
  refrigeration cycles, 327–332
Thermodynamic problem analysis
  entropy and consistency of, 228–230
  ideal gas mixtures, 363–370
  procedure for solution of, 59
  scientific laws for, 5
  software packages for, 60
  thermodynamic systems, 59
  treating gases separately, 368–370
  use of, 5–7
Thermodynamic systems
  analysis of, 59
  choice of type of, 10
  closed, 7, 8–9
  energy change in, 107
  isolated, 7–8
  open, 7, 9
  problem analysis, 59
  software packages for analysis of, 60
  solution procedure, 59
  surroundings, 6
  system boundary, 6
  thermodynamic processes in, 11–12
Thermodynamics, 1–27
  analysis, 5–6
  applications of, 4–5
  defined, 3, 5
  energy conversion, 2–3
  engineering accuracy and, 24
  equilibrium, 12–13
  first law of, 107–156
  force, 20–22
  importance and demand of energy, 1–2
  mass (m), 19–21
  molecular mass (M), 19–20
  phases of matter, 25–26
  pressure (P), 21–24
  process, 11–12, 24
  properties of, 13–14, 16–24
  scientific laws for, 5
  second law of, 173–193
  specific properties, 14, 21
  systems, 6–10
  temperature (T), 16–19
  third law of, 197
  units of, 15, 16–18, 20, 22
  volume (V), 21
  weight (W), 20–21
  zeroth law of, 24–25
Third law of thermodynamics, 197
Throttling devices, 129–130
Total energy (E), 36, 39
Total heat transfer, 41–48
  energy transfer, 56
Total work, 55–56
Transient systems, 140–144. *See also* Unsteady (transient) systems
Transition regions, 72–74
Transport of energy, 36, 41–59
  heat transfer (Q), 41–48
  mass transfer, 57–59
  work transfer (W), 48–56
Triple line, 71–72
Triple point, 19, 71–72
Turbines
  entropy of, 221–224
  isentropic efficiency of, 221–224
  single-inlet, single-outlet open systems, 121–124
  steady-state, steady-flow process of, 120–121

U

Units, 14–22
  conversion factor (g_c), 20
  energy, 36
  English Engineering (EE), 14–15
  force (F), 20
  International System (SI), 14–15
  mass (m), 19–20
  prefixes for SI, 15
  pressure (P), 22
  temperature (T), 16–19
  volume (V), 21
Unsteady (transient) systems, 110, 140–144
  air compressors, 140–141
  first law of thermodynamics in, 140–144

mass transfer in, 108–110
tank-filling processes, 141–144

V

van der Waals equation, 88–90
Vapor, 69–77
gas-phase distinction of, 72, 74
phase diagrams, 72–74
property comparison for phase determination, 92–93
saturated, 74–76, 93–94
superheated, 74–76, 93–94
Vapor dome, 74, 92–93
Vapor power cycle, 252–254, 255–287
Carnot, 253
Rankine, 255–287
working fluids in, 252–253
Vapor refrigeration cycle, 329
Vapor-compression refrigeration cycle, 330–336
cascade system, 336
coefficients of performance for, 331–332
cooling and dehumidifying process combined with, 404–405
cooling from, 332–333
processes of, 330
thermal efficiency (η) of, 331
Variable specific heats
entropy changes and, 203–205
equations of state, 85–86
isentropic efficiency and, 220
Volume (V), 14, 21
air mixture composition and, 386–387
Amagat law of partial volumes, 360–361
gas mixtures, 357–359
psychrometric, 389
specific, 14, 21, 70, 389
three-dimensional (3-D) diagrams for, 70–71
units of measurement, 21
Volume fraction, 360

W

Water properties, 92–94
compressed liquids data, 97
property determination for, 92–93
saturated property tables, 95
steam tables, 93
superheated vapor property tables, 95
vapor dome and, 92–93
Water vapor, 383–389
atmospheric content of, 383–384
comfort level and, 383–387
dew point temperature, 387–388
evaporation and, 384
partial pressure ratio, 385
relative humidity (ϕ), 385–388
saturation pressure and, 385–387
Water-gas shift reaction, 472–473
Weight, 20–21
Wet bulb temperature, 393–394
Wet cooling tower, 414
Wolfram Alpha software, 86, 93, 96
Work, 48
chemical work, 55
elastic solids, 54
electrical work, 53
energy transport by, 48–56
force acting through displacement, 48
magnetic work, 55
moving boundary work, 48–51
path function, 48
point function, 48
polytropic process for, 50
rotating shaft work, 51–53
sign convention for, 54–56
spring work, 54
surface tension, 55
total work, 55–56
Working fluids
air standard cycles (ASC), 287
combustion heat of reactions and, 456–458
gas mixtures, 351–352
gas power cycles, 287–288
phase change of, 252, 329
Rankine cycle, 255–256, 273
regeneration heat transfer and, 275–276
vapor power cycle, 251–253
Work transfer, 48–56

Z

Zeroth law of thermodynamics, 24–25